Lecture Notes in Computer Science 14226

Founding Editors

The series Lecture Notes in Computer Science (LNCS), including its subseries Lecture Notes in Artificial Intelligence (LNAI) and Lecture Notes in Bioinformatics (LNBI), has established itself as a medium for the publication of new developments in computer science and information technology research, teaching, and education.

LNCS enjoys close cooperation with the computer science R & D community, the series counts many renowned academics among its volume editors and paper authors, and collaborates with prestigious societies. Its mission is to serve this international community by providing an invaluable service, mainly focused on the publication of conference and workshop proceedings and postproceedings. LNCS commenced publication in 1973.

Hayit Greenspan · Anant Madabhushi ·
Parvin Mousavi · Septimiu Salcudean ·
James Duncan · Tanveer Syeda-Mahmood ·
Russell Taylor
Editors

Medical Image Computing and Computer Assisted Intervention – MICCAI 2023

26th International Conference
Vancouver, BC, Canada, October 8–12, 2023
Proceedings, Part VII

 Springer

Editors
Hayit Greenspan
Icahn School of Medicine, Mount Sinai,
NYC, NY, USA

Tel Aviv University
Tel Aviv, Israel

Parvin Mousavi
Queen's University
Kingston, ON, Canada

James Duncan ⓘ
Yale University
New Haven, CT, USA

Russell Taylor ⓘ
Johns Hopkins University
Baltimore, MD, USA

Anant Madabhushi ⓘ
Emory University
Atlanta, GA, USA

Septimiu Salcudean ⓘ
The University of British Columbia
Vancouver, BC, Canada

Tanveer Syeda-Mahmood ⓘ
IBM Research
San Jose, CA, USA

ISSN 0302-9743 ISSN 1611-3349 (electronic)
Lecture Notes in Computer Science
ISBN 978-3-031-43989-6 ISBN 978-3-031-43990-2 (eBook)
https://doi.org/10.1007/978-3-031-43990-2

This Springer imprint is published by the registered company Springer Nature Switzerland AG
The registered company address is: Gewerbestrasse 11, 6330 Cham, Switzerland

Paper in this product is recyclable.

Preface

We are pleased to present the proceedings for the 26th International Conference on Medical Image Computing and Computer-Assisted Intervention (MICCAI). After several difficult years of virtual conferences, this edition was held in a mainly in-person format with a hybrid component at the Vancouver Convention Centre, in Vancouver, BC, Canada October 8–12, 2023. The conference featured 33 physical workshops, 15 online workshops, 15 tutorials, and 29 challenges held on October 8 and October 12. Co-located with the conference was also the 3rd Conference on Clinical Translation on Medical Image Computing and Computer-Assisted Intervention (CLINICCAI) on October 10.

MICCAI 2023 received the largest number of submissions so far, with an approximately 30% increase compared to 2022. We received 2365 full submissions of which 2250 were subjected to full review. To keep the acceptance ratios around 32% as in previous years, there was a corresponding increase in accepted papers leading to 730 papers accepted, with 68 orals and the remaining presented in poster form. These papers comprise ten volumes of Lecture Notes in Computer Science (LNCS) proceedings as follows:

- Part I, LNCS Volume 14220: Machine Learning with Limited Supervision and Machine Learning – Transfer Learning
- Part II, LNCS Volume 14221: Machine Learning – Learning Strategies and Machine Learning – Explainability, Bias, and Uncertainty I
- Part III, LNCS Volume 14222: Machine Learning – Explainability, Bias, and Uncertainty II and Image Segmentation I
- Part IV, LNCS Volume 14223: Image Segmentation II
- Part V, LNCS Volume 14224: Computer-Aided Diagnosis I
- Part VI, LNCS Volume 14225: Computer-Aided Diagnosis II and Computational Pathology
- Part VII, LNCS Volume 14226: Clinical Applications – Abdomen, Clinical Applications – Breast, Clinical Applications – Cardiac, Clinical Applications – Dermatology, Clinical Applications – Fetal Imaging, Clinical Applications – Lung, Clinical Applications – Musculoskeletal, Clinical Applications – Oncology, Clinical Applications – Ophthalmology, and Clinical Applications – Vascular
- Part VIII, LNCS Volume 14227: Clinical Applications – Neuroimaging and Microscopy
- Part IX, LNCS Volume 14228: Image-Guided Intervention, Surgical Planning, and Data Science
- Part X, LNCS Volume 14229: Image Reconstruction and Image Registration

The papers for the proceedings were selected after a rigorous double-blind peer-review process. The MICCAI 2023 Program Committee consisted of 133 area chairs and over 1600 reviewers, with representation from several countries across all major continents. It also maintained a gender balance with 31% of scientists who self-identified as women. With an increase in the number of area chairs and reviewers, the reviewer load on the experts was reduced this year, keeping to 16–18 papers per area chair and about 4–6 papers per reviewer. Based on the double-blinded reviews, area chairs' recommendations, and program chairs' global adjustments, 308 papers (14%) were provisionally accepted, 1196 papers (53%) were provisionally rejected, and 746 papers (33%) proceeded to the rebuttal stage. As in previous years, Microsoft's Conference Management Toolkit (CMT) was used for paper management and organizing the overall review process. Similarly, the Toronto paper matching system (TPMS) was employed to ensure knowledgeable experts were assigned to review appropriate papers. Area chairs and reviewers were selected following public calls to the community, and were vetted by the program chairs.

Among the new features this year was the emphasis on clinical translation, moving Medical Image Computing (MIC) and Computer-Assisted Interventions (CAI) research from theory to practice by featuring two clinical translational sessions reflecting the real-world impact of the field in the clinical workflows and clinical evaluations. For the first time, clinicians were appointed as Clinical Chairs to select papers for the clinical translational sessions. The philosophy behind the dedicated clinical translational sessions was to maintain the high scientific and technical standard of MICCAI papers in terms of methodology development, while at the same time showcasing the strong focus on clinical applications. This was an opportunity to expose the MICCAI community to the clinical challenges and for ideation of novel solutions to address these unmet needs. Consequently, during paper submission, in addition to MIC and CAI a new category of "Clinical Applications" was introduced for authors to self-declare.

MICCAI 2023 for the first time in its history also featured dual parallel tracks that allowed the conference to keep the same proportion of oral presentations as in previous years, despite the 30% increase in submitted and accepted papers.

We also introduced two new sessions this year focusing on young and emerging scientists through their Ph.D. thesis presentations, and another with experienced researchers commenting on the state of the field through a fireside chat format.

The organization of the final program by grouping the papers into topics and sessions was aided by the latest advancements in generative AI models. Specifically, Open AI's GPT-4 large language model was used to group the papers into initial topics which were then manually curated and organized. This resulted in fresh titles for sessions that are more reflective of the technical advancements of our field.

Although not reflected in the proceedings, the conference also benefited from keynote talks from experts in their respective fields including Turing Award winner Yann LeCun and leading experts Jocelyne Troccaz and Mihaela van der Schaar.

We extend our sincere gratitude to everyone who contributed to the success of MICCAI 2023 and the quality of its proceedings. In particular, we would like to express our profound thanks to the MICCAI Submission System Manager Kitty Wong whose meticulous support throughout the paper submission, review, program planning, and proceeding preparation process was invaluable. We are especially appreciative of the effort and dedication of our Satellite Events Chair, Bennett Landman, who tirelessly coordinated the organization of over 90 satellite events consisting of workshops, challenges and tutorials. Our workshop chairs Hongzhi Wang, Alistair Young, tutorial chairs Islem

Rekik, Guoyan Zheng, and challenge chairs, Lena Maier-Hein, Jayashree Kalpathy-Kramer, Alexander Seitel, worked hard to assemble a strong program for the satellite events. Special mention this year also goes to our first-time Clinical Chairs, Drs. Curtis Langlotz, Charles Kahn, and Masaru Ishii who helped us select papers for the clinical sessions and organized the clinical sessions.

We acknowledge the contributions of our Keynote Chairs, William Wells and Alejandro Frangi, who secured our keynote speakers. Our publication chairs, Kevin Zhou and Ron Summers, helped in our efforts to get the MICCAI papers indexed in PubMed. It was a challenging year for fundraising for the conference due to the recovery of the economy after the COVID pandemic. Despite this situation, our industrial sponsorship chairs, Mohammad Yaqub, Le Lu and Yanwu Xu, along with Dekon's Mehmet Eldegez, worked tirelessly to secure sponsors in innovative ways, for which we are grateful.

An active body of the MICCAI Student Board led by Camila Gonzalez and our 2023 student representatives Nathaniel Braman and Vaishnavi Subramanian helped put together student-run networking and social events including a novel Ph.D. thesis 3-minute madness event to spotlight new graduates for their careers. Similarly, Women in MICCAI chairs Xiaoxiao Li and Jayanthi Sivaswamy and RISE chairs, Islem Rekik, Pingkun Yan, and Andrea Lara further strengthened the quality of our technical program through their organized events. Local arrangements logistics including the recruiting of University of British Columbia students and invitation letters to attendees, was ably looked after by our local arrangement chairs Purang Abolmaesumi and Mehdi Moradi. They also helped coordinate the visits to the local sites in Vancouver both during the selection of the site and organization of our local activities during the conference. Our Young Investigator chairs Marius Linguraru, Archana Venkataraman, Antonio Porras Perez put forward the startup village and helped secure funding from NIH for early career scientist participation in the conference. Our communications chair, Ehsan Adeli, and Diana Cunningham were active in making the conference visible on social media platforms and circulating the newsletters. Niharika D'Souza was our cross-committee liaison providing note-taking support for all our meetings. We are grateful to all these organization committee members for their active contributions that made the conference successful.

We would like to thank the MICCAI society chair, Caroline Essert, and the MICCAI board for their approvals, support and feedback, which provided clarity on various aspects of running the conference. Behind the scenes, we acknowledge the contributions of the MICCAI secretariat personnel, Janette Wallace, and Johanne Langford, who kept a close eye on logistics and budgets, and Diana Cunningham and Anna Van Vliet for including our conference announcements in a timely manner in the MICCAI society newsletters. This year, when the existing virtual platform provider indicated that they would discontinue their service, a new virtual platform provider Conference Catalysts was chosen after due diligence by John Baxter. John also handled the setup and coordination with CMT and consultation with program chairs on features, for which we are very grateful. The physical organization of the conference at the site, budget financials, fund-raising, and the smooth running of events would not have been possible without our Professional Conference Organization team from Dekon Congress & Tourism led by Mehmet Eldegez. The model of having a PCO run the conference, which we used at

MICCAI, significantly reduces the work of general chairs for which we are particularly grateful.

Finally, we are especially grateful to all members of the Program Committee for their diligent work in the reviewer assignments and final paper selection, as well as the reviewers for their support during the entire process. Lastly, and most importantly, we thank all authors, co-authors, students/postdocs, and supervisors for submitting and presenting their high-quality work, which played a pivotal role in making MICCAI 2023 a resounding success.

With a successful MICCAI 2023, we now look forward to seeing you next year in Marrakesh, Morocco when MICCAI 2024 goes to the African continent for the first time.

October 2023

Tanveer Syeda-Mahmood
James Duncan
Russ Taylor
General Chairs

Hayit Greenspan
Anant Madabhushi
Parvin Mousavi
Septimiu Salcudean
Program Chairs

Organization

General Chairs

Tanveer Syeda-Mahmood   IBM Research, USA
James Duncan   Yale University, USA
Russ Taylor   Johns Hopkins University, USA

Program Committee Chairs

Hayit Greenspan   Tel-Aviv University, Israel and Icahn School of
          Medicine at Mount Sinai, USA
Anant Madabhushi   Emory University, USA
Parvin Mousavi   Queen's University, Canada
Septimiu Salcudean   University of British Columbia, Canada

Satellite Events Chair

Bennett Landman   Vanderbilt University, USA

Workshop Chairs

Hongzhi Wang   IBM Research, USA
Alistair Young   King's College, London, UK

Challenges Chairs

Jayashree Kalpathy-Kramer   Harvard University, USA
Alexander Seitel   German Cancer Research Center, Germany
Lena Maier-Hein   German Cancer Research Center, Germany

Tutorial Chairs

Islem Rekik   Imperial College London, UK
Guoyan Zheng   Shanghai Jiao Tong University, China

Clinical Chairs

Curtis Langlotz   Stanford University, USA
Charles Kahn   University of Pennsylvania, USA
Masaru Ishii   Johns Hopkins University, USA

Local Arrangements Chairs

Purang Abolmaesumi   University of British Columbia, Canada
Mehdi Moradi   McMaster University, Canada

Keynote Chairs

William Wells   Harvard University, USA
Alejandro Frangi   University of Manchester, UK

Industrial Sponsorship Chairs

Mohammad Yaqub   MBZ University of Artificial Intelligence,
                            Abu Dhabi
Le Lu   DAMO Academy, Alibaba Group, USA
Yanwu Xu   Baidu, China

Communication Chair

Ehsan Adeli   Stanford University, USA

Publication Chairs

Ron Summers   National Institutes of Health, USA
Kevin Zhou   University of Science and Technology of China,
                                      China

Young Investigator Chairs

Marius Linguraru   Children's National Institute, USA
Archana Venkataraman   Boston University, USA
Antonio Porras   University of Colorado Anschutz Medical
                                      Campus, USA

Student Activities Chairs

Nathaniel Braman   Picture Health, USA
Vaishnavi Subramanian   EPFL, France

Women in MICCAI Chairs

Jayanthi Sivaswamy   IIIT, Hyderabad, India
Xiaoxiao Li   University of British Columbia, Canada

RISE Committee Chairs

Islem Rekik   Imperial College London, UK
Pingkun Yan   Rensselaer Polytechnic Institute, USA
Andrea Lara   Universidad Galileo, Guatemala

Submission Platform Manager

Kitty Wong   The MICCAI Society, Canada

Virtual Platform Manager

John Baxter   INSERM, Université de Rennes 1, France

Cross-Committee Liaison

Niharika D'Souza   IBM Research, USA

Program Committee

Sahar Ahmad	University of North Carolina at Chapel Hill, USA
Shadi Albarqouni	University of Bonn and Helmholtz Munich, Germany
Angelica Aviles-Rivero	University of Cambridge, UK
Shekoofeh Azizi	Google, Google Brain, USA
Ulas Bagci	Northwestern University, USA
Wenjia Bai	Imperial College London, UK
Sophia Bano	University College London, UK
Kayhan Batmanghelich	University of Pittsburgh and Boston University, USA
Ismail Ben Ayed	ETS Montreal, Canada
Katharina Breininger	Friedrich-Alexander-Universität Erlangen-Nürnberg, Germany
Weidong Cai	University of Sydney, Australia
Geng Chen	Northwestern Polytechnical University, China
Hao Chen	Hong Kong University of Science and Technology, China
Jun Cheng	Institute for Infocomm Research, A*STAR, Singapore
Li Cheng	University of Alberta, Canada
Albert C. S. Chung	University of Exeter, UK
Toby Collins	Ircad, France
Adrian Dalca	Massachusetts Institute of Technology and Harvard Medical School, USA
Jose Dolz	ETS Montreal, Canada
Qi Dou	Chinese University of Hong Kong, China
Nicha Dvornek	Yale University, USA
Shireen Elhabian	University of Utah, USA
Sandy Engelhardt	Heidelberg University Hospital, Germany
Ruogu Fang	University of Florida, USA

Aasa Feragen   Technical University of Denmark, Denmark
Moti Freiman   Technion - Israel Institute of Technology, Israel
Huazhu Fu   IHPC, A*STAR, Singapore
Adrian Galdran   Universitat Pompeu Fabra, Barcelona, Spain
Zhifan Gao   Sun Yat-sen University, China
Zongyuan Ge   Monash University, Australia
Stamatia Giannarou   Imperial College London, UK
Yun Gu   Shanghai Jiao Tong University, China
Hu Han   Institute of Computing Technology, Chinese
                             Academy of Sciences, China
Daniel Hashimoto   University of Pennsylvania, USA
Mattias Heinrich   University of Lübeck, Germany
Heng Huang   University of Pittsburgh, USA
Yuankai Huo   Vanderbilt University, USA
Mobarakol Islam   University College London, UK
Jayender Jagadeesan   Harvard Medical School, USA
Won-Ki Jeong   Korea University, South Korea
Xi Jiang   University of Electronic Science and Technology
                             of China, China
Yueming Jin   National University of Singapore, Singapore
Anand Joshi   University of Southern California, USA
Shantanu Joshi   UCLA, USA
Leo Joskowicz   Hebrew University of Jerusalem, Israel
Samuel Kadoury   Polytechnique Montreal, Canada
Bernhard Kainz   Friedrich-Alexander-Universität
                             Erlangen-Nürnberg, Germany and Imperial
                             College London, UK
Davood Karimi   Harvard University, USA
Anees Kazi   Massachusetts General Hospital, USA
Marta Kersten-Oertel   Concordia University, Canada
Fahmi Khalifa   Mansoura University, Egypt
Minjeong Kim   University of North Carolina, Greensboro, USA
Seong Tae Kim   Kyung Hee University, South Korea
Pavitra Krishnaswamy   Institute for Infocomm Research, Agency for
                             Science Technology and Research (A*STAR),
                             Singapore
Jin Tae Kwak   Korea University, South Korea
Baiying Lei   Shenzhen University, China
Xiang Li   Massachusetts General Hospital, USA
Xiaoxiao Li   University of British Columbia, Canada
Yuexiang Li   Tencent Jarvis Lab, China
Chunfeng Lian   Xi'an Jiaotong University, China

Jianming Liang	Arizona State University, USA
Jianfei Liu	National Institutes of Health Clinical Center, USA
Mingxia Liu	University of North Carolina at Chapel Hill, USA
Xiaofeng Liu	Harvard Medical School and MGH, USA
Herve Lombaert	École de technologie supérieure, Canada
Ismini Lourentzou	Virginia Tech, USA
Le Lu	Damo Academy USA, Alibaba Group, USA
Dwarikanath Mahapatra	Inception Institute of Artificial Intelligence, United Arab Emirates
Saad Nadeem	Memorial Sloan Kettering Cancer Center, USA
Dong Nie	Alibaba (US), USA
Yoshito Otake	Nara Institute of Science and Technology, Japan
Sang Hyun Park	Daegu Gyeongbuk Institute of Science and Technology, South Korea
Magdalini Paschali	Stanford University, USA
Tingying Peng	Helmholtz Munich, Germany
Caroline Petitjean	LITIS Université de Rouen Normandie, France
Esther Puyol Anton	King's College London, UK
Chen Qin	Imperial College London, UK
Daniel Racoceanu	Sorbonne Université, France
Hedyeh Rafii-Tari	Auris Health, USA
Hongliang Ren	Chinese University of Hong Kong, China and National University of Singapore, Singapore
Tammy Riklin Raviv	Ben-Gurion University, Israel
Hassan Rivaz	Concordia University, Canada
Mirabela Rusu	Stanford University, USA
Thomas Schultz	University of Bonn, Germany
Feng Shi	Shanghai United Imaging Intelligence, China
Yang Song	University of New South Wales, Australia
Aristeidis Sotiras	Washington University in St. Louis, USA
Rachel Sparks	King's College London, UK
Yao Sui	Peking University, China
Kenji Suzuki	Tokyo Institute of Technology, Japan
Qian Tao	Delft University of Technology, Netherlands
Mathias Unberath	Johns Hopkins University, USA
Martin Urschler	Medical University Graz, Austria
Maria Vakalopoulou	CentraleSupelec, University Paris Saclay, France
Erdem Varol	New York University, USA
Francisco Vasconcelos	University College London, UK
Harini Veeraraghavan	Memorial Sloan Kettering Cancer Center, USA
Satish Viswanath	Case Western Reserve University, USA
Christian Wachinger	Technical University of Munich, Germany

Hua Wang	Colorado School of Mines, USA
Qian Wang	ShanghaiTech University, China
Shanshan Wang	Paul C. Lauterbur Research Center, SIAT, China
Yalin Wang	Arizona State University, USA
Bryan Williams	Lancaster University, UK
Matthias Wilms	University of Calgary, Canada
Jelmer Wolterink	University of Twente, Netherlands
Ken C. L. Wong	IBM Research Almaden, USA
Jonghye Woo	Massachusetts General Hospital and Harvard Medical School, USA
Shandong Wu	University of Pittsburgh, USA
Yutong Xie	University of Adelaide, Australia
Fuyong Xing	University of Colorado, Denver, USA
Daguang Xu	NVIDIA, USA
Yan Xu	Beihang University, China
Yanwu Xu	Baidu, China
Pingkun Yan	Rensselaer Polytechnic Institute, USA
Guang Yang	Imperial College London, UK
Jianhua Yao	Tencent, China
Chuyang Ye	Beijing Institute of Technology, China
Lequan Yu	University of Hong Kong, China
Ghada Zamzmi	National Institutes of Health, USA
Liang Zhan	University of Pittsburgh, USA
Fan Zhang	Harvard Medical School, USA
Ling Zhang	Alibaba Group, China
Miaomiao Zhang	University of Virginia, USA
Shu Zhang	Northwestern Polytechnical University, China
Rongchang Zhao	Central South University, China
Yitian Zhao	Chinese Academy of Sciences, China
Tao Zhou	Nanjing University of Science and Technology, USA
Yuyin Zhou	UC Santa Cruz, USA
Dajiang Zhu	University of Texas at Arlington, USA
Lei Zhu	ROAS Thrust HKUST (GZ), and ECE HKUST, China
Xiahai Zhuang	Fudan University, China
Veronika Zimmer	Technical University of Munich, Germany

Reviewers

Alaa Eldin Abdclaal
John Abel
Kumar Abhishek
Shahira Abousamra
Mazdak Abulnaga
Burak Acar
Abdoljalil Addeh
Ehsan Adeli
Sukesh Adiga Vasudeva
Seyed-Ahmad Ahmadi
Euijoon Ahn
Faranak Akbarifar
Alireza Akhondi-asl
Saad Ullah Akram
Daniel Alexander
Hanan Alghamdi
Hassan Alhajj
Omar Al-Kadi
Max Allan
Andre Altmann
Pablo Alvarez
Charlems Alvarez-Jimenez
Jennifer Alvén
Lidia Al-Zogbi
Kimberly Amador
Tamaz Amiranashvili
Amine Amyar
Wangpeng An
Vincent Andrearczyk
Manon Ansart
Sameer Antani
Jacob Antunes
Michel Antunes
Guilherme Aresta
Mohammad Ali Armin
Kasra Arnavaz
Corey Arnold
Janan Arslan
Marius Arvinte
Muhammad Asad
John Ashburner
Md Ashikuzzaman
Shahab Aslani

Mehdi Astaraki
Angélica Atehortúa
Benjamin Aubert
Marc Aubreville
Paolo Avesani
Sana Ayromlou
Reza Azad
Mohammad Farid
    Azampour
Qinle Ba
Meritxell Bach Cuadra
Hyeon-Min Bae
Matheus Baffa
Cagla Bahadir
Fan Bai
Jun Bai
Long Bai
Pradeep Bajracharya
Shafa Balaram
Yaël Balbastre
Yutong Ban
Abhirup Banerjee
Soumyanil Banerjee
Sreya Banerjee
Shunxing Bao
Omri Bar
Adrian Barbu
Joao Barreto
Adrian Basarab
Berke Basaran
Michael Baumgartner
Siming Bayer
Roza Bayrak
Aicha BenTaieb
Guy Ben-Yosef
Sutanu Bera
Cosmin Bercea
Jorge Bernal
Jose Bernal
Gabriel Bernardino
Riddhish Bhalodia
Jignesh Bhatt
Indrani Bhattacharya

Binod Bhattarai
Lei Bi
Qi Bi
Cheng Bian
Gui-Bin Bian
Carlo Biffi
Alexander Bigalke
Benjamin Billot
Manuel Birlo
Ryoma Bise
Daniel Blezek
Stefano Blumberg
Sebastian Bodenstedt
Federico Bolelli
Bhushan Borotikar
Ilaria Boscolo Galazzo
Alexandre Bousse
Nicolas Boutry
Joseph Boyd
Behzad Bozorgtabar
Nadia Brancati
Clara Brémond Martin
Stéphanie Bricq
Christopher Bridge
Coleman Broaddus
Rupert Brooks
Tom Brosch
Mikael Brudfors
Ninon Burgos
Nikolay Burlutskiy
Michal Byra
Ryan Cabeen
Mariano Cabezas
Hongmin Cai
Tongan Cai
Zongyou Cai
Liane Canas
Bing Cao
Guogang Cao
Weiguo Cao
Xu Cao
Yankun Cao
Zhenjie Cao

Jaime Cardoso
M. Jorge Cardoso
Owen Carmichael
Jacob Carse
Adrià Casamitjana
Alessandro Casella
Angela Castillo
Kate Cevora
Krishna Chaitanya
Satrajit Chakrabarty
Yi Hao Chan
Shekhar Chandra
Ming-Ching Chang
Peng Chang
Qi Chang
Yuchou Chang
Hanqing Chao
Simon Chatelin
Soumick Chatterjee
Sudhanya Chatterjee
Muhammad Faizyab Ali
    Chaudhary
Antong Chen
Bingzhi Chen
Chen Chen
Cheng Chen
Chengkuan Chen
Eric Chen
Fang Chen
Haomin Chen
Jianan Chen
Jianxu Chen
Jiazhou Chen
Jie Chen
Jintai Chen
Jun Chen
Junxiang Chen
Junyu Chen
Li Chen
Liyun Chen
Nenglun Chen
Pingjun Chen
Pingyi Chen
Qi Chen
Qiang Chen

Runnan Chen
Shengcong Chen
Sihao Chen
Tingting Chen
Wenting Chen
Xi Chen
Xiang Chen
Xiaoran Chen
Xin Chen
Xiongchao Chen
Yanxi Chen
Yixiong Chen
Yixuan Chen
Yuanyuan Chen
Yuqian Chen
Zhaolin Chen
Zhen Chen
Zhenghao Chen
Zhennong Chen
Zhihao Chen
Zhineng Chen
Zhixiang Chen
Chang-Chieh Cheng
Jiale Cheng
Jianhong Cheng
Jun Cheng
Xuelian Cheng
Yupeng Cheng
Mark Chiew
Philip Chikontwe
Eleni Chiou
Jungchan Cho
Jang-Hwan Choi
Min-Kook Choi
Wookjin Choi
Jaegul Choo
Yu-Cheng Chou
Daan Christiaens
Argyrios Christodoulidis
Stergios Christodoulidis
Kai-Cheng Chuang
Hyungjin Chung
Matthew Clarkson
Michaël Clément
Dana Cobzas

Jaume Coll-Font
Olivier Colliot
Runmin Cong
Yulai Cong
Laura Connolly
William Consagra
Pierre-Henri Conze
Tim Cootes
Teresa Correia
Baris Coskunuzer
Alex Crimi
Can Cui
Hejie Cui
Hui Cui
Lei Cui
Wenhui Cui
Tolga Cukur
Tobias Czempiel
Javid Dadashkarimi
Haixing Dai
Tingting Dan
Kang Dang
Salman Ul Hassan Dar
Eleonora D'Arnese
Dhritiman Das
Neda Davoudi
Tareen Dawood
Sandro De Zanet
Farah Deeba
Charles Delahunt
Herve Delingette
Ugur Demir
Liang-Jian Deng
Ruining Deng
Wenlong Deng
Felix Denzinger
Adrien Depeursinge
Mohammad Mahdi
    Derakhshani
Hrishikesh Deshpande
Adrien Desjardins
Christian Desrosiers
Blake Dewey
Neel Dey
Rohan Dhamdhere

Maxime Di Folco
Songhui Diao
Alina Dima
Hao Ding
Li Ding
Ying Ding
Zhipeng Ding
Nicola Dinsdale
Konstantin Dmitriev
Ines Domingues
Bo Dong
Liang Dong
Nanqing Dong
Siyuan Dong
Reuben Dorent
Gianfranco Doretto
Sven Dorkenwald
Haoran Dou
Mitchell Doughty
Jason Dowling
Niharika D'Souza
Guodong Du
Jie Du
Shiyi Du
Hongyi Duanmu
Benoit Dufumier
James Duncan
Joshua Durso-Finley
Dmitry V. Dylov
Oleh Dzyubachyk
Mahdi (Elias) Ebnali
Philip Edwards
Jan Egger
Gudmundur Einarsson
Mostafa El Habib Daho
Ahmed Elazab
Idris El-Feghi
David Ellis
Mohammed Elmogy
Amr Elsawy
Okyaz Eminaga
Ertunc Erdil
Lauren Erdman
Marius Erdt
Maria Escobar

Hooman Esfandiari
Nazila Esmaeili
Ivan Ezhov
Alessio Fagioli
Deng-Ping Fan
Lei Fan
Xin Fan
Yubo Fan
Huihui Fang
Jiansheng Fang
Xi Fang
Zhenghan Fang
Mohammad Farazi
Azade Farshad
Mohsen Farzi
Hamid Fehri
Lina Felsner
Chaolu Feng
Chun-Mei Feng
Jianjiang Feng
Mengling Feng
Ruibin Feng
Zishun Feng
Alvaro Fernandez-Quilez
Ricardo Ferrari
Lucas Fidon
Lukas Fischer
Madalina Fiterau
Antonio
    Foncubierta-Rodríguez
Fahimeh Fooladgar
Germain Forestier
Nils Daniel Forkert
Jean-Rassaire Fouefack
Kevin François-Bouaou
Wolfgang Freysinger
Bianca Freytag
Guanghui Fu
Kexue Fu
Lan Fu
Yunguan Fu
Pedro Furtado
Ryo Furukawa
Jin Kyu Gahm
Mélanie Gaillochet

Francesca Galassi
Jiangzhang Gan
Yu Gan
Yulu Gan
Alireza Ganjdanesh
Chang Gao
Cong Gao
Linlin Gao
Zeyu Gao
Zhongpai Gao
Sara Garbarino
Alain Garcia
Beatriz Garcia Santa Cruz
Rongjun Ge
Shiv Gehlot
Manuela Geiss
Salah Ghamizi
Negin Ghamsarian
Ramtin Gharleghi
Ghazal Ghazaei
Florin Ghesu
Sayan Ghosal
Syed Zulqarnain Gilani
Mahdi Gilany
Yannik Glaser
Ben Glocker
Bharti Goel
Jacob Goldberger
Polina Golland
Alberto Gomez
Catalina Gomez
Estibaliz
    Gómez-de-Mariscal
Haifan Gong
Kuang Gong
Xun Gong
Ricardo Gonzales
Camila Gonzalez
German Gonzalez
Vanessa Gonzalez Duque
Sharath Gopal
Karthik Gopinath
Pietro Gori
Michael Götz
Shuiping Gou

Maged Goubran
Sobhan Goudarzi
Mark Graham
Alejandro Granados
Mara Graziani
Thomas Grenier
Radu Grosu
Michal Grzeszczyk
Feng Gu
Pengfei Gu
Qiangqiang Gu
Ran Gu
Shi Gu
Wenhao Gu
Xianfeng Gu
Yiwen Gu
Zaiwang Gu
Hao Guan
Jayavardhana Gubbi
Houssem-Eddine Gueziri
Dazhou Guo
Hengtao Guo
Jixiang Guo
Jun Guo
Pengfei Guo
Wenzhangzhi Guo
Xiaoqing Guo
Xueqi Guo
Yi Guo
Vikash Gupta
Praveen Gurunath Bharathi
Prashnna Gyawali
Sung Min Ha
Mohamad Habes
Ilker Hacihaliloglu
Stathis Hadjidemetriou
Fatemeh Haghighi
Justin Haldar
Noura Hamze
Liang Han
Luyi Han
Seungjae Han
Tianyu Han
Zhongyi Han
Jonny Hancox

Lasse Hansen
Degan Hao
Huaying Hao
Jinkui Hao
Nazim Haouchine
Michael Hardisty
Stefan Harrer
Jeffry Hartanto
Charles Hatt
Huiguang He
Kelei He
Qi He
Shenghua He
Xinwei He
Stefan Heldmann
Nicholas Heller
Edward Henderson
Alessa Hering
Monica Hernandez
Kilian Hett
Amogh Hiremath
David Ho
Malte Hoffmann
Matthew Holden
Qingqi Hong
Yoonmi Hong
Mohammad Reza
   Hosseinzadeh Taher
William Hsu
Chuanfei Hu
Dan Hu
Kai Hu
Rongyao Hu
Shishuai Hu
Xiaoling Hu
Xinrong Hu
Yan Hu
Yang Hu
Chaoqin Huang
Junzhou Huang
Ling Huang
Luojie Huang
Qinwen Huang
Sharon Xiaolei Huang
Weijian Huang

Xiaoyang Huang
Yi-Jie Huang
Yongsong Huang
Yongxiang Huang
Yuhao Huang
Zhe Huang
Zhi-An Huang
Ziyi Huang
Arnaud Huaulmé
Henkjan Huisman
Alex Hung
Jiayu Huo
Andreas Husch
Mohammad Arafat
   Hussain
Sarfaraz Hussein
Jana Hutter
Khoi Huynh
Ilknur Icke
Kay Igwe
Abdullah Al Zubaer Imran
Muhammad Imran
Samra Irshad
Nahid Ul Islam
Koichi Ito
Hayato Itoh
Yuji Iwahori
Krithika Iyer
Mohammad Jafari
Srikrishna Jaganathan
Hassan Jahanandish
Andras Jakab
Amir Jamaludin
Amoon Jamzad
Ananya Jana
Se-In Jang
Pierre Jannin
Vincent Jaouen
Uditha Jarayathne
Ronnachai Jaroensri
Guillaume Jaume
Syed Ashar Javed
Rachid Jennane
Debesh Jha
Ge-Peng Ji

Luping Ji
Zexuan Ji
Zhanghexuan Ji
Haozhe Jia
Hongchao Jiang
Jue Jiang
Meirui Jiang
Tingting Jiang
Xiajun Jiang
Zekun Jiang
Zhifan Jiang
Ziyu Jiang
Jianbo Jiao
Zhicheng Jiao
Chen Jin
Dakai Jin
Qiangguo Jin
Qiuye Jin
Weina Jin
Baoyu Jing
Bin Jing
Yaqub Jonmohamadi
Lie Ju
Yohan Jun
Dinkar Juyal
Manjunath K N
Ali Kafaei Zad Tehrani
John Kalafut
Niveditha Kalavakonda
Megha Kalia
Anil Kamat
Qingbo Kang
Po-Yu Kao
Anuradha Kar
Neerav Karani
Turkay Kart
Satyananda Kashyap
Alexander Katzmann
Lisa Kausch
Maxime Kayser
Salome Kazeminia
Wenchi Ke
Youngwook Kee
Matthias Keicher
Erwan Kerrien

Afifa Khaled
Nadieh Khalili
Farzad Khalvati
Bidur Khanal
Bishesh Khanal
Pulkit Khandelwal
Maksim Kholiavchenko
Ron Kikinis
Benjamin Killeen
Daeseung Kim
Heejong Kim
Jaeil Kim
Jinhee Kim
Jinman Kim
Junsik Kim
Minkyung Kim
Namkug Kim
Sangwook Kim
Tae Soo Kim
Younghoon Kim
Young-Min Kim
Andrew King
Miranda Kirby
Gabriel Kiss
Andreas Kist
Yoshiro Kitamura
Stefan Klein
Tobias Klinder
Kazuma Kobayashi
Lisa Koch
Satoshi Kondo
Fanwei Kong
Tomasz Konopczynski
Ender Konukoglu
Aishik Konwer
Thijs Kooi
Ivica Kopriva
Avinash Kori
Kivanc Kose
Suraj Kothawade
Anna Kreshuk
AnithaPriya Krishnan
Florian Kromp
Frithjof Kruggel
Thomas Kuestner

Levin Kuhlmann
Abhay Kumar
Kuldeep Kumar
Sayantan Kumar
Manuela Kunz
Holger Kunze
Tahsin Kurc
Anvar Kurmukov
Yoshihiro Kuroda
Yusuke Kurose
Hyuksool Kwon
Aymen Laadhari
Jorma Laaksonen
Dmitrii Lachinov
Alain Lalande
Rodney LaLonde
Bennett Landman
Daniel Lang
Carole Lartizien
Shlomi Laufer
Max-Heinrich Laves
William Le
Loic Le Folgoc
Christian Ledig
Eung-Joo Lee
Ho Hin Lee
Hyekyoung Lee
John Lee
Kisuk Lee
Kyungsu Lee
Soochahn Lee
Woonghee Lee
Étienne Léger
Wen Hui Lei
Yiming Lei
George Leifman
Rogers Jeffrey Leo John
Juan Leon
Bo Li
Caizi Li
Chao Li
Chen Li
Cheng Li
Chenxin Li
Chnegyin Li

Dawei Li
Fuhai Li
Gang Li
Guang Li
Hao Li
Haofeng Li
Haojia Li
Heng Li
Hongming Li
Hongwei Li
Huiqi Li
Jian Li
Jieyu Li
Kang Li
Lin Li
Mengzhang Li
Ming Li
Qing Li
Quanzheng Li
Shaohua Li
Shulong Li
Tengfei Li
Weijian Li
Wen Li
Xiaomeng Li
Xingyu Li
Xinhui Li
Xuelu Li
Xueshen Li
Yamin Li
Yang Li
Yi Li
Yuemeng Li
Yunxiang Li
Zeju Li
Zhaoshuo Li
Zhe Li
Zhen Li
Zhenqiang Li
Zhiyuan Li
Zhjin Li
Zi Li
Hao Liang
Libin Liang
Peixian Liang

Yuan Liang
Yudong Liang
Haofu Liao
Hongen Liao
Wei Liao
Zehui Liao
Gilbert Lim
Hongxiang Lin
Li Lin
Manxi Lin
Mingquan Lin
Tiancheng Lin
Yi Lin
Zudi Lin
Claudia Lindner
Simone Lionetti
Chi Liu
Chuanbin Liu
Daochang Liu
Dongnan Liu
Feihong Liu
Fenglin Liu
Han Liu
Huiye Liu
Jiang Liu
Jie Liu
Jinduo Liu
Jing Liu
Jingya Liu
Jundong Liu
Lihao Liu
Mengting Liu
Mingyuan Liu
Peirong Liu
Peng Liu
Qin Liu
Quan Liu
Rui Liu
Shengfeng Liu
Shuangjun Liu
Sidong Liu
Siyuan Liu
Weide Liu
Xiao Liu
Xiaoyu Liu

Xingtong Liu
Xinwen Liu
Xinyang Liu
Xinyu Liu
Yan Liu
Yi Liu
Yihao Liu
Yikang Liu
Yilin Liu
Yilong Liu
Yiqiao Liu
Yong Liu
Yuhang Liu
Zelong Liu
Zhe Liu
Zhiyuan Liu
Zuozhu Liu
Lisette Lockhart
Andrea Loddo
Nicolas Loménie
Yonghao Long
Daniel Lopes
Ange Lou
Brian Lovell
Nicolas Loy Rodas
Charles Lu
Chun-Shien Lu
Donghuan Lu
Guangming Lu
Huanxiang Lu
Jingpei Lu
Yao Lu
Oeslle Lucena
Jie Luo
Luyang Luo
Ma Luo
Mingyuan Luo
Wenhan Luo
Xiangde Luo
Xinzhe Luo
Jinxin Lv
Tianxu Lv
Fei Lyu
Ilwoo Lyu
Mengye Lyu

Qing Lyu
Yanjun Lyu
Yuanyuan Lyu
Benteng Ma
Chunwei Ma
Hehuan Ma
Jun Ma
Junbo Ma
Wenao Ma
Yuhui Ma
Pedro Macias Gordaliza
Anant Madabhushi
Derek Magee
S. Sara Mahdavi
Andreas Maier
Klaus H. Maier-Hein
Sokratis Makrogiannis
Danial Maleki
Michail Mamalakis
Zhehua Mao
Jan Margeta
Brett Marinelli
Zdravko Marinov
Viktoria Markova
Carsten Marr
Yassine Marrakchi
Anne Martel
Martin Maška
Tejas Sudharshan Mathai
Petr Matula
Dimitrios Mavroeidis
Evangelos Mazomenos
Amarachi Mbakwe
Adam McCarthy
Stephen McKenna
Raghav Mehta
Xueyan Mei
Felix Meissen
Felix Meister
Afaque Memon
Mingyuan Meng
Qingjie Meng
Xiangzhu Meng
Yanda Meng
Zhu Meng

Martin Menten
Odyssée Merveille
Mikhail Milchenko
Leo Milecki
Fausto Milletari
Hyun-Seok Min
Zhe Min
Song Ming
Duy Minh Ho Nguyen
Deepak Mishra
Suraj Mishra
Virendra Mishra
Tadashi Miyamoto
Sara Moccia
Marc Modat
Omid Mohareri
Tony C. W. Mok
Javier Montoya
Rodrigo Moreno
Stefano Moriconi
Lia Morra
Ana Mota
Lei Mou
Dana Moukheiber
Lama Moukheiber
Daniel Moyer
Pritam Mukherjee
Anirban Mukhopadhyay
Henning Müller
Ana Murillo
Gowtham Krishnan
    Murugesan
Ahmed Naglah
Karthik Nandakumar
Venkatesh
    Narasimhamurthy
Raja Narayan
Dominik Narnhofer
Vishwesh Nath
Rodrigo Nava
Abdullah Nazib
Ahmed Nebli
Peter Neher
Amin Nejatbakhsh
Trong-Thuan Nguyen

Truong Nguyen
Dong Ni
Haomiao Ni
Xiuyan Ni
Hannes Nickisch
Weizhi Nie
Aditya Nigam
Lipeng Ning
Xia Ning
Kazuya Nishimura
Chuang Niu
Sijie Niu
Vincent Noblet
Narges Norouzi
Alexey Novikov
Jorge Novo
Gilberto Ochoa-Ruiz
Masahiro Oda
Benjamin Odry
Hugo Oliveira
Sara Oliveira
Arnau Oliver
Jimena Olveres
John Onofrey
Marcos Ortega
Mauricio Alberto
    Ortega-Ruíz
Yusuf Osmanlioglu
Chubin Ou
Cheng Ouyang
Jiahong Ouyang
Xi Ouyang
Cristina Oyarzun Laura
Utku Ozbulak
Ece Ozkan
Ege Özsoy
Batu Ozturkler
Harshith Padigela
Johannes Paetzold
José Blas Pagador
    Carrasco
Daniel Pak
Sourabh Palande
Chengwei Pan
Jiazhen Pan

Jin Pan
Yongsheng Pan
Egor Panfilov
Jiaxuan Pang
Joao Papa
Constantin Pape
Bartlomiej Papiez
Nripesh Parajuli
Hyunjin Park
Akash Parvatikar
Tiziano Passerini
Diego Patiño Cortés
Mayank Patwari
Angshuman Paul
Rasmus Paulsen
Yuchen Pei
Yuru Pei
Tao Peng
Wei Peng
Yige Peng
Yunsong Peng
Matteo Pennisi
Antonio Pepe
Oscar Perdomo
Sérgio Pereira
Jose-Antonio
    Pérez-Carrasco
Mehran Pesteie
Terry Peters
Eike Petersen
Jens Petersen
Micha Pfeiffer
Dzung Pham
Hieu Pham
Ashish Phophalia
Tomasz Pieciak
Antonio Pinheiro
Pramod Pisharady
Theodoros Pissas
Szymon Płotka
Kilian Pohl
Sebastian Pölsterl
Alison Pouch
Tim Prangemeier
Prateek Prasanna

Raphael Prevost
Juan Prieto
Federica Proietto Salanitri
Sergi Pujades
Elodie Puybareau
Talha Qaiser
Buyue Qian
Mengyun Qiao
Yuchuan Qiao
Zhi Qiao
Chenchen Qin
Fangbo Qin
Wenjian Qin
Yulei Qin
Jie Qiu
Jielin Qiu
Peijie Qiu
Shi Qiu
Wu Qiu
Liangqiong Qu
Linhao Qu
Quan Quan
Tran Minh Quan
Sandro Queirós
Prashanth R
Febrian Rachmadi
Daniel Racoceanu
Mehdi Rahim
Jagath Rajapakse
Kashif Rajpoot
Keerthi Ram
Dhanesh Ramachandram
João Ramalhinho
Xuming Ran
Aneesh Rangnekar
Hatem Rashwan
Keerthi Sravan Ravi
Daniele Ravì
Sadhana Ravikumar
Harish Raviprakash
Surreerat Reaungamornrat
Samuel Remedios
Mengwei Ren
Sucheng Ren
Elton Rexhepaj

Mauricio Reyes
Constantino
    Reyes-Aldasoro
Abel Reyes-Angulo
Hadrien Reynaud
Razieh Rezaei
Anne-Marie Rickmann
Laurent Risser
Dominik Rivoir
Emma Robinson
Robert Robinson
Jessica Rodgers
Ranga Rodrigo
Rafael Rodrigues
Robert Rohling
Margherita Rosnati
Łukasz Roszkowiak
Holger Roth
José Rouco
Dan Ruan
Jiacheng Ruan
Daniel Rueckert
Danny Ruijters
Kanghyun Ryu
Ario Sadafi
Numan Saeed
Monjoy Saha
Pramit Saha
Farhang Sahba
Pranjal Sahu
Simone Saitta
Md Sirajus Salekin
Abbas Samani
Pedro Sanchez
Luis Sanchez Giraldo
Yudi Sang
Gerard Sanroma-Guell
Rodrigo Santa Cruz
Alice Santilli
Rachana Sathish
Olivier Saut
Mattia Savardi
Nico Scherf
Alexander Schlaefer
Jerome Schmid

Adam Schmidt
Julia Schnabel
Lawrence Schobs
Julian Schön
Peter Schueffler
Andreas Schuh
Christina
  Schwarz-Gsaxner
Michaël Sdika
Suman Sedai
Lalithkumar Seenivasan
Matthias Seibold
Sourya Sengupta
Lama Seoud
Ana Sequeira
Sharmishtaa Seshamani
Ahmed Shaffie
Jay Shah
Keyur Shah
Ahmed Shahin
Mohammad Abuzar
  Shaikh
S. Shailja
Hongming Shan
Wei Shao
Mostafa Sharifzadeh
Anuja Sharma
Gregory Sharp
Hailan Shen
Li Shen
Linlin Shen
Mali Shen
Mingren Shen
Yiqing Shen
Zhengyang Shen
Jun Shi
Xiaoshuang Shi
Yiyu Shi
Yonggang Shi
Hoo-Chang Shin
Jitae Shin
Keewon Shin
Boris Shirokikh
Suzanne Shontz
Yucheng Shu

Hanna Siebert
Alberto Signoroni
Wilson Silva
Julio Silva-Rodríguez
Margarida Silveira
Walter Simson
Praveer Singh
Vivek Singh
Nitin Singhal
Elena Sizikova
Gregory Slabaugh
Dane Smith
Kevin Smith
Tiffany So
Rajath Soans
Roger Soberanis-Mukul
Hessam Sokooti
Jingwei Song
Weinan Song
Xinhang Song
Xinrui Song
Mazen Soufi
Georgia Sovatzidi
Bella Specktor Fadida
William Speier
Ziga Spiclin
Dominik Spinczyk
Jon Sporring
Pradeeba Sridar
Chetan L. Srinidhi
Abhishek Srivastava
Lawrence Staib
Marc Stamminger
Justin Strait
Hai Su
Ruisheng Su
Zhe Su
Vaishnavi Subramanian
Gérard Subsol
Carole Sudre
Dong Sui
Heung-Il Suk
Shipra Suman
He Sun
Hongfu Sun

Jian Sun
Li Sun
Liyan Sun
Shanlin Sun
Kyung Sung
Yannick Suter
Swapna T. R.
Amir Tahmasebi
Pablo Tahoces
Sirine Taleb
Bingyao Tan
Chaowei Tan
Wenjun Tan
Hao Tang
Siyi Tang
Xiaoying Tang
Yucheng Tang
Zihao Tang
Michael Tanzer
Austin Tapp
Elias Tappeiner
Mickael Tardy
Giacomo Tarroni
Athena Taymourtash
Kaveri Thakoor
Elina Thibeau-Sutre
Paul Thienphrapa
Sarina Thomas
Stephen Thompson
Karl Thurnhofer-Hemsi
Cristiana Tiago
Lin Tian
Lixia Tian
Yapeng Tian
Yu Tian
Yun Tian
Aleksei Tiulpin
Hamid Tizhoosh
Minh Nguyen Nhat To
Matthew Toews
Maryam Toloubidokhti
Minh Tran
Quoc-Huy Trinh
Jocelyne Troccaz
Roger Trullo

Chialing Tsai
Apostolia Tsirikoglou
Puxun Tu
Samyakh Tukra
Sudhakar Tummala
Georgios Tziritas
Vladimír Ulman
Tamas Ungi
Régis Vaillant
Jeya Maria Jose Valanarasu
Vanya Valindria
Juan Miguel Valverde
Fons van der Sommen
Maureen van Eijnatten
Tom van Sonsbeek
Gijs van Tulder
Yogatheesan Varatharajah
Madhurima Vardhan
Thomas Varsavsky
Hooman Vaseli
Serge Vasylechko
S. Swaroop Vedula
Sanketh Vedula
Gonzalo Vegas
 Sanchez-Ferrero
Matthew Velazquez
Archana Venkataraman
Sulaiman Vesal
Mitko Veta
Barbara Villarini
Athanasios Vlontzos
Wolf-Dieter Vogl
Ingmar Voigt
Sandrine Voros
Vibashan VS
Trinh Thi Le Vuong
An Wang
Bo Wang
Ce Wang
Changmiao Wang
Ching-Wei Wang
Dadong Wang
Dong Wang
Fakai Wang
Guotai Wang

Haifeng Wang
Haoran Wang
Hong Wang
Hongxiao Wang
Hongyu Wang
Jiacheng Wang
Jing Wang
Jue Wang
Kang Wang
Ke Wang
Lei Wang
Li Wang
Liansheng Wang
Lin Wang
Ling Wang
Linwei Wang
Manning Wang
Mingliang Wang
Puyang Wang
Qiuli Wang
Renzhen Wang
Ruixuan Wang
Shaoyu Wang
Sheng Wang
Shujun Wang
Shuo Wang
Shuqiang Wang
Tao Wang
Tianchen Wang
Tianyu Wang
Wenzhe Wang
Xi Wang
Xiangdong Wang
Xiaoqing Wang
Xiaosong Wang
Yan Wang
Yangang Wang
Yaping Wang
Yi Wang
Yirui Wang
Yixin Wang
Zeyi Wang
Zhao Wang
Zichen Wang
Ziqin Wang

Ziyi Wang
Zuhui Wang
Dong Wei
Donglai Wei
Hao Wei
Jia Wei
Leihao Wei
Ruofeng Wei
Shuwen Wei
Martin Weigert
Wolfgang Wein
Michael Wels
Cédric Wemmert
Thomas Wendler
Markus Wenzel
Rhydian Windsor
Adam Wittek
Marek Wodzinski
Ivo Wolf
Julia Wolleb
Ka-Chun Wong
Jonghye Woo
Chongruo Wu
Chunpeng Wu
Fuping Wu
Huaqian Wu
Ji Wu
Jiangjie Wu
Jiong Wu
Junde Wu
Linshan Wu
Qing Wu
Weiwen Wu
Wenjun Wu
Xiyin Wu
Yawen Wu
Ye Wu
Yicheng Wu
Yongfei Wu
Zhengwang Wu
Pengcheng Xi
Chao Xia
Siyu Xia
Wenjun Xia
Lei Xiang

Tiange Xiang
Deqiang Xiao
Li Xiao
Xiaojiao Xiao
Yiming Xiao
Zeyu Xiao
Hongtao Xie
Huidong Xie
Jianyang Xie
Long Xie
Weidi Xie
Fangxu Xing
Shuwei Xing
Xiaodan Xing
Xiaohan Xing
Haoyi Xiong
Yujian Xiong
Di Xu
Feng Xu
Haozheng Xu
Hongming Xu
Jiangchang Xu
Jiaqi Xu
Junshen Xu
Kele Xu
Lijian Xu
Min Xu
Moucheng Xu
Rui Xu
Xiaowei Xu
Xuanang Xu
Yanwu Xu
Yanyu Xu
Yongchao Xu
Yunqiu Xu
Zhe Xu
Zhoubing Xu
Ziyue Xu
Kai Xuan
Cheng Xue
Jie Xue
Tengfei Xue
Wufeng Xue
Yuan Xue
Zhong Xue

Ts Faridah Yahya
Chaochao Yan
Jiangpeng Yan
Ming Yan
Qingsen Yan
Xiangyi Yan
Yuguang Yan
Zengqiang Yan
Baoyao Yang
Carl Yang
Changchun Yang
Chen Yang
Feng Yang
Fengting Yang
Ge Yang
Guanyu Yang
Heran Yang
Huijuan Yang
Jiancheng Yang
Jiewen Yang
Peng Yang
Qi Yang
Qiushi Yang
Wei Yang
Xin Yang
Xuan Yang
Yan Yang
Yanwu Yang
Yifan Yang
Yingyu Yang
Zhicheng Yang
Zhijian Yang
Jiangchao Yao
Jiawen Yao
Lanhong Yao
Linlin Yao
Qingsong Yao
Tianyuan Yao
Xiaohui Yao
Zhao Yao
Dong Hye Ye
Menglong Ye
Yousef Yeganeh
Jirong Yi
Xin Yi

Chong Yin
Pengshuai Yin
Yi Yin
Zhaozheng Yin
Chunwei Ying
Youngjin Yoo
Jihun Yoon
Chenyu You
Hanchao Yu
Heng Yu
Jinhua Yu
Jinze Yu
Ke Yu
Qi Yu
Qian Yu
Thomas Yu
Weimin Yu
Yang Yu
Chenxi Yuan
Kun Yuan
Wu Yuan
Yixuan Yuan
Paul Yushkevich
Fatemeh Zabihollahy
Samira Zare
Ramy Zeineldin
Dong Zeng
Qi Zeng
Tianyi Zeng
Wei Zeng
Kilian Zepf
Kun Zhan
Bokai Zhang
Daoqiang Zhang
Dong Zhang
Fa Zhang
Hang Zhang
Hanxiao Zhang
Hao Zhang
Haopeng Zhang
Haoyue Zhang
Hongrun Zhang
Jiadong Zhang
Jiajin Zhang
Jianpeng Zhang

Jiawei Zhang
Jingqing Zhang
Jingyang Zhang
Jinwei Zhang
Jiong Zhang
Jiping Zhang
Ke Zhang
Lefei Zhang
Lei Zhang
Li Zhang
Lichi Zhang
Lu Zhang
Minghui Zhang
Molin Zhang
Ning Zhang
Rongzhao Zhang
Ruipeng Zhang
Ruisi Zhang
Shichuan Zhang
Shihao Zhang
Shuai Zhang
Tuo Zhang
Wei Zhang
Weihang Zhang
Wen Zhang
Wenhua Zhang
Wenqiang Zhang
Xiaodan Zhang
Xiaoran Zhang
Xin Zhang
Xukun Zhang
Xuzhe Zhang
Ya Zhang
Yanbo Zhang
Yanfu Zhang
Yao Zhang
Yi Zhang
Yifan Zhang
Yixiao Zhang
Yongqin Zhang
You Zhang
Youshan Zhang

Yu Zhang
Yubo Zhang
Yue Zhang
Yuhan Zhang
Yulun Zhang
Yundong Zhang
Yunlong Zhang
Yuyao Zhang
Zheng Zhang
Zhenxi Zhang
Ziqi Zhang
Can Zhao
Chongyue Zhao
Fenqiang Zhao
Gangming Zhao
He Zhao
Jianfeng Zhao
Jun Zhao
Li Zhao
Liang Zhao
Lin Zhao
Mengliu Zhao
Mingbo Zhao
Qingyu Zhao
Shang Zhao
Shijie Zhao
Tengda Zhao
Tianyi Zhao
Wei Zhao
Yidong Zhao
Yiyuan Zhao
Yu Zhao
Zhihe Zhao
Ziyuan Zhao
Haiyong Zheng
Hao Zheng
Jiannan Zheng
Kang Zheng
Meng Zheng
Sisi Zheng
Tianshu Zheng
Yalin Zheng

Yefeng Zheng
Yinqiang Zheng
Yushan Zheng
Aoxiao Zhong
Jia-Xing Zhong
Tao Zhong
Zichun Zhong
Hong-Yu Zhou
Houliang Zhou
Huiyu Zhou
Kang Zhou
Qin Zhou
Ran Zhou
S. Kevin Zhou
Tianfei Zhou
Wei Zhou
Xiao-Hu Zhou
Xiao-Yun Zhou
Yi Zhou
Youjia Zhou
Yukun Zhou
Zongwei Zhou
Chenglu Zhu
Dongxiao Zhu
Heqin Zhu
Jiayi Zhu
Meilu Zhu
Wei Zhu
Wenhui Zhu
Xiaofeng Zhu
Xin Zhu
Yonghua Zhu
Yongpei Zhu
Yuemin Zhu
Yan Zhuang
David Zimmerer
Yongshuo Zong
Ke Zou
Yukai Zou
Lianrui Zuo
Gerald Zwettler

Outstanding Area Chairs

Mingxia Liu	University of North Carolina at Chapel Hill, USA
Matthias Wilms	University of Calgary, Canada
Veronika Zimmer	Technical University Munich, Germany

Outstanding Reviewers

Kimberly Amador	University of Calgary, Canada
Angela Castillo	Universidad de los Andes, Colombia
Chen Chen	Imperial College London, UK
Laura Connolly	Queen's University, Canada
Pierre-Henri Conze	IMT Atlantique, France
Niharika D'Souza	IBM Research, USA
Michael Götz	University Hospital Ulm, Germany
Meirui Jiang	Chinese University of Hong Kong, China
Manuela Kunz	National Research Council Canada, Canada
Zdravko Marinov	Karlsruhe Institute of Technology, Germany
Sérgio Pereira	Lunit, South Korea
Lalithkumar Seenivasan	National University of Singapore, Singapore

Honorable Mentions (Reviewers)

Kumar Abhishek	Simon Fraser University, Canada
Guilherme Aresta	Medical University of Vienna, Austria
Shahab Aslani	University College London, UK
Marc Aubreville	Technische Hochschule Ingolstadt, Germany
Yaël Balbastre	Massachusetts General Hospital, USA
Omri Bar	Theator, Israel
Aicha Ben Taieb	Simon Fraser University, Canada
Cosmin Bercea	Technical University Munich and Helmholtz AI and Helmholtz Center Munich, Germany
Benjamin Billot	Massachusetts Institute of Technology, USA
Michal Byra	RIKEN Center for Brain Science, Japan
Mariano Cabezas	University of Sydney, Australia
Alessandro Casella	Italian Institute of Technology and Politecnico di Milano, Italy
Junyu Chen	Johns Hopkins University, USA
Argyrios Christodoulidis	Pfizer, Greece
Olivier Colliot	CNRS, France

Lei Cui	Northwest University, China
Neel Dey	Massachusetts Institute of Technology, USA
Alessio Fagioli	Sapienza University, Italy
Yannik Glaser	University of Hawaii at Manoa, USA
Haifan Gong	Chinese University of Hong Kong, Shenzhen, China
Ricardo Gonzales	University of Oxford, UK
Sobhan Goudarzi	Sunnybrook Research Institute, Canada
Michal Grzeszczyk	Sano Centre for Computational Medicine, Poland
Fatemeh Haghighi	Arizona State University, USA
Edward Henderson	University of Manchester, UK
Qingqi Hong	Xiamen University, China
Mohammad R. H. Taher	Arizona State University, USA
Henkjan Huisman	Radboud University Medical Center, the Netherlands
Ronnachai Jaroensri	Google, USA
Qiangguo Jin	Northwestern Polytechnical University, China
Neerav Karani	Massachusetts Institute of Technology, USA
Benjamin Killeen	Johns Hopkins University, USA
Daniel Lang	Helmholtz Center Munich, Germany
Max-Heinrich Laves	Philips Research and ImFusion GmbH, Germany
Gilbert Lim	SingHealth, Singapore
Mingquan Lin	Weill Cornell Medicine, USA
Charles Lu	Massachusetts Institute of Technology, USA
Yuhui Ma	Chinese Academy of Sciences, China
Tejas Sudharshan Mathai	National Institutes of Health, USA
Felix Meissen	Technische Universität München, Germany
Mingyuan Mcng	University of Sydney, Australia
Leo Milecki	CentraleSupelec, France
Marc Modat	King's College London, UK
Tiziano Passerini	Siemens Healthineers, USA
Tomasz Pieciak	Universidad de Valladolid, Spain
Daniel Rueckert	Imperial College London, UK
Julio Silva-Rodríguez	ETS Montreal, Canada
Bingyao Tan	Nanyang Technological University, Singapore
Elias Tappeiner	UMIT - Private University for Health Sciences, Medical Informatics and Technology, Austria
Jocelyne Troccaz	TIMC Lab, Grenoble Alpes University-CNRS, France
Chialing Tsai	Queens College, City University New York, USA
Juan Miguel Valverde	University of Eastern Finland, Finland
Sulaiman Vesal	Stanford University, USA

Wolf-Dieter Vogl	RetInSight GmbH, Austria
Vibashan VS	Johns Hopkins University, USA
Lin Wang	Harbin Engineering University, China
Yan Wang	Sichuan University, China
Rhydian Windsor	University of Oxford, UK
Ivo Wolf	University of Applied Sciences Mannheim, Germany
Linshan Wu	Hunan University, China
Xin Yang	Chinese University of Hong Kong, China

Contents – Part VII

Clinical Applications – Dermatology

Clinical Applications – Fetal Imaging

Clinical Applications – Oncology

Clinical Applications – Ophthalmology

Clinical Applications – Vascular

Clinical Applications – Abdomen

Source-Free Domain Adaptation for Medical Image Segmentation via Prototype-Anchored Feature Alignment and Contrastive Learning

Qinji Yu[1], Nan Xi[2], Junsong Yuan[2], Ziyu Zhou[1], Kang Dang[3],
and Xiaowei Ding[1,3(✉)]

[1] Shanghai Jiao Tong University, Shanghai, China
dingxiaowei@sjtu.edu.cn
[2] State University of New York at Buffalo, New York, USA
[3] VoxelCloud, Inc., Los Angeles, USA

Abstract. Unsupervised domain adaptation (UDA) has increasingly gained interests for its capacity to transfer the knowledge learned from a labeled source domain to an unlabeled target domain. However, typical UDA methods require concurrent access to both the source and target domain data, which largely limits its application in medical scenarios where source data is often unavailable due to privacy concern. To tackle the source data-absent problem, we present a novel two-stage source-free domain adaptation (SFDA) framework for medical image segmentation, where only a well-trained source segmentation model and unlabeled target data are available during domain adaptation. Specifically, in the prototype-anchored feature alignment stage, we first utilize the weights of the pre-trained pixel-wise classifier as source prototypes, which preserve the information of source features. Then, we introduce the bi-directional transport to align the target features with class prototypes by minimizing its expected cost. On top of that, a contrastive learning stage is further devised to utilize those pixels with unreliable predictions for a more compact target feature distribution. Extensive experiments on a cross-modality medical segmentation task demonstrate the superiority of our method in large domain discrepancy settings compared with the state-of-the-art SFDA approaches and even some UDA methods. Code is available at: https://github.com/CSCYQJ/MICCAI23-ProtoContra-SFDA.

Keywords: Domain adaptation · Source-free · Contrastive learning

---

K. Dang—Co-Corresponding Author.

Supplementary Information The online version contains supplementary material available at https://doi.org/10.1007/978-3-031-43990-2_1.

1   Introduction

The inception of deep neural networks has revolutionized the landscape of medical image segmentation [14,24]. This tremendous success, however, is conditioned on the assumption that the training and testing data are drawn from the same distribution. Unfortunately, in real-world clinical scenarios, due to different acquisition protocols or various imaging modalities, domain shift is widespread between training (*i.e.*, source domain) and testing (*i.e.*, target domain) datasets [15]. This distribution gap usually degenerates the model performance on the target domain. To achieve reliable performance across different domains, a straightforward way is manually labeling some target data and fine-tuning the pretrained model on them [13]. However, obtaining expert-level annotation data in the medical imaging domain incurs significant time and expense [22]. Recently, unsupervised domain adaptation (UDA) has been widely investigated to reduce domain gap through transferring the knowledge learned from a rich-labeled source domain to an unlabeled target domain [4,7,17,19]. Existing UDA methods typically require sharing source data during adaptation, and enforce distribution alignment to diminish the domain discrepancy between source and target domains. This requirement limits the application of UDA methods when source domain data are not accessible. Hence, some very recent works have started to explore a more practical setting, source-free domain adaptation (SFDA), that adapts a pre-trained source model to unlabeled target domains without accessing any source data [1,5,6,12,20,21].

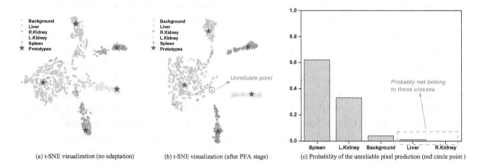

(a) t-SNE visualization (no adaptation)   (b) t-SNE visualization (after PFA stage)   (c) Probability of the unreliable pixel prediction (red circle point)

Fig. 1. (a→b) t-SNE visualization of target feature distributions in embedding space before and after Prototype-anchored Feature Alignment (PFA). (c) Category-wise probability of the unreliable pixel in (b).

Among these methods, [5] and [20] focus on generating reliable pseudo labels for target domain data by developing various denoising strategies. Unavoidably, these self-training methods depends heavily on initial probability maps produced by the source model, which are considerably unreliable when the domain discrepancy is large (*e.g.*, CT and MRI). To relieve the issues caused by noisy pseudo labels, Bateson et al. [1] proposed a prior-aware entropy minimization method to minimize the label-free entropy loss for target predictions. Furthermore, unlike the above self-adaption methods, Yang et al. [21] utilized the statistic information stored in the batch normalization layer of the source model and mutual Fourier Transform to synthesize the source-like image. However, the quality of the generated image is still influenced by the domain discrepancy.

In this work, we propose a novel SFDA framework for cross-modality medical image segmentation. Our framework contains two sequentially conducted stages, *i.e.*, Prototype-anchored Feature Alignment (PFA) stage and Contrastive Learning (CL) stage. As previous works [12] noted, the weights of the pre-trained classifier (*i.e.*, projection head) can be employed as the source prototypes during domain adaptation. That means we can characterize the features of each class with a source prototype and align the target features with them instead of the inaccessible source features. To that end, during the PFA stage, we first provide a target-to-prototype transport to ensure the target features get close to the corresponding prototypes. Then, considering the trivial solution that all target features are assigned to the dominant class prototype (*e.g.*, background), we add a reverse prototype-to-target transport to encourage diversity. However, although most target features have been assigned to the correct class prototype after PFA, some hard samples with high prediction uncertainty still exist in the decision boundary (see Fig. 1(a→b)). Moreover, we observe that those unreliable predictions usually get confused among only a few classes instead of all classes [18]. Taking the unreliable pixel in Fig. 1(b, c) for example, though it achieves similar high probabilities on the spleen and left kidney, the model is pretty sure about this pixel not belonging to the liver and right kidney. Inspired by this, we use confusing pixels as the negative samples for those unlikely classes, and then introduce the CL stage to pursue a more compact target feature distribution. Finally, we conduct experiments on a cross-modality abdominal multi-organ segmentation task. With only a source model and unlabeled target data, our method outperforms the state-of-the-art SFDA and even achieves comparable results with some classical UDA approaches.

2 Methods

We are first provided a segmentation model $\mathcal{M}^s$ trained on N_s labeled samples $\{(x_n^s, y_n^s)\}_{n=1}^{N_s}$ from the source domain $\mathcal{D}^s$, and an unlabeled dataset with N_t samples $\{x_m^t\}_{m=1}^{N_t}$ from the target domain $\mathcal{D}^t$, where $x^s, x^t \in \mathbb{R}^{H \times W \times D}$, $y_n^s \in \mathbb{R}^{H \times W}$, H and W are the height and width of the samples. The goal of SFDA is to adapt the source model $\mathcal{M}^s$ with only unlabeled x^t to predict pixel-wise label y^t for the target domain data. In general, the segmentation model consists of two parts: 1) a feature extractor $F_\theta : x_i \to f_i \in \mathbb{R}^{D_f}$, parameterized by θ, mapping each pixel $i \in \{1, \cdots, H \times W\}$ in image x to the feature f_i in the embedding space; 2) a one-layer pixel-wise classifier $\phi : f_i \to p_i \in \mathbb{R}^C$, that projects pixel feature into the semantic label space with C classes.

In the SFDA task, the source classifier ϕ^s encounters a domain shift problem when classifying the target domain feature. To tackle this challenge, we propose a novel SFDA framework mainly including two stages, shown in Fig. 2. We will elaborate on the details in the following.

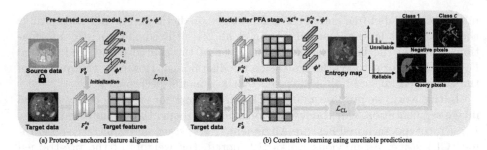

(a) Prototype-anchored feature alignment   (b) Contrastive learning using unreliable predictions

Fig. 2. An overview of the proposed two-stage SFDA framework. (a) is the first PFA stage. We freeze the classifier ϕ^s and use its weights for prototype-anchored feature alignment. (b) is the following CL stage. Given a target image, we first use $\mathcal{M}^{t_0}$ to make a prediction, and separate the pixels into query one and negative ones for each class based on their reliability (entropy). Then, features of query pixels come from F_θ^t (query samples), while features of negative pixels are from $F_\theta^{t_0}$ (negative samples), when minimizing $\mathcal{L}_{\mathrm{CL}}$.

2.1 Prototype-Anchored Feature Alignment

Since source data is not available, explicit feature alignment that directly minimizes the domain gap between the source and target data like many UDA methods [4,8] is inoperative. As shown by previous methods [12], the weights $[\boldsymbol{\mu}_1, \boldsymbol{\mu}_2, \cdots, \boldsymbol{\mu}_C] \in \mathbb{R}^{D_f \times C}$ of the source domain classifier ϕ^s can be interpreted as the source prototypes, which characterize the features of each class. Thus, we introduce a bi-directional transport cost to align the target features with these prototypes instead of the unaccessible source features.

Following [23], given a mini-batch $\{x_m^t\}_{m=1}^M$ with M images, we first adopt the cosine distance $d(\boldsymbol{\mu}_c, \boldsymbol{f}_{m,i}^t) = 1 - \langle \boldsymbol{\mu}_c, \boldsymbol{f}_{m,i}^t \rangle$ to define a point-to-point transport cost between $\boldsymbol{f}_{m,i}^t$ and $\boldsymbol{\mu}_c$, where $\langle \cdot, \cdot \rangle$ is the cosine similarity. Then, a conditional distribution $\pi_\theta\left(\boldsymbol{\mu}_c \mid \boldsymbol{f}_{m,i}^t\right)$ specifying the probability of transporting from $\boldsymbol{f}_{m,i}^t$ to $\boldsymbol{\mu}_c$ can be constructed as,

$$\pi_\theta\left(\boldsymbol{\mu}_c \mid \boldsymbol{f}_{m,i}^t\right) = \frac{\hat{p}\left(\boldsymbol{\mu}_c\right) \exp\left(\boldsymbol{\mu}_c^T \boldsymbol{f}_{m,i}^t / \tau\right)}{\sum_{c'=1}^C \hat{p}\left(\boldsymbol{\mu}_{c'}\right) \exp\left(\boldsymbol{\mu}_{c'}^T \boldsymbol{f}_{m,i}^t / \tau\right)} \tag{1}$$

where τ is the temperature parameter, and $\hat{p}\left(\boldsymbol{\mu}_c\right)$ is the prior distribution (*i.e.*, class proportion) over the C classes for the target domain. As the true class distribution is unavailable in the target domain, we use the EM algorithm to infer $\hat{p}\left(\boldsymbol{\mu}_c\right)$ instead of using a uniform prior distribution (see more details in [16]). Note that in Eq. 1, a target point is more likely to be transported to the class prototypes closer to it or those with higher class propotion.

With the conditional distribution and point-to-point transport cost, we can derive the target-to-prototype (T2P) expected cost of moving the target features in this mini-batch to source prototypes,

$$\mathcal{L}_{\mathrm{T2P}} = \frac{1}{M \times H \times W} \sum_{m=1}^{M} \sum_{i=1}^{H \times W} \sum_{c=1}^{C} d(\boldsymbol{\mu}_c, \boldsymbol{f}_{m,i}^t) \pi_\theta \left(\boldsymbol{\mu}_c \mid \boldsymbol{f}_{m,i}^t \right) \tag{2}$$

In this target-to-prototype direction, we assign each target pixel to the prototypes according to their similarities and the class distribution. However, like many entropy minimization methods [1,2], optimizing target-to-prototype cost alone may result in degenerate trivial solutions, biasing the prediction towards a single dominant class [16]. To avoid mapping most of the target features to only a few prototypes, we add a prototype-to-target (P2T) transport cost in the opposite direction, which ensures that each prototype can be assigned to some target features. Similarly, we have:

$$\mathcal{L}_{\mathrm{P2T}} = \sum_{c=1}^{C} \hat{p}(\boldsymbol{\mu}_c) \sum_{m=1}^{M} \sum_{i=1}^{H \times W} d(\boldsymbol{\mu}_c, \boldsymbol{f}_{m,i}^t) \frac{\exp\left(\boldsymbol{\mu}_c^T \boldsymbol{f}_{m,i}^t / \tau\right)}{\sum_{m'=1}^{M} \sum_{i'=1}^{H \times W} \exp\left(\boldsymbol{\mu}_c^T \boldsymbol{f}_{m',i'}^t / \tau\right)} \tag{3}$$

Then, combining the conditional transport cost in these two directions, we define the total prototype-anchored feature alignment (PFA) loss:

$$\mathcal{L}_{\mathrm{PFA}} = \mathcal{L}_{\mathrm{T2P}} + \mathcal{L}_{\mathrm{P2T}} \tag{4}$$

Similar to [6], we initialize the adaptation model $\mathcal{M}^{t_0}$ with the pre-trained source model $\mathcal{M}^s$ and fix the weights of the classifier during adaptation.

2.2   Contrastive Learning Using Unreliable Predictions

After the PFA stage, the clusters of target features are shifted towards their corresponding source prototypes, which brings remarkable improvements for the initial noisy prediction (see Fig. 3(b)). To further improve the compactness of the target feature distribution, previous self-training methods mainly focus on strengthening the reliability of pseudo labels by developing denoising strategies [5,20], but discard those low-confidence predictions. However, such contempt for unreliable predictions may result in information loss. For example, in Fig. 1(c), the probability of the unreliable pixel hovers between spleen and left kidney, yet is confident enough to indicate the categories it does not belong to.

With this intuition, we denote $\boldsymbol{p}_{m,i}^t$ as the softmax probabilities generated by model $\mathcal{M}^{t_0}$ for the target data $x_{m,i}^t$. Then, for each class c, we construct three components, named query samples, positive prototypes, and negative samples, to explore those unreliable predictions as [18].

Query Samples. During training, we employ the per-pixel entropy as uncertainty metric [18], and sample the pixels with low entropy (reliable pixel) in the current mini-batch as query candidates. We denote the set of features of all query pixels for class c as $\mathcal{P}_c$,

$$\mathcal{P}_c = \{\boldsymbol{f}_{m,i}^t \mid \mathcal{H}(\boldsymbol{p}_{m,i}^t) \leq \gamma_c, \ \arg\max_{c'} \boldsymbol{p}_{m,i}^t = c\} \tag{5}$$

where $\mathcal{H}(\cdot)$ is the entropy of the input probabilities and γ_c is the entropy threshold for class c. Here we set γ_c as the α_c-th percentile of all the entropy values of pixels assigned a pseudo label c.

Positive Prototypes. The positive prototype is the same for all query pixels from the same class. Instead of using the center of query samples like [18], we set them the same as the previous source prototype, which is denoted as $z_c^+ = \mu_c$.

Negative Samples. For a query sample from class c, its qualified negative samples should satisfy: 1) unreliable; 2) highly probable not belong to class c. Therefore, we introduce the pixel-level category order $\mathcal{O}_{m,i}^t = \mathrm{argsort}(\boldsymbol{p}_{m,i}^t)$. For example, we have $\mathcal{O}_{m,i}^t(\arg\max \boldsymbol{p}_{m,i}^t) = 1$ and $\mathcal{O}_{m,i}^t(\arg\min \boldsymbol{p}_{m,i}^t) = C$. Thus, we can use $\mathcal{O}_{m,i}^t(c)$ to define the set of all negative samples:

$$\mathcal{N}_c = \{\boldsymbol{f}_{m,i}^t \mid \mathcal{H}(\boldsymbol{p}_{m,i}^t) > \gamma_c, \ \mathcal{O}_{m,i}^t(c) \geq r_l\} \tag{6}$$

where r_l is the low rank threshold and is set to 3 in our task.

With the above definition, we have the pixel-level contrastive loss as:

$$\mathcal{L}_{\mathrm{CL}} = -\frac{1}{C \times K} \sum_{c=1}^{C} \sum_{k=1}^{K} \log \left[\frac{e^{\langle z_{c,k}, z_c^+ \rangle / \tau}}{e^{\langle z_{c,k}, z_c^+ \rangle / \tau} + \sum_{j=1}^{N} e^{\langle z_{c,k}, z_{c,k,j}^- \rangle / \tau}} \right] \tag{7}$$

where K is the number of query samples, and $z_{c,k} \in \mathcal{P}_c$ denotes the k-th query sample from class c. Each query sample is paired with a positive prototype z_c^+ and N negative samples $z_{c,k,j}^- \in \mathcal{N}_c$.

3   Experiments and Results

3.1   Experimental Setup

Datasets and Evaluation Metrics. We evaluate our SFDA approach on a cross-modality abdominal multi-organ segmentation task. For the abdominal datasets, we obtain 20 MRI volumes from the 2019 CHAOS Challenge [10] and 30 CT volumes from MICCAI 2015 [11], respectively. Both datasets are under the Creative Commons Attribution 4.0 International license and involve segmentation masks for the following abdominal organs: liver, right kidney, left kidney and spleen. We complete adaptation experiments both in the "MRI to CT" direction and in the "CT to MRI" direction. For the "MRI to CT" direction, we take the MRI modality to train the source model and vice verse. Both modalities are randomly divided into 80% for domain adaptation training and 20% for evaluation. For both datasets, we discard the axial slices that do not contain foreground and crop out the non-body region [3]. The value range in CT volumes is first clipped to $[-125, 275]$. Then min-max normalization has been performed on both datasets to normalize the intensity value to $[0, 1]$. After that, all the MRI and CT volumes are uniformly resized to 256×256 in axial plane. Due to the large variance in the slice thickness of CT and MRI modality, we split the volume into slices for the model training.

For the evaluation, two main metrics, dice similarity coefficient (Dice) and average symmetric surface distance (ASSD) are used to quantitatively evaluate the segmentation results [4,15].

Implementation Details. We adopt classic U-Net structure for the segmentation model as the previous work [1]. The source segmentation model is trained in a fully-supervised manner for 10k iterations. During adaptation, we use Adam optimizer with the learning rate 1×10^{-4} and a weight decay of 5×10^{-4}. The temperature τ and batch size is set as 0.1 and 16, respectively. In PFA stage, we freeze the classifier and optimize F_θ^{to} for 200 iterations. In CL stage, we empirically set hyper-parameters $\alpha_c = 80$, $K = 64$, and $N = 256$ for all classes. All experiments are conducted with PyTorch on a single NVIDIA RTX 3090 GPU of 24 GB memory. Data augmentation such as random cropping, rotation, and brightness are adopted for source domain training and target domain adaptation.

3.2 Results of Source-Free Domain Adaptation

Comparision with Other Methods. In our experiments, "no adaptation" lower bound denotes learning a model on the source domain and directly test on the target domain without adaptation. And "supervised" upper bound means training and testing in the same target domain. We compared our methods with recent SFDA methods all designed for medical image segmentation scenarios, including a denoised pseudo-labeling approach (DPL) [5], a prior-aware entropy minimization approach (AdaMI) [1], a fourier style mining approach (FSM) [21], and a feature map statistics-guided approach [9]. We also considered top-performing UDA methods (*i.e.*, SIFA [4], DAG-Net [19]). For a fair comparison, we utilized the same backbone for these methods [1,4,5,21] and reimplemented them according to their official codes. Note that we reported the results of methods [9,19] from papers, since their official codes were not released.

The quantitative evaluation results are presented in Table 1. Compared to the upper and lower bounds in both directions, a huge performance gap can be observed due to the severe domain shifts between MRI and CT modalities. In "MRI to CT" direction, our method remarkably outperforms all other SFDA approaches on the right kidney and spleen, achieving the highest average Dice

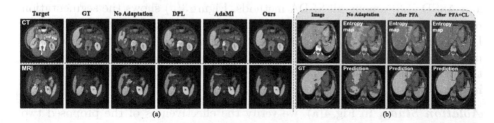

Fig. 3. (a) Qualitative segmentation results of different methods for abdominal images. (b) Visualized evolution of the model uncertainty and predictions in different stages.

Table 1. Comparision with other methods on abdominal multi-organ datasets.

Method	Source-free	Dice (%, mean ± std) ↑					ASSD (voxel, mean ± std) ↓				
		Liver	R. Kidney	L. Kidney	Spleen	Avg	Liver	R. Kidney	L. Kidney	Spleen	Avg
Abdominal MRI → Abdominal CT											
No Adaptation	–	49.2±9.6	43.4±24.9	65.4±18.1	62.4±20.9	55.1±14.3	8.7±2.0	10.8±4.6	7.4±3.1	4.5±2.9	7.9±2.5
Supervised	–	93.5±1.2	91.3±1.2	92.1±2.6	91.1±5.6	92.0±2.7	0.7±0.1	0.6±0.1	0.6±0.2	0.5±0.2	0.6±0.2
SIFA [4]	✗	89.0±3.2	83.8±4.0	82.7±5.8	84.6±8.0	85.0±5.9	1.2±0.5	1.3±0.6	1.5±0.7	1.6±0.9	1.4±0.8
DAG-Net [19]	✗	84.8±4.6	85.9±3.9	86.7±3.6	88.1±7.4	86.4±4.9	1.6±0.6	1.1±0.5	1.2±0.8	0.9±0.7	1.2±0.7
DPL [5]	✔	70.1±6.9	52.9±14.2	65.7±12.5	70.9±13.2	64.9±8.8	4.6±2.0	7.9±3.1	7.5±2.8	3.2±2.6	5.8±2.0
FSM [21]	✔	83.2±3.8	74.5±4.5	75.1±4.6	76.2±9.8	77.3±7.0	2.8±0.8	4.2±1.6	5.0±1.9	2.6±1.4	3.7±1.3
AdaMI [1]	✔	**90.2±1.0**	81.4±3.3	82.6±4.7	80.2±7.1	83.6±6.6	**1.0±0.5**	1.8±0.9	1.6±1.0	2.4±1.2	1.7±0.9
Hong et al [9]	✔	88.1	80.8	**88.1**	79.2	84.1	–	–	–	–	–
Ours	✔	89.9±2.7	**84.5±6.8**	84.9±4.0	**85.2±7.8**	**86.1±6.3**	1.3±0.5	**1.4±0.4**	**1.3±0.6**	**1.2±0.8**	**1.4±0.7**
Abdominal CT → Abdominal MRI											
No Adaptation	–	66.5±6.8	81.6±14.6	78.8±19.5	70.1±11.6	74.3±11.0	3.5±1.6	1.9±1.0	3.0±1.6	5.1±2.3	3.4±2.1
Supervised	–	93.6±3.5	94.2±1.8	90.4±3.7	92.2±2.6	92.6±2.1	0.5±0.2	0.3±0.1	0.4±0.1	0.4±0.2	0.4±0.2
SIFA [4]	✗	87.1±4.6	89.1±2.8	84.2±3.9	88.3±2.4	87.2±3.8	1.6±0.6	0.8±0.2	1.5±0.8	1.7±0.9	1.5±0.8
DAG-Net [19]	✗	86.3±3.3	89.0±2.0	89.9±2.0	90.6±2.9	89.0±2.6	1.8±0.6	0.9±0.2	0.9±0.4	1.3±1.0	1.2±0.6
DPL [5]	✔	77.2±2.0	83.5±12.5	80.3±13.6	83.3±7.2	81.1±8.3	3.0±0.6	2.3±1.6	2.5±1.4	3.1±1.7	2.7±1.4
FSM [21]	✔	84.3±3.1	83.5±6.6	82.1±9.0	84.2±5.6	83.5±4.3	2.1±0.5	1.5±0.6	1.6±0.7	2.2±0.6	1.9±0.6
AdaMI [1]	✔	85.5±2.4	86.4±5.1	82.1±7.6	89.9±3.2	86.0±3.1	**1.9±0.4**	1.2±0.5	2.2±0.6	**1.3±0.7**	1.7±0.5
Hong et al [9]	✔	**88.4**	89.1	86.4	**91.1**	88.8	–	–	–	–	–
Ours	✔	86.1±0.5	**91.7±5.1**	**88.6±8.0**	90.4±2.2	**89.2±3.3**	2.0±0.3	**0.7±0.2**	**1.0±0.4**	1.5±0.5	**1.3±0.5**

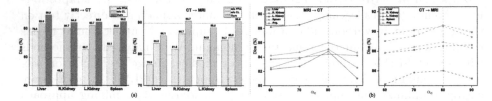

Fig. 4. (a) Ablation analysis of proposed two SFDA stages. "w/o CL" denotes only the PFA is performed; "w/o PFA" denotes directly optimizing the contrastive loss according to the source model prediction. (b) Effect of different uncertainty percentile α_c on the adaptation performance.

of 86.1% and the lowest average ASSD of 1.4. Moreover, compared with recent UDA methods, our method obtains competitive results on average Dice and ASSD, which may be due to the use of unreliable predictions. As for "CT to MRI" direction, our method similarly shows great superiority on most organs as well, achieving the best performance in terms of both the average Dice (89.2%) and ASSD (1.3) among all SFDA methods. Figure 3(a) shows the segmentation results obtained by existing and our methods in both modalities. As observed, DPL is prone to amplify the initial noisy regions since it directly discards the unreliable pixels in self-training. For comparison, our method substantially rectificate the uncertain regions from the initial prediction, and details are shown in Fig. 3(b).

Ablation Study. In Fig. 4(a), we verify the effectiveness of the proposed two SFDA stages by removing each stage while keeping the other. The consecutive two stage adaptation leads to the best performance, while the drop in Dice is more significant if we remove the PFA stage. This result is not surprising because, without PFA, the source model prediction is too noisy to sample the qualified query and negative pixels for contrastative learning. We also study the impact of different uncertainty percentile α_c in Fig. 4(b). This parameter has a certain impact on performance, and we find $\alpha_c = 80\%$ achieves the best performance for most organs. Large α_c may introduce low-confidence query samples for supervision, and small α_c will drop some informative negative samples.

4   Conclusion

In this paper, we propose a novel two-stage framework to address the source-free domain adaptation problem in medical image segmentation. We first introduce a bi-directional transport cost to encourage the alignment between target features and source class prototypes in the prototype-anchored feature alignment stage. Also, a contrastive learning stage using unreliable predictions is further devised to learn a more compact target feature distribution. Sufficient experiments on the cross-modality abdominal multi-organ segmentation task validate the effectiveness and superiority of our method against other strong SFDA baselines, even some classical UDA approaches.

References

1. Bateson, M., Kervadec, H., Dolz, J., Lombaert, H., Ayed, I.B.: Source-free domain adaptation for image segmentation. Med. Image Anal. **82**, 102617 (2022)
2. Bateson, M., Lombaert, H., Ben Ayed, I.: Test-time adaptation with shape moments for image segmentation. In: Wang, L., Dou, Q., Fletcher, P.T., Speidel, S., Li, S. (eds.) MICCAI 2022. LNCS, vol. 13434, pp. 736–745. Springer, Cham (2022). https://doi.org/10.1007/978-3-031-16440-8_70
3. Bian, C., Yuan, C., Ma, K., Yu, S., Wei, D., Zheng, Y.: Domain adaptation meets zero-shot learning: an annotation-efficient approach to multi-modality medical image segmentation. IEEE Trans. Med. Imaging **41**(5), 1043–1056 (2021)
4. Chen, C., Dou, Q., Chen, H., Qin, J., Heng, P.A.: Unsupervised bidirectional cross-modality adaptation via deeply synergistic image and feature alignment for medical image segmentation. IEEE Trans. Med. Imaging **39**(7), 2494–2505 (2020)
5. Chen, C., Liu, Q., Jin, Y., Dou, Q., Heng, P.-A.: Source-free domain adaptive fundus image segmentation with denoised pseudo-labeling. In: de Bruijne, M., et al. (eds.) MICCAI 2021. LNCS, vol. 12905, pp. 225–235. Springer, Cham (2021). https://doi.org/10.1007/978-3-030-87240-3_22
6. Ding, N., Xu, Y., Tang, Y., Xu, C., Wang, Y., Tao, D.: Source-free domain adaptation via distribution estimation. In: Proceedings of the IEEE/CVF Conference on Computer Vision and Pattern Recognition, pp. 7212–7222 (2022)
7. Dou, Q., Ouyang, C., Chen, C., Chen, H., Heng, P.A.: Unsupervised cross-modality domain adaptation of convnets for biomedical image segmentations with adversarial loss. In: Proceedings of the 27th International Joint Conference on Artificial Intelligence, pp. 691–697 (2018)
8. Han, X., et al.: Deep symmetric adaptation network for cross-modality medical image segmentation. IEEE Trans. Med. Imaging **41**(1), 121–132 (2021)

9. Hong, J., Zhang, Y.D., Chen, W.: Source-free unsupervised domain adaptation for cross-modality abdominal multi-organ segmentation. Knowl.-Based Syst. 109155 (2022)

10. Kavur, A.E., et al.: Chaos challenge-combined (CT-MR) healthy abdominal organ segmentation. Med. Image Anal. **69**, 101950 (2021)

11. Landman, B., Xu, Z., Igelsias, J., Styner, M., Langerak, T., Klein, A.: MICCAI multi-atlas labeling beyond the cranial vault-workshop and challenge. In: Proceedings of MICCAI Multi-Atlas Labeling Beyond Cranial Vault-Workshop Challenge, vol. 5, p. 12 (2015)

12. Liu, Y., Chen, Y., Dai, W., Gou, M., Huang, C.T., Xiong, H.: Source-free domain adaptation with contrastive domain alignment and self-supervised exploration for face anti-spoofing. In: Avidan, S., Brostow, G., Cissé, M., Farinella, G.M., Hassner, T. (eds.) ECCV 2022. LNCS, vol. 13672, pp. 511–528. Springer, Cham (2022). https://doi.org/10.1007/978-3-031-19775-8_30

13. Motiian, S., Jones, Q., Iranmanesh, S., Doretto, G.: Few-shot adversarial domain adaptation. In: Advances in Neural Information Processing Systems, vol. 30 (2017)

14. Ronneberger, O., Fischer, P., Brox, T.: U-Net: convolutional networks for biomedical image segmentation. In: Navab, N., Hornegger, J., Wells, W.M., Frangi, A.F. (eds.) MICCAI 2015. LNCS, vol. 9351, pp. 234–241. Springer, Cham (2015). https://doi.org/10.1007/978-3-319-24574-4_28

15. Stan, S., Rostami, M.: Privacy preserving domain adaptation for semantic segmentation of medical images. arXiv preprint arXiv:2101.00522 (2021)

16. Tanwisuth, K., et al.: A prototype-oriented framework for unsupervised domain adaptation. Adv. Neural. Inf. Process. Syst. **34**, 17194–17208 (2021)

17. Tzeng, E., Hoffman, J., Saenko, K., Darrell, T.: Adversarial discriminative domain adaptation. In: Proceedings of the IEEE Conference on Computer Vision and Pattern Recognition, pp. 7167–7176 (2017)

18. Wang, Y., et al.: Semi-supervised semantic segmentation using unreliable pseudo-labels. In: Proceedings of the IEEE/CVF Conference on Computer Vision and Pattern Recognition, pp. 4248–4257 (2022)

19. Xian, J., et al.: Unsupervised cross-modality adaptation via dual structural-oriented guidance for 3D medical image segmentation. IEEE Trans. Med. Imaging (2023)

20. Xu, Z., et al.: Denoising for relaxing: unsupervised domain adaptive fundus image segmentation without source data. In: Wang, L., Dou, Q., Fletcher, P.T., Speidel, S., Li, S. (eds.) MICCAI 2022. LNCS, vol. 13435, pp. 214–224. Springer, Cham (2022). https://doi.org/10.1007/978-3-031-16443-9_21

21. Yang, C., Guo, X., Chen, Z., Yuan, Y.: Source free domain adaptation for medical image segmentation with fourier style mining. Med. Image Anal. **79**, 102457 (2022)

22. Yu, Q., Dang, K., Tajbakhsh, N., Terzopoulos, D., Ding, X.: A location-sensitive local prototype network for few-shot medical image segmentation. In: 2021 IEEE 18th International Symposium on Biomedical Imaging (ISBI), pp. 262–266. IEEE (2021)

23. Zheng, H., Zhou, M.: Exploiting chain rule and Bayes' theorem to compare probability distributions. Adv. Neural. Inf. Process. Syst. **34**, 14993–15006 (2021)

24. Zhou, Z., Rahman Siddiquee, M.M., Tajbakhsh, N., Liang, J.: UNet++: a nested U-net architecture for medical image segmentation. In: Stoyanov, D., et al. (eds.) DLMIA/ML-CDS -2018. LNCS, vol. 11045, pp. 3–11. Springer, Cham (2018). https://doi.org/10.1007/978-3-030-00889-5_1

Segmentation of Kidney Tumors on Non-Contrast CT Images Using Protuberance Detection Network

Taro Hatsutani[✉], Akimichi Ichinose, Keigo Nakamura, and Yoshiro Kitamura

Medical Systems Research and Development Center,
FUJIFILM Corporation, Tokyo, Japan
taro.hatsutani@fujifilm.com

Abstract. Many renal cancers are incidentally found on non-contrast CT (NCCT) images. On contrast-enhanced CT (CECT) images, most kidney tumors, especially renal cancers, have different intensity values compared to normal tissues. However, on NCCT images, some tumors called isodensity tumors, have similar intensity values to the surrounding normal tissues, and can only be detected through a change in organ shape. Several deep learning methods which segment kidney tumors from CECT images have been proposed and showed promising results. However, these methods fail to capture such changes in organ shape on NCCT images. In this paper, we present a novel framework, which can explicitly capture protruded regions in kidneys to enable a better segmentation of kidney tumors. We created a synthetic mask dataset that simulates a protuberance, and trained a segmentation network to separate the protruded regions from the normal kidney regions. To achieve the segmentation of whole tumors, our framework consists of three networks. The first network is a conventional semantic segmentation network which extracts a kidney region mask and an initial tumor region mask. The second network, which we name protuberance detection network, identifies the protruded regions from the kidney region mask. Given the initial tumor region mask and the protruded region mask, the last network fuses them and predicts the final kidney tumor mask accurately. The proposed method was evaluated on a publicly available KiTS19 dataset, which contains 108 NCCT images, and showed that our method achieved a higher dice score of 0.615 (+0.097) and sensitivity of 0.721 (+0.103) compared to 3D-UNet. To the best of our knowledge, this is the first deep learning method that is specifically designed for kidney tumor segmentation on NCCT images.

Keywords: Renal Cancer · Tumor Segmentation · Non-Contrast CT

1 Introduction

Over 430,000 new cases of renal cancer were reported in 2020 in the world [1] and this number is expected to rise [22]. When the tumor size is large (greater

H. Greenspan et al. (Eds.): MICCAI 2023, LNCS 14226, pp. 13–22, 2023.
https://doi.org/10.1007/978-3-031-43990-2_2

than 7 cm) often the whole kidney is removed, however, when the tumor size is small (less than 4 cm), partial nephrectomy is the preferred treatment [20] as it could preserve kidney's function. Thus, early detection of kidney tumors can help to improve patient's prognosis. However, early-stage renal cancers are usually asymptomatic, therefore they are often incidentally found during other examinations [19], which includes non-contrast CT (NCCT) scans.

Segmentation of kidney tumors on NCCT images adds challenges compared to contrast-enhanced CT (CECT) images, due to low contrast and lack of multiphase images. On CECT images, the kidney tumors have different intensity values compared to the normal tissues. There are several works that demonstrated successful segmentation of kidney tumors with high precision [13,21]. However, on NCCT images, as shown in Fig. 1b, some tumors called isodensity tumors, have similar intensity values to the surrounding normal tissues. To detect such tumors, one must compare the kidney shape with tumors to the kidney shape without the tumors so that one can recognize regions with protuberance.

3D U-Net [3] is the go-to network for segmenting kidney tumors on CECT images. However, convolutional neural networks (CNNs) are biased towards texture features [5]. Therefore, without any intervention, they may fail to capture the protuberance caused by isodensity tumors on NCCT images.

In this work, we present a novel framework that is capable of capturing the protuberances in the kidneys. Our goal is to segment kidney tumors including isodensity types on NCCT images. To achieve this goal, we create a synthetic dataset, which has separate annotations for normal kidneys and protruded regions, and train a segmentation network to separate the protruded regions from the normal kidney regions. In order to segment whole tumors, our framework consists of three networks. The first is a base network, which extracts kidneys and an initial tumor region masks. The second protuberance detection network receives the kidney region mask as its input and predicts a protruded region mask. The last fusion network receives the initial tumor mask and the protruded region mask to predict a final tumor mask. This proposed framework enables a better segmentation of isodensity tumors and boosts the performance of segmentation of kidney tumors on NCCT images. The contribution of this work is summarized as follows:

1. Present a pioneering work for segmentation of kidney tumors on NCCT images.
2. Propose a novel framework that explicitly captures protuberances in a kidney to enable a better segmentation of tumors including isodensity types on NCCT images. This framework can be extended to other organs (e.g. adrenal gland, liver, pancreas).
3. Verify that the proposed framework achieves a higher dice score compared to the standard 3D U-Net using a publicly available dataset.

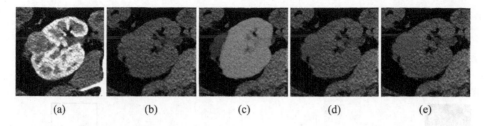

Fig. 1. Example CECT and NCCT images. Some kidney tumors have similar intensity values to its surrounding tissues. a) CECT image. b) NCCT image. c) Output of the protuberance detection network. d) Output of our model. e) Ground truth mask.

2   Related Work

The release of two public CT image datasets with kidney and tumor masks from the 2019/2021 Kidney and Kidney Tumor Segmentation challenge [8] (KiTS19, KiTS21) attracted researchers to develop various methods for segmentation.

Looking at the top 3 teams from each challenge [6,11,13,17,21], all teams utilized 3D U-Net [3] or V-Net [16], which bears a similar architecture. The winner of KiTS19 [13] added residual blocks [7] to 3D U-Net and predicted kidney and tumor regions directly. However, the paper notes that modifying the architecture resulted in only slight improvement. The other 5 teams took a similar approach to nnU-Net's coarse-to-fine cascaded network [12], where it predicts from a low-resolution image in the first stage and then predicts kidneys and tumors from a high-resolution image in the second stage. Thus, although other attempts were made, using 3D U-Net is the go-to method for predicting kidneys and tumors. In our work, we also make use of 3D U-Net, but using this network alone fails to learn some isodensity tumors. To overcome this issue, we developed a framework that specifically incorporates protuberances in kidneys, allowing for an effective segmentation of tumors on NCCT images.

In terms of focusing on protruded regions in kidneys, our work is close to [14,15]. [14] developed a computer-aided diagnosis system to detect exophytic kidney tumors on NCCT images using belief propagation and manifold diffusion to search for protuberances. An exophytic tumor is located on the outer surface of the kidney that creates a protrusion. While this method demonstrated high sensitivity (95%), its false positives per patient remained high (15 false positives per patient). In our work, we will not only segment protruded tumors but also other tumors as well.

3   Proposed Method

To capture the protuberances in kidneys, we specifically train a protuberance detection network, which receives a kidney region mask as an input and separates protruded regions from it. This enables us to extract a part of tumors that forms protuberance, but our goal is segmenting all visible kidney tumors on NCCT images. Thus, we make use of three networks as shown in Fig. 2.

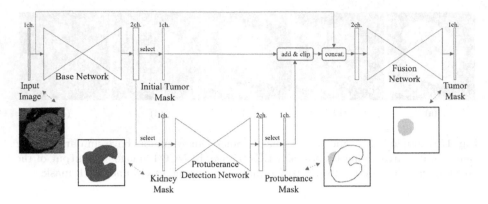

Fig. 2. Overview of our framework.

The first base network is responsible for predicting kidney and tumor region masks. Our architecture is based on 3D U-Net, which has an encoder-decoder style architecture, with few modifications. To reduce the required size of GPU memory, we only use the encoder that has only 16 channels at the first resolution, but instead we make the architecture deeper by having 1 strided convolution and 4 max-pooling layers. In the decoder, we replace the up-convolution layers with a bilinear up-sampling layer and a convolution layer. In addition, by only having a single convolution layer instead of two in the original architecture at each resolution, we keep the decoder relatively small. Throughout this paper, we refer this architecture as our 3D U-Net.

The second protuberance detection network is the same as the base network except it starts from 8 channels instead of 16. We train this network using synthetic datasets. The details of the dataset and training procedures are described in Sect. 3.2.

The last fusion network combines the outputs from the base network and the protuberance detection network and makes the final tumor prediction. In detail, we perform a summation of the initial tumor mask and the protruded region mask, and then concatenate the result with the input image. This is the input of the last fusion network, which also has the same architecture as the base network with an exception of having two input channels. This fusion network do not just combine the outputs but also is responsible for removing false positives from the base network and the protuberance detection network.

Our combined three network is fully differentiable, however, to train efficiently, we train the model in 3 steps.

3.1   Step1: Training Base Network

In the first step, we train the base network, which is a standard segmentation network, to extract kidney and tumor masks from the images. We use a sigmoid function for the last layer. And as a loss function, we use the dice loss [16] and the cross-entropy loss equally.

3.2 Step2: Training Protuberance Detection Network

In the second step, we train the protuberance detection network alone to separate protruded regions from the normal kidney masks. Here, we only use the cross-entropy loss and label smoothing with a smoothing factor of $\epsilon = 0.01$.

Synthetic Dataset. To enable a segmentation of protruded regions only, a separate annotation of each region is usually required. However, annotating such areas is time-consuming and preparing a large number of data is challenging. Alternatively, we create a synthetic dataset that mimics a kidney with protrusions. The synthetic dataset is created through the following steps:

1. Randomly sample a kidney mask without protuberance and a tumor mask.
2. Apply random rotation and scaling to the tumor mask.
3. Randomly insert the tumor mask into the kidney mask.
4. If both of the following conditions are met, append to the dataset.

$$\frac{\sum_i k_i t_i}{\sum_i k_i} < 0.3, \tag{1}$$

$$\frac{\sum_i k_i t_i}{\sum_i t_i} < 0.95, \tag{2}$$

where k_i is a voxel value (0 or 1) in the kidney mask and t_i is a voxel value in the tumor mask. Equation 1 ensures that only up to 30% of the kidney is covered with a tumor. Equation 2 ensures that not all tumors are covered by the kidney (at least 5% of the tumor is protruded from the kidney).

3.3 Step3: End-to-End Training with Fusion Network

In the final step, we train the complete network jointly. Although our network is fully differentiable, since there is no separate annotation for protruded regions other from the synthetic dataset, we freeze the parameters in protuberance detection network.

The output of the protuberance detection network will likely have more false positives than the base network since it has no access to the input image. Thus, when the output of the protuberance detection network is concatenated with the output of the base network, the fusion network can easily reduce the loss by ignoring the protuberance detection network's output, which is suboptimal. To avoid this issue, we perform summation not concatenation to avoid the model from ignoring all output from the protuberance detection network. We then clip the value of the mask to the range of 0 and 1. As a result, the input to the fusion network has two channels. The first channel is the input image, and the second channel is the result of summation of the initial tumor mask and the protruded region mask. We concatenate the input image so that the last network can remove false positives from the predicted masks as well as predicting the missing tumor regions from the protuberance detection network.

We use the dice loss and the cross-entropy loss as loss functions for the fusion network. We also keep the loss functions in the base network for predicting kidneys and tumors. The loss function for tumors in the base network acts like an intermediate supervision. Our network shares some similarities with the stacked hourglass network [18] where the network consists of multiple U-Net like hourglass modules and has intermediate supervision at the end of each hourglass module. By having multiple modules in this manner, the network can fix the initial mistakes in early modules and corrects in later modules.

4 Experiments

No prior work exists that uses NCCT images from KiTS19 [8,9]. Thus, we first created our baseline model and compared the performance with existing methods on CECT images. This allows us to ensure that our baseline model and training procedure is appropriate. We then trained the model using NCCT images and compared with our proposed method.

4.1 Datasets and Preprocessing

We used a dataset from KiTS19 [8] which contains both CECT and NCCT images. For CECT images, there are 210 images for training and validation and, 90 images for testing. For NCCT images, there are 108 images, which are different series of the 210 images. The ground truth masks are only available for the 210 CECT images. Thus, we transfer the masks to NCCT images. This is achieved by extracting kidney masks and adjusting the height of each kidney. The ground truth mask contains a kidney label and a kidney tumor label. Cysts are not annotated separately and included in the kidney label on this dataset. The data can be downloaded from The Cancer Imaging Archive (TCIA) [4,9].

The images were first clipped to the intensity value range of $[-90, 210]$ and normalized from -1 to 1. The voxel spacings were normalized to 1 mm. During the training, the images were randomly cropped to a patch size of $128 \times 128 \times 128$ voxels. We applied random rotation, random scaling and random noise addition as data augmentation.

During the $Step2$ phase of the training, where we used the synthetic dataset, we created 10,000 masks using the method from Sect. 3.2. We applied some augmentations during training to input masks to simulate the incoming inputs from the base network. The output of the base network is not binarized to keep gradient from flowing, so the values are in the range $[0, 1]$ and the edge of kidneys are usually smooth. Therefore, we applied gaussian blurring, gaussian noise addition and intensity value shifting.

Table 1. Dice performance of existing method and our baseline model. Evaluated using CECT images from KiTS19. Composite dice is an average dice between kidney and tumor dice. The results were obtained by submitting our predicted masks to the Grand Challenge page.

Method	Composite Dice	Kidney Dice	Tumor Dice
Isensee and Maier-Hein [13]	0.9123	0.9737	0.8509
Our baseline model	0.8832	0.9728	0.7935

4.2   Training Details and Evaluation Metrics

Our model was trained using SGD with a 0.9 momentum and a weight decay of $1e-7$. We employed a learning rate scheduler, which we warm-up linearly from 0.0001 to 0.1 during the first 30% (for $Step1$ and $Step3$) or 10% (for $Step2$) of the total training steps and decreased following the cosine decay learning rate. A mini-batch size of 8, 16 and 4 were used, and trained for 250k, 100k and 100k steps during $Step1$ to 3 respectively. We conducted our experiments using JAX (v.0.4.1) [2] and Haiku (v.0.0.9) [10]. We trained the model using a single NVIDIA RTX A5000 GPU.

For the experiment on CECT images, we used the dice score as our evaluation metrics following the same formula from KiTS19. For the experiment on NCCT images, we also evaluated the sensitivity and false positives per image (FPs/image). We calculated as true positive when the predicted mask has the dice score greater than 0.5, otherwise we calculated as false negative. On the other hand, false positives were counted when the predicted mask did not overlap with any ground truth masks.

5   Results

5.1   Performance on CECT Images

To show that our model is properly tuned, we compare our baseline model with an existing method using CECT images. As can be seen from Table 1, our model showed comparable scores to the winner of KiTS19 challenge. We used this baseline model as our base network for the experiments on NCCT images.

5.2   Performance on NCCT Images

Table 2 shows our experimental results and ablation studies on NCCT images. The proposed method (Table 2-bottom) outperformed the baseline model (Table 2-top). The ablation studies show that adding each component (CECT images and the protuberance detection network) resulted in an increase in the performance. While adding CECT images contributed the most for the increase in tumor dice and sensitivity, adding the protuberance detection network further pushed the performance. However, the false positives per image (FPs/image)

Table 2. Result of our proposed method on NCCT images from KiTS19. The values are average values of a five-fold cross-validation.

Protuberance Detection Network	with CECT images	Tumor Dice	Sensitivity	FPs/Image
✗	✗	0.518	0.618	**0.283**
✗	✓	0.585	0.686	0.340
✓	✓	**0.615**	**0.721**	0.421

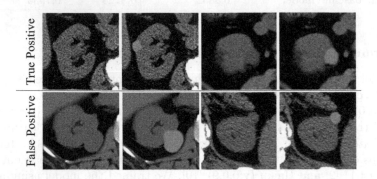

Fig. 3. Output examples from our model. The first row shows true positive results and the second row shows false positive results.

increased from 0.283 to 0.421. The protuberance detection network cannot distinguish the protrusions that were caused by tumors or cysts, so the output from this network has many FPs at this stage. Thus, the fusion network has to eliminate cysts by looking again the input image, however, it may have failed to eliminate some cysts (Fig. 3 second row).

6   Conclusion

In this paper, we proposed a novel framework for kidney tumor segmentation on NCCT images. To cope with isodensity tumors, which have similar intensity values to their surrounding tissues, we created a synthetic dataset to train a network that extracts protuberance from the kidney masks. We combined this network with the base network and fusion network. We evaluated our method using the publicly available KiTS19 dataset, and showed that the proposed method can achieve a higher sensitivity than existing approach. Our framework is not limited to kidney tumors but can also be extended to other organs (e.g., adrenal gland, liver, pancreas).

References

1. Estimated number of new cases in 2020, world, both sexes, all ages (excl. nmsc). https://gco.iarc.fr/today/online-analysis-table/. Accessed 27 Feb 2023

2. Bradbury, J., et al.: JAX: composable transformations of Python+NumPy programs (2018). http://github.com/google/jax
3. Çiçek, Ö., Abdulkadir, A., Lienkamp, S.S., Brox, T., Ronneberger, O.: 3D U-Net: learning dense volumetric segmentation from sparse annotation. In: Ourselin, S., Joskowicz, L., Sabuncu, M.R., Unal, G., Wells, W. (eds.) MICCAI 2016. LNCS, vol. 9901, pp. 424–432. Springer, Cham (2016). https://doi.org/10.1007/978-3-319-46723-8_49
4. Clark, K., et al.: The cancer imaging archive (TCIA): maintaining and operating a public information repository. J. Digit. Imaging **26**(6), 1045–1057 (2013)
5. Geirhos, R., Rubisch, P., Michaelis, C., Bethge, M., Wichmann, F.A., Brendel, W.: Imagenet-trained CNNs are biased towards texture; increasing shape bias improves accuracy and robustness. In: Proceedings of International Conference on Learning Representations (2019)
6. Golts, A., Khapun, D., Shats, D., Shoshan, Y., Gilboa-Solomon, F.: An ensemble of 3D U-net based models for segmentation of kidney and masses in CT scans. In: Heller, N., Isensee, F., Trofimova, D., Tejpaul, R., Papanikolopoulos, N., Weight, C. (eds.) KiTS 2021. LNCS, vol. 13168, pp. 103–115. Springer, Cham (2022). https://doi.org/10.1007/978-3-030-98385-7_14
7. He, K., Zhang, X., Ren, S., Sun, J.: Deep residual learning for image recognition. In: Proceedings of the IEEE Conference on Computer Vision and Pattern Recognition, pp. 770–778 (2016)
8. Heller, N., et al.: The state of the art in kidney and kidney tumor segmentation in contrast-enhanced CT imaging: results of the kits19 challenge. Med. Image Anal. **67**, 101821 (2021)
9. Heller, N., et al.: C4KC kits challenge kidney tumor segmentation dataset (2019)
10. Hennigan, T., Cai, T., Norman, T., Babuschkin, I.: Haiku: Sonnet for JAX (2020). http://github.com/deepmind/dm-haiku
11. Hou, X., Chunmei, X., Li, F., Yang, N.: Cascaded semantic segmentation for kidney and tumor. Submissions to the 2019 Kidney Tumor Segmentation Challenge: KiTS19 (2019)
12. Isensee, F., Jaeger, P.F., Kohl, S.A.A., Petersen, J., Maier-Hein, K.H.: nnU-net: a self-configuring method for deep learning-based biomedical image segmentation. Nat. Methods **18**(2), 203–211 (2020)
13. Isensee, F., Maier-Hein, K.: An attempt at beating the 3D U-NET. Submissions to the 2019 Kidney Tumor Segmentation Challenge: KiTS19 (2019)
14. Liu, J., Wang, S., Linguraru, M.G., Yao, J., Summers, R.M.: Computer-aided detection of exophytic renal lesions on non-contrast CT images. Med. Image Anal. **19**(1), 15–29 (2015)
15. Liu, J., Wang, S., Yao, J., Linguraru, M.G., Summers, R.M.: Manifold diffusion for exophytic kidney lesion detection on non-contrast CT images. In: Mori, K., Sakuma, I., Sato, Y., Barillot, C., Navab, N. (eds.) MICCAI 2013. LNCS, vol. 8149, pp. 340–347. Springer, Heidelberg (2013). https://doi.org/10.1007/978-3-642-40811-3_43
16. Milletari, F., Navab, N., Ahmadi, S.A.: V-net: fully convolutional neural networks for volumetric medical image segmentation. In: 2016 Fourth International Conference on 3D Vision (3DV), pp. 565–571. IEEE (2016)
17. Mu, G., Lin, Z., Han, M., Yao, G., Gao, Y.: Segmentation of kidney tumor by multi-resolution VB-Nets. Submissions to the 2019 Kidney Tumor Segmentation Challenge: KiTS19 (2019)

18. Newell, A., Yang, K., Deng, J.: Stacked hourglass networks for human pose esti-mation. In: Leibe, B., Matas, J., Sebe, N., Welling, M. (eds.) ECCV 2016. LNCS, vol. 9912, pp. 483–499. Springer, Cham (2016). https://doi.org/10.1007/978-3-319-46484-8_29
19. Pinsky, P.F., et al.: Incidental renal tumours on low-dose CT lung cancer screening exams. J. Med. Screen. **24**(2), 104–109 (2017)
20. Touijer, K., et al.: The expanding role of partial nephrectomy: a critical analysis of indications, results, and complications. Eur. Urol. **57**(2), 214–222 (2010)
21. Zhao, Z., Chen, H., Wang, L.: A coarse-to-fine framework for the 2021 kidney and kidney tumor segmentation challenge. In: Heller, N., Isensee, F., Trofimova, D., Tejpaul, R., Papanikolopoulos, N., Weight, C. (eds.) KiTS 2021. LNCS, vol. 13168, pp. 53–58. Springer, Cham (2022). https://doi.org/10.1007/978-3-030-98385-7_8
22. Znaor, A., Lortet-Tieulent, J., Laversanne, M., Jemal, A., Bray, F.: International variations and trends in renal cell carcinoma incidence and mortality. Eur. Urol. **67**(3), 519–530 (2015)

Eye-Guided Dual-Path Network for Multi-organ Segmentation of Abdomen

Chong Wang, Daoqiang Zhang, and Rongjun Ge[✉]

College of Computer Science and Technology,
Nanjing University of Aeronautics and Astronautics, Nanjing, China
rongjun.ge@nuaa.edu.cn

Abstract. Multi-organ segmentation of the abdominal region plays a vital role in clinical such as organ quantification, surgical planning, and disease diagnosis. Due to the dense distribution of abdominal organs and the close connection between each organ, the accuracy of the label is highly required. However, the dense and complex structure of abdominal organs necessitates highly professional medical expertise to manually annotate the organs, leading to significant costs in terms of time and effort. We found a cheap and easily accessible form of supervised information. Recording the areas by the eye tracker where the radiologist focuses while reading abdominal images, gaze information is able to force the network model to focus on relevant objects or features required for the segmentation task. Therefore how to effectively integrate image information with gaze information is a problem to be solved. To address this issue, we propose a novel network for abdominal multi-organ segmentation, which incorporates radiologists' gaze information to boost high-precision segmentation and weaken the demand for high-cost manual labels. Our network includes three special designs: 1) a dual-path encoder to further integrate gaze information; 2) a cross-attention transformer module (CATM) that embeds human cognitive information about the image into the network model; and 3) multi-feature skip connection (MSC), which combines spatial information during down-sampling to offset the internal details of segmentation. Additionally, our network utilizes discrete wavelet transform (DWT) to further provide information on organ location and edge in different directions. Extensive experiments performed on the publicly available Synapse dataset demonstrate that our proposed method can integrate effectively gaze information and achieves Dice similarity coefficient (DSC) up to 81.87% and Hausdorff distance (HD) reduction to 11.96%, as well as gain high-quality readable visualizations. Code will be available at https://github.com/code-Porunacabeza/gaze_seg/.

Keywords: Eye-tracking · Multiorgan segmentation ·
Computer-Aided Diagnosis

H. Greenspan et al. (Eds.): MICCAI 2023, LNCS 14226, pp. 23–32, 2023.
https://doi.org/10.1007/978-3-031-43990-2_3

1   Introduction

The automatic segmentation of abdominal multiple organs is clinically significant in extremely that can significantly reduce clinical resource costs. However,
the task of abdominal organ segmentation is difficult. The number of abdominal organs is large, and these multiple organs show diverse characteristics among
themselves. For example, the shape of the stomach varies greatly even in the same
individual at different times, making precise pixel segmentation extremely challenging. Accurate and automatic segmentation of readable results from abdominal multiple organs can provide accurate evidence of reality for surgical navigation, visual enhancement, radiation therapy, and biomarker measurement systems. Therefore, how to accurately make the segmentation results more readable
in the case of multiple organs influencing each other has a great contribution to
clinical examination and diagnosis.

Abdominal multi-organ network models based on deep neural networks
(DNN) are difficult to train. Training such a good enough model usually requires
a large amount of labeled data, or the model performance is likely to meet a heavy
drop. However, manual annotation of organs requires doctors to make accurate
judgments based on their professional knowledge and rich experience, this leads
to making manual labeling both expensive and time-consuming. In addition to
pixel-level annotated datasets, deep neural networks can also benefit from other
types of supervision. For example, boundary-level annotation can provide more
detailed boundary information. In addition, weakly supervised [6,12,14] learning
techniques can be used, such as training with pixel-level labels and unlabeled
data. Additionally, visual perceptual [1,7] supervision can be employed by utilizing visual perceptual theory in the training of deep networks to increase their
sensitivity to image features. Furthermore, pre-trained models can be utilized for
transfer learning, which allows the model to learn features from previous tasks
and improve its performance. In summary, deep neural networks can benefit from
various types of supervision, which can improve their performance in a variety of
visual tasks. These studies have demonstrated that incorporating finer-grained
additional supervision can enhance the accuracy of deep neural networks and
improve the interpretability of network models.

However, the practical process of collecting additional annotations remains
challenging, as it may require clinicians to repeatedly provide specific and refined
annotations to fine-tune the network model. There is a need to minimize the
impact of the annotation process on clinical work. To address this, we investigate novel annotation information that can be used for abdominal multi-organ
segmentation. In the context of medical image analysis, it has been observed that
radiologists tend to focus their attention on specific regions of interest (ROIs) or
lesions when interpreting medical images. Specifically, our method utilizes eye
gaze information collected by an eye-tracker during radiologists' image interpretation as a source of additional supervision. In clinical practice, experienced
radiologists can usually quickly locate specific organs when reading abdominal
images. In this process, the doctor's eye movement information can reflect the
location information of organs to a certain extent. Compared with manual label-

ing, this information is cheap and fast and can be used as effective supervision information to assist the localization and segmentation of each organ. The literature studies have implied that the potential of the radiologist's gaze data can be high in improving disease diagnosis [2,17]. Recently, Wang et al. [16] applied eye-tracking technology to diagnose knee osteoarthritis, while Men et al. [9] used eye-trackers to provide visual guidance to sonographers during ultrasound scanning. It can be seen that the use of eye movement attention information has great value and potential in automated auxiliary diagnosis.

In this paper, we propose a novel eye-guided multi-organ segmentation network for diverse abdominal organ images. The network model is forced to focus on relevant objects or features required for the segmentation task by fully and synergistically utilizing the radiologist's cognitive information about the abdominal image. This method of information collection is convenient and can make the positioning of each organ more accurate. The overall architecture is shown in Fig. 1. The proposed network has three special designs: 1) a dual-path encoder that integrates human cognitive information; 2) a cross-attention transformer module (CATM) that communicates information in network semantic perception and human semantic perception; and 3) multi-feature skip connection (MSC), which effectively combines spatial information during down-sampling to offset the internal details of segmentation.

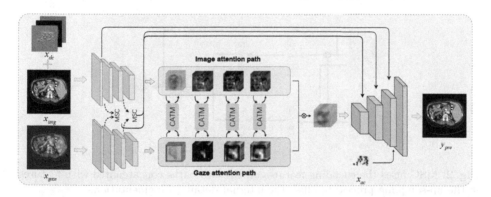

Fig. 1. Overview of the network architecture, detail coefficients x_{dc}, image x_{img} and gaze information x_{gaze} are input into the network for segmentation. In the decoding phase, approximation coefficients x_{ac} are fused to compensate the global information. MSC: multi-feature skip connection. CATM: cross-attention transformer module

2   Methodology

As shown in Fig. 1. The proposed network adopts an encoder-decoder structure, where the encoder part consists of parallel dual paths that utilize multi-feature skip connection (MSC) to combine spatial information during down-sampling to offset the internal details of segmentation. A cross-attention transformer module

(CATM) is designed at the bottleneck stage to effectively communicate information in network perception and human perception.

2.1   Wavelet Transform for Composite Information

Wavelet transform is able to obtain global information and edge information in different directions in gaze attention heatmaps so that the network can effectively fuse the composite information in the heatmaps. In the clinic, when radiologists read abdominal images, the more important location, the longer the radiologists' gaze. We convert this information into a heatmap representation. The heatmap reflects the rough position information of the target to be segmented. The single heatmap is unable to reflect the composite information it contains, therefore DWT is utilized for extracting it. Discrete wavelet transform [8] (DWT) is applied to decompose the approximation coefficients and detail coefficients of abdominal organ distribution information on the gaze attention heatmap to locate the position and edge of multiple organs in the abdomen. In the decoding phase, we fuse the approximation coefficient in the gaze heatmap so that compensates for the global topological information of decoding features at the final segmentation. The detail coefficients are input into the image encoder together with the original image, which is used to guide the dual-path encoder to reserve detailed information.

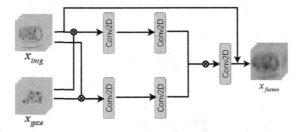

Fig. 2. MSC fuses the encoding features on the two paths concatenated with features in the up-sampling process to offset the internal details of segmentation.

2.2   Multi-feature Skip Connection

Multi-feature skip connection (MSC) comprehensively utilizes multiple features to guide the segmentation results of each abdominal organ toward accurate internal details. As shown in Fig. 2, we choose to integrate the encoding features on the two paths concatenated with features in the up-sampling process to offset the internal details of segmentation. Instead of using a simple concatenated and fusion strategy, we use multiple composite splicing and fusion to obtain matching features. The residual connection can aggregate the features of different levels to avoid additional noise and thus improve the network performance. The MSC can be expressed as follows:

$$F_{feat} = F_i + Conv(Conv(Conv(F_i \oplus F_g)) \oplus Conv(Conv(F_g \oplus F_i))), \qquad (1)$$

where F_i and F_g represent the output features from the down-sampling layers of the two paths, and F_{fusion} denotes the final multiple fusion features. Following the MSC, the dimension of the concatenated multiple features remains the same as the dimension of the upsampled features.

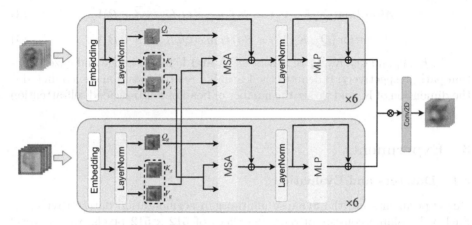

Fig. 3. CATM utilizes cross-attention efficiently enhances information interactive collaboration between network semantic perception and human semantic perception. The convolution operation fuses the feature information from two different paths.

2.3 Cross-Attention Transformer Module

The cross-attention transformer module (CATM) creatively enables the communication between network semantic perception and human semantic perception. Different from the traditional self-attention mechanism in transformer block [15], by using CATM, information interactive collaboration on two paths is enabled effectively. CATM, which is a multi-path structure, is embedded in the bottleneck layer between the encoder and decoder. As shown in Fig. 3, it consists of two paths: the image attention path and the gaze attention path. In our work, CATM is composed of L (we set $L = 6$) cross-attention transformer blocks (CTB) and Conv2D. The expression of CATM can be represented as:

$$F_{out} = Conv((CTB(F_i) \oplus CTB(F_g))_i), \qquad (2)$$

where F_i and F_g represent the final output features of the encoder on the image attention and gaze attention encoding pathways. Our experimental results demonstrate that the cross-attention operation within the CATM design efficiently enhances the communication of information between these two paths.

The convolution operation of CATM is used to fuse the feature information from two paths, which makes up for the possible information shortage in the decoding process. The CTB is the core design of the CATM, which is a variant of the Transformer [15] that exchanges the network semantic perception and human semantic perception on the two different paths of the image and gaze attention. The key of cross-attention is to exchange Q, K and V of respective features between different path features and fuse them. As shown in Fig. 3, the K and V in the image attention path exchange with the gaze attention path, and each original Q fuse with the exchanged K and V. It is represented by a formula:

$$Attention(Q_i, K_g, V_g) = Softmax(Q_i K_g / \sqrt{d} + B)V_g, \qquad (3)$$

$$Attention(Q_g, K_i, V_i) = Softmax(Q_g K_i / \sqrt{d} + B)V_i, \qquad (4)$$

Q_i, K_i, V_i and Q_g, K_g, V_g represent Q, K, and V in the image and gaze attention path, respectively; B denotes the learnable relative positional encoding; d is the dimension of K, and we set the number of head of multi-headed self-attention is 12.

3 Experiments

3.1 Datasets and Evaluation

Our experiments use the Synapse multi-organ segmentation dataset(Synapse). Each CT volume consists of $85 - 198$ slices of 512×512 pixels, with a voxel spatial resolution of $([0.54 - 0.54] \times [0.98 - 0.98] \times [2.5 - 5.0]) \, mm^3$. We use the 30 abdominal CT scans and split it 18 training cases and 12 testing cases randomly. Following [3,4], all 3D volumes are inferenced in a slice-by-slice fashion and the predicted 2D slices are stacked together to reconstruct the 3D prediction. We use the average Dice-Similarity coefficient(DSC) and average Hausdorff distance (HD) as the evaluation metric to evaluate our method on the full resolution of the original slice.

Table 1. Comparison on the Synapse multi-organ CT dataset (average dice score % and average Hausdorff distance in mm, and dice score % for each organ).

Methods	DSC↑	HD↓	Aorta	Gallbladder	Kidney(L)	Kidney(R)	Liver	Pancreas	Spleen	Stomach
V-Net [10]	68.81	-	75.34	51.87	77.10	80.75	87.84	40.05	80.56	56.98
DARR [5]	69.77	-	74.74	53.77	72.31	73.24	94.08	54.18	89.90	45.96
R50 U-Net [4]	74.68	36.87	87.74	63.66	80.60	78.19	93.74	56.90	85.87	74.16
U-Net [13]	76.85	39.70	89.07	**69.72**	77.77	68.60	93.43	53.98	86.67	75.58
U-Net$_{gaze}$	77.94	35.50	89.66	65.04	81.25	75.91	93.57	59.94	84.66	73.53
R50 Att-UNet [4]	75.57	36.97	55.92	63.91	79.20	72.71	93.56	49.37	87.19	74.95
Att-UNet [11]	77.77	36.02	89.55	68.88	77.98	71.11	93.57	58.04	87.30	75.75
R50 ViT [4]	71.29	32.87	73.73	55.13	75.80	72.20	91.51	45.99	81.99	73.95
TransUnet [4]	77.48	31.69	87.23	63.13	81.87	77.02	94.08	55.86	85.08	75.62
TransUnet$_{gaze}$	78.29	27.00	88.57	61.89	83.07	75.99	94.56	59.44	85.54	77.27
SwinUnet [3]	79.13	21.55	85.47	66.53	83.28	79.61	94.29	56.58	**90.66**	76.60
SwinUnet$_{gaze}$	80.02	19.47	86.58	66.97	84.05	81.80	94.22	61.33	89.22	76.00
Ours	**81.87**	**11.96**	**89.88**	64.16	**86.62**	**83.89**	**94.77**	**66.97**	87.53	**81.16**

3.2   Results and Analysis

Overall Performance. As the last row shown in Table 1. Our experimental results demonstrate that leveraging gaze attention as an auxiliary supervision mechanism for network training achieves superior segmentation performance, as evidenced by segmentation accuracies of 81.87% (Dice similarity coefficient) and 11.96% (Hausdorff distance).

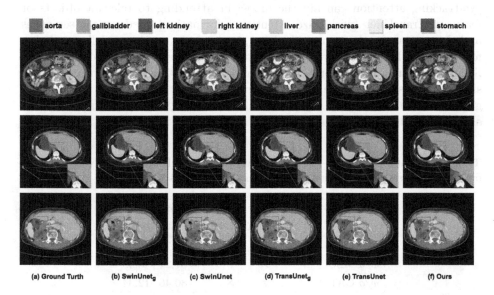

Fig. 4. 1). 1st row of the figure highlights that the approach leveraging gaze attention as auxiliary supervision yields fewer erroneous labels in comparison to the other methods; 2). TransUnet and SwinUnet (without gaze attention information) predict coarser edges and shapes compared to our method; 3). In the 3rd row, our method correctly identifies the stomach, while SwinUnet (with and without gaze attention information) failed to predict the shape of the stomach.

Comparison with Existing Methods Performance. As shown in Table 1, We also train $Unet_{gaze}$, $TransUnet_{gaze}$, and $SwinUnet_{gaze}$ networks with gaze attention information by simply concatenating the gaze attention information with the input image. Our experimental results reveal that this approach of introducing gaze attention information as an auxiliary supervision mechanism leads to an appreciable improvement in network segmentation performance. Specifically, Unet, TransUnet, and SwinUnet have an improvement of approximately 1% in terms of the DSC evaluation metric and 2–4% in terms of the HD evaluation metric by concatenating the gaze attention. Furthermore, our method surpasses the SwinUnet (without gaze attention information) by approximately 3% in terms of the DSC evaluation metric and approximately 10% in terms of the HD evaluation metric. Comparing our method with U-Net, TransUnet, and SwinUnet (with gaze attention information), we also observe a significant improvement of approximately 2–4% in terms of the DSC evaluation metric and 7–24% in terms of the HD evaluation metric, respectively.

Qualitative Visualization Results. As shown in Fig. 4. 1st row of the figure highlights that the approach leveraging gaze attention as auxiliary supervision yields fewer erroneous labels in comparison to the other methods, implying that eye-tracking attention can aid the model in attending to relevant objects or features. TransUnet (with gaze attention information) predicts coarser edges and shapes compared to our method (e.g. in the 2nd row, the model's prediction for the liver). In the 3rd row, our method correctly identifies the stomach, while SwinUnet (with and without gaze attention information) failed to predict the shape of the stomach. The results demonstrate that our method can leverage gaze attention as auxiliary supervision and better segment and retain edge and shape information.

Table 2. Ablation study on the different variant network under the Synapse multi-organ CT dataset (average dice score % and average Hausdorff distance in mm).

Variant	Modules			Metrics	
	DWT	MSC	CATM	DSC	HD
w/o DWT	✗	✓	✓	75.75	32.35
w/o MSC	✓	✗	✓	81.21	8.99
w/o CATM	✓	✓	✗	80.46	12.34
Full Version	✓	✓	✓	81.87	11.96

Ablation Study. We verified the DWT, MSC, and CATM separately using three different network configurations. We summarize the experimental results in Table 2. It can be seen that the w/o DWT performs the worst, indicating that the detail coefficients extracted from eye-tracking heatmaps can effectively locate organs in the image and provide strong support for edge segmentation. The w/o CATM does not effectively fuse the image features and eye-tracking attention features using CATM, resulting in a less-than-ideal improvement in segmentation results. In the results of w/o MSC and the full version of the network, we observed that MSC can further improve segmentation results by integrating down-sampling information from both paths.

4   Conclusion

In this paper, we propose a novel network that can realize the interactive communication between network semantic perception and human semantic perception, and apply it to the task of abdominal multi-organ segmentation for information interactive collaboration. The network is innovatively built with 1) a dual-path encoder that integrates human cognitive information; 2) a cross-attention transformer module (CATM) that communicates information in network semantic perception and human semantic perception; and 3) multi-feature skip connection (MSC), which effectively combines spatial information during down-sampling to offset the internal details of segmentation. Extensive experiments with promising results reveal gaze attention has great clinical value and potential in multi-organ segmentation.

Acknowledgements. This study was supported by the National Natural Science Foundation (No. 62101249 and No. 62136004), the Natural Science Foundation of Jiangsu Province (No. BK20210291), and the China Postdoctoral Science Foundation (No. 2021TQ0149 and No. 2022M721611).

References

1. Bertram, R., et al.: Eye movements of radiologists reflect expertise in CT study interpretation: a potential tool to measure resident development. Radiology **281**(3), 805–815 (2016)
2. Brunyé, T.T., Drew, T., Weaver, D.L., Elmore, J.G.: A review of eye tracking for understanding and improving diagnostic interpretation. Cogn. Res. Princ. Implic. **4**(1), 1–16 (2019). https://doi.org/10.1186/s41235-019-0159-2
3. Cao, H., et al.: Swin-Unet: Unet-like pure transformer for medical image segmentation. In: Karlinsky, L., Michaeli, T., Nishino, K. (eds.) ECCV 2022. LNCS, vol. 13803, pp. 205–218. Springer, Cham (2023). https://doi.org/10.1007/978-3-031-25066-8_9
4. Chen, J., et al.: Transunet: transformers make strong encoders for medical image segmentation. arXiv preprint arXiv:2102.04306 (2021)
5. Fu, S., et al.: Domain adaptive relational reasoning for 3D multi-organ segmentation. In: Martel, A.L., et al. (eds.) MICCAI 2020. LNCS, vol. 12261, pp. 656–666. Springer, Cham (2020). https://doi.org/10.1007/978-3-030-59710-8_64
6. Herzig, J., Nowak, P.K., Müller, T., Piccinno, F., Eisenschlos, J.M.: Tapas: weakly supervised table parsing via pre-training. arXiv preprint arXiv:2004.02349 (2020)
7. Kundel, H.L., Nodine, C.F., Krupinski, E.A., Mello-Thoms, C.: Using gaze-tracking data and mixture distribution analysis to support a holistic model for the detection of cancers on mammograms. Acad. Radiol. **15**(7), 881–886 (2008)
8. Li, G., Lyu, J., Wang, C., Dou, Q., Qin, J.: WavTrans: synergizing wavelet and cross-attention transformer for multi-contrast MRI super-resolution. In: Wang, L., Dou, Q., Fletcher, P.T., Speidel, S., Li, S. (eds.) MICCAI 2022. LNCS, vol. 13436, pp. 463–473. Springer, Cham (2022). https://doi.org/10.1007/978-3-031-16446-0_44
9. Men, Q., Teng, C., Drukker, L., Papageorghiou, A.T., Noble, J.A.: Multimodal-guidenet: gaze-probe bidirectional guidance in obstetric ultrasound scanning. In: Wang, L., Dou, Q., Fletcher, P.T., Speidel, S., Li, S. (eds.) MICCAI 2022. LNCS, vol. 13437, pp. 94–103. Springer, Cham (2022). https://doi.org/10.1007/978-3-031-16449-1_10
10. Milletari, F., Navab, N., Ahmadi, S.A.: V-net: fully convolutional neural networks for volumetric medical image segmentation. In: 2016 Fourth International Conference on 3D Vision (3DV), pp. 565–571. IEEE (2016)

11. Oktay, O., et al.: Attention U-net: learning where to look for the pancreas. arXiv preprint arXiv:1804.03999 (2018)
12. Ouyang, X., et al.: Learning hierarchical attention for weakly-supervised chest X-ray abnormality localization and diagnosis. IEEE Trans. Med. Imaging **40**(10), 2698–2710 (2020)
13. Ronneberger, O., Fischer, P., Brox, T.: U-Net: convolutional networks for biomedical image segmentation. In: Navab, N., Hornegger, J., Wells, W.M., Frangi, A.F. (eds.) MICCAI 2015. LNCS, vol. 9351, pp. 234–241. Springer, Cham (2015). https://doi.org/10.1007/978-3-319-24574-4_28
14. Shu, R., Chen, Y., Kumar, A., Ermon, S., Poole, B.: Weakly supervised disentanglement with guarantees. arXiv preprint arXiv:1910.09772 (2019)
15. Vaswani, A., et al.: Attention is all you need. In: Advances in Neural Information Processing Systems, vol. 30 (2017)
16. Wang, S., Ouyang, X., Liu, T., Wang, Q., Shen, D.: Follow my eye: using gaze to supervise computer-aided diagnosis. IEEE Trans. Med. Imaging **41**(7), 1688–1698 (2022)
17. Wu, C.C., Wolfe, J.M.: Eye movements in medical image perception: a selective review of past, present and future. Vision **3**(2), 32 (2019)

Scribble-Based 3D Multiple Abdominal Organ Segmentation via Triple-Branch Multi-Dilated Network with Pixel- and Class-Wise Consistency

Meng Han[1], Xiangde Luo[1], Wenjun Liao[2], Shichuan Zhang[2], Shaoting Zhang[1,3], and Guotai Wang[1,3(✉)]

[1] School of Mechanical and Electrical Engineering, University of Electronic Science and Technology of China, Chengdu, China
guotai.wang@uestc.edu.cn
[2] Department of Radiation Oncology, Sichuan Cancer Hospital and Institute, University of Electronic Science and Technology of China, Chengdu, China
[3] Shanghai Artificial Intelligence Laboratory, Shanghai, China

Abstract. Multi-organ segmentation in abdominal Computed Tomography (CT) images is of great importance for diagnosis of abdominal lesions and subsequent treatment planning. Though deep learning based methods have attained high performance, they rely heavily on large-scale pixel-level annotations that are time-consuming and labor-intensive to obtain. Due to its low dependency on annotation, weakly supervised segmentation has attracted great attention. However, there is still a large performance gap between current weakly-supervised methods and fully supervised learning, leaving room for exploration. In this work, we propose a novel 3D framework with two consistency constraints for scribble-supervised multiple abdominal organ segmentation from CT. Specifically, we employ a Triple-branch multi-Dilated network (TDNet) with one encoder and three decoders using different dilation rates to capture features from different receptive fields that are complementary to each other to generate high-quality soft pseudo labels. For more stable unsupervised learning, we use voxel-wise uncertainty to rectify the soft pseudo labels and then supervise the outputs of each decoder. To further regularize the network, class relationship information is exploited by encouraging the generated class affinity matrices to be consistent across different decoders under multi-view projection. Experiments on the public WORD dataset show that our method outperforms five existing scribble-supervised methods.

Keywords: Weakly-supervised learning · Scribble annotation · Uncertainty · Consistency

1 Introduction

Abdominal organ segmentation from medical images is an essential work in clinical diagnosis and treatment planning of abdominal lesions [17]. Recently, deep

H. Greenspan et al. (Eds.): MICCAI 2023, LNCS 14226, pp. 33–42, 2023.
https://doi.org/10.1007/978-3-031-43990-2_4

learning methods based on Convolution Neural Network (CNN) have achieved impressive performance in medical image segmentation tasks [2,24]. However, their success relies heavily on large-scale high-quality pixel-level annotations that are too expensive and time-consuming to obtain, especially for multiple organs in 3D volumes. Weakly supervised learning with a potential to reduce annotation costs has attracted great attention. Commonly-used weak annotations include dots [6,11], scribbles [1,11,13,15], bounding boxes [5], and image-level tags [20,25]. Compared with the other weak annotations, scribbles can provide more location information about the segmentation targets, especially for objects with irregular shapes [1]. Therefore, this work focuses on exploring high-performance models for multiple abdominal organ segmentation based on scribble annotations.

Training CNNs for segmentation with scribble annotations has been increasingly studied recently. Existing methods are mainly based on pseudo label learning [11,15], regularized losses [10,18,22] and consistency learning [7,13,26]. Pseudo label learning methods deal with unannotated pixels by generating fake semantic labels for learning. For example, Luo et al. [15] introduced a network with two slightly different decoders that generate dynamically mixed pseudo labels for supervision. Liang et al. [11] proposed to leverage minimum spanning trees to generate low-level and high-level affinity matrices based on color information and semantic features to refine the pseudo labels. Arguing that the pseudo label learning may be unreliable, Tang et al. [22] introduced the Conditional Random Field (CRF) regularization loss for image segmentation directly. Obukhov et al. [18] proposed to incorporate the gating function with CRF loss considering the directionality of unsupervised information propagation. Recently, consistency strategies that encourage consistent outputs of the network for the same input under different perturbations have achieved increasing attentions. Liu et al. [13] introduced transformation-consistency based on an uncertainty-aware mean teacher [4] model. Zhang et al. [26] proposed a framework composed of mix augmentation and cycle consistency. Although these scribble-supervised methods have achieved promising results, their performance is still much lower than that of fully-supervised training, leaving room for improvement.

Differently from most existing weakly supervised methods that are designed for 2D slice segmentation with a single or few organs, we propose a highly optimized 3D triple-branch network with one encoder and three different decoders, named TDNet, to learn from scribble annotations for segmentation of multiple abdominal organs. Particularly, the decoders are assigned with different dilation rates [25] to learn features from different receptive fields that are complementary to each other for segmentation, which also improves the robustness of dealing with organs at different scales as well as the feature learning ability of the shared encoder. Considering the features at different scales learned in these decoders, we fuse these multi-dilated predictions to obtain more accurate soft pseudo labels rather than hard labels [15] that tend to be over-confidence predictions. For more stable unsupervised learning, we use voxel-wise uncertainty to rectify the soft pseudo labels and then impose consistency constraints on the output of each branch. In addition, we extend the consistency to the class-related information

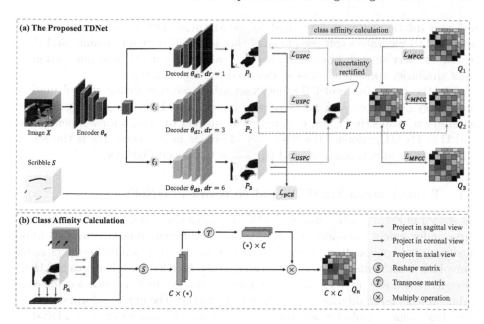

Fig. 1. Overview of the proposed Triple-branch multi-Dilated Network (TDNet) that uses different dilation rates at three decoders. The TDNet is optimized by Uncertainty-weighted Soft Pseudo label Consistency (USPC) using the mixed soft pseudo labels and Multi-view Projection-based Class-similarity Consistency (MPCC). The class affinity calculation process is shown in (b). Best viewed in color.

level [23] to constrain inter-class affinity for better distinguishing them. Specifically, we generate the class affinity matrices in different decoders and encourage them to be consistent after projection in different views.

The contributions of this paper are summarized as follows: 1) We propose a novel 3D Triple-branch multi-Dilated network called TDNet for scribble-supervised segmentation. By equipping with varying dilation rates, the network can better leverage multi-scale context for dealing with organs at different scales. 2) We propose two novel consistency loss functions, i.e., Uncertainty-weighted Soft Pseudo label Consistency (USPC) loss and Multi-view Projection-based Class-similarity Consistency (MPCC) loss, to regularize the prediction from the pixel-wise and class-wise perspectives respectively, which helps the segmentation network obtain reliable predictions on unannotated pixels. 3) Experiments results show our proposed method outperforms five existing scribble-supervised methods on the public dataset WORD [17] for multiple abdominal organ segmentation.

2   Method

Figure 1 shows the proposed framework for scribble-supervised medical image segmentation. We introduce a network with one encoder and three decoders with different dilation rates to learn multi-scale features. The decoders' outputs are averaged to generate a soft pseudo label that is rectified by uncertainty and then used to supervise each branch. To better deal with multi-class segmentation, a class similarity consistency loss is also used for regularization.

For the convenience of following description, we first define several mathematical symbols. Let X, S be a training image and the corresponding scribble annotation, respectively. Let C denote the number of classes for segmentation, and $\Omega = \Omega_S \cup \Omega_U$ denote the whole set of voxels in X, where Ω_S is the set of labeled pixels annotated in S, and Ω_U is the unlabeled pixel set.

2.1 Triple-Branch Multi-Dilated Network (TDNet)

As shown in Fig. 1(a), the proposed TDNet consists of a shared encoder (θ_e) and three independent decoders ($\theta_{d1}, \theta_{d2}, \theta_{d3}$) with different dilation rates to mine unsupervised context from different receptive fields. Specifically, decoders using convolution with small dilation rates can extract detailed local features but their receptive fields are small for understanding a global context. Decoders using convolution with large dilation rates can better leverage the global information but may lose some details for accurate segmentation. In this work, our TDNet is implemented by introducing two auxiliary decoders into a 3D UNet [3]. The dilation rate in the primary decoder and the two auxiliary decoders are 1, 3 and 6 respectively, with the other structure parameters (e.g., kernel size, channel number etc.) being the same in the three decoders. To further introduce perturbations for obtaining diverse outputs, the three branches are initialized with Kaiming initialization, Xavier and Normal initialization methods, respectively. In addition, the bottleneck's output features are randomly dropped out before sending into the auxiliary decoders. The probability prediction maps obtained by the three decoders are denoted as P_1, P_2 and P_3, respectively.

2.2 Pixel-Wise and Class-Wise Consistency

Uncertainty-Weighted Soft Pseudo Label Consistency (USPC). As the three decoders capture features at different scales that are complementary to each other, an ensemble of them would be more robust than a single branch. Therefore, we take an average of P_1, P_2, P_3 to get a better soft pseudo label $\bar{P} = (P_1 + P_2 + P_3)/3$ that is used to supervise each branch during training. However, $\bar{P}$ may also contain noises and be inaccurate, and it is important to highlight reliable pseudo labels while suppressing unreliable ones. Thus, we propose a regularization term named Uncertainty-weighted Soft Pseudo label Consistency (USPC) between P_n ($n = 1, 2, 3$) and $\bar{P}$:

$$\mathcal{L}_{USPC} = \frac{1}{3} \sum_{n=1,2,3} \frac{\sum_i w_i KL(P_{n,i} \| \bar{P}_i)}{\sum_i w_i} \qquad (1)$$

where $\bar{P}_i$ refers to the prediction probability at voxel i in $\bar{P}$, and $\bar{P}_{n,i}$ is the corresponding prediction probability at voxel i in $\bar{P}_n$. $KL()$ is the Kullback-

Leibler divergence. w_i is the voxel-wise weight based on uncertainty estimation:

$$w_i = e^{\sum_c \bar{P}_i^c log(\bar{P}_i^c)} \tag{2}$$

where the uncertainty is estimated by entropy. c is the class index, and $\bar{P}_i^c$ means the probability for class c at voxel i in the pseudo label. Note that a higher uncertainty leads to a lower weight. With the uncertainty-based weighting, the model will be less affected by unreliable pseudo labels.

Multi-view Projection-Based Class-Similarity Consistency (MPCC). For multi-class segmentation tasks, it is important to learn inter-class relationship for better distinguishing them. In addition to using $\mathcal{L}_{USPC}$ for pixel-wise supervision, we consider making consistency on class relationship across the outputs of the decoders as illustrated in Fig. 1. In order to save computing resources, we project the soft pseudo labels along each dimension and then calculate the affinity matrices, which also strengthens the class relationship information learning. We first project the soft prediction map of the n-th decoder $P_n \in \mathbb{R}^{C \times D \times H \times W}$ in axial view to a tensor with the shape of $C \times 1 \times H \times W$. It is reshaped into $C \times (WH)$ and multiplied by its transposed version, leading to a class affinity matrix $Q_n'^{axial} \in \mathbb{R}^{C \times C}$. A normalized version of $Q_n'^{axial}$ is denoted as $Q_n^{axial} = Q_n'^{axial}/||Q_n'^{axial}||$. Similarly, P_n is projected in the sagittal and coronal views, respectively, and the corresponding normalized class affinity matrices are denoted as $Q_n^{sagittal}$ and $Q_n^{coronal}$, respectively. Here, the affinity matrices represents the relationship between any pair of classes along the dimensions. Then we constraint the consistency among the corresponding affinity matrices by Multi-view Projection-based Class-similarity Consistency (MPCC) loss:

$$\mathcal{L}_{MPCC} = \frac{1}{3 \times 3} \sum_{v} \sum_{n=1,2,3} KL(Q_n^v || \bar{Q}^v) \tag{3}$$

where $v \in \{axial, sagittal, coronal\}$ is the view index, and $\bar{Q}^v$ is the average class affinity matrix in a certain view obtained by the three decoders.

2.3   Overall Loss Function

To learn from the scribbles, the partially Cross-Entropy (pCE) loss is used to train the network, where the labeled pixels are considered to calculate the gradient and the other pixels are ignored [21]:

$$\mathcal{L}_{sup} = -\frac{1}{3 |\Omega_S|} \sum_{n=1,2,3} \sum_{i \in \Omega_S} \sum_{c} S_i^c \log P_{n,i}^c \tag{4}$$

where S represents the one-hot scribble annotation, and Ω_S is the set of labeled pixels in S. The total object function is summarized as:

$$\mathcal{L}_{total} = \mathcal{L}_{sup} + \alpha_t \mathcal{L}_{USPC} + \beta_t \mathcal{L}_{MPCC} \tag{5}$$

where α_t and β_t are the weights for the unsupervised losses. Following [13], we define α_t based on a ramp-up function: $\alpha_t = \alpha \cdot e^{\left(-5(1-t/t_{\max})^2\right)}$, where t denotes the current training step and t_{max} is the maximum training step. We define $\beta_t = \beta \cdot e^{\left(-5(1-t/t_{\max})^2\right)}$ in a similar way. In this way, the model can learn accurate information from scribble annotations, which also avoids getting stuck in a degenerate solution due to low-quality pseudo labels at an early stage.

3 Experiments and Results

3.1 Dataset and Implementation Details

We used the publicly available abdomen CT dataset WORD [17] for experiments, which consists of 150 abdominal CT volumes from patients with rectal cancer, prostate cancer or cervical cancer before radiotherapy. Each CT volume contains 159–330 slices of 512×512 pixels, with an in-plane resolution of 0.976×0.976 mm and slice spacing of 2.5–3.0 mm. We aimed to segment seven organs: the liver, spleen, left kidney, right kidney, stomach, gallbladder and pancreas. Following the default settings in [17], the dataset was split into 100 for training, 20 for validation and 30 for testing, respectively, where the scribble annotations for foreground organs and background in the axial view of the training volumes had been provided and were used in model training. For pre-processing, we cut off the Hounsfield Unit (HU) values with a fixed window/level of 400/50 to focus on the abdominal organs, and normalized it to $[0, 1]$. We used the commonly-adopted Dice Similarity Coefficient (DSC), 95% Hausdorff Distance (HD$_{95}$) and the Average Surface Distance (ASD) for quantitative evaluation.

Our framework was implemented in PyTorch [19] on an NVIDIA 2080Ti with 11 GB memory. We employed the 3D UNet [3] as the backbone network for all experiments, and extended it with three decoders by embedding two auxiliary decoders with different dilation rates, as detailed in Sect. 2.1. To introduce perturbations, different initializations were applied to each decoder, and random perturbations (ratio $= (0, 0.5)$) were introduced in the bottleneck before the auxiliary decoders. The Stochastic Gradient Descent (SGD) optimizer with momentum of 0.9 and weight decay of 10^{-4} was used to minimize the overall loss function formulated in Eq. 5, where α=10.0 and β=1.0 based on the best performance on the validation set. The poly learning rate strategy [16] was used to decay learning rate online. The batch size, patch size and maximum iterations $t_{\max}$ were set to 1, $[80, 96, 96]$ and 6×10^4 respectively. The final segmentation results were obtained by using a sliding window strategy. For a fair comparison, we used the primary decoder's outputs as the final results during the inference stage and did not use any post-processing methods. Note that all experiments were conducted in the same experimental setting. The existing methods are implemented with the help of open source codebase from [14].

Table 1. Quantitative comparison between our method and existing weakly supervised methods on WORD testing set. * denotes p-value < 0.05 (paired t-test) when comparing with the second place method [15]. The best values are highlighted in bold.

Organ	FullySup [3]	pCE	TV [9]	USTM [13]	EM [8]	DMPLS [15]	Ours
Liver	$96.37_{\pm0.74}$	$86.53_{\pm3.61}$	$87.22_{\pm3.07}$	$80.57_{\pm3.87}$	$89.60_{\pm1.72}$	$88.89_{\pm2.54}$	$\mathbf{93.31_{\pm0.85}}^*$
spleen	$95.42_{\pm1.55}$	$86.81_{\pm5.75}$	$82.95_{\pm7.30}$	$86.49_{\pm4.71}$	$87.76_{\pm5.65}$	$89.19_{\pm3.93}$	$\mathbf{91.77_{\pm3.19}}^*$
kidney(L)	$94.95_{\pm1.58}$	$86.25_{\pm4.46}$	$83.78_{\pm4.60}$	$87.15_{\pm4.21}$	$87.29_{\pm3.88}$	$90.14_{\pm2.98}$	$\mathbf{92.34_{\pm2.30}}^*$
kidney(R)	$95.33_{\pm1.34}$	$89.41_{\pm2.97}$	$89.38_{\pm3.15}$	$89.26_{\pm2.35}$	$81.32_{\pm4.79}$	$90.93_{\pm2.15}$	$\mathbf{92.54_{\pm1.95}}^*$
stomach	$90.08_{\pm4.42}$	$61.09_{\pm12.20}$	$62.64_{\pm12.74}$	$77.33_{\pm6.19}$	$77.74_{\pm7.34}$	$77.06_{\pm7.39}$	$\mathbf{85.82_{\pm4.19}}^*$
gallbladder	$75.33_{\pm13.21}$	$56.61_{\pm20.12}$	$44.06_{\pm19.92}$	$63.94_{\pm17.17}$	$65.83_{\pm16.80}$	$\mathbf{70.25_{\pm15.57}}$	$69.01_{\pm16.53}$
pancreas	$80.90_{\pm7.67}$	$61.55_{\pm10.45}$	$65.31_{\pm9.20}$	$67.24_{\pm9.49}$	$73.52_{\pm7.54}$	$75.02_{\pm7.53}$	$\mathbf{75.40_{\pm7.51}}$
avg DSC(%)	$89.77_{\pm4.36}$	$75.46_{\pm8.51}$	$73.62_{\pm8.57}$	$78.85_{\pm6.86}$	$80.44_{\pm6.82}$	$83.07_{\pm6.01}$	$\mathbf{85.74_{\pm5.22}}^*$
avg ASD(mm)	$1.60_{\pm1.56}$	$25.11_{\pm11.59}$	$31.01_{\pm12.41}$	$18.24_{\pm9.29}$	$16.17_{\pm8.35}$	$7.77_{\pm6.33}$	$\mathbf{2.33_{\pm1.76}}^*$
avg HD$_{95}$(mm)	$5.71_{\pm5.36}$	$77.56_{\pm36.80}$	$98.61_{\pm39.93}$	$61.43_{\pm34.79}$	$50.90_{\pm29.92}$	$24.00_{\pm20.08}$	$\mathbf{7.84_{\pm5.84}}^*$

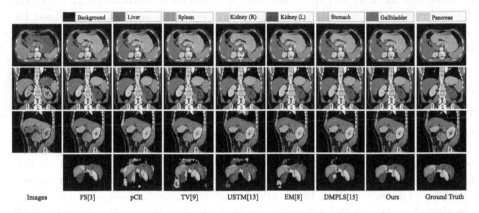

Fig. 2. Visual comparison between our method and other weakly supervised methods. Best view in color.

3.2 Comparison with Other Methods

We compared our method with five weakly supervised segmentation methods with the same set of scribbles, including pCE only [12], Total Variation Loss (TV) [9], Uncertainty-aware Self-ensembling and Transformation-consistent Model (USTM) [13], Entropy Minimization (EM) [8] and Dynamically Mixed Pseudo Labels Supervision (DMPLS) [15]. They were also compared with the upper bound by using dense annotation to train models (FullySup) [3]. The results in Table 1 show that our method leads to the best DSC, ASD and HD$_{95}$. Compared with the second best method DMPLS [15], the average DSC was increased by 2.67 percent points, and the average ASD and HD$_{95}$ were decreased by 5.44 mm and 16.16 mm, respectively. It can be observed that TV [9] obtained a worse performance than pCE, which is mainly because that method classifies pixels by minimizing the intra-class intensity variance, making it difficult to achieve good segmentation due to the low contrast. Figure 2 shows a visual comparison

Table 2. Ablation study of our proposed method on WORD validation set. N(s) and N(d) means N decoders with the same and different dilation rates, respectively. $\mathcal{L}_{sup}$ is used by default. The best values are highlighted in bold.

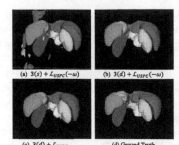

(a) $3(s) + \mathcal{L}_{USPC}(-\omega)$ (b) $3(d) + \mathcal{L}_{USPC}(-\omega)$

Decoder	Loss	DSC(%)	ASD(mm)	HD$_{95}$(mm)
1(s)		74.70$_{\pm 8.68}$	25.51$_{\pm 10.12}$	79.98$_{\pm 30.39}$
3(s)	$\mathcal{L}_{USPC}(-\omega)$	81.92$_{\pm 8.04}$	9.40$_{\pm 6.79}$	31.11$_{\pm 20.92}$
3(d)	$\mathcal{L}_{USPC}(-\omega)$	82.57$_{\pm 7.28}$	3.34$_{\pm 2.67}$	9.26$_{\pm 7.26}$
3(d)	$\mathcal{L}_{USPC}$	84.21$_{\pm 6.99}$	2.82$_{\pm 2.71}$	8.25$_{\pm 6.36}$
3(d)	$\mathcal{L}_{USPC} + \mathcal{L}_{MPCC}$	**84.75**$_{\pm 7.01}$	**2.64**$_{\pm 2.46}$	**7.91**$_{\pm 5.93}$
2(d)	$\mathcal{L}_{USPC} + \mathcal{L}_{MPCC}$	84.18$_{\pm 6.84}$	2.85$_{\pm 2.38}$	8.56$_{\pm 6.36}$
4(d)	$\mathcal{L}_{USPC} + \mathcal{L}_{MPCC}$	83.51$_{\pm 7.01}$	2.88$_{\pm 2.49}$	8.58$_{\pm 6.53}$

(c) $3(d) + \mathcal{L}_{USPC}$ (d) Ground Truth

Fig. 3. Visualization of the improvement obtained by using different dilation rates and uncertainty rectifying. Best viewed in color.

between our method and the other weakly supervised methods on the WORD dataset (word_0014.nii). It can be obviously seen that the results obtained by our method are closer to the ground truth, with less mis-segmentation in both slice level and volume level.

3.3   Ablation Experiment

We then performed ablation experiments to investigate the contribution of each part of our method, and the quantitative results on the validation set are shown in Table 2, where $\mathcal{L}_{USPC}(-\omega)$ means using $\mathcal{L}_{USPC}$ without pixel-wise uncertainty rectifying. Baseline refers to a triple-branch model with different initializations and random feature-level dropout in the bottleneck, supervised by pCE only. It can be observed that by using $\mathcal{L}_{USPC}(-\omega)$ with mutiple decoders, the model segmentation performance is greatly enhanced with average DSC increasing by 7.70%, ASD and HD$_{95}$ decreasing by 16.11 mm and 48.87 mm, respectively. By equipping each decoders with different dilation rates, the model's performance is further improved, especially in terms of ASD and HD$_{95}$, which proves our hypothesis that learning features from different scales can improve the segmentation accuracy. Replacing $\mathcal{L}_{USPC}(-\omega)$ with $\mathcal{L}_{USPC}$ further improved the DSC to 84.21%, and reduced the ASD and HD$_{95}$ by 0.52 mm and 1.01 mm through utilizing the uncertainty information. Visual comparison in Fig. 3 demonstrates that over-segmentation can be mitigated by using different dilation rates in the three decoders, and using the uncertainty-weighted pseudo labels can further improve the segmentation accuracy with small false positive regions removing.

Additionally, Table 2 shows that combining $\mathcal{L}_{USPC}$ and $\mathcal{L}_{MPCC}$ obtained the best performance, where the average DSC, ASD and HD$_{95}$ were 84.75%, 2.64 mm and 7.91 mm, respectively, which demonstrates the effectiveness of the proposed class similarity consistency. In order to find the optimal number of decoders, we set the decoder number to 2, 3 and 4 respectively. The quantitative results in the last three rows of Table 2 show that using three decoders outperformed using two and four decoders.

4   Conclusion

In this paper, we proposed a scribble-supervised multiple abdominal organ segmentation method consisting of a 3D triple-branch multi-dilated network with two-level consistency constraints. By equipping each decoder with different dilation rates, the model leverages features at different scales to obtain high-quality soft pseudo labels. In addition to mine knowledge from unannotated pixels, we also proposed USPC Loss and MPCC Loss to learn unsupervised information from the uncertainty-rectified soft pseudo labels and class affinity matrix information respectively. Experiments on a public abdominal CT dataset WORD demonstrated the effectiveness of the proposed method, which outperforms five existing scribble-based methods and narrows the performance gap between weakly-supervised and fully-supervised segmentation methods. In the future, we will explore the effect of our method on sparser labels, such as a volumetric data with scribble annotations on one or few slices.

Acknowledgements. This work was supported by the National Natural Science Foundation of China (No. 62271115), Science and Technology Department of Sichuan Province, China (2022YFSY0055) and Radiation Oncology Key Laboratory of Sichuan Province Open Fund (2022ROKF04).

References

1. Chen, Q., Hong, Y.: Scribble2D5: weakly-supervised volumetric image segmentation via scribble annotations. In: Wang, L., Dou, Q., Fletcher, P.T., Speidel, S., Li, S. (eds.) MICCAI 2022. LNCS, vol. 13438, pp. 234–243. Springer, Cham (2022). https://doi.org/10.1007/978-3-031-16452-1_23
2. Chen, X., et al.: A deep learning-based auto-segmentation system for organs-at-risk on whole-body computed tomography images for radiation therapy. Radiother. Oncol. **160**, 175–184 (2021)
3. Çiçek, Ö., Abdulkadir, A., Lienkamp, S.S., Brox, T., Ronneberger, O.: 3D U-Net: learning dense volumetric segmentation from sparse annotation. In: Ourselin, S., Joskowicz, L., Sabuncu, M.R., Unal, G., Wells, W. (eds.) MICCAI 2016. LNCS, vol. 9901, pp. 424–432. Springer, Cham (2016). https://doi.org/10.1007/978-3-319-46723-8_49
4. Cui, W., et al.: Semi-supervised brain lesion segmentation with an adapted mean teacher model. In: Chung, A.C.S., Gee, J.C., Yushkevich, P.A., Bao, S. (eds.) IPMI 2019. LNCS, vol. 11492, pp. 554–565. Springer, Cham (2019). https://doi.org/10.1007/978-3-030-20351-1_43
5. Dai, J., He, K., Sun, J.: Boxsup: exploiting bounding boxes to supervise convolutional networks for semantic segmentation. In: ICCV, pp. 1635–1643 (2015)
6. En, Q., Guo, Y.: Annotation by clicks: a point-supervised contrastive variance method for medical semantic segmentation. arXiv preprint arXiv:2212.08774 (2022)
7. Gao, F., et al.: Segmentation only uses sparse annotations: unified weakly and semi-supervised learning in medical images. Med. Image Anal. **80**, 102515 (2022)
8. Grandvalet, Y., Bengio, Y.: Semi-supervised learning by entropy minimization. In: NeurIPS, pp. 1–17 (2004)

9. Javanmardi, M., Sajjadi, M., Liu, T., Tasdizen, T.: Unsupervised total variation loss for semi-supervised deep learning of semantic segmentation. arXiv preprint arXiv:1605.01368 (2016)
10. Kim, B., Ye, J.C.: Mumford-shah loss functional for image segmentation with deep learning. IEEE Trans. Image Process. **29**, 1856–1866 (2019)
11. Liang, Z., Wang, T., Zhang, X., Sun, J., Shen, J.: Tree energy loss: towards sparsely annotated semantic segmentation. In: CVPR, pp. 16907–16916 (2022)
12. Lin, D., Dai, J., Jia, J., He, K., Sun, J.: Scribblesup: scribble-supervised convolutional networks for semantic segmentation. In: CVPR, pp. 3159–3167 (2016)
13. Liu, X., et al.: Weakly supervised segmentation of COVID19 infection with scribble annotation on CT images. Pattern Recogn. **122**, 108341 (2022)
14. Luo, X.: WSL4MIS (2021). https://github.com/Luoxd1996/WSL4MIS
15. Luo, X., et al.: Scribble-supervised medical image segmentation via dual-branch network and dynamically mixed pseudo labels supervision. In: Wang, L., Dou, Q., Fletcher, P.T., Speidel, S., Li, S. (eds.) MICCAI 2022. LNCS, vol. 13431, pp. 528–538. Springer, Cham (2022). https://doi.org/10.1007/978-3-031-16431-6_50
16. Luo, X., et al.: Efficient semi-supervised gross target volume of nasopharyngeal carcinoma segmentation via uncertainty rectified pyramid consistency. In: de Bruijne, M., et al. (eds.) MICCAI 2021. LNCS, vol. 12902, pp. 318–329. Springer, Cham (2021). https://doi.org/10.1007/978-3-030-87196-3_30
17. Luo, X., et al.: WORD: a large scale dataset, benchmark and clinical applicable study for abdominal organ segmentation from CT image. Med. Image Anal. **82**, 102642 (2022)
18. Obukhov, A., Georgoulis, S., Dai, D., Van Gool, L.: Gated CRF loss for weakly supervised semantic image segmentation. arXiv preprint arXiv:1906.04651 (2019)
19. Paszke, A., et al.: Pytorch: an imperative style, high-performance deep learning library. In: NeurIPS, vol. 32 (2019)
20. Ru, L., Zhan, Y., Yu, B., Du, B.: Learning affinity from attention: end-to-end weakly-supervised semantic segmentation with transformers. In: CVPR, pp. 16846–16855 (2022)
21. Tang, M., Djelouah, A., Perazzi, F., Boykov, Y., Schroers, C.: Normalized cut loss for weakly-supervised CNN segmentation. In: CVPR, pp. 1818–1827 (2018)
22. Tang, M., Perazzi, F., Djelouah, A., Ben Ayed, I., Schroers, C., Boykov, Y.: On regularized losses for weakly-supervised CNN segmentation. In: ECCV, pp. 507–522 (2018)
23. Tung, F., Mori, G.: Similarity-preserving knowledge distillation. In: ICCV, pp. 1365–1374 (2019)
24. Wang, Y., Zhou, Y., Shen, W., Park, S., Fishman, E.K., Yuille, A.L.: Abdominal multi-organ segmentation with organ-attention networks and statistical fusion. Med. Image Anal. **55**, 88–102 (2019)
25. Wei, Y., Xiao, H., Shi, H., Jie, Z., Feng, J., Huang, T.S.: Revisiting dilated convolution: a simple approach for weakly-and semi-supervised semantic segmentation. In: CVPR, pp. 7268–7277 (2018)
26. Zhang, K., Zhuang, X.: Cyclemix: a holistic strategy for medical image segmentation from scribble supervision. In: CVPR, pp. 11656–11665 (2022)

Geometry-Adaptive Network for Robust Detection of Placenta Accreta Spectrum Disorders

Zailiang Chen[1], Jiang Zhu[1], Hailan Shen[1(✉)], Hui Liu[2], Yajing Li[1], Rongchang Zhao[1], and Feiyang Yu[1]

[1] School of Computer Science and Engineering,
Central South University, Changsha, China
hn_shl@126.com

[2] Xiangya Hospital, Central South University, Changsha, China

Abstract. Placenta accreta spectrum (PAS) is a high-risk obstetric disorder associated with significant morbidity and mortality. Since the abnormal invasion usually occurs near the uteroplacental interface, there is a large geometry variation in the lesion bounding boxes, which considerably degrades the detection performance. In addition, due to the confounding visual representations of PAS, the diagnosis highly depends on the clinical experience of radiologists, which easily results in inaccurate bounding box annotations. In this paper, we propose a geometry-adaptive network for robust PAS detection. Specifically, to deal with the geometric prior missing problem, we design a Geometry-adaptive Label Assignment (GA-LA) strategy and a Geometry-adaptive RoI Fusion (GA-RF) module. The GA-LA strategy dynamically selects positive PAS candidates (RoIs) for each lesion according to its shape information. The GA-RF module aggregates the multi-scale RoI features based on the geometry distribution of proposals. Moreover, we develop a Lesion-aware Detection Head (LA-Head) to leverage high-quality predictions to iteratively refine inaccurate annotations with a novel multiple instance learning paradigm. Experimental results under both clean and noisy labels indicate that our method achieves state-of-the-art performance and demonstrate promising assistance for PAS diagnosis in clinical applications.

Keywords: PAS detection · Geometric information · Inaccurate annotations · Magnetic resonance imaging

1 Introduction

Placenta accreta spectrum (PAS) refers to the abnormal invasion of trophoblast cells into the myometrium at different depths of infiltration [10]. With the rising trend of advanced maternal age and cesarean section delivery, the incidence of PAS has increased steadily in recent years [5]. Undiagnosed PAS may lead to

H. Greenspan et al. (Eds.): MICCAI 2023, LNCS 14226, pp. 43–53, 2023.
https://doi.org/10.1007/978-3-031-43990-2_5

massive obstetric hemorrhage and hysterectomy [21]. Therefore, accurate detection of PAS prenatally is critical for appropriate treatment planning. Magnetic resonance imaging (MRI) provides valuable position information of placenta and can be an important complementary imaging method when ultrasound (US) diagnosis is inconclusive [1,15]. Since manually identifying PAS on MR images is time-consuming and labor-intensive, automated detection of PAS is significant in clinical practice.

However, due to the lack of open-source dataset, the research on computer-aided diagnosis of PAS is very limited. Previous studies [17,18,20,29] mainly focused on the classification task but accurate location cannot be provided. Moreover, existing object detection methods are designed for natural images [3,19,23, 30] or specific diseases [16,22,27], with no consideration for the characteristics of PAS. Given that the abnormal invasion usually occurs near the uteroplacental interface, the geometric information of lesion regions is highly correlated with the shape of placenta, thereby causing significant variation in the aspect ratios of PAS bounding boxes. Furthermore, MRI in most cases is used after suspecting an abnormality on US, this by itself raises the difficulty of accurate labeling [19].

To address the above issues, we propose a novel geometry-adaptive PAS detection method, which utilizes the shape prior of placenta to predict high-quality PAS bounding boxes and further refines inaccurate annotations. The prior knowledge is mainly reflected in the aspect ratio of lesion boxes. Specifically, to take advantage of the geometry prior, we design a Geometry-adaptive Label Assignment (GA-LA) strategy and a Geometry-adaptive RoI Fusion (GA-RF) module. The GA-LA strategy dynamically selects positive proposals by calculating the optimal IoU threshold for each lesion. The GA-RF module fuses multi-scale RoI features from different pyramid layers. To reduce the impact of lesions with large aspect ratio, the module generates fusion weights through the geometry distribution of proposals. Furthermore, in order to alleviate the reliance on accurate annotations, we construct a Lesion-aware Detection Head (LA-Head), which leverages the geometry-guided predictions to iteratively refine bounding box labels by multiple instance learning. To the best of our knowledge, this is the first work to automatically detect PAS disorders on MR images. The contributions of this paper can be summarized as follows: (1) A Lesion-aware Detection Head (LA-Head) is designed, which employs a new multiple instance learning approach to improve the robustness to inaccurate annotations. (2) A flexible Geometry-adaptive Label Assignment (GA-LA) strategy is proposed to select positive PAS candidates according to the shape of lesions. (3) A statistic-based Geometry-adaptive RoI Fusion (GA-RF) module is developed for aggregating multi-scale features based on the geometry distribution of proposals.

2   Method

The central idea of our geometry-adaptive network is to refine inaccurate bounding boxes with geometry-guided predictions. To this end, we propose a Geometry-adaptive Label Assignment (GA-LA) strategy and a Geometry-adaptive RoI

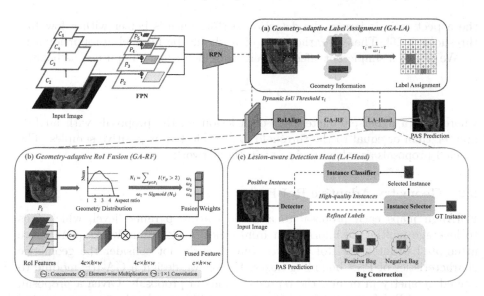

Fig. 1. Overview of our geometry-adaptive network, with novel (a) GA-LA, (b) GA-RF and (c) LA-Head. To solve the geometric prior missing problem, GA-LA strategy dynamically selects PAS RoIs according to the shape of placenta, and then GA-RF module fuses multi-scale features based on their geometry distribution. With the high-quality predictions, LA-Head refines inaccurate annotations by multiple instance learning.

Fusion (GA-RF) module to introduce geometry prior to detection. Then we develop a Lesion-aware Detection Head (LA-Head) to achieve the refinement of inaccurate annotations using multiple instance learning.

The overview of our method is illustrated in Fig. 1. Given an input image, Feature Pyramid Network (FPN) [12] extracts image features and then Region Proposal Network (RPN) [19] generates a set of region proposals. After obtaining the candidate boxes, RoIAlign [7] maps them to feature pyramid levels and extracts each proposal as a fixed-size feature map. Then the GA-RF module aggregates the multi-scale representations of each level through the statistical geometric information. Based on the fused RoI feature map, the LA-Head predicts and refines the classification and localization of lesions. During training, the GA-LA strategy assigns label to each proposal according to the shape of corresponding ground-truth box, making the model directly trainable.

2.1   Geometry-Adaptive Label Assignment

The static label assignment strategy adopted by Faster R-CNN predefines the IoU threshold τ to match objects and background to each anchor, but ignores the different shapes of PAS regions. To relieve the problem, we propose a GA-LA strategy to dynamically calculate the IoU threshold τ_i for each lesion bounding box g_i according to its aspect ratio r_i. A previous study demonstrated that the aspect ratio is larger, the detection performance is better with a low IoU threshold [9, 26]. Therefore, the value of τ_i should be inversely proportional to r_i. We design a simple but effective weighting factor and compute τ_i as below:

$$\tau_i = \frac{1}{\alpha r_i} \cdot \tau \tag{1}$$

where α is a hyper-parameter. For each lesion g_i, the proposals with an IoU greater than or equal to the threshold τ_i are selected as positive samples. The labeled proposals are then used to train the network.

2.2   Geometry-Adaptive RoI Fusion

RoIAlign maps each proposal to a single feature pyramid level, which fails to leverage multi-scale information from other levels. Some works [6, 14] have attempted to integrate RoI features, but they do not consider the geometric characteristics of PAS bounding boxes. Hence, we design a GA-RF module to embed geometry prior into the representations of proposals. Given a proposal, the mapped RoI features $\{f_i^l \in \mathbb{R}^{c \times h \times w}\}_{l=1}^L$ in different levels are concatenated in channel dimension, where L is the number of levels. We define the proposal whose aspect ratio is greater than R as a hard sample. The number of hard samples distributed to each level is formulated as geometry prior knowledge $s_i \in \mathbb{R}^{1 \times 4}$, which is used to generate fusion factor $\omega \in \mathbb{R}^{1 \times 4}$. The fused feature $f_i \in \mathbb{R}^{cL \times h \times w}$ is the weighted sum of all feature levels and expressed as follows:

$$N_l = \sum_{p \in P_l} \mathbb{I}(r_p \geq R), \ \omega = sigmoid(N_1, N_2, \cdots, N_L), \ f_i = \sum_{l=1}^L \omega_l \cdot f_i^l \tag{2}$$

where $\{N_l\}_{l=1}^L$ is the number of proposals with large aspect ratio in each layer. In this way, we take full advantage of the multi-scale information and the geometry distribution prior to enrich the feature representation for PAS prediction.

2.3   Lesion-Aware Detection Head

Motivated by [2, 25], we present a Lesion-aware Detection Head (LA-Head) to use high-quality proposals to iteratively refine the lesion box labels.

In this work, we regard PAS detection as a multiple instance learning (MIL) problem. In the standard paradigm for object detection, a MIL method treats an image as a bag and proposals in the image as instances [13, 28]. Different from previous studies, we treat an object as a bag B_i and proposals corresponding to the object as instances $P_i = \{p_i^j\}_{j=1}^N$, where N is the number of instances. Each bag has a label $y_i \in \{1, -1\}$, where $y_i = 1$ denotes an inaccurate ground-truth box containing at least one lesion candidate and $y_i = -1$ denotes a background box without lesions.

As shown in Fig. 1(c), our lesion-aware MIL method can be separated into three alternative parts: detector $\mu(\theta^d)$, instance selector $\phi(\theta^s)$ and instance classifier $\psi(\theta^c)$, where θ^d, θ^s and θ^c are parameters to be learned. First, the detector generates lesion instances based on proposal features. Then for each instance p_i^j, the instance selector computes the confidence score $\phi(p_i^j; \theta^s)$. Considering that the classification scores are not strongly correlated with localization quality [11], we select the most positive instance $p_i^{j^*}$ as follows:

$$j^* = \arg\max_j \phi(p_i^j; \theta^s) \tag{3}$$

where j^* is the index of the instance with the highest score. To leverage the classification information of predicted bounding boxes, we fuse the most positive instance $p_i^{j^*}$ and the ground-truth instance g_i to obtain a high-quality positive instance for training. The final selected instance is defined as below:

$$p_i^* = \delta_s(\phi(p_i^{j^*}; \theta^s)) \cdot p_i^{j^*} + (1 - \delta_s(\phi(p_i^{j^*}; \theta^s))) \cdot g_i \tag{4}$$

where δ_s is the weighting factor. Intuitively, the weight assigned to p_i^* should be higher when $\phi(p_i^{j^*}; \theta^s)$ is larger, and the weight assigned to g_i should have a lower bound γ to ensure that g_i can provide prior knowledge. Thus δ_s is calculated as:

$$\delta_s(x) = \min(e^{\beta \cdot x}, \gamma) \tag{5}$$

where β and γ are hyper-parameters. With the selected instances, instance classifier is trained to classify other instances as positive or negative. Then these proposals can serve as refined bounding box annotations for the detector. Furthermore, the detector generates instances to train the detection head and high-quality proposals can improve the detection performance. Thus this enables the self-feedback relationship between the detector and the LA-Head.

During training, instance selector, instance classifier and detector are jointly optimized based on the loss function:

$$L = \sum_i (L_s(B_i, \varphi(\theta^s)) + L_c(B_i, p_i^*, \phi(\theta^c)) + L_d(B_i, p_i^*, \mu(\theta^d)) \tag{6}$$

where L_s is the loss of instance selector which is defined as hinge loss:

$$L_s(B_i, \phi(\theta^c)) = \max(0, 1 - y_i \max_j \phi(p_i^j; \theta^c)) \tag{7}$$

The second term L_c is the loss of instance classifier and denoted as follows:

$$L_c(B_i, p_i^*, \psi(\theta^c)) = \sum_j \log(y_i^j \cdot (\psi(p_i^j; \theta^c) - \frac{1}{2}) + \frac{1}{2}) \tag{8}$$

where $\psi(p_i^j; \theta^c)$ is the probability that p_i^j contains lesions, and y_i^j is the label of p_i^j and calculated as follows:

$$y_i^j = \begin{cases} 1, y_i = 1 \text{ and } IoU(p_i^j, p_i^*) \geq \tau^* \\ -1, y_i = 1 \text{ and } IoU(p_i^j, p_i^*) < \tau^* \\ -1, y_i = -1 \end{cases} \tag{9}$$

where τ_i^* is the dynamic IoU threshold of p_i^j. The third term L_d is defined as:

$$L_d(B_i, p_i^*, \mu(\theta^d)) = \sum_j \mathbb{1}(y_i) \cdot \mathbb{1}(y_i^j) \cdot L_{reg}(p_i^j, p_i^*) \qquad (10)$$

where $\mathbb{1}(x)$ is the indicator function, set to 1 if $x = 1$; otherwise, set to 0. We adopt the smooth L_1 loss as the loss function L_{reg} for regression.

3   Experiments and Results

3.1   Dataset, Evaluation Metrics and Implementation Details

Dataset. Owing to the lack of open-source PAS dataset, our experiments are performed on a private dataset. We collected 110 placenta MRI scans of different patients upon the approval of Xiangya Hospital of Central South University. All T2-weighted image volumes were sliced along the sagittal plane. Two experienced radiologists with 20 and 14 years of clinical experience in medical imaging and PAS diagnosis selected the slices with PAS and manually annotated the lesion bounding boxes using LabelImg [24]. We finally obtain 312 2D MR images with a resolution of 640×640 px. To verify the robustness of our network, we simulate inaccurate annotations by perturbing clean bounding box labels. With the noise rate λ, we randomly shift and scale the ground-truth box $\{x_i, y_i, w_i, h_i\}$ as follows:

$$x_i' = x_i + \Delta_x w_i, y_i' = y_i + \Delta_y h_i, w_i' = w_i + \Delta_w w_i, h_i' = h_i + \Delta_h h_i \qquad (11)$$

where Δx, Δy, Δw, and Δh follow the uniform distribution $U(-\lambda, \lambda)$.

Evaluation Metrics. We adopt Average Precision (AP) and Sensitivity to evaluate the detection performance. In detail, AP is calculated over IoU threshold ranges from 0.25 to 0.95 at an interval of 0.05. Sensitivity denotes the proportion of correct prediction results in all ground-truths and is computed with an IoU threshold of 0.25 at 1 false positive per image.

Implementation Details. The proposed network is implemented with MMDetection 2.4.0 [4] and Pytorch 1.6 on NVIDIA GeForce RTX 1080Ti. We took FPN with ResNet50 [8] as backbone. The framework was trained using the SGD optimizer, where the initial learning rate, momentum, and weight decay were $5e-4$, 0.9, and $1e-3$, respectively. We adopt a batch size of 1 and the epochs of 120. The hyper-parameters are set as $\alpha = 2$, $\beta = 0.2$, $\gamma = 0.8$ and $R = 2$.

3.2   Ablation Study

The results of ablative experiments are listed in Table 1. Faster R-CNN is set as the baseline. We first explore the impact of GA-LA and GA-RF under accurate annotations. Compared with the baseline, two components bring 3.4% and

Table 1. Ablation study of the proposed method.

GA-LA	GA-RF	LA-Head	AP$_{25}$	AP$_{50}$	mAP	Sensitivity
Data with clean labels						
×	×	×	65.1	35.6	25.4	79.5
✓	×	×	68.7	41.8	28.8 (+3.4)	90.4 (+10.9)
×	✓	×	68.9	47.7	30.4 (+5.0)	82.2 (+2.7)
✓	✓	×	**79.2**	**51.1**	**34.1(+8.7)**	**91.8(+12.3)**
Data with noisy labels ($\lambda = 0.1$)						
×	×	×	65.2	30.2	24.2	83.6
×	×	✓	69.5	30.9	26.4 (+2.2)	84.9 (+1.3)
✓	✓	✓	**74.0**	**40.4**	**30.3(+6.1)**	**86.3(+2.7)**
Data with noisy labels ($\lambda = 0.2$)						
×	×	×	60.2	25.9	21.0	78.1
×	×	✓	61.4	34.3	22.3 (+1.3)	80.8 (+2.7)
✓	✓	✓	**71.1**	**39.5**	**27.3(+6.3)**	**83.6(+5.5)**

5.0% mAP gains separately. Another 3.7% mAP improvement is achieved when applied together. This finding indicates that geometry information of lesions can behave as effective prior for PAS detection. We then add LA-Head and conduct experiments under inaccurate annotations. It outperforms the baseline by at least 1.3% mAP, which demonstrates that the classification information of predictions is beneficial to alleviate the impact of noisy bounding box labels. We subsequently combine all key designs and obtain the optimal 30.3% and 26.8% mAP under 10% and 20% noise levels, outperforming the baseline by 6.1% and 5.8% respectively. The results reveal that high-quality predictions with geometry guidance can provide precise supervision for LA-Head. In addition, our method brings more obvious improvement under high annotation noise, which further proves its robustness to inaccurate annotations.

3.3   Comparison with State-of-the-Art Methods

We compare the geometry-adaptive network with seven object detection methods on both clean and noisy datasets. Faster R-CNN, Cascade R-CNN [3] and Dynamic R-CNN [30] are anchor-based methods, while FCOS is an anchor-free detector. ATSS [31] and SA-S [9] are dynamic label assignment strategies. Note that SA-S also uses the object shape information to select samples. AugFPN [6] is a variant of FPN and contains a soft RoI selection module.

The quantitative results are reported in Table 2. We first analyze the results of experiments under clean data. Compared with the general anchor-based and anchor-free methods, our method outperforms them by a large margin, thereby verifying that higher-quality predictions are generated under the guidance of geometry information. Compared with dynamic label assignment strategies, the

Table 2. Comparison with state-of-the-art methods.

Methods	Data with clean labels				Data with noisy labels			
	AP_{25}	AP_{50}	mAP	Sensitivity	AP_{25}	AP_{50}	mAP	Sensitivity
Faster R-CNN [19]	65.1	35.6	25.4	79.5	65.2	30.2	24.2	83.6
Cascade R-CNN [3]	62.6	38.7	27.2	76.7	57.0	30.1	22.7	79.5
Dynamic R-CNN [30]	71.5	40.2	30.3	89.0	69.2	39.8	28.3	84.9
FCOS [23]	70.8	36.0	27.4	86.3	68.9	38.7	26.6	90.4
ATSS [31]	68.3	40.1	28.2	82.2	65.1	28.5	23.8	90.4
SA-S [9]	71.3	46.1	31.5	90.4	66.4	33.9	26.0	**91.8**
AugFPN [6]	70.7	40.1	29.0	87.8	59.1	36.1	24.0	79.5
Ours	**79.2**	**51.1**	**34.1**	**91.8**	**74.0**	**40.4**	**30.3**	86.3

improvement of mAP by 5.9%, 2.6% and sensitivity by 9.6%, 1.4% demonstrated that our GA-LA strategy is simple but effective. Although SA-S also considers the shape of objects, the proposed method still achieves superior performance, likely because our model benefits from the two-stage structure. Compared with the feature fusion method, our approach performs favorably against heuristic-guided RoI selection of AugFPN. An intuitive explanation is that the optimal feature may be difficult to obtain using heuristic-guided method. Meanwhile our GA-RF module can adaptively generate representation according to the geometry prior of PAS lesions. We then analyze experimental results under 10% noise level data. Our geometry-adaptive network achieves consistent performance improvements compared with other state-of-the-art detectors. The result reveals that the high-quality predictions can serve as precise supervision signals for learning on inaccurate annotations. Figure 2 provides the visualization results of Faster R-CNN, Dynamic R-CNN, SA-S and AugFPN. The examples show that our method can generate PAS bounding boxes with more accurate shape and localization, which is consistent with the previous analysis.

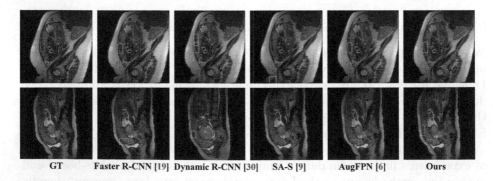

GT   Faster R-CNN [19] Dynamic R-CNN [30]   SA-S [9]   AugFPN [6]   Ours

Fig. 2. Visualization of the detection results. Red bounding boxes represent the ground-truth while green bounding boxes represent predictions. The first row and second row are the results under clean and noisy labels respectively. (Color figure online)

4   Conclusion

In this paper, we present a geometry-adaptive network for robust PAS detection. We point out that the geometry prior missing problem and inaccurate annotations could deteriorate the performance of detectors. To solve the problem, a Geometry-adaptive Label Assignment strategy (GA-LA) and a Geometry-adaptive RoI Fusion (GA-RF) module are proposed to fully utilize the geometry prior of lesions to predict high-quality proposals. Moreover, a Lesion-aware Detection Head (LA-Head) is developed to alleviate the impact of inaccurate annotations by leveraging the classification information of predicted boxes. The experimental results under both clean and noisy labels demonstrate that our method surpasses other state-of-the-art detectors.

Acknowledgements. This work was supported by the National Natural Science Foundation of China (No. 61972419), and the Natural Science Foundation of Hunan Province, China (No. 2021JJ30865 and 2023JJ30865).

References

1. Baughman, W.C., Corteville, J.E., Shah, R.R.: Placenta accreta: spectrum of us and MR imaging findings. Radiographics **28**(7), 1905–1916 (2008)
2. Bilen, H., Vedaldi, A.: Weakly supervised deep detection networks. In: Proceedings of the IEEE Conference on Computer Vision and Pattern Recognition, pp. 2846–2854 (2016)
3. Cai, Z., Vasconcelos, N.: Cascade R-CNN: delving into high quality object detection. In: Proceedings of the IEEE Conference on Computer Vision and Pattern Recognition, pp. 6154–6162 (2018)
4. Chen, K., et al.: Mmdetection: open mmlab detection toolbox and benchmark. arXiv preprint arXiv:1906.07155 (2019)
5. El Gelany, S., et al.: Placenta accreta spectrum (PAS) disorders: incidence, risk factors and outcomes of different management strategies in a tertiary referral hospital in Minia, Egypt: a prospective study. BMC Pregnancy Childbirth **19**, 1–8 (2019)
6. Guo, C., Fan, B., Zhang, Q., Xiang, S., Pan, C.: AugFPN: improving multi-scale feature learning for object detection. In: Proceedings of the IEEE/CVF Conference on Computer Vision and Pattern Recognition, pp. 12595–12604 (2020)
7. He, K., Gkioxari, G., Dollár, P., Girshick, R.: Mask R-CNN. In: Proceedings of the IEEE International Conference on Computer Vision, pp. 2961–2969 (2017)
8. He, K., Zhang, X., Ren, S., Sun, J.: Deep residual learning for image recognition. In: Proceedings of the IEEE Conference on Computer Vision and Pattern Recognition, pp. 770–778 (2016)
9. Hou, L., Lu, K., Xue, J., Li, Y.: Shape-adaptive selection and measurement for oriented object detection. In: Proceedings of the AAAI Conference on Artificial Intelligence, vol. 36, pp. 923–932 (2022)
10. Jauniaux, E., Chantraine, F., Silver, R., Langhoff-Roos, J.: Figo placenta accreta diagnosis and management expert consensus panel. figo consensus guidelines on placenta accreta spectrum disorders: epidemiology. Int. J. Gynaecol. Obstet. **140**(3), 265–273 (2018)

11. Jiang, B., Luo, R., Mao, J., Xiao, T., Jiang, Y.: Acquisition of localization confidence for accurate object detection. In: Proceedings of the European Conference on Computer Vision (ECCV), pp. 784–799 (2018)
12. Lin, T.Y., Dollár, P., Girshick, R., He, K., Hariharan, B., Belongie, S.: Feature pyramid networks for object detection. In: Proceedings of the IEEE Conference on Computer Vision and Pattern Recognition, pp. 2117–2125 (2017)
13. Liu, C., Wang, K., Lu, H., Cao, Z., Zhang, Z.: Robust object detection with inaccurate bounding boxes. In: Avidan, S., Brostow, G., Cissé, M., Farinella, G.M., Hassner, T. (eds.) European Conference on Computer Vision. LNCS, vol. 13670, pp. 53–69. Springer, Cham (2022). https://doi.org/10.1007/978-3-031-20080-9_4
14. Liu, S., Qi, L., Qin, H., Shi, J., Jia, J.: Path aggregation network for instance segmentation. In: Proceedings of the IEEE Conference on Computer Vision and Pattern Recognition, pp. 8759–8768 (2018)
15. Masselli, G., et al.: Magnetic resonance imaging in the evaluation of placental adhesive disorders: correlation with color doppler ultrasound. Eur. Radiol. 18, 1292–1299 (2008)
16. Mathai, T.S., et al.: Detection of lymph nodes in T2 MRI using neural network ensembles. In: Lian, C., Cao, X., Rekik, I., Xu, X., Yan, P. (eds.) MLMI 2021. LNCS, vol. 12966, pp. 682–691. Springer, Cham (2021). https://doi.org/10.1007/978-3-030-87589-3_70
17. Qi, H.: Prenatal assessment of placenta accreta spectrum disorders from ultrasound images using deep learning. Ph.D. thesis, University of Oxford (2019)
18. Qi, H., Collins, S., Noble, J.A.: Knowledge-guided pretext learning for uteroplacental interface detection. In: Martel, A.L., et al. (eds.) MICCAI 2020. LNCS, vol. 12261, pp. 582–593. Springer, Cham (2020). https://doi.org/10.1007/978-3-030-59710-8_57
19. Ren, S., He, K., Girshick, R., Sun, J.: Faster R-CNN: towards real-time object detection with region proposal networks. In: Advances in Neural Information Processing Systems, vol. 28 (2015)
20. Shao, Q., et al.: Deep learning and radiomics analysis for prediction of placenta invasion based on T2WI. Math. Biosci. Eng. 18(5), 6198–6215 (2021)
21. Silver, R.M., Branch, D.W.: Placenta accreta spectrum. N. Engl. J. Med. 378(16), 1529–1536 (2018)
22. Swinburne, N.C., et al.: Semisupervised training of a brain MRI tumor detection model using mined annotations. Radiology 303(1), 80–89 (2022)
23. Tian, Z., Shen, C., Chen, H., He, T.: FCOS: fully convolutional one-stage object detection. In: Proceedings of the IEEE/CVF International Conference on Computer Vision, pp. 9627–9636 (2019)
24. Tzutalin: Labelimg (2015). https://github.com/heartexlabs/labelImg
25. Wan, F., Liu, C., Ke, W., Ji, X., Jiao, J., Ye, Q.: C-mil: continuation multiple instance learning for weakly supervised object detection. In: Proceedings of the IEEE/CVF Conference on Computer Vision and Pattern Recognition, pp. 2199–2208 (2019)
26. Wan, Z., Chen, Y., Deng, S., Chen, K., Yao, C., Luo, J.: Slender object detection: diagnoses and improvements. arXiv preprint arXiv:2011.08529 (2020)
27. Wang, S., et al.: Global-local attention network with multi-task uncertainty loss for abnormal lymph node detection in MR images. Med. Image Anal. 77, 102345 (2022)
28. Xu, Y., Zhu, L., Yang, Y., Wu, F.: Training robust object detectors from noisy category labels and imprecise bounding boxes. IEEE Trans. Image Process. 30, 5782–5792 (2021)

29. Xuan, R., Li, T., Wang, Y., Xu, J., Jin, W.: Prenatal prediction and typing of placental invasion using MRI deep and radiomic features. Biomed. Eng. Online **20**(1), 56 (2021)

30. Zhang, H., Chang, H., Ma, B., Wang, N., Chen, X.: Dynamic R-CNN: towards high quality object detection via dynamic training. In: Vedaldi, A., Bischof, H., Brox, T., Frahm, J.-M. (eds.) ECCV 2020. LNCS, vol. 12360, pp. 260–275. Springer, Cham (2020). https://doi.org/10.1007/978-3-030-58555-6_16

31. Zhang, S., Chi, C., Yao, Y., Lei, Z., Li, S.Z.: Bridging the gap between anchor-based and anchor-free detection via adaptive training sample selection. In: Proceedings of the IEEE/CVF Conference on Computer Vision and Pattern Recognition, pp. 9759–9768 (2020)

Clinical Applications – Breast

Clinical Applications – Breast

DisAsymNet: Disentanglement of Asymmetrical Abnormality on Bilateral Mammograms Using Self-adversarial Learning

Xin Wang[1,2], Tao Tan[3,1(✉)], Yuan Gao[1,2], Luyi Han[1,4], Tianyu Zhang[1,2,4], Chunyao Lu[1,4], Regina Beets-Tan[1,2], Ruisheng Su[5], and Ritse Mann[1,4]

[1] Department of Radiology, Netherlands Cancer Institute (NKI),
1066 CX Amsterdam, The Netherlands
taotanjs@gmail.com
[2] GROW School for Oncology and Development Biology, Maastricht University,
6200 MD Maastricht, The Netherlands
[3] Faculty of Applied Sciences, Macao Polytechnic University, Macao 999078, China
[4] Department of Radiology and Nuclear Medicine,
Radboud University Medical Centre, Nijmegen, The Netherlands
[5] Erasmus Medical Center, Erasmus University,
3015 GD Rotterdam, The Netherlands

Abstract. Asymmetry is a crucial characteristic of bilateral mammograms (Bi-MG) when abnormalities are developing. It is widely utilized by radiologists for diagnosis. The question of *"what the symmetrical Bi-MG would look like when the asymmetrical abnormalities have been removed ?"* has not yet received strong attention in the development of algorithms on mammograms. Addressing this question could provide valuable insights into mammographic anatomy and aid in diagnostic interpretation. Hence, we propose a novel framework, DisAsymNet, which utilizes asymmetrical abnormality transformer guided self-adversarial learning for disentangling abnormalities and symmetric Bi-MG. At the same time, our proposed method is partially guided by randomly synthesized abnormalities. We conduct experiments on three public and one in-house dataset, and demonstrate that our method outperforms existing methods in abnormality classification, segmentation, and localization tasks. Additionally, reconstructed normal mammograms can provide insights toward better interpretable visual cues for clinical diagnosis. The code will be accessible to the public.

Keywords: Bilateral mammogram · Asymmetric transformer · Disentanglement · Self-adversarial learning · Synthesis

1 Introduction

Breast cancer (BC) is the most common cancer in women and incidence is increasing [14]. With the wide adoption of population-based mammography

H. Greenspan et al. (Eds.): MICCAI 2023, LNCS 14226, pp. 57–67, 2023.
https://doi.org/10.1007/978-3-031-43990-2_6

screening programs for early detection of BC, millions of mammograms are conducted annually worldwide [23]. Developing artificial intelligence (AI) for abnormality detection is of great significance for reducing the workload of radiologists and facilitating early diagnosis [21]. Besides using the data-driven manner, to achieve accurate diagnosis and interpretation of the AI-assisted system output, it is essential to consider mammogram domain knowledge in a model-driven fashion.

Authenticated by the BI-RADS lexicon [12], the asymmetry of bilateral breasts is a crucial clinical factor for identifying abnormalities. In clinical practice, radiologists typically compare the bilateral craniocaudal (CC) and mediolateral oblique (MLO) projections and seek the asymmetry between the right and left views. Notably, the right and the left view would not have pixel-level symmetry differences in imaging positions for each breast and biological variations between the two views. Leveraging bilateral mammograms (Bi-MG) is one of the key steps to detect asymmetrical abnormalities, especially for subtle and non-typical abnormalities. To mimic the process of radiologists, previous studies only extracted simple features from the two breasts and used fusion techniques to perform the classification [6,20,22,24,25]. Besides these simple feature-fusion methods, recent studies have demonstrated the powerful ability of transformer-based methods to fuse information in multi-view (MV) analysis (CC and MLO view of unilateral breasts) [1,16,26]. However, most of these studies formulate the diagnosis as an MV analysis problem without dedicated comparisons between the two breasts.

The question of *"what the Bi-MG would look like if they were symmetric?"* is often considered when radiologists determine the symmetry of Bi-MG. It can provide valuable diagnostic information and guide the model in learning the diagnostic process akin to that of a human radiologist. Recently, two studies explored generating healthy latent features of target mammograms by referencing contralateral mammograms, achieving state-of-the-art (SOTA) classification performance [18,19]. None of these studies is able to reconstruct a normal pixel-level symmetric breast in the model design. Image generation techniques for generating symmetric Bi-MG have not yet been investigated. Visually, the remaining parts after the elimination of asymmetrical abnormalities are the appearance of symmetric Bi-MG. A more interpretable and pristine strategy is disentanglement learning [9,17] which utilizes synthetic images to supervise the model in separating asymmetric anomalies from normal regions at the image level.

In this work, we present a novel end-to-end framework, DisAsymNet, which consists of an *asymmetric transformer-based classification (AsyC) module* and an *asymmetric abnormality disentanglement (AsyD) module*. The *AsyC* emulates the radiologist's analysis process of checking unilateral and comparing Bi-MG for abnormalities classifying. The *AsyD* simulates the process of disentangling the abnormalities and normal glands on pixel-level. Additionally, we leverage a self-adversarial learning scheme to reinforce two modules' capacity, where the feedback from the *AsyC* is used to guide the *AsyD*'s disentangling, and the *AsyD*'s output is used to refine the *AsyC* in detecting subtle abnormalities. To

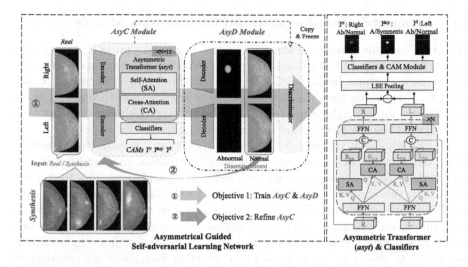

Fig. 1. The schematic overview of the proposed DisAsymNet.

facilitate the learning of semantic symmetry, we also introduce *Synthesis*, combining randomly created synthetic asymmetrical Bi-MG with real mammograms to supervise the learning process. **Our contributions are summarized as follows:** *(1)* We propose a framework comprising the *AsyC* and *AcyD* modules for exploiting clinical classification and localization of asymmetry in an interpretable way. *(2)* We propose *Synthesis* to simulate the asymmetry using normal pair of views to guide the model for providing normal symmetric breast views and to indicate the abnormal regions in an accurate supervised fashion. *(3)* We demonstrate the robustness of our approach on four mammogram datasets for classification, segmentation, and localization tasks.

2   Methodology

In this study, the "asymmetric" refers to the visual differences on perception level that can arise between the left and right breasts due to any abnormality, including both benign and malignant lesions. Thus, a paired Bi-MG is considered symmetrical only if both sides are normal and the task is different from the malignancy classification study [13]. The paired Bi-MG of the same projection is required, which can be formulated as $\mathcal{I} = \{\mathbf{x}^r, \mathbf{x}^l, \mathbf{y}^{asy}, \mathbf{y}^r, \mathbf{y}^l\}$. Here, $\mathbf{x} \in \mathbb{R}^{H \times W}$ represents a mammogram with the size of $H \times W$, $\mathbf{x}^r$ and $\mathbf{x}^l$ correspond to the right and left view respectively. $\mathbf{y}^r, \mathbf{y}^l, \mathbf{y}^{asy} \in \{0,1\}$ are binary labels, indicating abnormality for each side, and the asymmetry of paired Bi-MG. The framework of our DisAsymNet is illustrated in Fig. 1. Specifically, the *AsyC* module takes a pair of Bi-MG as input and predicts if it is asymmetric and if any side is abnormal. We employ an online Class Activation Mapping (CAM) module [10,11] to generate heatmaps for segmentation and localization. Subsequently, the *AsyD* module disentangles the abnormality from the normal part of the Bi-MG through the self-adversarial learning and *Synthesis* method.

2.1   Asymmetric Transformer-Based Classification Module

The *AsyC* module consists of shared encoders ψ_e and asymmetric transformer layers ψ_{asyt} to extract features and learn bilateral-view representations from the paired mammograms. In this part, we first extract the starting features f of each side (f^r, f^l represent the right and left features respectively) through ψ_e in the latent space for left-right inspection and comparison, which can be denoted as $f = \psi_e(\mathbf{x})$. Then the features are fed into the ψ_{asyt}.

Unlike other MV transformer methods [1,16] that use only cross-attention (CA), our asymmetric transformer employs self-attention (SA) and CA in parallel to aggregate information from both self and contralateral sides to enhance the side-by-side comparison. This is motivated by the fact that radiologists commonly combine unilateral (identifying focal suspicious regions according to texture, shape, and margin) and bilateral analyses (comparing them with symmetric regions in the contralateral breasts) to detect abnormalities in mammography [6]. As shown in the right of Fig. 1, starting features f are transformed into query (f_Q), key (f_K), and value (f_V) vectors through feed-forward network (FFN) layers. The SA and CA modules use multi-head attention (MHA), $\psi_{mha}^{h=8}(f_Q, f_K, f_V)$ with the number of heads $h = 8$, which is a standard component in transformers and has already gained popularity in medical image fields [1,16,26]. In the SA, the query, key, and value vectors are from the same features, $f_{SA} = \psi_{mha}^{h=8}(f_Q, f_K, f_V)$. While in the CA, we replace the key and value vectors with those from the contralateral features, $f_{CA}^l = \psi_{mha}^{h=8}(f_Q^l, f_K^r, f_V^r)$ or $f_{CA}^r = \psi_{mha}^{h=8}(f_Q^r, f_K^l, f_V^l)$. Then, the starting feature f, and the attention features f_{SA} and f_{CA} are concatenated in the channel dimension and fed into the FFN layers to fuse the information and maintain the same size as f. The transformer block is repeated $N = 12$ times to iteratively integrate information from Bi-MG, resulting in the output feature $f_{out}^r, f_{out}^l = \psi_{asyt}^{N=12}(f^r, f^l)$.

To predict the abnormal probability $\hat{y}$ of each side, the output features f_{out} are fed into the abnormal classifier. For the asymmetry classification of paired mammograms, we compute the absolute difference of the output features between the right and left sides ($f_{out}^{asy} = abs(f_{out}^r - f_{out}^l)$, which for maximizing the difference between the two feature) and feed it into the asymmetry classifier. We calculate the classification loss using the binary cross entropy loss (BCE) $\mathcal{L}_{bce}$, denoted as $\mathcal{L}_{diag} = \mathcal{L}_{cls}(\mathbf{y}^{asy}, \mathbf{y}^r, \mathbf{y}^l, \mathbf{x}^r, \mathbf{x}^l) = \mathcal{L}_{bce}(\mathbf{y}^{asy}, \hat{\mathbf{y}}^{asy}) + \mathcal{L}_{bce}(\mathbf{y}, \hat{\mathbf{y}})$.

2.2   Disentangling via Self-adversarial Learning

What would the Bi-MG look like when the asymmetrical abnormalities have been removed? Unlike previous studies [18,19], which only generated normal features in the latent space, our *AsyD* module use weights shared U-Net-like decoders ψ_g, to generate both abnormal ($\mathbf{x}_{ab}$) and normal ($\mathbf{x}_n$) images for each side through a two-channel separation, as $\mathbf{x}_n, \mathbf{x}_{ab} = \psi_g(f_{out})$. We constrain the model to reconstruct images realistically using L1 loss ($\mathcal{L}_{l1}$) with the guidance of CAMs (M), as follows, $\mathcal{L}_{rec} = \mathcal{L}_{l1}((1 - M)\mathbf{x}, (1 - M)\mathbf{x}_n) + \mathcal{L}_{l1}(M\mathbf{x}, \mathbf{x}_{ab})$. However, it is difficult to train the generator in a supervised manner due to the lack of annotations of the location for asymmetrical pairs. Inspired by previous self-adversarial learning work [10], we introduce a frozen discriminator ψ_d to impose constraints on the generator to address this challenge. The frozen discriminator comprises the same components as $AsyC$. In each training step, we update the discriminator parameters by copying them from the $AsyC$ for leading ψ_g to generate the symmetrical Bi-MG. The ψ_d enforces symmetry in the paired Bi-MG, which can be denoted as $\mathcal{L}_{dics} = \mathcal{L}_{cls}(\mathbf{y}^{asy} = 0, \mathbf{y}^r = 0, \mathbf{y}^l = 0, \mathbf{x}_n^r, \mathbf{x}_n^l)$. Furthermore, we use generated normal Bi-MG to reinforce the ability of $AsyC$ to recognize subtle asymmetry and abnormal cues, as $\mathcal{L}_{refine} = \mathcal{L}_{cls}(\mathbf{y}^{asy}, \mathbf{y}^r, \mathbf{y}^l, \mathbf{x}_n^r, \mathbf{x}_n^l)$.

2.3 Asymmetric Synthesis for Supervised Reconstruction

To alleviate the lack of annotation pixel-wise asymmetry annotations, in this study, we propose a random synthesis method to supervise disentanglement. Training with synthetic artifacts is a low-cost but efficient way to supervise the model to better reconstruct images [15,17]. In this study, we randomly select the number $n \in [1, 2, 3]$ of tumors t from a tumor set $\mathcal{T}$ inserting into one or both sides of randomized selected symmetric Bi-MG ($\mathbf{x}^r, \mathbf{x}^l | y^{asy} = 0$). For each tumor insertion, we randomly select a position within the breast region. The tumors and symmetrical mammograms are combined by an alpha blending-based method [17], which can be denoted by $\mathbf{x}|fake = \mathbf{x} \prod_{k=1}^n (1 - \alpha_k) + \sum_{k=1}^n t_k \alpha_k, t \in \mathcal{T}$. The alpha weights α_k is a 2D Gaussian distribution map, in which the co-variance is determined by the size of k-th tumor t, representing the transparency of the pixels of the tumor. Some examples are shown in Fig. 1. The tumor set $\mathcal{T}$ is collected from real-world datasets. Specifically, to maintain the rule of weakly-supervised learning of segmentation and localization tasks, we collect the tumors from the DDSM dataset as $\mathcal{T}$ and train the model on the INBreast dataset. When training the model on other datasets, we use the tumor set collected from the INBreast dataset. Thus, the supervised reconstruction loss is $\mathcal{L}_{syn} = \mathcal{L}_{l1}(\mathbf{x}|real, \mathbf{x}_n|fake)$, where $\mathbf{x}|real$ is the real image before synthesis and $\mathbf{x}_n|fake$ is the disentangled normal image from the synthesised image $\mathbf{x}|fake$.

2.4 Loss Function

For each training step, there are two objectives, training $AsyC$ and $AsyD$ module, and then is the refinement of $AsyC$. For the first, the loss function can be denoted by $\mathcal{L} = \lambda_1 \mathcal{L}_{diag} + \lambda_2 \mathcal{L}_{rec} + \lambda_3 \mathcal{L}_{dics} + \lambda_4 \mathcal{L}_{syn}$. The values of weight terms $\lambda_1, \lambda_2, \lambda_3$, and λ_4 are experimentally set to be 1, 0.1, 1, and 0.5, respectively. The loss of the second objective is $\mathcal{L}_{refine}$ as aforementioned.

3 Experimental

3.1 Datasets

This study reports experiments on four mammography datasets. The INBreast dataset [7] consists of 115 exams with BI-RADS labels and pixel-wise annotations, comprising a total of 87 normal (BI-RADS = 1) and 342 abnormal (BI-RADS ≠ 1) images. The DDSM dataset [3] consists of 2,620 cases, encompassing 6,406 normal and 4,042 (benign and malignant) images with outlines generated by an experienced mammographer. The VinDr-Mammo dataset [8] includes 5,000 cases with BI-RADS assessments and bounding box annotations, consisting of 13,404 normal (BI-RADS = 1) and 6,580 abnormal (BI-RADS ≠ 1) images. The In-house dataset comprises 43,258 mammography exams from 10,670 women between 2004–2020, collected from a hospital with IRB approvals. In this study, we randomly select 20% women of the full dataset, comprising 6,000 normal (BI-RADS = 1) and 28,732 abnormal (BI-RADS ≠ 1) images. Due to a lack of annotations, the In-house dataset is only utilized for classification tasks. Each dataset is randomly split into training, validation, and testing sets at the patient level in an 8:1:1 ratio, respectively (except for that INBreast which is split with a ratio of 6:2:2, to keep enough normal samples for the test).

Table 1. Comparison of asymmetric and abnormal classification tasks on four mammogram datasets. We report the AUC results with 95% CI. Note that, when ablating the "*AsyC*", we only drop the "*asyt*" and keep the encoders and classifiers.

Methods	INBreast Asymmetric	INBreast Abnormal	DDSM Asymmetric	DDSM Abnormal	VinDr-Mammo Asymmetric	VinDr-Mammo Abnormal	In-house Asymmetric	In-house Abnormal
ResNet18 [2]	NA	0.667 (0.460-0.844)	NA	0.768 (0.740-0.797)	NA	0.776 (0.750-0.798)	NA	0.825 (0.808-0.842)
HAM [11]	NA	0.680 (0.438-0.884)	NA	0.769 (0.742-0.796)	NA	0.780 (0.778-0.822)	NA	0.828 (0.811-0.845)
Late-fusion [4]	0.778 (0.574-0.947)	0.718 (0.550-0.870)	0.931 (0.907-0.954)	0.805 (0.779-0.830)	0.782 (0.753-0.810)	0.803 (0.781-0.824)	0.887 (0.867-0.904)	0.823 (0.807-0.841)
CVT [16]	0.801 (0.612-0.952)	0.615 (0.364-0.846)	0.953 (0.927-0.973)	0.790 (0.765-0.815)	0.803 (0.774-0.830)	0.797 (0.775-0.819)	0.886 (0.867-0.903)	0.821 (0.803-0.839)
Wang et al. [20]	0.850 (0.685-0.979)	0.755 (0.529-0.941)	0.950 (0.925-0.970)	0.834 (0.810-0.857)	0.798 (0.769-0.825)	0.796 (0.773-0.820)	0.885 (0.867-0.903)	0.865 (0.850-0.880)
Ours	0.907 (0.792-0.990)	0.819 (0.670-0.937)	0.958 (0.938-0.975)	0.845 (0.822-0.868)	0.823 (0.796-0.848)	0.841 (0.821-0.860)	0.898 (0.880-0.915)	0.884 (0.869-0.898)
AsyC	0.865 (0.694-1.000)	0.746 (0.522-0.913)	0.958 (0.937-0.974)	0.825 (0.800-0.850)	0.811 (0.783-0.839)	0.809 (0.788-0.831)	0.891 (0.8721-0.909)	0.860 (0.844-0.877)
AsyD	0.803 (0.611-0.947)	0.669 (0.384-0.906)	0.951 (0.927-0.971)	0.802 (0.777-0.828)	0.812 (0.786-0.840)	0.827 (0.806-0.848)	0.892 (0.871-0.910)	0.832 (0.815-0.849)
AsyD &Syn	0.866 (0.722-0.975)	0.754 (0.558-0.916)	0.955 (0.936-0.971)	0.824 (0.798-0.849)	0.818 (0.791-0.843)	0.824 (0.803-0.846)	0.907 (0.890-0.923)	0.846 (0.830-0.863)
AsyC &AsyD	0.862 (0.709-0.973)	0.798 (0.654-0.922)	0.953 (0.933-0.970)	0.841 (0.816-0.866)	0.820 (0.794-0.846)	0.833 (0.813-0.854)	0.894 (0.876-0.911)	0.881 (0.867-0.895)

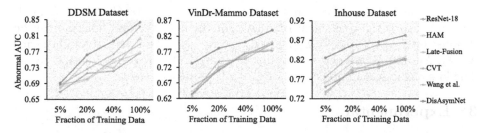

Fig. 2. Abnormality classification performance of DisAsymNet in terms of AUC trained on different sizes of training sets.

3.2 Experimental Settings

The mammogram pre-processing is conducted following the pipeline proposed by [5]. Then we standardize the image size to 1024 × 512 pixels. For training models, we employ random zooming and random cropping for data augmentation. We employ the ResNet-18 [2] with on ImageNet pre-trained weights as the common backbone for all methods. The Adam optimizer is utilized with an initial learning rate (LR) of 0.0001, and a batch size of 8. The training process on the INBreast dataset is conducted for 50 epochs with a LR decay of 0.1 every 20 epochs. For the other three datasets, the training is conducted separately on each one with 20 epochs and a LR decay of 0.1 per 10 epochs. All experiments are implemented in the Pytorch framework and an NVIDIA RTX A6000 GPU (48 GB). The training takes 3–24 h (related to the size of the dataset) on each dataset.

To assess the performance of different models in *classification tasks*, we calculate the area under the receiver operating characteristic curve (AUC) metric. The 95% confidence interval (CI) of AUC is estimated using bootstrapping (1,000 times) for each measure. For the *segmentation task*, we utilize Intersection over Union (IoU), Intersection over Reference (IoR), and Dice coefficients. For the *localization task*, we compute the mean accuracies of IoU or IoR values above a given threshold, following the approach [11]. Specifically, we evaluated the mean accuracy with thresholds for IoU at 0.1, 0.2, 0.3, 0.4, 0.5, 0.6, and 0.7, while the thresholds for IoR are 0.1, 0.25, 0.5, 0.75, and 0.9.

3.3 Experimental Results

We compare our proposed DisAsymNet with single view-based baseline ResNet18, attention-driven method HAM [11], MV-based late-fusion method [4], current SOTA MV-based methods cross-view-transformer (CVT) [16], and attention-based MV methods proposed by Wang et al., [20] on classification, segmentation, and localization tasks. We also conduct an ablation study to verify the effectiveness of "*AsyC*", "*AsyD*", and "*Synthesis*". Note that, the asymmetric transformer (*asyt*) is a core component of our proposed "*AsyC*". Thus, when ablating the "*AsyC*", we only drop the *asyt* and keep the encoders and classifiers. The features from the Bi-MG are simply concatenated and passed to the classifier.

Comparison of Performance in Different Tasks: For the classification task, the AUC results of abnormal classification are shown in Table 1. Our method outperforms all the single-based and MV-based methods in these classification tasks across all datasets. Furthermore, the ablation studies demonstrate the effectiveness of each proposed model component. In particular, our "*AsyC*" only method already surpasses the CAT method, indicating the efficacy of the proposed combination of SA and CA blocks over using CA alone. Additionally, our "*AsyD*" only method improves the performance compared to the late-fusion method, demonstrating that our disentanglement-based self-adversarial learning strategy can refine classifiers and enhance the model's ability to classify anomalies and asymmetries. The proposed "*Synthesis*" method further enhances

Table 2. Comparison of weakly supervised abnormalities segmentation and localization tasks on public datasets.

| | Segmentation Task | | | Localization Task | | | | | |
| | INbreast | | | INbreast | | DDSM | | VinDr-Mammo | |
Methods	IoU	IoR	Dice	mean TIoU	mean TIoR	mean TIoU	mean TIoR	mean TIoU	mean TIoR
ResNet18 [2]	0.193	0.344	0.283	21.4%	32.0%	7.9 %	21.2 %	15.3 %	33.8 %
HAM [11]	0.228	0.361	0.320	25.7%	34.0%	12.6 %	16.5 %	19.5 %	29.0 %
Late-fusion [4]	0.340	0.450	0.452	42.9%	42.0%	13.6 %	17.0 %	23.7 %	37.4 %
CVT [16]	0.047	0.072	0.072	4.3%	6.0%	8.8 %	14.6 %	20.7 %	40.7 %
Wang et al. [20]	0.301	0.398	0.406	34.3%	36.0%	15.4 %	16.6 %	26.4 %	44.4 %
Ours	0.461	0.594	0.601	60.0%	58.0%	19.2 %	26.5 %	27.2 %	48.8 %
AsyC	0.384	0.443	0.489	45.7%	42.0%	15.0 %	20.9 %	24.6 %	42.2 %
AsyD	0.336	0.524	0.465	42.9%	50.0%	16.6 %	25.9 %	23.2 %	43.5 %
AsyD & Syn	0.310	0.484	0.438	37.1%	48.0%	17.4 %	29.9 %	18.9 %	45.1 %
AsyC & AsyD	0.385	0.452	0.500	50.0%	42.0%	16.5 %	23.6 %	24.1 %	39.4 %

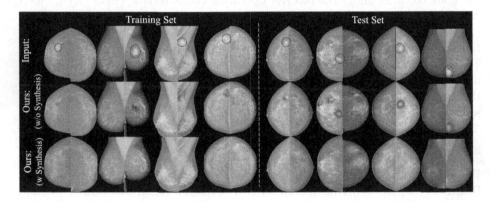

Fig. 3. Eight representative visualizations of normal mammogram reconstruction. The red circles indicate where the asymmetric abnormalities are. (Color figure online)

the performance of our proposed method. Moreover, we investigate the ability of different methods to classify abnormalities under various percentages of DDSM, VinDr, and In-house datasets. The INBreast dataset was excluded from this experiment due to its small size. Figure 2 illustrates the robustness of our method's advantage and our approach consistently outperformed the other methods, regardless of the size of the training data used and data sources. For the weakly supervised segmentation and localization tasks, results are shown in Table 2. The results demonstrate that our proposed framework achieves superior segmentation and localization performance compared to other existing methods across all evaluation metrics. The results of the ablation experiment also reveal that all modules incorporated in our framework offer improvements for the tasks.

Visualization: Figure 3 displays multiple disentangled normal Bi-MG cases. Our model achieves the efficient removal of asymmetrical abnormalities while retaining normal symmetric tissue. Without using pixel-level asymmetry ground truth from the *"Synthesis"* method, our generator tends to excessively remove asymmetric abnormalities at the cost of leading to the formation of black holes or areas that are visibly darker than the surrounding tissue because of the limitation of our discriminator and lack of pixel-level supervision. The incorporation of proposing synthetic asymmetrical Bi-MG during model training can lead to more natural symmetric tissue generation.

4   Conclusion

We present, DisAsymNet, a novel asymmetrical abnormality disentangling-based self-adversarial learning framework based on the image-level class labels only. Our study highlights the importance of considering asymmetry in mammography diagnosis in addition to the general multi-view analysis. The incorporation of pixel-level normal symmetric breast view generation boosts the classification of Bi-MG and also provides the interpretation of the diagnosis. The extensive experiments on four datasets demonstrate the robustness of our DisAsymNet framework for improving performance in classification, segmentation, and localization tasks. The potential of leveraging asymmetry can be further investigated in other clinical tasks such as BC risk prediction.

Acknowledgment. Xin Wang is funded by Chinese Scholarship Council scholarship (CSC) and this work was also funded by the Science and Technology Development Fund of Macau SAR (Grant number 0105/2022/A).

References

1. Chen, Y., et al.: Multi-view local co-occurrence and global consistency learning improve mammogram classification generalisation. In: Wang, L., Dou, Q., Fletcher, P.T., Speidel, S., Li, S. (eds.) MICCAI 2022, Part III. LNCS, vol. 13433, pp. 3–13. Springer, Cham (2022). https://doi.org/10.1007/978-3-031-16437-8_1
2. He, K., Zhang, X., Ren, S., Sun, J.: Deep residual learning for image recognition. In: Proceedings of the IEEE Conference on Computer Vision and Pattern Recognition, pp. 770–778 (2016)
3. Heath, M., et al.: Current status of the digital database for screening mammography. In: Karssemeijer, N., Thijssen, M., Hendriks, J., van Erning, L. (eds.) Digital Mammography. CIVI, vol. 13, pp. 457–460. Springer, Dordrecht (1998). https://doi.org/10.1007/978-94-011-5318-8_75
4. Li, C., et al.: Multi-view mammographic density classification by dilated and attention-guided residual learning. IEEE/ACM Trans. Comput. Biol. Bioinf. **18**(3), 1003–1013 (2020)
5. Liu, Y., Azizpour, H., Strand, F., Smith, K.: Decoupling inherent risk and early cancer signs in image-based breast cancer risk models. In: Martel, A.L., et al. (eds.) MICCAI 2020, Part VI. LNCS, vol. 12266, pp. 230–240. Springer, Cham (2020). https://doi.org/10.1007/978-3-030-59725-2_23
6. Liu, Y., et al.: From unilateral to bilateral learning: detecting mammogram masses with contrasted bilateral network. In: Shen, D., et al. (eds.) MICCAI 2019, Part VI. LNCS, vol. 11769, pp. 477–485. Springer, Cham (2019). https://doi.org/10.1007/978-3-030-32226-7_53

7. Moreira, I.C., Amaral, I., Domingues, I., Cardoso, A., Cardoso, M.J., Cardoso, J.S.: INbreast: toward a full-field digital mammographic database. Acad. Radiol. **19**(2), 236–248 (2012)
8. Nguyen, H.T., et al.: VinDr-Mammo: a large-scale benchmark dataset for computer-aided diagnosis in full-field digital mammography. MedRxiv (2022)
9. Ni, H., et al.: Asymmetry disentanglement network for interpretable acute ischemic stroke infarct segmentation in non-contrast CT scans. In: Wang, L., Dou, Q., Fletcher, P.T., Speidel, S., Li, S. (eds.) MICCAI 2022, Part VIII. LNCS, vol. 13438, pp. 416–426. Springer, Cham (2022). https://doi.org/10.1007/978-3-031-16452-1_40
10. Ouyang, X., et al.: Self-adversarial learning for detection of clustered microcalcifications in mammograms. In: de Bruijne, M., et al. (eds.) MICCAI 2021, Part VII. LNCS, vol. 12907, pp. 78–87. Springer, Cham (2021). https://doi.org/10.1007/978-3-030-87234-2_8
11. Ouyang, X., et al.: Learning hierarchical attention for weakly-supervised chest X-ray abnormality localization and diagnosis. IEEE Trans. Med. Imaging **40**(10), 2698–2710 (2020)
12. Spak, D.A., Plaxco, J., Santiago, L., Dryden, M., Dogan, B.: BI-RADS® fifth edition: a summary of changes. Diagn. Interv. Imaging **98**(3), 179–190 (2017)
13. Stadnick, B., et al.: Meta-repository of screening mammography classifiers. arXiv preprint arXiv:2108.04800 (2021)
14. Sung, H., et al.: Global cancer statistics 2020: GLOBOCAN estimates of incidence and mortality worldwide for 36 cancers in 185 countries. CA: A Canc. J. Clin. **71**(3), 209–249 (2021)
15. Tardy, M., Mateus, D.: Looking for abnormalities in mammograms with self-and weakly supervised reconstruction. IEEE Trans. Med. Imaging **40**(10), 2711–2722 (2021)
16. van Tulder, G., Tong, Y., Marchiori, E.: Multi-view analysis of unregistered medical images using cross-view transformers. In: de Bruijne, M., et al. (eds.) MICCAI 2021, Part III. LNCS, vol. 12903, pp. 104–113. Springer, Cham (2021). https://doi.org/10.1007/978-3-030-87199-4_10
17. Wang, C.R., Gao, F., Zhang, F., Zhong, F., Yu, Y., Wang, Y.: Disentangling disease-related representation from obscure for disease prediction. In: International Conference on Machine Learning, pp. 22652–22664. PMLR (2022)
18. Wang, C., Zhang, F., Yu, Y., Wang, Y.: BR-GAN: bilateral residual generating adversarial network for mammogram classification. In: Martel, A.L., et al. (eds.) MICCAI 2020, Part II. LNCS, vol. 12262, pp. 657–666. Springer, Cham (2020). https://doi.org/10.1007/978-3-030-59713-9_63
19. Wang, C., et al.: Bilateral asymmetry guided counterfactual generating network for mammogram classification. IEEE Trans. Image Process. **30**, 7980–7994 (2021)
20. Wang, X., Gao, Y., Zhang, T., Han, L., Beets-Tan, R., Mann, R.: Looking for abnormalities using asymmetrical information from bilateral mammograms. In: Medical Imaging with Deep Learning (2022)
21. Wang, X., Moriakov, N., Gao, Y., Zhang, T., Han, L., Mann, R.M.: Artificial intelligence in breast imaging. In: Fuchsjäger, M., Morris, E., Helbich, T. (eds.) Breast Imaging: Diagnosis and Intervention, pp. 435–453. Springer, Cham (2022). https://doi.org/10.1007/978-3-030-94918-1_20
22. Wu, N., et al.: Deep neural networks improve radiologists' performance in breast cancer screening. IEEE Trans. Med. Imaging **39**(4), 1184–1194 (2019)
23. Yala, A., et al.: Toward robust mammography-based models for breast cancer risk. Sci. Transl. Med. **13**(578), eaba4373 (2021)

24. Yang, Z., et al.: MommiNet-v2: mammographic multi-view mass identification networks. Med. Image Anal. **73**, 102204 (2021)
25. Zhao, X., Yu, L., Wang, X.: Cross-view attention network for breast cancer screening from multi-view mammograms. In: ICASSP 2020–2020 IEEE International Conference on Acoustics, Speech and Signal Processing (ICASSP), pp. 1050–1054. IEEE (2020)
26. Zhao, Z., Wang, D., Chen, Y., Wang, Z., Wang, L.: Check and link: pairwise lesion correspondence guides mammogram mass detection. In: Avidan, S., Brostow, G., Cissé, M., Farinella, G.M., Hassner, T. (eds.) ECCV 2022, Part XXI. LNCS, vol. 13681, pp. 384–400. Springer, Cham (2022). https://doi.org/10.1007/978-3-031-19803-8_23

Mammo-Net: Integrating Gaze Supervision and Interactive Information in Multi-view Mammogram Classification

Changkai Ji[1,2], Changde Du[2], Qing Zhang[3], Sheng Wang[1,4,5], Chong Ma[6],
Jiaming Xie[7], Yan Zhou[3], Huiguang He[1,2(✉)], and Dinggang Shen[1,5,8(✉)]

[1] School of Biomedical Engineering, ShanghaiTech University, Shanghai, China
{jichk,dgshen}@shanghaitech.edu.cn, huiguang.he@ia.ac.cn
[2] State Key Laboratory of Multimodal Artificial Intelligence Systems, Institute of
Automation, Chinese Academy of Sciences, Beijing, China
[3] Department of Radiology, Renji Hospital Shanghai Jiao Tong University School of
Medicine, Shanghai, China
[4] Institute for Medical Imaging Technology, School of Biomedical Engineering,
Shanghai Jiao Tong University, Shanghai, China
[5] Shanghai United Imaging Intelligence Co., Ltd., Shanghai, China
[6] School of Automation, Northwestern Polytechnical University, Xi'an, China
[7] Department of Computer Science,
The University of Hong Kong, Hong Kong, China
[8] Shanghai Clinical Research and Trial Center, Shanghai, China

Abstract. Breast cancer diagnosis is a challenging task. Recently, the application of deep learning techniques to breast cancer diagnosis has become a popular trend. However, the effectiveness of deep neural networks is often limited by the lack of interpretability and the need for significant amount of manual annotations. To address these issues, we present a novel approach by leveraging both gaze data and multi-view data for mammogram classification. The gaze data of the radiologist serves as a low-cost and simple form of coarse annotation, which can provide rough localizations of lesions. We also develop a pyramid loss better fitting to the gaze-supervised process. Moreover, considering many studies overlooking interactive information relevant to diagnosis, we accordingly utilize transformer-based attention in our network to mutualize multi-view pathological information, and further employ a bidirectional fusion learning (BFL) to more effectively fuse multi-view information. Experimental results demonstrate that our proposed model significantly improves both mammogram classification performance and interpretability through incorporation of gaze data and cross-view interactive information.

Keywords: Mammogram classification · Gaze · Multi-view interaction · Bidirectional fusion learning

Supplementary Information The online version contains supplementary material available at https://doi.org/10.1007/978-3-031-43990-2_7.

1   Introduction

Breast cancer is the most prevalent form of cancer among women and can have serious physical and mental health consequences if left unchecked [5]. Early detection through mammography is critical for early treatment and prevention [19]. Mammograms provide images of breast tissue, which are taken from two views: the cranio-caudal (CC) view, and the medio-lateral oblique (MLO) view [4]. By identifying breast cancer early, patients can receive targeted treatment before the disease progresses.

Deep neural networks have been widely adopted for breast cancer diagnosis to alleviate the workload of radiologists. However, these models often require a large number of manual annotations and lack interpretability, which can prevent their broader applications in breast cancer diagnosis. Radiologists typically focus on areas with breast lesions during mammogram reading [11,22], which provides valuable guidance. We propose using real-time eye tracking information from radiologists to optimize our model. By using gaze data to guide model training, we can improve model interpretability and performance [24].

Radiologists' eye movements can be automatically and unobtrusively recorded during the process of reading mammograms, providing a valuable source of data without the need for manual labeling. Previous studies have incorporated radiologists' eye-gaze as a form of weak supervision, which directs the network's attention to the regions with possible lesions [15,23]. Leveraging gaze from radiologists to aid in model training *not only* increases efficiency and minimizes the risk of errors linked to manual annotation, *but also* can be seamlessly implemented without affecting radiologists' normal clinical interpretation of mammograms.

Mammography primarily detects two types of breast lesions: masses and microcalcifications [16]. The determination of the benign or malignant nature of masses is largely dependent on the smoothness of their edges [13]. The gaze data can guide the model's attention towards the malignant masses. Microcalcifications are small calcium deposits which exhibit irregular boundaries on mammograms [9]. This feature makes them challenging to identify, often leading to missed or false detection by models. Radiologists need to magnify mammograms to differentiate between benign scattered calcifications and clustered calcifications, the latter of which are more likely to be malignant and necessitate further diagnosis. Leveraging gaze data can guide the model to locate malignant calcifications.

In this work, we propose a novel diagnostic model, namely Mammo-Net, which integrates radiologists' gaze data and interactive information between CC-view and MLO-view to enhance diagnostic performance. To the best of our knowledge, this is the first work to integrate gaze data into multi-view mammography classification. We utilize class activation map (CAM) [18] to calculate the attention maps for the model. Additionally, we apply pyramid loss to maintain consistency between radiologists' gaze heat maps and the model's attention maps at multiple scales of the pyramid [1]. Our model is designed for single-breast cases. Mammo-Net extracts multi-view features and utilizes transformer-based attention to mutualize information [21]. Furthermore, there are differences between multi-view mammograms of the same patient, arising from variations in breast shape and density. Capturing these multi-view shared features can be a challenge for models. To address this issue, we develop a novel method called bidirectional fusion learning (BFL) to extract shared features from multi-view mammograms.

Our contributions can be summarized as follows:

- We emphasize the significance of low-cost gaze to provide weakly-supervised positioning and visual interpretability for the model. Additionally, we develop a pyramid loss that adapts to the supervised process.
- We propose a novel breast cancer diagnosis model, namely Mammo-Net. This model employs transformer-based attention to mutualize information and uses BFL to integrate task-related information to make accurate predictions.
- We demonstrate the effectiveness of our approach through experiments using mammography datasets, which show the superiority of Mammo-Net.

2 Proposed Method

2.1 Overall Architecture

The pipeline of Mammo-Net is illustrated in Fig. 1. Mammo-Net feeds two-view mammograms of the same breast into two ResNet-style [7] CNN branch networks. We use several ResNet blocks pre-trained on ImageNet [3] to process mammograms. Then, we use global average pooling (GAP) and fully connected layers to compute the feature vectors produced by the model. Before the final residual block, we employ cross-view attention to mutualize multi-view information. Our proposed method employs BFL to effectively fuse multi-view information to improve diagnostic accuracy. Additionally, by integrating gaze data from radiologists, our proposed model is able to generate more precise attention maps. The fusion network combines multi-view feature representations using a stack of linear-activation layers and a fully connected layer, resulting in a classification output.

2.2 Gaze Supervision

In this module, we utilize CAM to calculate the attention map for the network by examining gradient-based activations in back-propagation. After that, we employ pyramid loss to make the network attention being consistent with the supervision of radiologists' gaze heat maps, guiding the network to focus on the same lesion areas as the radiologists. This module guides the network to accurately extract pathological features.

Class Activation Map. At the final convolutional layer of our model, the activation of the ith feature map $f_i(x, y)$ at coordinates (x, y) is associated with a weight w_i^k for class k. This allows us to generate the attention map H^k for class k as:

$$H^k = \sum_i w_i^k f_i(x, y). \tag{1}$$

Pyramid Loss. To enhance the learning of important attention areas, we propose a pyramid loss constraint that requires consistency between the network and gaze attention maps. The pyramid loss is based on using a pyramid representation of the attention map:

$$\mathcal{L}_{Pyramid} = \sum_{l}^{L} ||(Z(G_l(H)))^+ - (Z(G_l(R)))^+||_2, \qquad (2)$$

where H is the network attention map generated by the CAM and R is the radiologist's gaze heat map. $G_l(\cdot)$ represents the feature map at the lth level of the Gaussian pyramid, obtained by downsampling $G_{l-1}(\cdot)$ using a Gaussian kernel, where $G_1(R) = R$. Z means to perform Layernorm and ReLU activation on each feature map. This focuses the consistency loss on the more important pathological regions. The positive part of the normalized $Z(R)$, denoted as $Z(R)^+$, indicates the network focuses on the lesions where the radiologist spent most time reading. The minimization of the pyramid loss involves calculating the mean

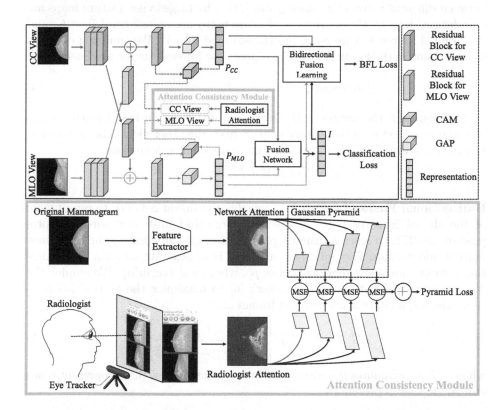

Fig. 1. Mammo-Net consists of two components: a multi-view classification network (upper half) and an attention consistency module (lower half). The classification network interacts multi-view information, while the attention consistency module provides positional supervision.

square error (MSE) between the attention maps generated by the radiologist and the model at each level of the Gaussian pyramid. This allows the model to mimic the attention of radiologists and enhance diagnostic performance.

Moreover, the pyramid representation enables the model to learn from the important pathological regions on which radiologists are focusing, without the need for precise pixel-level information. Layernorm is also employed to address the issue of imprecise gaze data. This reduces noise in the consistency process by performing consistency loss only in the regions where radiologist spent most time.

2.3   Interactive Information

Transformer-Based Mutualization Model. We use transformer-based attention to mutualize information from the two views at the level of the spatial feature map. For each attention head, we compute embeddings for the source and target pixels. Our model does not utilize positional encoding, as it encodes the relative position of each pixel and is not suitable for capturing information between different views of mammograms [21]. The target view feature maps are transformed into Q, the source view feature maps are transformed into K, and the original source feature maps are transformed into V. We can then obtain a weighted sum of the features from the source view for each target pixel using [21]:

$$Attention(Q, K, V) = softmax(\frac{QK^T}{\sqrt{d_k}})V. \tag{3}$$

Subsequently, the output is transformed into attention-based feature maps X and mutualized with the feature maps Y from the other view. The mutualized feature maps are normalized and used for subsequent calculations:

$$Z = Norm(Y + Linear(X)). \tag{4}$$

Bidirectional Fusion Learning. To enable the fusion network to retain more of the shared features between the two views and filter out noise, we propose to use BFL to learn a fusion representation that maximizes the cross-view mutual information. The optimization target is to generate a fusion representation I from multi-view representations p_v, where $v \in \{cc, mlo\}$. We employ the Noise-Contrastive Estimation framework [6] to maximize the mutual information, which is a contrastive learning framework:

$$\mathcal{L}(I, \boldsymbol{P}_v) = -\mathbb{E}_P \left[log \frac{s(I, p_v^i)}{\sum_{p_v^j \in \boldsymbol{P}_v} s(I, p_v^j)} \right], \tag{5}$$

where $s(I, p_v)$ evaluates the correlation between multi-view fused representations and single-view representations [17]:

$$s(I, p_v) = exp \left(\overline{p_v} \left(\overline{N(I)} \right)^T \right),$$
$$\overline{p_v} = \frac{p_v}{||p_v||_2}, \qquad \overline{N(I)} = \frac{N(I)}{||N(I)||_2}, \tag{6}$$

where $N(I)$ is a reconstruction of p_v generated by a fully connected network N from I and the Euclidean norm $||\cdot||_2$ is applied to obtain unit-length vectors. In contrastive learning, we consider the same patient mammograms as positive samples and those from different patient mammograms in the same batch $\tilde{P}_v^i = P_v \backslash \{p_v^i\}$ as negative samples [17]. Minimizing the similarity between the same patient mammograms enables the model to learn shared features. Maximizing the dissimilarity between different patient mammograms enhances the model's robustness.

In short, we require the fusion representation I to reversely reconstruct multi-view representations p_v so that more view-invariant information can be passed to I. By aligning the prediction $N(I)$ to p_v, we enable the model to decide how much information it should receive from each view.

The overall loss function for this module is the sum of the losses defined for each view:

$$\mathcal{L}_{BFL} = \mathcal{L}_I^{cc} + \mathcal{L}_I^{mlo}. \tag{7}$$

2.4 Loss Function

We use binary cross entropy loss (BCE) between the network prediction and the ground-truth as the classification loss. In conclusion, we have proposed a total of three loss functions to guide the model training: $\mathcal{L}_{BCE}$, $\mathcal{L}_{BFL}$, and $\mathcal{L}_{Pyramid}$. The overall loss function is defined as the sum of these three loss functions, with coefficients λ and μ used to adjust their relative weights:

$$\mathcal{L}_{overall} = \mathcal{L}_{BCE} + \lambda \mathcal{L}_{Pyramid} + \mu \mathcal{L}_{BFL}. \tag{8}$$

3 Experiments and Results

3.1 Datasets

Mammogram Dataset. Our experiments were conducted on CBIS-DDSM [12] and INbreast [16]. The CBIS-DDSM dataset contains 1249 exams that have been divided based on the presence or absence of masses, which we used to perform mass classification. The INbreast dataset contains 115 exams with both masses and micro-calcifications, on which we performed benign and malignant classification. We split the INbreast dataset into training and testing sets in a 7:3 ratio. It is worth noting that the official INbreast dataset does not provide image-level labels, so we obtained these labels following Shen et al. [20].

Eye Gaze Dataset. Eye movement data was collected by reviewing all cases in INbreast using a Tobii Pro Nano eye tracker. The scenario is shown in Appendix and can be accessed at https://github.com/JamesQFreeman/MicEye. Participated radiologist has 11 years of experience in mammography screening.

3.2 Implementation Details

We trained our model using the Adam optimizer [10] with a learning rate of 10^{-4} (partly implemented by MindSpore). To overcome the problem of limited data, we employed various data augmentation techniques, including translation, rotation, and flipping. To address the problem of imbalanced classes, we utilized a weighted loss function that assigns higher weights to malign cases in order to balance the number of benign and malign cases. The coefficients λ and μ of $\mathcal{L}_{overall}$ were set to 0.5 and 0.2, respectively, based on 5-fold cross validation on the training set. The network was trained for 300 epochs. We used Accuracy (ACC) and the Area Under the ROC Curve (AUC) [25] as our evaluation metrics, and we selected the final model based on the best validation AUC. Considering the relatively small size of our dataset, we used ResNet-18 as the backbone of our network.

3.3 Results and Analysis

Table 1. Ablation study of key components of Mammo-Net, and comparison of different models in terms of AUC and ACC. "BFL" denotes "Bidirectional Fusion Learning", and "RA" denotes "Radiologist Attention".

Dataset	Model	AUC	ACC
CBIS-DDSM	Lopez et al. [14]	0.739	0.754
	Tulder et al. [2]	0.802	0.811
	Xian et al. [26]	0.812	0.735
	MLO-view	0.701	0.763
	CC-view	0.721	0.754
	Cross-view	0.809	0.838
	Cross-view+BFL	**0.821**	**0.864**
INbreast	Wang et al. [23]	0.806	0.756
	Jiang et al. [8]	0.819	0.793
	Lopez et al. [14]	0.793	0.830
	Xian et al. [26]	0.859	0.791
	MLO-view	0.663	0.716
	CC-view	0.650	0.704
	Cross-view	0.762	0.755
	Cross-view+BFL	0.786	0.812
	Cross-view+RA	0.864	0.830
	Cross-view+BFL+RA (Mammo-Net)	**0.889**	**0.849**

Performance Comparison. As shown in Table 1, we compare our model to other methods and find that our model performs better. Lopez et al. [14] proposed the use of hypercomplex networks to mimic radiologists. By leveraging the properties of hypercomplex algebra, the model is able to continually process two mammograms together. Lee et al. [26] proposed a 2-channel approach that utilizes a Gaussian model to capture the spatial correlation between lesions across two views, and an LT-GAN to achieve a robust mammography classification.

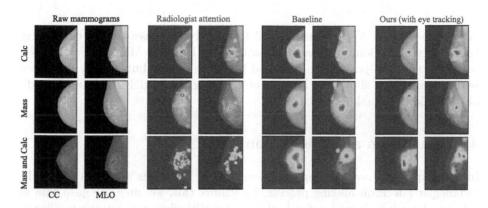

Fig. 2. Comparative visualization of mammography diagnosis with and without gaze supervision. After integrating gaze supervision, the model's capability in localizing lesions becomes more precise.

We also compare our model with other methods that use eye movement supervision as shown in Table 1. The GA-Net [23] proposed a ResNet-based model with class activation mapping guided by eye gaze data. We developed a multi-view model using this approach for a fair comparison, and found that our method performed better. We believe that one possible reason for the inferior performance of GA-Net compared to Mammo-Net might be the use of a simple MSE loss by GA-Net, which neglects the coarse nature of the gaze data. Jiang et al. [8] proposed a Double-model that fuses gaze maps with original images before training. However, this model did not consider the gap between research and clinical workflow. This model requires gaze input during both the training and inference stages, which limits its practical use in hospitals without eye-trackers. In contrast, our method does not rely on gaze input during inference stage.

Visualization. Figure 2 illustrates the visualization of our proposed model on three representative exams from the INbreast dataset that includes masses, calcifications, and a combination of both. For each exam, we present gaze heat maps generated from eye movement data. The preprocessing process is shown in Fig. 5 (see Appendix). To make an intuitive comparison, we exhibit attention maps generated by the model under both unsupervised and gaze-supervised cases. Each exam is composed of two views, i.e., the CC-view and the MLO-view. More exams can be found in Fig. 6 (see Appendix).

The results of the visualization demonstrate that the model's capability in localizing lesions becomes more precise when radiologist attention is incorporated in the training stage. The pyramid loss improves the model's robustness even when the radiologist's gaze data is not entirely focused on the breast. This intuitively demonstrates the effectiveness of training the model with eye-tracking supervision.

Ablation Study. We perform an ablation analysis to assess each component (radiologist attention, cross-view attention and BFL) in Mammo-Net. Table 1 suggests that each part of the proposed framework contributes to the increased performance. This shows the benefits of adapting the model to mimic the radiologist's decision-making process.

4   Conclusion and Discussion

In this paper, we have developed a breast cancer diagnosis model to mimic the radiologist's decision-making process. To achieve this, we integrate gaze data as a form of weak supervision for both lesion positioning and interpretability of the model. We also utilize transformer-based attention to mutualize multi-view information and further develop BFL to fully fuse multi-view information. Our experimental results on mammography datasets demonstrate the superiority of our proposed model. In future work, we intend to explore the use of scanning path analysis as a means of obtaining insights into the pathology-relevant regions of lesions.

Acknowledgements. This work was supported in part by The Key R&D Program of Guangdong Province, China (grant number 2021B0101420006), National Natural Science Foundation of China (grant numbers 62131015, 82272072), Science and Technology Commission of Shanghai Municipality (STCSM) (grant number 21010502600), and the CAAI-Huawei MindSpore Open Fund.

References

1. Adelson, E.H., Anderson, C.H., Bergen, J.R., Burt, P.J., Ogden, J.M.: Pyramid methods in image processing. RCA Eng. **29**(6), 33–41 (1984)
2. Cheng, K., Ma, Y., Sun, B., Li, Y., Chen, X.: Depth estimation for colonoscopy images with self-supervised learning from videos. In: de Bruijne, M., et al. (eds.) MICCAI 2021. LNCS, vol. 12906, pp. 119–128. Springer, Cham (2021). https://doi.org/10.1007/978-3-030-87231-1_12
3. Deng, J., Dong, W., Socher, R., Li, L.J., Li, K., Fei-Fei, L.: Imagenet: a large-scale hierarchical image database. In: 2009 IEEE Conference on Computer Vision and Pattern Recognition, pp. 248–255. IEEE (2009)

4. Frazer, H.M., Qin, A.K., Pan, H., Brotchie, P.: Evaluation of deep learning-based artificial intelligence techniques for breast cancer detection on mammograms: results from a retrospective study using a breastscreen victoria dataset. J. Med. Imaging Radiat. Oncol. **65**(5), 529–537 (2021)
5. Giaquinto, A.N., Miller, K.D., Tossas, K.Y., Winn, R.A., Jemal, A., Siegel, R.L.: Cancer statistics for African American/black people 2022. CA Cancer J. Clin. **72**(3), 202–229 (2022)
6. Gutmann, M., Hyvärinen, A.: Noise-contrastive estimation: a new estimation principle for unnormalized statistical models. In: Proceedings of the Thirteenth International Conference on Artificial Intelligence and Statistics, pp. 297–304. JMLR Workshop and Conference Proceedings (2010)
7. He, K., Zhang, X., Ren, S., Sun, J.: Deep residual learning for image recognition. In: Proceedings of the IEEE Conference on Computer Vision and Pattern Recognition, pp. 770–778 (2016)
8. Jiang, H., et al.: Eye tracking based deep learning analysis for the early detection of diabetic retinopathy: a pilot study. Available at SSRN 4247845 (2023)
9. Jørgensen, K.J., et al.: Breast-cancer screening-viewpoint of the iarc working group. New Engl. J. Med. **373**, 1478 (2015)
10. Kingma, D.P., Ba, J.: Adam: a method for stochastic optimization. arXiv preprint arXiv:1412.6980 (2014)
11. Kundel, H.L., Nodine, C.F., Krupinski, E.A., Mello-Thoms, C.: Using gaze-tracking data and mixture distribution analysis to support a holistic model for the detection of cancers on mammograms. Acad. Radiol. **15**(7), 881–886 (2008)
12. Lee, R.S., Gimenez, F., Hoogi, A., Miyake, K.K., Gorovoy, M., Rubin, D.L.: A curated mammography data set for use in computer-aided detection and diagnosis research. Sci. Data **4**(1), 1–9 (2017)
13. Li, Z., et al.: Domain generalization for mammography detection via multi-style and multi-view contrastive learning. In: de Bruijne, M., et al. (eds.) MICCAI 2021. LNCS, vol. 12907, pp. 98–108. Springer, Cham (2021). https://doi.org/10.1007/978-3-030-87234-2_10
14. Lopez, E., Grassucci, E., Valleriani, M., Comminiello, D.: Multi-view breast cancer classification via hypercomplex neural networks. arXiv preprint arXiv:2204.05798 (2022)
15. Ma, C., et al.: Eye-gaze-guided vision transformer for rectifying shortcut learning. IEEE Trans. Med. Imaging (2023)
16. Moreira, I.C., Amaral, I., Domingues, I., Cardoso, A., Cardoso, M.J., Cardoso, J.S.: Inbreast: toward a full-field digital mammographic database. Acad. Radiol. **19**(2), 236–248 (2012)
17. Oord, A.V.d., Li, Y., Vinyals, O.: Representation learning with contrastive predictive coding. arXiv preprint arXiv:1807.03748 (2018)
18. Ouyang, X., et al.: Learning hierarchical attention for weakly-supervised chest x-ray abnormality localization and diagnosis. IEEE Trans. Med. Imaging **40**(10), 2698–2710 (2020)
19. Selvi, R.: Breast Diseases: Imaging and Clinical Management. Springer, Heidelberg (2014). https://doi.org/10.1007/978-81-322-2077-0
20. Shen, L., Margolies, L.R., Rothstein, J.H., Fluder, E., McBride, R., Sieh, W.: Deep learning to improve breast cancer detection on screening mammography. Sci. Rep. **9**(1), 12495 (2019)
21. Vaswani, A., et al.: Attention is all you need. Adv. Neural Inf. Process. Syst. **30**, 1–11 (2017)

22. Voisin, S., Pinto, F., Xu, S., Morin-Ducote, G., Hudson, K., Tourassi, G.D.: Investigating the association of eye gaze pattern and diagnostic error in mammography. In: Medical Imaging 2013: Image Perception, Observer Performance, and Technology Assessment, vol. 8673, p. 867302. SPIE (2013)

23. Wang, S., Ouyang, X., Liu, T., Wang, Q., Shen, D.: Follow my eye: using gaze to supervise computer-aided diagnosis. IEEE Trans. Med. Imaging 41(7), 1688–1698 (2022)

24. Wu, C.C., Wolfe, J.M.: Eye movements in medical image perception: a selective review of past, present and future. Vision 3(2), 32 (2019)

25. Wu, N., et al.: Deep neural networks improve radiologists' performance in breast cancer screening. IEEE Trans. Med. Imaging 39(4), 1184–1194 (2019)

26. Xian, J., Wang, Z., Cheng, K.-T., Yang, X.: Towards robust dual-view transformation via densifying sparse supervision for mammography lesion matching. In: de Bruijne, M., et al. (eds.) MICCAI 2021. LNCS, vol. 12905, pp. 355–365. Springer, Cham (2021). https://doi.org/10.1007/978-3-030-87240-3_34

Synthesis of Contrast-Enhanced Breast MRI Using T1- and Multi-b-Value DWI-Based Hierarchical Fusion Network with Attention Mechanism

Tianyu Zhang[1,2,3], Luyi Han[1,3], Anna D'Angelo[4], Xin Wang[1,2,3],
Yuan Gao[1,2,3], Chunyao Lu[1,3], Jonas Teuwen[5], Regina Beets-Tan[1,2],
Tao Tan[1,6(✉)], and Ritse Mann[1,3]

[1] Department of Radiology, Netherlands Cancer Institute (NKI),
Plesmanlaan 121, 1066 CX Amsterdam, The Netherlands
[2] GROW School for Oncology and Development Biology, Maastricht University,
P. O. Box 616, 6200 MD Maastricht, The Netherlands
[3] Department of Diagnostic Imaging, Radboud University Medical Center,
Geert Grooteplein 10, 6525 GA Nijmegen, The Netherlands
[4] Dipartimento di diagnostica per immagini, Radioterapia, Oncologia ed ematologia,
Fondazione Universitaria A. Gemelli, IRCCS Roma, Roma, Italy
[5] Department of Radiation Oncology, Netherlands Cancer Institute (NKI),
Plesmanlaan 121, 1066 CX Amsterdam, The Netherlands
[6] Faculty of Applied Sciences, Macao Polytechnic University, Macao 999078, China
taotanjs@gmail.com

Abstract. Magnetic resonance imaging (MRI) is the most sensitive technique for breast cancer detection among current clinical imaging modalities. Contrast-enhanced MRI (CE-MRI) provides superior differentiation between tumors and invaded healthy tissue, and has become an indispensable technique in the detection and evaluation of cancer. However, the use of gadolinium-based contrast agents (GBCA) to obtain CE-MRI may be associated with nephrogenic systemic fibrosis and may lead to bioaccumulation in the brain, posing a potential risk to human health. Moreover, and likely more important, the use of gadolinium-based contrast agents requires the cannulation of a vein, and the injection of the contrast media which is cumbersome and places a burden on the patient. To reduce the use of contrast agents, diffusion-weighted imaging (DWI) is emerging as a key imaging technique, although currently usually complementing breast CE-MRI. In this study, we develop a multi-sequence fusion network to synthesize CE-MRI based on T1-weighted MRI and DWIs. DWIs with different b-values are fused to efficiently utilize the difference features of DWIs. Rather than proposing a pure data-driven approach, we invent a multi-sequence attention module to obtain

T. Zhang and L. Han—Equal contribution.

Supplementary Information The online version contains supplementary material available at https://doi.org/10.1007/978-3-031-43990-2_8.

refined feature maps, and leverage hierarchical representation information fused at different scales while utilizing the contributions from different sequences from a model-driven approach by introducing the weighted difference module. The results show that the multi-b-value DWI-based fusion model can potentially be used to synthesize CE-MRI, thus theoretically reducing or avoiding the use of GBCA, thereby minimizing the burden to patients. Our code is available at https://github.com/Netherlands-Cancer-Institute/CE-MRI.

Keywords: Contrast-enhanced MRI · Diffusion-weighted imaging · Deep learning · Multi-sequence fusion · Breast cancer

1   Introduction

Breast cancer is the most common cancer and the leading cause of cancer death in women [18]. Early detection of breast cancer allows patients to receive timely treatment, which may have less burden and a higher probability of survival [6]. Among current clinical imaging modalities, magnetic resonance imaging (MRI) has the highest sensitivity for breast cancer detection [12]. Especially, contrast-enhanced MRI (CE-MRI) can identify tumors well and has become an indispensable technique for detecting and defining cancer [13]. However, the use of gadolinium-based contrast agents (GBCA) requires iv-cannulation, which is a burden to patients, time consuming and cumbersome in a screening situation. Moreover, contrast administration can lead to allergic reactions and finaly CE-MRI may be associated with nephrogenic systemic fibrosis and lead to bioaccumulation in the brain, posing a potential risk to human health [4,9,14–16]. In 2017, the European Medicines Agency concluded its review of GBCA, confirming recommendations to restrict the use of certain linear GBCA used in MRI body scans and to suspend the authorization of other contrast agents, albeit macrocyclic agents can still be freely used [10].

With the development of computer technology, artificial intelligence-based methods have shown potential in image generation and have received extensive attention. Some studies have shown that some generative models can effectively perform mutual synthesis between MR, CT, and PET [19]. Among them, synthesis of CE-MRI is very important as mentioned above, but few studies have been done by researchers in this area due to its challenging nature. Li et al. analyzed and studied the feasibility of using T1-weighted MRI and T2-weighted MRI to synthesize CE-MRI based on deep learning model [11]. Their results showed that the model they developed could potentially synthesize CE-MRI and outperform other cohort models. However, MRI source data of too few sequences (only T1 and T2) may not provide enough valuable informative to effectively synthesize CE-MRI. In another study, Chung et al. investigated the feasibility of using deep learning (a simple U-Net structure) to simulate contrast-enhanced breast MRI of invasive breast cancer, using source data including T1-weighted non-fat-suppressed MRI, T1-weighted fat-suppressed MRI, T2-weighted fat-suppressed

MRI, DWI, and apparent diffusion coefficient [5]. However, obtaining a complete MRI sequence makes the examination costly and time-consuming. On the other hand, the information provided by multi-sequences may be redundant and may not contain the relevant information of CE-MRI. Therefore, it is necessary to focus on the most promising sequences to synthesize CE-MRI.

Diffusion-weighted imaging (DWI) is emerging as a key imaging technique to complement breast CE-MRI [3]. DWI can provide information on cell density and tissue microstructure based on the diffusion of tissue water. Studies have shown that DWI could be used to detect lesions, distinguish malignant from benign breast lesions, predict patient prognosis, etc [1,3,7,8,17]. In particular, DWI can capture the dynamic diffusion state of water molecules to estimate the vascular distribution in tissues, which is closely related to the contrast-enhanced regions in CE-MRI. DWI may be a valuable alternative in breast cancer detection in patients with contraindications to GBCA [3]. Inspired by this, we develop a multi-sequence fusion network based on T1-weighted MRI and multi-b-value DWI to synthesize CE-MRI. Our contributions are as follows:

i From the perspective of method, we innovatively proposed a multi-sequence fusion model, designed for combining T1-weighted imaging and multi-b-value DWI to synthesize CE-MRI for the first time.
ii We invented hierarchical fusion module, weighted difference module and multi-sequence attention module to enhance the fusion at different scale, to control the contribution of different sequence and maximising the usage of the information within and across sequences.
iii From the perspective of clinical application, our proposed model can be used to synthesize CE-MRI, which is expected to reduce the use of GBCA.

2   Methods

2.1   Patient Collection and Pre-processing

This study was approved by Institutional Review Board of our cancer institute with a waiver of informed consent. We retrospectively collected 765 patients with breast cancer presenting at our cancer institute from January 2015 to November 2020, all patients had biopsy-proven breast cancers (all cancers included in this study were invasive breast cancers, and ductal carcinoma in situ had been excluded). The MRIs were acquired with Philips Ingenia 3.0-T scanners, and overall, three sequences were present in the in-house dataset, including T1-weighted fat-suppressed MRI, contrast-enhanced T1-weighted MRI and DWI. DWI consists of 4 different b-values (b = 0, b = 150, b = 800 and b = 1500). All MRIs were resampled to 1 mm isotropic voxels and uniformly sized, resulting in volumes of 352×352 pixel images with 176 slices per MRI, and subsequent registration was performed based on Advanced Normalization Tools (ANTs) [2].

2.2 Model

Figure 1 illustrates the structure of the proposed model. First, the reconstruction module is used to automatically encode and decode each input MRI sequence information to obtain the latent representation of different MRI sequences at multi-scale levels. Then, the hierarchical fusion module is used to extract the hierarchical representation information and fuse them at different scales.

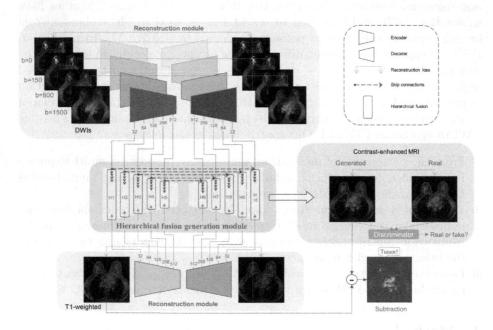

Fig. 1. Model details and flowchart for this study.

In each convolutional layer group of the reconstruction module, we use two 3×3 filters (same padding) with strides 1 and 2, respectively. The filters are followed by batch normalization, and after batch normalization, the activation functions *LeakyReLU* (with a slope of 0.2) and *ReLU* are used in the encoder and decoder, respectively. The l_1-norm is used as a reconstruction loss to measure the difference between the reconstructed image and the ground truth.

Figure 2 shows the detailed structure of the hierarchical fusion module, which includes two sub-modules, a weighted difference module and a multi-sequence attention module. The calculation of the apparent diffusion coefficient (ADC) map is shown in Eq. 1, which provides a quantitative measure of observed diffusion restriction in DWIs. Inspired by ADC, a weighted difference module is designed, in which the neural network is used to simulate the dynamic analysis of the ln function, and the element-wise subtraction algorithm is used to extract the differentiation features between DWIs with different b-values, and finally the features are weighted to obtain weighted feature maps (F_{DWI}, Eq. 2).

$$\text{ADC} = -\ln(S_h/S_l)/(b_h - b_l) = [\ln(S_l) - \ln(S_h)]/(b_h - b_l) \tag{1}$$

$$F_{\text{DWI}} = [f_{\theta^l}(S_l) - f_{\theta^h}(S_h)]/(b_h - b_l) \tag{2}$$

where S_l and S_h represent the image signals obtained from lower b value b_l and higher b_h, f_{θ^l} and f_{θ^h} represent the corresponding neural networks for DWI with a lower and higher b value.

In the multi-sequence attention module, a channel-based attention mechanism is designed to automatically apply weights (A_s) to feature maps (F_{concat}) from different sequences to obtain a refined feature map (F'_{concat}), as shown in Eq. 3. The input feature maps (F_{concat}) go through the maximum pooling layer and the average pooling layer respectively, and then are added element-wise after passing through the shared fully connected neural network, and finally the weight map A_s is generated after passing through the activation function, as shown in Eq. 4.

$$F'_{concat} \in \mathbb{R}^{C \times H \times W} = F_{concat} \otimes A_s \tag{3}$$

$$A_s = \sigma(f_{\theta^{fc}}(AvgPool(F_{concat})) \oplus f_{\theta^{fc}}(MaxPool(F_{concat}))) \tag{4}$$

where $\otimes$ represents element-wise multiplication, $\oplus$ represents element-wise summation, σ represents the sigmoid function, θ^{fc} represents the corresponding network parameters of the shared fully-connected neural network, and $AvgPool$ and $MaxPool$ represent average pooling and maximum pooling operations, respectively.

In the synthesis process, the generator $\mathcal{G}$ tries to generate an image according to the input multi-sequence MRI (d_1, d_2, d_3, d_4, t_1), and the discriminator $\mathcal{D}$ tries to distinguish the generated image $G(d_1, d_2, d_3, d_4, t_1)$ from the real image y, and at the same time, the generator tries to generate a realistic image to mislead the discriminator. The generator's objective function is as follows:

$$\mathcal{L}_G(G, D) = \mathbb{E}_{d_1,d_2,d_3,d_4,t_1 \sim pro_{data}(d_1,d_2,d_3,d_4,t_1)}$$
$$[\log(1 - D(d_1, d_2, d_3, d_4, t_1, (G(d_1, d_2, d_3, d_4, t_1))))] \tag{5}$$
$$+ \lambda_1 \mathbb{E}_{d_1,d_2,d_3,d_4,t_1,y}[\|y - G(d_1, d_2, d_3, d_4, t_1)\|_1]$$

and the discriminator's objective function is as follows:

$$\mathcal{L}_D(G, D) = \mathbb{E}_{y \sim pro_{data}(y)}[\log D(y)]$$
$$+ \mathbb{E}_{d_1,d_2,d_3,d_4,t_1 \sim pro_{data}(d_1,d_2,d_3,d_4,d_1)} \tag{6}$$
$$[\log(1 - D(G(d_1, d_2, d_3, d_4, t_1)))]$$

where $pro_{data}(d_1, d_2, d_3, d_4, t_1)$ represents the empirical joint distribution of inputs d_1 (DWI_{b0}), d_2 (DWI_{b150}), d_3 (DWI_{b800}), d_4 (DWI_{b1500}) and t_1 (T1-weighted MRI), λ_1 is a non-negative trade-off parameter, and l_1-norm is used to measure the difference between the generated image and the corresponding ground truth. The architecture of the discriminator includes five convolutional layers, and in each convolutional layer, 3×3 filters with stride 2 are used. Each filter is followed by batch normalization, and after batch normalization, the activation function *LeakyReLU* (with a slope of 0.2) is used. The numbers of filters are 32, 64, 128, 256 and 512, respectively.

2.3   Visualization

The T1-weighted images and the contrast-enhanced images were subtracted to obtain a difference MRI to clearly reveal the enhanced regions in the CE-MRI. If the CE-MRI was successfully synthesized, the enhanced region would be highlighted in the difference MRI, otherwise it would not.

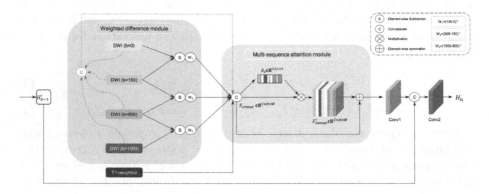

Fig. 2. Detailed structure of the hierarchical fusion module.

2.4   Experiment Settings

Based on the ratio of 8:2, the training set and independent test set of the in-house dataset have 612 and 153 cases, respectively. The trade-off parameter λ_1 was set to 100 during training, and the trade-off parameter of the reconstruction loss in the reconstruction module is set to 5. Masks for all breasts were used (weighted by a factor of 100 during the calculation of the loss between generated and real CE-MRI) to reduce the influence of signals in the thoracic area. The batch was set to 8 for 100 epochs, the initial learning rate was 1e−3 with a decay factor of 0.8 every 5 epochs (total run time is about 60 h). *Adam* optimizer was applied to update the model parameters. MMgSN-Net [11] and the method of Chung et al. [5] were used as cohort models, and all models were trained on NVIDIA RTX A6000 48 GB GPU.

2.5   Evaluation Metrics

Results analysis was performed by Python 3.7. Structural Similarity Index Measurement (SSIM), Peak Signal-to-Noise Ratio (PSNR) and Normalized Mean

Squared Error (NMSE) were used as metrics, all formulas as follows:

$$SSIM = \frac{(2\mu_{y(x)}\mu_{G(x)} + c_1)(2\sigma_{y(x)G(x)} + c_2)}{(\mu_{y(x)}^2 + \mu_{G(x)}^2 + c_1)(\sigma_{y(x)}^2 + \sigma_{G(x)}^2 + c_2)} \tag{7}$$

$$PSNR = 10\log_{10}\frac{\max^2(y(x), G(x))}{\frac{1}{N}\|y(x) - G(x)\|_2^2} \tag{8}$$

$$NMSE = \frac{y(x) - G(x)}{\|y(x)\|_2^2} \tag{9}$$

Table 1. Results for synthesizing breast ceT1 MRI for different models.

Method	Sequence	SSIM↑	PSNR↑	NMSE↓
MMgSN-Net [11]	T1+DWI(0)	86.61 ± 2.52	26.39 ± 1.38	0.0982 ± 0.038
Chung et al. [5]	T1+DWIs	87.58 ± 2.68	27.80 ± 1.56	0.0692 ± 0.035
Proposed	**T1+DWIs**	**89.93 ± 2.91**	**28.92 ± 1.63**	**0.0585 ± 0.026**

where $G(x)$ represents a generated image, $y(x)$ represents a ground-truth image, $\mu_{y(x)}$ and $\mu_{G(x)}$ represent the mean of $y(x)$ and $G(x)$, respectively, $\sigma_{y(x)}$ and $\sigma_{G(x)}$ represent the variance of $y(x)$ and $G(x)$, respectively, $\sigma_{y(x)G(x)}$ represents the covariance of $y(x)$ and $G(x)$, and c_1 and c_2 represent positive constants used to avoid null denominators.

3 Results

First, we compare the performance of different existing methods on synthetic CE-MRI using our source data, The quantitative indicators used include PSNR, SSIM and NMSE. As shown in Table 1, the SSIM of MMgSN-Net [11] and the method of Chung et al. [5] in synthesizing ceT1 MRI is 86.61 ± 2.52 and 87.58 ± 2.68, respectively, the PSNR is 26.39 ± 1.38 and 27.80 ± 1.56, respectively, and the NMSE is 0.0982 ± 0.038 and 0.0692 ± 0.035, respectively. In contrast, our proposed multi-sequence fusion model achieves better SSIM of 89.93 ± 2.91, better PSNR of 28.92 ± 1.63 and better NMSE of 0.0585 ± 0.026 in synthesizing ceT1 MRI, outperforming existing cohort models.

MMgSN-Net [11] combined T1-weighted and T2-weighted MRI in their work to synthesize CE-MRI. Here we combined T1-weighted MRI and DWI with a b-value of 0 according to their method, but the model did not perform well. It may be because their model can only combine bi-modality and cannot integrate the features of all sequences, so it cannot mine the difference features between multiple b-values, which limits the performance of the model. In addition, although the method of Chung et al. [5] used full-sequence MRI to synthesize CE-MRI, it would be advantageous to obtain synthetic CE-MRI images using as little data as possible, taking advantage of the most contributing sequences. They did not take advantage of multi-b-value DWI, nor did they use the hierarchical fusion module to fully fuse the hierarchical features of multi-sequence MRI.

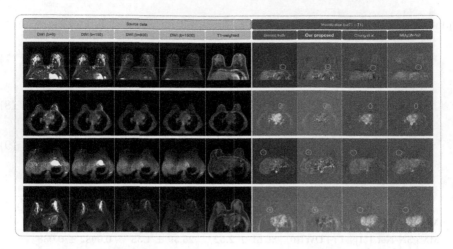

Fig. 3. Some examples of source data and visualization results

As described in Methods, the proposed model consists of several key components, including a hierarchical fusion generation module, a weighted difference module, and a multi-sequence attention module. Therefore, ablation studies were performed to demonstrate the importance and effectiveness of our three key components. Several network structures were selected for comparison, as follows: (1) Input-level fusion network without other modules (called IF-Net), (2) Hierarchical fusion generation network combined with reconstruction module, without weighted difference module and multi-sequence attention module (called HF-Net), (3) Hierarchical fusion generation network with weighted difference module (called HFWD-net), (4) Hierarchical fusion generation network with weighted difference module and multi-sequence attention module (proposed model). As shown in Table 2, IF-Net achieves SSIM of 87.25 ± 2.62 and PSNR of 26.51 ± 1.52 in synthesizing ceT1 MRI. HF-Net achieves SSIM of 88.32 ± 2.70 and PSNR of 27.95 ± 1.59 in synthesizing ceT1 MRI. After adding the weighted difference module, SSIM and PSNR were improved to 89.18 ± 2.73 and 28.45 ± 1.61, respectively. Finally, the addition of the multi-sequence attention module further improved the performance of the model, with SSIM of 89.93 ± 2.91, PSNR of 28.92 ± 1.63, and NMSE of 0.0585 ± 0.026.

Table 2. Ablation results for synthesizing breast ceT1 MRI.

Method	Sequence	SSIM↑	PSNR↑	NMSE↓
IF-Net	T1+DWIs	87.25 ± 2.62	26.51 ± 1.52	0.0722 ± 0.032
HF-Net	T1+DWIs	88.32 ± 2.70	27.95 ± 1.59	0.0671 ± 0.033
HFWD-Net	T1+DWIs	89.18 ± 2.73	28.45 ± 1.61	0.0622 ± 0.030
Proposed	**T1+DWIs**	$\mathbf{89.93 \pm 2.91}$	$\mathbf{28.92 \pm 1.63}$	$\mathbf{0.0585 \pm 0.026}$

The visualization results of random samples are shown in Fig. 3. It can be seen from the visualization results that after the difference between the generated CE-MRI and the original T1-weighted MRI, the lesion position of the breast is highlighted, the red circle represents the highlighted area, which proves that our method can effectively synthesize contrast-enhanced images, highlighting the same parts as the real enhanced position. See Supplementary Material for more visualization results, including visualizations of breast CE-MRI synthesized in axial, coronal, and sagittal planes.

4 Conclusion

We have developed a multi-sequence fusion network based on multi-b-value DWI to synthesize CE-MRI, using source data including DWIs and T1-weighted fat-suppressed MRI. Compared to existing methods, we avoid the challenges of using full-sequence MRI and aim to be selective on valuable source data DWI. Hierarchical fusion generation module, weighted difference module, and multi-sequence attention module have all been shown to improve the performance of synthesizing target images by addressing the problems of synthesis at different scales, leveraging differentiable information within and across sequences. Given that current research on synthetic CE-MRI is relatively sparse and challenging, our study provides a novel approach that may be instructive for future research based on DWIs. Our further work will be to conduct reader studies to verify the clinical value of our research in downstream applications, such as helping radiologists on detecting tumors. In addition, synthesizing dynamic contrast-enhanced MRI at multiple time points will also be our future research direction. Our proposed model can potentially be used to synthesize CE-MRI, which is expected to reduce or avoid the use of GBCA, thereby optimizing logistics and minimizing potential risks to patients.

Acknowledgments. The authors are thankful for the support from the Guangzhou Elite Project (TZ-JY201948) and Chinese Scholarship Council.

References

1. Amornsiripanitch, N., Bickelhaupt, S., Shin, H.J., Dang, M., Rahbar, H., Pinker, K., Partridge, S.C.: Diffusion-weighted MRI for unenhanced breast cancer screening. Radiology **293**(3), 504–520 (2019)
2. Avants, B.B., Tustison, N.J., Song, G., Cook, P.A., Klein, A., Gee, J.C.: A reproducible evaluation of ants similarity metric performance in brain image registration. Neuroimage **54**(3), 2033–2044 (2011)
3. Baltzer, P., et al.: Diffusion-weighted imaging of the breast-a consensus and mission statement from the EUSOBI international breast diffusion-weighted imaging working group. Eur. Radiol. **30**, 1436–1450 (2020)
4. Broome, D.R., Girguis, M.S., Baron, P.W., Cottrell, A.C., Kjellin, I., Kirk, G.A.: Gadodiamide-associated nephrogenic systemic fibrosis: why radiologists should be concerned. Am. J. Roentgenol. **188**(2), 586–592 (2007)

5. Chung, M., et al.: Deep learning to simulate contrast-enhanced breast MRI of invasive breast cancer. Radiology **306**, 213199 (2022)
6. Goldhirsch, A., et al.: Personalizing the treatment of women with early breast cancer: highlights of the St Gallen international expert consensus on the primary therapy of early breast cancer 2013. Ann. Oncol. **24**(9), 2206–2223 (2013)
7. van der Hoogt, K.J.J., et al.: Factors affecting the value of diffusion-weighted imaging for identifying breast cancer patients with pathological complete response on neoadjuvant systemic therapy: a systematic review. Insights Imaging **12**(1), 1–22 (2021). https://doi.org/10.1186/s13244-021-01123-1
8. Iima, M., Honda, M., Sigmund, E.E., Ohno Kishimoto, A., Kataoka, M., Togashi, K.: Diffusion MRI of the breast: current status and future directions. J. Magn. Reson. Imaging **52**(1), 70–90 (2020)
9. Kanda, T., Ishii, K., Kawaguchi, H., Kitajima, K., Takenaka, D.: High signal intensity in the dentate nucleus and globus pallidus on unenhanced t1-weighted MR images: relationship with increasing cumulative dose of a gadolinium-based contrast material. Radiology **270**(3), 834–841 (2014)
10. Kleesiek, J., et al.: Can virtual contrast enhancement in brain MRI replace gadolinium?: a feasibility study. Invest. Radiol. **54**(10), 653–660 (2019)
11. Li, W., et al.: Virtual contrast-enhanced magnetic resonance images synthesis for patients with nasopharyngeal carcinoma using multimodality-guided synergistic neural network. Int. J. Radiat. Oncol.* Biol.* Phys. **112**(4), 1033–1044 (2022)
12. Mann, R.M., Cho, N., Moy, L.: Breast MRI: state of the art. Radiology **292**(3), 520–536 (2019)
13. Mann, R.M., Kuhl, C.K., Moy, L.: Contrast-enhanced MRI for breast cancer screening. J. Magn. Reson. Imaging **50**(2), 377–390 (2019)
14. Marckmann, P., et al.: Nephrogenic systemic fibrosis: suspected causative role of gadodiamide used for contrast-enhanced magnetic resonance imaging. J. Am. Soc. Nephrol. **17**(9), 2359–2362 (2006)
15. Nguyen, N.C., Molnar, T.T., Cummin, L.G., Kanal, E.: Dentate nucleus signal intensity increases following repeated gadobenate dimeglumine administrations: a retrospective analysis. Radiology **296**(1), 122–130 (2020)
16. Olchowy, C., et al.: The presence of the gadolinium-based contrast agent depositions in the brain and symptoms of gadolinium neurotoxicity-a systematic review. PLoS ONE **12**(2), e0171704 (2017)
17. Partridge, S.C., Newitt, D.C., Chenevert, T.L., Rosen, M.A., Hylton, N.M., Team, A.T., Investigators, I.S.T.: Diffusion-weighted MRI in multicenter trials of breast cancer. Radiology **291**(2), 546–546 (2019)
18. Sung, H., et al.: Global cancer statistics 2020: GLOBOCAN estimates of incidence and mortality worldwide for 36 cancers in 185 countries. CA: Cancer J. Clin. **71**(3), 209–249 (2021)
19. Yi, X., Walia, E., Babyn, P.: Generative adversarial network in medical imaging: a review. Med. Image Anal. **58**, 101552 (2019)

Developing Large Pre-trained Model for Breast Tumor Segmentation from Ultrasound Images

Meiyu Li[1], Kaicong Sun[1], Yuning Gu[1], Kai Zhang[1], Yiqun Sun[1], Zhenhui Li[2], and Dinggang Shen[1,3(✉)]

[1] School of Biomedical Engineering, ShanghaiTech University, Shanghai, China
dgshen@shanghaitech.edu.cn
[2] Department of Radiology, Yunnan Cancer Hospital, Kunming, China
[3] Shanghai United Imaging Intelligence Co., Ltd., Shanghai, China

Abstract. Early detection and diagnosis of breast cancer using ultrasound images are crucial for timely diagnostic decision and treatment in clinical application. However, the similarity between tumors and background and also severe shadow noises in ultrasound images make accurate segmentation of breast tumor challenging. In this paper, we propose a large pre-trained model for breast tumor segmentation, with robust performance when applied to new datasets. Specifically, our model is built upon UNet backbone with deep supervision for each stage of the decoder. Besides using Dice score, we also design discriminator-based loss on each stage of the decoder to penalize the distribution dissimilarity from multi-scales. Our proposed model is validated on a large clinical dataset with more than 10000 cases, and shows significant improvement than other representative models. Besides, we apply our large pretrained model to two public datasets without fine tuning, and obtain extremely good results. This indicates great generalizability of our large pre-trained model, as well as robustness to multi-site data. The code is publicly available at https://github.com/limy-ulab/US-SEG.

Keywords: Large-scale clinical dataset · Deep-supervision · Multi-scale segmentation · Breast ultrasound images

Registration Number: 4319

1 Introduction

Breast cancer is a serious health problem with high incidence and wide prevalence for women throughout the world [1, 2]. Regular screening and early detection are crucial for effective diagnosis and treatment, and hence for improved prognosis and survival rate [3, 4]. Clinical researches have shown that ultrasound imaging is an effective tool for screening breast cancer, due to its critical characteristics of non-invasiveness, non-radiation and inexpensiveness [5–7]. In clinical diagnosis, delineating tumor regions from background is a crucial step for quantitative analysis [8]. Manual delineation always depends on the experience of radiologists, which tends to be subjective and time-consuming [9, 10].

H. Greenspan et al. (Eds.): MICCAI 2023, LNCS 14226, pp. 89–96, 2023.
https://doi.org/10.1007/978-3-031-43990-2_9

Therefore, there is a high demand for automatic and robust methods to achieve accurate breast tumor segmentation. However, due to speckle noise and shadows in ultrasound images, breast tumor boundaries tend to be blurry and are difficult to be distinguished from background. Furthermore, the boundary and size of breast tumors are always variable and irregular [11, 12]. These issues pose challenges and difficulties for accurate breast tumor segmentation in ultrasound images.

Various approaches based on deep learning have been developed for tumor segmentation with promising results [13–19]. Su et al. [13] designed a multi-scale U-Net to extract more semantic and diverse features for medical image segmentation, using multiple convolution sequences and convolution kernels with different receptive fields. Zhou et al. [14] raised a deeply-supervised encoder-decoder network, which is connected through a series of nested and dense skip pathways to reduce semantic gap between feature maps. In [15], a multi-scale selection and multi-channel fusion segmentation model was built, which gathers global information from multiple receptive fields and integrates multi-level features from different network positions for accurate pancreas segmentation. Oktay et al. [16] proposed an attention gate model, which is capable of suppressing irrelevant regions while highlighting useful features for a specific task. Huang et al. [17] introduced a UNet 3+ for medical image segmentation, which incorporates low-level and high-level feature maps in different scales and learns full-scale aggregated feature representations. Liu et al. [18] established a convolution neural network optimized by super-pixel and support vector machine, segmenting multiple organs from CT scans to assist physicians diagnosis. Pei et al. [19] introduced channel and position attention module into deep learning neural network to obtain contextual information for colorectal tumors segmentation in CT scans. However, although these proposed models have achieved satisfactory results in different medical segmentation tasks, their performances are limited for breast tumor segmentation in ultrasound images due to the low image contrast and blurry tissue boundary.

To address these challenges, we present, to the best of our knowledge, the first work to adopt multi-scale features collected from large set of clinical ultrasound images for breast tumor segmentation. The main contributions of our work are as follows: (1) we propose a well-pruned simple but effective network for breast tumor segmentation, which shows remarkable and solid performance on large clinical dataset; (2) our large pretrained model is evaluated on two additional public datasets without fine-tuning and shows extremely stabilized improvement, indicating that our model has outstanding generalizability and good robustness against multi-site data data.

2   Method

We demonstrate the architecture of our proposed network in Fig. 1. It is based on the UNet [20] backbone. In order to collect multi-scale rich information for tumor tissues, we propose to use GAN [21]-based multi-scale deep supervision. In particular, we apply similarity constraint for each stage of the UNet decoder to obtain consistent and stable segmentation maps. Instead of using Dice score in the final layer of UNet, we also use Dice loss on each of the decoder stages. Besides, we integrate an adversarial loss as additional constraint to penalize the distribution dissimilarity between the predicted segmentation map and the ground truth. In the framework of GAN, we take our segmentation network as the generator and a convolutional neural network as the discriminator. The discriminator consists of five convolution layers with the kernel sizes of 7×7, 5×5, 4×4, 4×4 and 4×4. Therefore, we formulate the overall loss for the generator, namely the segmentation network, as

$$
\begin{aligned}
L_{overall} = \sum_{n=1}^{4} & \alpha_n \cdot (1 - 2\left|P^{(n)} \cap G\right| \cdot \left(\left|P^{(n)}\right| + |G|\right)^{-1}) \\
& + \beta_n \cdot E_{P^{(n)} \sim P_{SN}(proba)} \log(1 - CN_{\theta_{CN}^n}(SN_{\theta_{SN}}(P^{(n)})))
\end{aligned}
\tag{1}
$$

where SN represents the segmentation network and CN represents the involved convolutional network. θ_{SN} and θ_{CN} refer to the parameters in the segmentation and convolutional network, respectively. $P^{(n)}$ represents the segmentation maps obtained from the n-th stage in the segmentation network, and G refers to the corresponding ground truth. $p_{SN}(proba)$ denotes the distribution of probability maps. $CN_{\theta_{CN}}(SN_{\theta_{SN}}(P^{(2)}))$ represents the probability for the input of CN coming from the predicted maps rather than the real ones. The parameters α_1 is empirically set as 1, α_2, α_3, α_4 are set as 0.1, and β_1, β_2, β_3 and β_4 are set as 0.05. It should be noted that, in UNet, there are 4 stages and hence we employ 4 CNNs for each of them without sharing their weights.

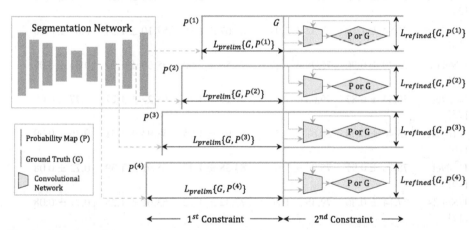

Fig. 1. Overview of the proposed architecture. The first constraint is used to enhance prediction similarity to the standard ones, for preliminary segmentation. The second constraint is used to capture data distribution to maintain consistency in high-dimensional space for map refinement.

Meanwhile, the adversarial loss for each of the CNN is defined as:

$$
\begin{aligned}
L_{CN}\left\{G, P^{(2)}\right\} = & -E_{G \sim p_{CN}(truth)} \log(CN_{\theta_{CN}}(G)) \\
& - E_{P^{(2)} \sim p_{SN}(proba)} \log(1 - CN_{\theta_{CN}}(SN_{\theta_{SN}}(P^{(2)})))
\end{aligned}
\tag{2}
$$

where $p_{CN}(truth)$ denotes the distribution of original samples. $CN_{\theta_{CN}}(G)$ represents the probability for the input of CN coming from the original dataset.

In the implementation, we update the segmentation network and all the discriminators alternatingly in each iteration until both the generator and discriminators are converged.

3 Experiments

3.1 Dataset and Implementation Details

We collected 10927 cases for this research from Yunnan Cancer Hospital. Each scan is with resolution of 1×1 mm^2 and size of 512×480. The breast tumors of each case are delineated by three experienced experts. Five-fold cross validation is performed on the dataset in all experiments to verify our proposed network. For external validation, we further test our model on two independent publicly-available datasets collected by STU-Hospital (Dataset 1) [22] and SYU-University (Dataset 2) [23]. In order to comprehensively evaluate segmentation efficiency of our model, Dice Similarity Coefficient (DSC), Precision, Recall, Jaccard, and Root Mean Squared Error (RMSE) are used as evaluation metrics in this work.

Table 1. Quantitative comparison with state-of-the-art segmentation methods on large-scale clinical breast ultrasound dataset.

Method	DSC (%)↑	Precision (%)↑	Recall (%)↑	Jaccard (%)↑	RMSE (mm)↓
DeepRes [24]	75.16 ± 1.13	80.04 ± 1.33	73.52 ± 2.18	61.24 ± 1.53	0.81 ± 0.08
MSUNet [13]	71.84 ± 1.19	67.32 ± 2.04	81.67 ± 1.85	57.10 ± 1.54	0.82 ± 0.10
UNet++ [14]	73.20 ± 0.96	75.98 ± 1.42	73.58 ± 2.00	58.59 ± 1.27	0.84 ± 0.09
SegNet [25]	76.84 ± 0.87	79.08 ± 1.33	77.29 ± 1.73	63.23 ± 1.27	0.77 ± 0.08
AttU-Net [16]	73.46 ± 0.98	80.81 ± 1.33	69.93 ± 1.95	58.94 ± 1.31	0.86 ± 0.09
U^2-Net [26]	78.78 ± 0.91	75.90 ± 1.52	**84.58 ± 1.43**	65.91 ± 1.39	0.72 ± 0.08
UNet 3+ [17]	77.14 ± 0.86	79.45 ± 1.20	77.32 ± 1.72	63.61 ± 1.25	0.77 ± 0.08
Ours	**80.97 ± 0.88**	**81.64 ± 1.30**	82.19 ± 1.61	**68.83 ± 1.38**	**0.68 ± 0.06**

Our proposed algorithm is conducted on PyTorch, and all experiments are performed on NVIDIA Tesla A100 GPU. We use Adam optimizer to train the framework with an initial learning rate of 10^{-4}. The epochs to train all models are 100 and the batch size in training process is set as 4.

3.2 Comparison with State-of-the-Art Methods

To verify the advantages of our proposed model for breast tumor segmentation in ultrasound images, we compare our deep-supervised convolutional network with the state-of-the-art tumor segmentation methods, including DeepRes [24], MSUNet [13], UNet++

[14], SegNet [25], AttU-Net [16], U^2-Net [26] and UNet 3+ [17]. The comparison experiments are carried on a large-scale clinical breast ultrasound dataset, and the quantitative results are reported in Table 1. It is obvious that our proposed model achieves the optimal performance compared with other segmentation models. For example, our method obtains a DSC score of 80.97%, which is 5.81%, 9.13%, 7.77%, 4.13%, 7.51%, 2.19%, and 3.83% higher than other seven models. These results indicate the effectiveness of the proposed model in delineating breast tumors in ultrasound images.

Representative segmentation results using different methods are provided in Fig. 2. The probability maps predicted by our model are more consistent with the ground truth, especially in the tiny structures which are difficult to capture. This verifies the superior ability of the proposed model in maintaining detailed edge information compared with state-of-the-art methods.

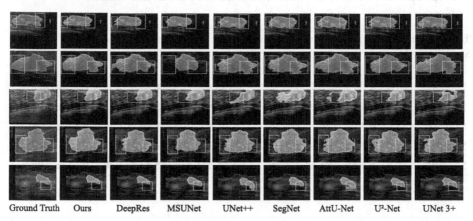

Ground Truth   Ours   DeepRes   MSUNet   UNet++   SegNet   AttU-Net U^2-Net   UNet 3+

Fig. 2. Segmentation results of five subjects obtained from different models. Red boxes are used to highlight the boundaries which are difficult to segment.

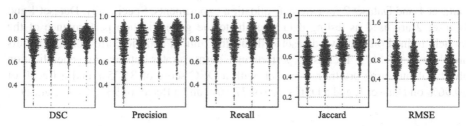

DSC   Precision   Recall   Jaccard   RMSE

Fig. 3. Five evaluation criteria for total cases from different frameworks: Stage I, Stage II, Stage III, and Stage IV (From left to right in each index).

3.3 Ablation Studies

We demonstrate the efficacy of the deep supervision strategy using ablation studies. Four groups of frameworks (Stage I, Stage II, Stage III and Stage IV) are designed, with the numerals denoting the level of deep supervision counting from the last deconvolutional layer.

We test these four frameworks on the in-house breast ultrasound dataset, and verify their segmentation performance using the same five evaluation criteria. The evaluation metrics from all cases are presented by the scatter plots in Fig. 3. The obtained quantitative results are shown in Table 2, where Stage IV model achieves the optimal DSC, Precision, Recall, and Jaccard. All these results draw a unanimous conclusion on the relationship between these four frameworks. That is, the segmentation ability of the proposed Stage IV is ameliorated from every possible perspective. Moreover, Stage IV obtains minimal RMSE compared with other three models (0.68 mm vs 0.84 mm, 0.82 mm, 0.75 mm), which means better matching of the predicted maps from Stage IV with the corresponding ground truth. All these comparison results verify the superiority of deep supervision for breast tumor segmentation in ultrasound images.

Table 2. Quantitative results among four groups of segmentation frameworks Stage I, Stage II, Stage III, and Stage IV.

Method	DSC (%)↑	Precision (%)↑	Recall (%)↑	Jaccard (%)↑	RMSE (mm)↓
Stage I	70.52 ± 1.25	70.01 ± 1.77	76.16 ± 2.04	55.72 ± 1.53	0.84 ± 0.09
Stage II	73.43 ± 1.07	76.10 ± 1.30	74.19 ± 2.30	58.99 ± 1.37	0.82 ± 0.08
Stage III	77.67 ± 1.06	79.63 ± 1.43	78.17 ± 1.98	64.42 ± 1.47	0.75 ± 0.07
Stage IV	**80.97 ± 0.88**	**81.64 ± 1.30**	**82.19 ± 1.61**	**68.83 ± 1.38**	**0.68 ± 0.06**

3.4 Performance on Two External Public Datasets

In order to evaluate the generalizability of the proposed model, we introduce external Dataset 1 and Dataset 2 for external validation experiments. Specifically, Dataset 1 and Dataset 2 are used as testing data to evaluate the generalization performance of the models trained on our own dataset without fine tuning, and the corresponding results are shown in Table 3. Promising performance demonstrates outstanding generalization ability of our large pre-trained model, with a DSC score of 81.35% and a Recall of 80.96% on Dataset 1, and a DSC score of 77.16% and a Recall of 93.22% on Dataset 2.

Table 3. External validation performance on two independent publicly-available Dataset 1 and Dataset 2.

Dataset	DSC (%)	Precision (%)	Recall (%)	Jaccard (%)	RMSE (mm)
Dataset 1	81.35 ± 2.04	85.75 ± 2.97	80.96 ± 2.09	70.65 ± 3.07	0.19 ± 0.01
Dataset 2	77.16 ± 2.51	68.34 ± 4.03	93.22 ± 0.65	65.26 ± 3.67	0.42 ± 0.01

4   Conclusion

In this paper, we have developed a large pre-trained model for breast tumor segmentation from ultrasound images. In particular, two constraints are proposed to exploit both image similarity and space correlation information for refining the prediction maps. Moreover, our proposed deep supervision strategy is used for quality control at each decoding stage, optimizing prediction maps layer-by-layer for overall performance improvement. Using a large clinical dataset, our proposed model demonstrates *not only* state-of-the-art segmentation performance, *but also* the outstanding generalizability to new ultrasound data from different sites. Besides, our large pre-trained model is general and robust in handling various tumor types and shadow noises in our acquired clinical ultrasound images. This also shows the potential of directly applying our model in real clinical applications.

Acknowledgment. This work was supported in part by The Key R&D Program of Guangdong Province, China (grant number 2021B0101420006), National Natural Science Foundation of China (grant number 62131015), and Science and Technology Commission of Shanghai Municipality (STCSM) (grant number 21010502600).

References

1. Siegel, R.L., Miller, K.D., Fuchs, H.E., Jemal, A.: Cancer statistics. CA: Cancer J. Clin. **72**(1), 7–33 (2022)
2. Arnold, M., Morgan, E., Rumgay, H., Mafra, A., Singh, D., Laversanne, M., et al.: Current and future burden of breast cancer: global statistics for 2020 and 2040. Breast **66**, 15–23 (2022)
3. Berg, W.A., Zhang, Z., Lehrer, D., Jong, R.A., Pisano, E.D., Barr, R.G., et al.: Detection of breast cancer with addition of annual screening ultrasound or a single screening MRI to mammography in women with elevated breast cancer risk. J. Am. Med. Assoc. (JAMA) **307**(13), 1394–1404 (2012)
4. Kalager, M., Haldorsen, T., Bretthauer, M., Hoff, G., Thoresen, S.O., Adami, H.O.: Improved breast cancer survival following introduction of an organized mammography screening program among both screened and unscreened women: a population-based cohort study. Breast Cancer Res. BCR **11**(4), 1–9 (2009)
5. Cheng, H.D., Shan, J., Ju, W., Guo, Y., Zhang, L.: Automated breast cancer detection and classification using ultrasound images: a survey. Pattern Recogn. **43**(1), 299–317 (2010)
6. Kratkiewicz, K., Pattyn, A., Alijabbari, N., Mehrmohammadi, M.: Ultrasound and photoacoustic imaging of breast cancer: clinical systems, challenges, and future outlook. J. Clin. Med. **11**(5), 1165 (2022)
7. Ragab, M., Albukhari, A., Alyami, J., Mansour, R.F.: Ensemble deep-learning-enabled clinical decision support system for breast cancer diagnosis and classification on ultrasound images. Biology (Basel) **11**(3), 439 (2022)
8. Xian, M., Zhang, Y., Cheng, H.-D., Xu, F., Zhang, B., Ding, J.: Automatic breast ultrasound image segmentation: a survey. Pattern Recogn. **79**, 340–355 (2018)
9. Calas, M.J.G., Almeida, R.M.V.R., Gutfilen, B., Pereira, W.C.A.: Intraobserver interpretation of breast ultrasonography following the BI-RADS classification. Eur. J. Radiol. **74**(3), 525–528 (2010)

10. Yap, M.H., Edirisinghe, E.A., Bez, H.E.: Processed images in human perception: a case study in ultrasound breast imaging. Eur. J. Radiol. **73**(3), 682–687 (2010)
11. Jalalian, A., Mashohor, S.B., Mahmud, H.R., Saripan, M.I.B., Ramli, A.R.B., Karasfi, B.: Computer-aided detection/diagnosis of breast cancer in mammography and ultrasound: a review. Clin. Imaging **37**(3), 420–426 (2013)
12. Zhou, Y., Chen, H., Li, Y., Cao, X., Wang, S., Shen, D.: Cross-model attention-guided tumor segmentation for 3D automated breast ultrasound (ABUS) images. IEEE J. Biomed. Health Inform. **26**(1), 301–311 (2021)
13. Su, R., Zhang, D., Liu, J., Cheng, C.: Msu-net: Multi-scale u-net for 2D medical image segmentation. Front. Genet. **12**, 639930 (2021)
14. Zhou, Z., Rahman Siddiquee, M.M., Tajbakhsh, N., Liang, J.: Unet++: a nested U-net architecture for medical image segmentation. In: Stoyanov, D., et al. (eds.) DLMIA/ML-CDS -2018. LNCS, vol. 11045, pp. 3–11. Springer, Cham (2018). https://doi.org/10.1007/978-3-030-00889-5_1
15. Li, M., Lian, F., Guo, S.: Multi-scale selection and multi-channel fusion model for pancreas segmentation using adversarial deep convolutional nets. J. Digit. Imaging **35**, 47–55 (2022)
16. Oktay, O., et al.: Attention U-net: learning where to look for the pancreas. arXiv preprint arXiv:180403999 (2018)
17. Huang, H., et al. (eds.) Unet 3+: a full-scale connected unet for medical image segmentation. In: ICASSP 2020–2020 IEEE International Conference on Acoustics, Speech and Signal Processing (ICASSP). IEEE (2020)
18. Liu, X., Guo, S., Yang, B., Ma, S., Zhang, H., Li, J., et al.: Automatic organ segmentation for CT scans based on super-pixel and convolutional neural networks. J. Digit. Imaging **31**, 748–760 (2018)
19. Pei, Y., Mu, L., Fu, Y., He, K., Li, H., Guo, S., et al.: Colorectal tumor segmentation of CT scans based on a convolutional neural network with an attention mechanism. IEEE Access. **8**, 64131–64138 (2020)
20. Ronneberger, O., Fischer, P., Brox, T.: U-net: convolutional networks for biomedical image segmentation. In: Navab, N., Hornegger, J., Wells, W.M., Frangi, A.F. (eds.) MICCAI 2015. LNCS, vol. 9351, pp. 234–241. Springer, Cham (2015). https://doi.org/10.1007/978-3-319-24574-4_28
21. Goodfellow, I.J., et al.: Generative adversarial nets. In: Proceedings of the 27th International Conference on Neural Information Processing Systems, Montreal, Canada, vol. 2, pp. 2672–2680. MIT Press, Cambridge (2014)
22. Negi, A., Raj, A.N.J., Nersisson, R., Zhuang, Z., Murugappan, M.: RDA-UNET-WGAN: an accurate breast ultrasound lesion segmentation using Wasserstein generative adversarial networks. Arab. J. Sci. Eng. **45**(8), 6399–6410 (2020)
23. Huang, Q., Huang, Y., Luo, Y., Yuan, F., Li, X.: Segmentation of breast ultrasound image with semantic classification of superpixels. Med. Image Anal. **61**, 101657 (2020)
24. Zhang, Z., Liu, Q., Wang, Y.: Road extraction by deep residual u-net. IEEE Geosci. Remote Sens. Lett. **15**(5), 749–753 (2018)
25. Badrinarayanan, V., Kendall, A., Cipolla, R.: Segnet: a deep convolutional encoder-decoder architecture for image segmentation. IEEE Trans. Pattern Anal. Mach. Intell. **39**(12), 2481–2495 (2017)
26. Qin, X., Zhang, Z., Huang, C., Dehghan, M., Zaiane, O.R., Jagersand, M.: U2-Net: going deeper with nested U-structure for salient object detection. Pattern Recogn. **106**, 107404 (2020)

Clinical Applications – Cardiac

Clinical Applications – Cardiac

Towards Expert-Amateur Collaboration: Prototypical Label Isolation Learning for Left Atrium Segmentation with Mixed-Quality Labels

Zhe Xu[1], Jiangpeng Yan[3], Donghuan Lu[2(✉)], Yixin Wang[4], Jie Luo[5],
Yefeng Zheng[2], and Raymond Kai-yu Tong[1(✉)]

[1] Department of Biomedical Engineering, The Chinese University of Hong Kong,
Hong Kong, China
jackxz@link.cuhk.edu.hk , kytong@cuhk.edu.hk
[2] Tencent Healthcare Co., Jarvis Lab, Shenzhen, China
caleblu@tencent.com
[3] Department of Automation, Tsinghua University, Beijing, China
[4] Department of Bioengineering, Stanford University, Stanford, CA, USA
[5] Massachusetts General Hospital, Harvard Medical School, Boston, MA, USA

Abstract. Deep learning-based medical image segmentation usually requires abundant high-quality labeled data from experts, yet, it is often infeasible in clinical practice. Without sufficient expert-examined labels, the supervised approaches often struggle with inferior performance. Unfortunately, directly introducing additional data with low-quality cheap annotations (e.g., crowdsourcing from non-experts) may confuse the training. To address this, we propose a Prototypical Label Isolation Learning (PLIL) framework to robustly learn left atrium segmentation from scarce high-quality labeled data and massive low-quality labeled data, which enables effective expert-amateur collaboration. Particularly, PLIL is built upon the popular teacher-student framework. Considering the structural characteristics that the semantic regions of the same class are often highly correlated and the higher noise tolerance in the high-level feature space, the self-ensembling teacher model isolates clean and noisy labeled voxels by exploiting their relative feature distances to the class prototypes via multi-scale voting. Then, the student follows the teacher's instruction for adaptive learning, wherein the clean voxels are introduced as supervised signals and the noisy ones are regularized via perturbed stability learning, considering their large intra-class variation. Comprehensive experiments on the left atrium segmentation benchmark demonstrate the superior performance of our approach.

Keywords: Image Segmentation · Class Prototype · Label Noises

1 Introduction

Segmenting the left atrium (LA) from magnetic resonance images (MRI) is critical in treating atrial fibrillation. Recent success of deep learning (DL)-

© The Author(s), under exclusive license to Springer Nature Switzerland AG 2023
H. Greenspan et al. (Eds.): MICCAI 2023, LNCS 14226, pp. 99–109, 2023.
https://doi.org/10.1007/978-3-031-43990-2_10

based methods usually requires a large amount of high-quality (HQ) labeled data (termed as Set-HQ). However, since labeling medical images is expertise-demanding and laborious, acquiring massive HQ labeled data from experts is expensive and not always feasible. Without sufficient HQ labels, the DL approaches often struggle with inferior performance. Despite the recent success of semi-supervised learning (SSL) that leverages abundant unlabeled data [3,11,20,21], it is still difficult for SSL to accurately propagate label information at the voxel level especially when the HQ labeled data is extremely scarce. Thus, an intuitive cost-efficient alternative is to collect additional labels via cheaper ways, e.g., crowdsourcing from non-experts, as depicted in Fig. 1. Unfortunately, the quality of cheap labels is always unsatisfactory. Directly introducing additional data with low-quality (LQ) noisy labels (termed as Set-LQ) may mislead the model training, easily causing performance degradation [10,18]. Such a pervasive dilemma poses a challenging yet practical scenario: how to robustly learn segmentation from scarce HQ labeled data and abundant LQ noisy labeled data?

The existing works on mining LQ labeled data for medical image segmentation can be categorized by two distinct application scenarios: (i) **HQ-agnostic**, e.g., Set-HQ and Set-LQ are mixed as one dataset [6,9,24,26–28]. TriNet [24] uses a tri-network that integrates predictions from two peer networks to supervise the third network; PNL [28] introduces an image-level label quality evaluation module to identify clean labels to tune the network. (ii) **HQ-aware**, e.g., recruiting experts to obtain a reasonable amount of HQ labeled data and thus Set-HQ and Set-LQ are separate. Such scenario extends SSL [3,11,20,21] to further exploit the potentially useful information of LQ labels, which will be more beneficial when the HQ labeled data is extremely scarce (as detailed in Sect. 3). Luo et al. [10] proposed to implicitly decouple the learning processes for Set-HQ and Set-LQ using two separate decoders; KDEM [5] extends [10] with knowledge distillation and entropy minimization regularization. However, this implicit decoupling strategy is experimentally hard-to-control. Thus, MTCL [18] estimates the joint distribution matrix between observed and latent true labels to explicitly characterize mislabeled locations for smooth label refurbishment. However, MTCL is based on the class-conditional noise (CCN) assumption that the noise is independent of input features given the true label, which may be impractical [2]. Considering the clinical practice, we advocate the HQ-aware scenario because: (a) HQ/LQ labeled data can be separated since the sources of medical annotation are usually recorded and acquiring a reasonable amount of HQ labels from radiologists is feasible; (b) the separation may implicitly embed rewarding prior knowledge on discriminating HQ/LQ labeled data into training.

Tailoring for the HQ-aware scenario, in this work, we propose the Prototypical Label Isolation Learning (PLIL) framework for left atrium segmentation, enabling effective expert-amateur collaboration. Specifically, PLIL is built upon the popular teacher-student framework. Besides the prime supervised signals from HQ labeled data, PLIL robustly exploits the additional LQ labeled data via two steps: (i) Considering the structural characteristics that semantic regions of the same class are often highly correlated and the higher noise tolerance in

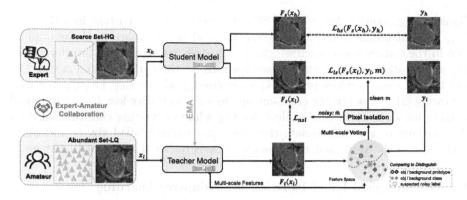

Fig. 1. Overview of our prototypical label isolation learning (PLIL) framework for robustly learning segmentation with scarce HQ labeled data and abundant LQ labeled data. m is the estimated clean-label selection mask; $\bar{m}$ is the noisy-label selection mask.

the high-level feature space [13,23], the self-ensembling teacher model isolates clean and noisy labeled voxels by exploiting their relative feature distances to the class prototypes via multi-scale voting. Besides the advantage of explicit spatial isolation, this strategy takes the input features into account, which is more realistic compared to [18] as the mislabeled voxels often present difficult and ambiguous regions in the image. (ii) Synergistically, the student follows the teacher's instruction for adaptive learning, wherein the clean voxels are further introduced as supervised signals and the noisy ones are especially regularized via perturbed stability learning, considering their vulnerable large intra-class variation in general. Comprehensive experiments on left atrium segmentation under extreme budget settings demonstrate the superior performance of our approach. The ablation study further verifies the effectiveness of each component.

2 Methods

2.1 Problem Formulation

Our PLIL framework is depicted in Fig. 1. Following the HQ-aware scenario, we have access to scarce expert-examined HQ labeled data $\mathcal{S}_h = \left\{ \left(x_{h(i)}, y_{h(i)} \right) \right\}_{i=1}^{M}$ that only contains M samples, and abundant non-expert LQ noisy labeled data $\mathcal{S}_l = \left\{ \left(x_{l(i)}, y_{l(i)} \right) \right\}_{i=M+1}^{N}$ that consists of $N - M$ (usually $\gg M$) samples, where $x_{h(i)}, x_{l(i)} \in \mathbb{R}^{\Omega_i}$ denote the images and $y_{h(i)}, y_{l(i)} \in \{0,1\}^{\Omega_i \times C}$ are the given HQ or LQ label (C denotes the class number). Our goal is to learn segmentation with scarce Set-HQ and abundant Set-LQ by optimizing the following loss function:

$$\mathcal{L} = \mathcal{L}_{HQ} + \lambda \mathcal{L}_{LQ}, \tag{1}$$

where $\mathcal{L}_{HQ}$ and $\mathcal{L}_{LQ}$ denote the guidance from HQ and LQ labeled data, respectively. λ is a trade-off weight for $\mathcal{L}_{LQ}$, scheduled by the time-dependent ramp-up

Gaussian function [4] $\lambda(t) = e^{-5\left(1-\frac{t}{t_{max}}\right)^2}$, where t is the current iteration and t_{max} is the maximal iteration. Since our method heavily relies on the manipulation in the feature space, such weighting schedule can reduce the interference of LQ labeled data to the feature space learning at the early training stage. The HQ labeled data provides prime HQ supervised guidance $\mathcal{L}_{hs}$, i.e., $\mathcal{L}_{HQ} = \mathcal{L}_{hs}$. Following [21], we adopt the cross-entropy loss $\mathcal{L}_{ce}$ and Dice loss $\mathcal{L}_{dice}$ with equal weights for $\mathcal{L}_{hs}$. To further exploit Set-LQ while alleviating confirmation bias [10], we aim to spatially isolate the *clean* and *noisy* labeled voxels and make better use of the suspected *noisy* labeled voxels rather than discarding them.

2.2 Prototypical Label Isolation for Adaptive Learning

Teacher-Student Architecture. Our framework is built upon the popular teacher-student architecture [15], where the student model F_s is updated by back-propagation and the teacher F_t is updated by the exponential moving average (EMA) weights of the student θ across training steps. Denoting the weights of the teacher model at step t as $\tilde{\theta}_t$, $\tilde{\theta}_t$ is updated by: $\tilde{\theta}_t = \alpha\tilde{\theta}_{t-1} + (1-\alpha)\theta_t$, where α is the EMA decay rate and empirically set to 0.99 [15]. As such, the teacher model owns the self-ensembling property [4], which can avoid sharp deterioration of the feature quality and thus suits our following prototypical label isolation strategy that appreciates high-quality and smooth embedding space.

Multi-scale Voting-Based Prototypical Label Isolation. Considering the structural characteristics that the targeted segmentation regions of the same class are often highly correlated and the higher noise tolerance in the high-level feature space [1,13,14,23], our label isolation strategy is inherently motivated by the assumption that for a *clean* labeled voxel, its features should lie closer to its corresponding class prototype (class-wise feature centroid); otherwise, a potential *noisy* labeled voxel is suspected. Specifically, we determine whether a voxel-wise label is a *clean* one by exploiting the relative feature distances to the class prototypes. Considering that different layers perceive the entire image with different perspectives, a multi-scale voting mechanism is introduced. Technically, given a medical scan x_l of Set-LQ and its noisy label y_l, we denote the last i-th feature map from the teacher model F_t as e_i^{temp}, which is then upsampled to $e_i \in \mathbb{R}^{H \times W \times Z \times L_i}$ (H, W and D denote height, width and depth of x_l, respectively, and L_i is the channel number) to be consistent with the size of segmentation mask via trilinear interpolation. Then, we resort to the pseudo label from the teacher model as the "mask" for the target class, which will be utilized to extract the class features. Denoting the teacher's prediction of x_l as $F_t(x_l)$, the pseudo label corresponds to the class with the maximal posterior probability. Since the HQ labeled data is scarce which makes it hard to obtain confident prediction, Monte Carlo dropout [7] based model uncertainty is leveraged to calibrate the pseudo label. For the LQ labeled image x_l, K stochastic forward inferences through F_t are performed with random dropout. Then, the normalized predictive entropy of the mean of the K softmax predictions is regarded as the uncertainty map u [21]. When the uncertainty u_v at voxel v is smaller than a threshold η, i.e., $u_v < \eta$, this voxel will be used as the final pseudo mask $\hat{F}_t(x_l)$. As such, at the i-th scale, the object prototype q_i^{obj} can be obtained via the masked average pooling [20, 25] as: $q_i^{obj} = \dfrac{\sum_v \hat{F}_{t(v)}^{obj} \cdot p_{t(v)}^{obj} \cdot e_{i(v)}}{\sum_v \hat{F}_{t(v)}^{obj} \cdot p_{t(v)}^{obj}}$, where the predicted probabilities of object $p_{t(v)}^{obj}$ from the teacher model weight the contribution of voxel v to prototype generation. Similarly, the background prototype q_i^{bg} can be also obtained. Then, the relative feature distances $d_{i(v)}^{obj}$ and $d_{i(v)}^{bg}$ between the feature vector of voxel v and the prototypes are defined as:

$$d_{i(v)}^{obj} = \left\| e_{i(v)} - q_i^{obj} \right\|_2 \quad \text{and} \quad d_{i(v)}^{bg} = \left\| e_{i(v)} - q_i^{bg} \right\|_2. \tag{2}$$

Intuitively, if the given label $y_{l(v)}$ at voxel v is object (background) yet its feature vector e_v lies closer to the background (object) prototype than the object (background) prototype, this voxel will be isolated to the *noisy* group. Otherwise, it will be selected as the *clean* labeled one. Formally, the i-th scale determines the *clean-label* selection mask m_i for image x_l as:

$$m_{i(v)} = \mathbb{1}[y_{l(v)} = 1] \cdot \mathbb{1}[d_{i(v)}^{obj} < d_{i(v)}^{bg}] + \mathbb{1}[y_{l(v)} = 0] \cdot \mathbb{1}[d_{i(v)}^{obj} > d_{i(v)}^{bg}]. \tag{3}$$

We select the last three scales of features from the teacher model to perform multi-scale voting. Thus, for the final *clean-label* selection mask, $m_v = 1$ if $\sum_i^3 m_{i(v)} \geq 2$. The *noisy-label* selection mask $\bar{m}$ is the negation of m.

Adaptive Learning Scheme for Isolated Voxels. As shown in Fig. 1, the additional supervised loss for Set-LQ ($\mathcal{L}_{ls}$) is applied to the isolated *clean* labeled voxels, which takes the form of m-masked cross-entropy loss and Dice loss as:

$$\mathcal{L}_{ls} = \sum_v (m_v \cdot \mathcal{L}_{ce,v} + m_v \cdot \mathcal{L}_{dice,v}). \tag{4}$$

For the noisy group, since it is extremely difficult to perfectly find out the noisy labels, we do not advocate label refinement as in [18] to avoid additional error propagation. Instead, we regularize the model behavior on these ambiguous noisy voxels via perturbed stability learning [15], i.e., encouraging consistent pre-softmax predictions between the student and teacher model for the same input with different perturbations ξ and ξ', formulated as:

$$\mathcal{L}_{nsl} = \dfrac{\sum_v \bar{m}_v \left\| F_{t(v)}(x_l + \xi) - F_{s(v)}(x_l + \xi') \right\|^2}{\sum_v \bar{m}_v}. \tag{5}$$

The design of $\bar{m}$-masked stability loss is motivated by the fact that the estimated noisy group correlates with the voxels with large intra-class variation, wherein these voxels often exhibit difficult and ambiguous nature, which potentially have serious instability problem. Besides, compared to [15], such a noise-selective stability learning avoids the distraction by the redundant easy regions, considering this loss takes the form of mean squared error (MSE) with the average nature. As such, the LQ loss $\mathcal{L}_{LQ}$ in Eq. 1 can be formulated as $\mathcal{L}_{LQ} = \mathcal{L}_{ls} + \beta\mathcal{L}_{nsl}$, where β is a tradeoff weight for the two learning manners. By combining $\mathcal{L}_{HQ}$ and $\mathcal{L}_{LQ}$, the model can not only receive HQ supervision from the scarce Set-HQ but also adaptively exploit different kinds of productive information in Set-LQ towards effective expert-amateur collaboration.

3   Experiments and Results

Materials. The left atrium (LA) segmentation dataset [17] provides 100 3D gadolinium-enhanced magnetic resonance images (GE-MRIs) with expert labels. The images have the isotropic resolution of $0.625 \times 0.625 \times 0.625$ mm^3. Following the same data preprocessing and split in [21], 80 samples are selected for training and the remaining 20 samples for testing. All the images are cropped to the center of the heart region and the intensities are normalized to zero mean and unit variance. We investigate the scenarios of scarce HQ labeled data, where only 4 (5%) or 6 (7.5%) samples are used as Set-HQ and the rest is utilized as non-expert Set-LQ, simulated by the commonly used label corruption scheme [22,28] including random erosion and dilation with 3–15 voxels.

Implementation and Evaluation Metrics. The framework is based on PyTorch using an NVIDIA GeForce RTX 3090 GPU. 3D V-Net [12] is adopted as the backbone, referring to [21]. We randomly crop patches of $112 \times 112 \times 80$ voxels as the input and use sliding window strategy with stride of $18 \times 18 \times 4$ voxels for inference. The batch size is set to 4 including 2 labeled samples and 2 unlabeled samples. $t_{\max}$ is set to 8,000. K, η and β are empirically set to 8, 0.1 and 0.1. The learning rate is initialized as 0.01 and decayed by multiplication with $(1.0 - t/t_{\max})^{0.9}$. Data augmentation, including random flip and rotation, is applied. Four metrics, including Dice, Jaccard, average surface distance (ASD) and 95% Hausdorff distance (95HD), are adopted for comprehensive evaluation. The code will be available at https://github.com/lemoshu/PLIL.

Comparison Study. The quantitative results are presented in Table 1. H-Sup denotes the supervised baseline that only Set-HQ is utilized, while HL-Sup denotes that Set-HQ and Set-LQ are mixed for supervised learning. We also include recent SSL methods (UAMT [21], CPS [3], CPCL [20] and URPC [11]), HQ-agnostic noisy label learning (NLL) methods (TriNet [24] and PNL [28]) and HQ-aware NLL methods (Decoupled [10] and MTCL [18]). All the methods are implemented with the same backbone and training protocols to ensure fairness. As observed, H-Sup performs poorly with scarce Set-HQ, yet, HL-Sup even further degrades, implying that our simulated LQ labels have led to serious confirmation bias. Relying on some model assumptions [16], SSL methods ignore the LQ labels and exploit the image information only from Set-LQ. Despite effectiveness, it is still difficult for SSL to accurately propagate voxel-level label

Table 1. Quantitative comparison study. Cross-subject standard deviations are shown in parentheses. ∗ indicates $p \leq 0.05$ from Wilcoxon signed rank test when comparing ours with the second best under the HQ-aware setting. The best results are in **bold**.

Methods		Settings			Metrics			
		Set-HQ	Set-LQ	HQ-aware?	Dice [%] ↑	Jaccard [%] ↑	95HQ [voxel] ↓	ASD [voxel] ↓
Sup	H-Sup (Upper Bound)	80	0	-	91.25 (1.93)	83.36 (3.24)	6.32 (6.45)	1.50 (0.61)
	H-Sup	4	0	-	77.84 (8.29)	64.02 (9.60)	22.60 (14.57)	5.03 (1.22)
	HL-Sup	4	76	-	77.21 (7.52)	63.47 (9.66)	21.59 (10.72)	6.35 (2.31)
SSL	UAMT [21]	4	76	-	79.88 (9.69)	67.46 (11.93)	24.11 (14.75)	3.21 (1.56)
	CPS [3]	4	76	-	81.54 (6.56)	69.33 (8.89)	24.54 (16.47)	3.86 (1.08)
	CPCL [20]	4	76	-	82.46 (8.15)	70.92 (11.02)	23.23 (14.49)	3.23 (1.32)
	URPC [11]	4	76	-	78.41 (8.53)	65.27 (14.36)	21.74 (15.03)	6.17 (2.33)
NLL	TriNet [24]	4	76	×	80.12 (7.11)	68.11 (8.36)	20.13 (11.74)	4.85 (1.24)
	PNL [28]	4	76	×	78.01 (7.23)	65.01 (8.22)	18.57 (13.05)	4.78 (1.33)
	Decoupled [10]	4	76	✓	80.74 (6.73)	68.23 (7.31)	16.55 (13.54)	4.90 (1.22)
	MTCL [18]	4	76	✓	83.36 (6.12)	71.53 (8.48)	16.81 (14.13)	3.44 (1.51)
	PLIL (ours)	4	76	✓	**84.91** (3.32)*	**73.93** (5.07)*	**15.49** (12.00)	**3.10** (1.31)
Sup	H-Sup	6	0	-	79.41 (4.86)	65.02 (5.11)	24.36 (14.02)	2.78 (1.01)
	HL-Sup	6	74	-	78.09 (4.45)	64.27 (5.78)	13.18 (9.63)	4.14 (0.76)
SSL	UAMT [21]	6	74	-	83.72 (7.15)	73.10 (9.77)	16.44 (14.07)	2.75 (1.10)
	CPS [3]	6	74	-	82.15 (6.89)	70.26 (9.27)	27.61 (15.07)	2.92 (0.90)
	CPCL [20]	6	74	-	83.99 (5.40)	73.55 (7.56)	19.60 (11.80)	2.79 (0.97)
	URPC [11]	6	74	-	80.52 (7.77)	68.14 (9.53)	22.81 (13.67)	6.18 (1.54)
NLL	TriNet [24]	6	74	×	84.82 (3.68)	74.04 (7.29)	15.37 (7.62)	3.01 (1.19)
	PNL [28]	6	74	×	80.05 (4.72)	68.08 (8.53)	17.02 (10.23)	3.58 (0.83)
	Decoupled [10]	6	74	✓	85.01 (3.76)	74.58 (6.43)	12.35 (8.36)	3.37 (1.02)
	MTCL [18]	6	74	✓	86.06 (4.78)	75.73 (7.18)	12.47 (10.06)	2.87 (1.09)
	PLIL (ours)	6	74	✓	**87.66** (2.61)	**78.12** (4.15)*	**10.93** (9.29)*	**2.41** (0.83)

information when the HQ labeled data is scarce. For the HQ-agnostic methods, TriNet and PNL show effectiveness in alleviating the negative effects brought by the agnostic LQ labels, yet, even fall behind some SSL methods given the violent simulated label noises, revealing that the HQ-agnostic setting may be sub-optimal. For the HQ-aware scenario, Decoupled [10] and MTCL [18] perform well under both labeling settings, demonstrating the benefits of HQ-aware strategy. Our PLIL relies on the manipulation in the feature space, more HQ labeled data will help the network learn more discriminative representations towards accurate isolation. As observed in the 6-HQ-sample setting, PLIL achieves the Dice of 87.66%, only 3.59% away from the upper bound trained with all 80 HQ labeled data. Despite less-discriminative features learned under the 4-HQ-sample setting, PLIL can still achieve respectable results, demonstrating its robustness. The impact of varying expert labeling budgets is further illustrated in Fig. 2(c) while Fig. 2(a) presents exemplar results of our PLIL and other approaches under the 6-HQ-sample setting. Consistently, the predicted mask of our PLIL fits more accurately with the ground truth. To better understand our method, we visualize the estimated noisy-label selection mask $\bar{m}$ for a dilated LQ label y_l of LA in Fig. 2(b), where it can be observed that most dilated regions are well-characterized, further demonstrating the efficacy of our label selection strategy.

Ablation Study and Discussions. To further investigate how our method works, we perform an ablation study under the 4-HQ-sample setting (as presented in Table 2) with the following variants: (i) **PLIL (HQ-agnostic)**: mixing Set-HQ and Set-LQ instead of the separate strategy; (ii) **w/o multi-scale voting**: only utilizing the features before the penultimate convolution for autocratic label isolation; (iii) **w/o** $\mathcal{L}_{ls}$: removing the m-masked supervised loss for the identified clean labeled group; (iv) **w/o** $\mathcal{L}_{nsl}$: removing the $\bar{m}$-masked (noise-selective) stability loss for the identified noisy labeled group. First, our PLIL is tailored for the HQ-aware scenario. As shown in Table 2, the HQ-agnostic input has interfered with the network training and led to obvious performance degradation, showing the efficacy of our separate strategy. We also observe that the arbitration-based multi-scale voting mechanism enables more reliable isolation due to the consideration of different perspectives of the images. When removing $\mathcal{L}_{ls}$, considerable performance degradation can be observed, revealing that our strategy effectively finds out the clean labeled voxels. Besides the productive guidance provided by the isolated clean labeled voxels, the noisy group exploited by the stability learning can further provide informative clues to boost the performance. Empirically, the noisy labeled regions often appear in the challenging areas, which are more sensitive to the perturbations and therefore exploring their perturbed stability during training is rewarding and can enhance the generalizability of the model [15,19]. However, as a methodological study, we only evaluated the methods with the commonly used simulated LQ noisy labels. As observed, the violent simulated noises lead to serious confirmation bias. Some existing NLL methods cannot handle such violent noises well, but some SSL methods, which discard LQ labels, achieve appealing performance. Thus, further clinical validation with real-world amateur noises is an important future work. Besides, to facilitate practical expert-amateur collaboration, we should further consider two intertwined problems in the future: (i) how to cost-efficiently edu-

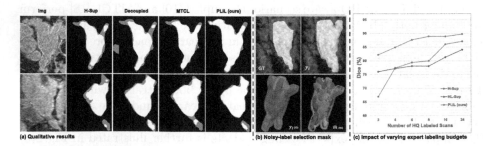

Fig. 2. (a) Examples of LA segmentation results with only 6 HQ labeled data. Grey color represents the inconsistency between the prediction and the ground truth (GT). (b) An example of the dilated LQ label y_l (white in 3D with fused red GT) and the estimated noisy-label selection mask $\bar{m}$ (yellow in 3D with fused red GT). (c) Segmentation performances (indicated by Dice score) with varying expert labeling budgets. (Color figure online)

Table 2. Ablation study with 5% HQ labeled data. The best mean results are in **bold**.

Methods	Metrics			
	Dice [%] ↑	Jaccard [%] ↑	95HQ [voxel] ↓	ASD [voxel] ↓
PLIL (HQ-aware)	**84.91** (3.32)	**73.93** (5.07)	**15.49** (12.00)	**3.10** (1.31)
PLIL (HQ-agnostic)	79.64 (7.46)	66.76 (9.50)	19.48 (11.01)	6.13 (2.55)
w/o multi-scale voting	83.36 (6.83)	72.11 (7.93)	17.36 (11.54)	3.65 (1.77)
w/o $\mathcal{L}_{ls}$	80.65 (9.36)	68.47 (11.53)	23.11 (13.66)	3.12 (1.11)
w/o $\mathcal{L}_{nsl}$	83.22 (7.36)	71.86 (9.59)	19.69 (14.26)	3.26 (1.67)

cate the amateurs on good medical annotation; (ii) how to automatically perform quality controls for the crowdsourced pixel-level labels [8].

4   Conclusion

In this work, we proposed a novel Prototypical Label Isolation Learning (PLIL) framework to robustly learn left atrium segmentation from scarce high-quality labeled data and massive low-quality labeled data. Taking advantage of our multi-scale voting-based prototypical label isolation and adaptive learning scheme for clean and suspected noisy labeled voxels, our approach can robustly exploit the additional low-quality labeled data (e.g., via cheap crowdsourcing), which enables effective expert-amateur collaboration. Comprehensive experiments on the left atrium segmentation benchmark demonstrated the superior performance of our method as well as the effectiveness of each proposed component.

Acknowledgement. This research was done with Tencent Jarvis Lab and Tencent Healthcare (Shenzhen) Co., LTD and supported by General Research Fund from Research Grant Council of Hong Kong (No. 14205419) and the National Key R&D Program of China (No. 2020AAA0109500 and No. 2020AAA0109501).

References

1. Chen, C., Liu, Q., Jin, Y., Dou, Q., Heng, P.-A.: Source-free domain adaptive fundus image segmentation with denoised pseudo-labeling. In: de Bruijne, M., et al. (eds.) MICCAI 2021. LNCS, vol. 12905, pp. 225–235. Springer, Cham (2021). https://doi.org/10.1007/978-3-030-87240-3_22
2. Chen, P., Ye, J., Chen, G., Zhao, J., Heng, P.A.: Beyond class-conditional assumption: A primary attempt to combat instance-dependent label noise. In: Proceedings of the AAAI Conference on Artificial Intelligence, vol. 35, pp. 11442–11450 (2021)
3. Chen, X., Yuan, Y., Zeng, G., Wang, J.: Semi-supervised semantic segmentation with cross pseudo supervision. In: Proceedings of the IEEE/CVF Conference on Computer Vision and Pattern Recognition, pp. 2613–2622 (2021)

4. Cui, W., et al.: Semi-supervised brain lesion segmentation with an adapted mean teacher model. In: Chung, A.C.S., Gee, J.C., Yushkevich, P.A., Bao, S. (eds.) IPMI 2019. LNCS, vol. 11492, pp. 554–565. Springer, Cham (2019). https://doi.org/10.1007/978-3-030-20351-1_43

5. Dolz, J., Desrosiers, C., Ayed, I.B.: Teach me to segment with mixed supervision: confident students become masters. In: Feragen, A., Sommer, S., Schnabel, J., Nielsen, M. (eds.) IPMI 2021. LNCS, vol. 12729, pp. 517–529. Springer, Cham (2021). https://doi.org/10.1007/978-3-030-78191-0_40

6. Guo, X., Yuan, Y.: Joint class-affinity loss correction for robust medical image segmentation with noisy labels. In: Wang, L., Dou, Q., Fletcher, P.T., Speidel, S., Li, S. (eds.) MICCAI 2022. LNCS, vol. 13434, pp. 588–598. Springer, Cham (2022). https://doi.org/10.1007/978-3-031-16440-8_56

7. Kendall, A., Gal, Y.: What uncertainties do we need in Bayesian deep learning for computer vision? arXiv preprint arXiv:1703.04977 (2017)

8. Kentley, J., et al.: Agreement between experts and an untrained crowd for identifying dermoscopic features using a gamified app: reader feasibility study. JMIR Med. Inform. 11(1), e38412 (2023)

9. Li, S., Gao, Z., He, X.: Superpixel-guided iterative learning from noisy labels for medical image segmentation. In: de Bruijne, M., et al. (eds.) MICCAI 2021. LNCS, vol. 12901, pp. 525–535. Springer, Cham (2021). https://doi.org/10.1007/978-3-030-87193-2_50

10. Luo, W., Yang, M.: Semi-supervised semantic segmentation via strong-weak dual-branch network. In: Vedaldi, A., Bischof, H., Brox, T., Frahm, J.-M. (eds.) ECCV 2020. LNCS, vol. 12350, pp. 784–800. Springer, Cham (2020). https://doi.org/10.1007/978-3-030-58558-7_46

11. Luo, X., et al.: Efficient semi-supervised gross target volume of nasopharyngeal carcinoma segmentation via uncertainty rectified pyramid consistency. In: de Bruijne, M., et al. (eds.) MICCAI 2021. LNCS, vol. 12902, pp. 318–329. Springer, Cham (2021). https://doi.org/10.1007/978-3-030-87196-3_30

12. Milletari, F., Navab, N., Ahmadi, S.A.: V-net: fully convolutional neural networks for volumetric medical image segmentation. In: Fourth International Conference on 3D Vision, pp. 565–571. IEEE (2016)

13. Qu, Y., Mo, S., Niu, J.: DAT: training deep networks robust to label-noise by matching the feature distributions. In: Proceedings of the IEEE/CVF Conference on Computer Vision and Pattern Recognition, pp. 6821–6829 (2021)

14. Snell, J., Swersky, K., Zemel, R.: Prototypical networks for few-shot learning. In: Advances in Neural Information Processing Systems, pp. 4080–4090 (2017)

15. Tarvainen, A., Valpola, H.: Mean teachers are better role models: weight-averaged consistency targets improve semi-supervised deep learning results. In: Advances in Neural Information Processing Systems, pp. 1195–1204 (2017)

16. Van Engelen, J.E., Hoos, H.H.: A survey on semi-supervised learning. Mach. Learn. 109(2), 373–440 (2020)

17. Xiong, Z., et al.: A global benchmark of algorithms for segmenting the left atrium from late gadolinium-enhanced cardiac magnetic resonance imaging. Med. Image Anal. 67, 101832 (2021)

18. Xu, Z., et al.: Noisy labels are treasure: mean-teacher-assisted confident learning for hepatic vessel segmentation. In: de Bruijne, M., et al. (eds.) MICCAI 2021. LNCS, vol. 12901, pp. 3–13. Springer, Cham (2021). https://doi.org/10.1007/978-3-030-87193-2_1

19. Xu, Z., et al.: Ambiguity-selective consistency regularization for mean-teacher semi-supervised medical image segmentation. Med. Image Anal. 88, 102880 (2023)

20. Xu, Z., et al.: All-around real label supervision: cyclic prototype consistency learning for semi-supervised medical image segmentation. IEEE J. Biomed. Health Inform. **26**, 3174–3184 (2022)
21. Yu, L., Wang, S., Li, X., Fu, C.-W., Heng, P.-A.: Uncertainty-aware self-ensembling model for semi-supervised 3D left atrium segmentation. In: Shen, D., et al. (eds.) MICCAI 2019. LNCS, vol. 11765, pp. 605–613. Springer, Cham (2019). https:// doi.org/10.1007/978-3-030-32245-8_67
22. Zhang, M., et al.: Characterizing label errors: confident learning for noisy-labeled image segmentation. In: Martel, A.L., et al. (eds.) MICCAI 2020. LNCS, vol. 12261, pp. 721–730. Springer, Cham (2020). https://doi.org/10.1007/978-3-030-59710-8_70
23. Zhang, P., Zhang, B., Zhang, T., Chen, D., Wang, Y., Wen, F.: Prototypical pseudo label denoising and target structure learning for domain adaptive semantic segmentation. In: Proceedings of the IEEE/CVF Conference on Computer Vision and Pattern Recognition, pp. 12414–12424 (2021)
24. Zhang, T., Yu, L., Hu, N., Lv, S., Gu, S.: Robust medical image segmentation from non-expert annotations with tri-network. In: Martel, A.L., et al. (eds.) MICCAI 2020. LNCS, vol. 12264, pp. 249–258. Springer, Cham (2020). https://doi.org/10. 1007/978-3-030-59719-1_25
25. Zhang, X., Wei, Y., Yang, Y., Huang, T.S.: SG-One: similarity guidance network for one-shot semantic segmentation. IEEE Trans. Cybern. **50**(9), 3855–3865 (2020)
26. Zhang, Z., Zhang, H., Arik, S.O., Lee, H., Pfister, T.: Distilling effective supervision from severe label noise. In: Proceedings of the IEEE/CVF Conference on Computer Vision and Pattern Recognition, pp. 9294–9303 (2020)
27. Zhou, X., Liu, X., Wang, C., Zhai, D., Jiang, J., Ji, X.: Learning with noisy labels via sparse regularization. In: Proceedings of the IEEE/CVF International Conference on Computer Vision, pp. 72–81 (2021)
28. Zhu, H., Shi, J., Wu, J.: Pick-and-learn: automatic quality evaluation for noisy-labeled image segmentation. In: Shen, D., et al. (eds.) MICCAI 2019. LNCS, vol. 11769, pp. 576–584. Springer, Cham (2019). https://doi.org/10.1007/978-3-030-32226-7_64

Conditional Physics-Informed Graph Neural Network for Fractional Flow Reserve Assessment

Baihong Xie[1], Xiujian Liu[1], Heye Zhang[1], Chenchu Xu[2,3], Tieyong Zeng[4], Yixuan Yuan[5], Guang Yang[6], and Zhifan Gao[1(✉)]

[1] School of Biomedical Engineering, Sun Yat-sen University, Shenzhen, China
gaozhifan@mail.sysu.edu.cn
[2] Institute of Artificial Intelligence, Hefei Comprehensive National Science Center, Hefei, China
[3] Anhui University, Hefei, China
[4] Department of Mathematics, The Chinese University of Hong Kong, Hong Kong, China
[5] Department of Electronic Engineering, The Chinese University of Hong Kong, Hong Kong, China
[6] Bioengineering Department and Imperial-X, Imperial College London, London, UK

Abstract. The assessment of fractional flow reserve (FFR) is significant for diagnosing coronary artery disease and determining the patients and lesions in need of revascularization. Deep learning has become a promising approach for the assessment of FFR, due to its high computation efficiency in contrast to computational fluid dynamics. However, it suffers from the lack of appropriate priors. The current study only considers adding priors into the loss function, which is insufficient to learn features having strong relationships with the boundary conditions. In this paper, we propose a conditional physics-informed graph neural network (CPGNN) for FFR assessment under the morphology and boundary condition information. Specially, CPGNN adds morphology and boundary conditions into inputs to learn the conditioned features and penalizes the residual of physical equations and the boundary condition in the loss function. Additionally, CPGNN consists of a multi-scale graph fusion module (MSGF) and a physics-informed loss. MSGF is to generate the features constrained by the coronary topology and better represent the different-range dependence. The physics-informed loss uses the finite difference method to calculate the residuals of physical equations. Our CPGNN is evaluated over 183 real-world coronary observed from 143 X-ray and 40 CT angiography. The FFR values of CPGNN correlate well with FFR measurements r = 0.89 in X-ray and r = 0.88 in CT.

Supplementary Information The online version contains supplementary material available at https://doi.org/10.1007/978-3-031-43990-2_11.

Keywords: Fractional Flow Reserve Assessment · Physics-Informed
Neural Networks · Coronary Angiography

1 Introduction

The assessment of fractional flow reserve (FFR) is significant for diagnosing coronary artery disease (CAD) and determining the patients and lesions in need of revascularization [1,2]. Although CAD is widely diagnosed by angiography technique in routine, the anatomical markers often underestimate or overestimate a lesion's functional severity [3]. FFR is the gold standard for the functional diagnosis of CAD and guides the revascularization strategy, due to its ability to assess the ischemic potential of a stenosis [4]. According to a multicenter trial, the use of FFR recalls an additional 30% of severe stenosis required revascularization [5]. Additionally, another trial demonstrates a 5.1% reduction of the 1-year adverse event rate by using FFR [3]. Therefore, it advocates the routine use of FFR in the clinical practice guidelines of the European Society and American Heart Association [1,6]. However, the widespread adoption of FFR is restricted by the risk associated with maneuvering a pressure wire down a coronary artery and the added time to assess multiple vessels [7]. Therefore, it is an eager need for assessing FFR derived from angiography.

Deep learning has become a promising approach for the assessment of FFR, due to its high computation efficiency in contrast to the computational fluid dynamics [8–11]. However, it suffers from the lack of appropriate priors. An appropriate prior is challenging to be found, due to FFR being ruled by the complex physical process (i.e. Navier-Stokes equations) [12].

Physics-informed neural networks (PINNs) have the potential to address the aforementioned challenge of lacking appropriate priors. PINNs add priors to the loss function by penalizing the residual of physical equations and the boundary conditions [13]. This prior guides the learning to a direction in compliance with the physical principles and boundary conditions. For example, PINNs add the Navier-Stokes equations and the boundary conditions to the loss function to generate a physically consistent prediction of blood pressure and velocity on the arterial networks [14–16]. However, only adding the prior to the loss function is insufficient, because the learned features are weakly related to boundary conditions in the loss function.

In this paper, we apply a prior to the inputs. Adding boundary conditions as prior to the inputs resulting in a strong relationship between learned features and boundary conditions. Therefore, the learned features contain more direct and powerful boundary condition information. Additionally, a graph network is introduced as a prior to enforce the coronary topology constraint, because the interaction between the nodes on the graph is similar to the interaction of FFR between the spatial points on the coronary. To this end, we propose a conditional physics-informed graph neural network (CPGNN) for FFR assessment under the constraint of the morphology and boundary conditions. CPGNN adds morphology and boundary conditions as priors to the inputs for learning the conditioned features, besides adding priors to the loss function by penalizing the residual of physical equations and the boundary conditions. Specially, CPGNN consists of a multi-scale graph fusion module (MSGF) and a physics-informed loss. The purpose of MSGF is to generate the features constrained by the coronary topology and better represents the different-range dependence. The physics-informed loss uses the finite difference method to calculate the residuals of physical equations.

The main contributions in the paper are three-fold:

(1) CPGNN provides FFR assessment under the condition of morphology and boundary.
(2) CPGNN introduces a prior by adding the information of the morphology and boundary into inputs and a multi-scale graph fusion module is designed to capture the conditional features related to those information.
(3) The extensive experiments on the 6600 synthetic coronary and 183 clinical angiography including 40 CT and 143 X-ray. The performance of our CPGNN demonstrates the advantages over six existing methods.

2   Method

2.1   Problem Statement

The purpose of CPGNN is to add the appropriate prior, which makes sure that the prediction of blood pressure and flow meet boundary conditions and the conservation of physical principle. The idea is to find a loss term describing the rules of blood pressure and flow. There exists the hemodynamic theory [12] to solve the pressure Q and flow P on spatial coordinate z, which is defined as:

$$\begin{cases} \mathcal{F}(Q(z), P(z); \theta) = 0 & z \in \Omega \\ \mathcal{B}(Q(z), P(z); \gamma) = 0 & z \in \partial\Omega \end{cases} \tag{1}$$

where Ω is the domain of the coronary, $\partial\Omega$ is the boundary, θ is the morphology parameters, γ is boundary condition, $\mathcal{F}$ and $\mathcal{B}$ are operators to describe the control equation and boundary constraints respectively.

A neural network with parameter ω is introduced to approximate the pressure and flow, defined as $Q_\omega(z)$ and $P_\omega(z)$. Then, the residuals of the Eq. (1) can be added to the loss function as prior. The prediction can be constrained by minimizing the residuals on coronary, namely

$$\arg\min_\omega \int_\Omega \|\mathcal{F}(Q_\omega(z), P_\omega(z); \theta)\|_2 \, dz + \int_{\partial\Omega} \|\mathcal{B}(Q_\omega(z), P_\omega(z); \gamma)\|_2 \, dz \tag{2}$$

Further, considering the feature learned by Eq. (2) have a weak relation with the γ and θ, both are added as prior conditional inputs to approximate the flow $Q_\omega(z|\gamma, \theta)$ and pressure $P_\omega(z|\gamma, \theta)$ by the process:

$$\arg\min_\omega \int_\Omega \|\mathcal{F}(Q_\omega(z|\gamma, \theta), P_\omega(z|\gamma, \theta))\|_2 \, dz$$
$$+ \int_{\partial\Omega} \|\mathcal{B}(Q_\omega(z|\gamma, \theta), P_\omega(z|\gamma, \theta))\|_2 \, dz \tag{3}$$

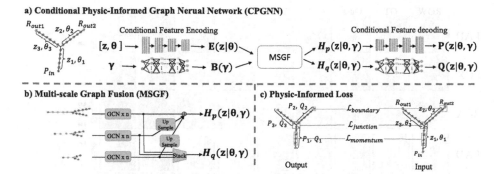

Fig. 1. The architecture of CPGNN. The morphology θ and boundary γ are added into the inputs. The multi-scale graph fusion is used to enforce the constraint from coronary topology and capture the different-range dependency. The physics-informed loss is used to guide conditional features encoding

2.2   Conditional Physics-Informed Graph Neural Network

As shown in the Fig. 1, CPGNN contains an encoding-decoding architecture to predict the pressure and flow, by the produce:

$$Q(z|\theta, \gamma), P(z|\theta, \gamma) = decoder(MSGF(embedding(\gamma), encoder(z, \theta)))   \quad (4)$$

According to hemodynamic theory [12], the morphology θ is the cross-sectional area, and boundary condition γ consists of the outlet resistances and inlet pressure. By sampling uniformly on coronary, the inputs of CPGNN the morphology $\theta \in R^{N_v \times N_d \times 1}$, the spatial coordinate $Z \in R^{N_v \times N_d \times 1}$ and the boundary condition $\gamma \in R^{N_b \times 1}$. The N_v, N_d, and N_b are the number of vessel branches, sampling points, and coronary boundary.

Conditional Feature Encoding. The θ is directly combined with Z, due to their corresponding relation. A serial of 1D convolution and down-sampling is used to generate the morphology features $M(z|\theta) \in R^{N_v \times N_d/8 \times C}$. There is a full connection layer to embed the vector γ into $B(\gamma) \in R^{N_b \times C}$.

Multi-scale Graph Fusion. The graph convolution operator is introduced to enforce the coronary topology constraint. The morphology features $M(z|\theta)$ compose the graph on coronary. The graph adds $B(\gamma)$ as new points at the corresponding boundary. Besides, a multi-scale mechanism is introduced to enhance the representation of the features, because it is beneficial to capture the different-range dependency. According to the hemodynamic theory [12], the flow is constant on the branch due to mass conservation under steady-state one-dimensional modeling, and the pressure varies associated with the position. The feature $H_p(z|\theta, \gamma) \in R^{N_v \times N_d/8 \times C}$ is obtained by the element-wise fusion on up-sampling

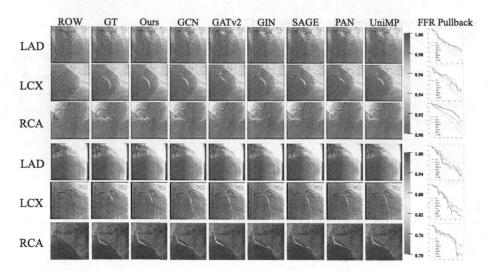

Fig. 2. Representative pressure predictions show our CPGNN is better than other state-of-the-art methods on the X-ray angiography data

feature graphs while the feature $H_q(z|\theta, \gamma) \in R^{N_v \times N_q \times C}$ is obtained by concatenating feature graphs at the branch channel. The N_q is the concatenated dimension according to the number of feature graphs used to fuse.

Conditional Feature Decoding. A serial of 1D convolution and up-sampling is used to decode the feature $H_p(z|\theta, \gamma)$ and generate the pressure prediction $P(z|\theta, \gamma) \in R^{N_v \times N_d \times 1}$. Considering the conversation of the mass, the points share the same prediction on branch. Thus, a full connection layer is used to decode the feature $H_q(z|\theta, \gamma)$ to generate the flow prediction $Q(z|\theta, \gamma) \in R^{N_v \times 1}$

Momentum Loss. The pointset $D = (Z, S, Q, P)$ contain the coordinate Z, cross-sectional area S, flow Q, and pressure P of the points at the coronary. The vessel wall is assumed to be rigid and blood is Newtonian fluid. The momentum loss term combines the hemodynamic equation and finite differences [12,17], defined as

$$\mathcal{L}_m = \frac{1}{N_v} \frac{1}{N_d} \sum_{i=1}^{N_v} \sum_{j=1}^{N_d} \left(\sum_{k=k_{min}}^{k_{max}} \frac{N_k}{Dh_i} \left(\frac{4}{3} \frac{Q_{i,j+k}^2}{S_{i,j+k}} + \frac{S_{i,j} P_{i,j+k}}{\rho} \right) + C \frac{Q_{i,j}}{S_{i,j}} \right)^2 \quad (5)$$

where (i, j) denote the j-th point at the i-th branch , N_d is point number on branch, N_v is branch number on coronary, h_i is the point interval on i-th branch defined as $h_i = (Z_{i,N_d} - Z_{i,1})/(N_d - 1)$, C is a coefficient describing the stenosis influence defined in [12].

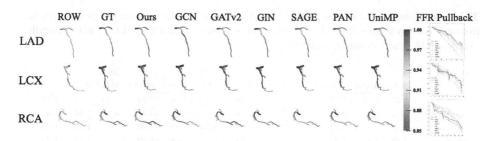

Fig. 3. Representative pressure predictions show our CPGNN is better than the other six state-of-the-art methods on the CT data.

Junction Loss. Given the pointset $J = (Q, P)$ containing flow Q and pressure P of the points at the branch junction, the conservation of momentum and mass is constrained by

$$\mathcal{L}_j = \frac{1}{N_j} \sum_{i=1}^{N_j} \left(\sum_{j=1}^{N_p^i} Q_{i,j} \right)^2 + \sum_{j=2}^{N_p^i} \left(P_{i,1} + \frac{\rho}{2} \left(\frac{Q_{i,1}}{S_{i,1}} \right)^2 - P_{i,j} - \frac{\rho}{2} \left(\frac{Q_{i,j}}{S_{i,j}} \right)^2 \right)^2 \quad (6)$$

where N_j is the number of junction on coronary tree, N_p^i is the number of points of the i-th junction, (i, j) denotes the j-th point at the i-th junction and the 1-st point is the nearest point to the coronary inlet.

Boundary Loss. Given the pointset $B = (Q, P, \gamma)$ containing flow Q, pressure P and condition γ of the points at the boundary, the boundary constraint is penalized by

$$\mathcal{L}_b = (P_1 - \gamma_1)^2 + \frac{1}{N_b} \sum_{i=2}^{N_b} (P_i - Q_i \gamma_i)^2 \quad (7)$$

where N_b is the number of boundary, the 1-st point is inlet, the rest points are outlet. Eventually, the objective of our CPGNN is to minimize the total loss $\mathcal{L}_{total}$:

$$\mathcal{L}_{total} = \mathcal{L}_m + \lambda_1 \mathcal{L}_j + \lambda_2 \mathcal{L}_b \quad (8)$$

where λ_1 and λ_2 are the trade-off parameters.

3   Experiments and Results

3.1   Materials and Experiment Setup

The experiment contains one synthetic dataset and two in-vivo datasets. (1) the synthetic data. We generate 6600 synthetic coronary trees for training. The parameters of geometry and boundary conditions are randomly set in the appropriate ranges, including vessel radius, the length and number of branches, inlet

Table 1. Comparison of our CPGNN and six state-of-the-arts methods for pressure prediction on synthetic data. LAD, left anterior descending artery; LCX, left circumflex artery; RCA, right coronary artery.

method	ALL			LAD			LCX			RCA		
	MAE	RMSE	MAPE	MAE	RMSE	MAPE	MAE	RMSE	MAPE	MAE	RMSE	MAPE
GCN [22]	1.17	1.32	1.25	0.94	1.08	1.00	0.91	1.08	0.97	1.65	1.77	1.77
GATv2 [23]	2.22	2.57	2.40	1.77	2.11	1.90	1.71	2.08	1.84	3.18	3.51	3.45
GIN [24]	1.22	1.35	1.31	1.04	1.16	1.11	1.01	1.15	1.07	1.62	1.72	1.73
SAGE [25]	1.15	1.29	1.22	0.92	1.06	0.98	0.91	1.08	0.97	1.60	1.73	1.72
PAN [26]	2.03	2.34	2.19	1.62	1.91	1.74	1.56	1.90	1.68	2.90	3.19	3.15
UniMP [27]	1.48	1.65	1.59	1.26	1.42	1.35	1.23	1.43	1.32	1.96	2.10	2.11
Our CPGNN	**0.74**	**0.86**	**0.79**	**0.67**	**0.79**	**0.71**	**0.65**	**0.76**	**0.69**	**0.90**	**1.01**	**0.96**

Table 2. Ablation of the number of graph scale levels and the conditional inputs on both synthetic data and in-vivo data. The S1, S2, and S3 represent the number of graphs are 1, 2, and 3

conditional inputs		graph number			synthetic data					X-ray	CT
θ	γ	S1	S2	S3	MAE	RMSE	MAPE	$\mathcal{L}_{mom}$	$\mathcal{L}_{junc}$	AUC	AUC
		✓			45.71	55.82	52	7.34	21.44	0.568	0.538
	✓	✓			41.45	50.38	45	7.14	17.36	0.63	0.549
✓		✓			19.89	23.95	21	6.20	9.44	0.742	0.632
✓	✓	✓			1.15	1.32	1.21	5.09	1.39	0.917	0.786
✓	✓		✓		1.06	1.18	1.11	3.15	1.60	0.93	0.84
✓	✓			✓	**0.74**	**0.86**	**0.79**	**2.59**	**1.17**	**0.97**	**0.93**

pressure, and outlet resistances [18]. The pressure ground truth (GT) is simulated by Simvascular [12]. (2) the in-vivo data. There are 143 X-ray and 40 CT angiography from 183 patients. The acquisition process obeys the standard clinical practice [19]. The setting of the boundary conditions is based on the TIMI count method [20] and PP-outlet strategy [21]. The FFR was performed using pressure guide-wire by the manufacturer Abbott with model HI-TORQUE for all patients in the clinic dataset.

All experiments run on a platform with NVIDIA RTX A6000 48 GB GPU. The Adam optimizer is used with 16 batch size per step. Initial learning rate is 0.001 and the decay rate is 0.95. The ratio of the training and verification of the synthetic dataset is 8:2. CPGNN is trained on the synthetic dataset and tested on both synthetic and clinical datasets.

CPGNN is compared with six state-of-the-art methods, GCN [22], GATv2 [23], GIN [24], SAGE [25], PAN [26] and UniMP [27]. The evaluation metrics are Root Mean Square Errors (RMSE), Mean Absolute Errors (MAE), Mean Absolute Percentage Errors (MAPE), the residual of momentum equation ($\mathcal{L}_{mom}$) and the residual of junction equation ($\mathcal{L}_{junc}$). The units of RMSE, MAE, $\mathcal{L}_{mom}$ and $\mathcal{L}_{junc}$ are mmHg, mmHg, e^{-6} and e^{-2}.

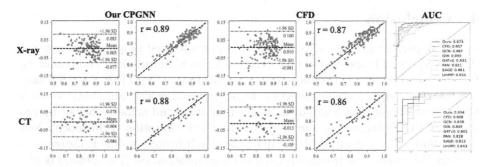

Fig. 4. Comparison of our CPGNN with computational fluid dynamics (CFD) for FFR assessment. The ground truth is in-vivo FFR measurement. The Bland-Altman analysis and Pearson correlation are performed for both CPGNN and CFD. The area under the curve (AUC) of CPGNN, CFD, and six state-of-the-art methods for the stenosis diagnosis (FFR<0.8) is performed

3.2   Results on Synthetic Data

Comparison of Pressure Prediction: The performance of CPGNN is closer to the GT with an overall MAE of 0.74, RMSE of 0.86, and MAPE of 0.79. Table 1 shows that CPGNN performs well than six state-of-the-art methods.

Ablation of Conditional Inputs and Multi-scale Mechanism: As shown in Table 2, the prediction of PINNs has a poor performance on unseen coronary without the morphology and boundary condition inputs. Thus, it is necessary to add both as inputs. The results gradually improve MAE, RMSE, and MAPE from 1.15 to 0.74, 1.32 to 0.86, and 1.21 to 0.79 when adding the number of graph scale-level. Thus, the multi-scale mechanism plays a key role.

3.3   Results on In-Vivo Data

Comparison of FFR Assessment: As shown in Fig. 4, Bland-Altman analysis and Pearson correlation are conducted to evaluate the differences between CPGNN and in-vivo FFR measurement based on X-ray and CT data. Compared to the computational fluid dynamics, CPGNN has a well consistency with in-vivo FFR in the Fig. 4. Figure 2 and Fig. 3 directly display FFR results of CPGNN and six state-of-the-art methods. Figure 4 presents the area under curve (AUC) of CPGNN is best for the stenosis diagnosis (FFR<0.8). Those results demonstrate that CPGNN outperforms the other six existing methods in the clinical FFR assessment.

Ablation of CPGNN Components: As shown in Table 2, the AUC of different settings is shown. The results also demonstrate that the conditional inputs and the multi-scale mechanism are important for FFR Assessment in the clinic.

4   Conclusion

In this paper, we propose a conditional physics-informed graph neural network (CPGNN) for FFR assessment under the condition of morphology and boundary. Compared to the current reduce-order computation method [12], CPGNN does not need to couple stenosis detection algorithm and enable automatic stenosis feature extraction, which avoids error accumulation and achieves better performance. CPGNN introduces a prior by adding the information of the morphology and boundary into inputs and a multi-scale graph fusion module is designed to capture the conditional features related to those information. The method is conducted on 143 X-ray and 40 CT subjects. The performance of CPGNN is higher than the six state-of-the-art methods. The FFR of CPGNN correlates well with FFR measurements ($r = 0.89$ in X-ray and $r = 0.88$ in CT). The computation speed of CPGNN is 0.03 s per case, which is $600\times$ faster than the SimVascular computation method [12]. Those results demonstrate that our CPGNN can aid in the clinical FFR assessment.

Acknowledgement. This work was supported in part by National Natural Science Foundation of China (62101606, 62276282, 62101610, 62106001, U1908211), Guangdong Basic and Applied Basic Research Foundation (2022A1515011384), the University Synergy Innovation Program of Anhui Province (GXXT-2021-007), and the Anhui Provincial Natural Science Foundation (2208085Y19).

References

1. Knuuti, J., Revenco, V., Saraste, A., et al.: 2019 ESC guidelines for the diagnosis and management of chronic coronary syndromes. Eur. Heart J. **41**(5), 407–477 (2020)
2. Neumann, F.-J., Sousa-Uva, M., Ahlsson, A., et al.: 2018 ESC/EACTS guidelines on myocardial revascularization. Eur. Heart J. **40**(2), 87–165 (2019)
3. Tonino, P.A.L., De Bruyne, B., Pijls, N.H.J., et al.: Fractional flow reserve versus angiography for guiding percutaneous coronary intervention. New England J. Med. **360**(3), 213–224 (2009)
4. Pijls, N.H., Van Son, J.A., Kirkeeide, R.L., et al.: Experimental basis of determining maximum coronary, myocardial, and collateral blood flow by pressure measurements for assessing functional stenosis severity before and after percutaneous transluminal coronary angioplasty. Circulation **87**(4), 1354–1367 (1993)
5. Tonino, P.A.L., Fearon, W.F., De Bruyne, B., et al.: Angiographic versus functional severity of coronary artery stenoses in the fame study: fractional flow reserve versus angiography in multivessel evaluation. J. Am. College Cardiol. **55**(25), 2816–2821 (2010)
6. Levine, G.N., Bates, E.R., Bittl, J.A., et al.: 2016 ACC/AHA guideline focused update on duration of dual antiplatelet therapy in patients with coronary artery disease: a report of the American college of cardiology/American heart association task force on clinical practice guidelines. Circulation **134**(10), e123–e155 (2016)
7. Tebaldi, M., Biscaglia, S., Fineschi, M., et al.: Evolving routine standards in invasive hemodynamic assessment of coronary stenosis: the nationwide Italian SICI-GISE cross-sectional ERIS study. JACC Cardiovasc. Interv. **11**(15), 1482–1491 (2018)

8. Li, Y., Qiu, H., Hou, Z., et al.: Additional value of deep learning computed tomographic angiography-based fractional flow reserve in detecting coronary stenosis and predicting outcomes. Acta Radiol. **63**(1), 133–140 (2022)
9. Gao, Z., Wang, X., Sun, S., et al.: Learning physical properties in complex visual scenes: an intelligent machine for perceiving blood flow dynamics from static CT angiography imaging. Neural Netw. **123**, 82–93 (2020)
10. Itu, L., Rapaka, S., Passerini, T., et al.: A machine-learning approach for computation of fractional flow reserve from coronary computed tomography. J. Appl. Physiol. **121**(1), 42–52 (2016)
11. Zhang, D., Liu, X., Xia, J., et al.: A physics-guided deep learning approach for functional assessment of cardiovascular disease in IoT-based smart health. IEEE Internet Things J. (2023)
12. Updegrove, A., Wilson, N.M., Merkow, J., et al.: SimVascular: an open source pipeline for cardiovascular simulation. Ann. Biomed. Eng. **45**, 525–541 (2017)
13. Raissi, M., Perdikaris, P., Karniadakis, G.E.: Physics-informed neural networks: a deep learning framework for solving forward and inverse problems involving non-linear partial differential equations. J. Comput. Phys. **378**, 686–707 (2019)
14. Fathi, M.F., Perez-Raya, I., Baghaie, A., et al.: Super-resolution and denoising of 4D-flow MRI using physics-informed deep neural nets. Comput. Methods Programs Biomed. **197**, 105729 (2020)
15. Sarabian, M., Babaee, H., Laksari, K.: Physics-informed neural networks for brain hemodynamic predictions using medical imaging. IEEE Trans. Med. Imaging **41**(9), 2285–2303 (2022)
16. Kissas, G., Yang, Y., Hwuang, E., et al.: Machine learning in cardiovascular flows modeling: predicting arterial blood pressure from non-invasive 4D flow MRI data using physics-informed neural networks. Comput. Methods Appl. Mech. Eng. **358**, 112623 (2020)
17. Eberly, D.: Derivative Approximation by Finite Differences. Magic Software Inc. (2008)
18. El Sayed, S., El Sawa, E.A., Atta-Alla, A.E.S., et al.: Morphometric study of the right coronary artery. Int. J. Anat. Res. **3**(3), 1362–1370 (2015)
19. Di Mario, C., Sutaria, N.: Coronary angiography in the angioplasty era: projections with a meaning. Heart **91**(7), 968–976 (2005)
20. Tu, S., Barbato, E., Köszegi, Z., et al.: Fractional flow reserve calculation from 3-dimensional quantitative coronary angiography and TIMI frame count: a fast computer model to quantify the functional significance of moderately obstructed coronary arteries. JACC Cardiovasc. Interv. **7**(7), 768–777 (2014)
21. Liu, X., Chuangye, X., Rao, S., et al.: Physiologically personalized coronary blood flow model to improve the estimation of noninvasive fractional flow reserve. Med. Phys. **49**(1), 583–597 (2022)
22. Kipf, T.N., Welling, M.: Semi-supervised classification with graph convolutional networks. In: International Conference on Learning Representations (2017)
23. Brody, S., Alon, U., Yahav, E.: How attentive are graph attention networks? In: International Conference on Learning Representations (2022)
24. Xu, K., Hu, W., Leskovec, J., et al.: How powerful are graph neural networks? In: International Conference on Learning Representations (2019)
25. Hamilton, W.L., Ying, R., Leskovec, J.: Inductive representation learning on large graphs. In: International Conference on Neural Information Processing Systems (2017)

26. Ma, Z., Xuan, J., Wang, Y.G., et al.: Path integral based convolution and pooling for graph neural networks. In: International Conference on Neural Information Processing Systems (2020)
27. Shi, Y., Huang, Z., Feng, S., et al.: Masked label prediction: unified message passing model for semi-supervised classification. In: International Joint Conference on Artificial Intelligence (2021)

Semantic Difference Guidance for the Uncertain Boundary Segmentation of CT Left Atrial Appendage

Xin You[1], Ming Ding[2], Minghui Zhang[1], Yangqian Wu[1], Yi Yu[2(✉)], Yun Gu[1], and Jie Yang[1(✉)]

[1] Institute of Medical Robotics, Shanghai Jiao Tong University, Shanghai, China
{sjtu_youxin,jieyang}@sjtu.edu.cn
[2] Shanghai Xinhua Hospital, Shanghai Jiaotong University, Shanghai, China

Abstract. Atrial fibrillation (AF) is one of the most common types of cardiac arrhythmia, which is closely relevant to anatomical structures including the left atrium (LA) and the left atrial appendage (LAA). Thus, a thorough understanding of the LA and LAA is essential for the AF treatment. In this paper, we have modeled relative relations between the LA and LAA via deep segmentation networks for the first time, and introduce a new LA & LAA CT dataset. To deal with uncertain boundaries between the LA and LAA, we propose the semantic difference module (SDM) based on diffusion theory to refine features with enhanced boundary information. Besides, disconnections between the LA and LAA are frequently observed in the segmentation results due to uncertain boundaries of the LAA region and CT imaging noise. To address this issue, we devise another connectivity-refined network with the connectivity loss. The loss function exerts a distance regularization on coarse predictions from the first-stage network. Experiments demonstrate that our proposed model can achieve state-of-the-art segmentation performance compared with classic convolutional-neural-networks (CNNs) and recent Transformer-based models on this new dataset. Specifically, SDM can also outperform existing methods on refining uncertain boundaries. Codes are available at https://github.com/AlexYouXin/LA-LAA-segmentation.

Keywords: Left atrial appendage · Difference operator · Uncertain boundary · Image segmentation

1 Introduction

Atrial fibrillation (AF) has been one of the most common types of cardiovascular diseases and is closely related to the left atrium (LA) [18]. Beside this chamber structure, there is a finer anatomy termed with the left atrial appendage

---

Supplementary Information The online version contains supplementary material available at https://doi.org/10.1007/978-3-031-43990-2_12.

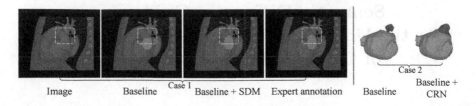

Fig. 1. Improved segmentation results of the baseline model after adding SDM on case 1 and after adding CRN on case 2. (Green Area: LA. Red Area: LAA). (Color figure online)

(LAA). The majority of strokes due to AF result from clots existing in the LAA [13]. A common measure for treatment is anticoagulant therapy. However, many patients have contraindications to this type of therapy. A more effective and feasible stroke prevention procedure is the left atrial appendage closure, which can avoid most of the drawbacks by anticoagulant therapy [12]. And the size of occlusion devices designed for patients is strongly associated with the anatomical interface between the LA and LAA [13]. Thus, enhancing the understanding for the structure of the LA and LAA is beneficial to carry out treatments for strokes due to AF. A normal pre-surgery imaging is Cardiac Computed Tomography (Cardiac CT), which is a popular physical inspection for diagnoses [18]. Thus, automatic and accurate segmentation of the LA and LAA from Cardiac CT images is essential to provide support for the diagnosis and treatment of various cardiovascular diseases.

Till now, many researches have focused on the automatic segmentation of the LA [24]. Compared with that, the LAA has not been sufficiently researched, particularly the relative relations between the LA and LAA [27]. In our work, we aim to design an automatic method for the correlation modeling between the LA and LAA, then give a quantitative and qualitative evaluation of segmentation performance. The LA and LAA both have large anatomical variations [9,27]. Besides, there are uncertain boundaries for the structure of the LAA, especially the interface between the LA and LAA. In contrast, cardiac tissues like the right atrium (RA), right ventricle (RV) and left ventricle (LV) have explicit boundaries, which can be easily and finely segmented [28]. As shown in Fig. 1, some cases show poor segmentation results on the uncertain boundary between the LA and LAA. There are many related works on the refinement for boundary segmentation, which can be grouped into three categories. The first strategy attempts to exert a strong loss constraint on boundaries via the multi-task learning paradigm [4]. Then some researchers apply a complex post-process to the segmentation of coarse boundaries, such as [26]. All the methods above mainly emphasize on refining predicted masks for high-quality images with clear boundaries, are not applicable for ambiguous or unclear boundaries [23] in the LA and LAA segmentation. Instead, the third strategy truly works, with the mechanism of enhancing deep features representing uncertain boundaries. Lee et al. [11] proposed a novel boundary-preserving block (BPB) with the ground-truth structure information indicated by experts. Xie et al. [23] used the confidence map to evaluate the uncertainty of each pixel to enhance the segmentation of ambiguous boundaries. Furthermore, some loss functions [10, 25] are specifically designed for the segmentation of uncertain boundaries.

Diffusion is a physical model aimed at minimizing the spatial concentration difference [15] and is widely used in computer vision [21, 22]. In our work, we detailedly explain the process of refining uncertain boundaries based on diffusion theory. Then we propose a semantic guidance module based on differential operators to refine features from ambiguous boundaries, which is called semantic difference module (SDM). Here we introduce semantic information from deeper layers to guide the diffusion process. As a result of fuzzy boundaries of the LAA region and CT imaging noise, there exists a disconnection between the LA and LAA as shown in Fig. 1. Thus, we design another connectivity-refined network (CRN) combined with the connectivity loss, to deal with the connectivity of two regions. The contributions of our work are listed as follows:

(1) We introduce a new LA & LAA CT dataset. And as far as we are concerned, this is the first work based on deep neural networks, to model relative relations between the LA and LAA.
(2) We propose a novel semantic difference module based on diffusion theory to deal with the segmentation of uncertain boundaries.
(3) We apply a connectivity-refined network with the connectivity loss to refine coarse masks, then achieve the connectivity between the LA and LAA.
(4) Our proposed network achieves state-of-the-art segmentation performance on the LA and LAA. Specifically, SDM outperforms other methods related to refining the segmentation of uncertain boundaries.

2   Methodology

2.1   Preliminaries

Diffusion is a physical phenomenon, in which molecules spread from regions with higher concentrations toward regions with lower concentrations [15, 16]. Then the whole system tends to be balanced. For a feature vector F to be smoothed, the diffusion process can be modeled as the following partial differential equation:

$$\frac{\partial F}{\partial t} = D \cdot \nabla^2 F \tag{1}$$

where D is the diffusivity function determining the diffusion speed along each direction, ∇ is the gradient operator. In our application, the stable state of $F(t)$ will show a more accurate localization for uncertain boundaries of the LAA.

Linear isotropic diffusion (D is equal to a constant) cannot be applied to complex scenes, because the diffusion velocity is the same in all directions. For a spatial-dependent function $D = D(x, y, z)$, the process is linear anisotropic. However, if we aim to extract refined boundary features, adopting linear diffusion processes will smooth both backgrounds and the edges. A more feasible solution is to devise complex diffusion functions $D = D(F)$ with nonlinear characteristics [21,22]. As a result, the diffusion process exerts more smoothing to regions parallel to boundaries compared to regions vertical to these edges.

Detailedly, given a feature F where regions of uncertain boundaries are not highlighted, it is updated by the diffusion process in infinite time. The diffusion adjacent to ambiguous boundaries should be restrained, while the diffusion far away from boundaries is promoted. And the final state of the diffused feature will accurately localize uncertain boundaries of the LAA.

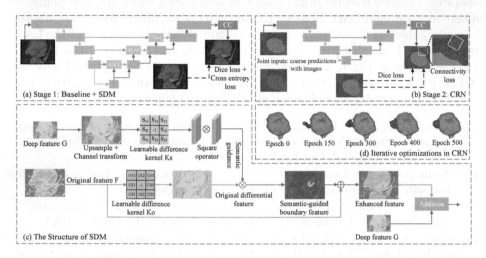

Fig. 2. Our proposed two-stage network. (a) Baseline model with the semantic difference module (CC: channel calibration). (b) Connectivity-refined network. (c) Boundary feature enhancement. (d) Iterative optimizations for the predicted mask of a specific testing case.

2.2   Semantic Difference Module

To localize fuzzy boundaries of the LAA, especially the interface between the LA and LAA, we propose the semantic difference module (SDM) to refine boundary features from these regions. Motivated by the diffusion process, we formulate the process of enhancing boundary features as solving a second-order partial differential equation. Due to the fact that semantic information is required to guide the localization of uncertain boundaries, we introduce the deep feature G from the precedent decoder layer to the diffusion process in each SDM.

Here we adopt ∇G as the semantic guidance map. And the square term $h(|\nabla G|^2)$ is deployed as function D to model nonlinear characteristics of the diffusion process, where h is a projection function. In terms of [15], Eq. 1 can be approximately solved via iterative updates as depicted by the following equations:

$$\hat{F}_p^{t+1} = \sum_{\widetilde{p} \in \delta_p} h(|G_{\widetilde{p}} - G_p|^2) \cdot (F_{\widetilde{p}}^t - F_p^t) \qquad (2)$$

$$F_p^{t+1} = \lambda \cdot F_p^t + \nu \cdot \hat{F}_p^{t+1} \qquad (3)$$

where p is the index of feature maps, δ_p is the local neighborhood centered at p, λ and ν are weighting coefficients. Indeed, $F_{\widetilde{p}}^t - F_p^t$ is the differential information of original feature F^t at point p, representing abundant boundary information, which contains complicated boundary features of anatomies as shown in Fig. 2.c, including the RA RV, etc. However, predicted boundaries between the LA and LAA are not accurate enough only with the diffusion process. Thus, the semantic difference guidance $|G_{\widetilde{p}} - G_p|^2$ is introduced to generate refined boundary feature $\hat{F}^{t+1}$. $\hat{F}^{t+1}$ will diffuse into the stable state as t increases, which can highlight boundaries between the LA and LAA, and suppress the activation on other boundary regions. And refined feature F^{t+1} is attained by fusing the original feature F^t with enhanced boundary feature $\hat{F}^{t+1}$.

We design the semantic difference module based on Eq. 3 as illustrated by Fig. 2. Here we make an improvement on the calculation of differential maps. Motivated by the fact that there exists an anisotropic distribution for our LA and LAA dataset in x, y and z dimensions, traditional edge operator cannot finely extract the differential map of feature F. Therefore, we propose a learnable boundary operator, which bears different values in each position of the kernel. As shown in Fig. 2, we fix the center value as -1 to maintain the difference attribute of the edge kernel. The revised description of enhanced boundary feature is calculated by Eq. 4.

$$\hat{F}_p^{t+1} = \sum_{\widetilde{p} \in \delta_p} \omega_{\widetilde{p}} \cdot |\alpha_{\widetilde{p}} G_{\widetilde{p}} - G_p|^2 \cdot (\beta_{\widetilde{p}} F_{\widetilde{p}}^t - F_p^t) \tag{4}$$

where $\alpha_{\widetilde{p}}$ and $\beta_{\widetilde{p}}$ refer to the learnable edge operator for feature F and semantic feature G respectively. And $\omega_{\widetilde{p}}$ means a vanilla $3 \times 3 \times 3$ convolution kernel.

2.3   Connectivity-Refined Network

Due to the CT imaging noise, some cases show the phenomenon that the LAA is separated from the LA. To deal with disconnections between the LA and LAA, we propose the second-stage network called connectivity-refined network (CRN). Inspired by the metric of 95% Hausdorff distance (HD_{95}) [7], we figure out that a poor connectivity between the LA and LAA will bring a large HD_{95} value for the LAA segmentation, which is not what we expected. Therefore, we adopt another distance constraint loss called connectivity loss $\mathcal{L}_c$, besides the per-pixel Dice loss $\mathcal{L}_d$ in the training process of CRN. Specifically, we choose coarse predictions from validation datasets from the first stage, concatenated with original images as the input of CRN. We firstly localize the predicted LAA region via its unique label. Then to make the region of LAA connected with the LA, we can only focus on the voxels most adjacent to the boundary interface. For the training efficiency, we locate the predicted LAA region with the approximately minimum external cube C (Please refer to **supplementary material** for more details about the algorithm). Four vertexes V_i (i=1, 2, 3, 4) of C neighbored with the LA are selected to calculate the connectivity loss, which is indeed an improved minimal distance from vertexes to the surface of LA. Here we note the point set in the LA surface as P, and each point from P is noted as P_j.

$$\mathcal{L}_c = \sigma \left\{ \frac{\sum_{i=1}^{4} \min_{P_j \in P} D(V_i, P_j)}{S} \right\} - 0.5 \tag{5}$$

$$\mathcal{L} = \mathcal{L}_d + \lambda \times \mathcal{L}_c \tag{6}$$

where D means the Euclidean distance, σ is the sigmoid function. S is a scaling coefficient, and we set it as 20 according to the ablation study on this hyper-parameter (Please refer to **supplementary material** for more quantitative results). Besides, λ is set as 1 if the training epoch reaches more than 300 epochs, or it is 0. When there is no LAA predictions in a cropped patch, $\mathcal{L}_c$ is equal to 0.

Table 1. (a) Results compared with other CNNs and Transformer-based models (**Values on both sides of '|' represent evaluation metrics for each stage of our model**). (b) Segmentation performance including our SDM and other methods related to deal with uncertain boundaries (LA: Left Atrium, LAA: Left Atrial Appendage. Bold: the best, Underlined numbers: the second best. '—' indicates that loss functions do not change Parameters and FLOPs of the baseline model).

(a) Segmentation benchmark on LA & LAA								
Model	Params(M)	FLOPs(T)	Dice score (%) ↑			HD_{95} (mm) ↓		
			LAA	LA	Average	LAA	LA	Average
3D UNet [2]	16.47	0.514	76.64	95.75	86.19±13.0	10.25	<u>6.94</u>	8.59±5.0
ResUNet [3]	32.46	0.813	77.45	95.86	86.66±13.1	9.04	9.79	9.41±6.9
V-Net [14]	45.72	2.391	74.21	92.68	83.44±14.7	12.54	8.82	10.68±7.9
TransBTS [20]	35.61	0.613	73.65	95.05	84.35±12.5	10.85	15.53	13.19±11.1
UNETR [5]	93.54	0.458	75.97	95.26	85.62±12.9	11.41	15.32	13.37±10.1
TransUNet (3D) [1]	82.41	0.220	78.43	95.45	86.94±11.3	8.95	8.47	8.71±5.3
Swin UNETR [17]	62.19	0.975	76.32	95.94	86.13±12.9	11.45	7.71	9.58±5.7
UNeXt [19]	4.02	0.035	79.37	95.34	87.36±9.7	**8.38**	8.22	8.31±2.3
nnUNet [8]	30.79	0.835	<u>81.16</u>	**96.29**	<u>88.72</u>±9.6	<u>8.67</u>	7.17	<u>7.92</u> ±2.8
Ours	23.47|1.92	0.666|0.130	**81.85**	<u>95.96</u>	**88.91** ±6.9	8.87	**6.63**	**7.75** ±2.4
(b) Comparison with other methods focused on uncertain boundaries								
Baseline	16.47	0.514	76.64	95.75	86.19±13.0	10.25	6.94	8.59±5.0
+ Boundary loss [10]	–	–	77.02	95.74	86.38 ±11.3	<u>9.82</u>	7.33	8.58 ±3.2
+ BU loss [25]	–	–	77.01	**96.16**	86.59 ±12.7	10.51	**6.78**	8.65 ±5.0
+ BPB [11]	27.97	0.717	77.49	96.02	86.76 ±12.8	9.96	6.97	8.46 ±4.7
+ CCM [23]	18.98	0.596	<u>77.71</u>	<u>96.05</u>	<u>86.88</u> ±11.8	9.86	7.04	<u>8.45</u> ±3.4
+ SDM (Ours)	23.47	0.666	**78.88**	95.91	**87.40** ±8.6	9.76	<u>6.93</u>	**8.35** ±3.5

On the ground that cases with a poor connectivity need to be refined in the training process of CRN, we increase sampling ratios of the whole LAA region from coarse masks of validation datasets. By cropping patches containing the boundary interface as network inputs, CRN will better learn the connectivity prior from the mapping between coarse predictions and expert annotations.

In the inference stage, final decoded features F_d are applied with the softmax operator to attain predicted masks. However, different channels of F_d bear different maximum values, which will affect segmentation performance. Thus, we propose the concept of channel calibration (CC). Before per-pixel selecting maximum values between channels, we uniform the maximum value of F_d channel by channel.

3   Experiment

3.1   Experimental Settings

Dataset. To evaluate the performance of our network, we conduct experiments on a new dataset, containing accurate annotations of LA and LAA provided by multiple experts. In detail, we collect 80 CT scans from 80 patients, which are split into 45/15/20 for training, validation and testing cases in Stage 1. In Stage 2, 50 predictions of the validation dataset in Stage 1 are generated by various models, in which there are 30 predicted masks with disconnections. Then we randomly split them as 35/15 for training and validation. Moreover, we choose the Dice score and HD_{95} as quantitative metrics.

Table 2. (a) Ablation study on the structure of the semantic difference module (SDM). DK refers to difference kernel) (b) Ablation study on the efficacy of our proposed key components, including SDM, connectivity-refined network (CRN), channel calibration (CC) and connectivity loss.

(a) Ablation study on SDM						
Model	Dice score (%) ↑			HD_{95} (mm) ↓		
	LAA	LA	Average	LAA	LA	Average
+ SDM	**78.88**	**95.91**	**87.40**	9.76	**6.93**	**8.35**
w/o learnable DK	76.60 (↓ 2.28)	95.77 (↓ 0.14)	86.18 (↓ 1.22)	10.53 (↑ 0.77)	7.30 (↑ 0.37)	8.92 (↑ 0.57)
w/o original feature F	77.15 (↓ 1.73)	95.71 (↓ 0.20)	86.43 (↓ 0.97)	**9.71** (↓ 0.05)	7.16 (↑ 0.23)	8.44 (↑ 0.09)
w/o semantic guidance	77.82 (↓ 1.06)	95.49 (↓ 0.42)	86.66 (↓ 0.74)	9.82 (↑ 0.06)	7.38 (↑ 0.45)	8.60 (↑ 0.25)
(b) Ablation study on key components						
Baseline	76.64	95.75	86.19	10.25	6.94	8.59
+ SDM	78.88	95.91	87.40	9.76	6.93	8.35
+ CC	77.44	**96.06**	86.75	9.70	6.84	8.27
+ SDM + CC	79.72	96.03	87.88	9.51	6.90	8.21
+ CRN	79.51	95.62	87.57	9.42	6.89	8.16
+ CRN w/o connectivity loss	78.24	95.65	86.95	9.93	6.95	8.44
+ SDM + CRN	81.32	95.64	88.48	9.08	6.78	7.93
+ SDM + CRN + CC	**81.85** (↑ 5.21)	95.96 (↑ 0.21)	**88.91** (↑ 2.72)	**8.87** (↓ 1.38)	**6.63** (↓ 0.31)	**7.75** (↓ 0.84)

Implementation Details. In the first stage, we choose the vanilla 3D UNet as our baseline model, trained for 2000 epochs. And we utilize a combination of cross entropy loss and Dice loss followed by [8]. For the second stage, CRN is trained for 500 epochs, which is a smaller 3D UNet with only $1.92M$ parameters.

And we utilize Dice loss and connectivity loss as illustrated by Eq. 6. We train all models using AdamW optimizer. With the linear warm-up strategy, the initial learning rate is set as $5e-4$ with a cosine learning rate decay scheduler, and weight decay is set as $1e-5$. The size of cropped patches is $160 \times 160 \times 192$. All models are implemented based on Pytorch and trained on 2 NVIDIA Tesla V100 GPUs.

3.2 Experimental Results

Table 1 consists of two sub-figures (a) and (b). According to Table 1.a, our proposed network outperforms classic CNNs and recent Transformer-based models. Specifically, our model shows superior to powerful nnUNet [8] on four metrics, and there exist 0.69% increase on the Dice score of LAA, 0.54 mm decrease on the HD_{95} of LA. Compared with nnUNet, our two-stage model requires less computational cost and is free from complicated multi-model ensembles. However, the Dice score of our model on LA is inferior to nnUNet. We argue that our model is aimed at improving the segmentation mask of LAA, which bears a different structure from LA. And the iterative optimization process of CRN in Fig. 2 shows the refinement for LAA segmentation. More results about the generalization of CRN can be found in **supplementary material**. Besides, Swin UNETR [17] takes the lead in Transformer-based models, which is not as good as our model because Transformer-based models are data-hungry [6]. Another phenomenon worth to mention is that our model shows an ordinary performance on the HD_{95} of LAA, which is owing to the appearance of outliers. And not only our model but other CNNs and Transformer-based models suffer from the existence of outliers in predictions (More visualization results can be found in **supplementary material**). We will address this issue in the future research. In Table 1.b, we choose 3D UNet as the baseline model. SDM shows better segmentation performance compared with other methods on improving the segmentation of uncertain boundaries. And Fig. 2 illustrated that SDM can enhance uncertain boundary features, especially the boundary interface between LA and LAA.

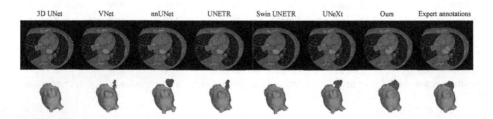

Fig. 3. Qualitative segmentation results of other CNNs and Transformer-based models. (Green Area: LA. Red Area: LAA). (Color figure online)

3.3   Ablation Studies

Table 2 shows ablation studies on key components and on the detailed structure of SDM. From Table 2.a, SDM, CRN and CC can all boost segmentation performance of the baseline model. Besides, we visualize qualitative results in Fig. 3. For the first row, our model give a more accurate localization for the boundary interface, which proves the effectiveness of SDM. The second row is a strong proof that CRN can effectively improve the connectivity between LA and LAA. Then we probe into the efficacy of connectivity loss $\mathcal{L}_c$ by removing it from CRN, which results in a $0.51mm$ increase on the HD_{95} of LAA, which is reasonable because $\mathcal{L}_c$ is indeed a distance regularization on LAA boundaries.

In Table 2.b, we investigate the significance of each component in SDM. (1) Without learnable difference kernels for differential maps of feature F and semantic feature G, the Dice score for LAA declines sharply, which reveals we need to focus on the anisotropic distribution of this dataset. (2) In SDM, we adopt a residual block by fusing the original feature F and the boundary feature. With F removed, there is some details and texture information missing, resulting in a performance drop. (3) Finally, semantic guidance from deeper features is deployed to guide the extraction for uncertain boundaries between LA and LAA. Besides, some false boundaries are restrained, which is illustrated in Fig. 2.

4   Conclusion

In this paper, we introduce a new CT dataset on LA and LAA, then carry out the segmentation task. Detailedly, we explain the refined process for the segmentation of uncertain boundaries via diffusion theory. Based on this, we apply SDM to successfully improve the segmentation for uncertain boundaries between LA and LAA. Then CRN with the connectivity loss can deal with the poor connectivity between two structures. Detailed quantitative and qualitative results have demonstrated the efficacy of two proposed elements.

Acknowledgement. This work is supported in part by National Key R&D Program of China (2019YFB1311503), the Shanghai Sailing Program (20YF1420800), the Shanghai Health and Family Planning Commission (202240110) and Xinhua Hospital affiliated with the School of Medicine (XHKC2021-07).

References

1. Chen, J., et al.: Transunet: transformers make strong encoders for medical image segmentation. arXiv preprint arXiv:2102.04306 (2021)
2. Çiçek, Ö., Abdulkadir, A., Lienkamp, S.S., Brox, T., Ronneberger, O.: 3D U-net: learning dense volumetric segmentation from sparse annotation. In: Ourselin, S., Joskowicz, L., Sabuncu, M.R., Unal, G., Wells, W. (eds.) MICCAI 2016. LNCS, vol. 9901, pp. 424–432. Springer, Cham (2016). https://doi.org/10.1007/978-3-319-46723-8_49

3. Diakogiannis, F.I., Waldner, F., Caccetta, P., Wu, C.: Resunet-A: a deep learning framework for semantic segmentation of remotely sensed data. ISPRS J. Photogrammetry Remote Sens. **162**, 94–114 (2020)
4. Hatamizadeh, A., Terzopoulos, D., Myronenko, A.: End-to-end boundary aware networks for medical image segmentation. In: Suk, H.-I., Liu, M., Yan, P., Lian, C. (eds.) MLMI 2019. LNCS, vol. 11861, pp. 187–194. Springer, Cham (2019). https://doi.org/10.1007/978-3-030-32692-0_22
5. Hatamizadeh, A., et al.: UnetR: transformers for 3D medical image segmentation. In: Proceedings of the IEEE/CVF Winter Conference on Applications of Computer Vision, pp. 574–584 (2022)
6. He, K., et al.: Transformers in medical image analysis: a review. Intell. Med. (2022)
7. Huttenlocher, D.P., et al.: Comparing images using the hausdorff distance. TPAMI **15**(9), 850–863 (1993)
8. Isensee, F., Jaeger, P.F., Kohl, S.A.A., Petersen, J., Maier-Hein, K.H.: NNU-net: a self-configuring method for deep learning-based biomedical image segmentation. Nat. Methods **18**(2), 203–211 (2021)
9. Jin, C., et al.: Left atrial appendage segmentation using fully convolutional neural networks and modified three-dimensional conditional random fields. IEEE J. Biomed. Health Inform. **22**(6), 1906–1916 (2018)
10. Kervadec, H., et al.: Boundary loss for highly unbalanced segmentation. Med. Image Anal. **67**, 101851 (2021)
11. Lee, H.J., Kim, J.U., Lee, S., Kim, H.G., Ro, Y.M.: Structure boundary preserving segmentation for medical image with ambiguous boundary. In: CVPR, pp. 4817–4826 (2020)
12. Leventić, H., Benčević, M., Babin, D., Habijan, M., Galić, I.: A survey of left atrial appendage segmentation and analysis in 3D and 4d medical images. arXiv preprint arXiv:2205.06486 (2022)
13. Leventić, H., et al.: Left atrial appendage segmentation from 3D CCTA images for occluder placement procedure. Comput. Biol. Med. **104**, 163–174 (2019)
14. Milletari, F., Navab, N., Ahmadi, S.-A.: V-net: fully convolutional neural networks for volumetric medical image segmentation. In: 2016 Fourth International Conference on 3D Vision (3DV), pp. 565–571. IEEE (2016)
15. Sapiro, G.: Geometric Partial Differential Equations and Image Analysis. Cambridge University Press, Cambridge (2006)
16. Tan, H., Sitong, W., Pi, J.: Semantic diffusion network for semantic segmentation. NeurIPS **35**, 8702–8716 (2022)
17. Tang, Y., et al.: Self-supervised pre-training of swin transformers for 3D medical image analysis. In: Proceedings of the IEEE/CVF Conference on Computer Vision and Pattern Recognition, pp. 20730–20740 (2022)
18. Tobon-Gomez, C., et al.: Benchmark for algorithms segmenting the left atrium from 3D CT and MRI datasets. IEEE Trans. Med. Imaging **34**(7), 1460–1473 (2015)
19. Valanarasu, J.M.J., Patel, V.M.: UNeXt: MLP-based rapid medical image segmentation network. In: Wang, L., Dou, Q., Fletcher, P.T., Speidel, S., Li, S. (eds.) MICCAI 2022. LNCS, vol. 13435, pp. 23–33. Springer, Cham (2022). https://doi.org/10.1007/978-3-031-16443-9_3
20. Wang, W., Chen, C., Ding, M., Yu, H., Zha, S., Li, J.: TransBTS: multimodal brain tumor segmentation using transformer. In: de Bruijne, M., et al. (eds.) MICCAI 2021. LNCS, vol. 12901, pp. 109–119. Springer, Cham (2021). https://doi.org/10.1007/978-3-030-87193-2_11

21. Weickert, J.: Coherence-enhancing diffusion filtering. Int. J. Comput. Vision **31**(2–3), 111 (1999)
22. Weickert, J., Ter Haar Romeny, B.M., Viergever, M.A.: Efficient and reliable schemes for nonlinear diffusion filtering. IEEE Trans. Image Process. **7**(3), 398–410 (1998)
23. Xie, Y., Liao, H., Zhang, D., Chen, F.: Uncertainty-aware cascade network for ultrasound image segmentation with ambiguous boundary. In: Wang, L., Dou, Q., Fletcher, P.T., Speidel, S., Li, S. (eds.) MICCAI 2022. LNCS, vol. 13434, pp. 268–278. Springer, Cham (2022). https://doi.org/10.1007/978-3-031-16440-8_26
24. Xiong, Z., et al.: A global benchmark of algorithms for segmenting the left atrium from late gadolinium-enhanced cardiac magnetic resonance imaging. Med. Image Anal. **67**, 101832 (2021)
25. Yeung, M., Yang, G., Sala, E., Schönlieb, C.-B., Rundo, L.: Incorporating boundary uncertainty into loss functions for biomedical image segmentation. arXiv preprint arXiv:2111.00533 (2021)
26. Yuan, Y., Xie, J., Chen, X., Wang, J.: SegFix: model-agnostic boundary refinement for segmentation. In: Vedaldi, A., Bischof, H., Brox, T., Frahm, J.-M. (eds.) ECCV 2020. LNCS, vol. 12357, pp. 489–506. Springer, Cham (2020). https://doi.org/10.1007/978-3-030-58610-2_29
27. Zheng, Y., Yang, D., John, M., Comaniciu, D.: Multi-part modeling and segmentation of left atrium in c-arm CT for image-guided ablation of atrial fibrillation. IEEE Trans. Med. Imaging **33**(2), 318–331 (2013)
28. Zhuang, X., Shen, J.: Multi-scale patch and multi-modality atlases for whole heart segmentation of MRI. Med. Image Anal. **31**, 77–87 (2016)

DMCVR: Morphology-Guided Diffusion Model for 3D Cardiac Volume Reconstruction

Xiaoxiao He[1]⬤, Chaowei Tan[2], Ligong Han[1], Bo Liu[3], Leon Axel[4], Kang Li[5], and Dimitris N. Metaxas[1](✉)

[1] Department of Computer Science, Rutgers University, Piscataway, USA
dnm@cs.rutgers.edu
[2] FocusAI Inc., Sunnyvale, USA
[3] Walmart Global Tech, San Bruno, USA
[4] School of Medicine, New York University, New York, USA
[5] West China Biomedical Big Data Center,
Sichuan University West China Hospital, Chengdu, China

Abstract. Accurate 3D cardiac reconstruction from cine magnetic resonance imaging (cMRI) is crucial for improved cardiovascular disease diagnosis and understanding of the heart's motion. However, current cardiac MRI-based reconstruction technology used in clinical settings is 2D with limited through-plane resolution, resulting in low-quality reconstructed cardiac volumes. To better reconstruct 3D cardiac volumes from sparse 2D image stacks, we propose a morphology-guided diffusion model for 3D cardiac volume reconstruction, DMCVR, that synthesizes high-resolution 2D images and corresponding 3D reconstructed volumes. Our method outperforms previous approaches by conditioning the cardiac morphology on the generative model, eliminating the time-consuming iterative optimization process of the latent code, and improving generation quality. The learned latent spaces provide global semantics, local cardiac morphology and details of each 2D cMRI slice with highly interpretable value to reconstruct 3D cardiac shape. Our experiments show that DMCVR is highly effective in several aspects, such as 2D generation and 3D reconstruction performance. With DMCVR, we can produce high-resolution 3D cardiac MRI reconstructions, surpassing current techniques. Our proposed framework has great potential for improving the accuracy of cardiac disease diagnosis and treatment planning. Code can be accessed at https://github.com/hexiaoxiao-cs/DMCVR.

Keywords: Diffusion model · 3D Reconstruction · Generative model

Supplementary Information The online version contains supplementary material available at https://doi.org/10.1007/978-3-031-43990-2_13.

1   Introduction

Medical imaging technology has revolutionized the field of cardiac disease diagnosis, enabling the assessment of both cardiac anatomical structures and motion, including the creation of 3D models of the heart [5]. Cardiac cine magnetic resonance imaging (cMRI) [16,20] is widely used in clinical diagnosis [14], allowing for non-invasive visualization of the heart in motion with detailed information on cardiac function and anatomy [17]. While cMRI has great potential in helping doctors understand and analyze cardiac function [9,15], the imaging technique has certain drawbacks including low through-plane resolution to accommodate for the limited scanning time, as visualized in Fig. 1. Recently, researchers have approached the problem of cardiac volume reconstruction with learning-based generative models [2]. However, most of the methods suffer from low generation quality, missing key cardiac structures and long generation times. This paper focuses on improving the cardiac model generation quality, while reducing the generation time, aiming to better reconstruct the missing structure of the cardiac model from low through-plane resolution cMRI.

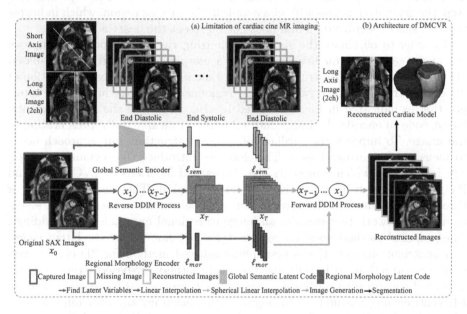

Fig. 1. (a) demonstrates the limitations of cardiac cMRI. The white line in the short axis (SAX) image is the location of 2 chamber (2ch) long axis (LAX) image slice and vice versa. The grey images indicate the missing slices which are not captured during the MRI scan. (b) is an overview of our DMCVR architecture. The SAX images x_0 are first encoded to global semantic ℓ_{sem}, regional morphology ℓ_{mor} and stochastic latent codes x_T, followed by interpolation in their respective latent space. The reconstructed images are sampled from a forward denoising diffusion implicit model (DDIM) process conditioned on the three latent codes. Finally, the 3D cardiac model is reconstructed via stacking the labels. The red, green, and blue regions represent the left ventricle cavity (LVC), left ventricle myocardium (LVM), and right ventricle cavity (RVC), respectively. (Color figure online)

Conventional 3D cardiac modeling [12] consists of 2D cardiac image segmentation followed by 3D cardiac volume reconstruction. Recent advances in deep learning methods have shown great success in medical image segmentation [4,6,11,23]. After obtaining 2D labels, the neighboring labels are stacked to reconstruct the 3D model. Nevertheless, due to the low inter-slice spatial cMRI resolution, a significant amount of structural information is lost in the resulting 3D volume. Thus, the interpolation between cMRI slices is necessary. Traditional intensity-based interpolation methods often yield blurring effects and unrealistic results. Conventional deformable model-based method [13] does not need consistency across images of the corresponding cardiac structures, but requires image-based structure segmentation which is nontrivial and hinders their ability to generalize. To overcome these limitations, an end-to-end pipeline based on generative adversarial networks (GANs), DeepRecon, was recently proposed in [2] that utilizes the latent space to interpolate the missing information between adjacent 2D slices. The generative network is first trained and a semantic image embedding in the $\mathcal{W}^+$ space [1] is computed. Evidently, the acquired semantic latent code is not optimal and needs iterative optimization with segmentation information for improving image qualities. However, even with the optimization step, the generated images still miss details in the cardiac region, which indicates the $\mathcal{W}^+$ space DeepRecon found does not represent the heart accurately.

In order to eliminate the step for optimizing the latent code and improve the image generation quality, we propose a morphology-guided diffusion-based 3D cardiac volume reconstruction method that improves the axial resolution of 2D cMRIs through global semantic and regional morphology latent code interpolation as indicated in Fig. 1. Inspired by [19], we utilize the global semantic latent code to encode the image into a high-level meaningful representation of the image. To improve the cardiac volume reconstruction, our approach needs to focus on the cardiac region. Therefore, we introduce the regional morphology latent code which represents the shapes and locations of LVC, LVM and RVC, which will help generating the cardiac region. The method consists of three parts: an implicit diffusion model, a global semantic encoder and a segmentation network that encodes an image to regional morphology embeddings. The proposed method does not require iteratively fine-tuning the latent codes. Our contributions are: 1) the first diffusion-based method for 3D cardiac volume reconstruction, 2) introducing the local morphology-based latent code for improved conditioning on the image generation process, 3) 8% improvement of left ventricle myocardium (LVM) segmentation accuracy and 35% improvement of structural similarity index compared to previous methods, and 4) improved efficiency by eliminating the iterative step for optimizing the latent code.

2   Methods

Figure 2 demonstrates the structure of our DMCVR approach that learns the global semantic, regional morphology, and stochastic latent spaces from MR images to yield a broad range of outcomes, including generation of high-quality

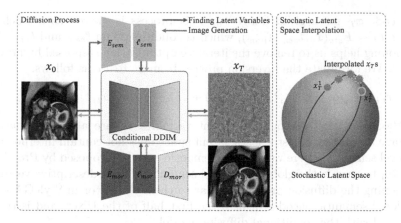

Fig. 2. On the left side, we demonstrates the network structure of the DMCVR, which consists of an global semantic encoder, a regional morphology encoder/decoder and a conditional DDIM. The right side shows the visualization of the stochastic latent space sampled from a high-dimensional Gaussian distribution $\mathcal{N}(0, I)$.

2D image and high-resolution 3D reconstructed volume. In this section, we will first describe the architecture of our DMCVR method and then elaborate on the latent space-based 3D volume generation which enables 3D volume reconstruction.

2.1 DMCVR Architecture

Our DMCVR is composed of a global semantic encoder E_{sem}, a regional moprphology network (E_{mor}, D_{mor}) and a diffusion-based generator G. The generating process G is defined as follows: given input $x_T, \ell_{sem}, \ell_{mor}$, which are the stochastic, global semantic and regional morphology latent codes, we want to reconstruct the image x_0 recursively as follows:

$$x_{t-1} = \sqrt{\alpha_{t-1}} f_\theta(x_t, t, \ell_{sem}, \ell_{mor}) + \sqrt{1 - \alpha_{t-1}} \epsilon_\theta(x_t, t, \ell_{sem}, \ell_{mor}), \quad (1)$$

where $\epsilon_\theta(x_t, t, \ell_{sem}, \ell_{mor})$ is the noise prediction network and f_θ is defined as removing the noise from x_t or Tweedie's formula [3]:

$$f_\theta(x_t, t, \ell_{sem}, \ell_{mor}) = \frac{1}{\sqrt{\alpha_t}}(x_t - \sqrt{1 - \alpha_t} \epsilon_\theta(x_t, t, \ell_{sem}, \ell_{mor})) \quad (2)$$

Here, the term α_t is a function of t affecting the sampling quality.

The forward diffusion process takes the noise x_T as input and produces x_0 the target image. Since the change in x_T will affect the details of the output images, we can treat x_T as the stochastic latent code. Therefore, finding the correct stochastic latent code is crucial for generating image details. Thanks to DDIM proposed by Song *et al.* [21], it is possible to get x_T in a deterministic fashion by running the generative process backwards to obtain the stochastic latent code x_T for a given image x_0. This process is viewed as a stochastic encoder $x_T = E_{sto}(x_0, \ell_{sem}, \ell_{mor})$, which is conditioned on ℓ_{sem} and ℓ_{mor}. This conditioning helps us to remove the iterative optimization step used by previous method. We formulate the inversion process from x_0 to x_T as follows:

$$x_{t+1} = \sqrt{\alpha_{t+1}} f_\theta(x_t, t, \ell_{sem}, \ell_{mor}) + \sqrt{1 - \alpha_{t+1}} \epsilon_\theta(x_t, t, \ell_{sem}, \ell_{mor}) \qquad (3)$$

Although using the stochastic latent variables we are able to reconstruct the image accurately, the stochastic latent space does not contain interpolatable high-level semantics. Here we utilize a semantic encoder proposed by Preechakul et al. [19] to encode the global high-level semantics into a descriptive vector for conditioning the diffusion process, similar to the style vector in StyleGAN [10]. The global semantic encoder utilizes the first half of the UNet, and is trained end-to-end with the conditional diffusion model.

One drawback of the global semantic encoder is that it encodes the general high-level features, but tends to pay little attention to the cardiac region. This is due to the relatively small area of LVC, LVM and RVC in the cMRI slice. However, the generation accuracy of the cardiac region is crucial for the cardiac reconstruction task. For this reason, we introduce the regional morphology encoder E_{mor} that embeds the image into the latent space containing necessary information to produce the segmentation map of the target cardiac tissues. With this extra morphology information, we are able to guide the generative model to focus on the boundary of the ventricular cavity and myocardium region, which will produce increased image accuracy in the cardiac region and the downstream segmentation task. Here, we do not assume any particular architecture for the segmentation network. However, in our experiments, we utilize the segmentation network MedFormer proposed by Gao et al. [4] for its excellent performance.

The training of DMCVR contains the training of the segmentation network and the training of the generative model. We first train the segmentation model with summation of focal loss and dice loss [4]. We utilize the simple loss introduced in [7] for training the conditional diffusion implicit model, where

$$L_{gen}(x) = \mathbb{E}_{t \sim \mathrm{Unif}(1,T), \epsilon \sim \mathcal{N}(0,I)} ||\epsilon_\theta(x_t, t, E_{sem}(x_0), E_{mor}(x_0)) - \epsilon||_2^2. \qquad (4)$$

2.2   3D Volume Reconstruction and Latent-Space-Based Interpolation

Due to various limitations, the gap between consecutive cardiac slices in cMRI is large, which results in an under-sampled 3D model. In order to output a smooth super-resolution cine image volume, we generate the missing slices by using the interpolated global semantic, regional morphology and stochastic latent codes. For global semantic and regional morphology latent code ℓ, since it is similar to the idea of latent code in StyleGAN, we utilize the same interpolation strategies as in the original paper between adjacent slices. Assume that $k < j - i, i < j$,

$$\ell^{i+k} = (1 - \frac{k}{j-i})\ell^i + \frac{k}{j-i}\ell^j. \qquad (5)$$

For interpolating the stochastic latent variable, it is important to consider that the distribution of stochastic noise is high-dimensional Gaussian, as shown in Eq. (4). Thus, our stochastic embedding is positioned on a sphere shown in Fig. 2. Using linear interpolation on the stochastic noise deviates from the underlying distribution assumption and causes the diffusion model to generate unrealistic images. Hence, to preserve the Gaussian property of the stochastic latent space, we interpolate the stochastic latent codes over a unit sphere, which can be written as follows: Let $k < j - i, i < j$ and $x_T^i \cdot x_T^j = \cos\theta$,

$$ x_T^{i+k} = \frac{\sin((1 - \frac{k}{j-i})\theta)}{\sin(\theta)} x_T^i + \frac{\sin(\frac{k}{j-i}\theta)}{\sin(\theta)} x_T^j. \tag{6} $$

Table 1. Quantitative comparison among the segmentation results of the original image (Original), DeepRecon with 1k optimization steps (DeepRecon$_{1k}$), Diffusion AutoEncoder [19] (DiffAE) and our DMCVR. We use a pretrained segmentation model on images generated by different methods. All metrics are evaluated against the ground truth based on 3D SAX images.

Cardiac Region	Method	DICE ↑	VOE↓	ASD↓	HD↓	ASSD↓
All labels	Original	0.943	10.730	0.229	4.056	0.229
	DeepRecon$_{1k}$	0.914	15.179	0.367	5.879	0.397
	DiffAE	0.919	14.913	0.322	4.654	0.326
	DMCVR	**0.935**	**12.153**	**0.261**	**4.093**	**0.266**
LVC	Original	0.937	11.579	0.221	3.156	0.224
	DeepRecon$_{1k}$	0.928	12.955	0.336	4.299	0.328
	DiffAE	0.910	16.049	0.330	3.710	0.320
	DMCVR	**0.929**	**12.940**	**0.250**	**3.236**	**0.254**
LVM	Original	0.875	22.082	0.226	3.140	0.237
	DeepRecon$_{1k}$	0.796	33.382	0.390	5.730	0.389
	DiffAE	0.825	29.333	0.351	4.032	0.338
	DMCVR	**0.865**	**23.636**	**0.282**	**3.519**	**0.267**
RVC	Original	0.898	18.187	0.273	4.458	0.267
	DeepRecon$_{1k}$	0.858	23.662	0.381	6.304	0.473
	DiffAE	0.857	24.518	0.346	5.217	0.382
	DMCVR	**0.884**	**20.467**	**0.273**	**4.460**	**0.308**

3 Experiments

3.1 Experimental Settings

In this study we use data from the publicly available UK Biobank cardiac MRI data [18], which contains SAX and LAX cine CMR images of normal subjects.

LVC, LVM and RVC are manually annotated on SAX images at the end-diastolic (ED) and end-systolic (ES) cardiac phases. We use 808 cases containing 484,800 2D SAX MR slices for training and 200 cases containing 120,000 2D images for testing. To evaluate the 3D volume reconstruction performance, we randomly choose 50 testing 2D LAX cases to evaluate the 3D volume reconstruction task. All models are implemented on PyTorch 1.13 and trained with 4×RTX8000.

3.2 Evaluation of the 2D Slice Generation Quality

We provide peak signal-to-noise ratio (PSNR) and structural similarity index measure (SSIM) [8] to evaluate the similarity between the generated images and the original images. In addition to image quality assessment, we want to consider the segmentation performance on the generated images by using a segmentation network trained on the real training data as the evaluator and segment the testing images generated by DeepRecon$_{1k}$, DiffAE which only uses the global semantic latent code as the condition on the DDIM model, and our DMCVR methods. The segmentation accuracy of the evaluator on the generated images can be viewed as a quantitative metric to represent the generation quality of the generated data compared to the cMRI data. We compare segmentation obtained based on three methods against ground truth on the SAX images in Table 1. The Dice coefficient (DICE), volumetric overlap error (VOE), average surface distance (ASD), Hausdorff distance (HD) and average symmetric surface distance (ASSD) [22] are reported for comparison.

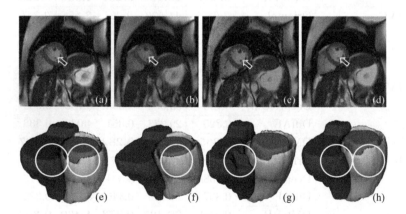

Fig. 3. 2D and 3D visualization results of the generated images and segmentation. (a, e) original image, (b, f) DeepRecon$_{1k}$, (c, g) DiffAE [19], (d, h) our proposed DMCVR. (Color figure online)

Our method achieves a PSNR score of **30.504** and SSIM score of **0.982**, which is a significant improvement (35% increase in SSIM) compared to Deep-Recon (PSNR: **27.684**, SSIM: **0.724**) with 1k optimization steps. This indicates that our method generates more realistic image compared to DeepRecon. The segmentation results on the original images in Table 1 provide an upper bound for other results. DMCVR outperforms all other methods in every metric with an 8% increase in LVM segmentation compared to $DiffRecon_{1k}$ Moreover, by comparing the DiffAE and DMCVR, the introduction of the regional morphology latent code drastically improves the generation results due to the extra information on the shape of LVC, LVM, and RVC. Figure 3 demonstrates the original image and corresponding synthetic images. The white arrow points towards the presence of cardiac papillary muscles. As indicated in the images, $DeepRecon_{1k}$ (b) cannot effectively recover the information of the papillary muscles from the latent space. However, both diffusion-based (c, d) methods accurately synthesize the information. Our method (d) generates a cleaner image with less artifacts than (c), especially around the LV and RV regions. By comparing the yellow

Table 2. Evaluation of 3D volumetric reconstruction from the DICE score of the intersection on each LAX plane against ground truth based on 2D LAX sampled images: mean (standard deviation). Nearest Neighbor, Image-based Linear Interpolation, $DeepRecon_{1k}$ and our DMCVR method are compared.

Method	Average DICE	2ch DICE	3ch DICE	4ch DICE
Nearest Neighbor	0.780 (0.111)	0.787 (0.091)	0.793 (0.105)	0.766 (0.128)
Linear Interpolation	0.781 (0.080)	0.797 (0.051)	0.773 (0.070)	0.768 (0.102)
$DeepRecon_{1k}$	0.817 (0.097)	**0.848** (0.056)	0.802 (0.141)	0.797 (0.091)
DMCVR	**0.836 (0.052)**	0.841 **(0.042)**	**0.809 (0.069)**	**0.854 (0.043)**

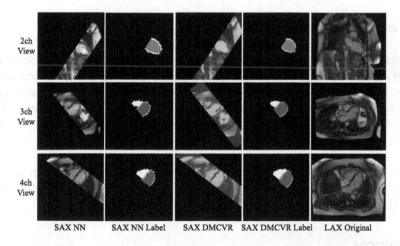

SAX NN   SAX NN Label   SAX DMCVR   SAX DMCVR Label   LAX Original

Fig. 4. Visual comparison of 3D volumetric reconstruction from SAX images to LAX. Each row from top to bottom are 2ch, 3ch and 4ch images. The column from left to right represents: resampled original images using nearest neighbour (NN), resampled original labels using NN, resampled DMCVR images, resampled DMCVR labels and the corresponding LAX images.

circled area, our method produces image closer to the ground truth compared to
DeepRecon$_{1k}$. Also, the white circle in Fig. 3 demonstrates the benefits of incor-
porating regional morphology information. Besides, the generative model used in
DeepRecon$_{1k}$ needs to be trained for 14 days with additional time to iteratively
optimize the latent code for each slice. Our method uses 4.8 days for training.
Since DDIM inversion does not have test-time optimization as DeepRecon does,
DMCVR generates images faster than DeepRecon.

3.3 Evaluation of the 3D Volume Reconstruction Quality Through Latent Space Interpolation

In this section, we exploit the relationship between SAX and LAX images and
leverage the LAX label to evaluate the volume reconstruction quality. In cardiac MRI, long axis (LAX) slices typically comprise 2-chamber (2ch), 3-chamber
(3ch), and 4-chamber (4ch) views. To evaluate the performance of different interpolation methods on LAX slices, we conducted the following experiments: 1)
Nearest Neighbor resampling of short-axis (SAX) volume to each LAX view, 2)
Image-based Linear Interpolation, 3) DeepRecon$_{1k}$, and 4) our DMCVR. Table 2
shows the computed 2D DICE score between the annotation of different LAX
views and the intersection between the corresponding LAX plane and 3D reconstructed volume. Our method outperforms other methods in three categories
and has only less than 1% performance degradation compared to DeepRecon$_{1k}$
but with more stable performance. Figure 4 presents three examples for each
LAX view, showing better reconstructed LAX results compared to the original
images.

4   Conclusion

Integrating analysis of cMRI holds significant clinical importance in understanding and evaluating cardiac function. We propose a diffusion-model-based volume
reconstruction method. Our finding shows that through an interpolatable latent
space, we are able to improve the spatial resolution and produce meaningful MR
images. In the future, we will consider incorporating LAX slices as part of the
generation process to help refine the latent space.

Acknowledgement. This research has been partially funded by research grants to D.
Metaxas through NSF: IUCRC CARTA 1747778, 2235405, 2212301, 1951890, 2003874,
and NIH-5R01HL127661.

References

1. Abdal, R., Qin, Y., Wonka, P.: Image2StyleGAN: how to embed images into the
StyleGAN latent space? In: Proceedings of the IEEE/CVF International Conference on Computer Vision, pp. 4432–4441 (2019)

2. Chang, Q., et al.: DeepRecon: joint 2D cardiac segmentation and 3D volume reconstruction via a structure-specific generative method. In: Wang, L., Dou, Q., Fletcher, P.T., Speidel, S., Li, S. (eds.) MICCAI 2022, Part IV. LNCS, vol. 13434, pp. 567–577. Springer, Cham (2022). https://doi.org/10.1007/978-3-031-16440-8_54

3. Efron, B.: Tweedie's formula and selection bias. J. Am. Stat. Assoc. **106**(496), 1602–1614 (2011)

4. Gao, Y., Zhou, M., Liu, D., Yan, Z., Zhang, S., Metaxas, D.N.: A data-scalable transformer for medical image segmentation: architecture, model efficiency, and benchmark. arXiv preprint arXiv:2203.00131 (2022)

5. van der Geest, R.J., Reiber, J.H.: Quantification in cardiac MRI. J. Magn. Reson. Imaging Off. J. Int. Soc. Magn. Reson. Med. **10**(5), 602–608 (1999)

6. He, X., Tan, C., Qiao, Y., Tan, V., Metaxas, D., Li, K.: Effective 3D humerus and scapula extraction using low-contrast and high-shape-variability MR data. In: Medical Imaging 2019: Biomedical Applications in Molecular, Structural, and Functional Imaging, vol. 10953, pp. 118–124. SPIE (2019)

7. Ho, J., Jain, A., Abbeel, P.: Denoising diffusion probabilistic models. In: Advances in Neural Information Processing Systems, vol. 33, pp. 6840–6851 (2020)

8. Hore, A., Ziou, D.: Image quality metrics: PSNR vs. SSIM. In: 2010 20th International Conference on Pattern Recognition, pp. 2366–2369. IEEE (2010)

9. Isensee, F., Jaeger, P.F., Full, P.M., Wolf, I., Engelhardt, S., Maier-Hein, K.H.: Automatic cardiac disease assessment on cine-MRI via time-series segmentation and domain specific features. In: Pop, M., et al. (eds.) STACOM 2017. LNCS, vol. 10663, pp. 120–129. Springer, Cham (2018). https://doi.org/10.1007/978-3-319-75541-0_13

10. Karras, T., Laine, S., Aila, T.: A style-based generator architecture for generative adversarial networks. In: Proceedings of the IEEE/CVF Conference on Computer Vision and Pattern Recognition, pp. 4401–4410 (2019)

11. Liu, D., et al.: TransFusion: multi-view divergent fusion for medical image segmentation with transformers. In: Wang, L., Dou, Q., Fletcher, P.T., Speidel, S., Li, S. (eds.) MICCAI 2022, Part V. LNCS, vol. 13435, pp. 485–495. Springer, Cham (2022). https://doi.org/10.1007/978-3-031-16443-9_47

12. Lopez-Perez, A., Sebastian, R., Ferrero, J.M.: Three-dimensional cardiac computational modelling: methods, features and applications. Biomed. Eng. Online **14**, 1–31 (2015). https://doi.org/10.1186/s12938-015-0033-5

13. Myronenko, A., Song, X.: Point set registration: coherent point drift. IEEE Trans. Pattern Anal. Mach. Intell. **32**(12), 2262–2275 (2010)

14. Patel, R., et al.: Diagnostic performance of cardiac magnetic resonance imaging and echocardiography in evaluation of cardiac and paracardiac masses. Am. J. Cardiol. **117**(1), 135–140 (2016)

15. Pattynama, P.M., De Roos, A., Van der Wall, E.E., Van Voorthuisen, A.E.: Evaluation of cardiac function with magnetic resonance imaging. Am. Heart J. **128**(3), 595–607 (1994)

16. Pelc, N.J., Herfkens, R.J., Shimakawa, A., Enzmann, D.R., et al.: Phase contrast cine magnetic resonance imaging. Magn. Reson. Q. **7**(4), 229–254 (1991)

17. Peng, P., Lekadir, K., Gooya, A., Shao, L., Petersen, S.E., Frangi, A.F.: A review of heart chamber segmentation for structural and functional analysis using cardiac magnetic resonance imaging. Magn. Reson. Mater. Phys., Biol. Med. **29**(2), 155–195 (2016). https://doi.org/10.1007/s10334-015-0521-4

18. Petersen, S.E., et al.: UK biobank's cardiovascular magnetic resonance protocol. J. Cardiovasc. Magn. Reson. **18**(1), 1–7 (2015)

19. Preechakul, K., Chatthee, N., Wizadwongsa, S., Suwajanakorn, S.: Diffusion autoencoders: toward a meaningful and decodable representation. In: Proceedings of the IEEE/CVF Conference on Computer Vision and Pattern Recognition, pp. 10619–10629 (2022)
20. Sechtem, U., Pflugfelder, P., Higgins, C.B.: Quantification of cardiac function by conventional and cine magnetic resonance imaging. Cardiovasc. Intervent. Radiol. **10**, 365–373 (1987). https://doi.org/10.1007/BF02577347
21. Song, J., Meng, C., Ermon, S.: Denoising diffusion implicit models. In: International Conference on Learning Representations (2021). https://openreview.net/forum?id=St1giarCHLP
22. Taha, A.A., Hanbury, A.: Metrics for evaluating 3D medical image segmentation: analysis, selection, and tool. BMC Med. Imaging **15**(1), 1–28 (2015)
23. Zhangli, Q., et al.: Region proposal rectification towards robust instance segmentation of biological images. In: Wang, L., Dou, Q., Fletcher, P.T., Speidel, S., Li, S. (eds.) MICCAI 2022, Part IV. LNCS, vol. 13434, pp. 129–139. Springer, Cham (2022). https://doi.org/10.1007/978-3-031-16440-8_13

A Conditional Flow Variational Autoencoder for Controllable Synthesis of Virtual Populations of Anatomy

Haoran Dou[1], Nishant Ravikumar[1], and Alejandro F. Frangi[1,2,3,4(✉)]

[1] Centre for Computational Imaging and Simulation Technologies in Biomedicine
(CISTIB), University of Leeds, Leeds, UK
n.ravikumar@leeds.ac.uk

[2] Division of Informatics, Imaging and Data Science, Schools of Computer Science
and Health Sciences, University of Manchester, Manchester, UK
alejandro.frangi@manchester.ac.uk

[3] Medical Imaging Research Center (MIRC), Electrical Engineering and
Cardiovascular Sciences Departments, KU Leuven, Leuven, Belgium

[4] Alan Turing Institute, London, UK

Abstract. The generation of virtual populations (VPs) of anatomy is essential for conducting in silico trials of medical devices. Typically, the generated VP should capture sufficient variability while remaining plausible and should reflect the specific characteristics and demographics of the patients observed in real populations. In several applications, it is desirable to synthesise virtual populations in a *controlled* manner, where relevant covariates are used to conditionally synthesise virtual populations that fit a specific target population/characteristics. We propose to equip a conditional variational autoencoder (cVAE) with normalising flows to boost the flexibility and complexity of the approximate posterior learnt, leading to enhanced flexibility for controllable synthesis of VPs of anatomical structures. We demonstrate the performance of our conditional flow VAE using a data set of cardiac left ventricles acquired from 2360 patients, with associated demographic information and clinical measurements (used as covariates/conditional information). The results obtained indicate the superiority of the proposed method for conditional synthesis of virtual populations of cardiac left ventricles relative to a cVAE. Conditional synthesis performance was evaluated in terms of generalisation and specificity errors and in terms of the ability to preserve clinically relevant biomarkers in synthesised VPs, that is, the left ventricular blood pool and myocardial volume, relative to the real observed population.

Keywords: Virtual Population · Generative Model · Normalizing Flow

N. Ravikumar and A. F. Frangi—Joint last authors.

Supplementary Information The online version contains supplementary material available at https://doi.org/10.1007/978-3-031-43990-2_14.

1   Introduction

In-silico trials (ISTs) use computational modelling and simulation techniques with virtual twin or patient models of anatomy and physiology to evaluate the safety and efficacy of medical devices virtually [22]. Virtual patient populations (VPs), distinct from virtual twin populations, comprise plausible instances of anatomy and physiology that do not represent any specific real patient's data (as in the case of the latter, viz. virtual twins). In other words, VPs comprise synthetic data that help expand/enrich the diversity of anatomical and physiological characteristics that can be investigated within an IST for a given medical device. A key aspect of patient recruitment in real clinical trials used to assess device performance and generate regulatory evidence for device approval is the clear definition of inclusion and exclusion criteria for the trial. These criteria define the target patient population considered appropriate/safe to assess the performance of the device of interest. Consequently, it is desirable to enable the *controlled* synthesis of VPs that may be used for device ISTs, in a manner that emulates the imposition of trial inclusion and exclusion criteria.

Virtual populations can be considered to be parametric representations of the anatomy sampled from a generative model. Traditional statistical shape models (SSMs), based on methods such as principal component analysis (PCA), have been widely explored in the past decade [8,9,15]. Recent studies focus on deep learning-based generative models due to their automatic and powerful hierarchical feature extraction [3,7]. For instance, Bonazzola *et al.* [3] used a graph convolutional variational auto-encoder (gcVAE) to learn latent representations of 3D left ventricular meshes and used the learnt representations as surrogates for cardiac phenotypes in genome-wide association studies. Dou *et al.* [7] proposed learning the shape representations of multiple cardiovascular anatomies using gcVAE independently and then assembling them into complete whole-heart anatomies termed virtual heart chimaeras. Other studies have investigated conditional-generative models for synthesis of VPs of anatomies. For example, Beetz *et al.* [1] employed a conditional VAE (cVAE), conditioned on gender and cardiac phase, to allow the synthesis of VPs from biventricular anatomies. In subsequent work [2,12], they extended their method to a multidomain VAE to model biventricular anatomies at multiple times (across the cardiac cycle), using patient-specific electrocardiogram (ECG) signals as additional conditioning information (in addition to patient demographic data and standard clinical measurements) to guide the synthesis. All aforementioned methods model the latent space in the VAEs/cVAEs as a multivariate Gaussian distribution with a diagonal covariance matrix. This limits the flexibility afforded to the cVAE, as the Gaussian distribution, being unimodal, is a poor approximation to multimodal latent posterior distributions. This in turn limits the overall variability in anatomical shape that can be captured by standard VAEs and cVAEs.

In this study, we address the limitations of the state-of-the-art conditional generative models used to synthesise VPs of anatomical structures. In particular, we propose a method to relax the constraint on modelling the latent distribution as a unimodal multivariate Gaussian, to boost the flexibility of the generative model, and to enable conditional synthesis of diverse and plausible VPs generation. Recent advances in normalising flows [14,16,21] introduce a new solution for this limitation by leveraging a series of invertible parameterized functions to transform the unimodal distribution to a multimodal one. Motivated by this technique, we propose the first conditional flow VAE (parameterised as a graph-convolutional network) for the task of *controllable* synthesis of VPs of anatomy. The contributions are as follows: (i) we introduce normalising flows to learn a multimodal latent posterior distribution by transforming the latent variables from a simple unimodal distribution. This helps the generative model capture greater anatomical variability from the observed real population, leading to the synthesis of more diverse VPs; (ii) we condition the flow-based VAE on patient demographic data and clinical measurements. This enables conditional synthesis of plausible VPs (given relevant covariates/conditioning information as inputs), which reflect the observed correlations between nonimaging patient information and anatomical characteristics in the real population.

2 Methodology

In this study, we propose a cVAE model equipped with normalising flows for controllable synthesis of VPs of cardiovascular anatomy. A schematic of the proposed conditional flow VAE network architecture is shown in Fig. 1. We employ normalising flows in the latent space of the cVAE to transform the initial Gaussian posterior to a complex multimodal distribution.

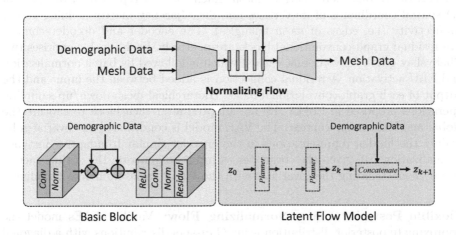

Fig. 1. Schematic illustration of our proposed conditional flow VAE

Conditional Variational Autoencoder: A VAE is a probabilistic generative model/network [11] that comprises an encoder and a decoder network branch. The encoder learns a mapping from the input data to a low-dimensional latent space that abstracts the semantic representations from the observations, and the decoder reconstructs the original data from the low-dimensional latent representation. The latent space from which the observed data is generated is given by approximating the posterior distribution of the latent variables using variational inference. The VAE network is trained by maximising the evidence lower bound (ELBO), which is a summation of the expected log-likelihood of the data and the Kullback-Leibler divergence between the approximate posterior and some assumed prior distribution over the latent variables (typically a multivariate Gaussian distribution). Despite its effectiveness in capturing some of the observed variability in the training population (e.g. of anatomical shapes or images), VAEs do not provide any control over the generation process and hence cannot guarantee that the generated population anatomical shapes are representative of target patient populations with specific inclusion/exclusion criteria. Controllable synthesis of anatomical VPs is essential for constructing meaningful cohorts for use in ISTs. Conditional VAE [18] is a VAE-variant that uses additional covariates/conditioning information in addition to the input data (e.g. anatomical shapes) to learn a conditional latent posterior distribution (conditioned on the covariates), enabling controllable synthesis of VPs during inference (given relevant covariates/conditioning information as input).

Our conditional flow VAE (cVAE-NF) is a graph-convolutional network which takes as input a triangular surface mesh representation of an anatomical structure of interest, i.e., the Left Ventricle (LV) in this study, and its associated covariates/conditioning variables, i.e., the patient demographic data and clinical measurements, such as gender, age, weight, blood cholesterol, etc., and outputs the reconstructed surface mesh. Each mesh is represented by a list of 3D spatial coordinates of its vertices and an adjacency matrix defining vertex connectivity (i.e. edges of mesh triangles). The encoder and decoder contain five residual graph-convolutional blocks, respectively. Each block comprises two Chebyshev graph convolutions, each of which is followed by batch normalisation and ELU activation. A residual connection is added between the input and the output of each graph-convolutional block. Hierarchical mesh down/up-sampling operations proposed in CoMA [13] are adopted after each block to capture the global and local shape context. The VAE model is conditioned on covariates by scaling the hidden representations in the encoder similar to adaptive instance normalization [10] given the covariates as input to generate the scaling factor, and by concatenating the covariates with the latent variables before decoding.

Flexible Posterior Using Normalizing Flow: Vanilla cVAEs model the approximate posterior distribution using Gaussian distributions with a diagonal covariance matrix. However, such a unimodal distribution is a poor approximation of the complex true latent posterior distribution in most real-world applications (e.g. for shapes of the LV observed across a population), limiting the anatomical variability captured by the model. In this study, we introduce normalising flows to construct a flexible multi-modal latent posterior distribution by applying a series of differentiable, invertible/diffeomorphic transformations iteratively to the initial simple unimodal latent distribution. As shown in Fig. 2, a two-dimensional Gaussian distribution can be transformed into a multi-modal distribution by applying several normalising flow steps to the former.

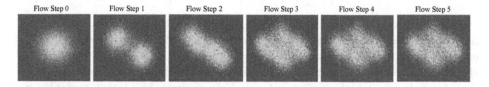

Fig. 2. Effect of normalising flow on Gaussian distribution. Step 0 is the initial two-dimensional Gaussian distribution, and step 1–5 represents the distribution of latent variables transformed by the normalising flow layers (i.e., planar flow).

Consider an invertible and smooth mapping function $f : \mathbb{R}^d \rightarrow \mathbb{R}^d$ with inverse $f^{-1} = g$, and a random variable $\mathbf{z}$ with distribution $q(\mathbf{z})$. The transformed variable $\mathbf{z}' = f(\mathbf{z})$ follows a distribution given by:

$$q(\mathbf{z}') = q(\mathbf{z}) \left| \det \frac{\partial f}{\partial \mathbf{z}} \right|^{-1} \tag{1}$$

where the $\det \frac{\partial f}{\partial \mathbf{z}}$ is the Jacobian determinant of f. Therefore, we can obtain a complex multi-modal density by composing multiple invertible mappings to transform the initial, simple and tractable density sequentially, as follows,

$$\mathbf{z}_i = f_i \circ \cdots \circ f_2 \circ f_1(\mathbf{z}_0) \tag{2}$$

$$\ln q_i(\mathbf{z}_i) = \ln q_0(\mathbf{z}_0) - \sum^{i} \ln \left| \det \frac{\partial f_i}{\partial \mathbf{z}_{i-1}} \right| \tag{3}$$

The specific mathematical formulation of the normalising flow function is important and must be chosen with care to allow for efficient gradient computation during training, scalable inference, and efficiency in computing the determinant of the Jacobian. In this study, we leverage the planar flow in [16] as a basic unit of our latent normalising flow net. Specifically, each transformation unit is given by,

$$f(\mathbf{z}) = \mathbf{z} + \mathbf{u}h(w^{\top}\mathbf{z} + b) \tag{4}$$

where $\mathbf{w} \in \mathbb{R}^d$, $\mathbf{u} \in \mathbb{R}^d$ and $b \in \mathbb{R}$ are learnable parameters; $h(\cdot)$ is a smooth element-wise non-linear function with derivative $h'(\cdot)$ (we use tanh in our study) and $\mathbf{z}$ denotes the latent variables sampled from the posterior distribution. Therefore, we could compute the log determinant of the Jacobian term in $O(D)$ time as follows:

$$\phi(\mathbf{z}) = h'(\mathbf{w}^{\top}\mathbf{z} + b)\mathbf{w} \tag{5}$$

$$\left| \det \frac{\partial f_i}{\partial \mathbf{z}_{i-1}} \right| = \left| \det(\mathbf{I} + \mathbf{u}\phi(\mathbf{z})^{\top}) \right| = \left| 1 + \mathbf{u}^{\top}\phi(\mathbf{z}) \right| \tag{6}$$

Finally, the network is trained by optimizing the modified ELBO based on Eq. 3:

$$\ln p(\mathbf{x}|\mathbf{c}) \geq \mathbb{E}_{q(\mathbf{z}_0|\mathbf{x},\mathbf{c})} \left[\ln p(\mathbf{x}|\mathbf{z}_i,\mathbf{c}) + \sum^{i} \ln \left| \det \frac{\partial f_i}{\partial \mathbf{z}_{i-1}} \right| \right] - \mathrm{KL}(q(\mathbf{z}_0|\mathbf{x},\mathbf{c})\|p(\mathbf{z}_i)) \quad (7)$$

where, $\ln p(\mathbf{x}|\mathbf{c})$ is the marginal log-likelihood of the observed data $\mathbf{x}$ (i.e. here $\mathbf{x}$ represents an LV graph/mesh), conditioned on the covariates of interest (i.e. patient demographics and clinical measurements) $\mathbf{c}$; i is the steps of the normalizing flows. $p(\mathbf{x}|\mathbf{z}_i,\mathbf{c})$ is the likelihood of data parameterised by the decoder network, which reconstructs/predicts $\mathbf{x}$ given the latent variables $\mathbf{z}_i$, transformed by latent (planar) normalising flows, and the conditioning variables $\mathbf{c}$; $\mathrm{KL}(q(\mathbf{z}_0|\mathbf{x})\|p(\mathbf{z}_i))$ is the Kullback-Leibler divergence of the approximate posterior initial $q(\mathbf{z}_0|\mathbf{x},\mathbf{c})$ from the prior, $p(z) = \mathcal{N}(z \mid 0, I)$.

3   Experimental Setup and Results

Data: In this study, we created a cohort of 2360 triangular meshes of the left ventricle (LV) based on a subset of cardiac cine-MR imaging data available from the UK Biobank (UKBB) by registering a cardiac LV atlas mesh [17] in manual contours (as described in [23]). We randomly split the data set into 422/59/1879 for training, validation, and testing, respectively. All meshes have the same and fixed graph topology, sharing the same edges and faces but differing in the position of vertices; i.e. there is pointwise correspondence across all shapes. We used 14 covariates available for the same subjects in UKBB as conditioning variables for our model, including, gender, age, height, weight, pulse, alcohol drinker status, smoking status, HbA1c, cholesterol, C-reactive protein, glucose, high-density lipoprotein cholesterol (HDL), insulin-like growth factor 1 (IGF-1), and low-density lipoprotein (LDL) cholesterol. These covariates were chosen because they are known cardiovascular risk factors.

Implementation Details: The framework was implemented using PyTorch on a standard PC with a NVIDIA RTX 2080Ti GPU. We trained our model using the AdamW optimizer with an initial learning rate of 1e–3 and batch size of 16 for 1000 epochs. The feature number for each graph convolutional block in the encoder was 16, 32, 32, 64, 64, and in reverse order in the decoder. The latent dimension was set at 16. The down/up-sampling factor was four, and we used a warm-up strategy [19] to the weight of the KL loss to prevent model collapse.

Evaluation Metrics: We compared our model (cVAE-NF) with a traditional PCA-based SSM, two generative models without conditioning information including a vanilla VAE and a VAE with normalising flow (VAE-NF) and the vanilla cVAE. Comparison of the vanilla cVAE can also validate the performance of existing approaches [1,2] because they are built on the cVAE with different covariates and basic units in the network. We evaluated the performance of all methods using three different metrics: 1) the reconstruction error, which evaluates the generalisability of the trained model to reconstruct/represent unseen shapes, using the distance between the reconstructed mesh with the ground truth/original mesh; 2) the specificity error, which measures the anatomical plausibility of the virtual cohorts synthesised, using the distance between the generated meshes and its nearest neighbour in the unseen real population [6]; and 3) the variability in the left ventricular volume in the synthesised cohorts, to assess the diversity of the instances generated in terms of a clinically relevant cardiac index. The variability in LV volume was quantified as the standard deviation of the volumes of LV blood pools (BPVols). The Euclidean distance was used to evaluate all three metrics. Additionally, we measured the activity of the latent dimension using the statistic $A = \mathrm{Cov}_{\mathbf{x}}(\mathbb{E}_{\mathbf{z}\sim q(\mathbf{z}|\mathbf{x})}[z])$ of the observations $\mathbf{x}$ [4]. A higher activity score indicates that a given latent dimension can capture greater population-wide shape variability.

Table 1. The quantitative results of the investigated methods in a hold-out test dataset. The bold values represent the results are significantly better than those of other methods.

Methods	Reconstruction Error↓	Specificity Error↓	Volume Variability↑
PCA	**0.82 ± 0.16**	1.48 ± 0.26	**32.74**
VAE	1.29 ± 0.21	**1.39 ± 0.98**	3.00
VAE-NF	0.90 ± 1.76	1.60 ± 0.34	16.03
cVAE	1.43 ± 0.26	**1.32 ± 0.21**	28.39
Ours	**1.23 ± 0.23**	1.38 ± 0.20	**29.91**

The results of our method are presented in Table 1. Our model outperforms the cVAE in terms of reconstruction error and the amount of volume variability captured in the synthesised VP (the reference volume variability for the real UKBB population was 33.38 mm^3). However, the cVAE achieved lower specificity errors than our model. This indicates that our method is better at capturing the population's shape variability, but it also creates some instances that are further away from the real population, resulting in higher specificity errors. We attribute this to the normalising flow's ability to learn a more flexible approximate posterior latent distribution of the observed shapes than the cVAE. This is also seen when comparing the performance of VAE and VAE-NF, where the latter can synthesise significantly more diverse VPs (e.g. it improves the volume variability from 3.00 to 16.03). Figure 3 shows the variability captured in each latent dimension. We observe that VAE-NF has higher activity scores in all latent dimensions compared to vanilla cVAE. The normalising flow allows for the approximation of multimodal latent distributions in the generative model, resulting in greater shape variability. Although PCA outperforms our method in terms of generalisation error and volume variability captured, it does not allow for controllable synthesis of VPs based on relevant patient demographic information and clinical measurements, making it less useful for our application of synthesising VPs for use in ISTs.

It is essential to capture the distribution of clinically relevant biomarkers (e.g. BPVol) in the synthesised virtual populations (VPs) based on the specified covariates/conditioning information available for real patients, in order to effectively replicate the inclusion/exclusion criteria used during trial design in ISTs. For example, the BPVol of women is known to be lower than that of men [20]. To verify this, we generated VPs using cVAE and our method, conditioned on real patient data (covariates) from the UK Biobank. Figure 3 summarises the BPVol distribution for both genders in the synthesised VPs and the real UKBB population, and the former accurately reflects the known trend of women having lower BPVol than men. Compared to cVAE, our model generates a VP that more closely matches the distribution of the volume of the LV blood pool observed in the real population. We also visualised the effect of manipulating individual attributes on two real patients in Fig. 4. We selected two representative attributes that are significantly associated with BPVol and myocardial volume (MyoVol): weight and age. We observe that BPVol and MyoVol of the

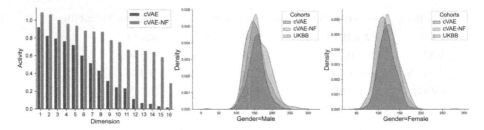

Fig. 3. Left: Comparison of the activity scores in different latent dimensions between the cVAE and cVAE-NF; right: Kernel density plots for BPVol from the VPs generated by cVAE and cVAE-NF and the real patient population (UKBB).

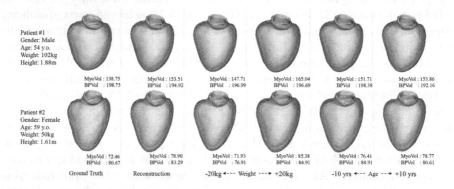

Fig. 4. Two representative examples of the reconstructed shapes and their variations through manipulation over two demographic attributes, i.e., weight and Age. MyoVol and BPVol are shown in the bottom right corner.

LV are positively correlated with the weight of the patients (as expected). On the other hand, increasing the individual's age results in a smaller BPVol, but an increased MyoVol (as visualised in Fig. 4), which is known to be due to cardiac hypertrophy caused by aging [5].

4 Conclusion

We proposed a conditional flow VAE model for the controllable synthesis of VPs of anatomy. Our approach was demonstrated to increase the flexibility of the learnt latent distribution, resulting in VPs that captured greater variability in the LV shape than the vanilla cVAE. Furthermore, our model was able to model the relationship between covariates/conditional variables and the shape of the LV, and synthesise target VPs that fit the desired criteria (in terms of demographics of the patient and clinical measurements) and closely matched the real population in terms of a clinically relevant biomarker (LV BPVol). These results suggest that our approach has potential for the controllable synthesis of diverse, yet plausible, VPs of anatomy. Future work will focus on modelling the whole heart and exploring the impact of individual covariates on VP synthesis in more detail.

Acknowledgement. This research was carried out using data from the UK Biobank (access application 11350). This work was supported by the Royal Academy of Engineering (INSILEX CiET1819/19), Engineering and Physical Sciences Research Council (EPSRC) UKRI Frontier Research Guarantee Programmes (INSIL-ICO, EP/Y030494/1), and the Royal Society Exchange Programme CROSSLINK IES\NSFC\201380.

References

1. Beetz, M., Banerjee, A., Grau, V.: Generating subpopulation-specific biventricular anatomy models using conditional point cloud variational autoencoders. In: Puyol Antón, E., et al. (eds.) STACOM 2021. LNCS, vol. 13131, pp. 75–83. Springer, Cham (2022). https://doi.org/10.1007/978-3-030-93722-5_9
2. Beetz, M., Banerjee, A., Grau, V.: Multi-domain variational autoencoders for combined modeling of mri-based biventricular anatomy and ecg-based cardiac electrophysiology. Front. Physiol. **991** (2022)
3. Bonazzola, R., Ravikumar, N., Attar, R., Ferrante, E., Syeda-Mahmood, T., Frangi, A.F.: Image-derived phenotype extraction for genetic discovery via unsupervised deep learning in CMR images. In: de Bruijne, M., et al. (eds.) MICCAI 2021. LNCS, vol. 12905, pp. 699–708. Springer, Cham (2021). https://doi.org/10.1007/978-3-030-87240-3_67
4. Burda, Y., Grosse, R., Salakhutdinov, R.: Importance weighted autoencoders. arXiv preprint arXiv:1509.00519 (2015)
5. Chiao, Y.A., Rabinovitch, P.S.: The aging heart. Cold Spring Harbor Perspect. Med. **5**(9), a025148 (2015)

6. Davies, R.H., Twining, C.J., Cootes, T.F., Taylor, C.J.: Building 3-d statistical shape models by direct optimization. IEEE Trans. Med. Imaging **29**(4), 961–981 (2009)
7. Dou, H., Virtanen, S., Ravikumar, N., Frangi, A.F.: A generative shape compositional framework: towards representative populations of virtual heart chimaeras. arXiv preprint arXiv:2210.01607 (2022)
8. Frangi, A.F., Rueckert, D., Schnabel, J.A., Niessen, W.J.: Automatic construction of multiple-object three-dimensional statistical shape models: application to cardiac modeling. IEEE Trans. Med. Imaging **21**(9), 1151–1166 (2002)
9. Gooya, A., Davatzikos, C., Frangi, A.F.: A bayesian approach to sparse model selection in statistical shape models. SIAM J. Imaging Sci. **8**(2), 858–887 (2015)
10. Huang, X., Belongie, S.: Arbitrary style transfer in real-time with adaptive instance normalization. In: Proceedings of the IEEE International Conference on Computer Vision, pp. 1501–1510 (2017)
11. Kingma, D.P., Welling, M.: Auto-encoding variational bayes. arXiv preprint arXiv:1312.6114 (2013)
12. Li, L., Camps, J., Banerjee, A., Beetz, M., Rodriguez, B., Grau, V.: Deep computational model for the inference of ventricular activation properties. In: Camara, O., et al. (eds.) Statistical Atlases and Computational Models of the Heart. Regular and CMRxMotion Challenge Papers: 13th International Workshop, STACOM 2022, Held in Conjunction with MICCAI 2022, Singapore, 18 September 2022, Revised Selected Papers, pp. 369–380. Springer, Heidelberg (2023). https://doi.org/10.1007/978-3-031-23443-9_34
13. Ranjan, A., Bolkart, T., Sanyal, S., Black, M.J.: Generating 3d faces using convolutional mesh autoencoders. In: Proceedings of the European Conference on Computer Vision (ECCV), pp. 704–720 (2018)
14. Rasal, R., Castro, D.C., Pawlowski, N., Glocker, B.: Deep structural causal shape models. In: European Conference on Computer Vision, pp. 400–432. Springer, Heidelberg (2022). https://doi.org/10.1007/978-3-031-25075-0_28
15. Ravikumar, N., Gooya, A., Çimen, S., Frangi, A.F., Taylor, Z.A.: Group-wise similarity registration of point sets using student's t-mixture model for statistical shape models. Med. Image Anal. **44**, 156–176 (2018)
16. Rezende, D., Mohamed, S.: Variational inference with normalizing flows. In: International Conference on Machine Learning, pp. 1530–1538. PMLR (2015)
17. Rodero, C., et al.: Linking statistical shape models and simulated function in the healthy adult human heart. PLoS Comput. Biol. **17**(4), e1008851 (2021)
18. Sohn, K., Lee, H., Yan, X.: Learning structured output representation using deep conditional generative models. Adv. Neural Inf. Process. Syst. **28**, 1–9 (2015)
19. Sønderby, C.K., Raiko, T., Maaløe, L., Sønderby, S.K., Winther, O.: Ladder variational autoencoders. Adv. Neural Inf. Process. Syst. **29**, 1–9 (2016)
20. St Pierre, S.R., Peirlinck, M., Kuhl, E.: Sex matters: a comprehensive comparison of female and male hearts. Front. Physiol. **13**, 303 (2022)
21. Tomczak, J.M., Welling, M.: Improving variational auto-encoders using householder flow. arXiv preprint arXiv:1611.09630 (2016)
22. Viceconti, M., Pappalardo, F., Rodriguez, B., Horner, M., Bischoff, J., Tshinanu, F.M.: In silico trials: verification, validation and uncertainty quantification of predictive models used in the regulatory evaluation of biomedical products. Methods **185**, 120–127 (2021)
23. Xia, Y., et al.: Automatic 3d+ t four-chamber CMR quantification of the UK biobank: integrating imaging and non-imaging data priors at scale. Med. Image Anal. **80**, 102498 (2022)

Radiomics-Informed Deep Learning for Classification of Atrial Fibrillation Sub-Types from Left-Atrium CT Volumes

Weihang Dai[1] , Xiaomeng Li[1]([✉]) , Taihui Yu[2,3], Di Zhao[4], Jun Shen[2,3],
and Kwang-Ting Cheng[1]

[1] The Hong Kong University of Science and Technology, Hong Kong, China
eexmli@ust.hk
[2] Sun Yat-Sen University, Guangzhou, China
[3] Sun Yat-Sen Memorial Hospital, Guangzhou, China
[4] Chinese Academy of Sciences, Beijing, China

Abstract. Atrial Fibrillation (AF) is characterized by rapid, irregular heartbeats, and can lead to fatal complications such as heart failure. The disease is divided into two sub-types based on severity, which can be automatically classified through CT volumes for disease screening of severe cases. However, existing classification approaches rely on generic radiomic features that may not be optimal for the task, whilst deep learning methods tend to over-fit to the high-dimensional volume inputs. In this work, we propose a novel radiomics-informed deep-learning method, RIDL, that combines the advantages of deep learning and radiomic approaches to improve AF sub-type classification. Unlike existing hybrid techniques that mostly rely on naïve feature concatenation, we observe that radiomic feature selection methods can serve as an information prior, and propose supplementing low-level deep neural network (DNN) features with locally computed radiomic features. This reduces DNN over-fitting and allows local variations between radiomic features to be better captured. Furthermore, we ensure complementary information is learned by deep and radiomic features by designing a novel feature de-correlation loss. Combined, our method addresses the limitations of deep learning and radiomic approaches and outperforms state-of-the-art radiomic, deep learning, and hybrid approaches, achieving 86.9% AUC for the AF sub-type classification task. Code is available at https://github.com/xmed-lab/RIDL.

Keywords: Atrial Fibrillation · Radiomics · CT Imaging Analysis

1 Introduction

Atrial fibrillation (AF) is a cardiac disease characterized by rapid, irregular heartbeats [4]. The disease can lead to stroke and heart failure, and has a mortal-

Supplementary Information The online version contains supplementary material available at https://doi.org/10.1007/978-3-031-43990-2_15.

ity rate of almost 20% [5,10,13]. AF is classified as either persistent atrial fibrillation (PeAF), where abnormal heart rhythms occur continuously for more than
seven days, or paroxysmal atrial fibrillation (PaAF), where the heart rhythm
returns to normal within seven days. Although AF can be treated through a
procedure called catheter ablation, PeAF cases have high recurrence rates and
often require re-intervention [8]. Accurate knowledge of the disease type is therefore highly valuable for treatment planning and has high prognostic value [22].

Clinical studies have discovered a strong relationship between AF and epicardial adipose tissue (EAT), a fat depot layer on the surface of the myocardium
that can cause inflammation and disrupt cardiac function [3,15]. Recent works
have shown that automatic classification of AF sub-types can be done using CT
volumes of the left atrium and surrounding EAT, which can be used to screen
for patients with high risk of PeAF. Huber *et al.* [7] showed that EAT volume,
approximated from left-atrium CT images, can be used as a predictor for AF
recurrence. Yang *et al.* [22] trained a random forest model to classify AF sub-
type based on radiomic features and volume measurements, achieving 85.3%
AUC. Although these methods demonstrate the usefulness of radiomic features
for AF sub-type classification, such features are generic and not specific to the
task, which can limit model performance [12]. Radiomic features also rely on
summary statistics such as entropy or homogeneity to obtain global descriptors,
and these have limited effectiveness when capturing local feature variations [16].

Deep learning has achieved outstanding results on medical imaging analysis
tasks, largely due to its ability to learn task-specific features and complex relations between them [17]. Naïvely using deep neural networks (DNNs) to predict
AF sub-types from CT volumes yields poor results however due to over-fitting
on high-dimensional volume inputs (see results for DNN in Table 1). Existing
works have attempted to combine deep and radiomic features through methods
such as direct concatenation [2,19], attention modules [14], or contrastive learning between feature types [24]. Although these methods propose different ways
of using both approaches, *they do not explicitly address the limitations of either
approach or explore ways to combine their complementary advantages.*

In this work, we propose a novel approach to atrial fibrillation sub-type classification from CT volumes by integrating radiomic and deep learning methods.
We note that textural radiomic features identified by feature selection methods
can serve as an information prior to supplement low-level features from DNNs,
since they are designed to capture low-level context and have predictive power
[23]. To this end, we *locally* calculate radiomic features based on patches surrounding each voxel, and perform feature fusion with low-level DNN features.
This provides the DNN with pre-defined features known to be relevant to the
task to reduce over-fitting, and also allows spatial relations between radiomic
features to be learned. Furthermore, we encourage the DNN to learn features
complementary to radiomic features to obtain more comprehensive signals and
design a novel feature de-correlation loss. The overall framework, which we term
Radiomics-**I**nformed **D**eep **L**earning (RIDL), is illustrated in Fig. 1. Unlike existing works, our method is designed to directly addresses the limitations of both
deep learning and radiomic approaches and achieves state-of-the-art performance
on AF sub-type classification. To summarize our key contributions:

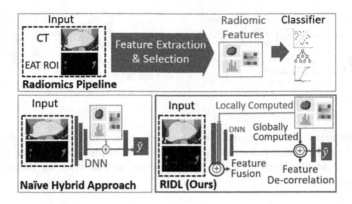

Fig. 1. Overview of our method compared with alternative approaches. Radiomic modeling uses feature-selection algorithms to select predictive features. Naïve hybrid approaches directly concatenates these radiomic with deep features. Our method also fuses *locally computed* radiomic features with low-level DNN features and encourages *complementary* deep and radiomic features to be learned.

- We propose a novel radiomics-informed deep learning (RIDL) method for AF sub-type classification from CT volumes, which achieves state-of-the-art results and can be used to screen for patients with high risk of PeAF.
- Our method uses a novel approach of fusing *locally computed* radiomic features with low-level DNN features to improve capturing of local context.
- Furthermore, we enforce feature de-correlation using a novel feature-bank design to ensure *complementary* deep and radiomic features are extracted.

2 Methodology

We combine radiomic and deep learning approaches using two novel components: 1) feature fusion of local radiomic features and low-level DNN features to improve local context, 2) encouraging complementary deep and radiomic features through feature de-correlation. These are illustrated in Fig. 2 and explained in detail below. Our dataset $\mathcal{D} := \{(x_i, y_i)\}_{i=1}^{N}$ includes N samples of input x_i and binary label y_i, where 0 indicates PaAF and 1 indicates PeAF. x_i has two channels, one consisting of the 3D CT volume centered around the left atrium and the other the binary region-of-interest (ROI) mask indicating EAT. The ROI is obtained through Hounsfield value thresholding such that all voxels valued between -250 and 0 are identified as EAT [7,22].

2.1 Feature Fusion of Locally Computed Radiomic Features with Low-Level DNN Features for Improved Local Context

Under the radiomics pipeline, a large set of features, typically more than a thousand, is first extracted by performing calculations over the volume and ROI input

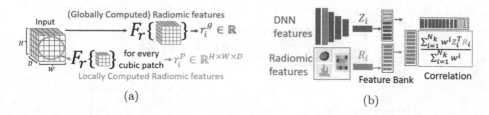

Fig. 2. Illustration of the two main components of our method. (a) Calculation of global *v.s.* local radiomic features. (b) Feature bank implementation for feature decorrelation. Light blue features indicate features in current training iteration (Color figure online)

x_i. Feature selection methodologies such as mutual information (MI), principal component analysis (PCA), or LASSO regularization, are then used to identify predictive features for classification [23]. Radiomic features are classified into shape, first-order statistical features, and texture features. Texture features are designed to capture local variations and use measures such as Gray-level Co-occurrence Matrices (GLCM) to reflect second-order textural distributions. Conventional statistics such as entropy and correlation are then used to summarize these measures [25], but these tend to be limited in their ability to capture local heterogeneity, such as the varying textures on the surface of a cancer tumor. Although DNN's are more effective at capturing local variations, they can overfit without sufficient data for training [17].

Unlike existing works that naïvely concatenate radiomic and deep features before the classification layer [2,19], we observe that textural features selected through radiomics feature selection algorithms are known to be predictive and *can be used as prior knowledge* to improve low-level DNN features. Given radiomic feature extractor F_r, the global radiomic feature, $r_i^g \in \mathcal{R}$, for input x_i is represented by:

$$r_i^g = F_r(x_i) . \tag{1}$$

Our method applies feature calculations *locally* to cubic patches centered around each voxel, such that features are obtained on a voxel basis and reflect the statistics of the neighbouring region. For a cubic patch with radius p and input x_i, the local feature at location (h, w, d), denoted by $r_{i,(h,w,d)}^p$, is obtained by performing R on the cubic patch in x_i centered around (h, w, d):

$$r_{i,(h,w,d)}^p = F_r(x_{i,[h-p:h+p,w-p:w+p,d-p:d+p]}) , \tag{2}$$

where the input of F_r is the cubic sub-volume. This process is illustrated in Fig. 2a. Local features can be calculated for multiple texture features and patch size p, which are then concatenated to obtain $r_i^l \in \mathcal{R}^{L \times H \times W \times D}$, where L is the total number of features used and H, W, and D are original input dimensions. We note that only texture radiomic features are used for local calculation since they are specifically intended to capture local context.

r_i^l is then concatenated with low-level DNN features, $z_i \in \mathcal{R}^{C \times H \times W \times D}$, to supplement the DNN with local radiomic features. To effectively fuse the features, we apply a channel attention module, $\mathcal{A}$, following the design in [20]:

$$z_i' = \mathcal{A}(r_i^l \oplus z_i) \otimes (r_i^l \oplus z_i) \tag{3}$$

where z_i' is the fused feature, $\oplus$ is channel concatenation, and $\otimes$ is element-wise multiplication. The learned attention tensor $\mathcal{A}(r_i^l \oplus z_i)$ has dimensions $(C + L) \times 1 \times 1 \times 1$ and is broadcasted along the volume dimension, such that attention is applied channel-wise and spatial feature distributions are preserved.

2.2 Encouraging Complementary Deep and Radiomic Features Through Feature De-Correlation

Global radiomic features are also included in our model by concatenation with high-level DNN features before the classification layer. Unlike existing approaches however, we encourage our DNN to learn features *complementary* to radiomic features by enforcing de-correlation between the two. This ensures that different variations are captured, which provides a more comprehensive signal to the classification layer.

Accurate approximation of correlation requires large batches sizes however, which requires large GPU memory and can affect model convergence [9]. We instead propose a novel feature-bank implementation with exponential weighting to estimate sample statistics. Every iteration, we save DNN and global radiomic features in feature-bank K, which holds up to N_k features in a first-in first-out queue. After a warm-up period, we calculate the sample correlation using an exponential weighting scheme. Given weight parameter $w < 1$, and the normalized deep feature Z_i and radiomic feature R_i from K, we calculate feature de-correlation loss $\mathcal{L}_{corr}$ as:

$$\mathcal{L}_{corr} = ||\frac{\sum_{i=1}^{N_k} w^i Z_i^T R_i}{\sum_{i=1}^{N_k} w^i}||_1 . \tag{4}$$

The first B samples, where B is the batch size, belong to the training sample of the current iteration, and their losses are back-propagated to encourage deep features to have zero correlation with radiomic features. This process is illustrated in Fig. 2b. Although feature banks have been used in techniques such as contrastive learning to address batch size limitations [6,21], we are the first to formulate this technique *for feature de-correlation*.

2.3 Overall Framework

The DNN model uses raw CT volumes concatenated with ROI masks as input. Global and local radiomic features are pre-computed for input into the feature layer. Binary cross-entropy is used for AF sub-type classification loss $\mathcal{L}_{cls}$:

$$\mathcal{L}_{cls} = \frac{1}{N} \sum_{i=1}^{N} [y_i \ln(\hat{y}_i) + (1 - y_i) \ln(1 - \hat{y}_i)] \tag{5}$$

where $\hat{y}_i$ is the model prediction for sample y_i. The model is trained together with feature de-correlation loss $\mathcal{L}_{corr}$ and its loss weighting, w_{corr}. To provide further regularization and prevent over-fitting, we perform an additional self-reconstruction task, using loss $\mathcal{L}_{rec}$, which we describe in more detail in the supplementary materials. The overall loss function is then:

$$\mathcal{L} = \mathcal{L}_{cls} + w_{corr}\mathcal{L}_{corr} + \mathcal{L}_{rec}. \tag{6}$$

3   Experiments

3.1   Implementation Details

Dataset. We use a dataset of 172 patients containing 94 PaAF and 78 PeAF cases collected from the Sun Yat-Sen Memorial Hospital in China. CT volumes are centered on the left atrium and normalized to between -1 and 1. ROI masks for EAT are obtained through Hounsfield value thresholding between -250 and 0. Volumes are resized to the same aspect ratio to ensure consistent dimensions across samples. We use an input size of $96 \times 128 \times 128$ voxels and apply zero padding for smaller volumes. We use five-fold cross-validation and report average test performance across folds. Cross-validation is implemented by splitting the dataset into five equal subsets and using three subsets for training, one subset for validation, and one subset for testing. A rolling scheme is used such that different validation and test subsets are used for each of the five folds. Data acquisition procedures and statistics are given in the supplementary materials.

Setup. We use the PyRadiomic package [18] to extract radiomic features from the input volumes and masks. Using the cross-validation splits, we perform feature selection and classification using LASSO regularized logistic regression. LASSO regularization consistently selects four radiomic features as the ones with the most significant predictive power: maximum 3D diameter, Maximum 2D Diameter, Maximum voxel value, and normalized inverse difference of GLCM (glcm_Idn). The texture feature glcm_Idn is calculated locally for $p \in \{1, 2, 5, 10\}$ to obtain local radiomic features $r_i^l \in \mathcal{R}^{4 \times 96 \times 128 \times 128}$.

For our DNN network, we use a modified 3D U-Net [1] (abbreviated as m3DUNet) with skip connections between the encoder and decoder removed to enhance bottle-neck feature compression. Bottle-neck features are averaged across spatial dimensions for classification, whilst decoder outputs are used for self-reconstruction regularization. The model is trained using the Adam optimizer with learning rate 10^{-4} for 100 epochs and 0.1 decay at 30 epochs. We use batch size $B = 1$, feature bank size $N_k = 25$, and warm-up period of one epoch. We use $w_{corr} = 2$ for de-correlation loss weighting, which was chosen based on the validation splits. Mean and standard deviation of ten runs are reported. Additional experiments and details are included in the supplementary materials.

3.2   Comparison with State-of-the-Art Methodologies

We compare our method with alternative state-of-the-art approaches based on radiomics, deep learning, and hybrid techniques. Deep and hybrid volume-based classification methods [11,14,24] are adapted to our task since there are no existing works for AF sub-type classification. We use the same encoder for all deep architectures for fair comparison, except for methods that are architecture specific. A naïve feature concatenation method is used as our baseline for the hybrid approach. Radiomic features for the hybrid approach are selected through LASSO regularization as it is the most effective. Results are shown in Table 1.

Table 1. Comparison with state-of-the-art methods for radiomic, deep learning, and hybrid approaches. Selector* refers to the feature selection method. Hybrid# methods use radiomic features selected by LASSO regularization, which is the most effective. DNN* is a naïve implementation using the m3DUNet model. Baseline† is a naïve hybrid implementation using simple feature concatenation.

Type	Selector*	Classifier	AUC (%)	MAP (%)	F1 (%)	Acc. (%)
Radiomic	Mutual Information	SVM	74.2	65.2	**71.1**	**74.7**
		Random Forest [22]	72.4	63.4	70.5	72.4
		Logistic Regression	68.7	59.1	65.7	69.3
	LASSO	Logistic Regression	**83.4**	**81.7**	69.1	73.9

Type	Method	Model	AUC (%)	MAP (%)	F1 (%)	Acc. (%)
Deep Learning	Lee *et al.* [11]	f-rMC5	63.3 ± 6.3	64.8 ± 3.9	31.5 ± 7.6	63.3 ± 4.3
	DNN*	m3DUNet	**77.2 ± 1.5**	**73.7 ± 1.4**	**68.4 ± 1.4**	**70.7 ± 1.8**

Type	Method	Model	AUC (%)	MAP (%)	F1 (%)	Acc. (%)
Hybrid#	Zhao *et al.* [24]	m3DUNet	85.4 ± 0.1	85.2 ± 0.3	70.6 ± 1.2	73.5 ± 1.0
	TMSS [14]	TMSS	84.0 ± 0.5	82.3 ± 0.2	71.6 ± 3.2	74.5 ± 1.2
	Baseline†	m3DUNet	85.8 ± 0.5	84.5 ± 0.6	73.9 ± 1.6	75.7 ± 1.4
	RIDL (ours)	m3DUNet	**86.9 ± 0.6**	**86.3 ± 0.6**	**74.7 ± 1.5**	**76.9 ± 1.0**

We can see that hybrid methods outperform radiomic and deep methods in general. Our method, RIDL, achieves the best results across all metrics however and improves AUC by 1.1% over the baseline method (86.9% *v.s.* 85.8%) and 3.5% over the best radiomics approach (86.9% *v.s.* 83.4%).

3.3   Ablation

Component Analysis. We perform ablation experiments to demonstrate improvements from using local radiomic features, global radiomic features, and feature de-correlation loss. Results are shown in Table 2.

We can see that including local radiomic features, improving AUC by up to 1.6% when included with a standard DNN (78.8% *v.s.* 77.2%). Using feature de-correlation further boosts performance and leads to the best overall results.

Table 2. Ablation study for proposed method. ✓ indicates use of global radiomic features (R_i^g), local radiomic features (r_i^l) and feature de-correlation loss ($\mathcal{L}_{corr}$). DNN* is a naïve implementation using the m3DUNet model. Baseline† is a naïve hybrid implementation using simple feature concatenation.

Method	R_i^g	r_i^l	$\mathcal{L}_{corr}$	AUC (%)	MAP (%)
DNN*				77.2 ± 1.5	73.7 ± 1.4
DNN* + Local Radiomic Features		✓		78.8 ± 1.1	74.9 ± 1.1
Baseline†	✓			85.8 ± 0.5	84.5 ± 0.6
Baseline† + Local Radiomic Features	✓	✓		86.3 ± 0.9	84.9 ± 8.0
RIDL (ours)	✓	✓	✓	$\mathbf{86.9 \pm 0.6}$	$\mathbf{86.3 \pm 0.6}$

Effectiveness of Radiomic Feature Selection. To demonstrate the effectiveness of radiomic feature selection as prior knowledge for feature fusion, we compare with results from using features discarded by radiomics feature selection. We randomly select three discarded features to generate local features r_i^l as input whilst keeping other components constant. Results are shown in Table 3.

Table 3. Results using different texture radiomic features for input r_i^l as displayed by their PyRadiomics key [18]. "Selected" indicates whether the feature was selected or discarded by the radiomic feature selection algorithm.

Feature used for local calculation	Selected	AUC (%)	MAP (%)
gldm_DependenceNonUniformityNormalized	✗	86.1 ± 0.8	85.5 ± 0.9
glrlm_LongRunEmphasis	✗	86.4 ± 0.8	85.4 ± 0.8
gldm_LargeDependenceEmphasis	✗	85.6 ± 0.7	84.6 ± 0.9
glcm_IDN (ours)	✓	$\mathbf{86.9 \pm 0.6}$	$\mathbf{86.3 \pm 0.6}$

We can see that using discarded features leads to worse performance in general. Given the large set of radiomic features, it is possible some discarded features may outperform selected features due to differences in global and local computation. Nevertheless, our results indicate that the radiomic feature selection process serves as an reasonable information prior. Our work is the first to propose *fusing locally computed radiomic features with low-level DNN features*, and we leave detailed local feature selection methods to future works.

4   Conclusion

In this work, we propose a new approach to atrial fibrillation sub-type classification from CT volumes by integrating radiomic and deep learning approaches through a radiomics-informed deep learning method, RIDL. Our method is based on two key ideas: feature fusion of locally computed radiomic features with low-level DNN features to improve local context, and encouraging complementary deep and radiomic features through feature de-correlation. Unlike existing hybrid approaches, our method specifically addresses the advantages and limitations of both techniques to improve feature extraction. We achieve state-of-the-art results on AF sub-type classification and outperform existing radiomic, deep learning, and hybrid methods.

Future improvements to RIDL can be made by introducing more sophisticated local radiomic features selection methods, given the large set features to choose from. Experiments on larger datasets or alternative tasks can also be done to provide more empirical support, since current results show only slight improvements over baseline. These issues may be addressed in future works. Overall, our method is a novel way of combining radiomic and deep learning approaches, and can be used to improve accuracy of PeAF screening from CT volumes for better preventive care of high-risk patients.

Acknowledgement. This work was supported in part by grants from Hong Kong Innovation and Technology Commission (Project no. ITS/030/21 & Project no. PRP/041/22FX), and by Foshan HKUST Projects under FSUST21-HKUST10E and FSUST21-HKUST11E.

References

1. Çiçek, Ö., Abdulkadir, A., Lienkamp, S.S., Brox, T., Ronneberger, O.: 3D U-Net: learning dense volumetric segmentation from sparse annotation. In: Ourselin, S., Joskowicz, L., Sabuncu, M.R., Unal, G., Wells, W. (eds.) MICCAI 2016. LNCS, vol. 9901, pp. 424–432. Springer, Cham (2016). https://doi.org/10.1007/978-3-319-46723-8_49

2. Cui, Y., et al.: A ct-based deep learning radiomics nomogram for predicting the response to neoadjuvant chemotherapy in patients with locally advanced gastric cancer: A multicenter cohort study. EClinicalMedicine **46**, 101348 (2022)

3. Gaeta, M., et al.: Is epicardial fat depot associated with atrial fibrillation? a systematic review and meta-analysis. Europace **19**(5), 747–752 (2017)

4. Go, A.S., et al.: Prevalence of diagnosed atrial fibrillation in adults: national implications for rhythm management and stroke prevention: the anticoagulation and risk factors in atrial fibrillation (atria) study. Jama **285**(18), 2370–2375 (2001)

5. Gomez-Outes, A., Lagunar-Ruiz, J., Terleira-Fernandez, A.I., Calvo-Rojas, G., Suárez-Gea, M.L., Vargas-Castrillon, E.: Causes of death in anticoagulated patients with atrial fibrillation. J. Am. Coll. Cardiol. **68**(23), 2508–2521 (2016)

6. He, K., Fan, H., Wu, Y., Xie, S., Girshick, R.: Momentum contrast for unsupervised visual representation learning. In: CVPR, pp. 9729–9738 (2020)

7. Huber, A.T., et al.: The relationship between enhancing left atrial adipose tissue at ct and recurrent atrial fibrillation. Radiology **305**(1), 56–65 (2022)

8. January, C.T., et al.: 2014 aha/acc/hrs guideline for the management of patients with atrial fibrillation: a report of the American college of cardiology/american heart association task force on practice guidelines and the heart rhythm society. J. Am. Coll. Cardiol. **64**(21), e1–e76 (2014)

9. Keskar, N.S., Mudigere, D., Nocedal, J., Smelyanskiy, M., Tang, P.T.P.: On large-batch training for deep learning: Generalization gap and sharp minima. arXiv preprint arXiv:1609.04836 (2016)

10. Lee, H.Y., et al.: Atrial fibrillation and the risk of myocardial infarction: a nationwide propensity-matched study. Sci. Rep. **7**(1), 12716 (2017)
11. Lee, J., et al.: Moving from 2d to 3d: volumetric medical image classification for rectal cancer staging. In: Wang, L., Dou, Q., Fletcher, P.T., Speidel, S., Li, S. (eds.) MICCAI 2022. LNCS, vol. 13433, pp. 780–790. Springer, Heidelberg (2022). https://doi.org/10.1007/978-3-031-16437-8_75
12. Li, Q., et al.: A fully-automatic multiparametric radiomics model: towards reproducible and prognostic imaging signature for prediction of overall survival in glioblastoma multiforme. Sci. Rep. **7**(1), 14331 (2017)
13. Pastori, D., et al.: Incidence of myocardial infarction and vascular death in elderly patients with atrial fibrillation taking anticoagulants: relation to atherosclerotic risk factors. Chest **147**(6), 1644–1650 (2015)
14. Saeed, N., Sobirov, I., Al Majzoub, R., Yaqub, M.: Tmss: An end-to-end transformer-based multimodal network for segmentation and survival prediction. In: Wang, L., Dou, Q., Fletcher, P.T., Speidel, S., Li, S. (eds.) MICCAI 2022. LNCS, vol. 13437, pp. 319–329. Springer, Heidelberg (2022). https://doi.org/10.1007/978-3-031-16449-1_31
15. Shamloo, A.S., et al.: Is epicardial fat tissue associated with atrial fibrillation recurrence after ablation? a systematic review and meta-analysis. IJC Heart Vascul. **22**, 132–138 (2019)
16. Sun, Q., et al.: Deep learning vs. radiomics for predicting axillary lymph node metastasis of breast cancer using ultrasound images: don't forget the peritumoral region. Front. Oncol. **10**, 53 (2020)
17. Truhn, D., Schrading, S., Haarburger, C., Schneider, H., Merhof, D., Kuhl, C.: Radiomic versus convolutional neural networks analysis for classification of contrast-enhancing lesions at multiparametric breast mri. Radiology **290**(2), 290–297 (2019)
18. Van Griethuysen, J.J., et al.: Computational radiomics system to decode the radiographic phenotype. Canc. Res. **77**(21), e104–e107 (2017)
19. Wang, S., et al.: A deep learning radiomics model to identify poor outcome in covid-19 patients with underlying health conditions: a multicenter study. IEEE J. Biomed. Health Inf. **25**(7), 2353–2362 (2021)
20. Woo, S., Park, J., Lee, J.Y., Kweon, I.S.: CBAM: convolutional block attention module. In: Proceedings of the European Conference on Computer Vision (ECCV), pp. 3–19 (2018)
21. Wu, Z., Xiong, Y., Yu, S.X., Lin, D.: Unsupervised feature learning via nonparametric instance discrimination. In: Proceedings of the IEEE Conference on Computer Vision and Pattern Recognition, pp. 3733–3742 (2018)
22. Yang, M., et al.: Development and validation of a machine learning-based radiomics model on cardiac computed tomography of epicardial adipose tissue in predicting characteristics and recurrence of atrial fibrillation. Front. Cardiovasc. Med. **9**, 813085 (2022)
23. Zhang, X., et al.: Deep learning with radiomics for disease diagnosis and treatment: challenges and potential. Front. Oncol. **12**, 773840 (2022)
24. Zhao, Z., Yang, G.: Unsupervised contrastive learning of radiomics and deep features for label-efficient tumor classification. In: de Bruijne, M., et al. (eds.) MICCAI 2021. LNCS, vol. 12902, pp. 252–261. Springer, Cham (2021). https://doi.org/10.1007/978-3-030-87196-3_24
25. Zwanenburg, A., et al.: The image biomarker standardization initiative: standardized quantitative radiomics for high-throughput image-based phenotyping. Radiology **295**(2), 328–338 (2020)

A Spatial-Temporally Adaptive PINN Framework for 3D Bi-Ventricular Electrophysiological Simulations and Parameter Inference

Yubo Ye[1], Huafeng Liu[1,3,4]($\boxtimes$), Xiajun Jiang[2], Maryam Toloubidokhti[2], and Linwei Wang[2]

[1] State Key Laboratory of Modern Optical Instrumentation, Department of Optical Engineering, Zhejiang University, Hangzhou 310027, China
liuhf@zju.edu.cn
[2] Rochester Institute of Technology, Rochester, NY 14623, USA
[3] Jiaxing Key Laboratory of Photonic Sensing and Intelligent Imaging, Jiaxing 314000, China
[4] Intelligent Optics and Photonics Research Center, Jiaxing Research Institute, Zhejiang University, Jiaxing 314000, China

Abstract. Physics-informed neural networks (PINNs) is a new paradigm for solving the forward and inverse problems of partial differential equations (PDEs). Its penetration into 3D bi-ventricular electrophysiology (EP) however has been slow, owing to its fundamental limitations to solve PDEs over large or complex solution domains with sharp transitions. In this paper, we propose a new PINN framework to overcome these challenges via three key innovations: 1) a weak-form PDE residual to bypass the challenges of high-order spatial derivatives over irregular spatial domains, 2) a spatial-temporally adaptive training strategy to mitigate the failure of PINN to propagate correct solutions and accelerate convergence, and 3) a sequential learning strategy to enable solutions over longer time domains. We experimentally demonstrated the effectiveness of the presented PINN framework to obtain the complete forward and inverse EP solutions over the 3D bi-ventricular geometry, which is otherwise not possible with vanilla PINN frameworks.

Keywords: Physics-informed neural network · Parameter Inference · Cardiac electrophysiology · Spatial-temporally adaptive training

1 Introduction

Virtual models of cardiac electrophysiology (EP) have demonstrated significant potential in various clinical tasks, such as stratifying the risk for lethal arrhythmias [2] and predicting responses to cardiac resynchronization therapy [18]. Numerical solutions to the governing partial differential equations (PDEs) for cardiac EP (forward problem), however, are difficult to obtain. The inverse

© The Author(s), under exclusive license to Springer Nature Switzerland AG 2023
H. Greenspan et al. (Eds.): MICCAI 2023, LNCS 14226, pp. 163–172, 2023.
https://doi.org/10.1007/978-3-031-43990-2_16

estimation of the parameters of these PDEs given observation data (inverse problem) is even more difficult, due to challenges such as the complex relation between the PDE parameters and the observations, and the need to embed a numerical PDE solver within the inverse optimization process.

Physics-informed neural networks (PINNs) is a new paradigm for solving both the forward and inverse problems of PDEs [16]. The general idea of PINNs is to train neural networks to satisfy physical laws as described by the PDEs, optimized via the so-called PDE residuals. When partial observations of the PDE solutions are available, this PDE residual can be combined with data-residuals to solve the forward and inverse problems at the same time [4,11,13,15,23]. Despite substantial attention and successes, however, the use of PINN in cardiac EP has been rather limited. To date, there have only been two attempts of PINN-based EP simulation that are limited to 1D/2D [9] or atrial surfaces [17].

What fundamental challenges does 3D bi-ventricular EP present for PINNs? First, 3D bi-ventricular EP simulation requires the PDE solutions to be obtained over a complex geometry domain in space and a long duration in time. Second, the PDE solution to bi-ventricular EP – in the form of the spatiotemporal propagation of transmembrane potential (TMP) activation – exhibits sharp spatial and temporal gradients. These characteristics of bi-ventricluar EP simulations present fundamental challenges to the state-of-the-art PINN framework, which has been shown to face potential failure modes when the PDEs are solved over large or complex solution domains with sharp transitions [8,10,14,21,22].

In this paper, we present a novel PINN framework to overcome these challenges and enable the forward and inverse solutions to 3D bi-ventricular EP simulations. This is achieved with three key innovations. First, to avoid dealing with higher-order spatial derivatives over the complex 3D geometry, we formulated the PINN over the meshfree representation of the 3D bi-ventricular geometry with a modified PDE residual incorporating the weak form of the original PDE. Second, to enable PINN solutions over the long temporal domain, we present a temporally adaptive and sequential training strategy to guide the PINN to respect the causality of the underlying physics of wave propagation. Finally, we introduce a spatially adaptive training strategy to guide the PINN to exploit the spatiotemporal sparsity of the sharp gradients exhibited in the PDE solution.

We experimentally demonstrated that the presented PINN framework was able to enable complete simulation of the bi-ventricular EP activation process that is otherwise not possible with vanilla PINN frameworks. We further conducted detailed ablation studies of the benefits of the presented spatial-temporally adaptive and sequential learning strategies, and demonstrated preliminary feasibility of the presented PINN framework for supporting the inverse parameter estimation of 3D bi-ventricular EP models. These represent an innovative first attempt to enable PINN solutions for 3D bi-ventricular EP applications.

2   Background: Bi-Ventricular EP Simulations

Existing ventricular EP models range from macroscopic-level two-variable PDEs to ionic models with tens of variables [5]. In this proof-of-concept study, we consider the two-variable diffusion-reaction Aliev-Panfilov (AP) PDE [1] for its ability to reproduce key excitation features without formidable computation:

$$\frac{\partial u}{\partial t} = \nabla(\mathbf{D}\nabla u) + f_1(u,v), \quad f_1(u,v) = cu(u-\gamma)(u-1) - uv \tag{1}$$

$$\frac{\partial v}{\partial t} = f_2(u,v), \quad f_2(u,v) = (e_0 + (\mu_1 v)/(u+\mu_2))(-v - cu(u-\gamma-1)) \tag{2}$$

$$\frac{\partial u}{\partial n} = 0: \text{natural boundary condition} \tag{3}$$

where $u \in [0,1]$ is the unit-less TMP and v is the recovery current. The diffusion tensor $\mathbf{D}$ describes local conductivity anisotropy determined by fiber structures. Parameters $\{\gamma, c, e_0, \mu_1, \mu_2\}$ control the temporal dynamics of u and v.

3   Methodology

We present a novel PINN framework to support forward and inverse 3D bi-ventricular EP simulations. As outlined in Fig. 1, it includes three key components: 1) a weak-form PDE residual to bypass the challenges of high-order spatial derivatives over irregular geometry; 2) a novel spatial-temporally adaptive strategy to mitigate PDE propagation failure over large solution domains and accelerate convergence; and 3) a sequential training strategy to enable solutions over the complete bi-ventricular EP activation process.

3.1   Weak-Form PDE Residual over Meshfree Representations

In a vanilla PINN framework, we will use a neural network $u_\theta(x,t)$ to approximate PDE solutions $u(x,t)$ to the AP model. The network will be optimized by the initial, boundary, and PDE residuals ($Loss_I$, $Loss_B$ and $Loss_R$) over a set of points $\{t_k\}_{k=1}^{N_T}$, $\{x_i\}_{i=1}^{N_\Omega}$, $\{x_j\}_{j=1}^{N_{\partial\Omega}}$ sampled in time, space, and boundary domains:

$$Res_{1,ik} = \frac{\partial u_\theta(x_i, t_k)}{\partial t} - \nabla(\mathbf{D}\nabla u_\theta(x_i, t_k)) - f_1(u_\theta(x_i, t_k), v_\theta(x_i, t_k)) \tag{4}$$

$$Res_{2,ik} = \frac{\partial v_\theta(x_i, t_k)}{\partial t} - f_2(u_\theta(x_i, t_k), v_\theta(x_i, t_k)) \tag{5}$$

$$Loss_R = \frac{1}{N_\Omega N_T} \sum_{i=1}^{N_\Omega} \sum_{k=1}^{N_T} |Res_{1,ik}|^2 + |Res_{2,ik}|^2 \tag{6}$$

$$Loss_I = \frac{1}{N_\Omega} \sum_{i=1}^{N_\Omega} |u(x_i, 0) - u_\theta(x_i, 0)|^2 \tag{7}$$

$$Loss_B = \frac{1}{N_{\partial\Omega} N_T} \sum_{j=1}^{N_{\partial\Omega}} \sum_{k=1}^{N_T} \left| \frac{\partial u_\theta(x_j, t_k)}{\partial n} \right|^2 \tag{8}$$

$$Loss = \lambda_I Loss_I + \lambda_B Loss_B + \lambda_R Loss_R \tag{9}$$

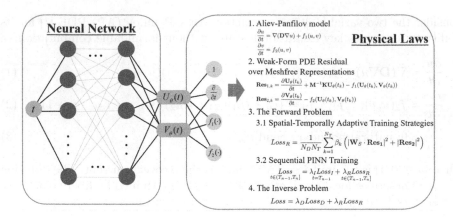

Fig. 1. The schematics of the presented PINN framework

where λ_I, λ_B and λ_R are the hyperparameters that balance these residuals during training. Because $u_\theta(x,t)$ lives on the 3D geometry of the ventricles, it is non-trivial to calculate the second-order spatial derivatives.

To address this, we utilize the weak form of PDE and spatially discretize it on a bi-ventricular mesh through the Mesh-free method [19] as:

$$\frac{\partial \mathbf{U}(t)}{\partial t} = -\mathbf{M}^{-1}\mathbf{K}\mathbf{U}(t) + f_1(\mathbf{U}(t),\mathbf{V}(t)), \quad \frac{\partial \mathbf{V}(t)}{\partial t} = f_2(\mathbf{U}(t),\mathbf{V}(t)) \quad (10)$$

where vectors $\mathbf{U} = [u_1, u_2, \ldots, u_{N_\Omega}]^T$ and $\mathbf{V} = [v_1, v_2, \ldots, v_{N_\Omega}]^T$ consist of u and v from all N_Ω mesh-free points inside the myocardium. Matrices $\mathbf{M}$ and $\mathbf{K}$ represent numerical approximations of the second-order spatial derivatives in Eq. (1), which also automatically incorporate the natural boundary condition [19,24]. We then let the PINN neural network describe the PDE solutions over the discrete ventricular mesh as $\mathbf{U}_\theta(t)$ and $\mathbf{V}_\theta(t)$ as a function of time t, and obtain a modified PDE residual as:

$$\mathbf{Res}_{1,k} = \frac{\partial \mathbf{U}_\theta(t_k)}{\partial t} + \mathbf{M}^{-1}\mathbf{K}\mathbf{U}_\theta(t_k) - f_1(\mathbf{U}_\theta(t_k),\mathbf{V}_\theta(t_k)) \quad (11)$$

$$\mathbf{Res}_{2,k} = \frac{\partial \mathbf{V}_\theta(t_k)}{\partial t} - f_2(\mathbf{U}_\theta(t_k),\mathbf{V}_\theta(t_k)) \quad (12)$$

$$Loss_R = \frac{1}{N_\Omega N_T} \sum_{k=1}^{N_T} |\mathbf{Res}_{1,k}|^2 + |\mathbf{Res}_{2,k}|^2 \quad (13)$$

With this modification, PINN no longer needs to deal directly with the second-order spatial derivatives over the 3D ventricular geometry.

3.2   Spatial-Temporally Adaptive Training Strategies

A common challenge in PINN training is the failure to propagate PDE solutions over a large solution domain [6,14,20]. Fundamentally different from conventional training, only labels of the initial PDE solutions are known in PINN training, from which the correct solutions need to be propagated to the entire solution domain. Before the correct solution arrives, the intermediate (and mostly trivial) solutions also propagate. If dominating the propagation, this will result in a chain reaction that prevents the propagation of correct solutions, leading to propagation failure. This problem is escalated if the propagation gradient is sharp.

Temporally Adaptive Training: We argue that the propagation failure is caused by the inability of vanilla PINNs to respect the causality underlying the physical laws of wave propagation. Therefore, we propose temporal weights $\{\beta_1, \beta_2, \ldots, \beta_{N_T}\}$ to guide PINNs to respect temporal causality during training. The intuition is that, when the correct solution is propagated to time t_c, the previous times $\{t_1, t_2, \ldots, t_{c-1}\}$ should all have lower PDE residuals. We thus propose the temporal weights to be:

$$\beta_k = \begin{cases} 1 & \text{if } \sum_{m=1}^{k-1} Loss_{R_m} < \epsilon_t \\ 0 & \text{else} \end{cases} \tag{14}$$

where ϵ_t is a threshold to be tuned. This will guide the propagation of correct solutions while avoiding obtaining and propagating trivial intermediate solutions.

Spatially Adaptive Training: While sharp gradients in PDE solutions are challenging for PINNs [12,14,22], they are sparse both in space and time in ventricular EP activation. We thus propose to exploit these sharp-gradients to focus PINN training on these sparse regions, thus accelerating PINN convergence. This is achieved by spatial weights $\mathbf{W}_S = [w_1, w_2, \ldots, w_{N_\Omega}]$ defined as:

$$w_i = \begin{cases} w_h & \text{if} \frac{\partial u_i}{\partial t}|_{t_c} > \epsilon_s \\ 1 & \text{else} \end{cases} \tag{15}$$

where $w_h > 1$ and ϵ_s is another threshold to be tuned.

With the above strategy, PDE residual is further modified to:

$$Loss_R = \frac{1}{N_\Omega N_T} \sum_{k=1}^{N_T} \beta_k \left(|\mathbf{W}_S \cdot \mathbf{Res}_{1,k}|^2 + |\mathbf{Res}_{2,k}|^2 \right) \tag{16}$$

Sequential PINN Training: Even when propagation causality and sparse regions of sharp gradients are respected, PINN cannot solve for arbitrarily long time domains because the loss landscape becomes increasingly complex as N_T increases. We thus further utilize a sequential learning method where we first uniformly discretize the time domain $[0, T]$ into n segments, and then train the PINN across these segments sequentially as:

$$\underset{t \in (T_{n-1}, T_n]}{Loss} = \lambda_I \underset{t=T_{n-1}}{Loss_I} + \lambda_R \underset{t \in (T_{n-1}, T_n]}{Loss_R} \tag{17}$$

where PDE solutions obtained from previous time segments become the initial residual for training the PINN for the next time segment. The complete PDE solution for the entire time domain is obtained at the end of the training.

3.3 Solving the Forward and Inverse Problems

In the forward problem, PINNs approximate the solution of the AP model by optimizing the initial residual $Loss_I$ and the PDE residual $Loss_R$:

$$Loss_I = \frac{1}{N_\Omega} |\mathbf{U}_\theta(0) - \mathbf{U}(0)|^2 + |\mathbf{V}_\theta(0) - \mathbf{V}(0)|^2 \tag{18}$$

$$Loss = \lambda_I Loss_I + \lambda_R Loss_R \tag{19}$$

In the inverse problem, PINNs utilize partially known solutions $\{\mathbf{U}(t_k)\}_{k=1}^{N_T}$ to simultaneously optimize the PINN parameters θ and unknown PDE parameters ϕ with an additional data loss $Loss_D$:

$$Loss_R = \frac{1}{N_\Omega N_T} \sum_{k=1}^{N_T} |\mathbf{Res}_{1,k}|^2_{\phi=\phi_\theta} + |\mathbf{Res}_{2,k}|^2_{\phi=\phi_\theta} \tag{20}$$

$$Loss_D = \frac{1}{N_\Omega N_T} \sum_{k=1}^{N_T} |\mathbf{U}_\theta(t_k) - \mathbf{U}(t_k)|^2 \tag{21}$$

$$Loss = \lambda_D Loss_D + \lambda_R Loss_R \tag{22}$$

4   Experiments and Results

In all experiments, the neural network we used has 5 quadratic residual layers [3] as hidden layers with 512 neurons in each layer, where the input is time t and the output is $N_\Omega(=1862)$ dimensional vectors $\mathbf{U}_\theta$ and $\mathbf{V}_\theta$. The quadratic residual layers have a stronger nonlinearity than the fully connected layers [3]. We used Adam optimizer and the learning rate is set at 1×10^{-3}. The parameters of AP model are fixed to standard values as documented in the literature [1]:$c = 8$, $e_0 = 0.002$, $\mu_1 = 0.2$ and $\mu_2 = 0.3$. The hyperparameters of our proposed PINN framework are set as follows: $\epsilon_t = 0.001$, $\epsilon_s = 0.01$, $w_h = 4$ and $T_n - T_{n-1} = 1$. In the forward problem $\lambda_I : \lambda_R = 10 : 1$, and in the inverse problem $\lambda_D : \lambda_R = 1 : 1000$. All experiments were run on NVIDIA Titan RTX and numerical solution generated through the method in the literature [19].

The Effectiveness of Spatial-Temporally Adaptive PINNs: To verify the effectiveness of the spatial-temporally adaptive strategies for PINN training, we solve the forward problem of the AP model on time domain $t \in [0,1]$, $t \in [0,5]$ and $t \in [0,10]$ considering three ablation models: vanilla PINN, PINN + temporal weights, PINN + temporal and spatial weights. Figure 2 shows MSE error between PINN output and numerical solution during the training process.

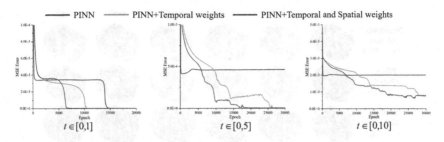

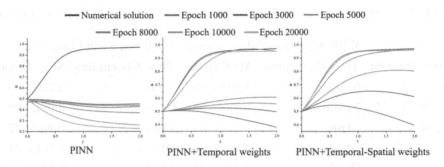

Fig. 2. Training MSE between PDE and PINN solutions for three ablation models. (Color figure online)

Fig. 3. Examples of PDE solutions obtained by PINN during training.

As shown, the vanilla PINN (black) can only solve for a short time domain ($t \in [0,1]$). Temporal weights (red) helped mitigate propagation failure over longer time domains, while spatial weights (blue) further accelerated convergence.

To understand why spatial-temporal adaptive training works, we further observed the PINN outputs during the training process. Figure 3 provides an example of the waveform of a PDE solution output by PINN (training on $t \in [0,5]$) during training. As shown, the propagation of the trivial solution trumped the propagation of the correct solution in the vanilla PINN. Temporal weights were able to guide the correct solution to propagate. The addition of spatial weights was able to further speed up this process.

Figure 4 provides examples of how the correct solution, PINN solutions, and spatial weights were propagated over space. As shown, the spatial weights were able to track the "wavefront" of the TMP activation – a sparse set of locations where high gradients exist both in space and time. This succeeded in guiding the optimizer to focus more on these regions, thus accelerating the convergence.

The Effectiveness of Sequential Training: We then compared the computation cost and the accuracy of the PDE solution achieved by the presented PINNs with and without sequence learning, across different lengths of time domains as summarized in Table 1. As shown, the sequential training was able to achieve a 1.3–1.8× speedup with increased accuracy. Also note that without sequence

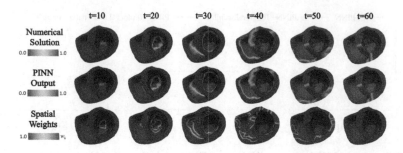

Fig. 4. Propagation of the correct solutions, PINN output, and spatial weights

Table 1. Comparison of time cost and accuracy with and without sequence learning

	Without Sequential Learning		With Sequential Learning	
Time Domain	Time Consuming	MSE Error	Time Consuming	MSE Error
$t \in [0, 2]$	5 min	1.80E−08	4 min	5.23E−09
$t \in [0, 5]$	14 min	6.77E−06	9 min	3.08E−08
$t \in [0, 7]$	24 min	7.17E−05	13 min	2.21E−07
$t \in [0, 10]$	26 min	6.18E−04	20 min	9.59E−07
$t \in [0, 20]$	59 min	1.39E−02	41 min	9.51E−06

learning, the MSE error increases rapidly as the time domain increases, making it difficult to scale to longer time domains. In comparison, sequence learning enables the PDE solution (training enough time) to preserve a low MSE error$(= 1.02 \times 10^{-5})$ throughout the complete ventricular activation ($t \in [0, 60]$), enabling simulation of the complete ventricular activation process as shown in Fig. 4 that is otherwise not possible with the other ablation models.

PINN for Supporting PDE Parameter Inference: Finally, we tested the feasibility of the presented PINNs to support parameter estimation of the AP model. We assumed measurements of TMP solutions to be available and considered unknown parameter γ in the AP model (Eq. 1) due to its relatively large influence on the PDE solution [7]. We considered spatially varying values of γ representing infarct tissues, and tested joint optimization of the PINN and the unknown γ for six different cases. The estimated parameter, as shown in Fig. 5, exhibited small MSEs in comparison to the ground truth.

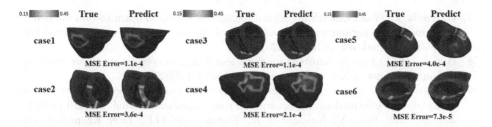

Fig. 5. True parameter γ and PINN predicted parameter $\hat{\gamma}$

5   Conclusion

We presented a spatial-temporally adaptive PINN framework to solve the forward and inverse problems of cardiac EP over 3D bi-ventricular models, overcoming the current limitations of PINNs in solving PDEs with sharp gradients/transitions over tricky spatial domains and long time domains. However, our PINN framework still has a much larger computational cost than traditional numerical methods, which is a drawback of PINN itself. Future works will pursue the use of this PINN framework for more complex PDEs of bi-ventricular EP, improve its computational efficiency as well as enable inverse EP using surface measurements such as electrocardiograms (ECGs) in real-data settings.

Acknowledgements. This work is supported in part by the National Key Research and Development Program of China(No: 2020AAA0109502); the National Natural Science Foundation of China (No: U1809204); the Talent Program of Zhejiang Province (No: 2021R51004); NIH/NHLBI under Award Numbers R01HL145590; and NSF grants OAC-2212548.

References

1. Aliev, R.R., Panfilov, A.V.: A simple two-variable model of cardiac excitation. Chaos Solitons Fract. **7**(3), 293–301 (1996)
2. Arevalo, H.J., et al.: Arrhythmia risk stratification of patients after myocardial infarction using personalized heart models. Nat. Commun. **7**(1), 11437 (2016)
3. Bu, J., Karpatne, A.: Quadratic residual networks: a new class of neural networks for solving forward and inverse problems in physics involving PDEs. In: Proceedings of the 2021 SIAM International Conference on Data Mining (SDM), pp. 675–683. SIAM (2021)
4. Chen, Y., Lu, L., Karniadakis, G.E., Dal Negro, L.: Physics-informed neural networks for inverse problems in nano-optics and metamaterials. Opt. Express **28**(8), 11618–11633 (2020)
5. Clayton, R., et al.: Models of cardiac tissue electrophysiology: progress, challenges and open questions. Prog. Biophys. Mol. Biol. **104**(1–3), 22–48 (2011)
6. Daw, A., Bu, J., Wang, S., Perdikaris, P., Karpatne, A.: Rethinking the importance of sampling in physics-informed neural networks. arXiv preprint arXiv:2207.02338 (2022)

7. Dhamala, J., et al.: Embedding high-dimensional Bayesian optimization via generative modeling: parameter personalization of cardiac electrophysiological models. Med. Image Anal. **62**, 101670 (2020)
8. Hao, Z., et al.: Physics-informed machine learning: a survey on problems, methods and applications. arXiv preprint arXiv:2211.08064 (2022)
9. Herrero Martin, C., et al.: Ep-pinns: cardiac electrophysiology characterisation using physics-informed neural networks. Front. Cardiovasc. Med. **8**, 2179 (2022)
10. Jagtap, A.D., Shin, Y., Kawaguchi, K., Karniadakis, G.E.: Deep Kronecker neural networks: a general framework for neural networks with adaptive activation functions. Neurocomputing **468**, 165–180 (2022)
11. Jin, X., Cai, S., Li, H., Karniadakis, G.E.: NSFnets (Navier-Stokes flow nets): physics-informed neural networks for the incompressible Navier-Stokes equations. J. Comput. Phys. **426**, 109951 (2021)
12. Karniadakis, G.E., Kevrekidis, I.G., Lu, L., Perdikaris, P., Wang, S., Yang, L.: Physics-informed machine learning. Nat. Rev. Phys. **3**(6), 422–440 (2021)
13. Kissas, G., Yang, Y., Hwuang, E., Witschey, W.R., Detre, J.A., Perdikaris, P.: Machine learning in cardiovascular flows modeling: predicting arterial blood pressure from non-invasive 4D flow MRI data using physics-informed neural networks. Comput. Methods Appl. Mech. Eng. **358**, 112623 (2020)
14. Krishnapriyan, A., Gholami, A., Zhe, S., Kirby, R., Mahoney, M.W.: Characterizing possible failure modes in physics-informed neural networks. In: Advances in Neural Information Processing Systems, vol. 34, pp. 26548–26560 (2021)
15. Mathews, A., Francisquez, M., Hughes, J.W., Hatch, D.R., Zhu, B., Rogers, B.N.: Uncovering turbulent plasma dynamics via deep learning from partial observations. Phys. Rev. E **104**(2), 025205 (2021)
16. Raissi, M., Perdikaris, P., Karniadakis, G.E.: Physics-informed neural networks: a deep learning framework for solving forward and inverse problems involving nonlinear partial differential equations. J. Comput. Phys. **378**, 686–707 (2019)
17. Sahli Costabal, F., Yang, Y., Perdikaris, P., Hurtado, D.E., Kuhl, E.: Physics-informed neural networks for cardiac activation mapping. Front. Phys. **8**, 42 (2020)
18. Sermesant, M., et al.: Patient-specific electromechanical models of the heart for the prediction of pacing acute effects in CRT: a preliminary clinical validation. Med. Image Anal. **16**(1), 201–215 (2012)
19. Wang, L., Zhang, H., Wong, K.C., Liu, H., Shi, P.: Physiological-model-constrained noninvasive reconstruction of volumetric myocardial transmembrane potentials. IEEE Trans. Biomed. Eng. **57**(2), 296–315 (2009)
20. Wang, S., Sankaran, S., Perdikaris, P.: Respecting causality is all you need for training physics-informed neural networks. arXiv abs/2203.07404 (2022)
21. Wang, S., Teng, Y., Perdikaris, P.: Understanding and mitigating gradient flow pathologies in physics-informed neural networks. SIAM J. Sci. Comput. **43**(5), A3055–A3081 (2021)
22. Wang, S., Yu, X., Perdikaris, P.: When and why PINNs fail to train: a neural tangent kernel perspective. J. Comput. Phys. **449**, 110768 (2022)
23. Xu, K., Darve, E.: Physics constrained learning for data-driven inverse modeling from sparse observations. J. Comput. Phys. **453**, 110938 (2022)
24. Zhang, H., Shi, P.: A meshfree method for solving cardiac electrical propagation. In: 2005 IEEE Engineering in Medicine and Biology 27th Annual Conference, pp. 349–352. IEEE (2006)

ModusGraph: Automated 3D and 4D Mesh Model Reconstruction from Cine CMR with Improved Accuracy and Efficiency

Yu Deng[1], Hao Xu[1], Sashya Rodrigo[1], Steven E. Williams[1,2],
Michelle C. Williams[2], Steven A. Niederer[1], Kuberan Pushparajah[1],
and Alistair Young[1(✉)]

[1] School of Biomedical Engineering and Imaging Sciences,
King's College London, London, UK
`alistair.young@kcl.ac.nz`
[2] University/BHF Centre for Cardiovascular Science,
University of Edinburgh, Edinburgh, UK

Abstract. Anatomical heart mesh models created from cine cardiac images are useful for the evaluation and monitoring of cardiovascular diseases, but require challenging and time-consuming reconstruction processes. Errors due to reduced spatial resolution and motion artefacts limit the accuracy of 3D models. We proposed ModusGraph to produce a higher quality 3D and 4D (3D+time) heart models automatically, employing i) a voxel processing module with Modality Handles and a super-resolution decoder to define low-resolution and high-resolution segmentations and correct motion artefacts with multi-modal unpaired data, ii) a Residual Spatial-temporal Graph Convolution Network to generate mesh models by controlled and progressive spatial-temporal deformation to better capture the cardiac motion, and iii) a Signed Distance Sampling process to bridge those two parts for end-to-end training. Modus-Graph was trained and evaluated on CT angiograms and cardiovascular MRI cines, showing superior performance compared to other mesh reconstruction methods. It creates well-defined meshes from sparse MRI cines, enabling vertex tracking across cardiac cycle frames. This process aids in analyzing myocardium function and conducting biomechanical analyses from imaging data https://github.com/MalikTeng/ModusGraph.

Keywords: Cardiovascular Magnetic Resonance · 3D heart model · Motion Artefacts · Super-resolution · Graph Neural Network

Supplementary Information The online version contains supplementary material available at https://doi.org/10.1007/978-3-031-43990-2_17.

1   Introduction

Multi-slice cine cardiovascular magnetic resonance (CMR) scanning is a common method for the accurate diagnosis and evaluation of cardiovascular diseases [16, 25]. Although this provides a series of images of the heart and blood vessels over time, it is often a lengthy process that obtains only certain slices of the heart which limit the visualization of certain structures, and has breath-hold motion artefacts resulting in misalignment between slices

To better evaluate heart disease, plan interventions, and monitor heart disease, a 3D heart model can be created from cine CMRs, which is a digitized heart object visualized as triangular meshes [9]. This reconstruction is accomplished in several steps: segmentation, registration, reconstruction, refinement, and visualisation. Because of the time cost and expert knowledge required for this task, it is desirable to create 3D or 4D (3D+time) heart models automatically for every patient [3,23]. However, it is challenging to create accurate 3D heart models, because of the impact of low spatial resolution and motion artefacts [11,18].

To this end, we proposed a Recurrent Graph Neural Network based method, ModusGraph, to fully automate the reconstruction of 4D heart models from cine CMR. This includes i) a voxel processing module with Modality Handles (Modhandle) and ResNet decoder for super-resolution and correction of motion artefacts from the acquired cine CMR, ii) a Residual Spatial-temporal Graph Convolution module (R-StGCN) for 4D mesh models generation by hierarchically spatial deformation and temporal motion estimation, and iii) a Signed Distance Sampling process bridge voxel features from segmentation and vertex features from deformation.

2   Related Works

Surface meshing involves constructing polygonal representations of geometric objects or surfaces, and creating high-quality and feature-aware surface meshes for medical imaging applications it is of particular interest.

With available large-volume training data and advanced computational resources, more studies harness the strength of deep learning and traditional methods to avoid user supervision. Aubert et al. [2] use convolutional neural networks to automatically detect anatomical landmarks for spine reconstruction. Ma et al. [22] propose a dense SLAM technique for colon surface reconstruction. Gopinath et al. [7] presented SegRecon, an end-to-end deep learning approach that for simultaneous reconstruction and segmentation of cortical surfaces directly from an MRI volume. Wang et al. [26] reconstructed 3D surfaces from 2D images using a neural network that learns a Signed Distance Function (SDF) representation from 2D images. Ma et al. [21] proposed a deep learning framework that uses neural ordinary differential equations (ODEs) for efficient cortical surface reconstruction from brain MRI scans. Similarly, Lebrat et al. [17] presents CorticalFlow, a geometric deep learning model that learns to deform a reference template mesh towards a targeted object in a 3D image, by solving ODEs from stationary velocity fields.

In contrast to those methods applying directly to the surface manifold, others, similar to our method, combine image segmentation with explicit surface representations and mesh deformation with coarse to fine controls. Wickramasinghe et al. [27] introduced Voxel2Mesh, a two-stage method that uses a CNN for voxel labelling and a GCN for mesh generation. Bongratz et al. [4] presented a deep learning algorithm that reconstructs explicit meshes of cortical surfaces from brain MRI scans using a convolutional and graph neural network, resulting in four meshes. Kong et al. [14] proposed a method that generates simulation-ready meshes of cardiac structures using atlas-based registration and shape-preserving interpolation. They also introduced a deep learning approach that constructs whole heart meshes by learning to deform a small set of deformation handles on a whole heart template [15]. Here, we utilize sparse CMRs to generate temporally coherent dynamic meshes for the cardiac cycle, leveraging unpaired high-resolution CT datasets (Fig. 1). These meshes are geometrically and topologically well-defined, with trackable vertices across consecutive frames. Such features enhance the analysis of myocardium function and enable biomechanical computational analysis (e.g., for stiffness or contractility estimation from finite element analysis of imaging data).

3   Method

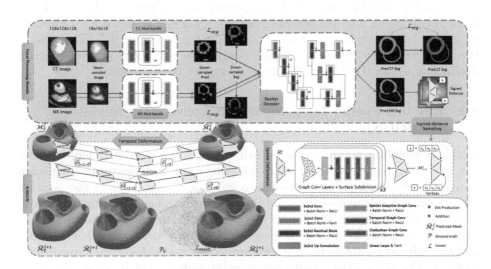

Fig. 1. Schematic of ModusGraph. It consists of two main components: a voxel processing module to generate high-resolution segmentations from image volumes, and an R-StGCN module based on a graph convolution network that deforms an initial coarse mesh progressively and dynamically by frames. A Signed Distance Sampling procedure bridges those two parts for end-to-end training.

3.1 Voxel Processing Module

Modality Handles. Compared to cine CMRs, CT imaging provides higher spatial resolution and enables easier high-resolution segmentation of the heart. The developed network to generate such segmentation is transferable to process cine CMRs, allowing for comparable cardiac structural information to be extracted from similar patient populations. The following module is thus proposed to estimate high-resolution segmentations from unpaired CT and cine CMR image volumes. An input image volume $X_{CT} \in \mathbb{R}^{H \times W \times D}$ is cropped around the anatomy of interest to the size of $128 \times 128 \times 128$, and then down-sampled following bilinear interpolation method to $X'_{CT} \in \mathbb{R}^{16 \times 16 \times 16}$, which enables a common low-resolution segmentation space for cine CMRs. Its predicted segmentation $\tilde{Y}'_{CT} \in \mathbb{R}^{16 \times 16 \times 16}$ is generated by a CT Modality Handle (CT Mod-handle, $h(X_{CT})$), using ResNet blocks followed by ReLU non-linear activity and a one-by-one convolution layer. A MR Modality Handle (MR Mod-handle, $h(X_{MR})$) generates predicted segmentation $\tilde{Y}_{MR}$ from cine CMRs in the same way.

Super-Resolution Decoder. $\tilde{Y}'_{CT}$ is passed to a decoder for super-resolution reconstruction to a size of $128 \times 128 \times 128$. The decoder ψ includes three layers of up convolution followed by ResNet blocks. The high-resolution segmentation of cine CMR is generated similarly through $\tilde{Y}_{MR} = \mathbf{W}_\psi \mathbf{W}_h X_{MR}$, where $\mathbf{W}_\psi$ and $\mathbf{W}_h$ are trainable weights of decoder and Mod-handle, respectively. To include heart morphological features in the graph convolution process, the signed distance is calculated from the decoder's output. This signed distance is computed as geodesic distances from each voxel to the surface boundary [1,5]. The mesh is then scaled to the output size and each mesh surface vertex is assigned a signed distance based on the output channel and vertex's coordinates.

3.2 Residual Spatial-Temporal Graph Convolution Module (R-StGCN)

Graph convolution networks can be utilized to reshape a heart mesh model, but regressing the surface near sharp edges or areas with aggressive Laplacian changes is challenging due to the networks' lack of awareness of the position relationships. We borrowed the idea of graph construction from human joints to overcome this issue [30]. The area of the 1-connected neighbourhood near sharp edges is divided into 3 subsets k: one subset includes vertices on the valves' edges, and the other two subsets include vertices on different sides of the myocardium surfaces. This defines the robust position relationships of sharp edges and other convex areas of the mesh, allowing us to use Adaptive Graph Convolutional (AGC) layers [24] to learn such relationships.

Spatial Deformation. Given a dynamic mesh at level l and frame t, i.e. $\mathcal{M}_t^l = (\mathcal{V}_t^l, \mathcal{E}_t^l)$, where $\mathcal{V}_t^l$ and $\mathcal{E}_t^l$ are N vertices and M edges, respectively. A dense adjacency matrix $\mathbf{A}_k \in \mathbb{R}^{N \times N}$ denotes the edges between every two vertices.

A data-dependent matrix $\mathbf{C}_k \in \mathbb{R}^{N \times N}$ determines the similarity of every two vertices as normalized Gaussian function, $\mathbf{C}_k = softmax(f_{in}^T \mathbf{W}_{\theta k}^T \mathbf{W}_{\phi k} f_{in})$. $\mathbf{W}_{\theta k}$ and $\mathbf{W}_{\phi k}$ are the parameters of the 1×1 convolution layer θ and ϕ, respectively. $f_{in} \in \mathbb{R}^{3 \times T \times N}$ is input feature matrix of the convolution layer, where T is the number of frames for each cardiac cycle. The sampling area of convolution is a 1-connected neighbourhood includes 3 subsets, which conforms with the aforementioned mesh topology. It is described as $f_{out} = \Sigma_k^{K=3} \mathbf{W}_k f_{in}(\mathbf{A}_k + \mathbf{C}_k)$. f_{out} is output feature of the convolution layer, $\mathbf{W}_k$ is trainable weights for the convolution. Following the AGC layer, we used graph convolution with first-order Chebyshev polynomial approximation. It is formalized as $f_{out} = \sigma(\mathbf{W}_{\theta 0} f_{in} + \mathbf{W}_{\theta 1} f_{in} \tilde{\mathbf{L}})$, where $\mathbf{W}_{\theta 0}, \mathbf{W}_{\theta 1}$ are trainable weights and $\tilde{\mathbf{L}} = 2\mathbf{L}_{norm}/\lambda_{max} - \mathbf{I}, \tilde{\mathbf{L}} \in \mathbb{R}^{N \times N}$ is the scaled and normalized Laplacian matrix [6]. The signed distance was added to mesh vertices prior to the graph convolution, and a straightforward Loop method [19] for surface subdivision is applied to refine the coarse mesh.

Temporal Deformation. The deformation field vector $\vec{\mathcal{V}}_{t-1 \to t}$ and $\vec{\mathcal{V}}_{t \to t-1}$ are learnt through temporal convolutions, where the sampling neighbourhood is defined as a vertex in consecutive frames. It is a $T \times 1$ convolution performed on the output feature matrix f_{out} in a bidirectional manner. $\vec{\mathcal{V}}$ is regularized following the principle of motion estimation. With vertices of meshes at consecutive frames $\mathcal{V}_0$ and $\mathcal{V}_1$, the vertices of intermediate mesh $\mathcal{V}_t, 0 < t < 1$ is approximated under symmetric assumption [28], as $\tilde{\mathcal{V}}_t = 0.5 \cdot (t \cdot (\mathcal{V}_0 + t \cdot \vec{\mathcal{V}}_{0 \to 1}) + (1 - t) \cdot (\mathcal{V}_1 + (1 - t) \cdot \vec{\mathcal{V}}_{1 \to 0}))$, and we measure the L1 difference between $\tilde{\mathcal{V}}_t$ and $\mathcal{V}_t$.

3.3 Training Scheme

Generally, the MR Mod-handle was trained on cine CMRs and down-sampled segmentation, described as $\mathcal{L}_{seg,\text{MR}}(h(\mathbf{X}_{\text{MR}}), \mathbf{Y'}_{\text{MR}})$. The CT Mod-handle and ResNet decoder were trained on CT image volumes, segmentation and their down-sampled counterparts using dice loss and cross-entropy loss, i.e. $\mathcal{L}_{seg,\text{CT}}(\psi \cdot h(\mathbf{X}_{\text{CT}}), \mathbf{Y}_{\text{CT}}, \mathbf{Y'}_{\text{CT}})$. Supervised by the ground-truth point clouds from CT segmentation, meshes were predicted from the R-StGCN module, where Chamfer distance is minimized together with surface regularization [27] and deformation field vector regularization as $\mathcal{L}_{mesh,\text{CT}} = \Sigma_{t=0}^{T=1} \Sigma_{l=0}^{L=2} \text{dCD}(\tilde{\mathcal{M}}_t^l, \mathcal{P}_t) + \lambda_{reg} \cdot (\mathcal{L}_{regular}(\mathcal{M}_t^l) + \|\tilde{\mathcal{V}}_t^l - \mathcal{V}_t^l\|_1)$. $\mathcal{P}_t$ was generated via Marching Cubes [20] and uniform surface sampling applied to the ground-truth segmentation. Similarly, pseudo point clouds $\tilde{\mathcal{P}}_t$ from super-resolved cine CMRs segmentation were used for fine-tuning the R-StGCN module, i.e. $\mathcal{L}_{mesh,\text{MR}}$. The total loss is $\mathcal{L}_{total} = \lambda_{seg} \cdot (\mathcal{L}_{seg,\text{CT}} + \mathcal{L}_{seg,\text{MR}}) + \lambda_{mesh} \cdot (\mathcal{L}_{mesh,\text{CT}} + \mathcal{L}_{mesh,\text{MR}})$, where $\lambda_{seg} = 0.5$, $\lambda_{mesh} = 1.0$ and $\lambda_{reg} = 0.1$ were selected by extensive experiments from $[0, 1]$. Find a detailed training/testing scheme in Appendix V. ModusGraph is implemented with PyTorch 1.12.1 and the experiment was conducted on an RTX 3090 GPU, with Adam optimizer and a learning rate of $1e-4$. The training and validation losses converge after 200 epochs in less than 2 h.

4 Results and Discussion

4.1 Datasets

The training and validation data consisted of CT image volumes from the SCOT-HEART study [12] and cine CMRs from the Cardiac Atlas Project (CAP) tetralogy of Fallot [8] database. CT data were included to provide high-resolution geometry information while tetralogy of Fallot CMR cases were used because functional analyses are important for these patients. The SCOT-HEART dataset provided 400 and 200 image volumes for training and testing, while the CAP dataset provided 84 and 48 time-series image volumes for training and testing. Data augmentation techniques included random intensity shifting, scaling, contrast adjustment, random rotation, and intensity normalization. Ground-truth segmentations of four heart chambers, left ventricle myocardium, and aorta artery from a previously validated method [29] was used for the whole heart meshing on the SCOT-HEART dataset while left and right ventricle and myocardium manual segmentations were used for the dynamic meshing with the CAP dataset.

4.2 Evaluation of Whole Heart Meshes Quality

Table 1. Comparison of accuracy of generated mesh on SCOT-HEART test cases. Dice (decimal) and INTersection (percentage) scores were derived on voxelized meshes and ASD (decimal) is evaluated on meshes. Compared methods include Voxel2Mesh (VM), CorticalFlow (CF), nnU-Net3D+Point2Mesh (NNP), ResNet Decoder+Marching Cubes (RES) and ModusGraph (MG).

Metrics	Methods	LV	LV-MYO	RV	LA	RA	AV
Dice	VM (6k)	0.66 ± 0.11	0.35 ± 0.13	0.63 ± 0.12	0.54 ± 0.16	0.58 ± 0.13	0.47 ± 0.17
	CF (6k)	0.75 ± 0.08	0.50 ± 0.12	0.73 ± 0.08	0.67 ± 0.12	0.72 ± 0.10	0.61 ± 0.13
	NNP (6k)	0.75 ± 0.05	0.49 ± 0.08	0.72 ± 0.07	0.66 ± 0.04	0.70 ± 0.05	0.58 ± 0.06
	RES (14k)	**0.92 ± 0.02**	**0.82 ± 0.05**	**0.91 ± 0.03**	**0.89 ± 0.03**	**0.90 ± 0.04**	**0.87 ± 0.05**
	MG (6k)	0.77 ± 0.08	0.51 ± 0.13	0.75 ± 0.09	0.68 ± 0.11	0.73 ± 0.09	0.65 ± 0.14
ASD	VM (6k)	3.20e-01	5.88e-01	1.14e-01	1.89e-01	3.21e-01	1.31e+00
	CF (6k)	1.13e-02	8.24e-03	1.17e-02	1.64e-02	1.27e-02	1.40e-02
	NNP (6k)	1.16e-02	9.78e-03	1.21e-02	1.75e-02	1.46e-02	1.57e-02
	RES (14k)	**1.01e-03**	**8.65e-04**	**1.45e-03**	**1.76e-03**	**1.81e-03**	**1.48e-03**
	MG (6k)	1.06e-02	8.60e-03	1.09e-02	1.64e-02	1.18e-02	1.16e-02
INT	VM (6k)	15.74 ± 7.62	16.25 ± 5.82	9.16 ± 8.09	6.00 ± 8.41	7.57 ± 9.95	5.16 ± 8.86
	CF (6k)	12.34 ± 5.45	15.29 ± 9.81	11.77 ± 10.22	9.35 ± 6.38	6.42 ± 6.47	8.61 ± 7.31
	NNP (6k)	1.85 ± 0.96	2.24 ± 1.27	0.62 ± 0.42	0.25 ± 0.86	0.20 ± 0.32	0.18 ± 0.13
	RES (14k)	**0.00 ± 0.00**	**0.00 ± 0.00**	**0.00 ± 0.00**	**0.00 ± 0.00**	**0.00 ± 0.00**	**0.00 ± 0.00**
	MG (6k)	1.79 ± 0.68	2.39 ± 0.49	1.17 ± 0.14	0.43 ± 0.29	0.18 ± 0.10	1.50 ± 1.80

We evaluated the quality of meshes generated by ModusGraph, by comparing them to those produced by other state-of-the-art methods using the SCOT-HEART dataset. ModusGraph, Voxel2Mesh [27], and CorticalFlow [17] started

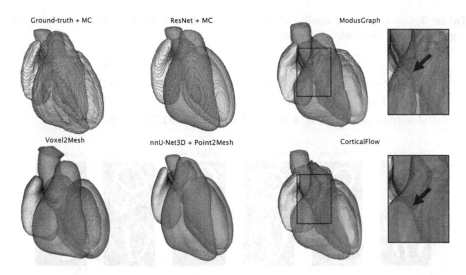

Fig. 2. Visualized meshes for a SCOT-HEART case. Arrows pointing to intersection structures in CorticalFlow

from the same template mesh and progressively deformed it to a finer mesh using $128 \times 128 \times 128$, while nnU-Net [13] segmentation and Point2Mcsh [10] reconstruction used a user controlled, differentiable refinement process to warp the segmentation boundary point clouds to meshes. These methods produce meshes with around 6,000 vertices for each anatomy. The Marching Cubes method was also applied to the ResNet decoder's segmentation to generate a finer mesh with around 14,000 vertices for evaluation. All methods for generating meshes had training times ranging from 2–5 hours.

Regarding the Dice score and Average Surface Distance (ASD) in Table 1, RES is a high benchmark since it was straightforward to reconstruct a high-resolution marching cubes mesh from the well-defined morphologies in the segmentations, but this is not suitable for tracking or computational models. ModusGraph's mesh accuracy is compromised due to information loss in the QuickHull-derived template mesh (Appendix I), alignment issues with deformed meshes and segmentation in different coordinate systems, and difficulties in capturing patient-specific geometric variations. However, refining the template mesh and registration process could potentially enhance the results. Figure 2 and the intersection scores in Table 1 that measure the ratio of surface collision show that ModusGraph can generate accurate whole-heart meshes with less surface distortion and collision when compared to its closest result from CorticalFlow.

4.3 Dynamic Mesh for Biomechanical Simulation

We evaluated ModusGraph and other methods on the task of creating dynamic ventricle myocardium meshes. When them to ground-truth short-axis slices segmentation, the Average Surface Distance (ASD) at end-diastole (ED) and

Table 2. Comparison of quality of generated mesh on CAP test cases. Numbers are decimal except for Angle in degree.

	ASD	Aspect Ratio	Angle (min, max)	Jacobian
Voxel2Mesh	1.08e-1	1.55 ± 0.08	(37.21, 91.01)	0.72 ± 0.03
CorticalFlow	8.21e-2	1.62 ± 0.15	(31.77, 94.57)	0.69 ± 0.04
ModusGraph	**5.14e-2**	**1.35 ± 0.00**	**(41.75, 81.21)**	**0.76 ± 0.00**

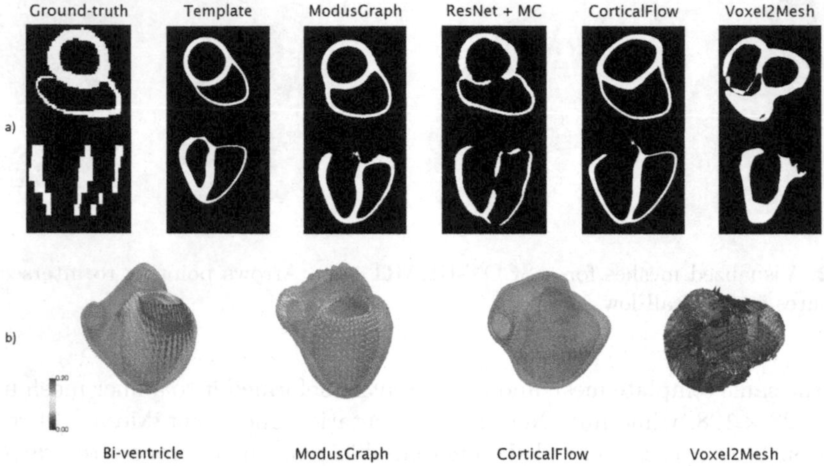

Fig. 3. Visualization of generated dynamic meshes' a) silhouette on short- and long-axis b) surface displacement between end-diastole and end-systole.

end-systole phases (ES) were used for evaluation since ground-truth dynamic meshes for each patient are challenging to obtain. ModusGraph more accurately deformed the template mesh towards true anatomy per patient, as shown by the silhouette of the generated meshes at ED in Fig 3-a. Generated meshes from nnU-Net + Point2Mesh or ResNet + MC are not suitable for mechanical simulations without proper post-processing and were thus not compared to ModusGraph, Voxel2Mesh, and CorticalFlow.

Mesh quality for biomechanical simulations, including aspect ratio, min and max angle, and Jacobian of surface mesh triangular cells, was also evaluated. ModusGraph showed less distorted cells, leading to faster convergence for mechanical simulations, as shown in Table 2. Changes in mesh surface from ED to ES phase were compared to a reference dynamic mesh, Bi-ventricle [8], and ModusGraph was found to more accurately describe the deformation in areas surrounding valves and the displacement of the myocardium surface, as shown in Fig 3-b. The dynamic mesh generated by ModusGraph is included in supplementary materials, along with details for creating template meshes for the two tasks and reference Bi-ventricle mesh.

5   Conclusion

Our proposed method, ModusGraph, automates 4D heart model reconstruction from cine CMR using a voxel processing module, a Residual Spatial-temporal Graph Convolution module and a Signed Distance Sampling process. Modus-Graph outperforms other state-of-the-art methods in reconstructing accurate 3D heart models from high-resolution segmentations on computed tomography images, and generates 4D heart models suitable for biomechanical analysis, which will aid in the understanding of congenital heart disease. This approach offers an efficient and automated solution for creating 3D and 4D heart models, with potential benefits for heart disease assessment, intervention planning, and monitoring.

Acknowledgements. YD was funded by the Kings-China Scholarship Council PhD Scholarship Program. HX was funded by Innovate UK (104691) London Medical Imaging & Artificial Intelligence Centre for Value Based Healthcare. SCOT-HEART was funded by The Chief Scientist Office of the Scottish Government Health and Social Care Directorates (CZH/4/588), with supplementary awards from Edinburgh and Lothian's Health Foundation Trust and the Heart Diseases Research Fund. AAY and KP acknowledge funding from the National Institutes of Health R01HL121754 and Welcome ESPCR Centre for Medical Engineering at King's College London WT203148/Z/16/Z.

References

1. Asad, M., Dorent, R., Vercauteren, T.: Fastgeodis: fast generalised geodesic distance transform. J. Open Sourc. Softw. **7**(79), 4532 (2022)
2. Aubert, B., Vazquez, C., Cresson, T., Parent, S., de Guise, J.A.: Toward automated 3d spine reconstruction from biplanar radiographs using CNN for statistical spine model fitting. IEEE Trans. Med. Imaging **38**(12), 2796–2806 (2019)
3. Banerjee, A., et al.: A completely automated pipeline for 3d reconstruction of human heart from 2d cine magnetic resonance slices. Phil. Trans. R. Soc. A **379**(2212), 20200257 (2021)
4. Bongratz, F., Rickmann, A.M., Pölsterl, S., Wachinger, C.: Vox2cortex: fast explicit reconstruction of cortical surfaces from 3d mri scans with geometric deep neural networks. In: Proceedings of the IEEE/CVF Conference on Computer Vision and Pattern Recognition, pp. 20773–20783 (2022)
5. Criminisi, A., Sharp, T., Blake, A.: GeoS: geodesic image segmentation. In: Forsyth, D., Torr, P., Zisserman, A. (eds.) ECCV 2008. LNCS, vol. 5302, pp. 99–112. Springer, Heidelberg (2008). https://doi.org/10.1007/978-3-540-88682-2_9
6. Defferrard, M., Bresson, X., Vandergheynst, P.: Convolutional neural networks on graphs with fast localized spectral filtering. In: Advances in Neural Information Processing Systems, vol. 29 (2016)
7. Gopinath, K., Desrosiers, C., Lombaert, H.: SegRecon: learning joint brain surface reconstruction and segmentation from images. In: de Bruijne, M., et al. (eds.) MICCAI 2021. LNCS, vol. 12907, pp. 650–659. Springer, Cham (2021). https://doi.org/10.1007/978-3-030-87234-2_61
8. Govil, S., et al.: A deep learning approach for fully automated cardiac shape modeling in tetralogy of Fallot. J. Cardiovasc. Magn. Reson. **25**(1), 15 (2023)

9. Guo, F., Li, M., Ng, M., Wright, G., Pop, M.: Cine and multicontrast late enhanced MRI registration for 3D heart model construction. In: Pop, M., et al. (eds.) STACOM 2018. LNCS, vol. 11395, pp. 49–57. Springer, Cham (2019). https://doi.org/10.1007/978-3-030-12029-0_6

10. Hanocka, R., Metzer, G., Giryes, R., Cohen-Or, D.: Point2mesh: a self-prior for deformable meshes. arXiv preprint arXiv:2005.11084 (2020)

11. Havsteen, I., Ohlhues, A., Madsen, K.H., Nybing, J.D., Christensen, H., Christensen, A.: Are movement artifacts in magnetic resonance imaging a real problem?-a narrative review. Front. Neurol. **8**, 232 (2017)

12. Investigators, S.H.: Coronary CT angiography and 5-year risk of myocardial infarction. N. Engl. J. Med. **379**(10), 924–933 (2018)

13. Isensee, F., Jaeger, P.F., Kohl, S.A., Petersen, J., Maier-Hein, K.H.: NNU-net: a self-configuring method for deep learning-based biomedical image segmentation. Nat. Methods **18**(2), 203–211 (2021)

14. Kong, F., Shadden, S.C.: Whole heart mesh generation for image-based computational simulations by learning free-from deformations. In: de Bruijne, M., et al. (eds.) MICCAI 2021. LNCS, vol. 12904, pp. 550–559. Springer, Cham (2021). https://doi.org/10.1007/978-3-030-87202-1_53

15. Kong, F., Shadden, S.C.: Learning whole heart mesh generation from patient images for computational simulations. IEEE Trans. Med. Imaging **42**, 533–545 (2022)

16. Kramer, C.M., Barkhausen, J., Flamm, S.D., Kim, R.J., Nagel, E.: Standardized cardiovascular magnetic resonance imaging (CMR) protocols, society for cardiovascular magnetic resonance: board of trustees task force on standardized protocols. J. Cardiovasc. Magn. Reson. **10**, 1–10 (2008)

17. Lebrat, L., et al.: CorticalFlow: a diffeomorphic mesh transformer network for cortical surface reconstruction. Adv. Neural. Inf. Process. Syst. **34**, 29491–29505 (2021)

18. Liao, J.R., Pauly, J.M., Brosnan, T.J., Pelc, N.J.: Reduction of motion artifacts in cine MRI using variable-density spiral trajectories. Magn. Reson. Med. **37**(4), 569–575 (1997)

19. Loop, C.: Smooth subdivision surfaces based on triangles (1987)

20. Lorensen, W.E., Cline, H.E.: Marching cubes: a high resolution 3d surface construction algorithm. ACM SIGGRAPH Comput. Graph. **21**(4), 163–169 (1987)

21. Ma, Q., Li, L., Robinson, E.C., Kainz, B., Rueckert, D., Alansary, A.: CortexODE: learning cortical surface reconstruction by neural odes. IEEE Trans. Med. Imaging **42**, 430–443 (2022)

22. Ma, R., Wang, R., Pizer, S., Rosenman, J., McGill, S.K., Frahm, J.-M.: Real-time 3D reconstruction of colonoscopic surfaces for determining missing regions. In: Shen, D., et al. (eds.) MICCAI 2019. LNCS, vol. 11768, pp. 573–582. Springer, Cham (2019). https://doi.org/10.1007/978-3-030-32254-0_64

23. Menchón-Lara, R.M., Simmross-Wattenberg, F., Casaseca-de-la Higuera, P., Martín-Fernández, M., Alberola-López, C.: Reconstruction techniques for cardiac cine MRI. Insights Imaging **10**, 1–16 (2019)

24. Shi, L., Zhang, Y., Cheng, J., Lu, H.: Two-stream adaptive graph convolutional networks for skeleton-based action recognition. In: Proceedings of the IEEE/CVF Conference on Computer Vision and Pattern Recognition, pp. 12026–12035 (2019)

25. Suinesiaputra, A., Gilbert, K., Pontre, B., Young, A.A.: Imaging biomarkers for cardiovascular diseases. In: Handbook of Medical Image Computing and Computer Assisted Intervention, pp. 401–428. Elsevier (2020)

26. Wang, P., Liu, L., Liu, Y., Theobalt, C., Komura, T., Wang, W.: NEUS: learning neural implicit surfaces by volume rendering for multi-view reconstruction. arXiv preprint arXiv:2106.10689 (2021)
27. Wickramasinghe, U., Remelli, E., Knott, G., Fua, P.: Voxel2Mesh: 3D mesh model generation from volumetric data. In: Martel, A.L., et al. (eds.) MICCAI 2020. LNCS, vol. 12264, pp. 299–308. Springer, Cham (2020). https://doi.org/10.1007/978-3-030-59719-1_30
28. Wolberg, G., Sueyllam, H., Ismail, M., Ahmed, K.: One-dimensional resampling with inverse and forward mapping functions. J. Graphics Tools **5**(3), 11–33 (2000)
29. Xu, H., et al.: Whole heart anatomical refinement from CCTA using extrapolation and parcellation. In: Ennis, D.B., Perotti, L.E., Wang, V.Y. (eds.) FIMH 2021. LNCS, vol. 12738, pp. 63–70. Springer, Cham (2021). https://doi.org/10.1007/978-3-030-78710-3_7
30. Yan, S., Xiong, Y., Lin, D.: Spatial temporal graph convolutional networks for skeleton-based action recognition. In: Proceedings of the AAAI Conference on Artificial Intelligence, vol. 32 (2018)

Twelve-Lead ECG Reconstruction from Single-Lead Signals Using Generative Adversarial Networks

Jinho Joo[1] , Gihun Joo[1] , Yeji Kim[2] , Moo-Nyun Jin[2] ,
Junbeom Park[2(✉)] , and Hyeonseung Im[1(✉)]

[1] Kangwon National University, Chuncheon 24341, Republic of Korea
{wnwlsgh111,joo9327,hsim}@kangwon.ac.kr
[2] Ewha Womans University Medical Center, Seoul 07985, Republic of Korea
{mnjin31,parkjb}@ewha.ac.kr

Abstract. Recent advances in wearable healthcare devices such as smartwatches allow us to monitor and manage our health condition more actively, for example, by measuring our electrocardiogram (ECG) and predicting cardiovascular diseases (CVDs) such as atrial fibrillation in real-time. Nevertheless, most smart devices can only measure single-lead signals, such as Lead I, while multichannel ECGs, such as twelve-lead signals, are necessary to identify more intricate CVDs such as left and right bundle branch blocks. In this paper, to address this problem, we propose a novel generative adversarial network (GAN) that can faithfully reconstruct 12-lead ECG signals from single-lead signals, which consists of two generators and one 1D U-Net discriminator. Experimental results show that it outperforms other representative generative models. Moreover, we also validate our method's ability to effectively reconstruct CVD-related characteristics by evaluating reconstructed ECGs with a highly accurate 12-lead ECG-based prediction model and three cardiologists.

Keywords: ECG reconstruction · Biosignal synthesis · Generative model

1 Introduction

Although these days smart healthcare devices such as smartwatches can be used to monitor a single-lead electrocardiogram (ECG) for Lead I and detect cardiovascular diseases (CVDs) such as atrial fibrillation (AF), multichannel ECGs such as twelve-lead signals are still required to diagnose more complex CVDs such as left and right bundle branch blocks (LBBBs and RBBBs) or myocardial

---

J. Joo and G. Joo—Contributed equally as first authors.

Supplementary Information The online version contains supplementary material available at https://doi.org/10.1007/978-3-031-43990-2_18.

infarction. To proactively deal with such intricate CVDs, therefore, one may need to undergo a 12-lead ECG measurement at a hospital and utilize 12-lead ECG-based deep learning algorithms for predicting CVDs [3,4,15,18], which can be a cumbersome process in everyday life. It is neither plausible to train a prediction model using only single-lead ECGs measured by smart devices as it is not possible to correctly label complex CVDs for them in the first place. To address this problem, in this paper, we propose a novel generative adversarial network (GAN) [13], called EKGAN, that can faithfully reconstruct 12-lead ECGs only from single-lead ones.

Although the ECG synthesis problem is not new, most previous studies have focused on utilizing it for data augmentation purpose as it is difficult to collect a sufficient amount of labeled ECGs (with CVDs) for developing prediction models. For example, many researchers have focused on synthesizing realistic ECGs using variants of autoencoders and GANs [5,6,9–11,14,22,23]. These methods can be useful for training prediction models, but it is unclear how they can be leveraged with commonly available wearable devices that can measure only single-lead ECGs. Meanwhile, another line of work has focused on reconstructing the corresponding (missing) ECGs from only a few actual lead signals [2,7,17,21], usually generating 12 leads from three leads including Lead I and II. We note here that if we know Lead I and II, then all six limb leads can be derived by Willem Einthoven's law [8] and Goldberger's law [12]. Thus, their reconstruction problem indeed reduces to the problem of reconstructing six precordial leads.

Our work differs significantly from previous studies, as we generate all 11 remaining leads simultaneously from only a single lead. Thus, our method can be used to bridge commonly available wearable devices that can measure only Lead I and high-performance deep learning-based prediction models using 12-lead ECGs. To the best of our knowledge, our work is the first to reconstruct all 12 leads simultaneously from a real single lead. By using the approach in [6], one can also generate all 12 leads, but only incrementally. Our proposed method EKGAN employs two generators and one 1D U-Net [19] discriminator to capture CVD-specific characteristics and correlation patterns between Lead I and the remaining leads from ECG training data. Experimental results show that the reconstruction performance of EKGAN outperforms other representative generative models such as Pix2Pix [16], CycleGAN [24] and CardioGAN [20] under various metrics. We also evaluated the practical applicability of our method by applying an existing 12-lead ECG-based CVD prediction model [18] to the reconstructed ECGs. To this end, we consider AF, LBBB, and RBBB, among which LBBB and RBBB require multichannel ECGs to detect. Moreover, three cardiologists examined reconstructed ECGs to see if they accurately reflected the important CVD-related characteristics of the original ECGs. All the results confirm the effectiveness and usefulness of our method, thus enabling preventive healthcare with smart wearable devices for CVDs. For reproducibility, the source code of EKGAN is available at https://github.com/knu-plml/ecg-recon.

2   Methods

In this section, we introduce a novel conditional GAN for ECG reconstruction, called EKGAN, which is based on Pix2Pix [16] but employs an additional label generator G_L and uses a 1D U-Net discriminator instead of a PatchGAN discriminator. Its overall structure is shown in Fig. 1. The inference generator G_I takes a Lead I signal and generates corresponding 12-lead signals, while the label generator G_L takes the same input as G_I and simply returns the same signal. The whole purpose of G_L is to enable G_I's encoder to learn the important characteristics of Lead I. To this end, the latent vector produced by G_I's encoder is approximated to that of G_L's encoder so that G_I's decoder can produce more detailed ECGs that are closely correlated with the input Lead I signals. The discriminator D distinguishes 12-lead ECGs generated by G_I and the original ECGs. The whole process is repeated, adversarially learning G_I and D.

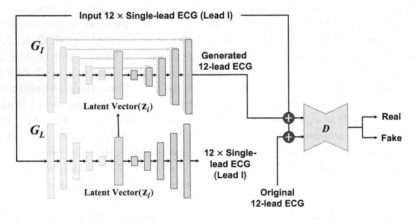

Fig. 1. Overall structure of EKGAN. G_I, inference generator; G_L, label generator; D, discriminator.

2.1   Generator

Inference Generator. The inference generator G_I is based on Pix2Pix's 2D U-Net generator and consists of an encoder and a decoder (Fig. 1). It takes an input of size $(16, 512, 1)$: 16 for 12 replicated signals of the input single lead plus 4 zero padding, 512 for the length of the ECG signal, and 1 for the channel size. More specifically, to generate 12-lead signals from single-lead signals, an input single-lead signal of length 512 is copied 12 times and two rows of zeros are added to both the top and bottom. Each original 12-lead ECG is arranged in the order of Lead I, II, III, aVR, aVL, aVF, and V1–V6, and is also zero padded. The encoder consists of five blocks each of which consists of convolution, batch normalization, and Leaky ReLU layers (an exception is the first block which excludes batch normalization). The numbers of convolution filters are

$64, 128, 256, 512, 1024$, while the kernel size is set to $(2, 4)$. The stride size is $(2, 2)$ except for the last block whose stride size is $(1, 2)$. The decoder is the inverse of the encoder and takes the encoder output as input. It also consists of five blocks each of which consists of concatenation, deconvolution, batch normalization, and ReLU layers (exceptions are the first block, which excludes concatenation, and the last block, which only uses concatenation and deconvolution). The numbers of deconvolution filters are $512, 256, 128, 64, 1$, while the kernel size is set to $(2, 4)$. The stride size is $(2, 2)$ except for the first block whose stride size is $(1, 2)$. From the second block, the output of the previous decoder block and that of the corresponding encoder block are concatenated and used as input.

Label Generator. An autoencoder-based model like U-Net [19] should create a latent vector in the encoder that represents the features of input data well. Then, the decoder should be learned to generate a target-like output from the latent vector. When training a U-Net, however, as we only use the reconstruction loss between the original and generated data, we cannot accurately determine how well the latent vector captures the essential features of the input (because there is no ground truth for the latent vector). In this paper, as a workaround, we use a label generator G_L which takes Lead I signals and returns the same signals. Accordingly, the latent vector produced by G_L's encoder would represent the features of the input accurately and thus can be used as ground truth for G_I's encoder. By doing so, G_I's decoder can produce a 12-lead ECG that is not only realistic but also closely associated with the input Lead I signal. The structure of G_L is similar to that of G_I, but it does not incorporate concatenation between the encoder and decoder. We experimentally validate the effectiveness of the label generator in Sect. 3.

2.2 Discriminator

In 12-lead ECGs, each lead has its own characteristics and thus it is important for a discriminator to analyze each lead signal individually at the pixel level. If a standard 2D convolution-based discriminator is used, as for image-to-image translation, which uses 2D patches, then the unique characteristics of each lead signal may be intermixed with others, which may in turn degrade the reconstruction quality. To prevent this problem, instead, we use a 1D U-Net discriminator D, which has the same layer architecture as G_L. D takes as input either G_I's output or an original 12-lead ECG, which is concatenated with G_I's input. In the encoder, the numbers of convolution filters are $32, 64, 128, 256, 512$, kernel sizes are $64, 32, 16, 8, 4$, and stride sizes are $4, 4, 4, 2, 2$. The decoder has the inverse structure of the encoder as in the two generators, except that it uses a sigmoid activation function at the last layer.

2.3 Loss

The objective of EKGAN is similar to that of other conditional GANs except that it uses the label generator G_L to train the inference generator G_I. Let us write e_i for 12 replicated ECG segments from the input single-lead ECG for G_I and e_o for the corresponding ground-truth 12-lead ECG segments. In addition, let z_i and z_l be the latent vectors produced by the encoders of G_I and G_L, respectively. We use the following adversarial loss and L1 losses:

$$\mathcal{L}_{\mathrm{adv}}(G_I, D) = \mathbb{E}_{e_i, e_o}[\log D(e_i, e_o)] + \mathbb{E}_{e_i}[\log(1 - D(e_i, G_I(e_i)))]$$
$$\mathcal{L}_{\mathrm{L1}}(G_I) = \mathbb{E}_{e_i, e_o}[\||e_o - G_I(e_i)\||_1]$$
$$\mathcal{L}_{\mathrm{L1}}(G_L) = \mathbb{E}_{e_i}[\||e_i - G_L(e_i)\||_1]$$
$$\mathcal{L}_{\mathrm{LV}}(G_I, G_L) = \mathbb{E}_{z_i, z_l}[\||z_i - z_l\||_1]$$

where the last one is the latent vector loss for the inference generator. Then, the objective of EKGAN is defined as:

$$G_I^* = \arg \min_{G_I, G_L} \max_D \{\mathcal{L}_{\mathrm{adv}}(G_I, D) + \lambda \mathcal{L}_{\mathrm{L1}}(G_I) + \alpha \mathcal{L}_{\mathrm{LV}}(G_I, G_L)\}$$

where λ and α control the relative importance of each function. Note that $\mathcal{L}_{\mathrm{L1}}(G_L)$ is used solely for training G_L and thus not included in the objective equation. Through a grid search, we determined $\lambda = 50$ and $\alpha = 1$, and these values have been used in the following experiments unless otherwise stated.

3   Evaluation

This section introduces our dataset and its preprocessing. Then, we extensively evaluate EKGAN in terms of its reconstruction performance. We also assess its applicability using an existing prediction model with proven performance [18] and with three cardiologists.

3.1   Experimental Setup

Datasets. To develop and evaluate EKGAN, we used about 326,000 ECGs collected from Ewha Womans University Mokdong and Seoul Hospitals between May 23, 2017, and November 30, 2022. Specifically, we first selected ECGs with LBBB, RBBB, and AF from the dataset and randomly selected normal sinus rhythm (NSR) ECGs in a 1:1 ratio to reduce the bias of the generative models. For the label information, we simply used the interpretation result of the ECG machine. Next, for a test set for CVD multi-label classification and cardiologists' examination, we randomly chose 100 ECGs for each disease and 300 NSR ECGs. Then, the remaining dataset was randomly divided into train and validation sets in an 8:2 ratio. Table 1 shows the configuration of the final dataset. We note here that since AF and NSR may coexist with LBBB and RBBB, the number of classes is different from the total number of data. The generative models were trained using the train set and evaluated using the validation set. The CVD prediction model was trained using the train and validation sets where the latter was used for hyperparameter tuning. Finally, 12-lead ECGs were generated by using both the validation and test sets, and their quality was evaluated by using the predictive model, *i.e.,* by comparing the classification performance with the original 12-lead ECGs and the generated ones, and examined by three cardiologists.

Table 1. Summary of the dataset. LBBB, Left Bundle Branch Block; RBBB, Right Bundle Branch Block; AF, Atrial Fibrillation; NSR, Normal Sinus Rhythm.

	LBBB	RBBB	AF	NSR	Total Classes	Total Data
Train	1,635	9,537	20,287	29,746	61,205	59,492
Validation	421	2,409	5,075	7,437	15,342	14,874
Test	102	108	140	460	810	600

Data Preparation. The 12-lead ECG data used in this study were measured for 10 s and the sampling rate is 500 Hz. To reduce the measurement noise, for each ECG signal, we remove the first 1 s and use only approximately the next 8.2 s (length of 4,096), while excluding the remaining part. In addition, for each 12-lead ECG, we normalize each lead signal individually, because if the amplitudes of 12-lead signals are significantly different from each other, the ones with small amplitudes may be ignored when predicting CVD. More specifically, for each signal, we first apply min-max normalization to $[-1, 1]$ and then a band-pass filter with lower and upper cutoff frequencies $[0.05, 150]$ as some fine details and important characteristics disappear when using a high lower cutoff frequency or a low upper cutoff frequency [1]. Finally, we downsample 4,096 lengths to 512 lengths.

Training and Test. All experiments were conducted on a workstation with an NVIDIA RTX 8000 and using TensorFlow 2.8. For comparisons, we implemented not only EKGAN but also Pix2Pix [16], CycleGAN [24], and CardioGAN [20] with minor modifications so that they can be applied to ECG data. In particular, for unpaired training of CycleGAN and CardioGAN, the input and label data were separated and shuffled independently. Moreover, due to the cycle consistency loss, Lead I of the label data was replaced with zero padding. All models were trained for 10 epochs, with the learning rate of 1e-4 until 5 epochs after which weight decay of 0.95 was applied per epoch. The kernel-initializer was sampled from a normal distribution $\mathcal{N}(0, 0.02^2)$.

3.2 Reconstruction Performance

Table 2 shows the reconstruction performance of EKGAN and other methods for the validation data in terms of root mean square error (RMSE), mean absolute error (MAE), percentage root mean square difference (PRD), maximum mean discrepancy (MMD), and mean absolute error for heart rates (MAE_{HR}). For all metrics, EKGAN significantly outperforms other methods, confirming the effectiveness of its label generator and 1D U-Net discriminator for pixel-level learning of ECG signals. Meanwhile, CycleGAN and CardioGAN, which are based on unpaired training, are not suitable for 12-lead ECG reconstruction. Figure 2 shows examples of reconstructed 12-lead ECGs by various methods from the Lead I, where EKGAN produces the most faithful reconstruction of the original ECG.

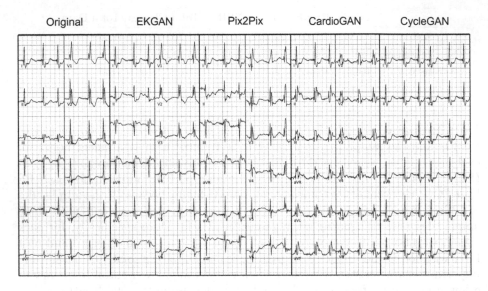

Fig. 2. Qualitative comparisons of various methods. The first column shows a sample original 12-lead ECG with RBBB and AF, while the rest was reconstructed from the original Lead I signal.

Table 2. Reconstruction performance of each method for the validation data.

Method	RMSE	MAE	PRD	MMD	MAE$_{HR}$
Pix2Pix	0.38	0.30	8.43	0.07×10^{-3}	27.68
CycleGAN	0.62	0.49	13.18	0.46×10^{-3}	24.16
CardioGAN	0.52	0.43	11.32	0.21×10^{-3}	24.09
EKGAN (proposed)	**0.32**	**0.25**	**6.95**	$\mathbf{0.04 \times 10^{-3}}$	**20.51**

Table 3. Performance of different variants of EKGAN for the validation data.

Method	RMSE	MAE	PRD	MMD	MAE$_{HR}$
Pix2Pix	0.38	0.30	8.43	0.07×10^{-3}	27.68
EKGAN w/o 1D discriminator	0.37	0.29	8.13	0.06×10^{-3}	24.06
EKGAN w/o label generator	0.35	0.28	7.73	0.05×10^{-3}	25.83
EKGAN (proposed)	**0.32**	**0.25**	**6.95**	$\mathbf{0.04 \times 10^{-3}}$	**20.51**

Table 3 shows the performance of different variants of EKGAN, that is, showing the effectiveness of our 1D discriminator and label generator. 'EKGAN w/o 1D discriminator' uses Pix2Pix's PatchGAN discriminator instead of our 1D U-Net discriminator. 'EKGAN w/o label generator' excludes a label generator and uses only an inference generator. We observe that the use of 1D discriminator is more effective than the use of a label generator, but using both results in a greater improvement over Pix2Pix for reconstructing 12-lead ECG signals.

Table 4. CVD multi-label classification results for the test data (F1-score). All results were obtained by using the algorithm proposed in [18].

Dataset	LBBB	RBBB	AF
Results reported in [18]	1.00	0.94	0.87
Original ECGs	0.92	0.97	0.92
Reconstructed ECGs by Pix2Pix	0.81	0.92	0.82
Reconstructed ECGs by EKGAN	0.94	0.96	0.88

Table 5. Concordance rate of three cardiologists for the test and reconstructed ECGs.

	LBBB	RBBB	AF	NSR
Cardiologist 1	96.32	92.31	93.65	93.65
Cardiologist 2	91.53	91.69	92.54	92.54
Cardiologist 3	94.77	92.24	91.06	91.06
Average	94.21	92.08	92.42	92.42

3.3   Reconstruction Quality Evaluation

To evaluate the applicability and quality of ECG signals reconstructed by EKGAN, we use a highly accurate prediction model for CVDs [18]. The model can predict six diseases by analyzing a 12-lead ECG, among which we choose LBBB, RBBB, and AF. We choose LBBB and RBBB as they require analysis of multi-lead ECGs and AF as it can be already predicted using commonly available wearable devices. This allows us to indirectly check if EKGAN is able to generate diverse ECG signals capturing different characteristics for each disease. Table 4 shows the multi-label classification results for the test set. Since the dataset used in [18] differs from ours, the performances using their datasets and ours are also slightly different, but both seem to perform well. We observe that the F1-scores when using the dataset reconstructed by EKGAN are comparable to those when using the original dataset and consistently better than those when using the ones reconstructed by Pix2Pix.

Table 5 shows the concordance rate between the test ECGs and the corresponding reconstructed ones by EKGAN, evaluated by three cardiologists. We randomly shuffled the original and reconstructed ECGs, and each cardiologist reviewed every ECG if it exhibited LBBB, RBBB, AF, or NSR. Then, for each case, the concordance rate is calculated as the ratio of pairs of original and corresponding reconstructed ECGs such that a cardiologist's read result coincides among all data pairs. We note here that since each ECG must exclusively include either AF or NSR, their concordance rates are the same. The results confirm that the reconstructed ECGs by EKGAN effectively capture the important characteristics of the original ones.

4   Conclusion

This paper studies a novel problem of reconstructing 12-lead ECGs from single-lead ECGs. To address this problem, we propose a novel conditional GAN, called EKGAN, based on Pix2Pix, which consists of two generators and one 1D U-Net discriminator. Experimental results show that EKGAN significantly outperforms other representative generative models such as Pix2Pix, CycleGAN, and CardioGAN, and is able to reconstruct 12-lead ECGs that faithfully capture the essential characteristics of the original 12-lead ECGs useful for predicting CVDs. Therefore, we expect that numerous deep learning models based on 12-lead ECGs with proven performance could be applied to smart healthcare devices that can measure only single-lead signals. It would be also interesting to investigate if our method is applicable to more complex CVDs such as acute myocardial infarction, which require a more detailed analysis of 12-lead ECGs by cardiologists.

Acknowledgments. This research was supported by "Regional Innovation Strategy (RIS)" through the National Research Foundation of Korea (NRF) funded by the Ministry of Education (MOE) (2022RIS-005). This work was also supported by the NRF grant funded by the Korea government (MSIT) (No. RS-2023-00208094 and RS-2023-00242528) and by Institute of Information & communications Technology Planning & Evaluation (IITP) grant funded by the Korea government (MSIT) (No. RS-2022-00155966, Artificial Intelligence Convergence Innovation Human Resources Development (Ewha Womans University)).

References

1. Philips DXL ECG Algorithm Physician's Guide. 2nd Edn. Publication number 453564106411 (2009)
2. Atoui, H., Fayn, J., Rubel, P.: A novel neural-network model for deriving standard 12-lead ECGs from serial three-lead ECGs: application to self-care. IEEE Trans. Inf. Technol. Biomed. **14**(3), 883–890 (2010)
3. Attia, Z.I., et al.: An artificial intelligence-enabled ECG algorithm for the identification of patients with atrial fibrillation during sinus rhythm: a retrospective analysis of outcome prediction. The Lancet **394**(10201), 861–867 (2019)
4. Bos, J.M., Attia, Z.I., Albert, D.E., Noseworthy, P.A., Friedman, P.A., Ackerman, M.J.: Use of artificial intelligence and deep neural networks in evaluation of patients with electrocardiographically concealed long QT syndrome from the surface 12-lead electrocardiogram. JAMA Cardiol. **6**(5), 532–538 (2021)

5. Chen, J., Liao, K., Wei, K., Ying, H., Chen, D.Z., Wu, J.: ME-GAN: learning panoptic electrocardio representations for multi-view ECG synthesis conditioned on heart diseases. In: Chaudhuri, K., Jegelka, S., Song, L., Szepesvari, C., Niu, G., Sabato, S. (eds.) Proceedings of the 39th International Conference on Machine Learning. Proceedings of Machine Learning Research, vol. 162, pp. 3360–3370. PMLR, 17–23 July 2022

6. Chen, J., Zheng, X., Yu, H., Chen, D.Z., Wu, J.: Electrocardio panorama: synthesizing new ECG views with self-supervision. In: Zhou, Z.H. (ed.) Proceedings of the Thirtieth International Joint Conference on Artificial Intelligence, IJCAI-21, pp. 3597–3605 (8 2021)

7. Cho, Y., et al.: Artificial intelligence algorithm for detecting myocardial infarction using six-lead electrocardiography. Sci. Rep. **10**, 20495 (2020)

8. Einthoven, W.: The different forms of the human electrocardiogram and their signification. The Lancet. **179**(4622), 853–861 (1912)

9. Golany, T., Lavee, G., Tejman Yarden, S., Radinsky, K.: Improving ECG classification using generative adversarial networks. Proc. AAAI Conf. Artif. Intell. **34**(08), 13280–13285 (2020)

10. Golany, T., Radinsky, K.: PGANs: personalized generative adversarial networks for ECG synthesis to improve patient-specific deep ECG classification. Proc. AAAI Conf. Artif. Intell. **33**(01), 557–564 (2019)

11. Golany, T., Radinsky, K., Freedman, D.: SimGANs: simulator-based generative adversarial networks for ECG synthesis to improve deep ECG classification. In: III, H.D., Singh, A. (eds.) Proceedings of the 37th International Conference on Machine Learning. Proceedings of Machine Learning Research, vol. 119, pp. 3597–3606. PMLR, 13–18 July 2020. https://proceedings.mlr.press/v119/golany20a.html

12. Goldberger, A.L., Goldberger, Z.D., Shvilkin, A.: Chapter 4 - ECG Leads. In: Goldberger, A.L., Goldberger, Z.D., Shvilkin, A. (eds.) Goldberger's Clinical Electrocardiography (Ninth Edition), pp. 21–31, 9th Edn. Elsevier (2018)

13. Goodfellow, I., et al.: Generative adversarial nets. In: Ghahramani, Z., Welling, M., Cortes, C., Lawrence, N., Weinberger, K. (eds.) Advances in Neural Information Processing Systems, vol. 27. Curran Associates, Inc. (2014)

14. Hossain, K.F., Kamran, S.A., Tavakkoli, A., Pan, L., Ma, X., Rajasegarar, S., Karmaker, C.: ECG-Adv-GAN: detecting ECG adversarial examples with conditional generative adversarial networks. In: 2021 20th IEEE International Conference on Machine Learning and Applications (ICMLA), pp. 50–56 (2021)

15. Hughes, J.W., et al.: Performance of a convolutional neural network and explainability technique for 12-lead electrocardiogram interpretation. JAMA Cardiol. **6**(11), 1285–1295 (2021)

16. Isola, P., Zhu, J.Y., Zhou, T., Efros, A.A.: Image-to-image translation with conditional adversarial networks. In: 2017 IEEE Conference on Computer Vision and Pattern Recognition (CVPR), pp. 5967–5976 (2017). https://doi.org/10.1109/CVPR.2017.632

17. Lee, J., Kim, M., Kim, J.: Reconstruction of precordial lead electrocardiogram from limb leads using the state-space model. IEEE J. Biomed. Health Inform. **20**(3), 818–828 (2016)

18. Ribeiro, A.H., et al.: Automatic diagnosis of the 12-lead ECG using a deep neural network. Nat. Commun. **11**, 1760 (2020)

19. Ronneberger, O., Fischer, P., Brox, T.: U-Net: convolutional networks for biomedical image segmentation. In: Navab, N., Hornegger, J., Wells, W.M., Frangi, A.F. (eds.) MICCAI 2015. LNCS, vol. 9351, pp. 234–241. Springer, Cham (2015). https://doi.org/10.1007/978-3-319-24574-4_28

20. Sarkar, P., Etemad, A.: Cardiogan: Attentive generative adversarial network with dual discriminators for synthesis of ECG from PPG. In: Thirty-Fifth AAAI Conference on Artificial Intelligence, AAAI 2021, Thirty-Third Conference on Innovative Applications of Artificial Intelligence, IAAI 2021, The Eleventh Symposium on Educational Advances in Artificial Intelligence, EAAI 2021, Virtual Event, February 2–9, 2021. pp. 488–496. AAAI Press (2021). https://ojs.aaai.org/index.php/AAAI/article/view/16126

21. Wang, L., Zhou, W., Xing, Y., Liu, N., Movahedipour, M., Zhou, X.: A novel method based on convolutional neural networks for deriving standard 12-lead ECG from serial 3-lead ECG. Front. Inf. Technol. Electron. Eng. **20**(3), 405–413 (2019)

22. Zhang, Y.H., Babaeizadeh, S.: Synthesis of standard 12-lead electrocardiograms using two-dimensional generative adversarial networks. J. Electrocardiol. **69**, 6–14 (2021)

23. Zhu, F., Fei, Y., Fu, Y., Liu, Q., Shen, B.: Electrocardiogram generation with a bidirectional LSTM-CNN generative adversarial network. Sci. Rep. **9**, 1–11, 6734 (2019)

24. Zhu, J.Y., Park, T., Isola, P., Efros, A.A.: Unpaired image-to-image translation using cycle-consistent adversarial networks. In: 2017 IEEE International Conference on Computer Vision (ICCV), pp. 2242–2251 (2017). https://doi.org/10.1109/ICCV.2017.244

Forward-Solution Aided Deep-Learning Framework for Patient-Specific Noninvasive Cardiac Ectopic Pacing Localization

Yashi Li, Huihui Ye, and Huafeng Liu[✉]

State Key Laboratory of Modern Optical Instrumentation, Zhejiang University,
Hangzhou 310027, China
liuhf@zju.edu.cn

Abstract. Accurate localization of the ectopic pacing is the key to effective catheter ablation for curing cardiac diseases such as premature ventricular contraction (PVC) and tachycardia. Invasive localization method can achieve high precision but has disadvantages of high risk, high cost, and time-consuming process, therefore, a non-invasive and convenient localization method is in demand. Noninvasive methods have been developed to utilize electrophysiological information provided by 12-lead electrocardiogram (ECG), and most of them are purely based on end-to-end data-driven architecture. This architecture generally needs a substantial and comprehensive labeled dataset, which is very difficult to obtain for whole ventricular ectopic beats in clinical setting. To address this issue, we propose a framework that combines cardiac forward-solution simulation and deep learning network for patient-specific noninvasive ectopic pacing localization. For each patient, it only requires his/her own CT images to establish a specific heart-torso model and to simulate various ECG data from different ectopic pacing locations and uses this simulated ECG data as the training dataset for our designed network. The network mainly contains time-frequency fusion module and local-global feature extraction module. Five PVC patient ECG data are tested with high precision and accuracy for ectopic pacing localization, which shows its high-potential in clinical setting.

Keywords: Ectopic pacing localization · Forward-solution · Deep Learning

1 Introduction

Catheter ablation surgery is a common operation that treats premature ventricular contraction arrhythmia effectively, and it mainly includes two steps: electrical physiological measurement and radiofrequency ablation. For electrical physiological measurement, it requires doctors to insert an electrode catheter in patient's heart, analyze the causes and parts of arrhythmia through the electrical signal obtained by the catheter, and finally determine the specific location of ectopic pacing and catheter ablation [1]. It highly depends on the doctors' experience and is very time-consuming [2]. Thus a noninvasive localization method that determines the ectopic pacing location in advance will significantly shorten the process of catheter ablation surgery and brings benefits for both doctors and patients.

H. Greenspan et al. (Eds.): MICCAI 2023, LNCS 14226, pp. 195–205, 2023.
https://doi.org/10.1007/978-3-031-43990-2_19

There is a trend to use computational tools based on 12-lead electrocardiogram (ECG) to analyze ablation location since ECG has been proven that can provide information about pacing areas in patients [3–5]. Earlier studies focused on extracting features of QRS axis through mathematical statistics and machine learning methods [6]. In recent years, deep learning has become a research hotspot due to its excellent feature extraction ability, which can automatically learn the mapping relationship between signal characteristics and pacing position [7–10]. Most methods are purely end-to-end data-driven architecture based on a large number of clinical databases. However, that will be greatly influenced by the comprehensiveness and size of the labeled data [11], especially when obtaining clinical data of whole ventricular ectopic beats is difficult.

Rapid developments in computer performance and theoretical knowledge have enabled detailed, physiologically realistic whole-heart simulations of arrhythmias and pacing [12, 13]. Based on this, a forward-solution computational mapping system for accurate localization of atrial and ventricular arrhythmias has been proposed, where a comprehensive arrhythmia simulation library is generated [14]. It can eliminate the impact of insufficient clinical trial data on the algorithm's accuracy. However, since the computational models are not specific, it will introduce errors in the ectopic pacing location process.

In this paper, inspired by work in [14], we propose a forward-solution aided deep-learning framework to realize noninvasive prediction of ectopic pacing from 12-lead ECG. Patient-specific heart-torso forward model is built, and a time-frequency fusion network based on the local-global feature extraction module is designed. The advantages of this paper can be summarized from three aspects:

1. Propose a framework that is trained based on ECG simulation data from the specific patient's CT for noninvasive cardiac ectopic pacing localization. It can eliminate the effect of insufficient clinical data and patient variance error on location accuracy.
2. Propose a network that combines time-frequency information and local-global information to achieve precise ectopic pacing location based on a small training data set.
3. Proposed method achieves great performance on PVC patient data, which demonstrates its potential for clinical cardiac treatment.

2   Methodology

2.1   Overview of Location Framework

Figure 1 outlines our framework for cardiac ectopic pacing localization. In brief, we construct a computational model of patient-specific anatomical structures, using a suitable cardiac source model and a transfer matrix H to obtain the solution of the electrocardiography forward problem for simulating whole-ventricle focal pacing as well as the corresponding ECGs. We then utilized this modeled dataset to train a deep-learning structure to locate the ectopic pacing. The structure is first evaluated on the simulated data and subsequently tested on clinical PVC patients' ECG data. We will describe the specific principles of our pipeline in detail in the following sections.

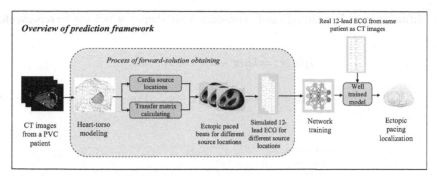

Fig. 1. Proposed forward-solution aided deep-learning framework for noninvasive cardiac ectopic pacing localization

2.2 Process of Forward-Solution Obtaining

Establishing a specific heart-torso model requires the geometric structure of both patient's heart and torso, which can be constructed from CT imaging data from the common preoperative scanning. The solution to the forward problem is obtained from the process of calculating torso surface potentials based on known cardiac source parameters [15], which can be mathematically represented as:

$$y = A(x) \tag{1}$$

where y denotes the torso surface and, x denotes the cardiac source, A is a transfer function dependent on the source model, which has been demonstrated in prior research that can be expressed as a transfer matrix $H \in R_{N*M}$. . This transfer matrix can be calculated through the resolution of the relationship between the cardiac and torso domains based on the finite element method (FEM) or boundary element method (BEM). In terms of the cardiac source, the two most common models are the activation-based model and the potential-based model [16]. One regards the arrival time of the depolarized wavefront as the main characteristic of the electrical activity of the heart; the latter uses the time-varying potential of the cardiac surface.

Compared to activation-based model, the potential-based model considers spatial distribution of cardiac potentials, offering insights into cardiac electric field and potential formation and propagation. Its comprehensive representation of cardiac electrical activity is advantageous for analyzing complex arrhythmias and cardiac disorders. In our experiment, we considered a simple two-variable potential-based model [17]; the process of cardiac excitation can be described as the follows:

$$\frac{\partial u}{\partial t} = \nabla(\boldsymbol{D}\nabla u) + f_1(u, v) \tag{2}$$

$$\frac{\partial v}{\partial t} = f_2(u, v) \tag{3}$$

where u is the transmembrane potentials, v is the conduction current, D is the diffusion tensor dependent on 3-D myocardial structure, and tissue conductive anisotropy, $\nabla(\boldsymbol{D}\nabla u)$

is the diffusion term. Functions f_1 and f_2 produce TMP shapes, which can be represented as:

$$f_1(u, v) = ku(u - a)(1 - u) - uv \tag{4}$$

$$f_2(u, v) = -e(v + ku(u - a - 1)) \tag{5}$$

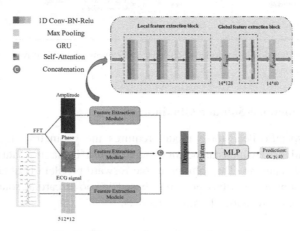

Fig. 2. Diagram of our proposed network

According to[17], parameter k, a, e is set as: $a = 0.15, k = 8, e = 0.01$. The longitudinal and transverse tensors of the diffusion tensor D are set to 4 and 1.

2.3 Network Structure

Figure 2 shows the structure of the proposed network. It mainly consists of the feature extraction part and the feature fusion part. Briefly, the 12-lead signal data (512 sampling points) is represented as a 512 * 12 matrix. Each lead signal undergoes Fourier transform to obtain frequency domain information, resulting in matrices of signal amplitude and signal phase with the same dimensions as the time domain matrix. These three matrices are input into the local-global feature extraction module to obtain encoded time and frequency information. An early fusion method is selected to concatenate these multi-level features, which is transformed into 3-dimensional coordinates of the predicted ectopic pacing location $\tilde{P}(x, y, z)$ by multilayer perceptron (MLP). The loss of the proposed network is defined as:

$$Loss = \frac{1}{N} \sum_{i=1}^{N} \|\tilde{P}(x, y, z) - P(x, y, z)\|_2^2 \tag{6}$$

where N is the number of samples and $P(x, y, z)$ is the labeled coordinates. To minimize the loss function, we can finally get a well-trained model. The training uses the Adam optimizer with a learning rate of 1e−4 and a batch size of 32.

Feature Fusion in Time-Frequency Domain. The electrocardiogram signal is a temporal signal, and therefore its analysis can be performed in both time and frequency domains.

The algorithm proposed in reference [18] was the first to convert the ECG signal to a Fourier spectrum to achieve discrimination of ventricular tachyarrhythmia. Currently, most algorithms process ECG signals in the time domain but ignore information in the frequency domain. Therefore, we choose the feature fusion method to combine time and frequency information for better accuracy localization. The Fourier transform breaks down a function into its constituent frequencies [19]. Given an ECG signal vector $\{\phi_t\}$ with $t \in [0, T-1]$ where T is the total number of samples, its Discrete Fourier transform (DFT) can be expressed as:

$$\Phi_k = \sum_{t=0}^{T-1} \phi_t e^{-\frac{2\pi i}{T} tk}, 0 \le k \le T - 1 \tag{7}$$

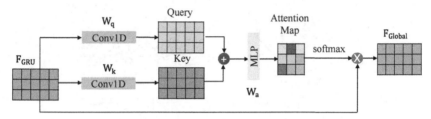

Fig. 3. Detailed illustration of our self-attention module

For each sample moment t, DFT algorithm generates a new representation Φ_k as the sum of all the original input ϕ_t. $\{\Phi_k\}$ is a complex matrix that can be represented as:

$$\{\Phi_k\} = \{r_k\}e^{i\{\theta_k\}}, 0 \le k \le T - 1 \tag{8}$$

where r_k and θ_k, respectively, denote amplitude and phase. We then separate amplitude matrix $\{r_0, r_1, \ldots, r_{T-1}\}$ and phase matrix $\{\theta_0, \theta_1, \ldots, \theta_{T-1}\}$ as new features that are sent to the network along with signal matrix $\{\phi_0, \phi_1, \ldots, \phi_{T-1}\}$ before normalization to $[-1, 1]$.

Local-Global Feature Extraction Module. Clinical diagnosis of heart disease based on 12-lead ECG mainly depends on its local features, such as P wave, T wave, or QRS complex. Convolutional neural network (CNN) has proven effective for extracting local information of ECG waveforms [20]. However, CNN's limited receptive field size makes it challenging to capture global feature representations, including dependence on long-distance signals and different leads. On the other hand, models like GRU [21] and attention mechanism [22], typical models in natural language processing, excel at capturing long-range dependencies but may compromise local feature details[23]. Inspired by these, our feature extraction module is divided into local feature extraction module and global feature extraction module, ensuring the acquisition of detailed and comprehensive information and improving the network's ability to encode ECG signals.

The local feature extraction block consists of three Conv1D blocks, each containing a 1DConv-BN-Relu layer and a max pooling layer. The Conv1D-BN-Relu layer comprises a one-dimensional convolutional layer, a batch normalization layer, and an activation layer with a Relu activation function. The kernel size of the convolutional layer is sequentially set to 7, 5, and 3; the stride is sequentially set to 2, 2, and 1.

Figure 2 shows the global feature extraction block. Gate Recurrent Unit (GRU)is a kind of recurrent neural network that occupies less memory and is more suitable for small data training [21], which is in line with our needs. The GRU layer is defined as:

$$H_{GRU}^t = GRU\left(\begin{matrix} H_{GRU}^{t-1}, F_{local}^t; \\ W_{GRU}^{t-1} \end{matrix}\right) \tag{9}$$

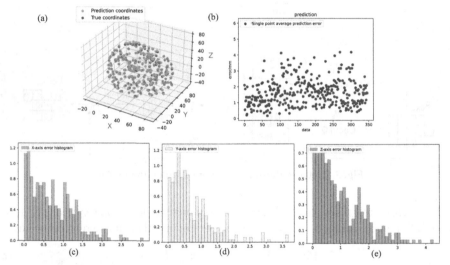

Fig. 4. Overall performance on testing data. (a) 3D display of predicted vs. labeled coordinates. (b) Location error point plot. (c) X-axis error histogram. (d) Y-axis error histogram. (e) Z-axis error histogram.

where $GRU(.)$ is the GRU network, H_{GRU} is the hidden state, F_{local} is the input and W_{GRU} represents the corresponding weight and biases.

In order to extract a broader dependence relationship, we introduced the additive self-attention mechanism after the GRU layer. Its network structure is shown in Fig. 3, which can be represented by the formula:

$$F_{Global} = softmax(MLP(W_q F_{GRU} + W_k F_{GRU}))F_{GRU} \tag{10}$$

where, F_{GRU} is the input of this part; more precisely, it denotes the output matrix from the previous GRU layer, with dimensions set to (14,40). W_* represents the weights to be learned and set to be (40,32); in this case, they are configured as a matrix with dimensions (40, 32). $MLP(.)$ means single layer MLP used for calculating the similarity of the

Query matrix and Key matrix. This operation yields an attention map that captures the correlation among the elements of the matrix. Finally, by applying the softmax function, we obtain the output F_{Global}, which encompasses the global feature.

3 Experimental Result

3.1 Overall Performance

The experimental data of this part are generated by ECGSIM software based on the UDL source model [24]. The solution of the electrocardiography forward problem is calculated to simulate 3485 sets of 12-lead ECG with 697 nodes of ectopic pacing on a 3-D heart model, and we set the active and resting TMP to 15 and 20, respectively. The training, validation, and test sets were divided into an 8:1:1 ratio.

Figure 4 shows the visual predictive performance of the test data. Figure 4(a) displays the coordinates of labeled data and the predicted location, showing that the predicted coordinate points nearly covered the actual value coordinate points. Figure 4(b) shows each test point's prediction error, indicating that the localization error was mainly below 2 mm. Figure 4(c)~(e) respectively displays the error in the x, y, and z-axis. The localization errors for each axis are mostly concentrated in the 1.5 mm range. Quantitative error data are presented in Table 1, which shows that the mean localization error for all test data is 1.76 mm, and the mean localization error in the x-axis, y-axis, and z-axis, respectively, is 0.84 mm, 0.95 mm, and 0.91 mm.

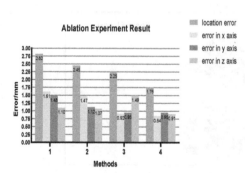

Fig. 5. Ablation study result.

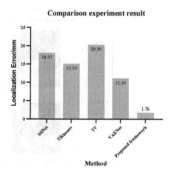

Fig. 6. Comparison experiment result.

Noise Robustness Experiment. In practical clinical measurement processes, noise is inevitably introduced; therefore, the prediction model needs robustness to noise.

In this section, we added Gaussian white noise with 5 dB, 15 dB, and 25 dB to the testing 12-lead data, respectively. From the results, we can see that network has certain robustness to different noise levels. At 25 dB noise, the prediction error is still only 2.97 mm. Though the error reached 10 mm at 5 dB, it is still within the clinically acceptable range considering the small signal-to-noise ratio (SNR).

Table 1. Localization error under different simulated levels of noise

SNR	Location Error/mm	Error in x axis/mm	Error in y axis/mm	Error in z axis/mm
Without noise	1.76	0.84	0.95	0.91
25DB	2.97	1.98	1.43	0.97
10DB	4.45	3.16	2.17	1.18
5DB	10.56	6.78	5.69	3.14

Ablation and Comparison Experiment. We compared four experiments' results to verify the effectiveness of the feature-fusion module and the local-global feature-extraction module: (1) CNN with feature fusion module: Remove the global feature extraction module; (2) GRU-Attention with feature fusion module: Remove the local feature extraction module; (3) Network without feature fusion module: only original ECGs sent to the local-global feature extraction module (4) Proposed Network. As the results shown in Fig. 5, single feature extraction networks, method (1) and method (2), are inferior compared to the proposed network. Method (3) only processes temporal signals and has lower error metrics than the above two methods, but the localization error is still 27.8% higher than that of the proposed network. These results demonstrate that combining temporal and frequency domain information of lead signals can improve the localization accuracy of ectopic pacing sites.

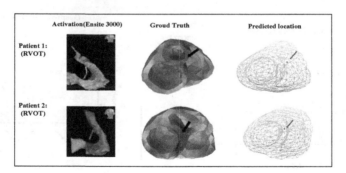

Fig. 7. Intuitive experiment result on clinical data.

To further demonstrate the capacity of the forward-solution aided method for ectopic pacing localization, we compared our method with the typical inverse solution methods shown in [25], which is also based on ECGSIM to simulate ectopic pacing data [26]. As is shown in Fig. 6, SSNet, Tikhonov, TV, and VAENet are inverse solution methods, and their localization errors are all >10 mm, while that of our method can reach within 2 mm.

Table 2. Quantitative results for the simulated patient pacing data

Patient ID	Location Error/mm	Error in x axis/mm	Error in y axis/mm	Error in z axis/mm
1	7.50	3.52	3.30	4.42
2	4.44	1.70	2.23	2.85
3	4.76	2.27	2.14	2.75
4	5.02	2.35	2.58	2.64
5	4.70	1.71	3.50	1.69

3.2 Clinical Data Experiment

Previous experiments confirmed the efficacy of our prediction framework on simulated cardiac data and the network's superior ability to learn the relationship between 12-lead data and ectopic pacing location. We now transfer the method to clinical PVC patients' data using the double variation cardiac source for calculating the forward solution of the specific heart-torso model and use them as the training data.

Table 2 shows the quantitative results of simulated patient pacing data. The location error ranges from 7.50 mm to 4.44 mm, with an average error of 5.28 mm. The network demonstrates precision and accuracy in localizing ectopic pacing in different cardiac models. Figure 7 shows the ectopic pacing location results for patient 1 and 2. The first column displays the activation of cardiomyocytes on the heart surface measured by the gold standard Ensite3000 system, with the red area indicating the earliest activation at the pacing location. The second column shows the activation mapping of the whole heart by Ensite3000. Our framework's localization results are consistent with the gold standard, with both pacing positions located in the posterior right ventricular outflow tract (RVOT), highlighting the practical significance of our study.

4 Discussion

In this paper, we developed a forward solution-aided deep learning framework for analyzing ectopic pacing from 12-lead ECG data. Only CT data is needed to establish the specific heart-torso model for simulating ECG data as the training set for the designed network. Time-frequency fusion module and local-global feature extraction module are the core component of the network. Experiments have shown that the framework performs well on both simulated and clinical data. In the future, to enhance the robustness of our proposed method, additional datasets and comprehensive simulations involving a broader spectrum of cardiac conditions should be incorporated. This paper primarily emphasizes the clinical potential of the proposed approach rather than extensively comparing actual and simulated data. The discrepancies observed could be attributed to noise, slight variations in physiological parameters, and consistent electrode placement in clinical practice. Further investigation and discussion on these aspects will be addressed in future research endeavors.

Acknowledgements. This work is supported in part by the National Natural Science Foundation of China (No: U1809204, 61525106) and by the Talent Program of Zhejiang Province (No: 2021R51004).

References

1. Soejima, K., Suzuki, M., Maisel, W.: Catheter ablation in patients with multiple and unstable ventricular tachycardias after myocardial infarction. Short ablation lines guided by reentry circuit isthmuses and sinus rhythm mapping. ACC Curr. J. Rev. **1**(11), 66–67 (2002)
2. Yamada, T.: Twelve-lead electrocardiographic localization of idiopathic premature ventricular contraction origins. J. Cardiovasc. Electrophysiol. **30**(11), 2603–2617 (2019)
3. Waxman, H.L., Josephson, M.E.: Ventricular activation during ventricular endocardial pacing: I. electrocardiographic patterns related to the site of pacing. Am. J. Cardiol. **50**(1), 1–10 (1982)
4. Miller, J.M., Marchlinski, F.E., Buxton, A.E., Josephson, M.E.: Relationship between the 12-lead electrocardiogram during ventricular tachycardia and endocardial site of origin in patients with coronary artery disease. Circulation **77**(4), 759–766 (1988)
5. Ito, S., et al.: Development and validation of an ECG algorithm for identifying the optimal ablation site for idiopathic ventricular outflow tract tachycardia. J. Cardiovasc. Electrophysiol. **14**(12), 1280–1286 (2003)
6. Feng, Q., Hu, H., Liu, H.: Multi-level information for non-invasive identification of exit site of ventricular tachycardia. In: 2020 Computing in Cardiology, pp. 1–4. IEEE (2020)
7. Gyawali, P.K., Horacek, B.M., Sapp, J.L., Wang, L.: Learning disentangled representation from 12-lead electrograms: application in localizing the origin of ventricular tachycardia. arXiv preprint arXiv:1808.01524 (2018)
8. Yang, T., Yu, L., Jin, Q., Wu, L., He, B.: Localization of origins of premature ventricular contraction by means of convolutional neural network from 12-lead ECG. IEEE Trans. Biomed. Eng. **65**(7), 1662–1671 (2017)
9. Monaci, S., et al.: Non-invasive localization of post-infarct ventricular tachycardia exit sites to guide ablation planning: a computational deep learning platform utilizing the 12-lead electrocardiogram and intracardiac electrograms from implanted devices. Europace **25**(2), 469–477 (2023)
10. Strodthoff, N., Strodthoff, C.: Detecting and interpreting myocardial infarction using fully convolutional neural networks. Physiol. Meas. **40**(1), 015001 (2019)
11. Sun, C., Shrivastava, A., Singh, S., Gupta, A.: Revisiting unreasonable effectiveness of data in deep learning era. In: Proceedings of the IEEE International Conference on Computer Vision, pp. 843–852 (2017)
12. Gonzales, M.J., Vincent, K.P., Rappel, W.J., Narayan, S.M., McCulloch, A.D.: Structural contributions to fibrillatory rotors in a patient-derived computational model of the atria. Europace **16**(suppl_4), iv3–iv10 (2014)
13. Villongco, C.T., Krummen, D.E., Stark, P., Omens, J.H., McCulloch, A.D.: Patient-specific modeling of ventricular activation pattern using surface ECG-derived vectorcardiogram in bundle branch block. Prog. Biophys. Mol. Biol. **115**(2–3), 305–313 (2014)
14. Krummen, D.E., et al.: Forward-solution noninvasive computational arrhythmia mapping: the VMAP study. Circulat. Arrhythmia Electrophysiol. **15**(9), e010857 (2022)
15. Bear, L.R., et al.: Forward problem of electrocardiography: is it solved? Circulat. Arrhythmia Electrophysiol. **8**(3), 677–684 (2015)
16. Langrill, D.M., Roth, B.J.: The effect of plunge electrodes during electrical stimulation of cardiac tissue. IEEE Trans. Biomed. Eng. **48**(10), 1207–1211 (2001)

17. Aliev, R.R., Panfilov, A.V.: A simple two-variable model of cardiac excitation. Chaos, Solitons Fract. **7**(3), 293–301 (1996)
18. Minami, K.I., Nakajima, H., Toyoshima, T.: Real-time discrimination of ventricular tachyarrhythmia with Fourier-transform neural network. IEEE Trans. Biomed. Eng. **46**(2), 179–185 (1999)
19. Lee-Thorp, J., Ainslie, J., Eckstein, I., Ontanon, S.: Fnet: mixing tokens with Fourier transforms. arXiv preprint arXiv:2105.03824 (2021)
20. Hannun, A.Y., et al.: Cardiologist-level arrhythmia detection and classification in ambulatory electrocardiograms using a deep neural network. Nat. Med. **25**(1), 65–69 (2019)
21. Chung, J., Gulcehre, C., Cho, K., Bengio, Y.: Empirical evaluation of gated recurrent neural networks on sequence modeling. arXiv preprint arXiv:1412.3555 (2014)
22. Vaswani, A., et al.: Attention is all you need. In: Advances in Neural Information Processing Systems, vol. 30 (2017)
23. Peng, Z., et al.: Local features coupling global representations for visual recognition. In: CVF International Conference on Computer Vision, ICCV, pp. 357–366. IEEE (2021)
24. Van Oosterom, A., Oostendorp, T.: ECGsim: an interactive tool for studying the genesis of QRST waveforms. Heart **90**(2), 165–168 (2004)
25. Huang, X., Yu, C., Liu, H.: Physiological model based deep learning framework for cardiac TMP recovery. In: Wang, L., Dou, Q., Fletcher, P.T., Speidel, S., Li, S. (eds.) MICCAI 2022. LNCS, vol. 13432, pp. 433–443. Springer, Cham (2022). https://doi.org/10.1007/978-3-031-16434-7_42
26. Ramanathan, C., Ghanem, R.N., Jia, P., Ryu, K., Rudy, Y.: Noninvasive electrocardiographic imaging for cardiac electrophysiology and arrhythmia. Nat. Med. **10**(4), 422–428 (2004)

Tensor-Based Multimodal Learning for Prediction of Pulmonary Arterial Wedge Pressure from Cardiac MRI

Prasun C. Tripathi[1]([envelope]), Mohammod N. I. Suvon[1], Lawrence Schobs[1], Shuo Zhou[1,2], Samer Alabed[3,4,5], Andrew J. Swift[3,4,5], and Haiping Lu[1,2,5]

[1] Department of Computer Science, University of Sheffield, Sheffield, UK
{p.c.tripathi,m.suvon,laschobs1,shuo.zhou,
h.lu,s.alabed,a.j.swift}@sheffield.ac.uk
[2] Centre for Machine Intelligence, University of Sheffield, Sheffield, UK
[3] Department of Infection, Immunity and Cardiovascular Disease,
University of Sheffield, Sheffield, UK
[4] Department of Clinical Radiology, Sheffield Teaching Hospitals, Sheffield, UK
[5] INSIGNEO, Institute for in Silico Medicine, University of Sheffield, Sheffield, UK

Abstract. Heart failure is a severe and life-threatening condition that can lead to elevated pressure in the left ventricle. Pulmonary Arterial Wedge Pressure (PAWP) is an important surrogate marker indicating high pressure in the left ventricle. PAWP is determined by Right Heart Catheterization (RHC) but it is an invasive procedure. A non-invasive method is useful in quickly identifying high-risk patients from a large population. In this work, we develop a tensor learning-based pipeline for identifying PAWP from multimodal cardiac Magnetic Resonance Imaging (MRI). This pipeline extracts spatial and temporal features from high-dimensional scans. For quality control, we incorporate an uncertainty-based binning strategy to identify poor-quality training samples. We leverage complementary information by integrating features from multimodal data: cardiac MRI with short-axis and four-chamber views, and cardiac measurements. The experimental analysis on a large cohort of 1346 subjects who underwent the RHC procedure for PAWP estimation indicates that the proposed pipeline has a diagnostic value and can produce promising performance with significant improvement over the baseline in clinical practice (i.e., ΔAUC = 0.10, ΔAccuracy = 0.06, and ΔMCC = 0.39). The decision curve analysis further confirms the clinical utility of our method. The source code can be found at: https://github.com/prasunc/PAWP.

Keywords: Cardiac MRI · Multimodal Learning · Pulmonary Arterial Wedge Pressure

1 Introduction

Heart failure is usually characterized by the inability of the heart to supply enough oxygen and blood to other organs of the body [4]. It is a major cause of

H. Greenspan et al. (Eds.): MICCAI 2023, LNCS 14226, pp. 206–215, 2023.
https://doi.org/10.1007/978-3-031-43990-2_20

mortality and hospitalization [14]. Elevated Pulmonary Arterial Wedge Pressure (PAWP) is indicative of raised left ventricular filling pressure and reduced contractility of the heart. In the absence of mitral valve or pulmonary vasculature disease, PAWP correlates with the severity of heart failure and risk of hospitalization [1]. While PAWP can be measured by invasive and expensive Right Heart Catheterization (RHC), simpler and non-invasive techniques could aid in better monitoring of heart failure patients. Cardiac Magnetic Resonance Imaging (MRI) is an effective tool for identifying various heart conditions and its ability to detect disease and predict outcome has been further improved by machine learning techniques [3]. For instance, Swift et al. [17] introduced a machine-learning pipeline for identifying Pulmonary Arterial Hypertension (PAH). Recently, Uthoff et al. [18] developed geodesically smoothed tensor features for predicting mortality in PAH.

Cardiac MRI scans contain high-dimensional spatial and temporal features generated throughout the cardiac cycle. The small number of samples compared to the high-dimensional features poses a challenge for machine learning classifiers. To address this issue, Multilinear Principal Component Analysis (MPCA) [11] utilizes a tensor-based approach to reduce feature dimensions while preserving the information for each mode, i.e. spatial and temporal information in cardiac MRI. Hence, the MPCA method is well-suited for analyzing cardiac MRI scans. The application of the MPCA method to predict PAWP might further increase the diagnostic yield of cardiac MRI in heart failure patients and help to establish cardiac MRI as a non-invasive alternative to RHC. Existing MPCA-based pipelines for cardiac MRI [2,17,18] rely on manually labeled landmarks that are used for aligning heart regions in cardiac MRI. The manual labeling of landmarks is a cumbersome task for physicians and impractical for analyzing large cohorts. Moreover, even small deviations in the landmark placement may significantly impact the classification performance of automatic pipelines [16]. To tackle this challenge, we leverage automated landmarks with uncertainty quantification [15] in our pipeline. We also extract complementary information from multimodal data from short-axis, four-chamber, and Cardiac Measurements (CM). We use CM features (i.e., left atrial volume and left ventricular mass) identified in the baseline work by Garg et al. [5] for PAWP prediction.

Our **main contributions** are summarized as follows: 1) **Methodology:** We developed a fully automatic pipeline for PAWP prediction using cardiac MRI data, which includes automatic landmark detection with uncertainty quantification, an uncertainty-based binning strategy for training sample selection, tensor feature learning, and multimodal feature integration. 2) **Effectiveness:** Extensive experiments on the cardiac MRI scans of 1346 patients with various heart diseases validated our pipeline with a significant improvement (ΔAUC = 0.1027, ΔAccuracy = 0.0628, and ΔMCC = 0.3917) over the current clinical baseline. 3) **Clinical utility:** Decision curve analysis indicates the diagnostic value of our pipeline, which can be used in screening high-risk patients from a large population.

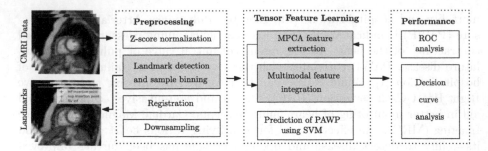

Fig. 1. The schematic overview of the PAWP prediction pipeline including preprocessing, tensor feature learning, and performance analysis. The blocks in gray color are explained in more detail in Sect. 2 (Color figure online).

2   Methods

As shown in Fig. 1, the proposed pipeline for PAWP prediction comprises three components: preprocessing, tensor feature learning, and performance analysis.

Cardiac MRI Preprocessing: The preprocessing of cardiac MRI contains (1) normalization of scans, (2) automatic landmark detection, (3) inter-subject registration, and (4) in-plane downsampling. We standardize cardiac MRI intensity levels using Z-score normalization [7] to eliminate inter-subject variations. Furthermore, we detect automatic landmarks which is explained in the next paragraph. We perform affine registration to align the heart regions of different subjects to a target image space. We then carry out in-plane scaling of scans by max-pooling at 2, 4, 8, and 16 times and obtain down-sampled resolutions of 128×128, 64×64, 32×32, and 16×16, respectively.

Landmark Detection and Uncertainty-based Sample Binning: We utilize supervised learning to automate landmark detection using an ensemble of Convolutional Neural Networks (CNNs) for each modality (short-axis and four-chamber). We use the U-Net-like architecture and utilize the same training regime implemented in [15]. We employ *Ensemble Maximum Heatmap Activation (E-MHA)* strategy [15] which incorporates an ensemble of five models for each modality. We utilize three landmarks for each modality, with the short-axis modality using the inferior hinge point, superior hinge point, and inferolateral inflection point of the right ventricular apex, and the four-chamber modality using the left ventricular apex and mitral and tricuspid annulus. E-MHA produces an associated uncertainty estimate for each landmark prediction, representing the model's epistemic uncertainty as a continuous scalar value.

A minor error in landmark prediction can result in incorrect image registration [16]. To address this issue, we hypothesize that incorrectly preprocessed samples resulting from inaccurate landmarks can introduce ambiguity during model training. For quality control, it is crucial to identify and effectively handle such samples. In this study, we leverage predicted landmarks and epistemic uncertainties to tackle this problem using uncertainty-based binning. To this end, we partition the training scans based on the uncertainty values of the landmarks. The predicted landmarks are divided into K quantiles, i.e., $Q = \{q_1, q_2, ..., q_K\}$, based on the epistemic uncertainty values. We then iteratively filter out training samples starting from the highest uncertain quantile. A sample is discarded if the uncertainty of any of its landmarks lies in quantile q_k where $k = \{1, 2, ..., K\}$. The samples are discarded iteratively until there is no improvement in the validation performance, as measured by the area under the curve (AUC), for two subsequent iterations.

Tensor Feature Learning: To extract features from processed cardiac scans, we employ tensor feature learning, i.e. Multilinear Principal Component Analysis (MPCA) [11], which learns multilinear bases from cardiac MRI stacks to obtain low-dimensional features for prediction. Suppose we have M scans as third-order tensors in the form of $\{\mathcal{X}_1, \mathcal{X}_2, .., \mathcal{X}_M \in \mathbb{R}^{I_1 \times I_2 \times I_3}\}$. The low-dimensional tensor features $\{\mathcal{Y}_1, \mathcal{Y}_2, .., \mathcal{Y}_M \in \mathbb{R}^{P_1 \times P_2 \times P_3}\}$ are extracted by learning three ($N = 3$) projection matrices $\{U^{(n)} \in \mathbb{R}^{I_n \times P_n}, n = 1, 2, 3\}$ as follows:

$$\mathcal{Y}_m = \mathcal{X}_m \times_1 U^{(1)^T} \times_2 U^{(2)^T} \times_3 U^{(3)^T}, m = 1, 2, ..., M, \tag{1}$$

where $P_n < I_n$, and $\times_n$ denotes a mode-wise product. Therefore, the feature dimensions are reduced from $I_1 \times I_2 \times I_3$ to $P_1 \times P_2 \times P_3$. We optimize the projection matrices $\{U^{(n)}\}$ by maximizing total scatter $\psi_\mathcal{Y} = \sum_{m=1}^{M} ||\mathcal{Y}_m - \bar{\mathcal{Y}}||_F^2$, where $\bar{\mathcal{Y}} = \frac{1}{M} \sum_{m=1}^{M} \mathcal{Y}_m$ is the mean tensor feature and $||.||_F$ is the Frobenius norm [10]. We solve this problem using an iterative projection method. In MPCA, $\{P_1, P_2, P_3\}$ can be determined by the explained variance ratio, which is a hyperparameter. Furthermore, we apply Fisher discriminant analysis to select the most significant features based on their Fisher score [8]. We select the top k-ranked features and employ Support Vector Machine (SVM) for classification.

Multimodal Feature Integration: To enhance performance, we perform multimodal feature integration using features extracted from the short-axis, four-chamber, and Cardiac Measurements (CM). We adopt two strategies for feature integration, namely the early and late fusion of features [6]. In early fusion, the features are fused at the input level without doing any transformation. We concatenate features from the short-axis and four-chamber to perform this fusion. We then apply MPCA [11] on the concatenated tensor, enabling the selection of multimodal features. In late fusion, the integration of features is performed at the common latent space that allows the fusion of features that have different dimensionalities. In this way, we can perform a late fusion of CM features with short-axis and four-chamber features. However, we can not perform an early fusion of CM features with short-axis and four-chamber features.

Performance Evaluation: In this paper, we use three primary metrics: Area Under Curve (AUC), accuracy, and Matthew's Correlation Coefficient (MCC), to evaluate the performance of the proposed pipeline. Decision Curve Analysis (DCA) is also conducted to demonstrate the clinical utility of our methodology.

Table 1. Baseline characteristics of included patients. p values were obtained using t-test [20].

	Low PAWP($\leq$ 15)	High PAWP($>$ 15)	p-value
Number of patients	940	406	-
Age (in years)	64.8 $\pm$ 14.2	70.5 $\pm$ 10.6	< 0.01
Body Surface Area (BSA)	1.88 $\pm$ 0.28	1.93 $\pm$ 0.24	< 0.01
Heart Rate (bpm)	73.9 $\pm$ 15.5	67.6 $\pm$ 15.9	< 0.01
Left Ventricle Mass (LVM)	92.3 $\pm$ 25	106 $\pm$ 33.1	< 0.01
Left Atrial Volume (ml^2)	72.2 $\pm$ 33.7	132.2 $\pm$ 56.7	< 0.01
PAWP (mmHg)	10.3 $\pm$ 3.1	21.7 $\pm$ 4.96	< 0.01

3 Experimental Results and Analysis

Study Population : Patients with suspected pulmonary hypertension were identified after institutional review board approval and ethics committee review. A total of 1346 patients who underwent Right Heart Catheterization (RHC) and cardiac MRI scans within 24 hours were included. Of these patients, 940 had normal PAWP ($\leq$ 15 mmHg), while 406 had elevated PAWP ($>$ 15 mmHg). Table 1 summarizes baseline patient characteristics. RHC was performed using a balloon-tipped 7.5 French thermodilution catheter.

Cardiac MRI and Measurement: MRI scans were obtained using a 1.5 Tesla whole-body GE HDx MRI scanner (GE Healthcare, Milwaukee, USA) equipped with 8-channel cardiac coils and retrospective electrocardiogram gating. Two cardiac MRI protocols, short-axis and four-chamber, were employed, following standard clinical protocols to acquire cardiac-gated multi-slice steady-state sequences with a slice thickness of 8 mm, a field of view of 48 $\times$ 43.2, a matrix size of 512 $\times$ 512, a bandwidth of 125 kHz, and TR/TE of 3.7/1.6 ms. Following [5], left ventricle mass and left atrial volume were selected as cardiac measurements.

Experimental Design: We conducted experiments on short-axis and four-chamber scans across four scales. To determine the optimal parameters, we performed 10-fold cross-validation on the training set. From MPCA, we selected the top 210 features. We employed early and late fusion on short-axis and four-chamber scans, respectively, while CM features were only fused using the late fusion strategy. We divided the data into a training set of 1081 cases and a testing set of 265 cases. To simulate a real testing scenario, we designed the experiments such that patients diagnosed in the early years were part of the training set, while patients diagnosed in recent years were part of the testing set. We also partitioned the test into 5 parts based on the diagnosis time to perform different runs of methods and report standard deviations of methods in comparison results. For SVM, we selected the optimal hyper-parameters from $\{0.001, 0.01, 0.1, 1\}$ using the grid search technique. The code for the experiments has been implemented in Python (version 3.9). We leveraged the cardiac MRI preprocessing pipeline and MPCA from the Python library PyKale [9] and SVM implementation is taken from scikit-learn [12].

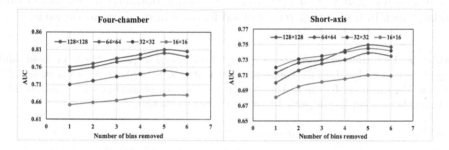

Fig. 2. Performance comparison of removing a different number of bins of training data on 10-fold cross-validation.

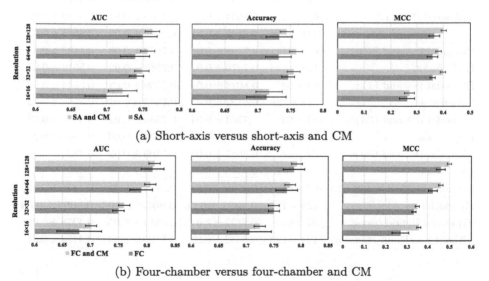

(a) Short-axis versus short-axis and CM

(b) Four-chamber versus four-chamber and CM

Fig. 3. The effect of combining CM features on short-axis and four-chamber. SA: Short-axis; FC: Four-chamber.

Uncertainty-Based Sample Binning: To improve the quality of training data, we used quantile binning to remove training samples with uncertain landmarks. The landmarks were divided into 50 bins, and then removed one bin at a time in the descending order of their uncertainties. Figure 2 depicts the results of binning using 10-fold cross-validation on the training set, where the performance improves consistently over the four scales when removed bins ≤ 5. Based on the results, we removed 5 bins (129 out of 1081 samples) from the training set, and used the remaining 952 training samples for the following experiments.

Unimodal Study: The performance of three models on single-modality is reported in Table 2, including short-axis (SA), four-chamber (FC), and cardiac measurements (CM), where the CM based unimodal is considered as the baseline. The results demonstrate an improvement of $\Delta AUC = 0.0800$ $\Delta Accuracy = 0.0527$, and $\Delta MCC = 0.3484$ over the baseline obtained by FC based unimodal, which indicates that tensor-based features have a diagnostic value.

Table 2. Performance comparison using three metrics (with **best** in bold and second best underlined). FC: Four-Chamber features; SA: Short-Axis features; CM: Cardiac Measurement features. The standard deviations of methods were obtained by dividing the test set into 5 parts based on the diagnosis time.

Modality	Resolution	AUC	Accuracy	MCC
Unimodal (CM) [5]	–	0.7300 ± 0.04	0.7400 ± 0.03	0.1182 ± 0.03
Unimodal (SA) [17]	64×64	0.7391 ± 0.05	0.7312 ± 0.07	0.3604 ± 0.02
	128×128	0.7495 ± 0.05	0.7321 ± 0.04	0.3277 ± 0.01
Unimodal (FC) [17]	64×64	0.8034 ± 0.02	0.7509 ± 0.04	0.4240 ± 0.02
	128×128	0.8100 ± 0.04	0.7925 ± 0.05	0.4666 ± 0.02
Bi-modal (SA and FC):	64×64	0.7998 ± 0.01	0.7698 ± 0.03	0.4185 ± 0.03
Early fusion	128×128	0.7470 ± 0.02	0.7283 ± 0.02	0.3512 ± 0.02
Bi-modal (SA and FC):	64×64	0.8028 ± 0.04	0.7509 ± 0.03	0.3644 ± 0.01
Late fusion	128×128	0.8122 ± 0.03	0.7547 ± 0.03	0.3594 ± 0.02
Bi-modal (SA and CM):	64×64	0.7564 ± 0.04	0.7585 ± 0.02	0.3825 ± 0.02
Late fusion	128×128	0.7629 ± 0.03	0.7434 ± 0.03	0.3666 ± 0.03
Bi-modal (FC and CM):	64×64	0.8061 ± 0.03	0.7709 ± 0.02	0.4435 ± 0.02
Late fusion	128×128	$\underline{0.8135 \pm 0.02}$	$\underline{0.7925 \pm 0.02}$	$\underline{0.4999 \pm 0.03}$
Tri-modal (FC, SA, and CM)	64×64	0.8146 ± 0.04	0.7774 ± 0.03	0.4460 ± 0.02
Hybrid fusion	128×128	$\mathbf{0.8327 \pm 0.06}$	$\mathbf{0.8038 \pm 0.05}$	$\mathbf{0.5099 \pm 0.04}$
Tri-modal Hybrid fusion	64×64	0.7892 ± 0.04	0.7513 ± 0.05	0.4278 ± 0.02
without uncertainty binning	128×128	0.8036 ± 0.03	0.7820 ± 0.04	0.4779 ± 0.01

Bi-modal Study: In this experiment, we compared the performance of bi-modal models. As shown in Table 2, bimodal (four-chamber and CM) produces superior performance (i.e., AUC = 0.8135, Accuracy=0.7925 and MCC = 0.4999) among bi-modal models. Next, we investigated the effect of fusing CM features with short-axis and four-chamber modalities in Fig. 3. It can be observed from these figures that the fusion of CM features enhances the diagnostic power of cardiac MRI modalities at all scales. The bi-modal (four-chamber and CM) model achieved the improvement in the performance (ΔAUC = 0.0035 and ΔMCC = 0.0333) over the unimodal (four-chamber) model.

Effectiveness of Tri-modal: In this experiment, we performed a fusion of CM features with the bi-modal models to create two tri-modal models. The first tri-modal is tri-modal late (CM with a late fusion of short-axis and four-chamber) and the second tri-modal is a tri-modal hybrid (CM with an early fusion of short-axis and four-chamber). As shown in Fig. 4, CM features enhance the performance of bi-modal models and tri-modal hybrid outperforms all. The tri-modal hybrid obtained the best performance (Table 2, where AUC = 0.8327, Accuracy = 0.8038, and MCC = 0.5099) and a significant improvement of ΔAUC = 0.1027, ΔAccuracy = 0.0628, and ΔMCC = 0.3917 over the baseline method.

Decision Curve Analysis (DCA) [13,19] on the performance suggests the potential clinical utility of the proposed method. As shown in Fig. 5, the Tri-modal model outperformed the baseline method for most possible benefit/harm preferences, where benefit indicates a positive net benefit (i.e. correct diagnosis) and harm indicates a negative net benefit (i.e. incorrect diagnosis). The tri-modal model (the best model) obtained a higher net benefit between decision threshold probabilities of 0.30 and 0.70 which implies that our method has a diagnostic value and can be used in screening high-risk patients from a large population.

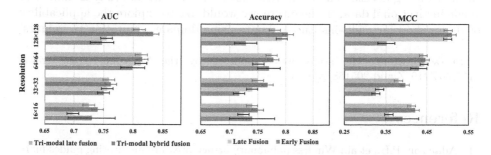

Fig. 4. The effect of combining CM features on the bi-modals including early and late fusion of four-chamber and short-axis. Early fusion: early fusion of short-axis and four-chamber; late fusion: late fusion of short-axis and four-chamber.

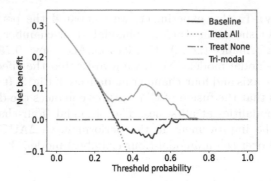

Fig. 5. Evaluating clinical utility of our method using Decision Curve Analysis (DCA) [19]. "Treat All" means treating all patients, regardless of their actual disease status, while "Treat None" means treating no patients at all. Our predictive model's net benefit is compared with the net benefit of treating everyone or no one to determine its overall utility.

Feature Contributions: Our model is interpretable. The highly-weighted features were detected in the left ventricle and interventricular septum in cardiac MRI. For cardiac measurements, left atrial volume (0.778/1) contributed more than left ventricular mass (0.222/1) to the prediction.

4   Conclusions

This paper proposed a tensor learning-based pipeline for PAWP classification. We demonstrated that: 1) tensor-based features have a diagnostic value for PAWP, 2) the integration of CM features improved the performance of unimodal and bi-modal methods, 3) the pipeline can be used to screen a large population, as shown using decision curve analysis. However, the current study is limited to single institutional data. In the future, we would like to explore the applicability of the method for multi-institutional data using domain adaptation techniques.

Acknowledgment. The study was supported by the Wellcome Trust grants 215799/Z/19/Z and 205188/Z/16/Z.

References

1. Adamson, P.B., et al.: Wireless pulmonary artery pressure monitoring guides management to reduce decompensation in heart failure with preserved ejection fraction. Circul. Heart Fail. **7**(6), 935–944 (2014)
2. Alabed, S., et al.: Machine learning cardiac-MRI features predict mortality in newly diagnosed pulmonary arterial hypertension. Eur. Heart J.-Digit. Health **3**(2), 265–275 (2022)

3. Assadi, H., et al.: The role of artificial intelligence in predicting outcomes by cardio-vascular magnetic resonance: a comprehensive systematic review. Medicina **58**(8), 1087 (2022)
4. Emdin, M., Vittorini, S., Passino, C., Clerico, A.: Old and new biomarkers of heart failure. Eur. J. Heart Fail. **11**(4), 331–335 (2009)
5. Garg, P., et al.: Cardiac magnetic resonance identifies raised left ventricular filling pressure: prognostic implications. Eur. Heart J. **43**(26), 2511–2522 (2022)
6. Huang, S.C., Pareek, A., Zamanian, R., Banerjee, I., Lungren, M.P.: Multimodal fusion with deep neural networks for leveraging CT imaging and electronic health record: a case-study in pulmonary embolism detection. Sci. Rep. **10**(1), 1–9 (2020)
7. Jain, A., Nandakumar, K., Ross, A.: Score normalization in multimodal biometric systems. Pattern Recogn. **38**(12), 2270–2285 (2005)
8. Li, J., et al.: Feature selection: a data perspective. ACM Comput. Surv. (CSUR) **50**(6), 94 (2018)
9. Lu, H., et al.: PyKale: knowledge-aware machine learning from multiple sources in python. In: Proceedings of the 31st ACM International Conference on Information & Knowledge Management, pp. 4274–4278 (2022)
10. Lu, H., Plataniotis, K.N., Venetsanopoulos, A.: Multilinear subspace learning: dimensionality reduction of multidimensional data. CRC Press (2013)
11. Lu, H., Plataniotis, K.N., Venetsanopoulos, A.N.: MPCA: multilinear principal component analysis of tensor objects. IEEE Trans. Neural Netw. **19**(1), 18–39 (2008)
12. Pedregosa, F., et al.: Scikit-learn: machine learning in Python. J. Mach. Learn. Res. **12**, 2825–2830 (2011)
13. Sadatsafavi, M., Adibi, A., Puhan, M., Gershon, A., Aaron, S.D., Sin, D.D.: Moving beyond AUC: decision curve analysis for quantifying net benefit of risk prediction models. Eur. Respirat. J. **58**(5), 2101186 (2021)
14. Savarese, G., Becher, P.M., Lund, L.H., Seferovic, P., Rosano, G.M., Coats, A.J.: Global burden of heart failure: a comprehensive and updated review of epidemiology. Cardiovasc. Res. **118**(17), 3272–3287 (2022)
15. Schöbs, L., Swift, A.J., Lu, H.: Uncertainty estimation for heatmap-based landmark localization. IEEE Transactions on Medical Imaging (2022)
16. Schobs, L., Zhou, S., Cogliano, M., Swift, A.J., Lu, H.: Confidence-quantifying landmark localisation for cardiac MRI. In: 2021 IEEE 18th International Symposium on Biomedical Imaging (ISBI), pp. 985–988. IEEE (2021)
17. Swift, A.J., et al.: A machine learning cardiac magnetic resonance approach to extract disease features and automate pulmonary arterial hypertension diagnosis. Eur. Heart J.-Cardiovascul. Imaging **22**(2), 236–245 (2021)
18. Uthoff, J., Alabed, S., Swift, A.J., Lu, H.: Geodesically smoothed tensor features for pulmonary hypertension prognosis using the heart and surrounding tissues. In: 23rd International Conference Medical Image Computing and Computer Assisted Intervention-MICCAI 2020, pp. 253–262 (2020)
19. Vickers, A.J., Elkin, E.B.: Decision curve analysis: a novel method for evaluating prediction models. Med. Decis. Making **26**(6), 565–574 (2006)
20. Welch, B.L.: The generalization of 'student's' problem when several different population varlances are involved. Biometrika **34**(1–2), 28–35 (1947)

Virtual Heart Models Help Elucidate the Role of Border Zone in Sustained Monomorphic Ventricular Tachycardia

Eduardo Castañeda[1,2]([✉]), Masahito Suzuki[3], Hiroshi Ashikaga[4], Èric Lluch[2], Felix Meister[2], Viorel Mihalef[5], Chloé Audigier[6], Andreas Maier[1], Henry Halperin[7], and Tiziano Passerini[5]

[1] Pattern Recognition Lab, Department of Computer Science, Friedrich-Alexander University Erlangen-Nürnberg, Erlangen, Germany
eduardo.castaneda@fau.de
[2] Siemens Healthineers, Digital Technologies and Innovation, Erlangen, Germany
[3] JA Toride Medical Ctr, Department of Cardiology, Ibaraki, Japan
[4] Pali Momi Heart Center, Aiea, Hawaii, USA
[5] Siemens Healthineers, Digital Technologies and Innovation, Princeton, USA
[6] Advanced Clinical Imaging Technology, Siemens Healthcare, Lausanne, Switzerland
[7] Johns Hopkins University, Department of Medicine, Division of Cardiology, Baltimore, Maryland, USA

Abstract. Post-ischemic Ventricular Tachycardia (VT) is sustained by a depolarization wave re-entry through channel-like structures within the post-ischemic scar. These structures are usually formed by partially viable tissue, called Border Zone (BZ). Understanding the anatomical and electrical properties of the BZ is crucial to guide ablation therapy to the right targets, reducing the likelihood of VT recurrence. Virtual Heart methods can provide ablation guidance non-invasively, but they have high computational complexity and have shown limited capability to accurately reproduce the specific mechanisms responsible for clinically observed VT. These outstanding challenges undermine the utility of Virtual Hearts for high precision ablation guidance in clinical practice. In this work, fast phenomenological models are developed to efficiently and accurately simulate the re-entrant dynamics of VT as observed in 12-lead ECG. Two porcine models of Myocardial Infarction (MI) are used to generate personalized bi-ventricular models from pre-operative LGE-MRI images. Myocardial conductivity and action potential duration are estimated using sinus rhythm ECG measurements. Multiple hypotheses for the BZ tissue properties are tested, and optimal values are identified. These allow the Virtual Heart model to produce VTs with good agreements with measurements in terms of ECG lead polarity and VT cycle length. Efficient GPU implementation of the cardiac electrophysiology model allows computation of sustained monomorphic VT in times compatible with the clinical workflow.

Supplementary Information The online version contains supplementary material available at https://doi.org/10.1007/978-3-031-43990-2_21.

Keywords: Cardiac modeling · VT · VAAT · ECG simulation

1   Introduction

Post-ischemic Ventricular Tachycardia (VT) is an arrhythmia that occurs after a myocardial infarction event. During Myocardial Infarction (MI), the blood flow to an area of the heart is blocked, producing tissue death and scarring [8]. After tissue healing, partially viable areas can appear within the scar tissue. These areas, usually referred to as Border Zone (BZ), contain a complex mixture of scar and viable myocardium [8]. These can produce channel-like structures that act as paths for an abnormal depolarization wave re-entry that sustains the VT [8,22].

Catheter ablation is recommended for ischemic heart disease patients, in the presence of recurrent monomorphic VT despite anti-arrhythmic drug therapy. Ablation therapy aims at destroying a part of the re-entry path (channel) that sustains the VT [22]. Precise identification of the target for ablation is paramount to maximize the efficacy of the procedure in terms of reduction of VT recurrence, while minimizing the amount and extension of ablation lesions. Both intracardiac electrograms and pre-operative Late Gadolinium Enhanced MRI have been adopted for channel identification [25], however the rate of success of catheter ablation is still unsatisfactory [22,25].

Virtual-Heart Arrhythmia Ablation Targeting (VAAT) methods have been developed to improve the utility of pre-operative imaging by using image-based computational models to reproduce the patient-specific re-entrant dynamics of the arrhythmia and thus increase precision in the localization of ablation targets [2,9,25]. These approaches generally suffer from two limitations. On the one hand, they have not shown the capability to consistently reproduce the same re-entries observed in-vivo and as measured by 12-lead ECG. On the other hand, they require a laborious process and time-intensive simulations incompatible with clinical use.

In this paper, we present a validation study of a novel VAAT method based on efficient phenomenological models of electrophysiology for high-fidelity simulation of VT. We successfully reproduce the inducibility of sustained monomorphic VT (more than 30 s) in two animal models of scar-related VT. In addition, we reproduce the ECG signature of each measured VT circuit, in terms of the polarity in the QRS complex of each lead as well as the VT cycle length. Finally, we demonstrate the feasibility of using this method in clinical practice by showing considerable speed up compared to the state-of-the art.

2   Methods

2.1   Data Description

Two porcine models of MI [11] were used for this study. The animal study was reviewed and approved by the Johns Hopkins University Animal Care and Use

218   E. Castañeda et al.

Committee (Baltimore, MD). High-resolution LGE-MRI (Aera, Siemens Healthineers) images of the subjects were acquired 6 weeks after the MI induction
procedure. One week after, the swine underwent the VT ablation procedure.
During the procedure, one VT was successfully induced in case 1, and three
different VTs were induced in case 2. The 12-lead ECG (CardioLab, GE Healthcare) was acquired pre-operatively and during VT for both cases. The QRS and
QT duration in the ECGs measured during sinus rhythm were annotated by
an electrophysiologist. Additionally, the ECG VT morphology was reviewed by
an electrophysiologist, which determined the VT cycle length and estimated the
approximate exit site location using the ECG lead polarity [1].

2.2 Anatomical Model Generation

The right ventricular (RV) endocardium, left ventricular (LV) endocardium, and
LV epicardium were manually contoured and reviewed by an electrophysiologist.
The computational domain for the electrophysiology model was defined as a
Cartesian grid of isotropic 0.5 mm resolution, obtained by rasterization of the
segmented surfaces [18]. A rule-based model of the myocardial fibers was also
included [3]. The complete anatomical generation process was applied as previously described [26]. A generic swine torso model was manually deformed and
oriented to fit the scout MRI images of each subject.

The scar and BZ were semi-automatically segmented by an expert and validated by an electrophysiologist. The full-width half-maximum method [23] was
applied to the stack of LGE-MRI images after resampling to an isotropic 3D
grid of 0.4 mm. The BZ segmentation was constrained to be within a maximum
distance of 4 mm from the closest scar following the assumption that BZ tissue
closely surrounds the scar core [23], which was consistent with measured voltage
maps. The tissue segmentation was then post-processed. First, we applied morphological closing with a spherical kernel of 4 voxel radius to close small holes
of BZ within the scar. Then we applied morphological opening with a spherical kernel of 1 voxel radius to remove small and isolated scar patches. Lastly,
we applied a dilation (max 2 mm) of the scar and BZ tissues in the transmural
direction to preserve transmurality after the domain rasterization. Preserving
scar transmurality prevents the generation of spurious endocardial or epicardial
re-entry pathways inconsistent with evidence from LGE-MRI and voltage maps.
See the Supplementary Materials for more details about the implementation.

2.3 Electrophysiology Simulation

The electrophysiology model is based on the monodomain equation using the
modified Mitchell-Schaeffer cellular model [6,19]. The numerical solver is an
efficient GPU implementation of the Lattice Boltzmann Method [17,21]. The
time-varying extracellular potential on the animal torso is computed following
the Boundary Element Method (BME) [26]. Virtual electrodes were placed on the
animal torso in the locations used in the clinical study, which were validated by

an electrophysiologist, and virtual 12-lead ECG signals were derived [17]. Each individual lead signal amplitude is then normalized to mitigate uncertainty on the torso anatomy.

Scar tissue was modeled as a non-conductive material. The BZ tissue was modeled as conductive, isotropic, and with a longer Action Potential Duration (APD) than the healthy tissue [16]. The transversal Conduction Velocity (CV) in the healthy myocardium was set to 0.45 times the longitudinal CV [7].

To reproduce the fast conduction produced by the Purkinje network, a Fast Endocardial Conductive (FEC) layer of 3 mm was added to the model to account for the increased transmurality of Purkinje networks in porcine models [24]. Longitudinal CV was set 4 times higher than in the healthy myocardium, while the transversal CV was unchanged [7,14]. The FEC layer was not placed over scar or BZ areas, which were measured as non-conductive by voltage mapping during the ablation procedure.

2.4   Parameter Personalization

The model parameters were tuned to match the ECG measurements during sinus rhythm. To simulate sinus rhythm, a one-second-long simulation was run with four pacing points placed in the basal and apical areas of the LV and RV septal walls to produce depolarization patterns consistent with previously published observations [10]. A grid-search approach was used to optimize the myocardial diffusivity and the closing time constant of the current gate (τ_{close}) of the Mitchell and Schaeffer cellular model, to match the measured QRS and QT interval respectively.

Myocardial diffusivity was sampled in a range of values corresponding to CV between 0.4 and 0.75 m/s, with 0.01 m/s steps, consistent with previous approaches [7,14]. After adjusting the CV, the τ_{close} parameter was sampled in a range of values corresponding to myocardial APD between 0.14 and 0.3 s with 0.005 s steps, consistent with reported values [13]. The APD was defined as the amount of time with voltage higher than 20% of its maximum value for each experiment.

The reference QRS and QT duration in the measured ECGs were annotated by an electrophysiologist in a digital trace plotted with a paper speed of 50 mm/s. The QRS and QT duration in the simulated ECGs were visually annotated by an expert and the set of parameters leading to the best match with the measured ECG were validated by an electrophysiologist. Due to the infarction features present in the ECG, such as ST elevation [12], the measurement procedure and thus the optimization procedure could not be fully automated, since automatic extraction of QRS and QT duration led to a relatively large uncertainty margin of up to 20 ms depending on the selected ECG lead and signal morphology.

Electrophysiology properties of the BZ were not personalized based on sinus rhythm ECG measurements, since we observed minimal effect on simulated sinus rhythm QRS and QT duration from variations of the BZ parameters within the range of previously reported values [7].

2.5 VT Induction Procedure

The simulation of a programmed stimulation experiment included three phases: model preconditioning, artificial stimulation delivery, and spontaneous activity simulation. First, the internal states of the model were pre-conditioned by simulating 5 s of sinus rhythm at 60 bpm. Then, 8 stimuli were delivered at a constant period S1, which ranged between 0.3 and 0.6 s, followed by three extra stimuli delivered with delays S2, S3, and S4. The pacing times S2, S3, and S4 were automatically reduced until VT was induced or no activation was produced, following the procedure described in the Supplementary Materials. The stimuli were delivered from one point at a time, which could be located in the RV outflow tract, RV apex, or each AHA region barycenter. Other points in proximity to areas that were visually identified as possible re-entry channels were additionally tested. For efficiency purposes, the internal states of the simulation were saved after the preconditioning phase and after the S1 stimuli train and were re-used for later simulations.

To elucidate the role of Electrophysiological (EP) properties of the BZ on the generation and maintenance of VT, multiple programmed stimulation experiments were conducted with varying BZ parameters. BZ diffusivity was sampled in a range of values corresponding to CV between 0.14 and 0.23 m/s, with 0.01 m/s steps, consistent with previous approaches [5,16]. BZ APD was increased from 0.015 s to 0.06 s over the myocardial APD with 0.015 s steps, consistent with reported values [5,16].

All simulations were performed on a high-performance GPU cluster with 24 GPUs (Tesla-V100-SXM2, NVIDIA Corporation) scheduled to perform multiple experiments in parallel. Additionally, to save computation time, the simulations were stopped 1 s after the last delivered stimulus if no spontaneous activity was present. If spontaneous activation was present, the simulations were computed for a maximum of 30 s after the last delivered stimulus. Each simulation was classified as inducible with sustained activity if the spontaneous activity lasted for 30 s after the last stimulus. For these simulations, the pseudo-ECG was computed and the intracardiac potentials sampled in each of the AHA region barycenters were saved for later analysis.

2.6 Analysis of VT Inducibility

The intracardiac potentials of each inducible simulation were analyzed by calculating the Cycle Length (CL) of spontaneous activity using a moving window of 4 s. The mean and standard deviation time between Action Potential (AP) peaks in the moving window was computed. This was done to study the evolution of the sustained VT during its duration. Additionally, this allowed us to determine whether the VT was monomorphic, in which case CL standard deviation is minimized.

3   Results

3.1   Personalization Results

The personalization resulted in a myocardial CV of 0.68 m/s and 0.64 m/s, which produced a QRSd consistent with the measurements of 70 ms and 80 ms, for case 1 and 2 respectively. The personalized APDs of 0.155 s for both cases produced a QTd of 370 ms and 380 ms for case 1 and case 2 respectively, which were in accordance with the measurements. The personalization was done with a BZ CV of 0.23 m/s and an APD increase of 0.03 s for both cases.

3.2   Inducibility of VT

The personalized values obtained in Sect. 3.1 were used for the VT inducibility simulations following the protocol described in Sect. 2.4.

For case 1, the artificial stimulation was delivered from 3 points as tested in the clinical scenario: the RV apex, AHA regions 1 and 2. An additional point in the LV apex was placed near a visually identified channel. A total of 7050 virtual programmed stimulation experiments were performed with an average compute time of 5 h and 31 min per GPU, resulting in 201 induced VTs. All VTs shared the same ECG signature, and were induced after LV apex pacing. The ECG signature was consistent with that of the measured VT. Additionally, one parameter combination produced VT with the same CL as the measured VT (0.27 s). This was achieved with a BZ CV of 0.21 m/s and BZ APD of 0.17 s. The 12-lead-ECG polarity matched in all leads. A qualitative comparison is shown in Fig. 1a. After inspection of the computed time-dependent transmembrane potential field (movie included in the Supplementary Materials), it was determined that the VT was a result of a re-entry through an LV apex channel. This was consistent with the measurements as confirmed by an electrophysiologist.

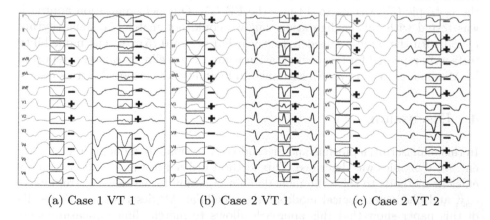

(a) Case 1 VT 1   (b) Case 2 VT 1   (c) Case 2 VT 2

Fig. 1. Computed VT morphology (left) vs. Measured VT morphology (right) for both cases.

For case 2, the artificial pacing was delivered from the RV outflow tract, RV Apex, and LV apex. Additionally, AHA regions 17, 15, 9, 8, 7, which surrounded the scar and BZ tissue were also tested. An extra point was placed inside the BZ tissue in the RV septum. Due to the more extensive BZ area and low CL of the measured VTs in this case, a larger span of parameter values for the BZ diffusivity was tested, corresponding to BZ CVs in the range [0.14, 0.32] m/s. In this case, 26646 inducibility experiments were done with an average computation time of 27 h and 39 min per GPU. 2414 experiments produced sustained monomorphic VT, with 4 distinct ECG signatures. Two of these ECG signatures were identified in VTs sustained by a LV apical re-entry pathway, and 2 in VTs sustained by a RV septal re-entry pathway. Both re-entry pathways are consistent with two of the VT exit sites estimated based on measured VT ECG. For these cases, one parameter combination produced optimal match between computed and measured VTs, in terms of CL and ECG lead polarity. This was achieved with BZ CV of 0.324 s and BZ APD of 0.1725 s. The best matching computed VT with LV apical exit site (VT 1) had a CL of 0.257 s (measured 0.240 s), and same lead polarity as the measured ECG. The best matching VT with RV septal exit site (VT 2) had a CL of 0.261 s (measured 0.190 s) and lead polarity consistent with the measured ECG except lead I. The qualitative comparison between the results can be observed in Figs. 1b and 1c. The re-entry movies for these cases can be found in the Supplementary Materials. The clinical agreement in both VTs was confirmed by an electrophysiologist.

3.3   Comparison with Previous Modeling Approaches

The modeling configuration reported by Mendonca Costa et al. [7] was reproduced in our study for Case 1. A FEC layer of 0.5 mm was placed in the endocardial layer, also covering BZ and Scar tissue. The FEC layer was configured with a longitudinal CV 6 times the healthy myocardial longitudinal CV, while the transversal CV was set equal. In the endocardium over the scar and BZ tissue the isotropic CV was set to 6 times the BZ CV. Our personalization approach was tested with this configuration, but we could not match the sinus rhythm QRSd with the largest considered myocardial CV of 0.75 m/s. The inducibility tests were repeated for this configuration using the baseline parameters reported by Mendonca-Costa et al. No sustained inducibility was achieved in this case.

4   Discussion

In this manuscript, a pipeline and methodology to simulate monomorphic VT reentrant dynamics was described. This method utilized fast and computationally efficient methods to solve the monodomain equation for cardiac electrophysiology, and phenomenological models for the cellular AP description. The results of this paper show that this approach allows to match clinical measurements of monomorphic VT, matching both the cycle length and ECG lead polarity, after myocardial parameter personalization and BZ parameter exploration. The optimal BZ tissue properties, maximizing the agreement between simulated and measured VT, could not be identified based on commonly adopted modeling assumptions (e.g. as a pre-defined ratio of myocardial tissue properties). Instead, we observed significant case-by-case variability, suggesting that improved methods for estimation of BZ properties are necessary to increase fidelity of Virtual Heart models for monomorphic VT simulation. Although this was only validated in 2 cases, we plan to add more in the continuation of the project. To our knowledge, this is the largest validation study of Virtual Heart models focusing on VT ECG. Lopez Perez et al. performed a similar study only using one case, showing a good agreement of VT morphology but not CL matching [15].

Our results show that BZ parameter exploration is necessary to reproduce the clinical VT. Accurately determining BZ tissue properties in a prospective setting, when VT ECG measurements are not available, requires future investigation. We hypothesize that it is possible to identify BZ features in sinus rhythm ECG measurements, possibly using more complex signal features than QRS and QT duration, although this has not been investigated in our study.

Additionally, we reproduced the alternative modeling approach described by Mendonca Costa et al. [7]. This approach showed a sub-optimal match of sinus rhythm ECG measurements. VT inducibility testing resulted in no sustained activity for the case considered in this study. We hypothesize that the lack of VT inducibility is related to incorrect selection of the BZ properties, and also to the assumption of a FEC covering the endocardial scar. This could produce spurious sub-endocardial re-entry pathways, preventing the maintenance of wave re-entry. Modeling assumptions can have a significant impact on the fidelity of the results, so it is important to evaluate them individually in the context of the available observations from each specific case.

The proposed pipeline is potentially capable of delivering results with times compatible with clinical practice: for a given parameter combination, a virtual programmed stimulation experiment requires 20 min of computation on a single GPU (i.e. 41 s per simulated second). This is faster than other approaches: 1 h per simulated second [20], or 17 min per simulated second (VARP) [4]. Nonetheless, absent a definitive way to characterize BZ parameters, the grid search requires extensive computational efforts, limiting direct application in a clinical setting. Other methods can identify plausible re-entry circuits in almost real-time using simplified electrophysiology models [4]. In contrast, our work focuses on identifying the circuits of clinical relevance, with the goal of proposing the minimal set of ablation targets with the maximal efficacy.

Clinical applicability will also require the full automation of several steps of this pipeline, including the tissue segmentation, parameter personalization, and unique VT signature characterization. At the current stage of our research, we have focused on careful manual curation of the input data to help reduce the potential impact of uncertainty due to data quality or algorithm performance.

An additional limitation of this study is that the virtual ECG model is not able to produce the full range of ECG morphologies observed in measured signals, in particular high frequency features in ECG leads including changes in slope, narrowing and notches within the QRS complex, as observed in Figs. 1a, 1b, and 1c. These signal characteristics might be produced by tissue properties heterogeneity which we do not include in our model. Nonetheless, the good agreement achieved in lead polarity suggests that the simulated re-entrant VTs follows a similar re-entry pathway as the corresponding measured VT.

The Virtual Heart model was not capable to match all measured VTs in case 2 in this study, characterized by a comparatively larger infarct extent, suggesting that additional VT exit sites may have been manifest in vivo, while not being produced in the model. As observed in previous computational modeling studies, we still have an incomplete understanding of the role of uncertainty on scar and BZ extent, as determined by the image processing and segmentation pipeline, on the model fidelity. This is a limitation of the current study to be addressed with larger validation studies including a wide variety of infarction presentations in pre-operative imaging.

Acknowledgements. The animal study was reviewed and approved by the Johns Hopkins University Animal Care and Use Committee (Baltimore, MD). Animal Welfare Assurance Number A3272-01.

References

1. Andreu, D., et al.: A QRS axis-based algorithm to identify the origin of scar-related ventricular tachycardia in the 17-segment American Heart Association model. Heart Rhythm **15**(10), 1491–1497 (2018)
2. Ashikaga, H., et al.: Feasibility of image-based simulation to estimate ablation target in human ventricular arrhythmia. Heart Rhythm **10**(8), 1109–1116 (2013)
3. Bayer, J.D., Blake, R.C., Plank, G., Trayanova, N.A.: A novel rule-based algorithm for assigning myocardial fiber orientation to computational heart models. Ann. Biomed. Eng. **40**, 2243–2254 (2012)
4. Campos, F.O., et al.: An automated near-real time computational method for induction and treatment of scar-related ventricular tachycardias. Med. Image Anal. **80**, 102483 (2022)
5. Campos, F.O., et al.: Factors promoting conduction slowing as substrates for block and reentry in infarcted hearts. Biophys. J. **117**(12), 2361–2374 (2019)
6. Corrado, C., Niederer, S.A.: A two-variable model robust to pacemaker behaviour for the dynamics of the cardiac action potential. Math. Biosci. **281**, 46–54 (2016)
7. Costa, C.M., et al.: Determining anatomical and electrophysiological detail requirements for computational ventricular models of porcine myocardial infarction. Comput. Biol. Med. **141**, 105061 (2022)
8. De Bakker, J., et al.: Reentry as a cause of ventricular tachycardia in patients with chronic ischemic heart disease: electrophysiologic and anatomic correlation. Circulation **77**(3), 589–606 (1988)
9. Deng, D., Prakosa, A., Shade, J., Nikolov, P., Trayanova, N.A.: Sensitivity of ablation targets prediction to electrophysiological parameter variability in image-based computational models of ventricular tachycardia in post-infarction patients. Front. Physiol. **10**, 628 (2019)
10. Durrer, D., Van Dam, R.T., Freud, G., Janse, M., Meijler, F., Arzbaecher, R.: Total excitation of the isolated human heart. Circulation **41**(6), 899–912 (1970)

11. Estner, H.L., et al.: The critical isthmus sites of ischemic ventricular tachycardia are in zones of tissue heterogeneity, visualized by magnetic resonance imaging. Heart Rhythm 8(12), 1942–1949 (2011)
12. Gard, J.J., Bader, W., Enriquez-Sarano, M., Frye, R.L., Michelena, H.I.: Uncommon cause of ST elevation. Circulation 123(9), e259–e261 (2011)
13. Kong, W., Fakhari, N., Sharifov, O.F., Ideker, R.E., Smith, W.M., Fast, V.G.: Optical measurements of intramural action potentials in isolated porcine hearts using optrodes. Heart Rhythm 4(11), 1430–1436 (2007)
14. Lee, A.W., et al.: A rule-based method for predicting the electrical activation of the heart with cardiac resynchronization therapy from non-invasive clinical data. Med. Image Anal. 57, 197–213 (2019)
15. Lopez-Perez, A., Sebastian, R., Izquierdo, M., Ruiz, R., Bishop, M., Ferrero, J.M.: Personalized cardiac computational models: from clinical data to simulation of infarct-related ventricular tachycardia. Front. Physiol. 10, 580 (2019)
16. Mendonca Costa, C., Plank, G., Rinaldi, C.A., Niederer, S.A., Bishop, M.J.: Modeling the electrophysiological properties of the infarct border zone. Front. Physiol. 9, 356 (2018)
17. Mihalef, V., Mansi, T., Rapaka, S., Passerini, T.: Implementation of a patient-specific cardiac model. In: Artificial Intelligence for Computational Modeling of the Heart, pp. 43–94. Elsevier (2020)
18. Mihalef, V., Passerini, T., Mansi, T.: Multi-scale models of the heart for patient-specific simulations. In: Artificial Intelligence for Computational Modeling of the Heart, pp. 3–42. Elsevier (2020)
19. Mitchell, C.C., Schaeffer, D.G.: A two-current model for the dynamics of cardiac membrane. Bull. Math. Biol. 65(5), 767–793 (2003)
20. Prakosa, A., et al.: Personalized virtual-heart technology for guiding the ablation of infarct-related ventricular tachycardia. Nat. Biomed. Eng. 2(10), 732–740 (2018)
21. Rapaka, S., et al.: LBM-EP: Lattice-Boltzmann method for fast cardiac electrophysiology simulation from 3D images. In: Ayache, N., Delingette, H., Golland, P., Mori, K. (eds.) MICCAI 2012. LNCS, vol. 7511, pp. 33–40. Springer, Heidelberg (2012). https://doi.org/10.1007/978-3-642-33418-4_5
22. Santangeli, P., et al.: Comparative effectiveness of antiarrhythmic drugs and catheter ablation for the prevention of recurrent ventricular tachycardia in patients with implantable cardioverter-defibrillators: a systematic review and meta-analysis of randomized controlled trials. Heart Rhythm 13(7), 1552–1559 (2016)
23. Schmidt, A., et al.: Infarct tissue heterogeneity by magnetic resonance imaging identifies enhanced cardiac arrhythmia susceptibility in patients with left ventricular dysfunction. Circulation 115(15), 2006–2014 (2007)
24. Tranum-Jensen, J., Wilde, A., Vermeulen, J.T., Janse, M.J.: Morphology of electrophysiologically identified junctions between purkinje fibers and ventricular muscle in rabbit and pig hearts. Circ. Res. 69(2), 429–437 (1991)
25. Trayanova, N.A., Doshi, A.N., Prakosa, A.: How personalized heart modeling can help treatment of lethal arrhythmias: a focus on ventricular tachycardia ablation strategies in post-infarction patients. Wiley Interdiscip. Rev. Syst. Biol. Med. 12(3), e1477 (2020)
26. Zettinig, O., et al.: Data-driven estimation of cardiac electrical diffusivity from 12-lead ECG signals. Med. Image Anal. 18(8), 1361–1376 (2014)

PCMC-T1: Free-Breathing Myocardial T1 Mapping with Physically-Constrained Motion Correction

Eyal Hanania[1]([✉]) [iD], Ilya Volovik[2] [iD], Lilach Barkat[3] [iD], Israel Cohen[1] [iD], and Moti Freiman[3] [iD]

[1] Faculty of Electrical and Computer Engineering, Technion - IIT, Haifa, Israel
eyalhan@campus.technion.ac.il
[2] Bnai Zion Medical Center, Haifa, Israel
[3] Faculty of Biomedical Engineering, Technion - IIT, Haifa, Israel

Abstract. T_1 mapping is a quantitative magnetic resonance imaging (qMRI) technique that has emerged as a valuable tool in the diagnosis of diffuse myocardial diseases. However, prevailing approaches have relied heavily on breath-hold sequences to eliminate respiratory motion artifacts. This limitation hinders accessibility and effectiveness for patients who cannot tolerate breath-holding. Image registration can be used to enable free-breathing T_1 mapping. Yet, inherent intensity differences between the different time points make the registration task challenging. We introduce PCMC-T1, a physically-constrained deep-learning model for motion correction in free-breathing T_1 mapping. We incorporate the signal decay model into the network architecture to encourage physically-plausible deformations along the longitudinal relaxation axis. We compared PCMC-T1 to baseline deep-learning-based image registration approaches using a 5-fold experimental setup on a publicly available dataset of 210 patients. PCMC-T1 demonstrated superior model fitting quality (R^2: 0.955) and achieved the highest clinical impact (clinical score: 3.93) compared to baseline methods (0.941, 0.946 and 3.34, 3.62 respectively). Anatomical alignment results were comparable (Dice score: 0.9835 vs. 0.984, 0.988). Our code and trained models are available at https://github.com/eyalhana/PCMC-T1.

Keywords: Quantitative T_1 mapping · Diffuse myocardial diseases · Motion correction

1 Introduction

Quantitative T_1 mapping is a magnetic resonance imaging (MRI) technique that allows for the precise measurement of intrinsic longitudinal relaxation time in myocardial tissue [13]. "Native" T_1 mapping, acquired without administration

This research was supported in part by a grant from the United States-Israel Binational Science Foundation (BSF), Jerusalem, Israel.

of a paramagnetic contrast agent, has been found to be sensitive to the presence of myocardial edema, iron overload, as well as myocardial infarcts and scarring [12]. It is increasingly recognized as an indispensable tool for the assessment of diffuse myocardial diseases such as diffuse myocardial inflammation, fibrosis, hypertrophy, and infiltration [13].

The derivation of accurate T_1 maps necessitates a sequential acquisition of registered images, where each pixel characterizes the same tissue at different timepoints (Fig. 1). However, the inherent motion of the heart, respiration, and spontaneous patient movements can introduce substantial distortions in the T_1 maps, ultimately impeding their reliability and clinical utility, and potentially leading to an erroneous diagnosis. [14]. Echo-triggering is a well-established approach to mitigate the effects of cardiac motion. Conversely, breath-hold sequences such as the Modified Look-Locker Inversion recovery (MOLLI) sequence and its variants [11] are commonly employed to suppress motion artifacts associated with respiration. However, the requirement for subjects to hold their breath places practical constraints on the number of images that can be acquired [11], as well as on the viability of the technique for certain patient populations who cannot tolerate breath-holding. Further, inadequate echo-triggering due to cardiac arrhythmia may lead to unreliable T_1 maps, compromising the diagnosis.

Alignment of the images obtained at different time-points via image registration can serve as a mitigation for residual motion and enable cardiac T_1 mapping with free-breathing sequences such as the slice-interleaved T_1 (STONE) sequence [15]. Yet, the intrinsic complexity of the image data, including contrast inversion, partial volume effects, and signal nulling for images acquired near the zero crossing of the T_1 relaxation curve, presents a daunting task in achieving registration for these images. Zhang et al. [18] proposed to perform motion correction in T_1 mapping by maximizing the similarity of normalized gradient fields in order to address the intensity differences across different time points. El-Rewaidy et al. [5] employed a segmentation-based approach in which the residual motion was computed by matching manually annotated contours of the myocardium to the different images. Xue et al. [16] and Tilborghs et al. [14] proposed an iterative approach in which the signal decay model parameters are estimated and synthetic images are generated. Then, image registration used the predicted images to register the acquired data. Van De Giessen et al. [6] used directly the error on the exponential curve fitting as the registration metric to spatially align images obtained from a Look-Locker sequence.

Deep-learning methods have been also proposed for motion correction by image registration as a pre-processing step in quantitative cardiac T_1 mapping [2,7,10]. A recent study by Yang et al. [17] introduced a sequential process to address the contrast differences between images. Initially, their approach aimed to separate intensity changes resulting from different inversion times from the fixed anatomical structure. However, this method heavily relied on the perfect disentanglement of the anatomical structure from the contrast. Moreover, the registration is performed exclusively between the disentangled anatomical images, overlooking the adherence of the signal along the inversion time axis to

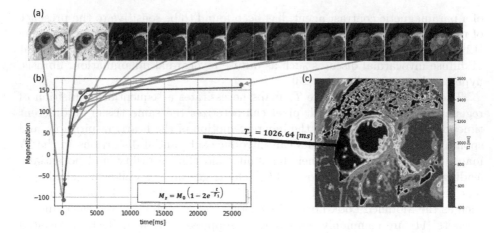

Fig. 1. Schematic description of T_1 mapping for a single voxel. (a) T_1-weighted myocardial images at 11 sequential time points. (b) Fitting an inversion recovery curve of the longitudinal magnetization M_z over different time points t, and extracting the corresponding T_1 and M_0 parameters. (c) Displaying the computed T_1 map.

the signal decay model. Nevertheless, these methods do not account directly for the signal decay model, therefore they may produce physically-unlikely deformations.

In this work, we introduce PCMC-T1, a physically-constrained deep-learning model for simultaneous motion correction and T_1 mapping from free-breathing acquisitions. Our network architecture combined an image registration module and an exponential T_1 signal decay model fitting module. The incorporation of the signal decay model into the network architecture encourages physically-plausible deformations along the longitudinal relaxation axis.

Our PCMC-T1 model has the potential to expand the utilization of quantitative cardiac T_1 mapping to patient populations who cannot tolerate breath-holding by enabling automatic motion-robust accurate T_1 parameter estimation without additional manual annotation of the myocardium.

2   Method

We formulate the simultaneous motion correction and signal relaxation model estimation for qMRI T_1 mapping as follows:

$$\widehat{T_1, M_0}, \Phi = \underset{T_1, M_0, \Phi}{\operatorname{argmin}} \sum_{i=0}^{N-1} \left\| M_0 \cdot (1 - 2 \cdot e^{\frac{-t_i}{T_1}}) - \phi_i \circ I_i \right\|^2 \qquad (1)$$

where N is the number of acquired images, M_0, T_1 are the exponential signal relaxation model parameters, ϕ_i is the i'th deformation field, $\Phi = \{\phi\}_{i=0}^{N-1}$, I_i is the i'th original image, and t_i is the i'th timestamp. However, direct optimization of this problem can be challenging and time-consuming [6].

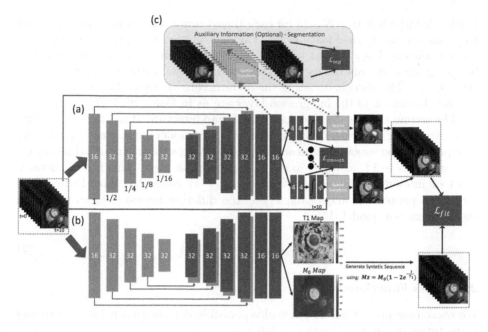

Fig. 2. Our PCMC-T1 model comprises two encoder-decoder components. (a) The first encoder-decoder extends the pair-wise VoxelMorph model to enable the registration of multiple images. (b) The second encoder-decoder generates parametric maps and motion-free synthetic images. The main goal of our network is to minimize the discrepancy between the registered images and the motion-free synthetic images, aiming for physically plausible deformations along the longitudinal relaxation axis. Additionally, we optionally promote anatomically consistent deformation fields by introducing a segmentation loss (c).

2.1 Model Architecture

To overcome this challenge, we propose PCMC-T1, a DNN architecture that simultaneously predicts the deformation fields and the exponential signal relaxation model parameters. Figure 2 summarizes the overall architecture of our model. It includes two U-Net-like encoder-decoder modules that are operating in parallel. Skip connections are connecting between the encoder and the decoder of each model. The first encoder-decoder module is a multi-image deformable image registration module based on the voxelmorph architecture [3], while the second encoder-decoder module is the qMRI signal relaxation model parameters prediction module.

The input of the DNN is a set of acquired images $\{I_i | i = 0 \ldots N-1\}$, stacked along the channel dimension. The first encoder-decoder is an extension of the pair-wise VoxelMorph model [3] for registration of multiple images. The encoder is a U-Net-like encoder consisting of convolutional and downsampling layers with an increasing number of filters. The decoder output splits into multiple separated heads of convolutional layers and integration layers that produce a specific deformation field $\{\phi_i | i = 0 \ldots N-1\}$ for each timestamp i. Skip connections are used to propagate the learned features into the deformation field prediction layers. A spatial warping layer is used to align the acquired images I_i to the synthetics images generated from the signal relaxation model parameters predictions (S_i): $R_i = I_i \circ \phi_i$. The specific details of the architecture are as in Balakrishnan et al. [3] and the details of the integration layer are as in Dalca et al. [4].

The second encoder-decoder has a similar architecture. It has two output layers representing the exponential signal relaxation model parameters: T_1 and M_0. The predicted parameters maps are then used, along with the input's timestamps $\{t_i | i = 0 \ldots N-1\}$, as input to a signal generation layer. This layer generates a set of motion-free images $\{S_i | i = 0 \ldots N-1\}$ computed directly from the estimated parametric maps (M_0, T_1) at the different inversion times using the signal relaxation model [15]:

$$S_i = M_0 \cdot (1 - 2 \cdot e^{\frac{-t_i}{T1}}) \tag{2}$$

2.2   Loss Functions

We encourage predictions of physically-plausible deformation fields by coupling three terms in our loss function as follows:

$$\mathcal{L}_{total} = \lambda_1 \cdot \mathcal{L}_{fit} + \lambda_2 \cdot \mathcal{L}_{smooth} + \lambda_3 \cdot \mathcal{L}_{seg} \tag{3}$$

The first term ($\mathcal{L}_{fit}$) penalizes for differences between the model-predicted images generated by the model-prediction decoder and the acquired images warped according to the deformation fields predicted by the registration decoder. Specifically, we use the mean-squared-error (MSE) between the registered images $\{R_i | i = 0 \ldots N-1\}$ and the synthetic images $\{S_i | i = 0 \ldots N-1\}$:

$$\mathcal{L}_{fit}(T_1, M_0, t_{i=0}^{N-1}, \Phi) = \sum_{i=0}^{N-1} (S_i - R_i)^2 \tag{4}$$

where S_i are the images generated with the signal model equation (Eq. 2), and the registered images are the output of the registration module. This term encourages deformation fields that are physically plausible by means of a signal relaxation that is consistent with the physical model of T_1 signal relaxation.

The second term ($\mathcal{L}_{smooth}$) encourages the model to predict realistic, smooth deformation fields Φ by penalizing for a large l_2 norm of the gradients of the velocity fields [3]:

$$\mathcal{L}_{smooth}(\Phi) = \sum_{i=0}^{N-1} \frac{1}{\Omega} \sum_{p \in \Omega} \|\nabla \phi_i(p)\|^2 \tag{5}$$

where Ω is the domain of the velocity field and p are the voxel locations within the velocity field. In addition, we encourage anatomically-consistent deformation fields by introducing a segmentation-based loss term ($\mathcal{L}_{seg}$) as a third term in the overall loss function [3]. This term can be used in cases where the left ventricle (LV)'s epicardial and endocardial contours are available during training. Specifically, the segmentation loss function is defined as follows:

$$\mathcal{L}_{seg}(r, Seg_{i=0}^{N-1}, \phi_{i=0}^{N-1}) = \sum_{i=0, i \neq r}^{N-1} DiceLoss(Seg_r, Seg_i \circ \phi_i) \qquad (6)$$

where $Seg_i, (i \in 0, \dots, N-1)$ is the $i'th$ binary segmentation mask of the myocardium, Seg_r is the binary segmentation mask of the fixed image, and r is the index of the fixed image. This term can be omitted in cases where the segmentations of the myocardium are not available.

2.3 Implementation Details

We implemented our models in PyTorch. We experimentally fixed the first time-point image, and predict deformation fields only for the rest of the time points. We optimized our hyperparameters using a grid search. The final setting for the loss function parameters were: $\lambda_1 = 1$, $\lambda_2 = 500$, $\lambda_3 = 70000$. We used a batch size of 8, ADAM optimizer with a learning rate of $2 \cdot 10^{-3}$. We trained the model for 300k iterations. We used the publicly available TensorFlow implementations of the diffeomorphic VoxelMorph [4] and SynthMorph [9] as baseline methods for comparison. We performed hyper-parameter optimization for baseline methods using a grid search. All experiments were run on an NVIDIA Tesla V100 GPU with 32G RAM.

3 Experiments and Results

3.1 Data

We used the publicly available myocardial T_1 mapping dataset [1,5]. The dataset includes 210 subjects, 134 males and 76 females aged 57 ± 14 years, with known or suspected cardiovascular diseases. The images were acquired with a 1.5T MRI scanner (Philips Achieva) and a 32-channel cardiac coil using the ECG-triggered free-breathing imaging slice-interleaved T_1 mapping sequence (STONE) [15]. Acquisition parameters were: field of view (FOV) $= 360 \times 351[\text{mm}^2]$, and voxel size of $2.1 \times 2.1 \times 8[\text{mm}^3]$. For each patient, 5 slices were acquired from base to apex in the short axis view at 11 time points. Additionally, manual expert segmentations of the myocardium were provided as part of the dataset [5]. We cropped the images to a size of 160×160 pixels for each time point. We normalized the images using a min-max normalization.

Table 1. Quantitative comparison between motion correction methods for myocardial T_1 mapping. All results are presented in mean±std.

	R^2	DSC	HD [mm]	clinical score
Original	0.911 ± 0.12	0.664 ± 0.23	14.93 ± 11.76	2.79 ± 0.99
SynthMorph	0.946 ± 0.09	0.88 ± 0.149	8.59 ± 9.98	3.62 ± 0.88
Voxelmorph-seg	0.941 ± 0.096	0.84 ± 0.188	9.39 ± 11.93	3.34 ± 0.79
Reg-MI	0.95 ± 0.08	0.73 ± 0.168	16.29 ± 11.43	3.68 ± 0.83
PCMC-T1 w.o $\mathcal{L}_{seg}$	0.971 ± 0.046	0.662 ± 0.172	21.5 ± 13.3	4 ± 0.83
PCMC-T1	0.955 ± 0.078	0.835 ± 0.137	9.34 ± 7.85	3.93 ± 0.78

3.2   Evaluation Methodology

Quantitative Evaluation: We used a 5-fold experimental setup. For each fold, we divided the 210 subjects into 80% as a training set and 20% as a test set. We conducted an ablation study to determine the added value of the different components in our model. Specifically, we compared our method using a few variations, including a multi-image registration model with a mutual-information-based loss function (REG-MI) [8], and our method (PCMC-T1) without the segmentation loss term. We used two state-of-the-art deep-learning algorithms for medical image registration including the pairwise probabilistic diffeomorphic VoxelMorph with a mutual-information-based loss [4], and pairwise SynthMorph [9], as well as with T_1 maps produced from the acquired images directly without any motion correction step. We quantitatively evaluated the T_1 maps produced by our PCMC-T1 model in comparison to T_1 maps produced after applying deep-learning-based image registration as a pre-processing step. We used the R^2 of the model fit to the observed data in the myocardium, the Dice score, and Hausdorff distance values of the myocardium segmentations as the evaluation metrics.

Clinical Impact: We further assessed the clinical impact of our method by conducting a semi-quantitative ranking of the T_1 maps for the presence of motion artifacts by an expert cardiac MRI radiologist (3 years of experience) who was blinded to the methods used to generate the maps. We randomly selected 29 cases (5 slices per case) from the test set with their associated T_1 maps. The radiologist was asked to rank each slice with 1 in case of a good quality map without visible motion artifacts and with 0 otherwise. We computed overall patient scores by summing the slice grades. The maximum grade per subject was 5 for cases in which no motion artifacts were present in all slices and 0 for cases in which motion artifacts were present in all slices. We assessed the statistical significance with the repeated measures ANOVA test; $p < 0.05$ was considered significant.

3.3   Results

Quantitative Evaluation: Table 1 summarizes our results for the test sets across all folds, encompassing a total of 210 patients. Our PCMC-T1 approach achieved the best result in terms of R^2 with the smallest variance. Although PCMC-T1 without the segmentation loss (L_{seg}) achieved a higher R^2 result compared to PCMC-T1 with the segmentation loss, it degraded the Dice value, representing over-fitted predictions. On the other hand, the slightly higher Dice score and Hausdorff distance values obtained by baseline methods compared to PCMC-T1 suggest bias of these methods toward the registration of the segmentation maps rather than producing deformation fields that are consistent with the signal relaxation model. The balanced result of PCMC-T1 indicates an improvement in the physical plausibility of the deformations produced by PCMC-T1 by means of signal relaxation and anatomical consistency.

Clinical Impact: Figure 3 presents several representative cases. Although the Dice score of the baseline methods is higher compared to this of PCMC-T1, the quality of the maps produced by PCMC-T1 is better. The rightmost column of

| Original | VoxelMorph | SynthMorph | PCMC-T1 |

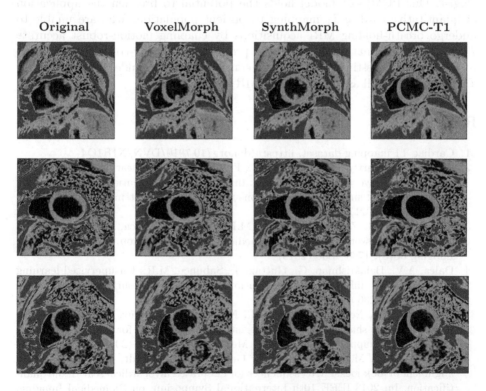

Fig. 3. Representative T_1 maps computed with the different approaches. Our approach (PCMC-T1) demonstrates a clearer delineation between the blood and the muscle with a reduced partial volume effect, resulting in a more homogeneous mapping of the myocardium.

Table 1 summarizes the results of the clinical impact assessment of our PCMC-T1 approach. Our PCMC-T1 received the highest quality score compared to the baseline methods. The difference in the radiologist grading was statistically significant ($p \ll 10e^{-5}$). The improvement in the radiological evaluation suggests that PCMC-T1 provides a balanced result that is not overly biased toward the segmentations or toward the signal relaxation model.

4   Conclusions

We presented PCMC-T1, a physically-constrained deep-learning model for motion correction in free-breathing T_1 mapping. Our main contribution is the incorporation of the signal decay model into the network architecture to encourage physically-plausible deformations along the longitudinal relaxation axis. We demonstrated a quantitative improvement by means of fit quality with comparable Dice score and Hausdorff distance. We further assessed the clinical impact of our method by conducting a qualitative evaluation of the T_1 maps produced by our method in comparison to baseline methods by an expert cardiac radiologist. Our PCMC-T1 model holds the potential to broaden the application of quantitative cardiac T_1 mapping to patient populations who are unable to undergo breath-holding MRI acquisitions by enabling motion-robust accurate T_1 parameter estimation. Further, the proposed physically-constrained motion robust parameter estimation approach can be directly extended to quantitative T2 mapping as well as to additional qMRI applications.

References

1. Cardiac T1 mapping dataset. https://doi.org/10.7910/DVN/N1R1Q4
2. Arava, D., Masarwy, M., Khawaled, S., Freiman, M.: Deep-learning based motion correction for myocardial T1 mapping. In: 2021 IEEE International Conference on Microwaves, Antennas, Communications and Electronic Systems (COMCAS), pp. 55–59. IEEE (2021)
3. Balakrishnan, G., Zhao, A., Sabuncu, M.R., Guttag, J., Dalca, A.V.: Voxelmorph: a learning framework for deformable medical image registration. IEEE Trans. Med. Imaging 38(8), 1788–1800 (2019)
4. Dalca, A.V., Balakrishnan, G., Guttag, J., Sabuncu, M.R.: Unsupervised learning of probabilistic diffeomorphic registration for images and surfaces. Med. Image Anal. 57, 226–236 (2019)
5. El-Rewaidy, H., Nezafat, M., Jang, J., Nakamori, S., Fahmy, A.S., Nezafat, R.: Nonrigid active shape model-based registration framework for motion correction of cardiac T1 mapping. Magn. Reson. Med. 80(2), 780–791 (2018)
6. van de Giessen, M., Tao, Q., van der Geest, R.J., Lelieveldt, B.P.: Model-based alignment of look-locker MRI sequences for calibrated myocardial scar tissue quantification. In: 2013 IEEE 10th International Symposium on Biomedical Imaging, pp. 1038–1041. IEEE (2013)
7. Gonzales, R.A., et al.: Moconet: robust motion correction of cardiovascular magnetic resonance T1 mapping using convolutional neural networks. Front. Cardiovasc. Med. 1689 (2021)

8. Hanania, E., Barkat, L., Cohen, I., Azhari, H., Freiman, M.: Deep-learning-based group-wise motion correction for myocardial T1 mapping. In: Proceedings of the ISMRM & SMRT Annual Meeting & Exhibition, Toronto, Canada (2023)
9. Hoffmann, M., Billot, B., Greve, D.N., Iglesias, J.E., Fischl, B., Dalca, A.V.: Synthmorph: learning contrast-invariant registration without acquired images. IEEE Trans. Med. Imaging 41(3), 543–558 (2021)
10. Li, Y., Wu, C., Qi, H., Si, D., Ding, H., Chen, H.: Motion correction for native myocardial T1 mapping using self-supervised deep learning registration with contrast separation. NMR Biomed. 35(10), e4775 (2022)
11. Roujol, S., et al.: Accuracy, precision, and reproducibility of four T1 mapping sequences: a head-to-head comparison of MOLLI, ShMOLLI, SASHA, and sapphire. Radiology 272(3), 683–689 (2014)
12. Schelbert, E.B., Messroghli, D.R.: State of the art: clinical applications of cardiac T1 mapping. Radiology 278(3), 658–676 (2016)
13. Taylor, A.J., Salerno, M., Dharmakumar, R., Jerosch-Herold, M.: T1 mapping: basic techniques and clinical applications. JACC Cardiovasc. Imaging 9(1), 67–81 (2016)
14. Tilborghs, S., et al.: Robust motion correction for cardiac T1 and ECV mapping using a T1 relaxation model approach. Med. Image Anal. 52, 212–227 (2019)
15. Weingärtner, S., Roujol, S., Akçakaya, M., Basha, T.A., Nezafat, R.: Free-breathing multislice native myocardial T1 mapping using the slice-interleaved T1 (stone) sequence. Magn. Reson. Med. 74(1), 115–124 (2015)
16. Xue, H., et al.: Motion correction for myocardial T1 mapping using image registration with synthetic image estimation. Magn. Reson. Med. 67(6), 1644–1655 (2012)
17. Yang, C., Zhao, Y., Huang, L., Xia, L., Tao, Q.: DisQ: disentangling quantitative MRI mapping of the heart. In: Wang, L., Dou, Q., Fletcher, P.T., Speidel, S., Li, S. (eds.) MICCAI 2022. LNCS, vol. 13436, pp. 291–300. Springer, Cham (2022). https://doi.org/10.1007/978-3-031-16446-0_28
18. Zhang, S., et al.: Cardiac magnetic resonance T1 and extracellular volume mapping with motion correction and co-registration based on fast elastic image registration. Magn. Reson. Mater. Phys., Biol. Med. 31, 115–129 (2018)

Machine Learning for Automated Mitral Regurgitation Detection from Cardiac Imaging

Ke Xiao[1]([✉]) [iD], Erik Learned-Miller[1] [iD], Evangelos Kalogerakis[1] [iD], James Priest[2] [iD], and Madalina Fiterau[1] [iD]

[1] University of Massachusetts Amherst, Amherst, MA 01003, USA
{kexiao,elm,kalo,mfiterau}@cs.umass.edu
[2] Stanford University, Stanford, CA 94305, USA
jpriest@stanford.edu

Abstract. Mitral regurgitation (MR) is a heart valve disease with potentially fatal consequences that can only be forestalled through timely diagnosis and treatment. Traditional diagnosis methods are expensive, labor-intensive and require clinical expertise, posing a barrier to screening for MR. To overcome this impediment, we propose a new semi-supervised model for MR classification called CUSSP. CUSSP operates on cardiac magnetic resonance (CMR) imaging slices of the 4-chamber view of the heart. It uses standard computer vision techniques and contrastive models to learn from large amounts of unlabeled data, in conjunction with specialized classifiers to establish *the first ever automated MR classification system using CMR imaging sequences*. Evaluated on a test set of 179 labeled – 154 non-MR and 25 MR – sequences, CUSSP attains an F1 score of 0.69 and a ROC-AUC score of 0.88, setting the first benchmark result for detecting MR from CMR imaging sequences.

1 Introduction

Mitral Regurgitation. Mitral regurgitation (MR) [7] is a valvular heart disease in which the mitral valve does not close completely during systole when the left ventricle contracts, causing regurgitation – leaking of blood backwards – from the left ventricle (LV), through the mitral valve, into the left atrium (LA) – Fig. 1. MR can be caused by either organic or functional mechanisms [6], with organic MR leading to atrial and annular enlargement and functional MR increasing atrial pressure. As MR progresses, it may cause arrhythmia, shortness of breath, heart palpitations and pulmonary hypertension [14]. Left undiagnosed and untreated, MR may cause significant hemodynamic instability and congestive heart failure which can lead to death [17], while acute MR usually necessitates immediate medical intervention [22]. Thus, *early detection and assessment of MR are crucial for optimal*

Supplementary Information The online version contains supplementary material available at https://doi.org/10.1007/978-3-031-43990-2_23.

treatment outcomes, with the best short-term and long-term results obtained in asymptomatic patients operated on in advanced repair centers with low operative mortality (<1%) and high repair rates (≥80–90%) [7].

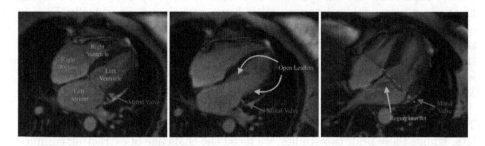

Fig. 1. Three cardiovascular magnetic resonance (CMR) images showing the long-axis four-chamber view of the heart. Left: a heart with normal mitral valve. Middle: a heart with normal mitral valve when the valve leaflets are open. Right: a heart with mitral regurgitation. The (red) dotted line denotes the mitral valve. (Color figure online)

MR Diagnosis. MR is often only detected following symptom onset. Among patients with asymptomatic MR, quantitative grading of mitral regurgitation is a powerful indicator for clinical treatment such as immediate cardiac surgery [8]. Clinically, MR is usually diagnosed with doppler echocardiography, with *cardiovascular magnetic resonance* (CMR) subsequently used to assess the MR severity and to accurately quantify the regurgitant volume, one of the indicators of severity [20]. Most studies that have evaluated CMR for assessing the mitral regurgitant volume use the difference between left ventricular stroke volume (LVSV) and forward stroke volume (FSV). LVSV is usually estimated with the short-axis (SA) view CMR – a 4-D tensor – while FSV is most commonly determined by aortic phase-contrast velocity-encoding images [20]. This diagnosis and assessment process requires continuous involvement from expert clinicians along with specific order and post-processing for the phase-contrast images of the proximal aorta or main pulmonary artery during the acquisition of the CMR data. The associated expense with this standard diagnostic procedure thus poses an obstacle to the large-scale screening for MR in the general population.

Towards Machine Learning for MR Diagnosis. Although quantitatively assessing mitral regurgitant volume requires specific CMR imaging sequences and expert analysis, four-chamber (4CH) CMR images provide a comprehensive view of all four heart chambers, including the mitral valve as it opens and closes, as shown in Fig. 1. Thus, we propose to train a model that uses 4CH CMR to automatically diagnose MR, making wide screening possible. As training data, we use the long axis 4CH CMR imaging data from the UK Biobank [1], from over 30,000 subjects, out of which N = 704 were labeled by an expert cardiologist. While the 4CH view has the potential to identify MR when the regurgitant jet is visible, the imaging is not accompanied by comprehensive annotations or diagnoses of diseases/conditions for individual patients. This is in contrast to Zhang et al. [27], where tens of thousands of annotated color doppler echocardiography images are available for MR assessments. To overcome this difficulty, we rely on weakly supervised and unsupervised methods. Weakly supervised deep learning has proved successful in detecting other heart pathologies. Specifically, Fries et al. [9] proposed a weakly supervised deep learning method (CNN-LSTM) to classify aortic valve malformation from the aortic valve cross section CMR present in the UK Biobank, wherein the critical feature of the aortic valve opening shape was easily extracted from the aortic valve cross section CMR imaging data. Meanwhile, Vimalesvaran et al. [21] proposed a deep learning based pipeline for detecting aortic valve pathology using 3CH CMR imaging from three hospitals. The data set was fully annotated with landmarks, stenotic jets and regurgitant jets. Unlike these prior two studies, we faced the challenge of extracting complex mitral valve regurgitant features from 4CH CMR images with *no annotations for landmarks, regurgitant jets or easily extractable features*, and only a small amount of binary MR labels. To the best of our knowledge, this is *the first study on identifying MR using the 4CH CMR imaging data in an automated pipeline*.

Our Approach. We propose an automated five stage pipeline named **C**ardio-vascular magnetic resonance **U**-Net localized **S**elf-**S**upervised **P**redictor (CUSSP). Our approach incorporates several different preexisting neural network architectures in the pipeline, discussed in Sect. 2.3, to address the challenges inherent to the MR classification task. Specifically, we use a U-Net [18] to perform segmentation of the heart chambers, which we then use to localize the area around the mitral valve. We apply histogram equalization to enhance the appearance of the valve. We then use a Barlow Twins [26] network to learn, without supervision, representations of the blood flow around the valve, and a Siamese network [25] to learn differences between instances of MR and non-MR. During training, CUSSP leverages a large amount of unlabeled CMR images, and *minimal supervision*, in the form of a comparatively small set of MR labels manually annotated by cardiologist. However, *at test time CUSSP is fully automated*.

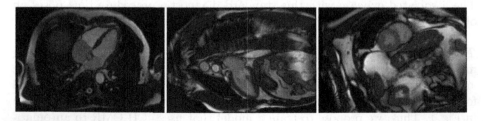

Fig. 2. Example of the segmentation outputs of the long axis 4CH (left), 2CH (middle) CMR view imaging data and the short axis (right) CMR imaging data.

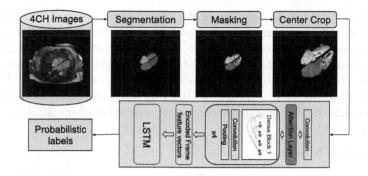

Fig. 3. Overview of the CNN-LSTM method pipeline for MR classification.

Contribution. Our work is the *first study on automated detection of mitral regurgitation (MR) from CMR*, providing a benchmark for the classification of MR in an automated pipeline from long axis 4CH CMR images. As a screening tool, it has the potential to support hospital diagnostics and improve patient care.

2 Methods

2.1 Segmentation of the Cardiac Magnetic Resonance Images

The CMR imaging data from the UK Biobank that is relevant to MR detection includes long-axis 2-chamber (2CH) view and long-axis 4-chamber (4CH) view, which are all shown in Fig. 2. In addition, the short-axis view CMR provides accurate description of the left ventricle. Both long-axis views and short-axis view are used to estimate heart measurements relevant to the MR detection task, while only the long-axis 4CH view is used for the deep learning models.

As a pre-processing step, we performed semantic segmentation on the CMR imaging data, using masks (Fig. 2) generated by a U-Net [18] segmentation model to highlight regions of interest to MR classification. U-Net is currently the leading model architecture for medical imaging segmentation, with various U-Net variants developed for different applications. TernausNet [12] is a U-Net variant that reshapes the U-Net encoder to match the VGG11 architecture, allowing it to use pre-trained VGG11 [19] model weights for faster convergence and improved segmentation results. While most medical imaging segmentation models are trained using supervised learning, weakly supervised segmentation methods such as VoxelMorph augmented segmentation [28], ACNN [16], CCNN [13], graph-based unsupervised segmentation [15], and GAN-based unsupervised segmentation [23,24] also produce comparable segmentation results. For the segmentation of the 4CH, 2CH, SA, and aorta view CMR imaging dataset from the UK Biobank, Bai et al. [2] offer a supervised segmentation model.

We manually labeled 100 CMR images for each view and trained a supervised segmentation model with the TernausNet [12] architecture. Then, segmentation outputs, shown in Fig. 2, are used to compute measurements of cardiac structure and function for the four chambers of the heart, as summarized in Table 1. The short-axis view CMR segmentation output is used to estimate the left ventricle and right ventricle measurements, while the long-axis 4CH view and 2CH view outputs are used to estimate the left atrium and right atrium measurements. Specifically, the left atrial volume is estimated using the biplane method with segmentation of both the 2CH and 4CH view, while the right atrial volume is estimated using single plane method with segmentation of the 4CH view.

2.2 Baseline Models

We first considered a random forest (RF) classifier [3] trained for MR classification on the tabular heart measurements in Table 1. We divided the 18 features by body surface area (BSA) prior to training the RF.

Next, we developed a deep learning baseline model for MR classification, a weakly supervised CNN-LSTM, following the principles in Fries et al. [9] and operating on the 4CH CMR imaging data. Fries et al. [9] used CMR imaging sequences from the UK Biobank, however, the objective of their work was the identification of aortic valve malformations. Their proposed deep learning architecture – CNN-LSTM – used DenseNet [11] as the CNN of choice to encode CMR imaging frames and the LSTM to encode embeddings of all frames within each sequence for a final classification of aortic valves into tricuspid (normal) and bicuspid (pathological). We point out that *our MR classification problem is considerably more challenging* due to the lack of direct view of the mitral valve in the CMR imaging data. Moreover, the flow information provided from the 4CH view CMR imaging data is difficult to learn and encode in the model, an issue which we alleviated in the CUSSP framework.

The CNN-LSTM pipeline, shown in Fig. 3, includes an image segmentation model, and an image classification model. It uses the 4CH CMR from the UK Biobank. The CMR data is center-cropped using the center of mass of the CMR imaging frames. The resulting sequence provided to the CNN-LSTM, which generates probabilistic labels of MR for the sample.

In the CNN-LSTM model architecture, the CNN serves as the frame encoder, which encodes each frame of each sequence into a representation vector. The model uses DenseNet-121 pre-trained on ImageNet as the CNN. To better learn

Table 1. Cardiac measurements derived from the semantic segmentation of the CMR.

Left Atrium	Right Atrium	Left Ventricle	Right Ventricle
Vol Max (mL)	Vol Max (mL)	End Systolic Vol (mL)	End Systolic Vol (mL)
Vol Min (mL)	Vol Min (mL)	End Diastolic Vol (mL)	End Diastolic Vol (mL)
Stroke Vol (mL)	Stroke Vol (mL)	Stroke Vol (mL)	Stroke Vol (mL)
Eject. fraction (%)	Eject. fraction (%)	Eject. fraction (%)	Eject. fraction (%)
		Cardiac Output (L/min)	
		Mass (g)	

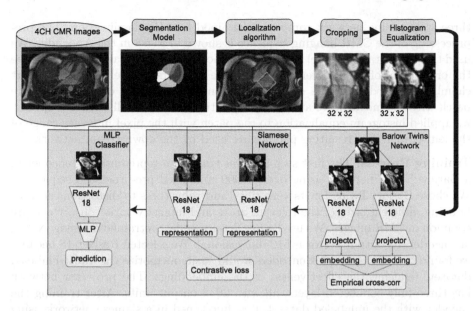

Fig. 4. Overview of the CUSSP pipeline for MR classification, with its 5 steps: (1) segmentation, (2) localization, (3) cropping, (4) equalization, and (5) prediction. In particular, the prediction stage of the CUSSP method contains three stages: (i) the feature encoder is trained in the Barlow-Twins network with unlabeled imaging data set, (ii) the feature encoder is fine-tuned in a siamese network with labeled imaging data set, and (iii) the feature encoder is assembled with a MLP, then trained with labeled imaging data set for the classification task of MR.

the attention span of the frame encoder, we added an attention layer to the DenseNet-121 after the first convolutional layer. After the bi-directional LSTM, a multi-layer perceptron (MLP) performs the final classification.

2.3   The CUSSP Framework

Conceptualization. To better encode the blood flow information relevant to MR classification from the 4CH CMR view, we investigated self-supervised representation learning methods which can leverage all the unlabeled CMR sequences present in the UK Biobank. Typically, self-supervised representation learning for visual data involves maximizing the similarity between representations of various distorted versions of a sample. Among the many self-supervised architectures, SimCLR [5], SwAV [4], and BYOL [10], we chose Barlow Twins [26], since it does not require large batches. With the labeled data, our siamese network compares the representation differences between classes by sampling two inputs from different classes as performed in [25]. Thus, our CUSSP MR classification pipeline takes advantage of both self-supervised and supervised representation learning.

Test-Time Pipeline. Our CUSSP method consists of five main steps, shown in Fig. 4, the first four for pre-processing, and the last one for prediction, with three stages using network components trained for MR classification. The pre-processing of the CMR imaging sequence is shown in Fig. 8 in the Appendix. We used the segmentation model in Sect. 2.1 to locate the mitral valve and determine the orientation of the left ventricle using the contours and centers of the heart chambers derived from the segmentation output. We then cropped a square patch with the mitral valve at its center positioned horizontally. After cropping, we applied histogram equalization to the patch with the pixel intensity range of the left atrium. The resulting patches are used by the downstream networks.

Training Process. The first step involves training a representation encoder in a Barlow Twins network using over 30,000 unlabeled pre-processed sequences. We chose Barlow Twins for its versatility and robustness to distortions such as blurring, different sizes of the relevant areas and intensity variations, which are common in CMR images. We used a ResNet-18 model pretrained on ImageNet as an encoder. Its output vector is 512-dimensional. We selected ResNet-18 because we found that more complex encoders would easily memorize the limited labeled dataset, reducing the effectiveness of the embeddings. The projector network has three fully connected layers, all with 2048 output units. After training the encoder with the unlabeled dataset, it is fine-tuned in a siamese network using a comparatively smaller labeled set, as indicated in Sect. 3. During training, two sequences are sampled from the labeled dataset, with the first being non-MR and the second being either MR or non-MR. The two sequences are passed through the representation encoder to obtain embeddings, which are then used to calculate the contrastive loss. The model is trained to maximize contrastive loss when the two samples are non-MR and MR and to minimize it when both are non-MR. Contrastive learning helps because it uses the very limited labels in our possession to obtain representations focused on encoding differences between classes. Once the representation encoder is fine-tuned in the siamese network, it is combined with a 3-layer multi-layer perceptron (MLP) network to form a classifier, which is trained on the same labeled dataset. To improve computation efficiency and training accuracy, we also tested the framework using a smaller window of 25 frames, since MR occurs between diastole and systole. The code is available at https://github.com/Information-Fusion-Lab-Umass/CUSSP_UKB_MR.

3   Experiments and Discussion

Experimental Setup. 4CH CMR images were used to conduct experiments with both the CNN-LSTM method and the CUSSP method. We used a total of 704 labeled sequences, with 525 sequences selected for the training set, including 452 labeled as non-MR and 73 labeled as MR. The remaining 179 sequences were used for testing, with 154 labeled as non-MR and 25 labeled as MR. Considering the substantial class imbalance, we opted to use F1 score as our primary evaluation metric, along with precision and recall.

Random Forest Classification Results. The random forest model is trained with 10-fold cross validation, with a random search over a parameter grid of

10–100 estimators, 2–16 depth, 2–8 min samples. The optimal hyper-parameter setting found is: 20 estimators, log2 max features, max depth of 9 and a minimum of 2 samples per leaf node. The best results obtained are presented in Table 2.

Table 2. Experimental results for Random Forest (RF) baseline, CNN-LSTM and CUSSP. CUSSP-1, CUSSP-2 and CUSSP-3 are trained with the BarlowTwins-MLP model without fine-tuning with the Siamese network. CUSSP-SIAM and CUSSP-SIAM-25 are trained with the BarlowTwins-Siamese-MLP model.

Model	Pos. Acc	Neg. Acc	Precision	Recall	F1	AUC
RF	0.09	0.99	0.43	0.09	0.14	0.58
CNN-LSTM	0.53	0.86	0.45	0.53	0.44	0.72
CUSSP-1	0.38	0.87	0.29	0.38	0.32	0.65
CUSSP-2	0.29	0.87	0.25	0.29	0.27	0.63
CUSSP-3	0.38	0.90	0.35	0.38	0.36	0.66
CUSSP-SIAM	0.55	0.96	0.66	0.55	0.60	0.80
CUSSP-SIAM-25	**0.62**	**0.96**	**0.8**	**0.62**	**0.69**	**0.88**

CNN-LSTM Classification Results. We conducted experiments on the DenseNet-LSTM classification model using various input image sizes, attention layer configurations, and masks. The best CNN-LSTM model attains a F1-score of 0.44, shown in Table 2, with further information on the performance under other settings summarized in the Appendix.

CUSSP Classification Results

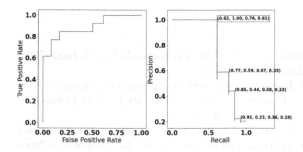

We evaluated various configurations of the CUSSP model, to determine the relative benefits of different components. In the first configuration, the ResNet18 model was combined with a 3-layer MLP to train a classifier using the labeled training set after being trained in the Barlow-Twins network with the unlabeled dataset. During the classifier training, the cross-correlation loss from the

Fig. 5. The ROC AUC curve and the precision-recall curve of CUSSP-SIAM-25 from Table 2. The annotated coordinates on the precision-recall curve plot are (recall, precision, F1-score, threshold).

Barlow-Twins network and the cross-entropy loss from the binary classification were weighted using three different configurations. For CUSSP-1 the cross-correlation loss has a weight of 0.9, while the cross-entropy loss has a weight of

0.1. For CUSSP-2, the weights are 0.5 and 0.5, while for CUSSP-3 they are 0.1 and 0.9, respectively. Both CUSSP-1 and CUSSP-3 outperform CUSSP-2, though the performance is low, indicating the importance of fine-tuning, described below.

In the second scenario, we fine-tuned the encoder with a siamese network to enhance the quality of the encoded representations after training the Barlow Twins network. To prevent overfitting of the model and to limit its capacity, we froze the parameters of all layers except the last block of the ResNet18 encoder when training the siamese network and the classifier. The resulting model, CUSSP-SIAM, showed a significant improvement in performance. In the final configuration CUSSP-SIAM-25, the number of frames in the training sequences was truncated from 50 frames to the 25 frames that correspond to the interval when mitral regurgitation occurs. The results are summarized in Table 2, while the ROC-AUC curve for CUSSP-SIAM-25 are shown in Fig. 5.

CUSSP attains an F1 score of 0.69, and an ROC AUC of 0.88, outperforming the CNN-LSTM approach. This is dues to CUSSP's focus on the area around the valve to capture the blood flow through the valve, combining the advantages of Barlow Twins and contrastive learning. Meanwhile, the CNN-LSTM relies on attention, which does not seem to work as well. Additionally, using only the frames relevant to the task reduces the number of parameters and makes the model sample-efficient. This is essential for attaining high performance in the low label setting. In the future, we aim to use the pipeline on the large unlabeled dataset to scan for and adjudicate more MR cases.

In conclusion, we present the first automated mitral regurgitation classification system using CMR imaging sequences. The CUSSP model we developed, trained with limited supervision, operates on 4CH CMR imaging sequences and attains an F1 score of 0.69 and an ROC AUC of 0.88, opening up the opportunity for large-scale screening for MR.

References

1. Allen, N.E., Sudlow, C., Peakman, T., Collins, R., UK Biobank: UK Biobank data: come and get it. Sci. Transl. Med. **6**(224) (2014)
2. Bai, W., et al.: Automated cardiovascular magnetic resonance image analysis with fully convolutional networks. J. Cardiovasc. Magn. Reson. **20**(1), 1–12 (2018)
3. Breiman, L.: Random forests. Mach. Learn. **45** (2001)
4. Caron, M., Misra, I., Mairal, J., Goyal, P., Bojanowski, P., Joulin, A.: Unsupervised learning of visual features by contrasting cluster assignments. In: Advances in Neural Information Processing Systems 33: Annual Conference on Neural Information Processing Systems 2020, NeurIPS 2020 (2020)
5. Chen, T., Kornblith, S., Norouzi, M., Hinton, G.E.: A simple framework for contrastive learning of visual representations. In: Proceedings of the 37th International Conference on Machine Learning, ICML 2020, 13–18 July 2020, Virtual Event. Proceedings of Machine Learning Research, vol. 119 (2020)
6. Dziadzko, V., et al.: Causes and mechanisms of isolated mitral regurgitation in the community: clinical context and outcome. Eur. Heart J. **40**(27) (2019)

7. Enriquez-Sarano, M., Akins, C.W., Vahanian, A.: Mitral regurgitation. Lancet **373**(9672) (2009)
8. Enriquez-Sarano, M., et al.: Quantitative determinants of the outcome of asymptomatic mitral regurgitation. N. Engl. J. Med. **352**(9) (2005)
9. Fries, J.A., et al.: Weakly supervised classification of aortic valve malformations using unlabeled cardiac MRI sequences. Nat. Commun. **10**(1) (2019)
10. Grill, J., et al.: Bootstrap your own latent - a new approach to self-supervised learning. In: Advances in Neural Information Processing Systems 33: Annual Conference on Neural Information Processing Systems 2020, NeurIPS 2020 (2020)
11. Huang, G., Liu, Z., van der Maaten, L., Weinberger, K.Q.: Densely connected convolutional networks. In: 2017 IEEE Conference on Computer Vision and Pattern Recognition, CVPR 2017 (2017)
12. Iglovikov, V., Shvets, A.: TernausNet: U-Net with VGG11 encoder pre-trained on ImageNet for image segmentation. CoRR abs/1801.05746 (2018)
13. Kervadec, H., Dolz, J., Tang, M., Granger, E., Boykov, Y., Ayed, I.B.: Constrained-CNN losses for weakly supervised segmentation. Med. Image Anal. **54** (2019)
14. Mirabel, M., et al.: What are the characteristics of patients with severe, symptomatic, mitral regurgitation who are denied surgery? Eur. Heart J. **28**(11) (2007)
15. Nian, Y., et al.: Graph-based unsupervised segmentation for lung tumor CT images. In: 2017 3rd IEEE International Conference on Computer and Communications (ICCC). IEEE (2017)
16. Oktay, O., et al.: Anatomically constrained neural networks (ACNNs): application to cardiac image enhancement and segmentation. IEEE Trans. Med. Imaging **37**(2) (2018)
17. Parcha, V., Patel, N., Kalra, R., Suri, S.S., Arora, G., Arora, P.: Mortality due to mitral regurgitation among adults in the United States: 1999–2018. In: Mayo Clinic Proceedings, vol. 95. Elsevier (2020)
18. Ronneberger, O., Fischer, P., Brox, T.: U-Net: convolutional networks for biomedical image segmentation. In: Navab, N., Hornegger, J., Wells, W.M., Frangi, A.F. (eds.) MICCAI 2015. LNCS, vol. 9351, pp. 234–241. Springer, Cham (2015). https://doi.org/10.1007/978-3-319-24574-4_28
19. Simonyan, K., Zisserman, A.: Very deep convolutional networks for large-scale image recognition. In: 3rd International Conference on Learning Representations, ICLR 2015, Conference Track Proceedings (2015)
20. Uretsky, S., Argulian, E., Narula, J., Wolff, S.D.: Use of cardiac magnetic resonance imaging in assessing mitral regurgitation: current evidence. J. Am. Coll. Cardiol. **71**(5) (2018)
21. Vimalesvaran, K., et al.: Detecting aortic valve pathology from the 3-chamber cine cardiac MRI view. In: Wang, L., Dou, Q., Fletcher, P.T., Speidel, S., Li, S. (eds.) MICCAI 2022. LNCS, vol. 13431, pp. 571–580. Springer, Cham (2022). https://doi.org/10.1007/978-3-031-16431-6_54
22. Watanabe, N.: Acute mitral regurgitation. Heart **105**(9) (2019)
23. Wu, X., Bi, L., Fulham, M.J., Feng, D.D., Zhou, L., Kim, J.: Unsupervised brain tumor segmentation using a symmetric-driven adversarial network. Neurocomputing **455**, 242–254 (2021)
24. Wu, X., Bi, L., Fulham, M.J., Kim, J.: Unsupervised positron emission tomography tumor segmentation via GAN based adversarial auto-encoder. In: 16th International Conference on Control, Automation, Robotics and Vision, ICARCV 2020, Shenzhen, China, 13–15 December 2020 (2020)

25. Xing, Z.J., Yin, F., Wu, Y.C., Liu, C.L.: Offline signature verification using convolution siamese network. In: Ninth International Conference on Graphic and Image Processing (ICGIP 2017), vol. 10615. SPIE (2018)
26. Zbontar, J., Jing, L., Misra, I., LeCun, Y., Deny, S.: Barlow twins: self-supervised learning via redundancy reduction. In: Proceedings of the 38th International Conference on Machine Learning, ICML 2021, 18–24 July 2021, Virtual Event. Proceedings of Machine Learning Research, vol. 139 (2021)
27. Zhang, Q., et al.: Automatic assessment of mitral regurgitation severity using the mask R-CNN algorithm with color Doppler echocardiography images. Comput. Math. Methods Med. **2021** (2021)
28. Zhao, A., Balakrishnan, G., Durand, F., Guttag, J.V., Dalca, A.V.: Data augmentation using learned transformations for one-shot medical image segmentation. In: IEEE Conference on Computer Vision and Pattern Recognition, CVPR 2019 (2019)

Clinical Applications – Dermatology

EPVT: Environment-Aware Prompt Vision Transformer for Domain Generalization in Skin Lesion Recognition

Siyuan Yan[1,2,3,6], Chi Liu[2,3,6], Zhen Yu[2,3,6], Lie Ju[1,2,3,6],
Dwarikanath Mahapatra[5,6], Victoria Mar[3,6], Monika Janda[4,6], Peter Soyer[4,6],
and Zongyuan Ge[2,6(✉)]

[1] Faculty of Engineering, Monash University, Melbourne, Australia
[2] AIM for Health Lab, Monash University, Victoria, Australia
zongyuan.ge@monash.edu
[3] Monash Medical AI, Monash University, Victoria, Australia
[4] Victorian Melanoma Service, Alfred Health, Victoria, Australia
[5] The University of Queensland Diamantina Institute, Dermatology Research Centre,
The University of Queensland, Brisbane, Australia
[6] Inception Institute of AI, Abu Dhabi, UAE

Abstract. Skin lesion recognition using deep learning has made remarkable progress, and there is an increasing need for deploying these systems in real-world scenarios. However, recent research has revealed that deep neural networks for skin lesion recognition may overly depend on disease-irrelevant image artifacts (*i.e.* dark corners, dense hairs), leading to poor generalization in unseen environments. To address this issue, we propose a novel domain generalization method called EPVT, which involves embedding prompts into the vision transformer to collaboratively learn knowledge from diverse domains. Concretely, EPVT leverages a set of domain prompts, each of which plays as a domain expert, to capture domain-specific knowledge; and a shared prompt for general knowledge over the entire dataset. To facilitate knowledge sharing and the interaction of different prompts, we introduce a domain prompt generator that enables low-rank multiplicative updates between domain prompts and the shared prompt. A domain mixup strategy is additionally devised to reduce the co-occurring artifacts in each domain, which allows for more flexible decision margins and mitigates the issue of incorrectly assigned domain labels. Experiments on four out-of-distribution datasets and six different biased ISIC datasets demonstrate the superior generalization ability of EPVT in skin lesion recognition across various environments. Code is available at https://github.com/SiyuanYan1/EPVT.

Keywords: Skin lesions · Prompt · Domain generalization · Debiasing

1 Introduction

Skin cancer is a serious and widespread form of cancer that requires early detection for successful treatment. Computer-aided diagnosis systems (CAD) using

---

Supplementary Information The online version contains supplementary material available at https://doi.org/10.1007/978-3-031-43990-2_24.

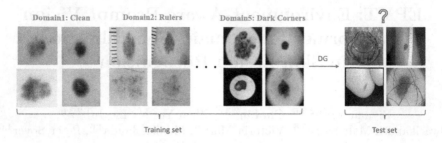

Fig. 1. The training data is split into five domains: clean, rulers, hairs, air pockets, and dark corners. Domain generalization aims to train the model to learn from these domains to generalize well in unseen domains.

deep learning models have shown promise in accurate and efficient skin lesion diagnosis. However, recent research has revealed that the success of these models may be a result of overly relying on "spurious cues" in dermoscopic images, such as rulers, gel bubbles, dark corners, and hairs [3–5,29], which leads to unreliable diagnoses. When a deep learning model overfits specific artifacts instead of learning the correct dermoscopic patterns, it may fail to identify skin lesions in real-world environments where the artifacts are absent or inconsistent.

To alleviate the artifact bias and enhance the model's generalization ability, we rethink the problem from the domain generalization (DG) perspective, where a model trained within multiple different but related domains are expected to perform well in unseen test domains. As illustrated in Fig. 1, we define the domain labels based on the types of artifacts present in the training images, which can provide environment-aware prior knowledge reflecting a range of noisy contexts. By doing this, we can develop a DG algorithm to learn the generalized and robust features from diverse domains.

Previous DG algorithms learning domain-invariant features from source domains have succeeded in natural image tasks [2,17,19], but cannot directly apply to medical images, in particular skin images, due to the vast cross-domain diversity of skin lesions in terms of shapes, colors, textures, etc. As each domain contains ad hoc intrinsic knowledge, learning domain-invariant features is highly challenging. One promising way is, as suggested in some recent works [7,24,32], exploiting multiple learnable domain experts (e.g., batch norm statistic, auxiliary classifiers, etc.) to capture domain-specific knowledge from different source domains individually. Still, two significant challenges remain. First, previous work only exploits some weak experts, like the batch norm, to capture knowledge, which naturally hampers the capability of capturing essential domain-specific knowledge. Second, previous methods such as [30] focused on learning domain knowledge independently while overlooking the rich cross-domain information that all domain experts can contribute collectively for the target domain prediction.

To overcome the above problems, we propose an environment-aware prompt vision transformer (EPVT) for domain generalization of skin lesion recognition.

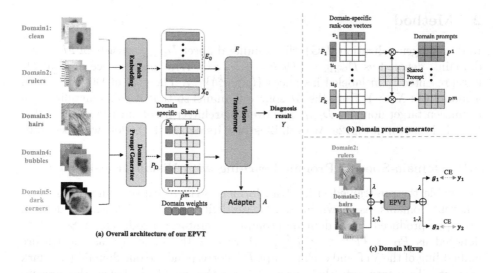

Fig. 2. The overview of our environment-aware prompt vision transformer (EPVT).

On the one hand, inspired by the emerging prompt learning techniques that embed prompts into a model for adaptation to diverse downstream tasks [12,26, 31], we construct different prompt vectors to strengthen the learning of domain-specific knowledge for adaptation to diverse domains. Then, the self-attention mechanism of the vision transformer (ViT) [8] is adopted to fully model the relationship between image tokens and prompt vectors. On the other hand, to encourage cross-domain information sharing while preserving the domain-specific knowledge of each domain prompt, we propose a domain prompt generator based on low-rank weights updating. The prompt generator enables multiple domain prompts to work collaboratively and benefit from each other for generalization to unknown domains. Additionally, we devise a domain mixup strategy to resolve the problem of co-occurring artifacts in dermoscopic images and mitigate the resulting noisy domain label assignments.

Our contributions can be summarized as: (1) We resolve an artifacts-derived biasing problem in skin cancer diagnosis using a novel environment-aware prompt learning-based DG algorithm, EPVT; (2) EPVT takes advantage of a ViT-based domain-aware prompt learning and a novel domain prompt generator to improve domain-specific and cross-domain knowledge learning simultaneously; (3) A domain mixup strategy is devised to reduce the co-artifacts specific to dermoscopic images; (4) Extensive experiments on four out-of-distribution skin datasets and six biased ISIC datasets demonstrate the outperforming generalization ability and robustness of EPVT under heterogeneous distribution shifts.

2   Method

In domain generalization (DG), the training dataset D_{train} consists of M source domains, denoted as $D_{train} = \{D_k | k = 1, ..., M\}$. Here, each source domain D_k is represented by n labeled instances $\{(x_j^k, y_j^k)\}_{j=1}^n$. The goal of DG is to learn a model $G : X \rightarrow Y$ from the M source domains so that it can generalize well in unseen target domains D_{test}. The overall architecture of our proposed model, EPVT, is shown in Fig. 2a. We will illustrate its details in the following sections.

2.1   Domain-Specific Prompt Learning with Vision Transformer

To enable the pre-trained vision transformer (ViT) to capture knowledge from different domains, as shown in Fig. 2a, we define a set of M learnable domain prompts produced by a domain prompt generator (introduced in Sect. 2.2), denoted as $P_D = \{P^m \in \mathbb{R}^d\}_{m=1}^M$, where d is the same size as the feature embedding of the ViT and each prompt P^m corresponds to one domain (*i.e.* dark corners). To incorporate these prompts into the model, we follow the conventional practice of visual prompt tuning [12], which prepends the prompts P_D into the first layer of the transformer. Particularly, for each prompt P^m in P_D, we extract the domain-specific features as:

$$F_m(x) = F([\ X_0, P^m, E_0\ ]) \tag{1}$$

where F is the feature encoder of the ViT, X_0 denotes the class token, E_0 is the image patch embedding, F_m is the feature extracted by ViT with the m-th prompt, and 0 is the index of the first layer. Domain prompts P_D are a set of learnable tokens, with each prompt P^m being fed into the vision transformer along with the image and corresponding class tokens from a specific domain. Through optimizing, each prompt becomes a domain expert only responsible for the images from its own domain. By the self-attention mechanism of ViT, the model can effectively capture domain-specific knowledge from the domain prompt tokens.

2.2   Cross-Domain Knowledge Learning

To facilitate effective knowledge sharing across different domains while maintaining its own parameters of each domain prompt, we propose a domain prompt generator, as depicted in Fig. 2b. Our approach is inspired by model adaptation and multi-task learning techniques used in natural language processing [13, 26]. Aghajanyan *et al.* [1] have shown that when adapting a model to a specific task, the updates to weights possess a low intrinsic rank. Similarly, each domain prompt P^m should also have a unique low intrinsic rank when learning knowledge from its own domain. To this end, we decompose each P^m into a Hadamard product between a randomly initialized shared prompt P^* and a rank-one matrix P_k obtained from two randomly initialized learnable vectors u_k and v_k, which is:

$$P^m = P^* \odot P_k \quad \text{where} \quad P_k = u_k \cdot v_k^T \tag{2}$$

where P^m represents the domain-specific prompt, computed by Hadamard product of P^* and P_k. Here, $P^* \in \mathbb{R}^{s \times d}$ is utilized to learn general knowledge, with s and d representing the dimensions of the prompt vector and feature embedding respectively. On the other hand, P_k is computed using domain-specific trainable vectors: $u_k \in \mathbb{R}^s$ and $v_k \in \mathbb{R}^d$. These vectors capture domain-specific information in a low-rank space. The decomposition of domain prompts into rank-one subspaces ensures that the model effectively encodes domain-specific information. By using the Hadamard product, the model can efficiently leverage cross-domain knowledge for target domain prediction.

2.3 Mitigating the Co-artifacts Issue

The artifacts-based domain labels can provide domain information for dermoscopic images. However, a non-trivial issue arises due to the possible co-occurrence of different artifacts from other domains within each domain. To address this issue, we employ a domain mixup strategy [27,28]. Instead of assigning a hard prediction label ("0" or "1") to each image, in each batch, we mix every image using two randomly selected images from two different domains. This allows us to learn a flexible margin relative to both domains. We then apply the cross-entropy loss to the corresponding labels of bot images, as shown in Fig. 2c and can be represented by the following equation:

$$\mathcal{L}_{mixup} = \lambda \mathcal{L}_{CE}(G(x_{mix}), y_i) + (1 - \lambda)\mathcal{L}_{CE}(G(x_{mix}), y_j) \tag{3}$$

where $x_{mix} = \lambda x_i^k + (1 - \lambda)x_j^q$; x_i^k and x_j^q are samples from two different domains k and q, and y_i^k and y_j^q are the corresponding labels. This strategy can overcome the challenge of ambiguous domain labels in dermoscopic images and improve the performance of our model.

2.4 Optimization

So far, we have introduced $\mathcal{L}_{mixup}$ in Eq. 3 for optimizing our model. However, since our goal is to generalize the model to unseen environments, we also need to take advantage of each domain prompt. Instead of assigning equal weights to each domain prompt, we employ an adapter [30] that learns the linear correlation between the domain prompts and the target image prediction. To obtain the adapted prompt for inference in the target domain, we define it as a linear combination of the source domain prompts:

$$P_{adapted} = A(F(x)) = \sum_{m=1}^{M} w_m \cdot P^m, \quad \text{s.t.} \quad \sum_{m=1}^{M} w_m = 1 \tag{4}$$

where A represents an adapter containing a two-layer MLP with a softmax layer, and w_m denotes the learned weights.

To train the adapter A, we simulate the inference process for each image in the source domain by treating it as an image from the pseudo-target domain.

Specifically, we first extract features from the ViT: $\hat{F}_m(x) = F([X_0, E_0])$. Then we calculated the adapted prompt $P_{adapted}$ for the pseudo-target environment image x using the adapter A: $P_{adapted} = A(\hat{F}_m(x))$. Next, we extract features from ViT using the adapted prompt: $\hat{F}_m(x) = F([\hat{F}_m(x), P_{adapted}, E_0])$. Finally, the classification head H is applied to predict the label y: $y = H(\hat{F}_m(x))$. Additionally, the inferece process is the same as the simulated inference process and our final prediction will be conditioned on the adapted prompt $P_{adapted}$.

To ensure that the adapter learns the correct linear correlation between the domain prompts and the target image, we use the domain label from source domains to directly supervise the weights w_m. We also use the cross-entropy loss to maintain the model performance with the adapted prompt:

$$\mathcal{L}_{adapted} = \mathcal{L}_{CE}(H(\hat{F}_m(x)), y) + \lambda(\frac{1}{M}\sum_{m=1}^{M}\frac{1}{M}(\mathcal{L}_{CE}(w_m^m, 1) + \sum_{t \neq m}\mathcal{L}_{CE}(w_t^m, 0))$$

(5)

where $\hat{F}_m(x)$ is the obtained feature map conditioned on the adapted prompt $P_{adapted}$, and H is the classification head. The total loss is then defined as $\mathcal{L}_{total} = \mathcal{L}_{mixup} + \mathcal{L}_{adapted}$.

3   Experiments

Experimental Setup: We consider two challenging melanoma-benign classification settings that can effectively evaluate the generalization ability of our model in different environments and closely mimic real-world scenarios. (1) Out-of-distribution evaluation: The task is to evaluate the model on test sets that contain different artifacts or attributes compared to the training set. We train and validate all algorithms on *ISIC2019* [6] dataset, following the split of [3]. We use the artifacts annotations from [3] and divide the training set of *ISIC2019* into five groups: *dark corner*, *hair*, *gel bubble*, *ruler*, and *clean*, with 2351, 4884, 1640, 672, and 2796 images, respectively. We evaluate models on four out-of-distribution (OOD) datasets, including *Derm7pt-Dermoscopic* [14], *Derm7pt-Clinical* [14], *PH2* [18], and *PAD-UFES-20* [21]. It's worth noting that *ISIC2019*, *Derm7pt-Dermoscopic*, and *PH2* are dermoscopic images, while *Derm7pt-Clinical* and *PAD* are clinical images. (2) Trap set debiasing: We train and test our EPVT with its baseline on six trap sets [3] with increasing bias levels, ranging from 0 (randomly split training and testing sets from the ISIC2019 dataset) to 1 (the highest bias level where the correlation between artifacts and class label is in the opposite direction in the dataset splits). More details about these datasets and splits are provided in the complementary material.

Implementation Details: For a fair comparison, we train all models using ViT-Base/16 [8] backbone pre-trained on Imagenet and report the ROC-AUC with five random seeds. Hyperparameter and model selection methods are crucial for domain generalization algorithms. We conduct a grid search over learning rate (from $3e^{-4}$ to $5e^{-6}$), weight decay (from $1e^{-2}$ to $1e^{-5}$), and the length of the

Table 1. The comparison on out-of-distribution datasets

Method	derm7pt_d	derm7pt_c	pad	ph2	Average
ERM	81.24 ± 1.6	71.61 ± 1.9	82.62 ± 1.6	83.06 ± 1.9	79.63 ± 1.5
DRO [23]	82.46 ± 1.7	72.88 ± 1.9	81.52 ± 1.2	84.64 ± 1.8	81.27 ± 1.6
CORAL [25]	81.42 ± 1.9	71.45 ± 1.3	**88.13 ± 1.2**	85.2 ± 2.2	81.55 ± 1.5
MMD [17]	82.08 ± 1.7	71.8 ± 1.5	85.89 ± 1.9	87.17 ± 1.4	81.73 ± 1.5
DANN [10]	81.79 ± 1.1	73.12 ± 1.6	84.12 ± 1.6	85.18 ± 1.9	81.87 ± 1.7
IRM [2]	79.07 ± 1.7	71.3 ± 1.8	77.82 ± 3.4	79.37 ± 1.2	76.64 ± 1.7
SagNet [20]	82.28 ± 1.8	73.19 ± 1.6	78.89 ± 4.5	88.79 ± 1.9	81.79 ± 1.8
MLDG [16]	81.06 ± 1.6	71.79 ± 1.6	83.41 ± 1.0	84.22 ± 1.8	79.87 ± 1.2
CAD [22]	82.72 ± 1.5	69.57 ± 1.6	81.36 ± 1.9	88.4 ± 1.5	81.51 ± 1.6
DoPrompt [30]	82.38 ± 1.0	71.61 ± 1.7	83.81 ± 1.4	91.33 ± 1.8	82.06 ± 1.6
SelfReg [15]	81.83 ± 1.9	73.29 ± 1.4	85.27 ± 1.3	85.16 ± 3.3	81.12 ± 1.0
EPVT (Ours)	**83.69 ± 1.4**	**73.96 ± 1.6**	86.67 ± 1.5	**91.91 ± 1.5**	**84.11 ± 1.4**

prompt (from 4 to 16, when available) and report the best performance of all models. We employ the training-domain validation set method [11] for model selection. After the grid search, we use the AdamW optimizer with a learning rate of $5e^{-6}$ and a weight decay of $1e^{-2}$. The batch size is 130, and the length of the prompt is 10. We resize the input image to a size of 224×224 and adopt the standard data augmentation like random flip, crop, rotation, and color jitter. An early stopping with the patience of 22 is set and with a total of 60 epochs for OOD evaluation and 100 epochs for trap set debiasing. All experiments are conducted on a single NVIDIA RTX 3090 GPU.

Out-of-Distribution Evaluation: Table 1 presents a comprehensive comparison of our EPVT algorithm with existing domain generalization methods. The results clearly demonstrate the superiority of our approach, with the best performance on three out of four OOD datasets and remarkable improvements over the ERM algorithm, especially achieving 4.1% and 8.9% improvement on the *PAD* and *PH2* datasets, respectively. Although some algorithms may perform similarly to our model on one of the four datasets, none can consistently match the performance of our method across all four datasets. Particularly, our approach showcases the highest average performance, with a 2.05% improvement over the second-best algorithm across all four datasets. These findings highlight the effectiveness of our algorithm in learning robust features and its strong generalization abilities across diverse environments.

Ablation Study: We perform ablation studies to analyze each component of our model, as shown in Table 2. We set our baseline as the Empirical Risk Minimization (ERM) algorithm, and we gradually add P (prompt [12]), A (Adapter), M (Mixup), and G (domain prompt generator) into the model. Firstly, we observe that the baseline model with prompt only improves the average performance by

Table 2. Ablation study on out-of-distribution datasets

Method	derm7pt_d	derm7pt_c	pad	ph2	Average
Baseline	81.24 ± 1.6	71.61 ± 1.9	82.62 ± 1.6	83.06 ± 1.9	79.63 ± 1.5
+P	82.13 ± 1.1	71.41 ± 1.3	82.15 ± 1.6	84.21 ± 1.4	79.73 ± 1.3
+P+A	82.55 ± 1.6	72.86 ± 1.1	81.02 ± 1.5	84.97 ± 1.8	81.10 ± 1.6
+P+A+M	81.43 ± 1.4	73.18 ± 1.5	85.78 ± 1.9	89.28 ± 1.3	82.42 ± 1.7
+P+A+M+G	**83.69 ± 1.4**	**73.96 ± 1.6**	**86.67 ± 1.5**	**91.91 ± 1.5**	**84.11 ± 1.4**

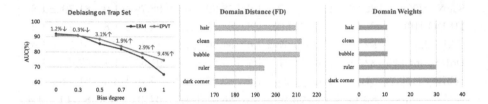

Fig. 3. (a) Deibiasing evaluation (b) domain distance (c) domain weights

0.1%, showing that simply combining prompt does not very helpful for domain generalization. When we combine the adapter, the model's average performance improves by 1.37%, but it performs worse than ERM on *PAD* dataset. Subsequently, we added domain mixup and domain prompt generator to the model, resulting in significant further improvements in the model's average performance by 1.32% and 1.69%, respectively. The consistently better performance than the baseline on all four datasets also highlights the importance of addressing co-artifacts and cross-domain learning for DG in skin lesion recognition.

Trap Set Debiasing: In Fig. 3a, we present the performance of the ERM baseline and our EPVT on six biased ISIC2019 datasets. Each point on the graph represents an algorithm that is trained and tested on a specific bias degree split. The graph shows that the ERM baseline performs better than our EPVT when the bias is low (0 and 0.3). However, this is because ERM relies heavily on spurious correlations between artifacts and class labels, leading to overfitting on the training set. As the bias degree increases, the correlation between artifacts and class labels decreases, and overfitting the train set causes the performance of ERM to drop dramatically on the test set with a significant distribution difference. In contrast, our EPVT exhibits greater robustness to different bias levels. Notably, our EPVT outperforms the ERM baseline by 9.4% on the bias 1 dataset.

Prompt Weights Analysis: To verify whether our model has learned the correct domain prompts for target domain prediction, we analyze and plot the results in Fig. 3b and 3c. Firstly, we extract the features of each domain from our training set and extract the feature from one target dataset, *Derm7pt-Clin*. We then calculate the Frechet distance [9] between each domain and the target dataset using the extracted feature, representing the domain distance between them. The results are recorded in Fig. 3b. Next, we record the learned weights of each domain prompt in Fig. 3c; it shows that our model assigns the highest weight to the "dark corner" group, as the domain distance between "dark corner" and *Derm7pt-Clin* is the closest, as shown in Fig. 3b. This suggests that they share the most similar domain information. Further, the "clean" group is assigned the smallest weight as the domain distance between them is the largest, indicating that their domains are significantly different and contain less useful information for target domain prediction. In summary, we observe a negative correlation between domain distance and the prompt's weights, indicating that our model can learn the correct knowledge from different domains precisely.

4   Conclusion

In this paper, we propose a novel DG algorithm called EPVT for robust skin lesion recognition. Our approach addresses the co-artifacts problem using a domain mixup strategy and cross-domain learning problems using a domain prompt generator. Compared to other competitive domain generalization algorithms, our method achieves outstanding results on three out of four OOD datasets and the second-best on the remaining one. Additionally, we conducted a debiasing experiment that highlights the shortcomings of conventional training using empirical risk minimization, which leads to overfitting in dermoscopic images due to artifacts. In contrast, our EPVT model effectively reduces overfitting and consistently performs better in different biased environments.

References

1. Aghajanyan, A., Zettlemoyer, L., Gupta, S.: Intrinsic dimensionality explains the effectiveness of language model fine-tuning. In: Annual Meeting of the Association for Computational Linguistics (2020)
2. Arjovsky, M., Bottou, L., Gulrajani, I., Lopez-Paz, D.: Invariant risk minimization. arXiv abs/1907.02893 (2019)
3. Bissoto, A., Barata, C., Valle, E., Avila, S.: Artifact-based domain generalization of skin lesion models. In: ECCV Workshops (2022)
4. Bissoto, A., Fornaciali, M., Valle, E., Avila, S.: (de) constructing bias on skin lesion datasets. In: 2019 IEEE/CVF Conference on Computer Vision and Pattern Recognition Workshops (CVPRW), pp. 2766–2774 (2019)
5. Bissoto, A., Valle, E., Avila, S.: Debiasing skin lesion datasets and models? Not so fast. In: 2020 IEEE/CVF Conference on Computer Vision and Pattern Recognition Workshops (CVPRW), pp. 3192–3201 (2020)
6. Combalia, M., et al.: Validation of artificial intelligence prediction models for skin cancer diagnosis using dermoscopy images: the 2019 international skin imaging collaboration grand challenge. Lancet Digit. Health 4(5), e330–e339 (2022)
7. Dai, Y., Li, X., Liu, J., Tong, Z., Duan, L.Y.: Generalizable person re-identification with relevance-aware mixture of experts. In: 2021 IEEE/CVF Conference on Computer Vision and Pattern Recognition (CVPR), pp. 16140–16149 (2021)
8. Dosovitskiy, A., et al.: An image is worth 16x16 words: transformers for image recognition at scale. In: ICLR (2021)

9. Dowson, D., Landau, B.: The fréchet distance between multivariate normal distributions. J. Multivar. Anal. **12**(3), 450–455 (1982)
10. Ganin, Y., et al.: Domain-adversarial training of neural networks. J. Mach. Learn. Res. **17**(1), 2096-2030 (2016)
11. Gulrajani, I., Lopez-Paz, D.: In search of lost domain generalization. In: International Conference on Learning Representations (2021). https://openreview.net/forum?id=lQdXeXDoWtI
12. Jia, M., et al.: Visual prompt tuning. In: Avidan, S., Brostow, G., Cissé, M., Farinella, G.M., Hassner, T. (eds.) ECCV 2022. LNCS, vol. 13693, pp. 709–727. Springer, Cham (2022). https://doi.org/10.1007/978-3-031-19827-4_41
13. Karimi Mahabadi, R., Henderson, J., Ruder, S.: Compacter: efficient low-rank hypercomplex adapter layers. Adv. Neural. Inf. Process. Syst. **34**, 1022–1035 (2021)
14. Kawahara, J., Daneshvar, S., Argenziano, G., Hamarneh, G.: Seven-point checklist and skin lesion classification using multitask multimodal neural nets. IEEE J. Biomed. Health Inform. **23**(2), 538–546 (2018)
15. Kim, D., Yoo, Y., Park, S., Kim, J., Lee, J.: Selfreg: self-supervised contrastive regularization for domain generalization. In: Proceedings of the IEEE/CVF International Conference on Computer Vision, pp. 9619–9628 (2021)
16. Li, D., Yang, Y., Song, Y.Z., Hospedales, T.: Learning to generalize: meta-learning for domain generalization. In: AAAI Conference on Artificial Intelligence (2018)
17. Li, H., Pan, S.J., Wang, S., Kot, A.C.: Domain generalization with adversarial feature learning. In: 2018 IEEE/CVF Conference on Computer Vision and Pattern Recognition, pp. 5400–5409 (2018)
18. Mendonça, T., Celebi, M., Mendonca, T., Marques, J.: PH2: a public database for the analysis of dermoscopic images. Dermoscopy Image Anal. (2015)
19. Motiian, S., Piccirilli, M., Adjeroh, D.A., Doretto, G.: Unified deep supervised domain adaptation and generalization. In: 2017 IEEE International Conference on Computer Vision (ICCV), pp. 5716–5726 (2017)
20. Nam, H., Lee, H., Park, J., Yoon, W., Yoo, D.: Reducing domain gap by reducing style bias. In: Proceedings of the IEEE/CVF Conference on Computer Vision and Pattern Recognition, pp. 8690–8699 (2021)
21. Pacheco, A.G., et al.: PAD-UFES-20: a skin lesion dataset composed of patient data and clinical images collected from smartphones. Data Brief **32**, 106221 (2020)
22. Ruan, Y., Dubois, Y., Maddison, C.J.: Optimal representations for covariate shift. In: International Conference on Learning Representations (2022). https://openreview.net/forum?id=Rf58LPCwJj0
23. Sagawa, S., Koh, P.W., Hashimoto, T.B., Liang, P.: Distributionally robust neural networks. In: International Conference on Learning Representations (2020). https://openreview.net/forum?id=ryxGuJrFvS
24. Seo, S., Suh, Y., Kim, D., Kim, G., Han, J., Han, B.: Learning to optimize domain specific normalization for domain generalization. In: Vedaldi, A., Bischof, H., Brox, T., Frahm, J.-M. (eds.) ECCV 2020. LNCS, vol. 12367, pp. 68–83. Springer, Cham (2020). https://doi.org/10.1007/978-3-030-58542-6_5
25. Sun, B., Saenko, K.: Deep CORAL: correlation alignment for deep domain adaptation. In: Hua, G., Jégou, H. (eds.) ECCV 2016. LNCS, vol. 9915, pp. 443–450. Springer, Cham (2016). https://doi.org/10.1007/978-3-319-49409-8_35
26. Wang, Z., Panda, R., Karlinsky, L., Feris, R., Sun, H., Kim, Y.: Multitask prompt tuning enables parameter-efficient transfer learning. In: International Conference on Learning Representations (2023). https://openreview.net/forum?id=Nk2pDtuhTq

27. Xu, M., Zhang, J., Ni, B., Li, T., Wang, C., Tian, Q., Zhang, W.: Adversarial domain adaptation with domain mixup. In: Proceedings of the AAAI Conference on Artificial Intelligence, vol. 34, pp. 6502–6509 (2020)
28. Yan, S., Song, H., Li, N., Zou, L., Ren, L.: Improve unsupervised domain adaptation with mixup training. arXiv preprint arXiv:2001.00677 (2020)
29. Yan, S., et al.: Towards trustable skin cancer diagnosis via rewriting model's decision. In: Proceedings of the IEEE/CVF Conference on Computer Vision and Pattern Recognition (CVPR), pp. 11568–11577 (2023)
30. Zheng, Z., Yue, X., Wang, K., You, Y.: Prompt vision transformer for domain generalization. arXiv abs/2208.08914 (2022)
31. Zhou, K., Yang, J., Loy, C.C., Liu, Z.: Learning to prompt for vision-language models. Int. J. Comput. Vision **130**, 2337–2348 (2021)
32. Zhou, K., Yang, Y., Qiao, Y., Xiang, T.: Domain adaptive ensemble learning. IEEE Trans. Image Process. **30**, 8008–8018 (2020)

Skin Lesion Correspondence Localization in Total Body Photography

Wei-Lun Huang[1]([✉]), Davood Tashayyod[5], Jun Kang[3], Amir Gandjbakhche[4], Michael Kazhdan[1], and Mehran Armand[1,2]

[1] Department of Computer Science, Johns Hopkins University, Baltimore, MD, USA
whuang44@jh.edu
[2] Department of Orthopaedic Surgery, Johns Hopkins University, Baltimore, MD, USA
[3] Department of Dermatology, Johns Hopkins School of Medicine, Baltimore, MD, USA
[4] Eunice Kennedy Shriver National Institute of Child Health and Human Development, Bethesda, MD, USA
[5] Lumo Imaging, Rockville, MD, USA

Abstract. Longitudinal tracking of skin lesions - finding correspondence, changes in morphology, and texture - is beneficial to the early detection of melanoma. However, it has not been well investigated in the context of full-body imaging. We propose a novel framework combining geometric and texture information to localize skin lesion correspondence from a source scan to a target scan in total body photography (TBP). Body landmarks or sparse correspondence are first created on the source and target 3D textured meshes. Every vertex on each of the meshes is then mapped to a feature vector characterizing the geodesic distances to the landmarks on that mesh. Then, for each lesion of interest (LOI) on the source, its corresponding location on the target is first coarsely estimated using the geometric information encoded in the feature vectors and then refined using the texture information. We evaluated the framework quantitatively on both a public and a private dataset, for which our success rates (at 10 mm criterion) are comparable to the only reported longitudinal study. As full-body 3D capture becomes more prevalent and has higher quality, we expect the proposed method to constitute a valuable step in the longitudinal tracking of skin lesions.

Keywords: Total body photography · Skin lesion longitudinal tracking · 3D correspondence

1 Introduction

Evolution, the change of pigmented skin lesions, is a risk factor for melanoma [1]. Therefore, longitudinal tracking of skin lesions over the whole body is beneficial for early detection of melanoma [5]. However, establishing skin lesion

M. Kazhdan and M. Armand—Co-senior authors.

Supplementary Information The online version contains supplementary material available at https://doi.org/10.1007/978-3-031-43990-2_25.

correspondences across multiple scans from different patient visits has not been well investigated in the context of full-body imaging.

Several techniques have been proposed to match skin lesions across pairs of 2D images [9–14,16,17,25]. Early work used geometric constraints imposed by initial matches of skin lesions (manual selection or automatic detection) to align images and further match other skin lesions [10,16,17,25]. Mirzaalian and colleagues published a series of works for establishing lesion correspondence in image space [11–14]. Li et al. [9] used a CNN to output a 2D vector field for pixel-wise correspondences between the two input images. Though effective at matching skin lesions across pairs of images, the extension of these methods to the context of total body photography (TBP) for longitudinal tracking remains a challenge.

Several works have been proposed for tackling the skin lesion tracking problem over the full body [7,8,21,22]. However, they are either only applicable in well-controlled environments or do not extend to the tracking of lesions across scans at different visits. Recently, the concept of finding lesion correspondence using a 3D representation of the human body has been explored in [26] and [2] by using a template mesh. However, accurately deforming a template mesh to fit varying body shapes is challenging when the scanned shape deviates from the template, leading to large errors in downstream tasks such as establishing shape correspondence. Additionally, [26] does not take advantage of texture, while [2] uses texture in a common UV map that may lead to failures when geodesically close locations on the surface are mapped to distant sites in the texture map (e.g. when the two locations are on opposite sides of a texture seam).

We propose a novel framework for finding skin lesion correspondence iteratively using geometric and texture information (Fig. 1). We demonstrate the effectiveness of the proposed method in localizing lesion correspondence across scans in a manner that is robust to changes in body pose and camera viewing directions. Our code is available at https://github.com/weilunhuang-jhu/LesionCorrespondenceTBP3D.

2 Methods

Given a set of lesions of interest (LOIs) X in the source mesh, we would like to find their corresponding positions Y in the target mesh. Formally, we assume we are given source and target meshes, $\mathcal{M}_0$ and $\mathcal{M}_1$, with vertex sets $\mathcal{V}_k \subset \mathcal{M}_k$ and corresponding landmark sets $L_k \subset \mathcal{V}_k$ with $|L_0| = |L_1| = S$. We achieve this by computing a dense correspondence map $\Phi_{L,\epsilon} : \mathcal{V}_0 \to \mathcal{V}_1$, initially defined using geometric information and refined using textural information. Then we use that to define a map taking lesions of interest on the source to positions on the target $\Phi : X \to \mathcal{V}_1$.

2.1 Landmark-Based Correspondences

We define an initial dense correspondence between source and target vertices by leveraging the sparse landmark correspondences [3,6]. We do this by mapping

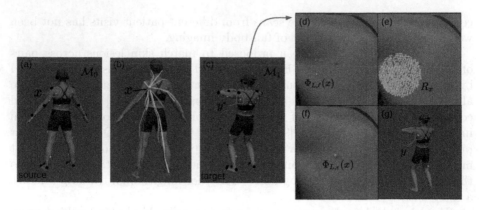

Fig. 1. Visualization of the correspondence localization workflow with geometric and texture information. (a) and (c) show a source and a target textured mesh with landmarks (black dots) and labeled lesions of interest (color dots). (b) shows the geodesic paths from a lesion x in (a) to all the landmarks. (d) shows the correspondence of the lesion $\Phi_{L,\ell}(x)$ derived from the geometric feature descriptors. (e) shows the local region R_x for refining the position of the correspondence. (f) shows the correspondence of the lesion $\Phi_{L,\epsilon}(x)$ from the local texture descriptors. (g) shows the lesion correspondences with confidence and anchored as new landmarks after one iteration.

source and target vertices into a high-dimensional space, based on their proximity to the landmarks, and then comparing positions in the high-dimensional space.

Concretely, we define maps $\ell_k : \mathcal{V}_k \to \mathbb{R}^S$, associating a vertex $v \in \mathcal{V}_k$ with an S-dimensional feature descriptor that describes the position of v relative to the landmarks:

$$\ell_k(v) = \left[\frac{1}{D_k(v, l_1)}, \cdots, \frac{1}{D_k(v, l_S)} \right]^\top \tag{1}$$

where $D_k : \mathcal{M}_k \times \mathcal{M}_k \to \mathbb{R}^{\geq 0}$ is the geodesic distance function on $\mathcal{M}_k$. We use the reciprocal of geodesic distance so that landmarks closer to v contribute more significantly to the feature vector.

Given this mapping, we create an initial dense correspondence between the source and target vertices, $\Phi_{L,\ell} : \mathcal{V}_0 \to \mathcal{V}_1$ by mapping a source vertex $v \in \mathcal{V}_0$ to the target vertex with the most similar feature descriptor (with similarity measured in terms of the normalized cross-correlation):

$$\Phi_{L,\ell}(v) = \arg\max_{v' \in \mathcal{V}_1} \left\{ C_{L,\ell}(v, v') = \frac{\langle \ell_0(v), \ell_1(v') \rangle}{||\ell_0(v)|| \cdot ||\ell_1(v')||} \right\}. \tag{2}$$

2.2   Texture-Based Refinement

While feature descriptors of corresponding vertices on the source and target mesh are identical when 1) the landmarks are in perfect correspondence, and 2) the source and target differ by an isometry, neither of these assumptions holds in real-world data. To address this, we use local texture to assign an additional feature descriptor to each vertex and use these texture-based descriptors to refine the coarse correspondence given by $\Phi_{L,\ell} : \mathcal{V}_0 \rightarrow \mathcal{V}_1$. Various texture descriptors have been proposed, e.g. SHOT [20,23,24], RoPS [4], and ECHO [15]. We selected the ECHO descriptor for its better descriptiveness and robustness to noise.

Letting $\epsilon_k(v) \in \mathbb{R}^N$ denote the ECHO descriptor of vertex $v \in \mathcal{V}_k$, our goal is to refine the dense correspondence so that corresponding source and target vertices also have similar descriptors. However, to avoid problems with repeating (local) textures, we would also like the correspondence to stay close to the correspondence defined by the landmarks.

We achieve this as follows: To every source vertex $v \in \mathcal{V}_0$ we associate a region $R_v \subset \mathcal{V}_1$ of target vertices that are either close to $\Phi_{L,\ell}(v)$ (the corresponding vertex on $\mathcal{V}_1$ as predicted by the landmarks) or have similar geometric feature descriptors:

$$R_v = \left\{ v' \in \mathcal{V}_1 \,\middle|\, D_1\left(v', \Phi_{L,\ell}(v)\right) < \varepsilon_1 \text{ or } C_{L,\ell}(v', \Phi_{L,\ell}(v)) > \varepsilon_2 \right\}. \tag{3}$$

Given this region, we define the target vertex corresponding to a source as the vertex within the region that has the most similar ECHO descriptor (using the normalized cross-correlation as before).

In practice, we compute the ECHO descriptor over three different radii, obtaining three descriptors for each vertex, $\epsilon_k^1(v), \epsilon_k^2(v), \epsilon_k^3(v) \in \mathbb{R}^N$. The selection of three different radii in ECHO descriptors is done to accommodate different sizes of lesions and their surrounding texture, and the values are empirically determined. This gives a mapping $\Phi_{L,\epsilon} : \mathcal{V}_0 \rightarrow \mathcal{V}_1$ defined in terms of the weighted sum of cross-correlations:

$$\Phi_{L,\epsilon}(v) = \underset{v' \in R_v}{\arg\max} \left\{ C_{L,\epsilon}(v, v') = \sum_{i=1}^{3} w_i \frac{\langle \epsilon_0^i(v), \epsilon_1^i(v') \rangle}{\|\epsilon_0^i(v)\| \cdot \|\epsilon_1^i(v')\|} \right\}, \tag{4}$$

where $C_{L,\epsilon}(v, v') \in [0,1]$ is the texture score of the target vertex v' and w_i is the weight of the cross-correlation between the ECHO descriptors computed at each radius.

2.3 Iterative Skin Lesion Correspondence Localization Framework

While each source LOI has a corresponding position on the target mesh as given by $\Phi_{L,\epsilon} : \mathcal{V}_0 \rightarrow \mathcal{V}_1$, not all correspondences are localized with high confidence when 1) the local texture is not well-preserved across scans and 2) the local region R_v does not include the true correspondence. To address this, we adapt our algorithm for computing the correspondence map $\Phi : X \rightarrow \mathcal{V}_1$ by iteratively growing the set of landmarks to include LOI correspondences about which we are confident, similar to the way in which a human annotator would label lesion correspondence (Fig. 2).

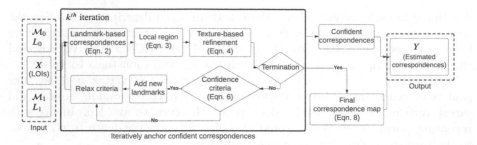

Fig. 2. Block diagram of the iterative lesion correspondence localization algorithm.

Iteratively Anchor Confident Correspondences. We iteratively compute correspondence maps $\Phi_{L,\ell}^k : \mathcal{V}_0 \to \mathcal{V}_1$ and $\Phi_{L,\epsilon}^k : \mathcal{V}_0 \to \mathcal{V}_1$, with the superscript denoting the kth iteration. For each map $\Phi_{L,\epsilon}^k$ and every LOI $x \in X$, we determine if we are confident in the correspondence $\{x, \Phi_{L,\epsilon}^k(x)\}$ by evaluating a binary function $\chi_L^k : X \to \{0,1\}$. Denoting by X' the subset of LOIs about which we are confident, we add the pairs $\{x', \Phi_{L,\epsilon}^k(x')\}$ to the landmark set L and remove the LOI $x' \in X'$ from X. We iterate this process until all the correspondences of LOIs are confidently found or a maximum number of iterations (K) have been performed.

Lesion correspondence confidence is measured using three criteria: i) texture similarity, ii) agreement between geometric and textural correspondences, and iii) the unique existence of a similar lesion within a region. To quantify uniqueness, we compute the set of target vertices whose textural descriptor is similar to that of the LOI:

$$S_x^\delta = \left\{ v' \in R_x \,\middle|\, C_{L,\epsilon}(x, v') > \delta \right\} , \tag{5}$$

and consider the diameter of the set (defined in terms of the mean of the distances of vertices in S_x^δ from the centroid of S_x^δ). Putting this together, we define confidence as

$$\chi_L^k(x) = C_{L,\epsilon}\big(x, \Phi_{L,\epsilon}^k(x)\big) > \varepsilon_3 \quad \vee \quad D_1\big(\Phi_{L,\ell}^k(x), \Phi_{L,\epsilon}^k(x)\big) < \varepsilon_4 \quad \vee \quad \varnothing(S_x^\delta) < \varepsilon_5, \tag{6}$$

where the initial values of thresholds ε_i are empirically chosen. To further support establishing correspondences, we relax the thresholds ε_i in subsequent iterations, allowing us to consider correspondences that are further away and about which we are less confident.

Final Correspondence Map. Having mapped every high-confidence LOI to a corresponding target vertex, we must complete the correspondence for the remaining low-confidence LOIs. We note that for a low-confidence LOI $x \in X$, the texture in the source mesh is not well-matched to the texture in the target, for any $v' \in R_x$. (Otherwise the first term in χ_L^k would be large.)

To address this, we would like to focus on landmark-based similarity. However, by definition of R_x, for all $v' \in R_x$, we know that the landmark descriptors of x and v' will all be similar, so that $C_{L,\ell}$ will not be discriminating. Instead, we use a standard transformation to turn distances into similarities. Specifically, we define geometric score between a source LOI x and a target vertex $v' \in R_x$ in terms of the geodesic distance between v' and the corresponding position of x in $\mathcal{V}_1$, as predicted by the landmark descriptors:

$$C_L(x, v') = e^{-\frac{1}{2\sigma^2} D_1^2\left(v',\ \Phi_{L,\ell}^K(x)\right)} \in [0, 1] \,, \tag{7}$$

where σ is the maximum geodesic distance from a vertex within R_x to $\Phi_{L,\ell}^K(x)$.

Therefore, for a remaining LOI, we define its corresponding target vertex as the vertex with the highest weighted sum of the geometric and texture scores:

$$\Phi(x) = \arg\max_{v' \in R_x} \left\{ w_1 \cdot C_L(x, v') + w_2 \cdot C_{L,\varepsilon}(x, v') \right\} \,, \tag{8}$$

where w_1 and w_2 are the weights for combining the scores.

3   Evaluation and Discussion

3.1   Dataset

We evaluated our methods on two datasets. The first dataset is from Skin3D [26] (annotated 3DBodyTex [18,19]). The second dataset comes from a 2D Imaging-Rich Total Body Photography system (IRTBP), from which the 3D textured meshes are derived from photogrammetry 3D reconstruction. The number of vertices is on average 300K and 600K for Skin3D and IRTBP datasets respectively. The runtime using 10 iterations is several minutes (on average) on an Intel i7-11857G7 processor. Example data of the two datasets can be found in the supplement.

3.2   Correspondence Localization Error and Success Rate

Average correspondence localization error (CLE) for individual subjects, defined as the geodesic distance between the ground-truth and the estimated lesion correspondence, is shown in Fig. 3. To interpret CLE in a clinical application, the localized correspondence is successful if its CLE is less than a threshold criterion. We measured the success rate as the percentage of the correctly localized skin lesions over the total number of skin lesion pairs in the dataset.

To compare our result to the existing method [26], we compute our success rates with the threshold criterion at 10 mm. As shown in Table 1, the performance of our method is comparable to the previously reported longitudinal accuracy. The qualitative result of the localized correspondence in the Skin3D dataset is shown in Fig. 4. A table of parameters used in the experiments can be found in the supplement.

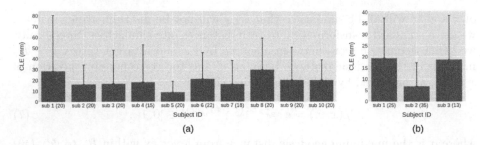

Fig. 3. Mean and standard deviation of correspondence localization error (CLE) for individual subjects in (a) Skin3D and (b) IRTBP datasets evaluated with the proposed methods. There are 10 and 3 subjects in the Skin3D and IRTBP datasets respectively. The number of LOIs for individual subjects is included in the parentheses.

Table 1. Comparison of the success rate on Skin3D dataset. Each metric is computed on a pair of meshes (for one subject) and averaged across paired meshes with the standard deviation shown in brackets. The method *Texture radius 50* and *Combined radius 50* are defined in Fig. 5.

	Skin3D [26]	Texture radius 50	Combined radius 50	Iterative algorithm
Success rate	0.5 (0.38)	0.48 (0.17)	0.45 (0.14)	0.57 (0.14)

3.3   Usage of Texture on 3D Surface

We believe that the geometric descriptor only provides a coarse correspondence while the local texture is more discriminating. Figure 5 shows the success rate under different threshold criteria for the proposed methods. Since we have two combinations of defining source and target for two scans, we measured the result in both to ensure consistency (Fig. 5(a) and (b)). As expected, we observed that using geometric information with body landmarks and geodesic distances is insufficient for localizing lesion correspondence accurately. However, the correspondence map $\Phi_{L,\epsilon}$ refined with local texture may lead to correspondences with large CLE when using a large region R_x. The figure shows the discriminating power and the large-error-prone property of using $\Phi_{L,\epsilon}$ to localize lesion correspondence with one iteration (relatively high success rates under strict criteria and relatively low success rates under loose criteria, compared to the correspondence map combining geometric and texture scores in Eq. 8). The figure also shows the effectiveness of the proposed algorithm when the iterative anchor mechanism is used to localize lesion correspondence, having consistently higher success rates with the criteria within 20 mm.

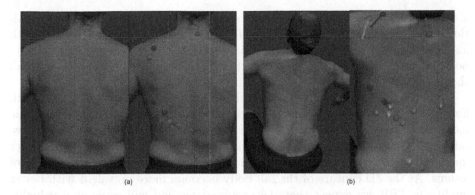

(a)   (b)

Fig. 4. The qualitative result on Skin3D dataset. (a) shows a source scan (left) and the scan with annotated LOIs (right). (b) shows a target scan (left) and the scan with annotated LOIs (transparent sphere) and the estimated correspondence of LOIs (solid dot) (right). The lesion correspondence pairs are shown in the same color. Large CLE occurs when the local texture is not well-preserved across scans.

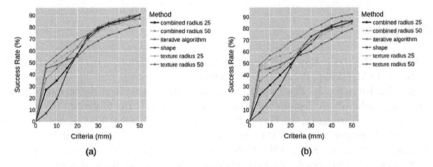

(a)   (b)

Fig. 5. Success rate under different criteria for the proposed algorithm on the Skin3D dataset. (a) and (b) show the results of two combinations of source and target meshes. *Iterative algorithm* is the proposed algorithm with the anchor mechanism. *Shape* is the method using $\Phi^1_{L,\ell}$. *Texture radius 25* and *texture radius 50* are the methods using $\Phi^1_{L,\epsilon}$ with ε_1 (Eq. 3) selected at 25 mm and 50 mm. *Combined radius 25* and *combined radius 50* are the methods using Eq. 8 with one iteration.

4   Conclusions and Limitations

The evolution of a skin lesion is an important sign of a potentially cancerous growth and total body photography is useful to keep track of skin lesions longitudinally. We proposed a novel framework that leverages geometric and texture information to effectively find lesion correspondence across TBP scans. The framework is evaluated on a private dataset and a public dataset with success rates that are comparable to those of the state-of-the-art method.

The proposed method assumes that the local texture enclosing the lesion and its surroundings should be similar from scan to scan. This may not hold when the appearance of the lesion changes dramatically (e.g. if the person acquires a tattoo). Also, the resolution of the mesh affects the precision of the positions of landmarks and lesions. In addition, the method may not work well with longitudinal data that has non-isometric deformation due to huge variations in body shape, inconsistent 3D reconstruction, or a dramatic change in pose and, therefore, topology, such as an open armpit versus a closed one.

In the future, the method needs to be evaluated on longitudinal data with longer duration and new lesions absent in the target. In addition, an automatic method to determine accurate landmarks is desirable. Note that although we rely on the manual selection of landmarks, the framework is still preferable over manually annotating lesion correspondences when a subject has hundreds of lesions. As the 3D capture of the full body becomes more prevalent with better quality in TBP, we expect that the proposed method will serve as a valuable step for the longitudinal tracking of skin lesions.

Acknowledgments. The research was in part supported by the Intramural Research Program (IRP) of the NIH/NICHD, Phase I of NSF STTR grant 2127051, and Phase I NIH/NIBIB STTR grant R41EB032304.

References

1. Abbasi, N.R., et al.: Early diagnosis of cutaneous melanoma: revisiting the ABCD criteria. JAMA **292**(22), 2771–2776 (2004)
2. Bogo, F., Romero, J., Peserico, E., Black, M.J.: Automated detection of new or evolving melanocytic lesions using a 3D body model. In: Golland, P., Hata, N., Barillot, C., Hornegger, J., Howe, R. (eds.) MICCAI 2014. LNCS, vol. 8673, pp. 593–600. Springer, Cham (2014). https://doi.org/10.1007/978-3-319-10404-1_74
3. Datar, M., Lyu, I., Kim, S.H., Cates, J., Styner, M.A., Whitaker, R.: Geodesic distances to landmarks for dense correspondence on ensembles of complex shapes. In: Mori, K., Sakuma, I., Sato, Y., Barillot, C., Navab, N. (eds.) MICCAI 2013. LNCS, vol. 8150, pp. 19–26. Springer, Heidelberg (2013). https://doi.org/10.1007/978-3-642-40763-5_3
4. Guo, Y., Sohel, F., Bennamoun, M., Lu, M., Wan, J.: Rotational projection statistics for 3D local surface description and object recognition. Int. J. Comput. Vision **105**(1), 63–86 (2013)
5. Halpern, A.C.: Total body skin imaging as an aid to melanoma detection. In: Seminars in Cutaneous Medicine and Surgery, vol. 22, pp. 2–8 (2003)
6. Kim, H., Kim, J., Kam, J., Park, J., Lee, S.: Deep virtual markers for articulated 3D shapes. In: Proceedings of the IEEE/CVF International Conference on Computer Vision, pp. 11615–11625 (2021)
7. Korotkov, K., et al.: An improved skin lesion matching scheme in total body photography. IEEE J. Biomed. Health Inform. **23**(2), 586–598 (2018)
8. Korotkov, K., Quintana, J., Puig, S., Malvehy, J., Garcia, R.: A new total body scanning system for automatic change detection in multiple pigmented skin lesions. IEEE Trans. Med. Imaging **34**(1), 317–338 (2014)
9. Li, Y., Esteva, A., Kuprel, B., Novoa, R., Ko, J., Thrun, S.: Skin cancer detection and tracking using data synthesis and deep learning. arXiv preprint arXiv:1612.01074 (2016)

10. Mcgregor, B.: Automatic registration of images of pigmented skin lesions. Pattern Recogn. **31**(6), 805–817 (1998)
11. Mirzaalian, H., Hamarneh, G., Lee, T.K.: A graph-based approach to skin mole matching incorporating template-normalized coordinates. In: 2009 IEEE Conference on Computer Vision and Pattern Recognition, pp. 2152–2159. IEEE (2009)
12. Mirzaalian, H., Lee, T.K., Hamarneh, G.: Uncertainty-based feature learning for skin lesion matching using a high order MRF optimization framework. In: Ayache, N., Delingette, H., Golland, P., Mori, K. (eds.) MICCAI 2012. LNCS, vol. 7511, pp. 98–105. Springer, Heidelberg (2012). https://doi.org/10.1007/978-3-642-33418-4_13
13. Mirzaalian, H., Lee, T.K., Hamarneh, G.: Spatial normalization of human back images for dermatological studies. IEEE J. Biomed. Health Inform. **18**(4), 1494–1501 (2013)
14. Mirzaalian, H., Lee, T.K., Hamarneh, G.: Skin lesion tracking using structured graphical models. Med. Image Anal. **27**, 84–92 (2016)
15. Mitchel, T.W., Rusinkiewicz, S., Chirikjian, G.S., Kazhdan, M.: Echo: extended convolution histogram of orientations for local surface description. In: Computer Graphics Forum, vol. 40, pp. 180–194. Wiley Online Library (2021)
16. Perednia, D.A., White, R.G.: Automatic registration of multiple skin lesions by use of point pattern matching. Comput. Med. Imaging Graph. **16**(3), 205–216 (1992)
17. Roning, J., Riech, M.: Registration of nevi in successive skin images for early detection of melanoma. In: Proceedings. Fourteenth International Conference on Pattern Recognition (Cat. No. 98EX170), vol. 1, pp. 352–357. IEEE (1998)
18. Saint, A., Ahmed, E., Cherenkova, K., Gusev, G., Aouada, D., Ottersten, B., et al.: 3DBodyTex: textured 3D body dataset. In: 2018 International Conference on 3D Vision (3DV), pp. 495–504. IEEE (2018)
19. Saint, A., Cherenkova, K., Gusev, G., Aouada, D., Ottersten, B., et al.: Bodyfitr: robust automatic 3D human body fitting. In: 2019 IEEE International Conference on Image Processing (ICIP), pp. 484–488. IEEE (2019)
20. Salti, S., Tombari, F., Di Stefano, L.: Shot: unique signatures of histograms for surface and texture description. Comput. Vis. Image Underst. **125**, 251–264 (2014)
21. Strakowska, M., Kociołek, M.: Skin lesion matching algorithm for application in full body imaging systems. In: Pietka, E., Badura, P., Kawa, J., Wieclawek, W. (eds.) ITIB 2022. Advances in Intelligent Systems and Computing, vol. 1429, pp. 222–233. Springer, Cham (2022). https://doi.org/10.1007/978-3-031-09135-3_19
22. Strzelecki, M.H., Strąkowska, M., Kozłowski, M., Urbańczyk, T., Wielowieyska-Szybińska, D., Kociołek, M.: Skin lesion detection algorithms in whole body images. Sensors **21**(19), 6639 (2021)
23. Tombari, F., Salti, S., Di Stefano, L.: Unique signatures of histograms for local surface description. In: Daniilidis, K., Maragos, P., Paragios, N. (eds.) ECCV 2010. LNCS, vol. 6313, pp. 356–369. Springer, Heidelberg (2010). https://doi.org/10.1007/978-3-642-15558-1_26
24. Tombari, F., Salti, S., Di Stefano, L.: A combined texture-shape descriptor for enhanced 3D feature matching. In: 2011 18th IEEE International Conference on Image Processing, pp. 809–812. IEEE (2011)
25. White, R.G., Perednia, D.A.: Automatic derivation of initial match points for paired digital images of skin. Comput. Med. Imaging Graph. **16**(3), 217–225 (1992)
26. Zhao, M., Kawahara, J., Abhishek, K., Shamanian, S., Hamarneh, G.: Skin3D: detection and longitudinal tracking of pigmented skin lesions in 3D total-body textured meshes. Med. Image Anal. **77**, 102329 (2022)

FEDD - Fair, Efficient, and Diverse Diffusion-Based Lesion Segmentation and Malignancy Classification

Héctor Carrión[1(✉)] and Narges Norouzi[2]

[1] University of California, Santa Cruz, Santa Cruz, USA
hcarrion@ucsc.edu
[2] University of California, Berkeley, Berkeley, USA

Abstract. Skin diseases affect millions of people worldwide, across all ethnicities. Increasing diagnosis accessibility requires fair and accurate segmentation and classification of dermatology images. However, the scarcity of annotated medical images, especially for rare diseases and underrepresented skin tones, poses a challenge to the development of fair and accurate models. In this study, we introduce a Fair, Efficient, and Diverse Diffusion-based framework for skin lesion segmentation and malignancy classification. FEDD leverages semantically meaningful feature embeddings learned through a denoising diffusion probabilistic backbone and processes them via linear probes to achieve state-of-the-art performance on Diverse Dermatology Images (DDI). We achieve an improvement in intersection over union of 0.18, 0.13, 0.06, and 0.07 while using only 5%, 10%, 15%, and 20% labeled samples, respectively. Additionally, FEDD trained on 10% of DDI demonstrates malignancy classification accuracy of 81%, 14% higher compared to the state-of-the-art. We showcase high efficiency in data-constrained scenarios while providing fair performance for diverse skin tones and rare malignancy conditions. Our newly annotated DDI segmentation masks and training code can be found on https://github.com/hectorcarrion/fedd.

Keywords: Lesion Segmentation · Classification · Fairness · Diffusion

1 Introduction and Related Work

Skin diseases are a major public health concern that impacts millions of people worldwide. The first step towards diagnosis and treatment of skin diseases often involves visual inspection and analysis of the lesion by dermatologists or other medical experts. However, this process is often subjective, time-consuming, costly, and inaccessible for many people, especially in low-resource communities or remote areas. It is estimated that around 3 billion people lack adequate access to dermatological care [1]. In the United States, only about one in three

Supplementary Information The online version contains supplementary material available at https://doi.org/10.1007/978-3-031-43990-2_26.

patients with skin disease are evaluated by a dermatologist, their average wait time exceeds 38 days while representing a cost of $75 billion on the health-care system [2,3]. Therefore, there exists a growing need for automated methods that can assist dermatologists, especially those in low-resource environments, in attending to skin lesions accurately and efficiently.

Skin lesion semantic segmentation and malignancy classification are essential for providing accurate and explainable diagnosis information for patients with skin diseases, and recently Artificial Intelligent (AI) systems have led the state-of-the-art for these tasks. However, these systems are commonly based on data and training methods that are prone to racial biases [4–6]. Some of the main challenges facing AI systems that can lead to bias are:

- **Data scarcity:** Annotated medical images are often scarce and expensive to obtain due to privacy issues, cost, and expert availability. This limits the amount of data available for training Deep Learning (DL) models, which may result in overfitting, especially in a medical context, as shown in [7].
- **Class imbalance:** The distribution of different types of skin lesions is often imbalanced in real-world datasets. For example, melanoma cases may be rare than basal cell carcinoma cases; this could then be exacerbated by datasets that are primarily sourced from light-skinned populations [8]. This class imbalance can introduce biases in modeling.
- **Data diversity:** The appearance and morphology of skin lesions can vary across different individuals due to factors such as age, gender, and ethnicity [9]. A dataset can be large but not necessarily diverse. This lack of diversity in the data may lead to poor generalization [4].
- **Base models:** Some recent works on dermatology images stem from transfer-learning models designed for ImageNet [10], which may be overly large for smaller dermatologic datasets [11,12]. Tuning these massive encoders could lead to overfitting.
- **Lack of diverse studies:** A recent review of 70 dermatological AI studies between 2015 and 2020 found that only 17 studies included ethnicity descriptors, and only 7 included skin tone descriptors [13]. This could lead to under-specification of model performance for different ethnicities.

Denoising Diffusion Probabilistic Models (DDPMs) have been introduced [14] as a new form of generative modeling. DDPMs have achieved state-of-the-art performance in image synthesis [15] and are effectively applied in colorization [16], super-resolution [17], segmentation [18], and other tasks. In the medical domain, recent work has presented results for DDPM-based anomaly detection [19] and segmentation [20], but these are limited to MRI, CT, and ultrasonography not natural smartphone-captured images of dermatology conditions. To our knowledge, none have explored segmentation and malignancy classification in this context from DDPM-based embeddings without re-training and evaluated performance on diverse dermatology images.

We introduce the FEDD framework, a denoising diffusion-based approach trained on small, skin tone-balanced, Diverse Dermatology Images (DDI) [4] subsets for skin lesion segmentation and malignancy classification that outperforms state-of-the-art across a diverse spectrum of skin tones and malignancy

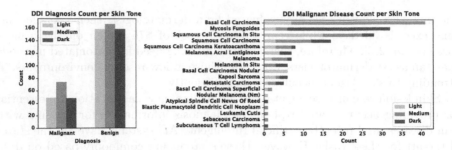

Fig. 1. The disease count per skin tone (left) shows a smaller amount of malignancy data in DDI but otherwise mostly balanced between light and dark skin tones. The distribution of malignant illnesses (right) shows high diversity and thus the morphological variation of lesions in DDI.

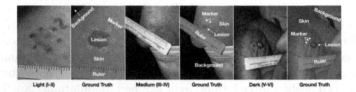

Fig. 2. Ground truth segmentation samples. We color code lesions as red, skin as green, markings as purple, and rulers as blue. Backgrounds are left to be transparent. (Color figure online)

conditions with very few training examples. FEDD leverages the highly semantically meaningful feature embeddings learned by DDPMs for image synthesis. Finally, linear probes predict per-pixel class or per-image malignancy, achieving state-of-the-art performance on the DDI dataset without fine-tuning the encoder.

2   Description of Data

The most commonly used dataset to train and evaluate fairness in dermatology AI is Fitzpatrick17k [8,21,22] thanks to its large size of nearly 17,000 images. However it contains a significant skin tone imbalance (3.6 times more light than dark skin toned samples) and greater than 30% disease label noise [8]. Further, the samples are not biopsied and visual inspection itself can be an unreliable way of diagnosis without the use of histopathological information [23]. Recently, DDI [4] was published as a dermatological image dataset with Fitzpatrick-scale [24] scores for all images, classifying them as light (I–II), medium (III–IV), or dark (V–VI) in skin tone. While at a lower skin tone resolution (3-point instead of 6-point), these labels are reviewed by two board-certified dermatologists. It also includes a mix of rare and common benign and malignant skin conditions, all of which are confirmed via biopsy. The dataset contains visually ambiguous lesions that would be difficult to visually diagnose but represent the kind of lesions that are seen in clinical practice. DDI is somewhat balanced between skin tones, with about 16% more information for medium skin tones; however, it is not balanced between malignant and benign classes. In total, the dataset contains 656 samples. Details and distribution of diagnosis are shown in Fig. 1.

We draw 4 balanced subsets of DDI for training, each representing approximately 5% (10 samples per skin tone), 10% (20 samples per skin tone), 15% (30 samples per skin tone), and 20% (40 samples per skin tone) of DDI. The smaller training sets are subsets of the larger ones, this is to say $5\% \subseteq 10\% \subseteq 15\% \subseteq 20\% \subseteq$ DDI. For classification we draw validation and test sets, each containing 30 samples (10 samples per skin tone). Further, we test model checkpoints trained on each DDI subset on all remaining DDI images (476 samples), accuracy results from this larger test set are reported on the paper text and on Table 3 of the supplementary materials. For segmentation, we test on 198 additionally annotated samples.

This is due to DDI including disease labels suitable for malignancy classification. For segmentation however, samples need to be semantically labeled and some samples may be difficult to correctly annotate, leading to discarding; for example if the target lesion is ambiguous, blurry, partially visible or occluded. We annotated the dataset and all masks underwent a secondary quality review. We define 5 classes: lesion, skin, marker, ruler, and background. We opted to label these classes as many images include a ruler or markings to denote the lesion of focus. Visualizations of our ground truth labels are shown in Fig. 2. Details on the annotation protocol, including skip criteria, can be found in the supplementary materials. We release our annotation work (a total of 378 annotated DDI images) as part of our contributions.

3   Approach

The UNet architecture was introduced for diffusion [14] and found to improve generative performance [25] over other denoising score-matching architectures. Recent work [15] has extensively ablated the diffusion UNet architecture by increasing depth while reducing the width, increasing the number of attention heads and applying it at different resolutions, applying BigGAN [26] residual blocks, and introducing AdaGN for injecting timestep and class embedding onto residual blocks, obtaining state-of-the-art for image synthesis. From this work, we designate the unconditional image generation model as our backbone and freeze its weights. This network is an ImageNet-trained DDPM with 256×256 input and output resolution.

3.1   Lesion, Marker, Ruler, Skin, and Background Segmentation

We obtain image encodings from blocks on the decoder side of the DDPM-UNet architecture, then apply bi-linear up-sampling up to some target output resolution (256×256 in this case) and concatenate them before feeding them to MLPs for per-pixel classification following [18]. However, we use fewer blocks, a single-time step, and 5 MLPs. This is to avoid overfitting and because results

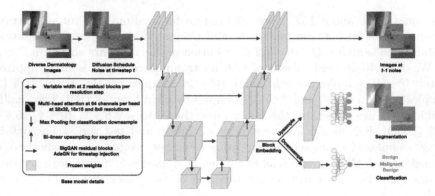

Fig. 3. Image noise is added according to the diffusion noise schedule for the selected timestep. The DDPM processes the image, and feature embeddings are obtained from the desired block levels. Embeddings are concatenated and either up-sampled for segmentation or down-sampled for classification. Finally, Multi-Layer Perceptrons (MLPs) predict per-pixel semantic class or whole image malignancy.

from [18] describe how different blocks at different timesteps perform differently depending on the target data. To understand which blocks are most promising for dermatology images, we obtain a sample of feature encodings at different blocks and perform K-Means clustering shown in the supplementary materials. We selected the blocks which clustered semantically meaningful areas (e.g. lesion, skin, and ruler). We identify block 6 and block 8 at timestep 100 as the most promising and use this setup for the rest of our segmentation experiments.

3.2   Malignancy Classification

For classification, we down-sample block encodings using a combination of 2D and 1D max pooling operations until the feature vectors are one-dimensional and of size 512. We note that the total number of pooling operations varies depending on the sampled block, as deeper blocks are smaller with more channels, while shallower blocks are larger with fewer channels. The vectors are then passed onto a 3-layer MLP of size 64, 32, and 1. We include batch normalization and dropout between each layer with 50% and 25%. This classification network is trained to predict the malignancy of the input image from the down-sampled feature vector. A summary of our approach is shown in Fig. 3.

4   Results and Discussion

4.1   Lesion, Marker, Ruler, Skin, and Background Segmentation

Our segmentation results are evaluated by Intersection over Union (IoU) performance and are compared against other architectures pre-trained on ImageNet: DenseNet121 [27], VGG16 [28], ResNet50 [29], and two other smaller networks,

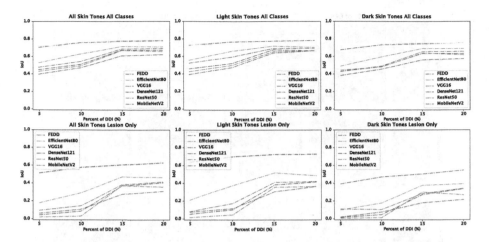

Fig. 4. Top Row: The test IoU score for all segmentation classes for all, light, and dark skin tones (left-to-right). **Bottom Row**: The test IoU score for the lesion segment for all, light, and dark skin tones (left-to-right).

EfficientNetB0 [30] and MobileNetV2 [31]. These architectures are configured as UNets and tasked with segmenting the input images. We observe FEDD outperforms all other architectures across all subsets of DDI on our validation and test sets. Importantly, other architectures (particularly the smaller networks) close the performance gap as the amount of training data increases, showcasing that FEDD's efficiency is most prevalent in very small data scenarios.

We further compare the FEDD's IoU performance on light and dark skin tone images to showcase algorithmic fairness. We note that all architectures show similar performance for both skin tones, suggesting that our balanced data subsets play a larger role in fairness than the choice of neural network. Finally, we plot test set performance when only considering the lesion class split between light and dark tones. We find that FEDD's performance is significantly better at segmenting the lesion class compared to other architectures. We believe this is due to the greater morphology variation of different skin lesions being harder to learn than other more consistent targets like the ruler. Since DDI contains a diversity of skin conditions, FEDD's efficiency becomes very useful for high-quality lesion segmentation of this morphologically changing class. These results are shown in Fig. 4.

In Fig. 5, we visualize predicted segmentation masks between FEDD and the next best IoU-performing architecture, EfficientNetB0. We observe that FEDD achieves significantly better segmentation masks at lower fractions of labeled data across all skin tones. With a larger percentage of labeled data, EfficientNet begins to produce similar results to FEDD, but FEDD comparatively outputs higher-quality segmentations with fewer segmentation artifacts and false positives. The skin lesions themselves, which appear in different sizes, locations, and morphologies, are also most accurately segmented by FEDD.

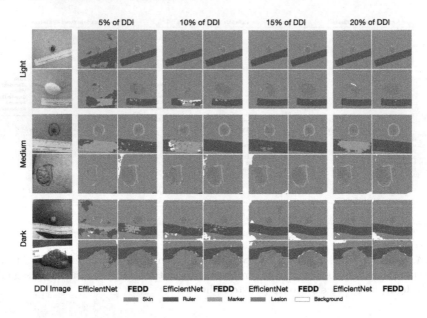

Fig. 5. Test set segmentation results for FEDD and EfficientNet.

4.2 Malignancy Classification

We ablate the performance of each individual block per timestep in the classification context as, to the best of our knowledge, it has not been done before. It is also not entirely intuitive which block depth and timestep combination would produce the best representations for classification, as well as how that performance varies as we introduce more data. We train FEDD's classifier on block embeddings produced between timestep 0 and 1000 of the backward diffusion process. We then record the accuracy of the classifier per block in increments of 50 timesteps. Figure 6 describes these results. While noisy, given the small amount of test data, the general pattern is that the earlier time-steps (later in the reverse diffusion de-noising process) allow for higher classification accuracy. This is likely due to the quality of the DDPM sample increasing as more noise is removed. Another finding is that as the classifier is shown more data, the shallower blocks begin to perform better. The best performing blocks are shown to be: block 4 at 5% DDI, block 6 at 10% DDI, block 12 at 15% DDI, and block 14 at 20% DDI. We attribute this to the fact that shallower blocks of the UNet decoder capture finer detail of the reconstructed image while deeper blocks capture lower-resolution detail. This coarse data is more generic and thus more generalizable than the finer features in later blocks. As we increase the amount of data, the classifier has enough information to learn from the finer details of later blocks, boosting performance.

We select the best-performing block and timestep combination at each fraction of data for the rest of our experiments. The previous classification state-of-

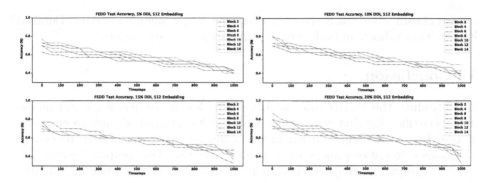

Fig. 6. Classification accuracy of each DDPM UNet decoder is shown. Later steps in the reverse diffusion process produce the highest quality embeddings. When less data is available (top row), earlier blocks of the UNet decoder perform best. When more data is available (bottom row), the later, shallower blocks of the decoder perform best.

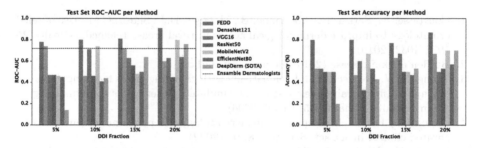

Fig. 7. ROC-AUC scores (left) show FEDD outperforms other methods, including an ensemble of dermatologists. Accuracy per method is also higher (right).

the-art on the DDI dataset is reported on [4] as the DeepDerm [11] architecture pre-trained on HAM10000 [32] and fine-tuned on DDI. We compare FEDD performance against this setup as well as other commonly used classifiers on each of our DDI subsets. We measure Receiver Operating Characteristic Area Under the Curve (ROC-AUC) at the best threshold for each method, F1 scores, and classification accuracy. We observe FEDD obtains a higher ROC-AUC than any other method at every level of data. It also surpasses the dermatologist ensemble performance reported by [4]. FEDD also reports the best accuracy, however, it does not see improvement at 10% or 15% of DDI compared to 5% and 20%, as shown in Fig. 7. When observing ROC curves for FEDD, we see it meets or exceeds the ensemble of dermatologists even at the smallest subset of DDI. We divide F1 scores between the light and dark skin tones finding that FEDD does not always obtain the best F1 performance at larger subsets of data, namely 15% and 20% of DDI. This result suggests that purpose-built classification networks could have a performance advantage over diffusion embeddings applied toward classification when allowed to train on larger amounts of data. Detailed F1 results are shown in table form on the supplementary materials.

5 Conclusion

We introduce the FEDD framework for skin lesion segmentation and malignancy classification that outperforms state-of-the-art methods and an ensemble of board-certified dermatologists across a diverse spectrum of skin tones and malignancy conditions under limited data scenarios. Our proposed methodology can improve the diagnosis and treatment of skin diseases while maintaining fair segmentation outcomes for under-represented skin tones and accurate malignancy predictions for rare malignancy conditions. We freely release our code and annotations to encourage further research around dermatological AI fairness.

References

1. Coustasse, A., Sarkar, R., Abodunde, B., Metzger, B.J., Slater, C.M.: Use of teledermatology to improve dermatological access in rural areas. Telemed. e-Health **25**, 1022–1032 (2019)
2. Burden of skin disease (2016). http://www.aad.org
3. Tsang, M.W., Resneck, J.S.: Even patients with changing moles face long dermatology appointment wait-times: a study of simulated patient calls to dermatologists. J. Am. Acad. Dermatol. **55**, 54–58 (2006)
4. Daneshjou, R., et al.: Disparities in dermatology AI performance on a diverse, curated clinical image set. Sci. Adv. **8**, 08 (2022)
5. Owens, K., Walker, A.: Those designing healthcare algorithms must become actively anti-racist. Nat. Med. **26**, 1327–1328 (2020)
6. Chen, I.Y., Pierson, E., Rose, S., Joshi, S., Ferryman, K., Ghassemi, M.: Ethical machine learning in healthcare. Ann. Rev. Biomed. Data Sci. **4**, 123–144 (2021)
7. Razzak, M.I., Naz, S., Zaib, A.: Deep learning for medical image processing: overview, challenges and the future. In: Dey, N., Ashour, A.S., Borra, S. (eds.) Classification in BioApps. LNCVB, vol. 26, pp. 323–350. Springer, Cham (2018). https://doi.org/10.1007/978-3-319-65981-7_12
8. Groh, M., et al.: Evaluating deep neural networks trained on clinical images in dermatology with the fitzpatrick 17k dataset. In: CVPRW (2021)
9. Adelekun, A., Onyekaba, G., Lipoff, J.B.: Skin color in dermatology textbooks: an updated evaluation and analysis. J. Am. Acad. Dermatol. **84**, 04 (2020)
10. Deng, J., Dong, W., Socher, R., Li, L. J., Li, K., Fei-Fei, L.: Imagenet: a large-scale hierarchical image database. In: 2009 IEEE Conference on Computer Vision and Pattern Recognition (2009)
11. Esteva, A., et al.: Dermatologist-level classification of skin cancer with deep neural networks. Nature **542**, 115–118 (2017)
12. Han, S.S., et al.: Augmented intelligence dermatology: deep neural networks empower medical professionals in diagnosing skin cancer and predicting treatment options for 134 skin disorders. J. Invest. Dermatol. **140**, 1753–1761 (2020)
13. Daneshjou, R., Smith, M.P., Sun, M.D., Rotemberg, V., Zou, J.: Lack of transparency and potential bias in artificial intelligence data sets and algorithms. JAMA Dermatol. **157**, 09 (2021)

14. Ho, J., Jain, A., Abbeel, P.: Denoising diffusion probabilistic models. In: NIPS (2020)
15. Dhariwal, P., Nichol, A.: Diffusion models beat GANs on image synthesis. In: NIPS (2021)
16. Song, Y., Sohl-Dickstein, J., Kingma, D.P., Kumar, A., Ermon, S., Poole, B.: Score-based generative modeling through stochastic differential equations, arXiv:2011.13456 (2021)
17. Saharia, C., Ho, J., Chan, W., Salimans, T., Fleet, D.J., Norouzi, M.: Image super-resolution via iterative refinement. IEEE Trans. Pattern Anal. Mach. Intell. **45**, 1–14 (2022)
18. Baranchuk, D., Rubachev, I., Voynov, A., Khrulkov, V., Babenko, A.: Label-efficient semantic segmentation with diffusion models. In: ICLR (2022)
19. Wolleb, J., Bieder, F., Sandkühler, R., Cattin, P.C.: Diffusion models for medical anomaly detection, arXiv:2203.04306 (2022)
20. Wu, J., Fu, R., Fang, H., Zhang, Y., Xu, Y.: MedSegDiff-V2: diffusion based medical image segmentation with transformer, arXiv:2301.11798 (2023)
21. Abid, A., Yuksekgonul, M., Zou, J.: Meaningfully debugging model mistakes using conceptual counterfactual explanations. In: proceedings.mlr.press, pp. 66–88 (2022)
22. Du, S., Hers, B., Bayasi, N., Hamarneh, G., Garbi, R.: FairDisCo: fairer AI in dermatology via disentanglement contrastive learning. In: Karlinsky, L., Michaeli, T., Nishino, K. (eds.) ECCV 2022. LNCS, vol. 13804, pp. 185–202. Springer, Cham (2023). https://doi.org/10.1007/978-3-031-25069-9_13
23. Daneshjou, R., et al.: Checklist for evaluation of image-based artificial intelligence reports in dermatology. JAMA Dermatol. **158**, 90 (2022)
24. Fitzpatrick, T.B.: The validity and practicality of sun-reactive skin types I through VI. Arch. Dermatol. **124**, 869–871 (1988)
25. Jolicoeur-Martineau, A., Piché-Taillefer, R., Combes, R.T.D., Mitliagkas, I.: Adversarial score matching and improved sampling for image generation, arXiv:2009.05475 (2020)
26. Brock, A., Donahue, J., Simonyan, K.: Large scale GAN training for high fidelity natural image synthesis, arXiv.org (2018)
27. Huang, G., Liu, Z., Weinberger, K.Q.: Densely connected convolutional networks, arXiv.org (2016)
28. Simonyan, K., Zisserman, A.: Very deep convolutional networks for large-scale image recognition, arXiv.org (2014)
29. He, K., Zhang, X., Ren, S., Sun, J.: Deep residual learning for image recognition, arXiv.org (2015)
30. Tan, M., Le, Q.V.: Efficientnet: rethinking model scaling for convolutional neural networks, arXiv.org (2019)
31. Sandler, M., Howard, A., Zhu, M., Zhmoginov, A., Chen, L.-C.: Mobilenetv2: inverted residuals and linear bottlenecks, arXiv.org (2018)
32. Tschandl, P., Rosendahl, C., Kittler, H.: The ham10000 dataset, a large collection of multi-source dermatoscopic images of common pigmented skin lesions. Sci. Data **5**, 1–9 (2018)

Clinical Applications – Fetal Imaging

Improving Automatic Fetal Biometry Measurement with Swoosh Activation Function

Shijia Zhou[1,2(✉)] , Euijoon Ahn[3] , Hao Wang[1] , Ann Quinton[4] ,
Narelle Kennedy[5] , Pradeeba Sridar[6] , Ralph Nanan[1,2,3,4,5,6] ,
and Jinman Kim[1]

[1] University of Sydney, School of Computer Science, Sydney, NSW 2000, Australia
szho6430@uni.sydney.edu.au
[2] James Cook University, Smithfield, QLD 4878, Australia
[3] Central Queensland University, Sydney, NSW 2000, Australia
[4] Liverpool Hospital, Liverpool, NSW 2170, Australia
[5] Indian Institute of Technology, Delhi, New Delhi, Delhi 110016, India
[6] University of Sydney, Sydney Medical School Nepean, Kingswood, NSW 2747,
Australia

Abstract. The measurement of fetal thalamus diameter (FTD) and fetal head circumference (FHC) are crucial in identifying abnormal fetal thalamus development as it may lead to certain neuropsychiatric disorders in later life. However, manual measurements from 2D-US images are laborious, prone to high inter-observer variability, and complicated by the high signal-to-noise ratio nature of the images. Deep learning-based landmark detection approaches have shown promise in measuring biometrics from US images, but the current state-of-the-art (SOTA) algorithm, BiometryNet, is inadequate for FTD and FHC measurement due to its inability to account for the fuzzy edges of these structures and the complex shape of the FTD structure. To address these inadequacies, we propose a novel Swoosh Activation Function (SAF) designed to enhance the regularization of heatmaps produced by landmark detection algorithms. Our SAF serves as a regularization term to enforce an optimum mean squared error (MSE) level between predicted heatmaps, reducing the dispersiveness of hotspots in predicted heatmaps. Our experimental results demonstrate that SAF significantly improves the measurement performances of FTD and FHC with higher intraclass correlation coefficient scores in FTD and lower mean difference scores in FHC measurement than those of the current SOTA algorithm BiometryNet. Moreover, our proposed SAF is highly generalizable and architecture-agnostic. The SAF's coefficients can be configured for different tasks, making it highly customizable. Our study demonstrates that the SAF activation function is a novel method that can improve measurement accuracy in fetal

This research was supported by Australian Research Council (ARC) DP200103748.

Supplementary Information The online version contains supplementary material available at https://doi.org/10.1007/978-3-031-43990-2_27.

biometry landmark detection. This improvement has the potential to contribute to better fetal monitoring and improved neonatal outcomes.

Keywords: 2-dimensional ultrasound · automatic measurement algorithm · fetal thalamus · activation function

1   Introduction

The thalamus is a critical brain region that relays and modulates information between different parts of the cerebral cortex, and plays a vital role in signal transmission and processing, including pain recognition and reaction [13]. Abnormal fetal thalamus development, which can disrupt serotonin receptor development, may contribute to the development of later neuropsychiatric disorders [12]. Thalamus development is influenced by maternal factors such as diabetes that suppresses thalamus development after 2 weeks of gestation [16]. To further investigate the relationship between maternal factors and fetal thalamus growth, a dataset containing maternal factors, fetal thalamus diameter (FTD), and fetal head circumference measurements (FHC) is needed. FHC measurement is necessary to normalize FTD against gestational age. However, measuring FTD and FHC manually from 2D-US scans is laborious, prone to high inter-observer variability, and complicated by 2D-US images' high signal-to-noise ratio (SNR) nature due to ultrasound wave's lack of penetration power compared to ionizing radiations such as the x-ray, acoustic shadows cast by highly echogenic objects, and unique noise characteristics due to ultrasound reverberation [3,4,9].

In recent years, landmark-based detection approaches based on deep learning have been employed to measure biometrics from fetal US images. They were used to detect measurement key points for brain structures in the fetal brains, and bony structures such as length of the femur or dimensions in the pelvic floors [2,7,14]. In these studies, the distance between a pair of landmarks represents the biometry being measured. The current state-of-the-art (SOTA) landmark detection algorithm for measuring biometry in fetal 2D-US images is the BiometryNet proposed by Avisdris et al., which was developed to detect landmarks for measuring the FHC and fetal femur length [2]. BiometryNet is based on High-Resolution Net (HRNet) and it has shown great performances in measuring dimensions of fetal skull and femur bone, outperforming other landmark-based methods [10]. BiometryNet uses dynamic orientation determination (DOD) to enforce a consistent orientation between detected landmarks', i.e. the first landmark is always the left/top measurement key point, and the second landmark is always the right/bottom key point [2].

However, BiometryNet cannot be directly used to measure FTD and FHC due to two specific difficulties. First being that the "guitar-shaped" structure (GsS) by Sridar et al. to measure FTD has similar echogenicity to surrounding brain tissues, resulting in fuzzy boundaries, especially around the wing-tips where measurement landmarks are located [9] (Fig. 1). This difficulty is also observed in 2D-US images of fetal skulls when they appear broken due to unfused bones

Fetal Head   Femur   Thalamus

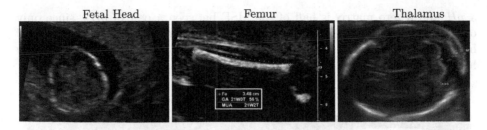

Fig. 1. Left: 2D-US image of fetal head from the HC18 dataset [11], note part of the skull is not mineralized and has similar echogenicty to the adjacent uterine tissue. Middle: 2D-US image of a fetal femur [1]. Right: 2D-US image of a GsS for measuring FTD, note the gaps in the skull due to unfused bones.

and acoustic shadows cast by the skull bones themselves (Fig. 1). Second being that the shape of the GsS resembles the silhouette of a guitar and more complex than that of a skull or femur bone (Fig. 2) [9]. These two difficulties causes uncertainties in the localization of measurement landmarks of FTD, resulting in inaccurate measurement of FTD.

To address the above difficulties, we present a novel Swoosh Activation Function (SAF) designed to enhance the regularization of heatmaps produced by landmark detection algorithms. The SAF takes its name from the Nike^{TM} swoosh logo, which resembles its shape (Supplementary Fig. 1). SAF can serve as a regularization term to enforce an optimum mean squared error (MSE) level between a pair of predicted heatmaps (pHs). By doing so, the landmark detection algorithm is compelled to highlight different areas. Additionally, SAF can enforce an optimum MSE between individual pH and a zero matrix to reduce hotspot dispersiveness. Moreover, because SAF does not grow exponentially when the input MSE is higher than the optimum MSE level, it does not hinder algorithm's learning. Consequently, we hypothesize that SAF can enhance landmark detection accuracy and overcome uncertainties arising from the fuzzy edges of GsS by promoting hotspot concentration in pHs.

2   Method

2.1   Swoosh Activation Function

SAF is introduced to optimize pHs by enforcing an optimum MSE between a pair of pHs and a secondary optimum MSE between a predicted heatmap (pH) and a zero matrix (O). We determine the optimum MSE by computing the MSE between a pair of ground truth heatmaps (gHs). Each ground truth heatmap (gH) represents a measurement landmark by a smaller matrix drawn from a Gaussian distribution that is centered at the landmark coordinates with the peak assigned at the value of 1 (Fig. 3.A). Since we determine that gHs represent optimum heatmaps, the MSE between a pair of pHs should approximate the MSE between a pair of gHs. In addition, the secondary optimum MSE between a *pH* and a zero matrix is half of the MSE between a pair of *gHs*, since only one Gaussian distribution is being compared. We demonstrate in Fig. 3.B and C how deviations from this optimum MSE value can lead to incorrect and noisy heatmaps.

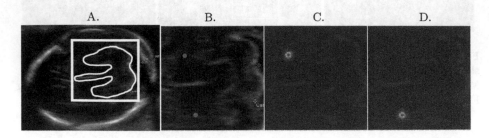

Fig. 2. A: a 2D-US fetal brain image that has the GsS (outlined with white curly lines) annotated with manually created bounding box (white box); B: input image of the GsS constrained by the bounding box. The red dots represent the ground truth landmarks of FTD; C: the first heatmap of one of the FTD landmarks with the hottest (red) spot representing the landmark; D: the second heatmap of the other FTD landmark with the hottest (red) spot representing the landmark (Color figure online)

To enforce this pre-determined optimum MSE, we defined SAF as:

$$f(x > 0) = \left(ax + \frac{1}{bx}\right)^c - Min \tag{1}$$

In Eq. 1, Min represents a function that ensures SAF minimizes to 0, and it is defined as $Min = f\left(\sqrt{\frac{1}{ab}}\right)$. The coefficient a determines the slope of SAF around the minimum point in Quadrant 1 of the Cartesian coordinate system (Supplementary Fig. 1). The slope of SAF determines its regularization strength. Coefficient b determines the x-axis coordinate of the minimum point where the x-coordinate of the minimum point correspond to the optimum MSE. Coefficient b is be deducted by Supplementary Eq. 1 and the coefficient c is deducted by Supplementary Eq. 2.

2.2 SAF Regularization

The loss function consists of the MSE between pH and gH, and three additional SAF regularization terms. We chose SAF as the activation function to control these regularization terms because SAF's output grows exponentially on either side of the minimum point (Supplementary Fig. 1) where the x-axis represents input values of $MSE(pH_1, pH_2)$, $MSE(pH_1, O)$, and $MSE(pH_2, O)$, while y-axis represents the output values of SAF which minimize to 0 when input values approximate the predetermined optimum MSE. SAF locks the input MSE to the pre-determined optimum value because deviation from this value would result in a fast increase in the gradient of SAF. The first SAF term regularizes the MSE between a pair of pHs. The next two SAF terms regularize the MSE between each pH and a zero matrix. The equation for the entire loss function is:

$$L(pH_1, pH_2, gH_1, gH_2, O) = MSE(pH_1, gH_1) + MSE(pH_2, gH_2) +$$
$$SAF(MSE(pH_1, pH_2)) + SAF(MSE(pH_1, O)) +$$
$$SAF(MSE(pH_2, O))$$

$$(2)$$

where pH_1 = the first predicted heatmap; pH_2 = the second predicted heatmap; gH_1 = the first ground truth heatmap; gH_2 = the second ground truth heatmap; O = a zero matrix where the elements are zero.

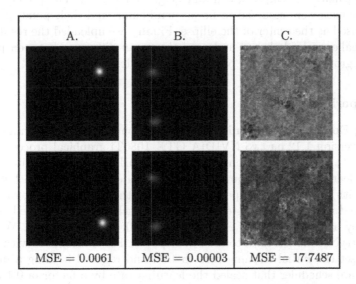

Fig. 3. A (left column): a pair of ground truth heatmaps with optimum MSE = 0.0061. B (middle column): a pair of predicted heatmaps with low MSE = 0.00003. C (right column): a pair of predicted heatmaps with high MSE = 17.7487

2.3   Datasets

FTD Dataset: The dataset used in this study consists of 1111 2D-US images acquired during the second trimester of pregnancies and confirmed by board-certified ultrasonographers to be suitable for measuring FTD [8]. No additional ethics approval was required. Spatial constraints were provided by manually added bounding boxes around the GsS, verified by the same ultrasonographers. Pycocotools generated two gHs for each pair of measurement landmarks, with hotspots representing landmarks [6]. 5-fold cross-validation was performed. During training, 100 training samples were randomly held out as the validation set.

The intraclass correlation coefficient (ICC) score was computed using IBM™ SPSS™ version 28, with the ICC configuration being Two-Way Random and Absolute Agreement [5].

HC18 Dataset: HC18 dataset is available on the Grand Challenge website [11]. We utilized least squared fitting of an ellipse to determine the center, width, height, and angle of rotation of the elliptical ground truth mask. We used trigonometry to determine the coordinates of the landmarks for the major and minor axes of the ellipse. Spatial constraints in the form of bounding boxes were built by the ground truth mask of head circumference published by the dataset author. Predicted landmark points were used for the major and minor axes to calculate the ellipse width and height for testing. The predicted major and minor axes were assumed to be perpendicular, and their point of intersection was used as the center of the ellipse. Finally, we uploaded the results to the Grand Challenge leaderboard and used the mean difference between predicted FHC and ground truth as the performance metric.

2.4   Experimental Setup

Training Epochs and Learning Rate. We conducted our experiments using PyTorch version 1.12 on two NVIDIA GTX-1080Ti graphical processing units, each with 11 GB of video memory. For landmark detection training, we used the same learning rate configuration as Avisdris et al. [2] to train both BiometryNet with/without SAF. Specifically, we set the initial learning rate to 10^{-5} and reduced it using a multi-step learning rate scheduler that scaled the learning rate by a factor of 0.2 at epoch numbers 10, 40, 90, and 150. We trained all BiometryNet models for a total of 200 epochs. For EfficientNet with/without SAF, we set the initial learning rate to 10^{-5} and reduced it using a multi-step learning rate scheduler that scaled the learning rate by a factor of 0.1 at epoch number 300 and 350. We trained both models for 400 epochs.

Pre-processing. Our pre-processing pipeline included random rotation of $\pm 180°$, random re-scaling of $\pm 5\%$, resizing to 384×384 pixels without preserving aspect ratio, and normalization using ImageNet-derived mean = (0.485, 0.456, 0.406) and standard deviation = (0.229, 0.224, 0.225) for each color channel.

SAF Configuration. Given our dataset configuration where each biometry landmark was represented by a 19×19 matrix with values derived from a Gaussian distribution with the center point's value peaked at 1, the MSE between a pair of gHs was 0.0061. This configuration followed the standard implementation used in human pose estimation landmark detection [15]. The secondary optimum MSE between a pH and a zero matrix was halved at 0.00305. We also predetermined the value of Min to be 0.001 to prevent SAF from overpowering the MSE loss between gHs and pHs. We experimented with different values of the coefficient a (1, 4, and 8) to evaluate coefficient a's influence on land-mark detection accuracy. We then determined the value of coefficient b using Supplementary Eq. 1, and the value of coefficient c using Supplementary Eq. 2.

We conditionally activated SAF when the average of $MSE(pH_1, gH_1)$ and $MSE(pH_2, gH_2)$ was less than 0.0009 because SAF is not bounded and early activation would hinder algorithm learning. The proposed SAF algorithm was evaluated using six model configurations, including Vanilla BiometryNet, Biom-etryNet with SAF with coefficient a values of 1, 4, and 8, an EfficientNet, and the EfficientNet with SAF configured with coefficient a value of 4. The model configurations were trained and tested to verify the usefulness of the proposed SAF using both FTD and HC18 datasets.

3   Results

The results of FTD dataset show that BiometryNet with SAF_a1 (Biome-tryNet_SAF_a1) achieved the highest ICC score at 0.737, surpassing the perfor-mance of the vanilla BiometryNet, which scored 0.684. Moreover, BiometryNet with SAF_a4 (BiometryNet_SAF_a4) and BiometryNet with SAF_a8 (Biome-tryNet_SAF_a8) also demonstrated superior ICC scores for FTD measurement, albeit to a lesser degree than BiometryNet_SAF_a1. The impact of SAF on per-formance was further observed in the modified EfficientNet, where Efficient-Net_SAF_a4 achieved a higher ICC score of 0.725, compared to the modified

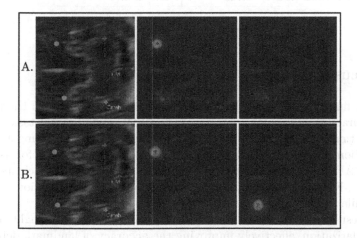

Fig. 4. Row A: landmarks predicted and heatmaps produced by BiometryNet. Left, input image overlaid with predicted landmarks (red spots). Middle, the first predicted heatmap, there are hotspots present near both upper and lower wing-tips. Right, the second predicted heatmap. Row B: landmarks predicted and heatmaps produced by BiometryNet_SAF_a1. Left input image overlaid with predicted landmarks (red spots). Middle, the first predicted heatmap. Right, the second predicted heatmap. (Color figure online)

EfficientNet without SAF, which only scored 0.688. We also observed that SAF reduced the similarities between a pair of pHs and the dispersiveness of hotspot in the pH. We display such heatmaps produced by BiometryNet in Fig .4.A and BiometryNet_SAF_a1 in Fig .4.B.

For the HC18 dataset, BiometryNet_SAF_a8 demonstrated the lowest measurement mean difference from the ground truth at 3.86 mm ± 7.74 mm. Additionally, all configurations outperformed the vanilla BiometryNet. The impact on FHC measurement was further observed in the modified EfficientNet, where EfficientNet_SAF_a4 achieved a lower measurement mean difference at 4.87 mm ± 5.79 mm compared to EfficientNet at 32.76 mm ± 21.01 mm. The FTD dataset ICC scores and HC18 dataset mean differences for each algorithm are presented in Table 1.

Table 1. ICC scores of the FTD dataset, and mean measurement differences ± confidence interval (CI) of the HC18 dataset achieved by all the algorithms.

Network Name	FTD ICC	HC18 Mean Difference ± CI (mm)
Vanilla BiometryNet	0.684	4.56 ± 7.41
BiometryNet_SAF_a1	0.737	4.02 ± 6.70
BiometryNet_SAF_a4	0.724	4.03 ± 7.73
BiometryNet_SAF_a8	0.719	3.86 ± 7.74
EfficientNet	0.688	32.76 ± 21.01
EfficientNet_SAF_a4	0.725	4.87 ± 5.79

4   Discussion

The main findings of this study are as follows: (1) SAF improved the measurement accuracy of algorithms in both FTD and FHC measurement tasks; (2) SAF regularization is architecture-agnostic, as it improved the measurement accuracy of both BiometryNet and EfficientNet compared to their vanilla forms that do not use SAF; and (3) the optimum configuration of SAF coefficients is task-dependent. For FTD measurement, the most optimum configuration was to use $a = 1$, while for FHC measurement was with coefficient $a = 8$.

The results demonstrate the performance improvement brought about by SAF regularization, effectively improving the accuracy of landmark detection of fetal biometries in 2D-US images. SAF regularization forces a pair of heatmaps to highlight different areas and reduce the dispersiveness of hotspots in pHs, which results in improved fetal biometry landmark detection accuracy. This is supported by the comparison of heatmaps produced by BiometryNet and BiometryNet_SAF_a1 displayed in Fig. 4. Moreover, SAF regularization is simple to implement and easy to configure, requiring no modification to the network architecture. Our results also suggest that SAF is highly generalizable as it is architecture-agnostic, improving the performance of both BiometryNet and EfficientNet. SAF is also highly configurable for different tasks via different coefficient configurations. Furthermore, we also suggest that SAF is generalizable to other imaging modalities that also require pair-wise landmark detection. For example, detecting mitral and aortic valves in 2D-US images of heart and detecting cranial sutures in CT images of skull.

As part of our future study, we will explore the effectiveness of SAF regularization in other fetal landmark detection tasks, especially those that suffer from similar issues with fuzzy edges and uncertain landmark locations. Additionally, the optimum configuration of SAF coefficients may vary depending on the specific dataset or imaging modality used.

5 Conclusion

Our study demonstrated the effectiveness of SAF as a novel activation function for regularizing heatmaps generated by fetal biometry landmark detection algorithms, resulting in improved measurement accuracy. SAF outperformed the previous state-of-the-art algorithm, BiometryNet, in both FTD and FHC measurement tasks. Importantly, our results showed that SAF is architecture-agnostic and highly configurable for different tasks through its coefficients, making it a generalizable solution for a wide range of landmark detection problems.

References

1. Apostolos Kolitsidakis: How to measure the femur length (2021)
2. Avisdris, N., et al.: Biometrynet: landmark-based fetal biometry estimation from standard ultrasound planes. In: Wang, L., Dou, Q., Fletcher, P.T., Speidel, S., Li, S. (eds.) MICCAI 2022, vol. 13434, pp. 279–289. Springer, Cham (2022). https://doi.org/10.1007/978-3-031-16440-8_27
3. Bethune, M., Alibrahim, E., Davies, B., Yong, E.: A pictorial guide for the second trimester ultrasound. Australas. J. Ultrasound Med. **16**(3), 98–113 (2013). https://doi.org/10.1002/j.2205-0140.2013.tb00106.x
4. Brickson, L.L., Hyun, D., Jakovljevic, M., Dahl, J.J.: Reverberation noise suppression in ultrasound channel signals using a 3D fully convolutional neural network. IEEE Trans. Med. Imaging **40**(4), 1184–1195 (2021). https://doi.org/10.1109/TMI.2021.3049307
5. IBM Corp: IBM SPSS Statistics for Windows (2020)
6. Lin, T.-Y., et al.: Microsoft COCO: common objects in context. In: Fleet, D., Pajdla, T., Schiele, B., Tuytelaars, T. (eds.) ECCV 2014. LNCS, vol. 8693, pp. 740–755. Springer, Cham (2014). https://doi.org/10.1007/978-3-319-10602-1_48
7. Shankar, H., et al.: Leveraging clinically relevant biometric constraints to supervise a deep learning model for the accurate caliper placement to obtain sonographic measurements of the fetal brain. In: Proceedings - International Symposium on Biomedical Imaging, vol. 2022-March. IEEE Computer Society (2022). https://doi.org/10.1109/ISBI52829.2022.9761493

8. Sridar, P., Kennedy, N.J., Quinton, A.E., Robledo, K., Kim, J., Nanan, R.: Normative ultrasound data of the fetal transverse thalamic diameter derived from 18 to 22 weeks of gestation in routine second-trimester morphology examinations. Australas. J. Ultrasound Med. **23**(1), 59–65 (2020). https://doi.org/10.1002/ajum.12196

9. Sridar, P., et al.: Automatic measurement of thalamic diameter in 2-D fetal ultrasound brain images using shape prior constrained regularized level sets. IEEE J. Biomed. Health Inform. **21**(4), 1069–1078 (2017). https://doi.org/10.1109/JBHI.2016.2582175

10. Sun, K., Xiao, B., Liu, D., Wang, J.: Deep high-resolution representation learning for human pose estimation. In: Proceedings of the IEEE/CVF Conference on Computer Vision and Pattern Recognition, pp. 5693–5703 (2019)

11. van den Heuvel, T.L., de Bruijn, D., de Korte, C.L., Ginneken, B.V.: Automated measurement of fetal head circumference using 2D ultrasound images (2018)

12. Wai, M.S., Lorke, D.E., Kwong, W.H., Zhang, L., Yew, D.T.: Profiles of serotonin receptors in the developing human thalamus. Psychiatry Res. **185**(1–2), 238–242 (2011). https://doi.org/10.1016/j.psychres.2010.05.003

13. Waxman, S.G., Waxman, S.G.: Chapter 9: Diencephalon: Thalamus and Hypothalamus. In: Clinical Neuroanatomy, chap. 9. McGraw-Hill Education, New York, 29th edn. (2020)

14. Xia, W., et al.: Automatic plane of minimal hiatal dimensions extraction from 3D female pelvic floor ultrasound. IEEE Trans. Med. Imaging **41**(12), 3873–3883 (2022). https://doi.org/10.1109/TMI.2022.3199968

15. Xiao, B., Wu, H., Wei, Y.: Simple baselines for human pose estimation and tracking. In: Proceedings of the European Conference on Computer Vision (ECCV), pp. 466–481 (2018)

16. You, L., Deng, Y., Li, D., Lin, Y., Wang, Y.: GLP-1 rescued gestational diabetes mellitus-induced suppression of fetal thalamus development. J. Biochem. Mol. Toxicol. (2022). https://doi.org/10.1002/jbt.23258

Robust Estimation of the Microstructure of the Early Developing Brain Using Deep Learning

Hamza Kebiri[1,2,3]([✉]), Ali Gholipour[3], Rizhong Lin[2,4], Lana Vasung[5],
Davood Karimi[3], and Meritxell Bach Cuadra[1,2]

[1] CIBM Center for Biomedical Imaging, Lausanne, Switzerland
hamza.kebiri@unil.ch
[2] Department of Radiology, Lausanne University Hospital (CHUV) and University of
Lausanne (UNIL), Lausanne, Switzerland
[3] Computational Radiology Laboratory, Department of Radiology, Boston Children's
Hospital and Harvard Medical School, Boston, MA, USA
[4] Signal Processing Laboratory 5 (LTS5), École Polytechnique Fédérale de Lausanne
(EPFL), Lausanne, Switzerland
[5] Department of Pediatrics, Boston Children's Hospital, and Harvard Medical
School, Boston, MA, USA

Abstract. Diffusion Magnetic Resonance Imaging (dMRI) is a powerful
non-invasive method for studying white matter tracts of the brain. However, accurate microstructure estimation with fiber orientation distribution (FOD) using existing computational methods requires a large number of diffusion measurements. In clinical settings, this is often not possible for neonates and fetuses because of increased acquisition times and
subject movements. Therefore, methods that can estimate the FOD from
reduced measurements are of high practical utility. Here, we exploited
deep learning and trained a neural network to directly map dMRI data
acquired with as low as six diffusion directions to FODs for neonates
and fetuses. We trained the method using target FODs generated from
densely-sampled multiple-shell data with the *multi-shell multi-tissue constrained spherical deconvolution* (MSMT-CSD). Detailed evaluations on
independent newborns' test data show that our method achieved estimation accuracy levels on par with the state-of-the-art methods while
reducing the number of required measurements by more than an order
of magnitude. Qualitative assessments on two out-of-distribution clinical
datasets of fetuses and newborns show the consistency of the estimated
FODs and hence the cross-site generalizability of the method.

D. Karimi and M. B. Cuadra—Equal contribution.

Supplementary Information The online version contains supplementary material
available at https://doi.org/10.1007/978-3-031-43990-2_28.

1   Introduction

Depiction of white matter fiber tracts is of paramount importance for brain characterization in health and disease. Diffusion-weighted magnetic resonance imaging (dMRI) is the method of choice to study axon bundles that connect different brain regions. Several models have been proposed to map the 4-dimensional diffusion signal to objects such as tensors or fiber orientation distribution functions (FODs) [23,32], which can be further processed to compute metrics such as tract orientation and apparent fiber density [16,26]. Model-based FODs are the mathematical frameworks of choice for microstructure estimation. In fact, FODs accurately describe the underlying microstructure by a deformed sphere in which different radii correspond to different intra-voxel fibers. Moreover, without FODs, tracking stops prematurely and favors shorter fiber tracts [8]. Standard FOD estimation methods [1,15,23,27,32] process voxels individually and thus do not exploit correlations between neighboring voxels. As a result, these methods demand dMRI measurements with multiple b-values and a high number of gradient directions to account for the response function of each tissue type [15].

These acquisitions require prolonged scans that are not affordable for newborn and fetal subjects because of the sensitivity of these cohorts and the increased risk of motion. Acquisitions have to be fast to freeze in-plane motion; yet data dropout rates are high in these cohorts because of motion artifacts. Reconstructing FODs in developing brains has been performed [4,5,7,30] using high-quality datasets and rich information including several gradient directions, higher and/or multiple b-values, and high signal-to-noise ratio (3 T magnetic field strength). Additionally, the datasets were acquired in a controlled and uniform research setting with healthy volunteers, which can hardly be reproduced in the clinical environment. Moreover, and in contrast to adult brains, anisotropy increases in white matter fibers of developing brains because of increased water volume and poor alignment of the fibers [6]. Gray matter on the other hand, during early gestational weeks, is highly anisotropic because of the complexity of the formation of cell bodies, glial cells, and the different neuronal structures [6]. This dynamic period for microstructure [2,9] makes FOD estimation a more challenging task. Adaptive learning-based methods can be leveraged to learn from high-quality datasets and exploit this knowledge in clinical routine acquisitions.

Deep learning models, first suggested in [12], promise to overcome the error accumulation of suboptimal processing steps that are characteristic of standard estimation techniques. This end-to-end learning paradigm has been then applied in dMRI for several purposes [13,18,22,24,25]. The authors in [24] have accurately predicted tensor maps with six diffusion measurements. In [22], a 2D convolutional neural network (CNN) was used to predict the orientation of the fibers in a classification approach whereas [25] deployed a 3D CNN to predict FODs using a small neighborhood of the diffusion signal. In [18], a feedforward neural network was used to predict the FODs and found that 44 directions can be sufficient. However, the network does not exploit neighboring voxels correlations. A more recent work [13] used a Transformer-CNN block to first map 200 to 60 directions and the latter to FODs. However, for uncooperative cohorts such as neonates or fetuses, this number of measurements is unrealistic to acquire.

To the best of our knowledge, no learning-based method to predict FODs has been reported for newborn and fetal brains. In this work, we demonstrate that a deep convolutional neural network with a large field of view (FOV) can accurately estimate FODs using only 6–12 diffusion-weighted measurements. Our contribution is three-fold. We first show that a deep learning method can achieve an accuracy level that is comparable with the agreement between the state-of-the-art methods, while drastically reducing the number of measurements, for developing brains. We then show a low agreement between state-of-the-art methods in terms of different metrics using data from a highly controlled setting, namely the developing Human Connectome Project (dHCP). This addresses the need to build reproducible and reliable pipelines for white matter characterization [29], particularly for the developing brain. Finally, we demonstrate the generalizability of our method on two clinical datasets of fetuses and newborns that were acquired in completely different settings than those used in the training data.

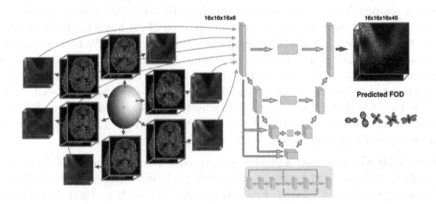

Fig. 1. The proposed framework to predict Fiber Orientation Distribution (FOD) functions in the spherical harmonics domain (SH-L_{max} order 8). The network takes 3D input patches from 6 diffusion measurements and outputs patches of SH coefficients. Network architecture details can be found in Figure S1 of Supplementary materials and code at (https://github.com/Medical-Image-Analysis-Laboratory/Perinatal_fODF_DL_estimation).

2   Methodology

2.1   Paradigm

The method is based on directly learning a mapping between the raw diffusion signal and the FOD in a supervised manner (Fig. 1). Our model inputs are respectively 6 single-shell (b = 1000 s/mm^2) measurements for the neonate network

(DL_n) and 12 single-shell (b = 400 s/mm^2) measurements for the fetal inference network (DL_f), trained on pre-term subjects (as in [17,19]). To be independent on gradient directions, projection of the signal onto spherical harmonics basis (SH) (SH-L_{max} order 2 and 4 respectively for the two networks) was performed to predict the FOD represented in the SH basis (SH-L_{max} order 8). To train the model, these target coefficients are estimated from 300 multi-shell measurements using MSMT-CSD [15]. These measurements are distributed over 3 shells of $\{400, 1000, 2600\}$ s/mm^2 with 64, 88 and 128 samples, respectively, and 20 b0 (b = 0 s/mm^2) images. The few measurements used as input to the model were based on the scheme in [31], whereby the gradient directions minimized the condition number of the diffusion tensor reconstruction matrix.

2.2 Data Processing

dHCP Newborns - We have selected two subsets from the developing Human Connectome Project (dHCP) dataset (1) 100 subjects (weeks: [32.1, 44.7], mean: 40, standard deviation: 2.4) and (2) 68 pre-term subjects (weeks: [29.3, 37.0], mean: 34.9, standard deviation: 1.8). The data was acquired with a 3T Philips Achieva scanner in a multi-shell scheme (b $\in \{0, 400, 1000, 2600\}$ s/mm^2) [14] and was denoised, motion and distortion corrected [3]. It has a final resolution of $1.17 \times 1.17 \times 1.5$ mm^3 in a field of view of $128 \times 128 \times 64$ voxels. We have upsampled the data to 1 mm isotropic resolution to account for network isotropic 3D patches. We have additionally normalized the input data by b0. A white matter mask was generated using the union of the *White Matter* and the *Brainstem* labels provided by the dHCP, and the voxels where Fractional Anisotropy (FA) was higher than 0.25. A resampling of the dHCP labels from T2-w resolution (0.5 mm^3 isotropic) to 1 mm^3 resolution was performed.

Clinical Newborns and Fetuses - Acquisitions of 8 neonates ([38.1, 39.4, 40.1, 40.4, 40.7, 40.9, 41.8, 42] weeks), were performed during natural sleep at 3T (Siemens Trio and Skyra). Five b0 images and 30 b = 1000 s/mm^2 were acquired. The TR-TE were 3700–104 ms and voxel size was 2 mm isotropic. Eight fetal subjects ([24, 25, 26.3, 26.6, 26.7, 26.9, 29.4, 38.7] gestational weeks, GW) were scanned using a 3T Siemens Skyra MRI scanner (TR = 3000–4000 ms, TE = 60 ms) with one b0 and 12 diffusion-sensitized images at b = 500 s/mm^2. All subjects were processed for noise [34] and bias field inhomogeneities [33]. Rigid registration to a T2 atlas [11] was performed and b-vectors were rotated accordingly for fetal data. The different volumes were upsampled to 1 mm^3 and normalized by b0. The studies were approved by the institutional review board committee.

2.3 Training

Two networks, DL_n and DL_f (see Subsect. 2.1 above), were trained using Adam optimizer [21] to minimize the ℓ_2 norm loss function between the predicted 45 SH coefficients and the ground truth FOD SH coefficients generated using the 300 directions and the 4 b-values ($\{0, 400, 1000, 2600\}$ s/mm^2), i.e.

$$minimize \sum_{i=1}^{45} \left\| FOD_i^{pred} - FOD_i^{GT} \right\|^2$$

We used 70% of the subjects for training, 15% for validation, and 15% for testing. We used the number of FOD peaks (extracted from Dipy [10]) to balance patch selection per batch. The central voxel of each patch was constrained to be in the generated white matter mask and to be 1 peak in $\frac{2}{3}$ of the batch and more than one peak in $\frac{1}{3}$ of the batch. This condition implicitly guarantees the non selection of empty patches. The patch size was empirically set to 16^3 voxels. The batch size was set to 27 for DL_n and 9 for DL_f, and the initial learning rate to 10^{-4} and was decreased by 0.9 whenever the validation loss did not improve after one epoch. The total number of training epochs was 10000 and a dropout rate of 0.1 was used in all layers to reduce overfitting and improve generalization. In DL_f, Gaussian noise injection (mean = 0, sigma = 0.025) was applied as well as small rotations (uniformly from $[-5°, +5°]$) to make the model robust to minor uncorrected movements due to small differences in FOV and fetal head motion.

2.4   Evaluation of dHCP Newborns

Comparison with State-of-the-Art Methods - In addition to comparing our network (DL_n) prediction with FODs estimated using MSMT-CSD of 300 directions (considered as ground truth, GT), we have assessed the agreement between two mutually exclusive subsets extracted from the ground truth (gold standards 1 and 2, respectively GS1 and GS2). Each subset contains 150 directions (b $\in \{0, 400, 1000, 2600\}$ s/mm^2) with respectively 10, 32, 44, and 64 measurements (half measurements of GT data). GS1 and GS2 subsets can be considered as independent high-quality scans, and differences in terms of subsequent metrics can be considered as an upper bound error for the different methods deployed. Furthermore, we have computed three state-of-the-art methods: (1) Constrained spherical deconvolution (CSD) [32] using the 128 gradient directions of the highest shell, i.e. b = 2600 s/mm^2 and 20 b0 images; (2) Constant Solid Angle ODF (Q-Ball) model [1] that we refer to as CSA and (3) the Sparse Fascicle Model (SFM) [27] model for which we have used the default regularization parameters. We also compared our method with the multilayer perceptron (MLP) in [18], which has been shown to outperform the method of [25].

Error Metrics - Quantitative validation was performed based on the number of peaks, the angular error and the apparent fiber density (AFD) [26]. The number of peaks was generated from the FOD predicted by the network and the ones estimated by the different methods (GT, GS1, GS2, CSD, CSA and SFM) using the same parameters (mean separation angle of 45°, a maximum number of 3 peaks and relative peak threshold of 0.5). The conservative choice of these parameters was guided by [28] which shows the limitations of current dMRI models at depicting multiple number of peaks and low angular crossing fibers. We have compared these models in terms of confusion matrices, and the *agreement rate* (AR). AR is defined for each number of peaks p as: $AR = \frac{A_p}{\Sigma D_p}$ where A_p is the percentage of voxels on which both methods agree on p number of peaks and D_p the percentage of voxels where at least one of the two methods predicts p and the other p' where $p \neq p'$. For the voxels containing the same number of peaks, we have computed the angular error with respect to the GT, as well as between GS1 and GS2. For voxels with multiple fibers, we have first extracted corresponding peaks between the two methods by computing the minimum angle between all configurations (4 for 2 peaks and 9 for 3 peaks); we then removed these peaks and recursively apply the same algorithm. We have also compared AFD, that is defined as the FOD amplitude. AFD was extensively demonstrated as a biologically plausible measure that is not only sensitive to the fiber partial volume fraction but also to fiber density or membrane permeability [26]. Statistical validation using paired t-test corrected for multiple comparisons with Bonferroni method was performed between the errors of the different methods with respect to GT and the difference between GS1 and GS2.

2.5   Evaluation of Clinical Datasets

DL_f was tested on fetal volumes whereas DL_n was tested on the clinical newborn dataset. Due to the lack of ground truth for both clinical datasets, we qualitatively assess the network predictions with 12 and 6 measurements, as compared to CSD using all available measurements (SH-L_{max} order 4 and 8), respectively for fetuses and newborns.

3   Results

The networks consistently learned a mapping between the six/twelve diffusion measurements and the ground truth FOD constructed with 300 measurements across 4 b-values, as evaluated on the independent test data (Figs. 2 and 3).

3.1   In-Domain Quantitative Evaluation in Newborns dHCP

Number of Peaks - We first observe a low agreement (AR) between the two gold standard acquisitions (GS1 vs. GS2), that is more pronounced for multiple fibers voxels. For instance, 1-peaks AR is 80.4%, 30.3% for 2-peaks and 27.9% for 3-peaks. SFM achieves a relatively high 1-peaks agreement with the GT of 83% and the lowest with multiple fibers voxels (10% and 3.5% for 2- and 3-peaks, respectively). In contrast, CSD estimates a high number of multiple fibers (16.5% and 5.9% for 1- and 2-peaks respectively) and achieves the lowest 1-peaks AR with 11.7%. In fact, the latter is biased towards multiple peaks estimation with more than 90% of the voxels modeled as either two or three peaks. This might be explained by the high b-value (b = 2600 s/mm^2) containing high levels of noise. Our method, DL_n, achieves an agreement for 1-, 2- and 3-peaks of respectively 79%, 16% and 3% that is globally the closest to the agreement between the gold standards when compared to other methods. We believe that the relatively low agreement for multiple intravoxel fiber orientations is due to their incongruence across GT subjects, and hence the absence of a

Table 1. Mean angular error, agreement rate on number of peaks and Apparent Fiber Density (AFD) error between GT (MSMT-CSD) and the different methods. ΔGS refers to GS1 and GS2 agreements. The number of measurements (N_m) and the b-values used are also reported. All results were statistically significant compared to ΔGS ($p \leq$9e^{-10} for angular error, except SFM three fibers, and $p \leq$4.5e^{-3} for AFD error).

Method	b-values (s/mm^2)	N_m	Angular error (Agreement rate in %)			AFD error
			Single fibers	Two fibers	Three fibers	
DL_n	{0, 1000}	7	12.6° (78.4%)	24.2° (15.8%)	33.3° (3.8%)	0.27 (±0.03)
CSD	{0, 2600}	148	7.5° (11.7%)	16.5° (16.5%)	27.2° (5.9%)	1.31 (±0.23)
CSA	{0, 400, 1000, 2600}	300	47.0° (27.7%)	41.4° (14.8%)	36.1° (7.9%)	3.46 (±0.46)
SFM	{0, 400, 1000, 2600}	300	51.4° (83%)	40.7° (10%)	35.4° (3.5%)	0.80 (±0.55)
ΔGS	{0, 400, 1000, 2600}	150	13.8° (80.4%)	29.1° (30.3%)	35.4° (27.9%)	0.2 (±0.025)

consistent pattern to be learned by the neural network. In fact, this is supported by the modest agreement between the two gold standards (ΔGS), in which both the subjects and the number of measurements are the same, only the gradient directions vary and already result in a drop of 70% in multiple fibers depiction. It is worth noting that the agreement between the different methods (CSD vs. CSA, SFM vs. CSA, CSD vs. DL_n, etc.) was also low. The confusion matrices for ΔGS agreement and the different methods can be found in Table 1 and the comparison with [18] in Section 3 of Supplementary materials.

Angular Error - The agreement in terms of the number of peaks does not guarantee that the fibers follow the same orientation. Table 1 shows the angular error and the agreement rate (AR) in numbers of peaks for the different configurations. In GS1 and GS2, the angular difference increases almost linearly for one, two and three fibers. Our learning model achieves an error rate that is comparable (although statistically different, $p \leq$9e^{-10}) to GS1 and GS2. SFM and CSA achieve a higher error rate for single and two fiber voxels, whereas CSD achieves the lowest. This is because of the low AR and hence the error is computed among a small subset of common voxels between the GT and CSD as shown in Table 1. It is worth mentioning that using 15 directions instead of 6 as input to the network did not improve the results; and in general, these angular errors are higher than those reported for adult data, such as the Human Connectome Project as in [13]. We hypothesize this can be due to immature and high variability of the developing brain anatomy.

Apparent Fiber Density - The last column in Table 1 shows the differences between AFD averaged over the 15 test subjects. Our model achieves the closest error rate of 0.27 (±0.03) to the GT compared with the gold standards difference of 0.2 (±0.025), in terms of mean and standard deviation. The other methods have an increased error rate compared to DL_n with factors of around 2.5, 4.5 and 9.5-fold for SFM, CSD and CSA respectively. Results were statistically significant ($p \leq$4.5e^{-3}) compared to the agreement between the gold standard models.

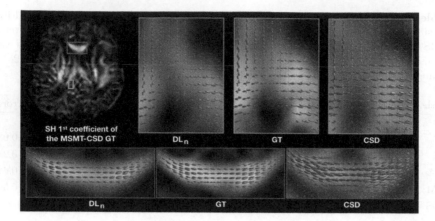

Fig. 2. Qualitative comparison between the deep learning method DL_n, the MSMT-CSD GT and CSD in two brain regions of a newborn dHCP subject.

3.2 Generalizability to Clinical Acquisitions (newborns and fetuses)

DL_f successfully generalized to fetal data as can be seen in Fig. 3 (right) for two subjects. Callossal fibers are clearly delineated on the top and bottom subjects. The radial coherence of cortical plate at early gestation [20] is also highlighted on the same panels. Similarly, DL_n generalized to the new newborn dataset (Fig. 3, left), despite differences in scanner and protocol. Both cortico-spinal tract and corpus callosum are shown in the bottom subject. As opposed to CSD that overestimated false positive crossing fibers, likely due to residual noise, the deep learning method trained on MSMT-CSD directly produced low amplitude FODs in isotropic or non-consistent regions. The results for six other subjects can be found in Figure S2 in Supplementary materials.

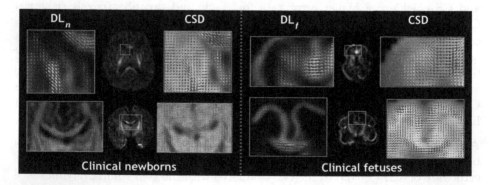

Fig. 3. The deep learning method compared to CSD in different brain regions for 2 newborn subjects (left) and 2 fetal subjects (right) of 25 (top) and 29.4 (bottom) weeks of gestation. FODs are superimposed to the first SH coefficient of the method used.

4   Conclusion

We have demonstrated how a deep neural network can successfully reconstruct high angular multi-shell FODs from a reduced number (6 to 12) of diffusion measurements. The substantially lower number of samples is compensated by learning from high-quality training data and by exploiting the spatial neighborhood information. The network was quantitatively evaluated on the dHCP dataset which was acquired in a highly controlled setting that cannot be reproduced in clinical settings. We showed that our method relying on six measurements can be leveraged to reconstruct plausible FODs of clinical newborn and fetal brains. We compared our model to commonly used methods such as CSD and MSMT-CSD between two gold standard datasets. The results exhibit low agreements between the different methods, particularly for multiple fiber orientations, despite using high angular multi-shell data. This highlights the need to build robust and reproducible methods for microstructure estimation in developing brains.

Acknowledgment. This work was supported by the Swiss National Science Foundation (project 205321-182602). We acknowledge the CIBM Center for Biomedical Imaging, a Swiss research center of excellence founded and supported by CHUV, UNIL, EPFL, UNIGE, HUG and the Leenaards and Jeantet Foundations. This research was also partly supported by the US National Institutes of Health (NIH) under awards R01NS106030 and R01EB032366; by the Office of the Director of the NIH under award S10OD0250111; and by NVIDIA Corporation; and utilized an NVIDIA RTX A6000 GPU.

References

1. Aganj, I., et al.: Reconstruction of the orientation distribution function in single- and multiple-shell q-ball imaging within constant solid angle. Magn. Reson. Med. **64**(2), 554–566 (2010)
2. Andescavage, N.N., et al.: Complex trajectories of brain development in the healthy human fetus. Cereb. Cortex **27**(11), 5274–5283 (2017)
3. Bastiani, M., et al.: Automated processing pipeline for neonatal diffusion MRI in the developing human connectome project. Neuroimage **185**, 750–763 (2019)
4. Chen, R., et al.: Deciphering the developmental order and microstructural patterns of early white matter pathways in a diffusion MRI based fetal brain atlas. Neuroimage **264**, 119700 (2022)
5. Christiaens, D., et al.: Multi-shell shard reconstruction from scattered slice diffusion MRI data in the neonatal brain. ISMRM (Paris) (2018)
6. Counsell, S.J., Arichi, T., Arulkumaran, S., Rutherford, M.A.: Fetal and neonatal neuroimaging. Handb. Clin. Neurol. **162**, 67–103 (2019)
7. Deprez, M., et al.: Higher order spherical harmonics reconstruction of fetal diffusion MRI with intensity correction. IEEE Trans. Med. Imaging **39**(4), 1104–1113 (2019)
8. Descoteaux, M.: High angular resolution diffusion MRI: from local estimation to segmentation and tractography. Ph.D. thesis, Univ. Nice Sophia Antipolis (2008)
9. Dubois, J., et al.: The early development of brain white matter: a review of imaging studies in fetuses, newborns and infants. Neuroscience **276**, 48–71 (2014)

10. Garyfallidis, E., et al.: Dipy, a library for the analysis of diffusion MRI data. Front. Neuroinform. **8**, 8 (2014)
11. Gholipour, A., et al.: A normative spatiotemporal MRI atlas of the fetal brain for automatic segmentation and analysis of early brain growth. Sci. Rep. **7**(1), 476 (2017)
12. Golkov, V., et al.: Q-space deep learning: twelve-fold shorter and model-free diffusion MRI scans. IEEE TMI **35**(5), 1344–1351 (2016)
13. Hosseini, S., et al.: CTtrack: a CNN+ transformer-based framework for fiber orientation estimation & tractography. Neurosci. Inform. **2**(4), 100099 (2022)
14. Hutter, J., et al.: Time-efficient and flexible design of optimized multishell HARDI diffusion. Magn. Reson. Med. **79**(3), 1276–1292 (2018)
15. Jeurissen, B., et al.: Multi-tissue constrained spherical deconvolution for improved analysis of multi-shell diffusion MRI data. Neuroimage **103**, 411–426 (2014)
16. Jeurissen, B., et al.: Diffusion MRI fiber tractography of the brain. NMR Biomed. **32**(4), e3785 (2019)
17. Karimi, D., et al.: Deep learning-based parameter estimation in fetal diffusion-weighted MRI. Neuroimage **243**, 118482 (2021)
18. Karimi, D., et al.: Learning to estimate the fiber orientation distribution function from diffusion-weighted MRI. Neuroimage **239**, 118316 (2021)
19. Kebiri, H., et al.: Slice estimation in diffusion MRI of neonatal and fetal brains in image and spherical harmonics domains using autoencoders. In: Cetin-Karayumak, S., et al. (eds.) CDMRI 2022. LNCS, vol. 13722, pp. 3–13. Springer, Cham (2022). https://doi.org/10.1007/978-3-031-21206-2_1
20. Khan, S., et al.: Fetal brain growth portrayed by a spatiotemporal diffusion tensor MRI atlas computed from in utero images. Neuroimage **185**, 593–608 (2019)
21. Kingma, D.P., Ba, J.: Adam: A method for stochastic optimization. arXiv preprint arXiv:1412.6980 (2014)
22. Koppers, S., Merhof, D.: Direct estimation of fiber orientations using deep learning in diffusion imaging. In: Wang, L., Adeli, E., Wang, Q., Shi, Y., Suk, H.-I. (eds.) MLMI 2016. LNCS, vol. 10019, pp. 53–60. Springer, Cham (2016). https://doi.org/10.1007/978-3-319-47157-0_7
23. Le Bihan, D., et al.: Diffusion tensor imaging: concepts and applications. J. Magn. Reson. Imaging **13**(4), 534–546 (2001)
24. Li, H., et al.: SuperDTI: ultrafast DTI and fiber tractography with deep learning. Magn. Reson. Med. **86**(6), 3334–3347 (2021)
25. Lin, Z., et al.: Fast learning of fiber orientation distribution function for MR tractography using convolutional neural network. Med. Phys. **46**(7), 3101–3116 (2019)
26. Raffelt, D., et al.: Apparent fibre density: a novel measure for the analysis of diffusion-weighted magnetic resonance images. Neuroimage **59**(4), 3976–3994 (2012)
27. Rokem, A., et al.: Evaluating the accuracy of diffusion MRI models in white matter. PLoS ONE **10**(4), e0123272 (2015)
28. Schilling, K.G., et al.: Histological validation of diffusion MRI fiber orientation distributions and dispersion. Neuroimage **165**, 200–221 (2018)
29. Schilling, K.G., et al.: Prevalence of white matter pathways coming into a single white matter voxel orientation: the bottleneck issue in tractography. Hum. Brain Mapp. **43**(4), 1196–1213 (2022)
30. Shen, K., et al.: A spatio-temporal atlas of neonatal diffusion MRI based on kernel ridge regression. In: International Symposium on Biomedical Imaging (ISBI) (2017)
31. Skare, S., et al.: Condition number as a measure of noise performance of diffusion tensor data acquisition schemes with MRI. J. Magn. Reson. **147**(2), 340–352 (2000)

32. Tournier, J.D., et al.: Direct estimation of the fiber orientation density function from diffusion-weighted MRI data using spherical deconvolution. Neuroimage **23**(3), 1176–1185 (2004)
33. Tustison, N.J., Avants, B.B., Cook, P.A., Zheng, Y., et al.: N4ITK: improved N3 bias correction. IEEE Trans. Med. Imaging **29**(6), 1310–1320 (2010)
34. Veraart, J., Fieremans, E., Novikov, D.S.: Diffusion MRI noise mapping using random matrix theory. Magn. Reson. Med. **76**(5), 1582–1593 (2016)

HACL-Net: Hierarchical Attention and Contrastive Learning Network for MRI-Based Placenta Accreta Spectrum Diagnosis

Mingxuan Lu[1], Tianyu Wang[2(✉)], Hao Zhu[3], and Mian Li[2]

[1] Columbia University in the City of New York, New York, USA
[2] Shanghai Jiao Tong University, Shanghai, China
gunnerwang27@sjtu.edu.cn
[3] Obstetrics and Gynecology Hospital affiliated to Fudan University, Shanghai, China

Abstract. Placenta Accreta Spectrum (PAS) can lead to high risks like excessive blood loss at the delivery. Therefore, prenatal screening with MRI is essential for delivery planning that ensures better clinical outcomes. For computer-aided PAS diagnosis, existing work mostly extracts radiomics features directly from ROI while ignoring the context information, or learns global semantic features with limited awareness of the focal area. Moreover, they usually select single or few MRI slices to represent the whole sequences, which can result in biased decisions. To deal with these issues, a novel end-to-end Hierarchical Attention and Contrastive Learning Network (HACL-Net) is proposed under the formulation of a multi-instance problem. Slice-level attention module is first designed to extract context-aware deep semantic features. These slice-wise features are then aggregated via the patient-level attention module into task-specific patient-wise representation for PAS prediction. A plug-and-play contrastive learning module is introduced to further improve the discriminating power of extracted features. Extensive experiments with ablation studies on a real clinical dataset show that HACL-Net can achieve state-of-the-art prediction performance with the effectiveness of each module.

Keywords: Placenta Accreta Spectrum · MRI · Multi-Instance Learning · Hierarchical Attention · Contrastive Learning

1 Introduction

Placenta Accreta Spectrum (PAS) occurs when the placenta becomes abnormally adherent to the myometrium rather than the uterine decidua [15]. The primary risk of PAS is hemorrhage and even life-threatening associated complications [15]. Particularly, it can lead to excessive blood loss and transfusions of blood products at the delivery [9]. Prenatal screening helps identify women

Supplementary Information The online version contains supplementary material available at https://doi.org/10.1007/978-3-031-43990-2_29.

with potential PAS and allows for appropriate delivery planning, which is critical for better clinical outcomes. Meanwhile, Magnetic Resonance Imaging (MRI) features high resolution and sensitivity, playing an increasingly important role in prenatal screening [12]. MRI is often used after an inconclusive ultrasonic examination to assess the invasive condition of PAS [9]. A detailed background illustration is included in Section A of supplementary material.

In the field of computer-aided PAS diagnosis for prenatal screening, some work based on traditional machine learning has been proposed [7,11–13] and achieved promising progress. In these works, radiomics features were first extracted from the Region of Interest (ROI, placenta) of raw images and a classifier was then fitted between the features and clinical outcomes. In [12], Random Forests (RF), K Nearest Neighbor (KNN), Naive Bayes (NB), and Multi-Layer Perception (MLP) were compared. Statistical analysis and multivariate logistic regression were employed in [13]. The features are mostly extracted only on placenta ROI with Pyradiomics [17] or predefined rules, ignoring the context information such as the location of the placenta and its interaction with nearby organs that are important in PAS diagnosis [10].

On the other hand, deep learning stands out in terms of automatic deep feature extraction [24] and end-to-end learning fashion [3]. Deep learning has been recently explored for image-based PAS diagnosis [20,23]. A deep dynamic convolution neural network with autoencoder training manner was proposed in [20] to extract deep features. In [23] the encoder of a placenta segmentation network was employed to capture the semantic features. To enhance such deep features with the awareness of the focal area, attention mechanisms [25] are considered in this paper. The ROI relevant to the prediction task can then be automatically localized, leading to more comprehensive features that capture both semantic and context information. Moreover, prior knowledge about the focal area can be incorporated by annotating a limited number of images with ROIs. That is more efficient than directly including the localization task in addition to the prediction task, which requires extra pixel-level annotation of each image.

Another gap to fill is the inherent diversity of MRI sequences. There are T1-weighted and T2-weighted MRI scans, upon which three planes with multiple slices are further present. Therefore, each patient corresponds to dozens of slice images. The physicians usually need to inspect a large portion of these images in order to identify the focus of suspected PAS and reach the patient-level diagnosis. For the existing work, a single (or few) slice image is selected to represent the information of a patient, which can lead to potentially biased results. Multi-Instance Learning (MIL), as a potential solution, has been widely adopted in lesion classification and localization [19,21,22]. Some recent works, such as slice attention transformer and convolutional transformer-based MIL, have studied the problem of patient as bag and images as instances [6,8,16]. It remains to explore how MIL can benefit the computer-aided PAS diagnosis.

To deal with the issues above, a novel end-to-end Hierarchical Attention and Contrastive Learning Network (HACL-Net) is proposed under the formulation of a multi-instance problem. Slice-level spatial residual attention with prior knowledge is designed to extract context-aware deep semantic features from individual

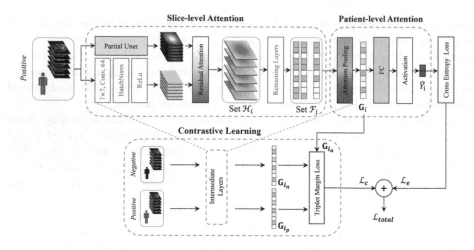

Fig. 1. An architecture overview of the proposed HACL-Net.

MRI slices of a patient. Patient-level attention-based pooling is then applied to aggregate these features into patient representation for PAS prediction. To make such features more discriminative, a plug-and-play contrastive loss is further included. Extensive experiments with ablation study show the proposed network can achieve state-of-the-art performance on a real clinical dataset involving 359 distinct patients.

2 Methodology

Consider each patient bag $\mathcal{P}_i$ includes K_i valid 2D MRI slices from axial, sagittal, and coronal views (possibly both T1-weighted and T2-weighted scans exist). In the training dataset, there is an associated binary label Y_i indicating whether PAS is reported at delivery. Denote $\mathcal{P}_i = \{\mathcal{X}_i, Y_i\}$, $\mathcal{X}_i = \{\mathbf{X}_i^{(1)}, \cdots, \mathbf{X}_i^{(K_i)}\}$, where $\mathbf{X}_i^{(j)}$ is the j-th MRI slice of i-th patient. Note that the number K_i can vary with different patients. The proposed HACL-Net aims to learn the relationship between $\mathcal{X}_i$ and Y_i in a weekly-supervised and end-to-end manner (Fig. 1). The slice-level attention module is designed to extract context-aware deep semantic features from each MRI slice. Then the slice-wise features are fused in a task-specific way via the patient-level attention module, after which being connected to FC layer for PAS prediction. A plug-and-play contrastive learning module is presented to facilitate discriminative patient-wise features between classes.

2.1 Slice-Level Spatial Residual Attention

Trunk Branch. The trunk branch aims to extract global semantic information from each MRI slice that is important for the PAS prediction task. It is composed of the top convolution layer, batch normalization layer, and ReLU layer from

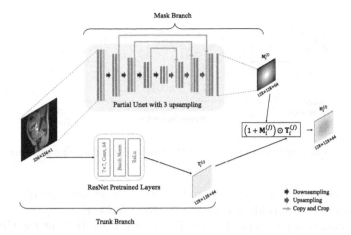

Fig. 2. The details of slice-level attention module with a two-branch structure.

ResNet18. For each grayscale MRI slice $\mathbf{X}_i^{(j)} \in \mathbb{R}^{256 \times 256 \times 1}$, a preliminary feature map $\mathbf{T}_i^{(j)} \in \mathbb{R}^{128 \times 128 \times 64}$ is generated. Note that to realize the spatial residual attention with the mask branch, shallow features are first derived since they can be aligned with the context features. Deep features will be extracted later after the spatial residual attention layer.

Mask Branch. The mask branch aims to extract the context information that indicates the location of ROI. Unlike the traditional implementation where a general bottom-up top-down network is employed, here the U-Net dedicated for medical image segmentation [14] is considered because the placenta area accounts for most of the ROI. Automatic segmentation of placenta with U-Net has proved effective in recent years [4]. The original network consists of 4 downsampling and 4 upsampling modules. To preserve the context information surrounding the placenta, a partial U-Net is constructed, as shown in Fig. 2. The last upsampling module to generate the single-channel mask ($\mathbb{R}^{256 \times 256 \times 1}$) is removed and the intermediate feature map $\mathbf{M}_i^{(j)} \in \mathbb{R}^{128 \times 128 \times 64}$ is obtained as the output of mask branch. Note that the entire U-Net is first pre-trained with a small number of placenta annotated MRI slices. Then the partial U-Net gets fine-tuned with a small learning rate in the end-to-end prediction task.

Spatial Residual Attention and Deep Feature Extraction. The mask branch output $\mathbf{M}_i^{(j)}$ has been aligned with the trunk branch output $\mathbf{T}_i^{(j)}$ in terms of the field of view. Therefore, $\mathbf{M}_i^{(j)}$ is multiplied in an element-wise manner with $\mathbf{T}_i^{(j)}$ as the attention mechanism. Besides, to avoid potential performance drop with attention, residual learning is further considered [18]. The shallow features after spatial residual attention become $\mathbf{H}_i^{(j)} = (1 + \mathbf{M}_i^{(j)}) \odot \mathbf{T}_i^{(j)}$. Note that the

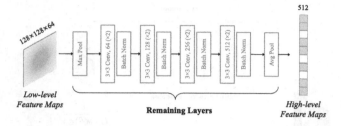

Fig. 3. Remaining layers of pre-trained ResNet18 except for the top three layers.

attention is not performed directly on deep features because the ROI considered has explicit physical meanings instead of being "hidden".

The remaining layers of pre-trained ResNet18 (excluding the top three layers used previously) are employed to further extract deep features, as shown in Fig. 3. These layers are fine-tuned in end-to-end learning so that task-specific deep features can be obtained. For each patient bag, a set of slice-wise deep features $\mathcal{F}_i = \{\mathbf{F}_i^{(1)}, \cdots, \mathbf{F}_i^{(K_i)}\}$ are eventually generated, where $\mathbf{F}_i^{(j)} \in \mathbb{R}^{512}$. These features capture the context-aware semantic information of MRI slices.

2.2 Patient-Level Attention-Based Pooling

Due to the MIL nature of the studied problem, MRI slices in a patient bag contribute unevenly to the PAS diagnosis. There are no detailed annotations to reflect the role of each slice. Therefore, patient-level attention mechanism is further proposed to aggregate the deep features from each slice in a task-specific manner, *i.e.*, deriving the patient-wise features $\mathbf{G}_i$ based on a set of slice-wise features $\mathcal{F}_i$. Some recent work [5,22] has proved that attention-based pooling is useful for feature aggregation in MIL problems. It shows certain similarities with channel attention which has been widely applied in image classification. Besides, such aggregators can be easily embedded in the network as a trainable layer. Concretely, the attention-based pooling layer works as follows:

$$\mathbf{G}_i = \sum_{j=1}^{K_i} a_j \mathbf{F}_i^{(j)}, \quad a_j = \frac{\exp\{\mathbf{w}^\top \tanh(\mathbf{V}\mathbf{F}_i^{(j)})\}}{\sum_{m=1}^{K_i} \exp\{\mathbf{w}^\top \tanh(\mathbf{V}\mathbf{F}_i^{(m)})\}}, \quad (1)$$

where the calculation of attention weight a_j involves trainable parameters $\mathbf{w} \in \mathbb{R}^{64 \times 1}$ and $\mathbf{V} \in \mathbb{R}^{64 \times 512}$. Such weight quantifies the degrees of "activation" toward the final prediction for each MRI slice. Note that the weight calculation resembles the softmax operation, and can deal with the varying number of slices for different patients unless the variation is not significant. The aggregated patient-wise features $\mathbf{G}_i$ are then mapped to an output vector $\hat{Y}_i$ with an FC layer. Binary Cross Entropy (BCE) is used as the classification loss function,

$$\mathcal{L}_e = -\frac{1}{N_b} \sum_{i=1}^{N_b} \Big(Y_i \log(\hat{Y}_i) + (1 - Y_i) \log(1 - \hat{Y}_i)\Big). \quad (2)$$

2.3 Contrastive Learning for Discriminative Deep Features

For the studied problem, the MRI slices of a positive patient (with PAS) that reflect the clues about PAS can account for only a small portion of all slices, while most ones look normal. Meanwhile, some slices of a negative patient (without PAS) can manifest suspected patterns. Therefore, the extracted patient-wise deep features should be more discriminative under the existence of inherent noises. In addition to aggregating the slice-wise features with attention, contrastive learning is further considered to realize that [2]. Specifically, the representation $\mathbf{G}_i$ needs to keep certain distances between positive and negative patients while getting relatively closer for patients in the same class. Since the proposed network does not require self-supervised learning, a plug-and-play contrastive loss is employed to avoid additional capacity.

Concretely, to make up each batch, a positive patient and a negative patient are first sampled from the dataset. Then another patient is randomly picked up. It is thus guaranteed that a matching pair and a non-matching pair exist, respectively. Assume the third patient is positive, the patient-wise features are denoted as $\mathbf{G}_{i_a}, \mathbf{G}_{i_n}, \mathbf{G}_{i_p}$. The triplet loss [1] is calculated between these feature maps ($\mathbf{G}_{i_a}$ is the anchor, $\mathbf{G}_{i_n}$ is the non-matching sample and $\mathbf{G}_{i_p}$ is the matching sample). The triplet loss is used as the contrastive loss:

$$\mathcal{L}_c = \max\left\{\|\mathbf{G}_{i_a} - \mathbf{G}_{i_p}\|_2 - \|\mathbf{G}_{i_a} - \mathbf{G}_{i_n}\|_2 + c, 0\right\} \tag{3}$$

where c is the positive margin constant that quantifies the distance. It is added to the BCE losses of the samples in a batch: $\mathcal{L}_{\text{total}} = \mathcal{L}_e + \mathcal{L}_c$.

3 Experiment

3.1 Dataset and Implementation Details

The experiments are conducted on a real clinical dataset collected from a large obstetrics and gynecology hospital. The dataset involves 359 distinct patient subjects with a total of 15,186 MRI slices. A 5-fold cross-validation is performed and the average metrics values with standard deviations over the folds are reported. To pre-train the U-Net in the mask branch, 185 MRI slices are annotated with placenta by the experts. The ResNet18 is pre-trained on ImageNet. HACL-Net is trained end to end with Adam optimizer and batch size $N_b = 3$ for 100 epochs. The learning rate of the slice-level attention module and final FC layer is set as 10^{-6} and 10^{-4} respectively, while that of the rest parts is set as 10^{-5}. PyTorch 1.10.1, and an NVIDIA GeForce RTX 2080 Ti GPU with CUDA 11.3 are used to train our method. More details can be found in Section B of supplementary. The code is publicly available on https://github.com/LouieBHLu/HACL-Net.

3.2 Performance Comparison of PAS Diagnosis

Section C of supplementary material illustrates the effect of different selections of trunk branch. Table 1 shows the performance of the pre-trained baseline U-Net

Table 1. Performance of pre-trained ROI networks. PA is pixel accuracy.

Method	Train			Test		
	DICE	IOU	PA	DICE	IOU	PA
U-Net	80.6 ± 0.5	89.2 ± 0.3	78.3 ± 0.9	77.1 ± 0.9	87.0 ± 0.5	72.4 ± 1.2
MaNet	82.2 ± 0.6	90.2 ± 0.3	79.8 ± 0.7	80.4 ± 1.0	89.1 ± 0.6	76.2 ± 1.0
FPN	85.4 ± 0.5	92.1 ± 0.4	81.7 ± 0.7	80.7 ± 0.9	89.3 ± 0.4	72.8 ± 0.9
LinkNet	79.5 ± 0.3	88.5 ± 0.2	77.0 ± 0.7	76.8 ± 0.6	86.9 ± 0.4	72.8 ± 1.1

Table 2. Performance comparison of HACL-Net and existing methods for PAS diagnosis. The results in bold denote the best performance.

Method	Accuracy	AUC	Sensitivity	Specificity
Radio+RF	70.1 ± 5.4	61.8 ± 4.3	15.4 ± 4.2	93.9 ± 4.2
Radio+NB	69.9 ± 4.0	48.1 ± 3.9	0.0 ± 0.0	**100.0 ± 0.0**
Radio+KNN	68.2 ± 5.7	60.2 ± 3.4	24.9 ± 8.3	87.0 ± 3.2
Radio+MLP	70.4 ± 6.4	61.7 ± 6.3	26.7 ± 8.8	89.4 ± 4.5
Xuan [20]	65.7 ± 1.2	51.4 ± 1.0	1.7 ± 2.4	91.3 ± 2.5
Ye [23]	55.7 ± 6.5	46.7 ± 1.4	35.0 ± 8.2	64.0 ± 12.3
HACL-Net	**84.2 ± 0.9**	**84.9 ± 0.6**	**79.2 ± 5.9**	86.9 ± 2.8

compared with other three common networks. It turns out such network choice does not have a significant impact. The placenta area can be accurately captured by U-Net, which justifies the usage of Partial U-Net in the mask branch. Visual ROI prediction can be found in Section D of supplementary material.

To evaluate the performance of machine learning methods with Pyradiomics features and deep learning methods [12,20,23], the slice-wise features are aggregated with simple mean to derive the patient-wise features. The accuracy, area under the ROC curve (AUC), sensitivity, and specificity are employed as the metrics. As shown in Table 2, HACL-Net achieves the best 84.9% AUC and 79.2% sensitivity with 23.1% and 44.2% improvement respectively over the second-best approach. The outcomes are reflected in Fig. 4, where the compared methods suffer from very low sensitivity and high specificity. Since only a small portion of MRI slices have clues about a patient's being positive, traditional aggregation of slices tends to dilute the effective information. As the features are extracted from a segmentation network in [23], the sensitivity is relatively improved but with highly degraded specificity. In general deep features even perform worse than ROI-based radiomics features, partially due to the increased complexity of input is not properly explained by the clinical outcome of PAS.

3.3 Ablation Study

Table 3 and Fig. 5 illustrate the effect of each module in HACL-Net. For experiments without patient-level attention, slice-wise features are also aggregated

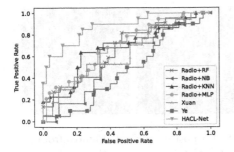

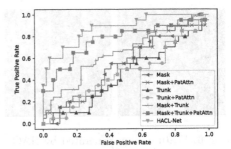

Fig. 4. ROC of HACL-Net and existing methods for PAS diagnosis.

Fig. 5. ROC of ablation study of each component of HACL-Net. PatAttn is patient-level attention module.

Table 3. Results of ablation study for each module in the proposed HACL-Net.

Mask Branch	Trunk Branch	Patient Attention	CL	Accuracy	AUC	Sensitivity	Specificity
✓				60.4 ± 13.5	50.1 ± 1.3	26.0 ± 32.2	74.1 ± 31.7
✓		✓		71.6 ± 0.2	50.0 ± 1.5	7.5 ± 2.5	94.1 ± 2.0
	✓			71.6 ± 1.4	51.5 ± 1.5	5.0 ± 3.5	98.0 ± 2.4
	✓	✓		70.7 ± 2.7	53.9 ± 5.2	12.5 ± 10.9	95.2 ± 4.4
✓	✓			77.7 ± 0.6	66.9 ± 1.5	35.0 ± 4.1	94.7 ± 1.0
✓	✓	✓		80.7 ± 4.1	79.2 ± 5.2	62.9 ± 1.9	**98.7 ± 1.3**
✓	✓	✓	✓	**84.2 ± 0.9**	**84.9 ± 0.6**	79.2 ± 5.9	86.9 ± 2.8
Sigmoid w/ 0.50 threshold				66.0 ± 9.6	60.9 ± 8.2	38.7 ± 4.1	84.8 ± 12.8
Sigmoid w/ 0.45 threshold				68.3 ± 9.8	68.1 ± 8.4	69.2 ± 6.6	67.0 ± 13.3
Sigmoid w/ 0.40 threshold				78.0 ± 8.0	82.1 ± 4.4	77.0 ± 10.1	81.3 ± 14.0
Sigmoid w/ 0.35 threshold				77.8 ± 8.4	81.6 ± 8.2	75.6 ± 0.9	78.5 ± 15.2
Sigmoid w/ 0.30 threshold				70.6 ± 3.1	66.3 ± 7.0	**87.8 ± 8.7**	49.7 ± 13.8

with simple mean to derive patient-level features. The combination of mask branch and trunk branch achieves 15.4% AUC improvement over a single trunk branch, indicating the gain of additional context information. A 12.3% AUC improvement is witnessed when patient-level attention is used as an aggregator instead of mean. Contrastive learning manifests 5.7% AUC and 16.3% sensitivity improvement with 11.8% specificity drop as a trade-off. It is believed that sensitivity can be more important from a clinical perspective, and the trade-off can be tuned according to specific demands. Moreover, the softmax activation is replaced with sigmoid, and different confidence thresholds are used. As the threshold gets lower, a non-monotonic trend of sens-spec trade-off is observed, while the optimal balance lies around 0.4. That reveals the potential effect of class imbalance, *i.e.*, fewer positive cases. To this end, contrastive learning can help separate the decision boundary between two classes. Sigmoid results differ from softmax due to the explicitly set decision threshold. The predicted scores between positive and negative patients can be close for hard samples, leading to large deviations in sigmoid results. Figure 6 shows feature maps of a positive

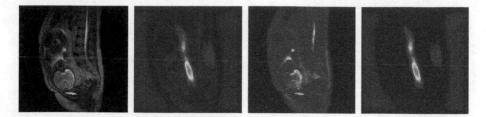

Fig. 6. Feature maps generated by HACL-Net of a positive patient. Figures from left to right are the MRI slice, the spatial residual attention feature map, the trunk branch feature map, and the mask branch feature map.

patient generated by HACL-Net. Spatial residual attention features capture both local focal area information without omitting global semantic context features compared to features of mask branch and trunk branch.

4   Conclusion

This paper presents a novel end-to-end HACL-Net for MRI-based computer-aided PAS diagnosis that facilitates more efficient prenatal screening. The proposed network utilizes slice-level spatial residual attention to extract context-aware deep semantic features, and aggregate them as a comprehensive representation of a patient with patient-level attention-based pooling. Moreover, a contrastive loss is added to improve the discriminating power of learned patient-wise features. Extensive experiments on a real clinical dataset show that HACL-Net achieves superior performance, with great potential for clinical applications.

Acknowledgement. This research was supported by Dr. Hao Zhu's team from Obstetrics and Gynecology Hospital affiliated to Fudan University in China.

References

1. Balntas, V., Riba, E., Ponsa, D., Mikolajczyk, K.: Learning local feature descriptors with triplets and shallow convolutional neural networks. In: Proceedings of the British Machine Vision Conference (BMVC), pp. 119.1-119.11 (2016)
2. Chopra, S., Hadsell, R., LeCun, Y.: Learning a similarity metric discriminatively, with application to face verification. In: 2005 IEEE Computer Society Conference on Computer Vision and Pattern Recognition (CVPR 2005), vol. 1, pp. 539–546 (2005)
3. Cummins, C., Petoumenos, P., Wang, Z., Leather, H.: End-to-end deep learning of optimization heuristics. In: 2017 26th International Conference on Parallel Architectures and Compilation Techniques (PACT), pp. 219–232. IEEE (2017)
4. Han, M., et al.: Automatic segmentation of human placenta images with u-net. IEEE Access 7, 180083–180092 (2019)
5. Ilse, M., Tomczak, J., Welling, M.: Attention-based deep multiple instance learning. In: International Conference on Machine Learning, pp. 2127–2136. PMLR (2018)

6. Jiao, M., Liu, H., Liu, J., Ouyang, H., Wang, X., Jiang, L., Yuan, H., Qian, Y.: Mal: Multi-modal attention learning for tumor diagnosis based on bipartite graph and multiple branches. In: Wang, L., Dou, Q., Fletcher, P.T., Speidel, S., Li, S. (eds.) MICCAI 2022. LNCS, vol. 13433, pp. 175–185. Springer, Cham (2022). https://doi.org/10.1007/978-3-031-16437-8_17

7. Kohli, M., Prevedello, L.M., Filice, R.W., Geis, J.R.: Implementing machine learning in radiology practice and research. Am. J. Roentgenol. **208**(4), 754–760 (2017)

8. Li, H., Chen, L., Han, H., Kevin Zhou, S.: Satr: slice attention with transformer for universal lesion detection. In: Wang, L., Dou, Q., Fletcher, P.T., Speidel, S., Li, S. (eds.) MICCAI 2022. LNCS, pp. 163–174. Springer Nature Switzerland, Cham (2022)

9. Liu, X., et al.: What we know about placenta accreta spectrum (PAS). Eur. J. Obstet. Gynecol. Reprod. Biol. **259**, 81–89 (2021)

10. Oyelese, Y., Smulian, J.C.: Placenta previa, placenta accreta, and vasa previa. Obstet. Gynecol. **107**(4), 927–941 (2006)

11. Ren, H., et al.: Prediction of placenta accreta spectrum using texture analysis on coronal and sagittal T2-weighted imaging. Abdom. Radiol. **46**, 5344–5352 (2021)

12. Romeo, V., et al.: Machine learning analysis of MRI-derived texture features to predict placenta accreta spectrum in patients with placenta previa. Magn. Reson. Imaging **64**, 71–76 (2019)

13. Romeo, V., et al.: Prediction of placenta accreta spectrum in patients with placenta previa using clinical risk factors, ultrasound and magnetic resonance imaging findings. Radiol. Med. (Torino) **126**(9), 1216–1225 (2021). https://doi.org/10.1007/s11547-021-01348-6

14. Ronneberger, O., Fischer, P., Brox, T.: U-net: convolutional networks for biomedical image segmentation. In: Navab, N., Hornegger, J., Wells, W.M., Frangi, A.F. (eds.) MICCAI 2015. LNCS, vol. 9351, pp. 234–241. Springer, Cham (2015). https://doi.org/10.1007/978-3-319-24574-4_28

15. Silver, R.M., Lyell, D.J.: Placenta accreta spectrum. Protocols for High-Risk Pregnancies: an evidence-based approach, pp. 571–580 (2020)

16. Tian, Y., et al.: Contrastive transformer-based multiple instance learning for weakly supervised polyp frame detection. In: Wang, L., Dou, Q., Fletcher, P.T., Speidel, S., Li, S. (eds.) MICCAI 2022. LNCS, pp. 88–98. Springer Nature Switzerland, Cham (2022). https://doi.org/10.1007/978-3-031-16437-8_9

17. Van Griethuysen, J.J., et al.: Computational radiomics system to decode the radiographic phenotype. Can. Res. **77**(21), e104–e107 (2017)

18. Wang, F., et al.: Residual attention network for image classification. In: Proceedings of the IEEE Conference on Computer Vision and Pattern Recognition, pp. 3156–3164 (2017)

19. Wang, S., et al.: RMDL: recalibrated multi-instance deep learning for whole slide gastric image classification. Med. Image Anal. **58**, 101549 (2019)

20. Xuan, R., Li, T., Wang, Y., Xu, J., Jin, W.: Prenatal prediction and typing of placental invasion using MRI deep and radiomic features. Biomed. Eng. Online **20**(1), 56 (2021)

21. Yao, J., Zhu, X., Huang, J.: Deep multi-instance learning for survival prediction from whole slide images. In: Shen, D., et al. (eds.) MICCAI 2019. LNCS, vol. 11764, pp. 496–504. Springer, Cham (2019). https://doi.org/10.1007/978-3-030-32239-7_55

22. Yao, J., Zhu, X., Jonnagaddala, J., Hawkins, N., Huang, J.: Whole slide images based cancer survival prediction using attention guided deep multiple instance learning networks. Med. Image Anal. **65**, 101789 (2020)

23. Ye, Z., Xuan, R., Ouyang, M., Wang, Y., Xu, J., Jin, W.: Prediction of placenta accreta spectrum by combining deep learning and radiomics using t2wi: a multi-center study. Abdominal Radiol. **47**(12), 4205–4218 (2022)
24. Zhang, D., Zou, L., Zhou, X., He, F.: Integrating feature selection and feature extraction methods with deep learning to predict clinical outcome of breast cancer. IEEE Access **6**, 28936–28944 (2018)
25. Zhang, Y., et al.: Spatiotemporal attention for early prediction of hepatocellular carcinoma based on longitudinal ultrasound images. In: Wang, L., Dou, Q., Fletcher, P.T., Speidel, S., Li, S. (eds.) MICCAI 2022. LNCS, vol. 13433, pp. 534–543. Springer, Cham (2022)

UM-CAM: Uncertainty-weighted Multi-resolution Class Activation Maps for Weakly-supervised Fetal Brain Segmentation

Jia Fu[1], Tao Lu[2], Shaoting Zhang[1,3], and Guotai Wang[1,3(✉)]

[1] University of Electronic Science and Technology of China, Chengdu 611731, China
[2] Sichuan Provincial People's Hospital, Chengdu 610072, China
[3] Shanghai Artificial Intelligence Laboratory, Shanghai 200030, China
guotai.wang@uestc.edu.cn

Abstract. Accurate segmentation of the fetal brain from Magnetic Resonance Image (MRI) is important for prenatal assessment of fetal development. Although deep learning has shown the potential to achieve this task, it requires a large fine annotated dataset that is difficult to collect. To address this issue, weakly-supervised segmentation methods with image-level labels have gained attention, which are commonly based on class activation maps from a classification network trained with image-level labels. However, most of these methods suffer from incomplete activation regions, due to the low-resolution localization without detailed boundary cues. To this end, we propose a novel weakly-supervised method with image-level labels based on semantic features and context information exploration. We first propose an Uncertainty-weighted Multi-resolution Class Activation Map (UM-CAM) to generate high-quality pixel-level supervision. Then, we design a Geodesic distance-based Seed Expansion (GSE) method to provide context information for rectifying the ambiguous boundaries of UM-CAM. Extensive experiments on a fetal brain dataset show that our UM-CAM can provide more accurate activation regions with fewer false positive regions than existing CAM variants, and our proposed method outperforms state-of-the-art weakly-supervised segmentation methods learning from image-level labels.

Keywords: Weakly-supervised segmentation · Class activation map · Geodesic distance · Fetal MRI

1 Introduction

Brain extraction is the first step in fetal Magnetic Resonance Image (MRI) analysis in advanced applications such as brain tissue segmentation [15] and quantitative measurement [22,24], which is essential for assessing fetal brain development and investigate the neuroanatomical correlation of cognitive impairments [14]. Current research based on Convolutional Neural Network (CNN) [5,19] has achieved promising performance for automatic fetal brain extraction from pixel-wise annotated fetal MRI. However, it is labor-intensive, time-consuming, and

H. Greenspan et al. (Eds.): MICCAI 2023, LNCS 14226, pp. 315–324, 2023.
https://doi.org/10.1007/978-3-031-43990-2_30

expensive to collect a large-scale pixel-wise annotated dataset, especially for images with poor quality and large variations. To address these issues, weakly-supervised segmentation methods with image-level supervision [21] are introduced due to their minimal annotation demand. However, learning from image-level supervision is extremely challenging since the image-level label only provides the existence of object class, but cannot indicate the information about location and shape that are essential for the segmentation task [1].

Prevailing methods learning from image-level labels for segmentation commonly produce a coarse localization of the objects based on Class Activation Maps (CAM) [27]. Due to the weak annotation, the CAMs from the classification network can only provide rough localization and coarse boundaries of objects. To alleviate the problem, a lot of approaches have been proposed, which can be categorized as one-stage and two-stage methods. One-stage methods aim to generate pixel-level segmentation by training a segmentation branch simultaneously with a classification network. For example, Reliable Region Mining (RRM) [26] comprises two parallel branches, in which pixel-level pseudo masks are produced from the classification branch and refined by Conditional Random Field (CRF) to supervise the segmentation branch. Despite their efficiency, one-stage methods commonly achieve inferior segmentation accuracy and incomplete activation of targets, owing to the failure to capture detailed contextual information from image-level labels [2,7].

In contrast to one-stage methods, two-stage methods can perform favourably, as they leverage dense labels generated by the classification network to train a segmentation network [12]. For instance, Discriminative Region Suppression (DRS) [9] suppresses the attention on discriminative regions and expands it to adjacent less activated regions. However, these methods leverage the CAMs from the deep layer of the classification network and raise the inherent drawback, i.e., low resolution, leading to limited localization and smooth boundaries of objects. Han et al. [8] proposed multi-layer pseudo supervision to reduce the false positive rate in segmentation results, while the weights for pseudo masks from different layers are constants that cannot be adaptive. Besides, though the quality of CAMs improves, they are still insufficient to provide accurate object boundaries for segmentation. Numerous methods [6,10,11] have been proposed to explore boundary information. For example, Kolesnikov et al. [10] proposed a joint loss function that constrains the global weighted rank pooling and low-level object boundary to expand activation regions. AffinityNet [1] trains another network to learn the semantic similarity between pixels and then propagates the semantics to adjacent pixels via random walk. Nevertheless, these methods use the initial seeds generated from the CAM method, resulting in limited performance when the object-related seeds from CAM are small and sparse. Thus, improving the initial prediction and exploring boundary information are both important for accurate object segmentation.

In this work, we propose a novel weakly-supervised method for accurate fetal brain segmentation using image-level labels. Our contribution can be summarized as follows: 1) We design an Uncertainty-weighted Multi-resolution CAM

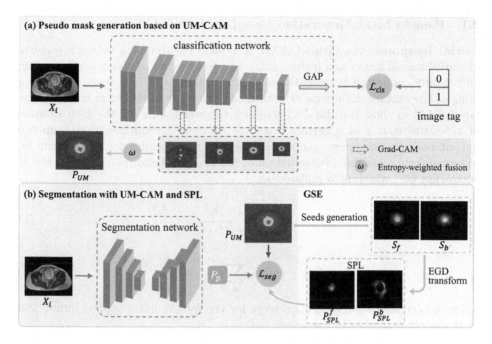

Fig. 1. Overview of the proposed method. (a) Uncertainty-weighted Multi-resolution CAM (UM-CAM) obtained by a classification network, (b) Segmentation model trained with UM-CAM and auxiliary supervision from Seed-derived Pseudo Label (SPL).

(UM-CAM) to integrate low- and high-resolution CAMs via entropy weighting, which can leverage the semantic features extracted from the classification network adaptively and eliminate the noise effectively. 2) We propose a Geodesic distance-based Seed Expansion (GSE) method to generate Seed-derived Pseudo Labels (SPLs) that can provide boundary cues for training a better segmentation model. 3) Extensive experiments conducted on a fetal brain dataset demonstrate the effectiveness of the proposed method, which outperforms several state-of-the-art approaches for learning from image-level labels.

2   Method

An overview of our method is presented in Fig. 1. First, to obtain high-quality pseudo masks, Uncertainty-weighted Multi-resolution CAM (UM-CAM) is produced by fusing low- and high-resolution CAMs from different layers of the classification network. Second, seed points are obtained from the UM-CAM automatically, and used to generate Seed-derived Pseudo Labels (SPL) via Geodesic distance-based Seed Expansion (GSE). The SPL provides more detailed context information in addition to UM-CAM for training the final segmentation model.

2.1 Psuedo Mask Generation Based on UM-CAM

Initial Response via Grad-CAM. A typical classification network consists of convolutional layers as a feature extractor, followed by global average pooling and a fully connected layer as the output classifier [3]. Given a set of training images, the classification network is trained with class labels. After training, the Grad-CAM method is utilized to compute the weights α_k for the k-th channel of a feature map f at a certain layer via gradient backpropagation from the output node for the foreground class. The foreground activation map A can be obtained from a weighted combination of feature maps and followed by a ReLU activation [20], which is formulated as:

$$\alpha_k = \frac{1}{N} \sum_{i=1}^{N} \frac{\partial y}{\partial f_k(i)}, \tag{1}$$

$$A(i) = ReLU(\sum_k \alpha_k f_k(i)), \tag{2}$$

where y is classification prediction score for the foreground. i is pixel index, and N is the pixel number in the image.

Multi-resolution Exploration and Integration. The localization map for each image typically provides discriminative object parts, which is insufficient to provide supervision for the segmentation task. As shown in Fig. 1(a), the activation maps generated from the shallow layers of the classification network contain high-resolution semantic features but suffer from noisy and dispersive localization. In contrast, activation maps generated from the deeper layers perform smoother localization but lack high-resolution information. To take advantage of activation maps from shallow and deep layers, we proposed UM-CAM to integrate the multi-resolution CAMs by uncertainty weighting. Let us denote a set of activation maps from M convolutional blocks as $\mathcal{A} = \{A_m\}_{m=0}^{M}$. Each activation map is interpolated to the input size and normalized by its maximum to the range of [0,1], and the normalized activation maps are $\hat{\mathcal{A}} = \{\hat{A}_m\}_{m=0}^{M}$. To minimize the uncertainty of pseudo mask, UM-CAM integrates the confident region of multi-resolution CAMs adaptively, which can be presented as the entropy-weighted combination of CAMs:

$$w_m(i) = 1 - (- \sum_{j=(b,f)} \hat{A}_m^j(i) log \hat{A}_m^j(i)), \tag{3}$$

$$P_{UM}(i) = \frac{\sum_m w_m(i) \hat{A}_m(i)}{\sum_m w_m(i)}, \tag{4}$$

where $\hat{A}_m^b$ and $\hat{A}_m^f$ represent the background and foreground probability of $\hat{A}_m$, respectively. w_m is the weight map for $\hat{A}_m$, and P_{UM} is the UM-CAM for the target.

2.2 Robust Segmentation with UM-CAM and SPL

Though UM-CAM is better than the CAM from the deep layer of the classification network, it is still insufficient to provide accurate object boundaries that are important for segmentation. Motivated by [13], we propose a Geodesic distance-based Seed Expansion (GSE) method to generate Seed-derived Pseudo Label (SPL) that contains more detailed context information. The SPL is combined with UM-CAM to supervise the segmentation model, as shown in Fig. 1 (b).

Concretely, we adopt the centroid and the corner points of the bounding box obtained from UM-CAM as the foreground seeds S_f and background seeds S_b, respectively. To efficiently leverage these seed points, SPL is generated via Exponential Geodesic Distance (EGD) transform of the seeds, leading to a foreground cue map P_{SPL}^f and a background cue map P_{SPL}^b. The values of P_{SPL}^b and P_{SPL}^f represent the similarity between each pixel and background/foreground seed points, which can be computed as:

$$P_{SPL}^b(i) = e^{-\alpha \cdot D_b(i)}, \quad P_{SPL}^f(i) = e^{-\alpha \cdot D_f(i)}, \tag{5}$$

$$D_b(i) = \min_{j \in S_b} D_{geo}(i, j, I), \quad D_f(i) = \min_{j \in S_f} D_{geo}(i, j, I), \tag{6}$$

$$D_{geo}(i, j, I) = \min_{p \in P_{i,j}} \int_0^1 \|\nabla I(p(n)) \cdot u(n)\| \, dn \tag{7}$$

where $P_{i,j}$ is the set of all paths between pixels i and j. $D_b(i)$ and $D_f(i)$ represent the minimal geodesic distance between target pixel i and background/foreground seed points, respectively. p is one feasible path and it is parameterized by $n \in [0, 1]$. $u(n) = p'(n) / \|p'(n)\|$ is a unit vector that is tangent to the direction of the path.

Based on the supervision from UM-CAM and SPL, the segmentation network can be trained by minimizing the following joint object function:

$$L_{seg} = \lambda L_{CE}(P_p, P_{UM}) + (1 - \lambda)L_{CE}(P_p, P_{SPL}). \tag{8}$$

where P_p is the prediction of the segmentation network, and λ is a weight factor to balance the supervision of UM-CAM and SPL. L_{CE} is the Cross-Entropy (CE) loss.

3 Experiments and Results

3.1 Experimental Details

Dataset. We collected clinical T2-weighted MRI data of 115 pregnant women in the second trimester with Single-shot Fast Spin-echo (SSFSE). The data were acquired in axial view with pixel size between 0.5547 mm × 0.5547 mm and 0.6719 mm × 0.6719 mm and slice thickness between 6.50 mm and 7.15 mm. Each slice was resampled to a uniform pixel size of 1 mm × 1 mm. In all experiments,

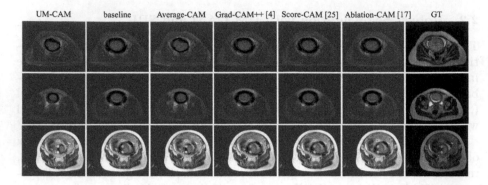

Fig. 2. Visual comparison between CAMs obtained by different methods.

we used 80 volumes with 976 positive and 3140 negative slices for training, 10 volumes with 116 positive and 408 negative slices for validation, and 25 volumes with 318 positive and 950 negative slices for testing. Positive and negative slice mean containing the brain and not, respectively. The ground truth was manually annotated by radiologists. Note that we used image-level labels for training and pixel-level ground truth for validation and testing.

Implementation Details. To boost the generalizability, we applied spatial and intensity-based data augmentation during the training stage, including gamma correction, random rotation, random flipping, and random cropping. For 2D classification, we employed VGG-16 [23] as backbone architecture, in which an additional convolutional layer is used to substitute the last three fully connected layers. The classification network was trained with 200 epochs using CE loss. Stochastic Gradient Descent (SGD) optimizer was used for training with batch size 32, momentum 0.99, and weight decay 5×10^{-4}. The learning rate was initialized to 1×10^{-3}. We used UNet [18] as the segmentation network. The learning rate was set as 0.01, and the SGD optimizer was used for training with 200 epochs, batch size 12, momentum 0.9, and weight decay 5×10^{-4}. The hyper-parameter M and λ were set as 4 and 0.1 based on the best results on the validation set, respectively. We used Dice Similarity Coefficient (DSC) and 95% Hausdorff Distance (HD_{95}) to evaluate the quality of 2D pseudo masks and the final segmentation results in 3D space.

3.2   Ablation Studies

Stage1: Quality of Pseudo Masks Obtained by UM-CAM. To evaluate the effectiveness of UM-CAM, we compared different pseudo mask generation strategies: 1) Grad-CAM (baseline): only using CAMs from the last layer of the classification network generated by using Grad-CAM method [20], 2) Average-CAM: fusing multi-resolution CAMs via averaging, 3) UM-CAM: fusing multi-resolution CAMs via uncertainty weighting. Table 1 lists the quantitative

Table 1. Ablation study on the validation set to validate the effectiveness of UM-CAM and SPL. * denotes p-value < 0.05 when comparing with the second place method.

Method		DSC (%)	HD_{95} (pixels)
Pseudo mask generation	Grad-CAM (baseline)	74.48 ± 13.26	39.88 ± 14.66
	Average-CAM	77.40 ± 13.28	38.70 ± 19.75
	UM-CAM	$\mathbf{79.00 \pm 12.83*}$	$\mathbf{35.33 \pm 16.98}$
Segmentation	Grad-CAM (baseline)	78.69 ± 10.02	22.17 ± 16.60
	UM-CAM	85.22 ± 6.62	5.26 ± 4.23
	SPL	89.05 ± 4.30	3.84 ± 4.07
	UM-CAM+SPL	$\mathbf{89.76 \pm 5.09*}$	$\mathbf{3.10 \pm 2.61}$

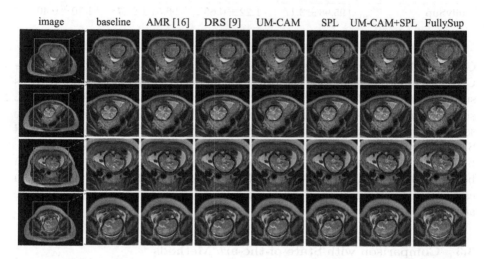

Fig. 3. Visual comparison of our method and other weakly-supervised segmentation methods. The green and red contours indicate the boundaries of ground truths and segmentation results, respectively. (Color figure online)

evaluation results of these methods, in which the segmentation is converted from CAMs using the optimal threshold found by grid search method. It can be seen that when fusing the information from multiple convolutional layers, the quality of pseudo masks improves. The proposed UM-CAM improves the average DSC by 4.52% and 1.60% compared with the baseline and Average-CAM, respectively. Figure 2 shows a visual comparison between CAMs obtained by the different methods. It can be observed that there are fewer false positive activation regions of UM-CAM compared with the other methods.

Stage2: Training Segmentation Model with UM-CAM and SPL. To investigate the effectiveness of SPL, we compared it with several segmentation models: 1) Grad-CAM (baseline): only using the pseudo mask generated from

Table 2. Comparison between ours and existing weakly-supervised segmentation methods. * denotes p-value < 0.05 when comparing with the second place weakly supervised method.

Method	Validation set		Test set	
	DSC (%)	HD_{95} (pixels)	DSC (%)	HD_{95} (pixels)
Grad-CAM++ [4]	74.52 ± 13.29	39.87 ± 14.69	76.60 ± 10.58	37.49 ± 11.72
Score-CAM [25]	74.49 ± 13.35	39.88 ± 14.83	76.58 ± 10.60	37.54 ± 11.73
Ablation-CAM [17]	74.56 ± 13.29	39.76 ± 14.64	76.55 ± 10.61	37.57 ± 11.82
AMR [16]	78.77 ± 8.83	11.53 ± 9.82	79.79 ± 6.86	11.35 ± 8.02
DRS [9]	84.98 ± 5.62	7.17 ± 8.01	83.79 ± 7.81	7.06 ± 5.13
UM-CAM+SPL (ours)	$\mathbf{89.76 \pm 5.09^*}$	$\mathbf{3.10 \pm 2.61^*}$	$\mathbf{90.22 \pm 3.75^*}$	$\mathbf{4.04 \pm 4.26^*}$
FullySup	95.98 ± 3.17	1.22 ± 0.65	96.51 ± 2.67	1.10 ± 0.40

Grad-CAM to train the segmentation model, 2) UM-CAM: only using UM-CAM as supervision for the segmentation model, 3) SPL: only using SPL as supervision, 4) UM-CAM+SPL: our proposed method using UM-CAM and SPL supervision for the segmentation model. Quantitative evaluation results in the second section of Table 1 show that the network trained with UM-CAM and SPL supervision achieves an average DSC score of 89.76%, improving the DSC by 11.07% compared to the baseline model. Figure 3 depicts a visual comparison between these models. It shows that SPL supervision with context information can better discriminate the fetal brain from the background, leading to more accurate boundaries for segmentation.

3.3   Comparison with State-of-the-art Methods

We compared three CAM invariants with our UM-CAM in pseudo mask generation stage, including Grad-CAM++ [4], Score-CAM [25], and Ablation-CAM [17]. Table 2 shows the quantitative results of these CAM variants. It can be seen that GradCAM++, Score-CAM, and Ablation-CAM achieve similar performance, which is consistent with the visualization results shown in Fig. 2. The proposed UM-CAM achieves higher accuracy than existing CAM variants, which generates more accurate boundaries that are closer to the ground truth.

We further compared the proposed method with fully supervised method (FullySup) and two state-of-the-art weakly-supervised methods, including DRS [9] that spreads the attention to adjacent non-discriminative regions by suppressing the attention on discriminative regions, and AMR [16] that incorporates a spotlight branch and a compensation branch to dig out more complete object regions. Table 2 lists the segmentation results of these methods. Our proposed method achieves an average DSC of 90.22% and an average HD_{95} of 4.04 pixels, which is at least 3.02 pixels lower than the other weakly-supervised methods on the testing set. It indicates that the proposed method can generate segmentation results with more accurate boundaries. Figure 3 shows some qualitative visualization results. The DRS and AMR predictions appear to be coarse and inaccurate in the boundary regions, while our proposed method generates more accurate segmentation results, even similar to those generated from the fully supervised model for some easy samples.

4 Conclusion

In this paper, we presented an uncertainty and context-based method for fetal brain segmentation using image-level supervision. An Uncertainty-weighted Multi-resolution CAM (UM-CAM) was proposed to integrate multi-resolution activation maps via uncertainty weighting to generate high-quality pixel-wise supervision. We proposed a Geodesic distance-based Seed Expansion (GSE) method to produce Seed-derived Pseudo Labels (SPL) containing detailed context information. The SPL is combined with UM-CAM for training the segmentation network. The proposed method was evaluated on the fetal brain segmentation task, and experimental results demonstrated the effectiveness of the proposed method and suggested the potential of our proposed method for obtaining accurate fetal brain segmentation with low annotation cost. In the future, it is of interest to validate our method with other segmentation tasks and apply it to other backbone networks.

Acknowledgement. This work was supported by the National Natural Science Foundation of China (62271115).

References

1. Ahn, J., Kwak, S.: Learning pixel-level semantic affinity with image-level supervision for weakly supervised semantic segmentation. In: CVPR, pp. 4981–4990 (2018)
2. Araslanov, N., Roth, S.: Single-stage semantic segmentation from image labels. In: CVPR, pp. 4253–4262 (2020)
3. Chang, Y.T., Wang, Q., Hung, W.C., Piramuthu, R., Tsai, Y.H., Yang, M.H.: Weakly-supervised semantic segmentation via sub-category exploration. In: CVPR, pp. 8991–9000 (2020)
4. Chattopadhay, A., Sarkar, A., Howlader, P., Balasubramanian, V.N.: Grad-CAM++: generalized gradient-based visual explanations for deep convolutional networks. In: WACV, pp. 839–847 (2018)
5. Ebner, M.: An automated framework for localization, segmentation and super-resolution reconstruction of fetal brain MRI. Neuroimage **206**, 116324 (2020)
6. Fan, J., Zhang, Z., Song, C., Tan, T.: Learning integral objects with intra-class discriminator for weakly-supervised semantic segmentation. In: CVPR, pp. 4283–4292 (2020)
7. Gao, W., et al.: TS-CAM: Token semantic coupled attention map for weakly supervised object localization. In: ICCV, pp. 2886–2895 (2021)
8. Han, C., et al.: Multi-layer pseudo-supervision for histopathology tissue semantic segmentation using patch-level classification labels. Med. Image Anal. **80**, 102487 (2022)

9. Kim, B., Han, S., Kim, J.: Discriminative region suppression for weakly-supervised semantic segmentation. In: AAAI. vol. 35, pp. 1754–1761 (2021)

10. Kolesnikov, A., Lampert, C.H.: Seed, expand and constrain: Three principles for weakly-supervised image segmentation. In: ECCV, pp. 695–711 (2016)

11. Lee, S., Lee, M., Lee, J., Shim, H.: Railroad is not a train: Saliency as pseudo-pixel supervision for weakly supervised semantic segmentation. In: CVPR, pp. 5495–5505 (2021)

12. Li, Y., Kuang, Z., Liu, L., Chen, Y., Zhang, W.: Pseudo-mask matters in weakly-supervised semantic segmentation. In: ICCV, pp. 6964–6973 (2021)

13. Luo, X., et al.: MIDeepSeg: minimally interactive segmentation of unseen objects from medical images using deep learning. Med. Image Anal. 72, 102102 (2021)

14. Makropoulos, A., Counsell, S.J., Rueckert, D.: A review on automatic fetal and neonatal brain MRI segmentation. Neuroimage 170, 231–248 (2018)

15. Makropoulos, A., et al.: Automatic whole brain MRI segmentation of the developing neonatal brain. IEEE Trans. Med. Imaging 33(9), 1818–1831 (2014)

16. Qin, J., Wu, J., Xiao, X., Li, L., Wang, X.: Activation modulation and recalibration scheme for weakly supervised semantic segmentation. In: AAAI, vol. 36, pp. 2117–2125 (2022)

17. Ramaswamy, H.G., et al.: Ablation-CAM: visual explanations for deep convolutional network via gradient-free localization. In: ICCV, pp. 983–991 (2020)

18. Ronneberger, O., Fischer, P., Brox, T.: U-Net: convolutional networks for biomedical image segmentation. In: MICCAI, pp. 234–241 (2015)

19. Salehi, S.S.M., et al.: Real-time automatic fetal brain extraction in fetal MRI by deep learning. In: ISBI, pp. 720–724 (2018)

20. Selvaraju, R.R., Cogswell, M., Das, A., Vedantam, R., Parikh, D., Batra, D.: Grad-CAM: visual explanations from deep networks via gradient-based localization. In: ICCV, pp. 618–626 (2017)

21. Shen, W., et al.: A survey on label-efficient deep image segmentation: bridging the gap between weak supervision and dense prediction. IEEE Trans. Pattern Anal. Mach. Intell. 45(8), 9284–9305 (2023)

22. Shi, W., et al.: Fetal brain age estimation and anomaly detection using attention-based deep ensembles with uncertainty. Neuroimage 223, 117316 (2020)

23. Simonyan, K., Zisserman, A.: Very deep convolutional networks for large-scale image recognition. In: ICLR, pp. 1–14 (2015)

24. Sridar, P., et al.: Automatic measurement of thalamic diameter in 2-D fetal ultrasound brain images using shape prior constrained regularized level sets. IEEE J. Biomed. Health Inform. 21(4), 1069–1078 (2016)

25. Wang, H., et al.: Score-CAM: Score-weighted visual explanations for convolutional neural networks. In: CVPR workshops, pp. 24–25 (2020)

26. Zhang, B., Xiao, J., Wei, Y., Sun, M., Huang, K.: Reliability does matter: An end-to-end weakly supervised semantic segmentation approach. In: AAAI, vol. 34, pp. 12765–12772 (2020)

27. Zhou, B., Khosla, A., Lapedriza, A., Oliva, A., Torralba, A.: Learning deep features for discriminative localization. In: CVPR, pp. 2921–2929 (2016)

ASC: Appearance and Structure Consistency for Unsupervised Domain Adaptation in Fetal Brain MRI Segmentation

Zihang Xu[1], Haifan Gong[1,2], Xiang Wan[1], and Haofeng Li[1(✉)]

[1] Shenzhen Research Institute of Big Data, Shenzhen, China
lhaof@sribd.cn
[2] The Chinese University of Hong Kong, Shenzhen, China

Abstract. Automatic tissue segmentation of fetal brain images is essential for the quantitative analysis of prenatal neurodevelopment. However, producing voxel-level annotations of fetal brain imaging is time-consuming and expensive. To reduce labeling costs, we propose a practical unsupervised domain adaptation (UDA) setting that adapts the segmentation labels of high-quality fetal brain atlases to unlabeled fetal brain MRI data from another domain. To address the task, we propose a new UDA framework based on Appearance and Structure Consistency, named ASC. We adapt the segmentation model to the appearances of different domains by constraining the consistency before and after a frequency-based image transformation, which is to swap the appearance between brain MRI data and atlases. Consider that even in the same domain, the fetal brain images of different gestational ages could have significant variations in the anatomical structures. To make the model adapt to the structural variations in the target domain, we further encourage prediction consistency under different structural perturbations. Extensive experiments on FeTA 2021 benchmark demonstrate the effectiveness of our ASC in comparison to registration-based, semi-supervised learning-based, and existing UDA-based methods.

Keywords: Unsupervised domain adaptation · Magnetic Resonance Imaging · Semantic segmentation · Fetal Brain · Consistency learning

Z. Xu and H. Gong—This work is supported by Chinese Key-Area Research and Development Program of Guangdong Province (2020B0101350001), and the National Natural Science Foundation of China (No.62102267), and the Guangdong Basic and Applied Basic Research Foundation (2023A1515011464), and the Shenzhen Science and Technology Program (JCYJ20220818103001002), and the Guangdong Provincial Key Laboratory of Big Data Computing, The Chinese University of Hong Kong, Shenzhen.
Z. Xu and H. Gong—Contribute equally to this work.

Supplementary Information The online version contains supplementary material available at https://doi.org/10.1007/978-3-031-43990-2_31.

1   Introduction

Magnetic resonance imaging (MRI) has emerged as an important tool for assessing brain development in utero [2,4,9,13]. Since manual segmentation is time-consuming [30] and suffers from high inter-rater variability in quantitative assessment, automatically segmenting brain tissue from MRI data becomes an urgent need [7,10,15]. However, the available annotated fetal brain datasets are limited in number and heterogeneity, hindering the development of automatic strategy.

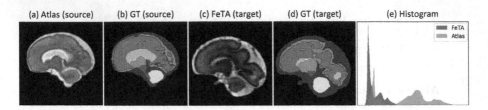

Fig. 1. Comparison of MR images from brain atlases (a) and FeTA dataset (c). (b) and (d) show the ground truth. (e) shows the gray-scale histograms of the samples of 30-week gestational age from the FeTA dataset and brain atlases. Horizontal and vertical coordinates denote intensity values and voxel numbers, respectively. (Color figure online)

To achieve the unsupervised fetal brain tissue segmentation, registration-based methods [18,22] use image registration and label fusion to obtain the segmentation result from a set of templates [19,24]. Still, the accuracy of these methods is not sufficient due to the complexity of registration, and they usually underperform on the abnormal fetal brain image. Recently, deep learning (DL) based unsupervised domain adaptation (UDA) methods have shown their advance on medical image segmentation tasks [1,3,8]. UDA methods usually narrow the distribution discrepancy between source and target domains by enforcing image-/feature-level alignment [11,21,25,26]. In addition to inter-domain knowledge transfer, some works [14,29] explored the knowledge from both intermediate domains.

However, existing UDA methods in medical imaging mainly focus on the gap from different modalities (e.g., CT and MR), and pay less attention to the domain gap from different centres. Due to motion artifacts [17], it is difficult to collect high-quality fetal brain MR images and is expensive to label the newly collected data voxel-wisely. The above observations motivate us to establish a new UDA problem setting that aims to transfer the segmentation knowledge from the publicly available atlases to unlabeled fetal brain MRIs from new centres.

To solve the above UDA task, we propose an Appearance and Structure Consistency (ASC) framework. Consider the fact [26,27] that swapping the low-level spectrum between images can exchange their style/color/brightness without changing semantic content, while swapping the higher spectrum introduces unwanted artifacts. Thus, we propose to align appearance by only swapping the

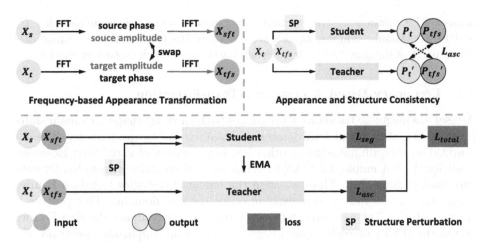

Fig. 2. Overview of the proposed Appearance and Structure Consistency framework. The student model learns from source data X_s and frequency-based transformed source data X_{sft} via the supervised loss L_{seg}. The appearance and structure consistency is achieved by the loss L_{asc}.

low-level spectrum. We develop an appearance consistency regularization based on a frequency-based appearance transformation, which is performed between labeled source data and unlabeled target data. Specifically, the source domain and the source data under the target appearance, are supervised with the same source labels. Then, the target data under the target appearance and source appearance are forced to maintain the same segmentation via dual unsupervised appearance consistency. Considering that significant variances in the shape of abnormal fetal brain tissue can cause difficulties in segmentation, we further constrain structure consistency under different perturbations in the target domain, besides aligning the inter-domain appearance gap. All the above consistency constraints are integrated with a teacher-student framework.

The contributions of this work are three-fold: (1) we propose a novel Appearance and Structure Consistency framework for UDA in fetal brain tissue segmentation; (2) we propose to address a practical UDA task adapting publicly available brain atlases to unlabeled fetal brain MR images; (3) experimental results on FeTA2021 benchmark [17] show that the proposed framework outperforms representative state-of-the-art methods.

2   Methodology

In the UDA setting, $D_s = \{X_s, Y_s\}_{s=1}^M$ denotes a set of source domain images (e.g., fetal brain atlases) and corresponding labels, respectively. $D_t = \{X_t\}_{t=1}^N$ denotes a set of target domain images (e.g., images from the FeTA benchmark). We aim to learn a semantic segmentation model for target domain data based on the labeled source and unlabeled target domain data. Usually, this goal is achieved by minimizing the domain gap between source domain samples D_s and target domain samples D_t. Figure 2 depicts the proposed Appearance and Structure Consistency (ASC) framework based on a teacher-student model.

2.1 Frequency-Based Appearance Transformation

Atlases are magnetic resonance fetal images with "average shape". Domain shifts between the atlases and fetal images are mainly due to the texture, different hospital sensors, illumination or other low-level sources of variability. However, traditional UDA employing GAN to synthetic style-transfer images hardly capture such domain shift. Thus, we align the low-level statistics based on Fourier transformation to narrow the distribution of the two domains. This process is shown in Fig. 3. Taking source data as an example, we compute the Fast Fourier transform (FFT) of each input image to obtain an amplitude spectrum $\mathcal{F}^A$ and a phase component $\mathcal{F}^P$, where the low-frequency part of the amplitude of the source image $\mathcal{F}^A(X_s)$ is swapped with the amplitude of the target image $\mathcal{F}^A(X_t)$. Then, the transformed spectral representation of X_s and the original phase $\mathcal{F}^P(X_s)$ are mapped back to the image X_{sft} by inverse FFT (iFFT). X_{sft} has the same content as X_s and similar appearance to X_t. The above process can be formally defined as:

$$\mathcal{F}^A(X_{sft}) = M \cdot \mathcal{F}^A(X_t) + (1 - M) \cdot \mathcal{F}^A(X_s), \tag{1}$$

$$X_{sft} = \mathcal{F}^{-1}([\mathcal{F}^A(X_{sft}), \mathcal{F}^P(X_s)]), \tag{2}$$

where the mask $M = \mathcal{I}_{(h,w,d)\in[-\beta H:\beta H,-\beta W:\beta W,-\beta D:\beta D]}$ controls the proportion of the swapped part over the whole amplitude by a parameter $\beta \in (0,1)$. Here we assume the center of the image is $(0, 0, 0)$. Then we can train a student network with domain alignment images X_{sft}, the original images X_s and the labels Y_s by minimizing the dice loss:

$$L_{seg} = L_{dice}(P_s, Y_s) + L_{dice}(P_{sft}, Y_s), \tag{3}$$

where P_s and P_{sft} are the prediction of X_s and X_{sft}, respectively.

2.2 Appearance Consistency

The above loss function imposes an implicit regularization before and after frequency-based transformation. In other words, source domain image X_s and its transformation image X_{sft} should predict the same segmentation. However, the label of images from the target domain is not available X_t in UDA settings. As a replacement, we propose a teacher model for keeping semantic consistency across domain transformation. Specifically, the target domain image X_t and its aligned image X_{tfs} are regarded as representations of an object under different domains. Given the inputs of X_t and X_{tfs} of teacher and student models, we expect their predictions to be consistent. Further, considering that appearance transformation may break certain semantic information and make the model

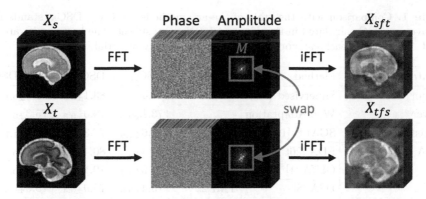

Fig. 3. Illustration of the frequency-based appearance transformation. The transformation exchanges the domain-specific appearance between two images by swapping their low-frequency components of the spectrum. FFT is the Fast Fourier Transform. Phase component and amplitude spectrum are denoted by $\mathcal{F}^P$ and $\mathcal{F}^A$ in Eq. (2), respectively.

learn the wrong mapping relationship, we employ a form of dual consistency, which directs the model to focus on invariant information between the two views. $f(\cdot)$ and $f'(\cdot)$ represent the outputs of the student model and the teacher model, respectively. Following the conventional consistency learning methods [20], we calculate the appearance consistency loss L_{con}^{app} between the teacher and student networks as:

$$L_{con}^{app} = \frac{1}{N} \sum_{i=1}^{N} ||f(X_{t,i}) - f'(X_{tfs,i})||^2 + \frac{1}{N} \sum_{i=1}^{N} ||f(X_{tfs,i}) - f'(X_{t,i})||^2. \quad (4)$$

2.3   Structure Consistency

Although frequency-based transformation and appearance consistency align the two domains' styles, the variance of tissue structure in pathological subjects still brings difficulty to domain alignment, which limits the model's gener alization ability. To this end, we utilise the teacher-student model [20] keeping prediction consistency L_{con}^{str} under structure perturbation [28] to alleviate the above problem. Here structure perturbation is *sp* for short. To achieve the structure perturbation, we first use a 3D cuboid mask consisting of a single box that randomly covers 25–50% of the image area at a random position, to blend two input images, which are sampled from the same batch. Then we blend the teacher predictions for the input images to produce a pseudo label for the student prediction of the blended image. Such an operation changes the original structural information, reduces the overfitting risk, and increases the robustness of the model to adapt to different structural variations. As appearance transformation doesn't affect the structure information, we add *sp* to both X_t and X_{tfs} to obtain $X_{t,sp}$ and $X_{tfs,sp}$, which are fed into the teacher-student model and expected their

Table 1. Comparison with the state-of-the-art methods. DSC_A / DSC_N stands for the average DSC calculated in the Abnormal / Normal subset. The best results are in **bold**. The DSC of each category is in the supplementary material.

Strategy	Method	Venue	DSC_A	DSC_N	Avg. DSC
Upper	Supervised (D_t)	-	$81.4_{\pm 0.2}$	$83.1_{\pm 0.3}$	$82.0_{\pm 0.1}$
Lower	W/o Adaptation	-	$73.1_{\pm 0.6}$	$79.0_{\pm 0.3}$	$75.3_{\pm 0.5}$
Registration-based	SCALE [19]	MIA'18	$63.6_{\pm 1.5}$	$77.3_{\pm 1.7}$	$68.7_{\pm 1.7}$
UDA	FDA [26]	CVPR'20	$74.4_{\pm 0.4}$	$80.4_{\pm 0.4}$	$76.7_{\pm 0.4}$
	OLVA [1]	MICCAI'21	$73.3_{\pm 0.6}$	$79.1_{\pm 0.3}$	$75.6_{\pm 0.3}$
	DSA [8]	TMI'22	$73.4_{\pm 0.6}$	$79.9_{\pm 0.4}$	$75.9_{\pm 0.5}$
SSL	CUTMIX [28]	ICCV'19	$74.1_{\pm 0.1}$	$79.7_{\pm 0.5}$	$76.2_{\pm 0.2}$
	ASE-NET [12]	TMI'22	$73.7_{\pm 0.3}$	$79.8_{\pm 0.3}$	$76.2_{\pm 0.2}$
UDA	ASC (ours)	-	$\mathbf{76.6_{\pm 0.2}}$	$\mathbf{81.7_{\pm 0.1}}$	$\mathbf{78.5_{\pm 0.1}}$

predictions to be consistent. Then, L_{con}^{str} and L_{con}^{app} are combined as L_{asc}:

$$L_{asc} = \frac{1}{N}\sum_{i=1}^{N}||f(X_{t,i,sp}) - f'(X_{tfs,i})||^2 + \frac{1}{N}\sum_{i=1}^{N}||f(X_{tfs,i,sp}) - f'(X_{t,i})||^2. \quad (5)$$

2.4 Overall Training Strategy

We calculate appearance and structure consistency using the same teacher model. Its model weight θ' is updated with the exponential moving average (EMA) of the student model $f(\theta)$, i.e., $\theta'_t = \alpha\theta'_{t-1} + (1-\alpha)\theta_t$, where α is the EMA decay rate that reflects the influence level of the current student model parameters.

Let λ control the trade-off between the supervised loss and the unsupervised regularization loss, the overall loss is:

$$L_{total} = L_{seg} + \lambda L_{asc}. \quad (6)$$

3 Experiment

3.1 Dataset and Pre-processing

We evaluated our method on the Fetal Brain Tissue Annotation and Segmentation Challenge (FeTA) 2021 benchmark dataset [17], which contains 80 3D T2 MRI volumes with manual segmentations annotation of external cerebrospinal fluid (eCSF), grey matter (GM), white matter (WM), lateral ventricles (LV), cerebellum (CBM), deep grey matter (dGM) and brainstem (BS). The dataset cohort consisted of two subgroups: 31 neurological fetuses and 49 fetuses with

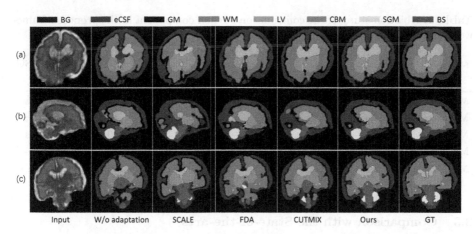

Fig. 4. Visual comparisons with existing methods. Due to space limitations, we only present those with the best average DSC in Registration-based, UDA and SSL methods. It can be seen that our predictions are visually closer to the ground truth than the other methods.

abnormal development. Following the general UDA setting [11], the target set was randomly divided into 40 scans for training and 40 scans for testing. A collection from three atlases was used as the source set, including 32 neurotypical fetal brain atlases [6,23] and 15 spina bifida fetal brain atlases [5]. Segmentations for all tissue types are available for all the atlas data. Some examples of slices, segmentations and histogram distribution are shown in Fig. 1. We cropped the foreground region of fetal volumes and reshaped them to $144 \times 144 \times 144$. Before being fed into the network, the input scans were normalized to a zero mean and unit variance.

3.2 Implementation Details

All models were implemented in PyTorch 1.12 and trained with NVIDIA A100 GPU with CUDA 11.3. Following the top-ranked method in the FeTA2021 competition [17], we use SegResNet [16] as the backbone for the teacher/student model. The network parameters were optimized with Adam with the initial learning rate of 1×10^{-4}. "Poly" learning rate policy is applied, where $lr = lr_{init} \times (1 - \frac{epoch}{epoch_{total}})^{0.9}$. The batch size and training epoch were set to 4 (2 from each domain) and 100, respectively. The EMA decay rate α of the teacher model was set to 0.99, and hyperparameters λ were ramped up individually with function $\lambda(t) = \gamma \times e^{(-5(1-\frac{t}{t_{max}})^2)}$, where t, t_{max} and γ were the current step, the last step and weight, respectively. β was set to 0.1. Cutmix [28] was used for target images as the structure perturbation for consistency regularization. We employed the student model prediction as the final result and used the Dice Similarity Coefficient (DSC) scores to evaluate the accuracy of the results. The average results of three runs were reported in all experiments.

Table 2. Ablation study of the contribution of each component in the proposed framework. The highest evaluation score is marked in **bold**.

Method	$L_{seg(X_s)}$	$L_{seg(X_{sft})}$	L_{asc}			DSC_A	DSC_N	Avg DSC
			$L_{con(X_t)}^{app}$	$L_{con(X_{tfs})}^{app}$	L_{con}^{str}			
M1	✓					$73.1_{\pm0.6}$	$79.0_{\pm0.3}$	$75.3_{\pm0.5}$
M2	✓	✓				$74.3_{\pm0.2}$	$80.4_{\pm0.3}$	$76.6_{\pm0.2}$
M3	✓	✓	✓			$75.6_{\pm0.5}$	$80.9_{\pm0.2}$	$77.3_{\pm0.3}$
M4	✓	✓	✓	✓		$75.8_{\pm0.1}$	$81.4_{\pm0.3}$	$77.9_{\pm0.2}$
M5	✓	✓	✓	✓	✓	$\mathbf{76.6_{\pm0.2}}$	$\mathbf{81.7_{\pm0.1}}$	$\mathbf{78.5_{\pm0.1}}$

3.3   Comparison with the State-of-the-arts

We implemented several state-of-the-art label-limited segmentation methods for comparison, including Registration-based (SCALE [19]), Unsupervised Domain Adaptatopm (UDA) (FDA [26], OLVA [1] and DSA [8]) and Semi-supervised Learning (CUTMIX [28], ASE-NET [12]) in Table 1. It reports the segmentation performance of UDA of adapting atlas to the fetal brain on FeTA2021, including average DSC for the full set, normal set and abnormal set. The upper bound is given by supervised training which uses fully-labeled target data for model training.

It is worth noting that registration-based [19] only successfully segments on the normal set, and GAN-based UDA approach [8] performs worse than the frequency-based approach [26] on the abnormal set. The proposed method is superior to the existing state-of-the-art methods [26,28] and achieves mean Dice of 78.5% over the seven tissue structures, reducing the Dice gap to supervised training to 3.5%. Compared with the baseline model (W/o adaptation), our proposed learning strategy further improves the performance by an average of 3.2% Dice. Visual results in Fig. 4 show our method can perform better in the junction areas of brain tissue.

3.4   Ablation Study and Sensitivity Analysis

The ablation study is shown in Table 2, and all the results boost our method's performance. "M1" represents the lower bound that only trains on the source domain data D_s. "M2" uses the aligned source images X_{sft} for training. The following component are based on L_{asc}, which is decoupled as $L_{con(X_t)}^{app}$, $L_{con(X_{tfs})}^{app}$ and L_{con}^{str}. "M3" denotes the appearance consistency loss $L_{con(X_t)}^{app}$ to align distribution from source to target. "M4" indicates the dual-view appearance consistency loss to constrain semantic invariance. "M5" denotes the structure consistency L_{con}^{str}.

We can see that appearance consistency can boost performance on normal and abnormal fetal MRIs, showing that minimizing the appearance gap between the source domain and target domain is effective. Further, we can obtain better

Table 3. Performance of our method with different values of hyperparameters γ, which are used to balance consistency loss and supervisory loss.

γ value	10	100	200	500	1000
Avg. DSC	$78.1_{\pm 0.1}$	$78.4_{\pm 0.1}$	$\mathbf{78.5_{\pm 0.1}}$	$78.4_{\pm 0.2}$	$78.2_{\pm 0.4}$

results on the abnormal samples by applying the structure consistency loss. Table 3 shows that the performance of our method grows with the increase of the hyper-parameter γ of consistency loss, and achieves the best when $\gamma = 200$. Besides, **efficiency analysis** is shown in the supplementary material.

4 Conclusion

In this paper, we present a novel UDA framework and an atlas-based UDA setting for fetal brain tissue segmentation. Our method integrates appearance consistency encouraging the model to adapt different domain styles to narrow the domain gap and structure consistency making the model robust against the anatomical variations in the target domain. Experiments on the FeTA2021 benchmark demonstrate that our method outperforms the state-of-the-art methods. The proposed novel setting of atlas-based UDA could provide accurate segmentation for the fetal brain MRI data without pixel-wise annotations, greatly reducing the labeling costs.

References

1. Al Chanti, D., Mateus, D.: OLVA: Optimal latent vector alignment for unsupervised domain adaptation in medical image segmentation. In: de Bruijne, M., et al. (eds.) Medical Image Computing and Computer Assisted Intervention – MICCAI 2021: 24th International Conference, Strasbourg, France, September 27–October 1, 2021, Proceedings, Part III, pp. 261–271. Springer, Cham (2021). https://doi.org/10.1007/978-3-030-87199-4_25
2. Benkarim, O.M., et al.: Toward the automatic quantification of in utero brain development in 3D structural MRI: a review: quantification of fetal brain development. Hum. Brain Mapp. **38**(5), 2772–2787 (2017)
3. Chen, C., Dou, Q., Chen, H., Qin, J., Heng, P.A.: Unsupervised bidirectional cross-modality adaptation via deeply synergistic image and feature alignment for medical image segmentation. IEEE Trans. Med. Imaging **39**(7), 2494–2505 (2020)
4. De Asis-Cruz, J., Andescavage, N., Limperopoulos, C.: Adverse prenatal exposures and fetal brain development: insights from advanced fetal magnetic resonance imaging. Biol. Psychiatry: Cogn. Neurosci. Neuroimaging **7**(5), 480–490 (2022)
5. Fidon, L., et al.: A spatio-temporal atlas of the developing fetal brain with spina bifida aperta. Open Res. Eur. **1**, 123 (2022)
6. Gholipour, A.: A normative spatiotemporal MRI atlas of the fetal brain for automatic segmentation and analysis of early brain growth. Sci. Rep. **7**(1), 476 (2017)

7. Gousias, I.S., et al.: Magnetic resonance imaging of the newborn brain: manual segmentation of labelled atlases in term-born and preterm infants. Neuroimage **62**(3), 1499–1509 (2012)
8. Han, X., et al.: Deep symmetric adaptation network for cross-modality medical image segmentation. IEEE Trans. Med. Imaging **41**(1), 121–132 (2021)
9. Hart, A.R., et al.: Accuracy of in-utero MRI to detect fetal brain abnormalities and prognosticate developmental outcome: postnatal follow-up of the meridian cohort. Lancet Child Adolesc. Health **4**(2), 131–140 (2020)
10. Huang, J., Li, H., Li, G., Wan, X.: Attentive symmetric autoencoder for brain MRI segmentation. In: Wang, L., Dou, Q., Fletcher, P.T., Speidel, S., Li, S. (eds.) Medical Image Computing and Computer Assisted Intervention – MICCAI 2022: 25th International Conference, Singapore, September 18–22, 2022, Proceedings, Part V, pp. 203–213. Springer, Cham (2022). https://doi.org/10.1007/978-3-031-16443-9_20
11. Huo, Y., et al.: SynSeg-Net: synthetic segmentation without target modality ground truth. IEEE Trans. Med. Imaging **38**(4), 1016–1025 (2018)
12. Lei, T., Zhang, D., Du, X., Wang, X., Wan, Y., Nandi, A.K.: Semi-Supervised medical image segmentation using adversarial consistency learning and dynamic convolution Network. IEEE Trans. Med. Imaging **42**(5), 1265–1277 (2023)
13. Li, H., et al.: View-disentangled transformer for brain lesion detection. In: 19th International Symposium on Biomedical Imaging (ISBI), pp. 1–5. IEEE (2022)
14. Li, K., Wang, S., Yu, L., Heng, P.-A.: Dual-Teacher: integrating Intra-domain and Inter-domain Teachers for Annotation-Efficient Cardiac Segmentation. In: Martel, A.L., Abolmaesumi, P., Stoyanov, D., Mateus, D., Zuluaga, M.A., Zhou, S.K., Racoceanu, D., Joskowicz, L. (eds.) Medical Image Computing and Computer Assisted Intervention – MICCAI 2020: 23rd International Conference, Lima, Peru, October 4–8, 2020, Proceedings, Part I, pp. 418–427. Springer International Publishing, Cham (2020). https://doi.org/10.1007/978-3-030-59710-8_41
15. Makropoulos, A., Counsell, S.J., Rueckert, D.: A review on automatic fetal and neonatal brain MRI segmentation. Neuroimage **170**, 231–248 (2018)
16. Myronenko, A.: 3D MRI brain tumor segmentation using autoencoder regularization. In: Crimi, A., Bakas, S., Kuijf, H., Keyvan, F., Reyes, M., van Walsum, T. (eds.) Brainlesion: Glioma, Multiple Sclerosis, Stroke and Traumatic Brain Injuries: 4th International Workshop, BrainLes 2018, Held in Conjunction with MICCAI 2018, Granada, Spain, September 16, 2018, Revised Selected Papers, Part II, pp. 311–320. Springer, Cham (2019). https://doi.org/10.1007/978-3-030-11726-9_28
17. Payette, K., et al.: An automatic multi-tissue human fetal brain segmentation benchmark using the fetal tissue annotation dataset. Sci. Data **8**(1), 167 (2021)
18. Sabuncu, M.R., Yeo, B.T., Van Leemput, K., Fischl, B., Golland, P.: A generative model for image segmentation based on label fusion. IEEE Trans. Med. Imaging **29**(10), 1714–1729 (2010)
19. Sanroma, G., et al.: Learning non-linear patch embeddings with neural networks for label fusion. Med. Image Anal. **44**, 143–155 (2018)
20. Tarvainen, A., Valpola, H.: Mean teachers are better role models: weight-averaged consistency targets improve semi-supervised deep learning results. In: Advances in Neural Information Processing Systems, vol. 30 (2017)
21. Tomar, D., Lortkipanidze, M., Vray, G., Bozorgtabar, B., Thiran, J.P.: Self-attentive spatial adaptive normalization for cross-modality domain adaptation. IEEE Trans. Med. Imaging **40**(10), 2926–2938 (2021)

22. Wang, H., Suh, J.W., Das, S.R., Pluta, J.B., Craige, C., Yushkevich, P.A.: Multi-atlas segmentation with joint label fusion. IEEE Trans. Pattern Anal. Mach. Intell. **35**(3), 611–623 (2012)

23. Wu, J., et al.: Age-specific structural fetal brain atlases construction and cortical development quantification for Chinese population. Neuroimage **241**, 118412 (2021)

24. Xie, L., et al.: Deep label fusion: a generalizable hybrid multi-atlas and deep convolutional neural network for medical image segmentation. Med. Image Anal. **83**, 102683 (2023)

25. Xie, Q., et al.: Unsupervised domain adaptation for medical image segmentation by disentanglement learning and self-training. IEEE Trans. Med. Imaging, 1 (2022). https://doi.org/10.1109/TMI.2022.3192303

26. Yang, Y., Soatto, S.: FDA: fourier domain adaptation for semantic segmentation. In: Proceedings of the IEEE/CVF Conference on Computer Vision and Pattern Recognition, pp. 4085–4095 (2020)

27. Yue, J., Li, H., Wei, P., Li, G., Lin, L.: Robust real-world image super-resolution against adversarial attacks. In: Proceedings of the 29th ACM International Conference on Multimedia, pp. 5148–5157 (2021)

28. Yun, S., Han, D., Oh, S.J., Chun, S., Choe, J., Yoo, Y.: CutMix: regularization strategy to train strong classifiers with localizable features. In: Proceedings of the IEEE/CVF International Conference on Computer Vision, pp. 6023–6032 (2019)

29. Zhao, Z., Xu, K., Li, S., Zeng, Z., Guan, C.: MT-UDA: towards unsupervised cross-modality medical image segmentation with limited source labels. In: de Bruijne, M., et al. (eds.) Medical Image Computing and Computer Assisted Intervention – MICCAI 2021: 24th International Conference, Strasbourg, France, September 27–October 1, 2021, Proceedings, Part I, pp. 293–303. Springer, Cham (2021). https://doi.org/10.1007/978-3-030-87193-2_28

30. Zhou, H.Y., et al.: SSMD: Semi-supervised medical image detection with adaptive consistency and heterogeneous perturbation. Med. Image Anal. **72**, 102117 (2021)

Simulation-Based Parameter Optimization for Fetal Brain MRI Super-Resolution Reconstruction

Priscille de Dumast[1,2], Thomas Sanchez[1,2]([✉]), Hélène Lajous[1,2],
and Meritxell Bach Cuadra[1,2]

[1] Department of Radiology, Lausanne University Hospital (CHUV) and University of
Lausanne (UNIL), Lausanne, Switzerland
[2] CIBM Center for Biomedical Imaging, Lausanne, Switzerland
thomas.sanchez@unil.ch

Abstract. Tuning the regularization hyperparameter α in inverse problems has been a longstanding problem. This is particularly true in the case of fetal brain magnetic resonance imaging, where an isotropic high-resolution volume is reconstructed from motion-corrupted low-resolution series of two-dimensional thick slices. Indeed, the lack of ground truth images makes challenging the adaptation of α to a given setting of interest in a quantitative manner. In this work, we propose a simulation-based approach to tune α for a given acquisition setting. We focus on the influence of the magnetic field strength and availability of input low-resolution images on the ill-posedness of the problem. Our results show that the optimal α, chosen as the one maximizing the similarity with the simulated reference image, significantly improves the super-resolution reconstruction accuracy compared to the generally adopted default regularization values, independently of the selected reconstruction pipeline. Qualitative validation on clinical data confirms the importance of tuning this parameter to the targeted clinical image setting. The simulated data and their reconstructions are available at https://zenodo.org/record/8123677.

Keywords: Magnetic resonance imaging (MRI) · fetal brain MRI · Super-resolution reconstruction (SRR) · parameter optimization · image synthesis

This work is supported by the Swiss National Science Foundation through grants 182602 and 141283, and by the Eranet Neuron MULTIFACT project (SNSF 31NE30_203977). We acknowledge access to the facilities and expertise of the CIBM Center for Biomedical Imaging, a Swiss research center of excellence founded and supported by CHUV, UNIL, EPFL, UNIGE and HUG.

Priscille de Dumast and Thomas Sanchez contributed equally to this work. Hélène Lajous and Mertixell Bach Cuadra share senior authroship.

Supplementary Information The online version contains supplementary material available at https://doi.org/10.1007/978-3-031-43990-2_32.

H. Greenspan et al. (Eds.): MICCAI 2023, LNCS 14226, pp. 336–346, 2023.
https://doi.org/10.1007/978-3-031-43990-2_32

1   Introduction

Magnetic resonance imaging (MRI) has become an increasingly important tool to investigate prenatal equivocal neurological situations, as it provides excellent anatomical details [1,2]. However, three-dimensional (3D) high-resolution (HR) imaging of the fetal brain is unfeasible due to the unpredictable fetal motion. In clinical practice, T2-weighted (T2w) fast spin echo (FSE) sequences are commonly used to minimize the effects of intra-slice random fetal movements, and multiple orthogonal series are acquired, resulting in several low-resolution (LR) series of two-dimensional (2D) thick slices [3,4]. Nevertheless, the strong anisotropy of the images leading to partial volume effects on small structures within the fetal brain and the remaining inter-slice motion hamper the accurate analysis of 3D imaging biomarkers.

Several post-processing techniques have been proposed to combine multiple motion-corrupted LR series and leverage the information redundancy from orthogonal orientations to reconstruct a single, 3D isotropic HR motion-free volume of the fetal brain [5–8]. These approaches all feature several pre-processing steps (e.g. brain extraction, intensity correction, and harmonization) leading to slice-to-volume registration (SVR) where inter-series inter-slice motion is estimated, followed by super-resolution reconstruction (SRR). This latter step can be framed as an inverse problem of the form

$$\min_{\mathbf{x}} \frac{1}{2} \|\mathbf{H}\mathbf{x} - \mathbf{x}^{\mathrm{LR}}\|^2 + \alpha R(\mathbf{x}), \tag{1}$$

where $\mathbf{x}$ is the target HR image, $\mathbf{x}^{\mathrm{LR}}$ the LR series, $\mathbf{H}$ an operator describing the motion, blurring and downsampling model estimated from the data, and R the regularization function (e.g., total-variation (TV) [6], first-order Tikhonov [7], etc.). α is a parameter that balances the strength of the regularization term compared to the data fidelity term.

Various applications in medical image computing are formulated as inverse problems, and the optimization of regularization parameters has been widely studied in this context [9,10]. Most strategies explicitly rely on reference data, which are not available in the context of fetal MRI, making the setting of the regularization parameter α in a principled and quantitative manner highly challenging. To circumvent the lack of HR data of the fetal brain, several works use HR MR images from newborns as ground truth data and downsample them to simulate the acquisition of LR series that are then reconstructed and compared to the HR image to set the default value of their regularization parameters [5,6]. Alternative approaches consider a leave-one-out approach where the left-out LR series serves as a reference for the quantitative evaluation of the SRR [5,6], or use a volume reconstructed from all available LR series as a reference to which SRR with fewer LR series can be compared [7]. However, all of these works rely on constructing surrogate ground truth images to study the influence of the regularization parameter on the quality of the SRR but do not provide insights on how to adapt it when new input acquisition setting has to be reconstructed. Furthermore, despite well-known differences in image acquisitions protocols, fetal brain

SRR MRI studies are still carried out using the default regularization values of the selected pipeline [6–8, 11].

This work proposes the first approach to optimize the setting of the regularization parameter α based on numerical simulations of imaging sequences tailored to clinical ones. We take advantage of a recent Fetal Brain MR Acquisition Numerical phantom (FaBiAN) [12] that provides a controlled environment to simulate the MR acquisition process of FSE sequences, and thus generates realistic T2w LR MR images of the fetal brain as well as corresponding HR volumes that serve as a reference to optimize the parameter α in a data-driven manner, considering both *acquisition setting-specific* and *subject-specific* strategies.

Our contributions are twofold. First, using synthetic, yet realistic data, we study the sensitivity of the regularization to three common variables in inverse SRR problem in fetal MRI: (i) the number of LR series used as input, (ii) the magnetic field strength which impacts also the in-plane through-plane spatial resolution ratio, and (iii) the gestational age (GA), which leads to substantial changes in brain anatomy. Secondly, we qualitatively illustrate the practical value of our framework, by translating our approach to clinical MR exams. We show that α^* estimated by our simulated framework echoes a substantial improvement of image quality in the clinical SRR. To generalize the validity of our findings, we perform our study using two state-of-the-art SRR pipelines, namely MIALSRTK [6] and NiftyMIC [7].

2   Materials and Methods

2.1   Simulated Acquisitions

We use FaBiAN [12, 13] to generate T2w MR images of the developing fetal brain derived from a normative spatiotemporal MRI atlas (STA) [14] that features 18 subjects from 21 to 38 weeks of GA. ypical FSE acquisitions are simulated using the extended phase graph (EPG) formalism [15], at either 1.5T or 3T, according to the MR protocol routinely performed at our local hospital for fetal brain examination. All sequence parameters are kept fixed at a given magnetic field strength (at $1.5/3T$: TR, $1200/1100ms$; TE, $90/101ms$; voxel size, $1.1 \times 1.1 \times 3/0.5 \times 0.5 \times 3$ mm ^{3}). Stochastic 3D rigid motion of little-to-moderate amplitude as well as random complex Gaussian noise (mean, 0; standard deviation, 0.15 at 1.5T, respectively 0.0025 at 3T) are applied during k-space sampling to simulate as closely as possible the MR acquisition process. Both realistic fetal movements and noise levels are qualitatively estimated from clinical LR series and set accordingly to match the characteristics of real scans [12, 16, 17]. More specifically, a maximum of 5% motion-corrupted slices is generated over the whole fetal brain volume with independent translation within a uniform distribution of $[-1,1]$ mm in every direction and 3D rotation within $[-2,2]°$ for little motion, respectively $[-3,3]$ mm and $[-5,5]°$ for moderate motion, to reproduce typical motion patterns. Multiple orthogonal LR series $\mathbf{x}^{LR} = \{\mathbf{x}^{LR,i}\}_i$ are simulated with a shift of the field-of-view of $1.6mm$ in the slice thickness direction for series in the same orientation. The amplitude of fetal motion and the number of simulated LR series are further detailed in the experimental settings (Sects. 2.3 and 2.4). A visual comparison between clinical LR series and the corresponding simulated data is available at Fig. 8 in the Supplementary material. A reference HR isotropic volume $\mathbf{x}^{HR}$ of the fetal brain is also simulated for each subject, without bias field or motion, to serve as a reference for the quantitative evaluation of the corresponding SRR.

2.2 Super-Resolution Reconstruction Methods

Two widely adopted reconstruction pipelines, MIALSRTK [6] and NiftyMIC [7], are used to reconstruct 3D isotropic HR images of the fetal brain from orthogonal LR series. For each pipeline, we perform a grid search approach of the regularization parameter space.

Remark. Contrary to NiftyMIC [7], MIALSRTK [6] places its regularization parameter λ on the data fidelity term. For the sake of consistency, we will only use the formulation of Eq. 1, with $\alpha = 1/\lambda$ in the case of MIALSRTK.

Quality Assessment. Solving Problem 1 yields a SR-reconstructed image $\hat{\mathbf{x}}^{HR}$ whose quality can be compared against the reference $\mathbf{x}^{HR}$ using various metrics. We use two common metrics for SRR assessment [5–7], namely the peak signal-to-noise ratio (PSNR) and the structural similarity index (SSIM) [18]. The best regularization parameter α is identified as the one maximizing a given performance metric.

2.3 Experiment 1 – Controlled in Silico Environment

In this first experiment, we study the sensitivity of the parameter α to common variations in the acquisition pipeline.

Dataset. For every STA subject, nine LR series (three per anatomical orientation) are simulated at 1.5T and 3T with little amplitude of stochastic 3D rigid motion.

Experimental Setting. We define four configurations based on the number of LR series given as input to the SRR pipeline (three or six series) and the magnetic field strength (1.5 or 3T). Note that the inter-magnetic field difference is especially captured in the image resolution, with a through-plane/in-plane ratio of $3.3/1.1 = 3$ at 1.5T and $3.3/0.5 = 6.6$ at 3T. In each configuration, individual brains are repeatedly reconstructed ($n = 3$) from a selection of different LR series among the nine series available per subject.

The grid of parameters searched for NiftyMIC consists of 10 values geometrically spaced between 10^{-3} and 2, plus the default parameter $\alpha_{def} = 0.01$. For MIALSRTK, we use $\alpha \in \{1/0.75, 1/1.0, 1/1.5, 1/2.0, 1/2.5, 1/3.0, 1/3.5, 1/5.0\}$ (8 values, with default parameter $\alpha_{def} = 1/0.75$). At the end of the experiment, the best parameter, for either of the pipelines, is referred to as α_1^*.

Statistical Analysis. The optimal regularization parameters evaluated for the different SRR configurations are compared using the Wilcoxon rank sum test. The difference between the metrics performance obtained with default or optimal parameters is tested with a paired Wilcoxon rank sum test. The p-value for statistical significance is set to 0.05.

2.4 Experiment 2 – Clinical Environment

Clinical MR fetal exams are prone to substantial inter-subject variation and heterogeneity. In particular, the number of LR series available for reconstruction, as well as the amplitude of fetal motion, greatly vary from one subject to the other [19]. Therefore, this second experiment has two purposes. First, we translate our findings from the first experiment to clinical data using the best value α_1^*. Secondly, we study an alternative approach to perform a tailored subject-wise regularization tuning by simulating synthetic data for each subject that mimic the clinical acquisitions available. We refer to the obtained value as α_2^*.

Dataset. Twenty fetal brain MR exams conducted upon medical indication were retrospectively collected from our institution. All brains were finally considered normal. Fetuses were aged between 21 and 34 weeks of GA (mean ± standard deviation (sd): 29.7 ± 3.6) at scan time. For each subject, at least three orthogonal series were acquired at 1.5T (voxel size: $1.125 \times 1.125 \times 3$ mm^3). After inspection, four to nine series (mean ± sd: 6.3 ± 1.5) were considered exploitable for SRR.

The local ethics committee approved the retrospective collection and analysis of MRI data and the prospective studies for the collection and analysis of the MRI data in presence of a signed form of either general or specific consent.

The same 20 subjects are simulated using exam-specific parameters to mimic as closely as possible the corresponding clinical acquisitions. In particular, we match the number and the orientation of the LR series, as well as the amplitude of fetal motion (from little to moderate), and the GA of each subject.

Experimental Setting. We consider the same regularization parameter space as in Experiment 1 (Sect. 2.3), and evaluate both clinical and simulated data on this parameter grid.

Statistical Analysis. We compare the similarity between the images reconstructed by MIALSRTK and NiftyMIC using both default and optimized parameters. In this experiment, no reference images are available. Statistical significance of the performance difference is tested using a paired Wilcoxon rank sum test ($p < 0.05$ for statistical significance).

3 Results

3.1 Experiment 1 – Controlled in Silico Environment

Optimal Regularization Parameter. Figure 1 shows the optimal regularization parameters α_1^* of SRR by MIALSRTK and NiftyMIC for each configuration.

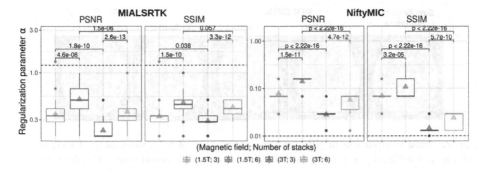

Fig. 1. Optimal regularization parameters α_1^* for MIALSRTK (left panel) and NiftyMIC (right panel) in the four different configurations studied. $\triangle$ indicates the mean optimal parameter. Inter-configurations p-values (Wilcoxon rank sum test, significance: $p < 0.05$) are indicated. The dashed line shows the value of the default parameter for each SRR technique.

Table 1. Mean metrics computed across all subjects on images reconstructed using the default regularization parameter α_{def} or the optimal parameter α_1^* respectively, compared to the simulated reference HR volume, for the four configurations studied. † indicates paired Wilcoxon rank sum test statistical significance ($p < 0.05$).

	MIALSRTK				NiftyMIC			
	PSNR (↑)		SSIM (↑)		PSNR (↑)		SSIM (↑)	
(Field strength; # LR)	α_{def}	α_1^*	α_{def}	α_1^*	α_{def}	α_1^*	α_{def}	α_1^*
(1.5T; 3)	18.9	**20.2**†	0.78	**0.80**†	17.3	**20.8**†	0.79	**0.82**†
(1.5T; 6)	20.1	**20.8**†	0.82	**0.83**†	17.0	**21.5**†	0.80	**0.84**†
(3T; 3)	19.9	**21.8**†	0.75	**0.77**†	20.5	**21.2**†	0.77	**0.77**†
(3T; 6)	21.0	**22.2**†	0.78	**0.80**†	20.9	**22.0**†	0.80	**0.80**†

Regardless of the magnetic field strength and the number of LR series used for reconstruction, the optimal regularization parameters that maximize the PSNR and SSIM compared to a synthetic HR volume greatly differ from the default values. For MIALSRTK, $\alpha_{\text{def}} = 1/0.75$, while the optimal range is found between $1/2.25$ and $1/4.5$. For NiftyMIC, $\alpha_{\text{def}} = 0.01$, whereas the optimal range is found between 0.015 and 0.15. We observe that for both the PSNR and the SSIM, the optimal regularization weight *increases* with the number of series used in the reconstruction, and decreases with the resolution. This is because changing the number of LR series or the magnetic field strength affects the magnitude of the data fidelity term with respect to the regularization term. When more series are combined in the SRR, a larger regularization parameter must be used to keep the ratio $\|\mathbf{Hx} - \mathbf{x}^{\text{LR}}\|^2 / \alpha R(\mathbf{x})$ constant.

Simulated HR   MIALSRTK   NiftyMIC
reference   Default α   α₁*   Default α   α₁*

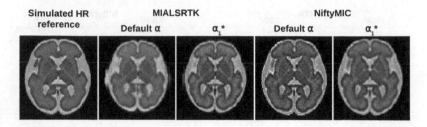

Fig. 2. Axial comparison of SRR of simulated cases of a 30-week old subject, using default and optimal parameters from Experiment 1 (α_1^*), reconstructed with three LR series.

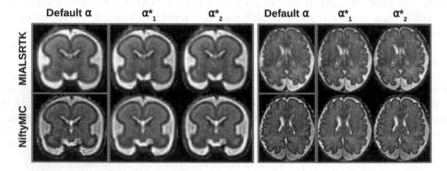

Fig. 3. Illustration of the domain shift between SRR techniques for two clinical subjects (left, GA: 23 weeks, 8 LR series and right, GA: 32 weeks, 4 LR series). The SR reconstructions using the default α (orange boxes) are much more different than the ones reconstructed with the optimal parameters (blue boxes).

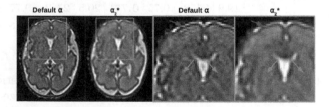

Fig. 4. Comparison of a clinical case (GA: 28 weeks, 8 LR series) reconstructed using NiftyMIC with different regularization parameters. We see that the optimal parameter α_2^* yields a smoother image with less ringing artifacts at the frontal lobe (red arrow) and a more clearly delineated deep gray matter (yellow arrows).

Quality Improvement. The corresponding mean PSNR and SSIM values, computed across all subjects both with default and optimal regularization parameters, and compared to the reference HR volume are displayed in Table 1. Overall, the quality metrics obtained from the SRR with optimal parameters are significantly improved compared to those obtained with default values. These results strongly suggest that the regularization parameters are highly

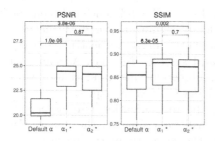

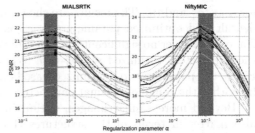

Fig. 5. Similarity (PSNR and SSIM) between the images reconstructed using MIALSRTK and NiftyMIC with default and optimal regularization parameters. Comparison is done between all 20 clinical exams; p-values from paired Wilcoxon rank sum test, statistical significance: $p < 0.05$).

Fig. 6. PSNR (left: MIALSRTK, right: NiftyMIC) for the simulated subjects of Experiment 2. Every light line represents an individual subject, and their average is the bold black curve. Dashed red (vertical) line is the default regularization for each method. Purple region highlights the range of optimal parameters determined in Experiment 1 at 1.5T. (Color figure online)

sub-optimally set for both SRR pipelines. This is further illustrated on Fig. 2, where we reconstruct simulated data and compare them to the corresponding ground truth image: using a default α, MIALSRTK tends to overly smooth the image, while NiftyMIC reconstructs images that are artificially sharp, enhancing edges beyond what is present on the reference image.

Gestational Age-Based Analysis. Since the human brain undergoes drastic morphological changes throughout gestation [20], one could expect to adjust α to GA. However, our experiments (cf. Supp. Figure 7) suggest that the α^* does not depend on GA, and is in line with the values reported on Fig. 1.

3.2 Experiment 2 – Clinical Environment

In this experiment, we compare two differently optimized regularization parameters. First, we use the optimal value α_1^* (from Fig. 1 at 1.5T, and rounded to the closest value on the grid of parameters). Second, we use the optimized regularization parameter α_2^* estimated from the subject-specific simulation.

Figure 3 shows the SRR of two subjects with default and optimal parameters using both pipelines. We observe that using the optimal parameters α_1^* and α_2^* makes the reconstructed images more similar. Indeed, the default parameters of MIALSRTK and NiftyMIC promote opposite behaviors, towards smoother (i.e., the default regularization is higher than the optimal one), respectively noisier (the default regularization is lower than the optimal one) images.

We quantitatively confirm the similarity of the optimized SRR by computing the PSNR and the SSIM between the reconstructed images from both methods for α_{def}, α_1^* and α_2^*. The results are shown in Fig. 5. The difference between the default and optimized parameters is statistically significant for both metrics. There is however no significant difference between the images reconstructed using the parameters optimized *setting-wise* (α_1^*) and *subject-wise* (α_2^*). As shown in Fig. 6, the parameters α_2^* optimized based on subject-specific simulations always lie within the range of optimal parameters α_1^* determined in Exp. 1.

Beyond more similar images, optimizing the regularization parameter can also matter in terms of the structures that will be visible on the image. On Fig. 4, using the optimal α_1^* allows to delineate the deep gray matter more clearly compared to the α_{def}.

4 Discussion and Conclusion

In this paper, we propose a novel simulation-based approach that addresses the need for automated, quantitative optimization of the regularization hyperparameter in ill-conditioned inverse problems, with a case study in the context of SRR fetal brain MRI. Our estimated regularization weight shows both qualitative and quantitative improvements over widely adopted default parameters. Our results also suggest that *subject-specific* parameter tuning – which is computationally expensive to run – might not be necessary, but that an *acquisition setting-specific* tuning, ran only once, might be sufficient in practice.

As such, the proposed methodology demonstrates a high practical value in a clinical setting where fetal MR protocols are not standardized, leading to heterogeneous acquisition schemes across centers and scanners. Besides, we show that our simulation-based optimization approach reduces the variability in image quality and appearance between the two SRR pipelines studied. We expect this behavior to contribute to mitigating the domain shift currently inherent to any reconstruction technique, a key challenge in the development of automated tissue segmentation methods [21,22]. Future work will address some limitations of this study. Indeed, we mostly focused on the influence of the main magnetic field strength and the number of available LR series, but the signal-to-noise ratio within LR series may also affect the regularization setting [12]. This aspect could also be tuned within the proposed MR acquisition simulation framework. Moreover, clinical assessment by radiologists of the different SRR would be important to further validate our method. Such an evaluation would allow to compare our approach to other techniques for parameter tuning, which cannot be done quantitatively due to the lack of HR ground truth data.

References

1. Griffiths, P., et al.: Use of MRI in the diagnosis of fetal brain abnormalities in utero (MERIDIAN): a multicentre, prospective cohort study. The Lancet **389**(10068), 538–546 (2017)
2. E. W. Group: Role of prenatal magnetic resonance imaging in fetuses with isolated mild or moderate ventriculomegaly in the era of neurosonography: international multicenter study. Ultrasound in Obstet. Gynecol. **56**(3), 340–347 (2020)
3. Gholipour, A., et al.: Fetal MRI: a technical update with educational aspirations. Concepts Magn. Reson. Part A **43**(6), 237–266 (2014)
4. Saleem, S.N.: Fetal MRI: an approach to practice: a review. J. Adv. Res. **5**(5), 507–523 (2014)
5. Kuklisova-Murgasova, M., Quaghebeur, G., Rutherford, M.A., Hajnal, J.V., Schnabel, J.A.: Reconstruction of fetal brain MRI with intensity matching and complete outlier removal. Med. Image Anal. **16**(8), 1550–1564 (2012). https://doi.org/10.1016/j.media.2012.07.004
6. Tourbier, S., Bresson, X., Hagmann, P., Thiran, J.-P., Meuli, R., Cuadra, M.B.: An efficient total variation algorithm for super-resolution in fetal brain MRI with adaptive regularization. NeuroImage **118**, 584–597 (2015). https://doi.org/10.1016/j.neuroimage.2015.06.018
7. Ebner, M., et al.: An automated framework for localization, segmentation and super-resolution reconstruction of fetal brain MRI, NeuroImage **206**, 116324 (2020)
8. Uus, A., et al.: Retrospective motion correction in foetal MRI for clinical applications: existing methods, applications and integration into clinical practice. Br. J. Radiol. **95**, 20220071 (2022)
9. Galatsanos, N.P., Katsaggelos, A.K.: Methods for choosing the regularization parameter and estimating the noise variance in image restoration and their relation. IEEE Trans. on Image Process. **1**(3), 322–336 (1992). https://doi.org/10.1109/83.148606
10. Afkham, B.M., et al.: Learning regularization parameters of inverse problems via deep neural networks. Inverse Prob. **37**(10), 105017 (2021)
11. Payette, K., et al.: An automatic multi-tissue human fetal brain segmentation benchmark using the fetal tissue annotation dataset. Sci. Data **8**(1), 167 (2021). https://doi.org/10.1038/s41597-021-00946-3
12. Lajous, H., et al.: A fetal brain magnetic resonance acquisition numerical phantom (FaBiAN). Sci. Rep. **12**(1), 8682 (2022)
13. Medical-Image-Analysis-Laboratory/FaBiAN: FaBiAN v2.0, Jul. (2023). https://doi.org/10.5281/zenodo.5471094
14. Gholipour, A., et al.: A normative spatiotemporal MRI atlas of the fetal brain for automatic segmentation and analysis of early brain growth. Sci. Rep. **7**(1), 476 (2017)
15. Weigel, M.: Extended phase graphs: dephasing, RF pulses, and echoes-pure and simple. J. Magn. Reson. Imaging **41**(2), 266–295 (2015)
16. Rousseau, F., et al.: Registration-based approach for reconstruction of high-resolution in utero fetal MR brain images. Acad. Radiol. **13**(9), 1072–1081 (2006)
17. Oubel, E., Koob, M., Studholme, C., Dietemann, J.-L., Rousseau, F.: Reconstruction of scattered data in fetal diffusion MRI. Med. Image Anal. **16**(1), 28–37 (2012)
18. Wang, Z., et al.: Image quality assessment: from error visibility to structural similarity. IEEE Trans. Image Process. **13**(4), 600–612 (2004)

19. Khawam, M., et al.: Fetal brain biometric measurements on 3D super-resolution reconstructed t2-weighted MRI: an intra-and inter-observer agreement study. Front. Pediatr. **9**, 639746 (2021)
20. Tierney, A., et al.: Brain development and the role of experience in the early years. Zero to three **30**(2), 9 (2009)
21. Payette, K., Kottke, R., Jakab, A.: Efficient multi-class fetal brain segmentation in high resolution MRI reconstructions with noisy labels. In: Hu, Y., et al. (eds.) Medical Ultrasound, and Preterm, Perinatal and Paediatric Image Analysis: First International Workshop, ASMUS 2020, and 5th International Workshop, PIPPI 2020, Held in Conjunction with MICCAI 2020, Lima, Peru, October 4-8, 2020, Proceedings, pp. 295–304. Springer, Cham (2020). https://doi.org/10.1007/978-3-030-60334-2_29
22. de Dumast, P., et al.: Synthetic magnetic resonance images for domain adaptation: application to fetal brain tissue segmentation. In: 2022 IEEE 19th International Symposium on Biomedical Imaging (ISBI) (2022)

TabAttention: Learning Attention Conditionally on Tabular Data

Michal K. Grzeszczyk[1(✉)], Szymon Płotka[1,2,3], Beata Rebizant[4],
Katarzyna Kosińska-Kaczyńska[4], Michał Lipa[5],
Robert Brawura-Biskupski-Samaha[4], Przemysław Korzeniowski[1],
Tomasz Trzciński[6,7,8], and Arkadiusz Sitek[9]

[1] Sano Centre for Computational Medicine, Cracow, Poland
m.grzeszczyk@sanoscience.org
[2] Informatics Institute, University of Amsterdam, Amsterdam, The Netherlands
[3] Amsterdam University Medical Center, Amsterdam, The Netherlands
[4] The Medical Centre of Postgraduate Education, Warsaw, Poland
[5] Medical University of Warsaw, Warsaw, Poland
[6] Warsaw University of Technology, Warsaw, Poland
[7] IDEAS NCBR, Warsaw, Poland
[8] Tooploox, Wroclaw, Poland
[9] Massachusetts General Hospital, Harvard Medical School, Boston, MA, USA

Abstract. Medical data analysis often combines both imaging and tabular data processing using machine learning algorithms. While previous studies have investigated the impact of attention mechanisms on deep learning models, few have explored integrating attention modules and tabular data. In this paper, we introduce TabAttention, a novel module that enhances the performance of Convolutional Neural Networks (CNNs) with an attention mechanism that is trained conditionally on tabular data. Specifically, we extend the Convolutional Block Attention Module to 3D by adding a Temporal Attention Module that uses multi-head self-attention to learn attention maps. Furthermore, we enhance all attention modules by integrating tabular data embeddings. Our approach is demonstrated on the fetal birth weight (FBW) estimation task, using 92 fetal abdominal ultrasound video scans and fetal biometry measurements. Our results indicate that TabAttention outperforms clinicians and existing methods that rely on tabular and/or imaging data for FBW prediction. This novel approach has the potential to improve computer-aided diagnosis in various clinical workflows where imaging and tabular data are combined. We provide a source code for integrating TabAttention in CNNs at https://github.com/SanoScience/Tab-Attention.

Keywords: Attention · Fetal Ultrasound · Tabular Data

1 Introduction

Many clinical procedures involve collecting data samples in the form of imaging and tabular data. New deep learning (DL) architectures fusing image and non-image data are being developed to extract knowledge from both sources of information and improve predictive capabilities [9]. While concatenation of tabular

H. Greenspan et al. (Eds.): MICCAI 2023, LNCS 14226, pp. 347–357, 2023.
https://doi.org/10.1007/978-3-031-43990-2_33

and imaging features in final layers is widely used [8,11], this approach limits the interaction between them. To facilitate better knowledge transfer between these modalities more advanced techniques have been proposed. Duanmu *et al.* [5] presented the Interactive network in which tabular features are passed through a separate branch and channel-wise multiplied with imaging features at different stages of Convolutional Neural Network (CNN). Pölsterl *et al.* [16] proposed a Dynamic Affine Feature Map Transform (DAFT) to shift and scale feature maps conditionally on tabular data. In [6], Guan *et al.* presented a method for transforming tabular data and processing them together with 3D feature maps via VisText self-attention module. The importance of the attention mechanism on DL models' performance has been extensively studied [23]. Convolutional Block Attention Module (CBAM) [25] has been shown to improve the performance of DL models on high dimensional data [24,26]. Despite these advances, few studies have explored the potential of incorporating attention maps with imaging and tabular data simultaneously.

We develop such a solution and as an example of application, we use fetal birth weight (FBW) prediction from ultrasound (US) data. It is a challenging task requiring clinicians to collect US videos of fetal body parts and fetal biometry measurements. Currently, abdominal circumference (AC), head circumference (HC), biparietal diameter (BPD), and femur length (FL) are used to estimate FBW with heuristic formulae [7]. The predicted weight is the indicator of perinatal health prognosis or complications in pregnancy and has an impact on the method of delivery (vaginal or Cesarean) [17]. Unfortunately, the current approach to FBW estimation is often imprecise and can lead to a mean absolute percentage error (MAPE) of 10%, even if performed by experienced sonographers [20]. An ensemble of Machine Learning algorithms was proposed by Lu *et al.* [12] for solving this task. CNNs are applied for fetal biometry measurements estimation from US standard planes [1] or US videos [15]. Tao *et al.* [21] approach this problem with a recurrent network utilizing temporal features of fetal weight changes over weeks concatenated with fetal parameters. Płotka *et al.* [14] developed BabyNet, a hybrid CNN with Transformer layers to estimate FBW directly from US videos. Recent studies show that there is a strong correlation between the image features of the abdominal plane and the estimated fetal weight, indicating that it can serve as a dependable indicator for evaluating fetal growth [3]. We utilize the US videos of the abdomen (imaging data) and biometry measurements with other numerical values (tabular data) during our experiments.

In this work, we introduce TabAttention, a novel module designed to enhance the performance of CNNs by incorporating tabular data. TabAttention extends the CBAM to the temporal dimension by adding a Temporal Attention Module (TAM) that leverages Multi-Head Self-Attention (MHSA) [23]. Our method utilizes pooled information from imaging feature maps and tabular data (represented as tabular embeddings) to generate attention maps through Channel Attention Module (CAM), Spatial Attention Module (SAM), and TAM. By incorporating tabular data, TabAttention enables the network to better identify *what*, *where*, and *when* to focus on, thereby improving performance. We evaluate our method on the task of estimating FBW from abdominal US

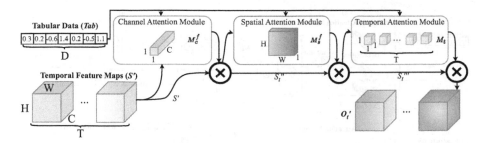

Fig. 1. The overview of proposed TabAttention module inspired by CBAM [25]. We add a Temporal Attention Module to the existing architecture to extend the method to 3D data processing. In our setting, channel, spatial and temporal attention maps are trained conditionally on tabular data. $\otimes$ indicates element-wise multiplication.

videos and demonstrate that TabAttention is at least on par with existing methods, including those based on tabular and/or imaging data, as well as clinicians. The main contributions of our work are: 1) the introduction of TabAttention, a module for conditional attention learning with tabular data, 2) the extension of CBAM to the temporal dimension via the TAM module, and 3) the validation of our method on the FBW estimation task, where we demonstrate that it is competitive with state-of-the-art methods.

2   Method

In this section, we introduce the fundamental components of the TabAttention module. We detail the development of CBAM augmented with a Temporal Attention Module. Then, we elaborate on how TabAttention leverages tabular embeddings to modulate the creation of attention maps and outline how the module can be seamlessly incorporated into the residual block of ResNet.

Figure 1 presents the overview of the TabAttention module. Given US video sequence $S \in \mathbb{R}^{T_0 \times 1 \times H_0 \times W_0}$ of height H_0, width W_0 and frame number T_0 as the input, 3D CNN produces intermediate temporal feature maps $S' \in \mathbb{R}^{T \times C \times H \times W}$ where C is the number of channels. In our setting, the CBAM block generates T 1D channel attention maps $M_c \in \mathbb{R}^{T \times C \times 1 \times 1}$ and T 2D spatial attention maps $M_s \in \mathbb{R}^{T \times 1 \times H \times W}$. We create attention maps separately for every temporal feature map as the information of *what* is meaningful and *where* it is important to focus on might change along the temporal dimension. To account for the temporal changes and focus on *when* is the informative part we add TAM which infers temporal attention map $M_t \in \mathbb{R}^{T \times 1 \times 1 \times 1}$. Intermediate temporal feature maps S' are refined with attention maps in the following way:

$$S'' = M_c(S') \otimes S' \qquad S''' = M_s(S'') \otimes S'' \qquad O' = M_t(S''') \otimes S''' \quad (1)$$

Here O' denotes the output of the module and $\otimes$ is an element-wise multiplication during which attention maps are broadcasted along all unitary dimensions.

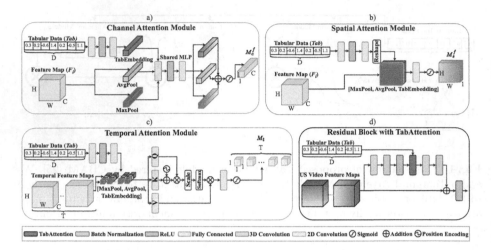

Fig. 2. Details of TabAttention components: CAM (a), SAM (b) and TAM (c) with tabular embeddings. In TAM, only one attention head is visualised. TabAttention is integrated with residual block as presented in (d). ⊗ indicates matrix multiplication.

In general, attention maps are computed based on information aggregated by average- and max-pooling along specified dimensions which are then passed through shared layers for refinement (Fig. 2). Then, these refined descriptors are passed through the sigmoid function to create final attention maps. To account for the tabular information during attention maps computing, we embed the input tabular data $Tab \in \mathbb{R}^D$, where D is the number of numerical features, with two linear layers and Rectified Linear Unit (ReLU) activation in between. The tabular data is embedded to the size of pooled feature maps. The embedding is passed through shared layers in the same way as pooled feature maps. Therefore, the attention maps are computed conditionally on tabular data. Thus, the output of TabAttention O'_t is computed as follows:

$$S''_t = M_c(S', Tab) \otimes S' \qquad S'''_t = M_s(S''_t, Tab) \otimes S''_t \qquad O'_t = M_t(S'''_t, Tab) \otimes S'''_t$$
(2)

Channel Attention Module. We follow the design of the original CBAM [25]. We split temporal feature maps into T feature maps F_i where $i \in 1, ..., T$ so that each of them is passed through CAM separately. To compute the channel attention (M_c), we aggregate the spatial information through average- and max-pooling to produce descriptors ($F^c_{avg_i}$, $F^c_{max_i} \in \mathbb{R}^{C \times 1 \times 1}$). We pass the tabular data through a multi-layer perceptron (MLP_{emb_c}) with one hidden layer (of size $\mathbb{R}^{\frac{C}{z}}$, where z is the reduction ratio set to 16) and ReLU activation to embed it into the same dimension as spatial descriptors. Then, both descriptors, with tabular embedding are passed through the shared network which is MLP with a hidden activation size of $\mathbb{R}^{\frac{C}{z}}$ and one ReLU activation. After the MLP is applied, the output vectors are element-wise summed to produce the attention map. We concatenate attention maps of all feature maps to produce M_c:

$$M_c(S', Tab) = [M_c^f(F_i, Tab)]_{i=1,...,T}$$
$$M_c^f(F_i, Tab) = \sigma(MLP(F_{max_i}^c) + MLP(F_{avg_i}^c) + MLP(MLP_{emb_c}(Tab))) \tag{3}$$

Spatial Attention Module. After splitting the temporal feature maps, we average- and max-pool them along channel dimension to produce feature descriptors ($F_{avg_i}^s$, $F_{max_i}^s \in \mathbb{R}^{1 \times H \times W}$). We pass the tabular data through MLP_{emb_s} with one hidden layer of size $\mathbb{R}^{\frac{H \times W}{2}}$ and ReLU activation to embed it into the same dimension as spatial descriptors. We reshape this embedding to the size of feature descriptors and concatenate it with them. We pass the following representation through a 2D convolution layer and the sigmoid activation:

$$M_s(S'', Tab) = [M_s^f(F_i, Tab)]_{i=1,...,T}$$
$$M_s^f(F_i, Tab) = \sigma(Conv([F_{max_i}^s, F_{avg_i}^s, Reshape(MLP_{emb_s}(Tab))])) \tag{4}$$

Temporal Attention Module. We create temporal descriptors by average- and max-pooling temporal feature maps along all non-temporal dimensions ($F_{avg_i}^t$, $F_{max_i}^t \in \mathbb{R}^{T \times 1 \times 1 \times 1}$). We embed tabular data with MLP_{emb_t} with one hidden layer of size $\mathbb{R}^{\frac{T}{2}}$ into the same dimension. We concatenate created vectors and treat them as the embedding of the US sequence which we pass to the MHSA layer (with 2 heads). We create the query (Q), key (K) and value (V) with linear layers and an output size of d (4). We add relative positional encodings [19] r to K. After passing through MHSA, we squash the refined representation with one MLP layer and sigmoid function to create a temporal attention map M_t:

$$MHSA(S_{emb}) = MLP \left(\left[softmax \left(\frac{Q_j(K_j + r)^T}{\sqrt{d}} \right) V_j \right]_{j=1,2} \right) \tag{5}$$
$$M_t(S''', Tab) = \sigma(MHSA([F_{max}^t, F_{avg}^t, MLP_{emb_t}(Tab)]))$$

TabAttention can be integrated within any 3D CNN (or 2D CNN in case TAM is omitted). As illustrated in Fig. 2, we add TabAttention between the first ReLU and the second convolution in the residual block to integrate our module with 3D ResNet-18.

3   Experiments and Results

This section describes the dataset used and provides implementation details of our proposed method. We benchmark the performance of TabAttention against several state-of-the-art methods and compare them to results obtained by clinicians. Additionally, we conduct an ablation study to demonstrate the significance of each key component utilized in our approach.

Dataset. This study was approved by the Ethics Committee of the Medical University of Warsaw (Reference KB.195/2021) and informed consent was obtained

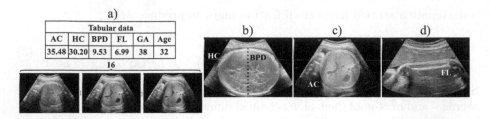

Fig. 3. Exemplary input to our method (a) of tabular data and frames of abdominal scans. Fetal US scans of the head (b), abdomen (c), and femur (d) were used to obtain AC, HC, BPD and FL utilized in tabular data.

for all subjects. The multi-site dataset was acquired using international standards approved by [18]. The dataset consists of 92 2D fetal US video scans captured in the standard abdominal plane view. These scans were collected from 92 pregnant women (31.89 ± 4.76 years), across three medical centers, and obtained as part of routine US examination done less than 24 h before delivery. This allowed us to obtain the real ground truth which was baby weight soon after birth. Five experienced sonographers (14.2 ± 4.02 years of experience) acquired the data using a single manufacturer device (General Electric) and several models (GE Voluson E6, S8, P8, E10, and S10). The abdominal fetal US videos (5–10 seconds, 13–37 frames per second) were saved in the DICOM file format. We resized the pixel spacing to 0.2 mm × 0.2 mm for all video clips. As tabular data, we used six numerical features: AC (34.51 ± 2.35 cm), HC (33.56 ± 1.41 cm), BPD (9.40 ± 0.46 cm), FL (7.28 ± 0.33 cm), GA (38.29 ± 1.47 weeks), and mother's age. The examples of how the measurements were obtained are presented in Fig. 3. The actual birth weight of the fetus obtained right post-delivery (3495 ± 507 g) was used as the target of the prediction.

Implementation Details. We use 3D ResNet-18 as our base model. We implement all experiments with PyTorch and train networks using NVIDIA A100 80GB GPU for 250 epochs with a batch size of 16 and an initial learning rate chosen with grid search from the set of $\{1 \times 10^{-2}, 1 \times 10^{-3}, 1 \times 10^{-4}\}$. To minimize the Mean Squared Error loss function, we employ the Adam [10] optimizer with L2 regularization of 1×10^{-4} and cosine annealing learning rate scheduler. To evaluate the reliability of the regression algorithm, we conduct five-fold cross-validation (CV) and ensure that each patient's data is present in only one fold. To ensure similar birth weight distribution in all folds, we stratify them based on the assignment of data samples into three bins: < 3000 g g, > 4000 g g, and in-between. The input frames are of size 128 × 128 pixels. We follow the approach presented in [14], we set the number of input frames to 16 and average per-patient predictions of all 16 frame segments from the single video. Throughout the training process, we employ various data augmentation techniques such as rotation, random adjustments to brightness and contrast, the addition of Gaussian noise, horizontal flipping, image compression, and motion blurring for every batch. We standardize all numerical features to a mean of 0

Table 1. Five-fold cross-validation results of state-of-the-art methods utilizing imaging (Img.) and/or tabular (Tab.) data. The mean of MAE, RMSE, and MAPE across all folds are presented. The best results are bolded.

Method	Img.	Tab.	mMAE [g]	mRMSE [g]	mMAPE [%]
BabyNet [14]	✓	✗	294 ± 30	386 ± 56	8.5 ± 1.0
3D ResNet-18 [22]	✓	✗	289 ± 38	373 ± 43	8.5 ± 1.1
XGBoost	✗	✓	259 ± 23	328 ± 26	7.6 ± 0.1
Linear Regression	✗	✓	207 ± 18	260 ± 19	6.0 ± 0.1
Clinicians	✗	✓	205 ± 14	253 ± 13	5.9 ± 0.0
DAFT [16]	✓	✓	175 ± 30	244 ± 42	5.3 ± 1.0
Interactive [5]	✓	✓	172 ± 27	230 ± 44	5.2 ± 0.9
TabAttention (ours)	✓	✓	**170 ± 26**	**225 ± 37**	**5.0 ± 0.8**

and a standard deviation of 1. We use Root Mean Square Error (RMSE), Mean Absolute Error (MAE), and MAPE to evaluate the regression performance.

Comparison with State-of-the-Art Methods. We compare TabAttention with several methods utilizing tabular data only (Linear Regression [13], XGBoost [4]), imaging data only (3D ResNet-18 [22], BabyNet [14]), both types of data (Interactive [5], DAFT [16]), and Clinicians. The predictions of Clinicians were achieved using Hadlock III [7] formula and AC, HC, BPD, FL measurements. The comparison of results from the five-fold CV is presented in Table 1. TabAttention achieves the lowest MAE, RMSE and MAPE (170 ± 26, 225 ± 37, 5.0 ± 0.8 respectively) among all tested methods. Our approach outperforms clinically utilized heuristic formulae, machine learning, and image-only DL methods (two-tailed paired t-test p-value < 0.05). Results of TabAttention are also best compared with all DL models utilizing tabular and imaging modalities, however, the difference does not reach statistical significance with a p-value around 0.11.

Ablation Study. We conduct ablation experiments to validate the effectiveness of key components of our proposed method (Table 2). We employ 3D ResNet-18 as the baseline model. The integration of TAM or CBAM with attention maps learned conditionally on tabular data into the 3D ResNet-18 architecture improves the predictive performance of the network. Subsequently, the incorporation of full TabAttention further enhances its capabilities.

Table 2. Five-fold cross-validation results of ablation study with key components of TabAttention. The first row is the result of the baseline method. The next rows refer to modules of TabAttention with or without tabular embeddings (Tab.) and the last one is full TabAttention.

Method	Img.	Tab.	mMAE [g]	mRMSE [g]	mMAPE [%]
3D ResNet-18 [22]	✓	✗	289 ± 38	373 ± 43	8.5 ± 1.1
+ TAM	✓	✗	288 ± 43	389 ± 65	8.4 ± 1.2
+ CBAM + Tab	✓	✓	271 ± 51	371 ± 99	7.7 ± 1.3
+ TAM + Tab	✓	✓	180 ± 32	237 ± 45	5.5 ± 0.1
+ **TabAttention (ours)**	✓	✓	$\mathbf{170 \pm 26}$	$\mathbf{225 \pm 37}$	$\mathbf{5.0 \pm 0.8}$

4   Discussion and Conclusions

In this work, we present a novel method, TabAttention, that can effectively compete with current state-of-the-art image and/or tabular-based approaches in estimating FBW. We found that it outperformed Clinicians achieving mMAPE of 5.0% vs. 5.9% (p-value < 0.05). A key advantage of our approach is that it does not require any additional effort from clinicians since the necessary data is already collected as part of standard procedures. This makes TabAttention an alternative to the heuristic formulas that are currently used in clinical practice. We should note that while TabAttention achieved the lowest metrics among the DL models we evaluated, the differences between our approach and other DL methods using tabular data were not statistically significant, partly due to the small performance change. This small difference in the performance is likely caused by the fact that the tabular features used in TabAttention are mainly derived from the same modality (i.e. US scans), so they do not carry additional information, but instead can be considered as refined features already present in the scans. To develop TabAttention, we used tabular data as a hint for the network to learn attention maps and gain additional knowledge about essential aspects presented in the scans. This approach significantly improved the performance of baseline methods and demonstrated its practical applicability.

Accurate estimation of FBW is crucial in determining the appropriate delivery method, whether vaginal or Cesarean. Low birth weight (less than 2500 g g) is a major risk factor for neonatal death, while macrosomia (greater than 4000 g) can lead to delivery traumas and maternal complications, such as birth canal injuries, as reported by Benacerraf et al. [2]. Thus, precise prediction of FBW is vital for very low and high weights. Notably, in this respect, our method is robust to outliers with high or low FBW since there is no correlation between true FBW and absolute prediction error (Pearson correlation coefficient of -0.029).

This study has limitations. Firstly, a relatively small study cohort was used, which may affect the accuracy and generalization of the results. To address this, future work will include a larger sample size by using additional datasets. Secondly, our dataset is limited to only Caucasian women and may not be representative of other ethnicities. It is important to investigate the performance of our method with datasets from different ethnic groups and US devices to obtain more robust and generalizable results. Lastly, our method relies on fetal biometry measurements that are subject to inter- and intra-observer variabilities. This variability could potentially affect the network's performance and influence the measurements' quality. Future studies should consider strategies to reduce measurement variabilities, such as standardized protocols or automated measurements, to improve the accuracy of the method.

To summarize, we have introduced TabAttention, a new module that enables the conditional learning of attention on tabular data and can be integrated with any CNN. Our method has many potential applications, including serving as a computer-aided diagnosis tool for various clinical workflows. We have demonstrated the effectiveness of TabAttention on the FBW prediction task, utilizing both US and tabular data, and have shown that it outperforms other methods, including clinically used ones. In the future, we plan to test the method in different clinical applications where imaging and tabular data are used together.

Acknowledgements. This work is supported by the European Union's Horizon 2020 research and innovation programme under grant agreement Sano No 857533 and the International Research Agendas programme of the Foundation for Polish Science, co-financed by the European Union under the European Regional Development Fund.

References

1. Bano, S., et al.: AutoFB: automating fetal biometry estimation from standard ultrasound planes. In: de Bruijne, M., et al. (eds.) MICCAI 2021. LNCS, vol. 12907, pp. 228–238. Springer, Cham (2021). https://doi.org/10.1007/978-3-030-87234-2_22
2. Benacerraf, B.R., Gelman, R., Frigoletto, F.D., Jr.: Sonographically estimated fetal weights: accuracy and limitation. Am. J. Obstet. Gynecol. **159**(5), 1118–1121 (1988)
3. Campbell, S., Wilkin, D.: Ultrasonic measurement of fetal abdomen circumference in the estimation of fetal weight. BJOG: Int. J. Obstet. Gynaecol. **82**(9), 689–697 (1975)
4. Chen, T., Guestrin, C.: XGBoost: a scalable tree boosting system. In: Proceedings of the 22nd ACM SIGKDD International Conference on Knowledge Discovery and Data Mining, pp. 785–794. KDD '16, ACM, New York, NY, USA (2016). https://doi.org/10.1145/2939672.2939785
5. Duanmu, H., et al.: Prediction of pathological complete response to neoadjuvant chemotherapy in breast cancer using deep learning with integrative imaging, molecular and demographic data. In: Martel, A.L., et al. (eds.) MICCAI 2020. LNCS, vol. 12262, pp. 242–252. Springer, Cham (2020). https://doi.org/10.1007/978-3-030-59713-9_24
6. Guan, Y., et al.: Predicting esophageal fistula risks using a multimodal self-attention network. In: de Bruijne, M., et al. (eds.) MICCAI 2021. LNCS, vol. 12905, pp. 721–730. Springer, Cham (2021). https://doi.org/10.1007/978-3-030-87240-3_69

7. Hadlock, F.P., Harrist, R., Sharman, R.S., Deter, R.L., Park, S.K.: Estimation of fetal weight with the use of head, body, and femur measurements-a prospective study. Am. J. Obstet. Gynecol. **151**(3), 333–337 (1985)

8. Holste, G., Partridge, S.C., Rahbar, H., Biswas, D., Lee, C.I., Alessio, A.M.: End-to-end learning of fused image and non-image features for improved breast cancer classification from MRI. In: Proceedings of the IEEE/CVF International Conference on Computer Vision, pp. 3294–3303 (2021)

9. Huang, S.C., Pareek, A., Seyyedi, S., Banerjee, I., Lungren, M.P.: Fusion of medical imaging and electronic health records using deep learning: a systematic review and implementation guidelines. NPJ Digital Med. **3**(1), 136 (2020)

10. Kingma, D.P., Ba, J.: Adam: a method for stochastic optimization. In: International Conference on Learning Representations (ICLR) (2015)

11. Liu, M., Zhang, J., Adeli, E., Shen, D.: Joint classification and regression via deep multi-task multi-channel learning for Alzheimer's disease diagnosis. IEEE Trans. Biomed. Eng. **66**(5), 1195–1206 (2018)

12. Lu, Y., Zhang, X., Fu, X., Chen, F., Wong, K.K.: Ensemble machine learning for estimating fetal weight at varying gestational age. In: Proceedings of the AAAI Conference on Artificial Intelligence, vol. 33, pp. 9522–9527 (2019)

13. Pedregosa, F., et al.: Scikit-learn: machine learning in Python. J. Mach. Learn. Res. **12**, 2825–2830 (2011)

14. Płotka, S., et al.: BabyNet: residual transformer module for birth weight prediction on fetal ultrasound video. In: Medical Image Computing and Computer Assisted Intervention-MICCAI 2022: 25th International Conference, Singapore, September 18–22, 2022, Proceedings, Part IV, pp. 350–359. Springer (2022). https://doi.org/10.1007/978-3-031-16440-8_34

15. Płotka, S., et al.: Deep learning fetal ultrasound video model match human observers in biometric measurements. Phys. Med. Biol. **67**(4), 045013 (2022)

16. Pölsterl, S., Wolf, T.N., Wachinger, C.: Combining 3D image and tabular data via the dynamic affine feature map transform. In: de Bruijne, M., et al. (eds.) MICCAI 2021. LNCS, vol. 12905, pp. 688–698. Springer, Cham (2021). https://doi.org/10.1007/978-3-030-87240-3_66

17. Pressman, E.K., Bienstock, J.L., Blakemore, K.J., Martin, S.A., Callan, N.A.: Prediction of birth weight by ultrasound in the third trimester. Obstet. Gynecol. **95**(4), 502–506 (2000)

18. Salomon, L., et al.: ISUOG practice guidelines: ultrasound assessment of fetal biometry and growth. Ultrasound Obstet. Gynecol. **53**(6), 715–723 (2019)

19. Shaw, P., Uszkoreit, J., Vaswani, A.: Self-attention with relative position representations. arXiv preprint arXiv:1803.02155 (2018)

20. Sherman, D.J., Arieli, S., Tovbin, J., Siegel, G., Caspi, E., Bukovsky, I.: A comparison of clinical and ultrasonic estimation of fetal weight. Obstet. Gynecol. **91**(2), 212–217 (1998)

21. Tao, J., Yuan, Z., Sun, L., Yu, K., Zhang, Z.: Fetal birthweight prediction with measured data by a temporal machine learning method. BMC Med. Informa. Decis. Making **21**(1), 1–10 (2021)

22. Tran, D., Wang, H., Torresani, L., Ray, J., LeCun, Y., Paluri, M.: A closer look at spatiotemporal convolutions for action recognition. In: Proceedings of the IEEE Conference on Computer Vision and Pattern Recognition, pp. 6450–6459 (2018)

23. Vaswani, A., et al.: Attention is all you need. In: Advances in Neural Information Processing Systems 30 (2017)

24. Wang, X., Liu, D., Zhang, Y., Li, Y., Wu, S.: A spatiotemporal multi-stream learning framework based on attention mechanism for automatic modulation recognition. Digit. Signal Process. **130**, 103703 (2022)
25. Woo, S., Park, J., Lee, J.Y., Kweon, I.S.: CBAM: convolutional block attention module. In: Proceedings of the European Conference on Computer Vision (ECCV), pp. 3–19 (2018)
26. Yadav, S., Rai, A.: Frequency and temporal convolutional attention for text-independent speaker recognition. In: ICASSP 2020–2020 IEEE International Conference on Acoustics, Speech and Signal Processing (ICASSP), pp. 6794–6798. IEEE (2020)

An Automated Pipeline for Quantitative T2* Fetal Body MRI and Segmentation at Low Field

Kelly Payette[1,2(✉)], Alena Uus[1,2], Jordina Aviles Verdera[1,2],
Carla Avena Zampieri[1,2], Megan Hall[1,3], Lisa Story[1,2], Maria Deprez[1,2],
Mary A. Rutherford[1], Joseph V. Hajnal[1,2], Sebastien Ourselin[1,2],
Raphael Tomi-Tricot[2,4], and Jana Hutter[1,2]

[1] Centre for the Developing Brain, School of Biomedical Engineering and Imaging
Sciences, King's College London, London, UK
[2] Department of Biomedical Engineering, School of Biomedical Engineering and
Imaging Sciences, King's College London, London, UK
kelly.m.payette@kcl.ac.uk
[3] Department of Women and Children's Health, King's College London, London, UK
[4] MR Research Collaborations, Siemens Healthcare Limited, Camberley, UK

Abstract. Fetal Magnetic Resonance Imaging at low field strengths is
emerging as an exciting direction in perinatal health. Clinical low field
(0.55T) scanners are beneficial for fetal imaging due to their reduced
susceptibility-induced artefacts, increased T2* values, and wider bore
(widening access for the increasingly obese pregnant population). However, the lack of standard automated image processing tools such as segmentation and reconstruction hampers wider clinical use. In this study,
we introduce a semi-automatic pipeline using quantitative MRI for the
fetal body at low field strength resulting in fast and detailed quantitative T2* relaxometry analysis of all major fetal body organs. Multi-echo
dynamic sequences of the fetal body were acquired and reconstructed into
a single high-resolution volume using deformable slice-to-volume reconstruction, generating both structural and quantitative T2* 3D volumes.
A neural network trained using a semi-supervised approach was created
to automatically segment these fetal body 3D volumes into ten different
organs (resulting in dice values > 0.74 for 8 out of 10 organs). The T2*
values revealed a strong relationship with GA in the lungs, liver, and
kidney parenchyma ($R^2 > 0.5$). This pipeline was used successfully for a
wide range of GAs (17–40 weeks), and is robust to motion artefacts. Low
field fetal MRI can be used to perform advanced MRI analysis, and is a
viable option for clinical scanning.

K. Payette and A. Uus—Authors contributed equally.

Supplementary Information The online version contains supplementary material
available at https://doi.org/10.1007/978-3-031-43990-2_34.

Keywords: Fetal MRI · Low field · T2*

1 Introduction

Fetal magnetic resonance imaging (MRI) is becoming increasingly common, supplementing ultrasound for clinical decision-making and planning. It has a wide range of functional contrasts, a higher resolution than ultrasound, and can be used from approximately 16 weeks of gestation until birth. The renewed interest in low-field MRI (0.55T) and the increasing availability of commercial low field scanners carries significant advantages for fetal MRI: Low-field MRIs allow a wider bore (widening access to this tool for the increasingly obese pregnant population) while maintaining field homogeneity at the lower field strength, and generally do not require helium for cooling. Low field MRI is especially advantageous for fetal functional imaging (often performed using Echo-Planar-Imaging) as the reduced susceptibility-induced artefacts and longer T2* times allow for longer read-outs and hence more efficient acquisitions. It therefore provides an excellent environment for fetal body T2* mapping [1].

T2* maps of the fetus provide an indirect measurement of blood oxygenation levels due to the differing relaxation times of deoxygenated and oxygenated hemoglobin [12,13]. Many fetal body organs including the lungs, kidneys, heart, liver, spleen have changing T2* values throughout gestation [2,15], indicating that quantitative measurements of fetal body T2* organs have the potential to be a clinically useful measurement.

However, the wider use of T2* relaxometry in the clinical setting is currently limited by significant methodological barriers such as quality of the data, fetal motion and time-consuming manual segmentations. As a consequence, it has mainly been used in research settings focused on the brain and placenta [3,6,14]. Studies to date have acquired only low-resolution images of the fetal body, which do not allow for proper imaging of small organs such as the adrenal gland, and they are still susceptible to motion artifacts. A lengthy echo time (TE) is required to acquire high-resolution multi-echo sequences. This is both impractical given the unpredictable motion and results in limited SNR on the later TEs. In structural fetal imaging, super-resolution algorithms have been used to transform several low-resolution scans into a single high-resolution volume [5,9,19]. High-resolution reconstructions of fetal T2* maps have only been done as proof of concept in phantoms [10] or in the brain [3]. Once they are acquired, the images require manual reorientation to a standard plane followed by manual segmentation in order to be clinically useful. The segmentations must also be very accurate, as inclusion of other organs or major vessels drastically changes the resulting T2* values. Overall, current state of T2* fetal body organ analysis involves a lengthy acquisition process followed by tedious manual image processing steps, resulting in a barrier to wider adoption of the technique in the clinical setting.

Here, we present an automated pipeline for quantitative mean T2* fetal body organs at low field MRI, resulting in normative growth curves from 17–40 weeks.

We use a low-resolution dynamic T2* acquisition framework, and then use a novel multi-channel deformable slice-to-volume reconstruction (dSVR) to generate a high-resolution 3D volume of the fetal body and its corresponding T2* map in the standard plane [18–20], followed by automatic segmentation of the fetal body organs across a wide range of gestational ages (GA; 17–40 weeks). We generate normative T2* growth curves of ten fetal body organs at low field MRI, which has not been done at higher field strengths in such detail. This pipeline (available: https://github.com/meerkat-tools/Fetal-T2star-Recon) will pave the way for advanced T2* fetal body mapping to become more prevalent in both research and the clinic, potentially allowing further insights into prenatal development, and better screening and diagnostic capacities.

2 Methods

Fetal MRI was acquired as part of an ethically approved study (MEERKAT, REC 21/LO/0742, Dulwich Ethics Committee, 08/12/2021) performed between May 2022 and February 2023 at St Thomas' Hospital in London, UK. Participants for this study were recruited prospectively, with inclusion criteria of a singleton pregnancy, maternal age over 18 years. Exclusion criteria were multiple pregnancies, maternal age <18 years, lack of ability to consent, weight >200 kg, and contraindications for MRI such as metal implants.

2.1 Image Acquisition

The subjects were scanned on a clinical 0.55T scanner (MAGNETOM Free.Max, Siemens Healthcare, Germany) using a 6-element blanket coil and a 9-element spine coil in the supine position. A multi-echo whole-uterus dynamic (time-resolved) gradient echo sequence was acquired in the maternal coronal orientation. The sequence parameters are as follows: Field of View (FOV): 400×400 mm; resolution: 3.125 mm $\times 3.125$ mm $\times 3$ mm; TE: [46, 120, 194] ms; TR: 10,420–18,400 ms; Number of slices: 28–85 Number of dynamics: 15–30; GRAPPA: 2; flip angle 90°. The acquisition time of the dynamic T2* relaxometry scan with 20 dynamics and 3 echoes was between 4–6 min.

2.2 Image Processing

The acquired images were reviewed to remove dynamics with excess motion. The remaining images were first denoised using MRTRIX3's dwidenoise tool [4,17]. Next, an in-house Python script was used to generate the T2* maps for each dynamic using mono-exponential decay fitting for each voxel [7].

Reconstructions. The image generated from the second echo was determined to have the best contract for the fetal body organs (an example of the echos can be seen in the Supplementary Material). This second echo image for each dynamic was then used to create a high-resolution 3D volume with isotropic resolution in a standard atlas space using deformable slice-to-volume (dSVR) registration, with the following non-default parameters: no intensity matching, no robust statistics, resolution = 1.2 mm, cp = [12 5], lastIter=0.015 [19,20]. The resulting transformations are then used to create a high resolution 3D T2* map [18].

Fetal Lung Reconstruction Validation. In order to validate that the 3D volume reconstructions do not alter the obtained mean T2* values, mean T2* values from 10 of the acquired dynamics were compared with mean lung T2* value of the corresponding 3D volume. The values were considered to be equal if the mean T2* determined from the 3D volume was within two standard deviations of the mean T2* values sampled from the 10 dynamic scans.

Segmentation. Initial segmentations of the second echo 3D volumes were generated with an existing in-house U-Net using the MONAI framework [11] that had been trained on dSVR reconstructions from T2-weighted (HASTE) images of the fetal body acquired at 1.5T and 3T and corresponding manual segmentations for the lungs, thymus, gall bladder, kidney pelvis, kidney parenchyma, spleen, adrenal gland, stomach, bladder, and liver (See Fig. 1). Seven of the initially generated label maps were corrected in detail to create the ground truth.

A 3D nnUNet [8] was trained on a Tesla V100 (5 folds, 250 epochs/fold, default nnUnet data augmentation, batch size: 2, kernel size: [3,3,3], Adam opti-

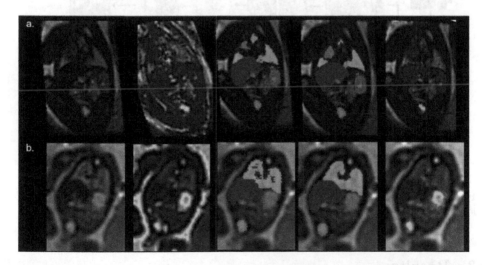

Fig. 1. From left to right: Example fetal deformable Slice-to-Volume Reconstructions (dSVR) of the second echo; the T2* map dSVR; the corresponding manual segmentations; label maps generated from the two-channel network; organ T2* maps overlaid on the second echo 3D volume reconstruction. Top row: 35.71 week, bottom row: 23.14 weeks

mizer, loss function: cross-entropy and Dice, 23 training cases, 4 validation cases, 7 testing cases) using these initial label maps as well as the second echo and the T2* 3D volumes. The final label maps were generated using a semi-automated refinement process, where at the end of each training cycle the cases with the best label maps were chosen, minor corrections were made by fetal anatomy specialists (Intra and inter-observer variability had previously been confirmed [16] before retraining. Two networks were trained, one with only the second echo 3D volumes as input (one channel), and one with both the second echo and the T2* map 3D volumes (two channel). Dice Similarity Coefficients (DSC) were calculated for both the one- and a two-channel networks and a two-sided t-test was performed.

Growth Curves. Organ-specific volume and T2* growth curves were created using control cases from the generated label maps and the high-resolution T2* volumes, excluding voxels where the T2* fitting failed. A linear regression analysis was performed in order to determine the relationship between the T2* values and GA.

The complete fetal body T2* 3D reconstruction and segmentation processing pipeline can be seen in Fig. 2.

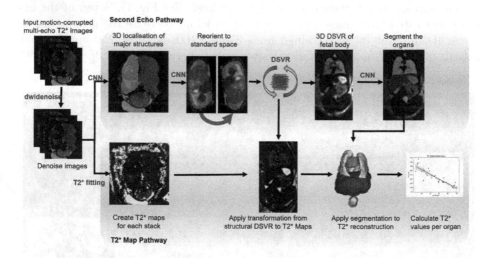

Fig. 2. Proposed pipeline for quantitative T2* fetal body MRI and segmentation at low field

3   Results

3.1   Fetal Body T2* Reconstruction

41 subjects had a multi-echo T2* dynamic scan (mean GA: 28.46 ± 6.78 weeks; mean body mass index: 29.0 ± 6.26). Dynamics with motion artifact were

removed (mean included: 13.6). All cases underwent 3D volume reconstruction. Both control and pathological cases were used during the iterative training process for the segmentation network. Nine cases were excluded from the creation of growth curves due to a pathology impacting the fetal body. A further two cases were excluded after reconstruction due to excessive motion.

Fetal Lung Reconstruction Validation. In the lung reconstruction validation experiment, all cases fell within the required range, and therefore are considered to be equal (Table 2). See Fig. 3 for an example of the individual dynamics and the final 3D volume reconstruction.

3.2   Fetal Body T2* Segmentation

The DSC scores of the one- and two-channel networks (Table 1) showed no significant difference. The two-channel network was used to generate the T2* values in the growth curves.

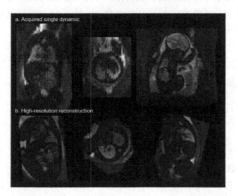

Fig. 3. Top row: Example of a single, motion-corrupted dynamic; Bottom Row: The reconstructed 3D volume of the same case

3.3   Growth Curves

The obtained T2* growth curves (Fig. 4) show a significant increase over gestation in the lungs and liver ($R^2 > 0.5$) and a significant decrease in the kidney parenchyma and kidney pelvis ($R^2 > 0.5$). A slight relationship with GA was also found in the spleen, adrenal glands, and gallbladder. No relationship with GA was found in the stomach, bladder, and thymus. The volumetric growth curves for all organs can be found in the Supplementary Material.

Table 1. Mean Dice Similarity Coefficients (DSC) for each label. One Channel: Trained using the second echo 3D volume; Two Channel: Trained using both the second echo and the T2* 3D volume. No significant difference was found between the two networks.

Label	One Channel	Two Channel	p-value
	DSC mean	DSC mean	
Lungs	0.893	0.892	0.97
Liver	0.910	0.910	0.99
Stomach	0.901	0.900	0.96
Spleen	0.7934	0.789	0.95
Kidney pelvis	0.772	0.774	0.96
Kidney Parenchyma	0.847	0.847	1.00
Bladder	0.918	0.917	0.97
Thymus	0.411	0.416	0.98
Gallbladder	0.744	0.747	0.98
Adrenal Glands	0.449	0.453	0.97

Table 2. Reconstruction robustness analysis: All mean lung T2* values were determined from 10 of the dynamics. All values from the reconstructed values fall within the 2σ range, and are therefore considered to be equal.

Case	Mean T2*	T2* 2σ range	T2* Reconstruction
1	194.05	188.9–199.2	189.07
2	299.49	271.9–327.1	312.25
3	225.93	207.9–244.0	236.09

4   Discussion and Conclusions

The data (including images, reconstructions, and segmentations) will be made available from the corresponding author upon reasonable academic request (REC 21/LO/0742).

The proposed pipeline overcomes barriers currently hindering wider clinical adaption such as burdensome manual image reorientation and segmentation. It requires users only to acquire the images and review the input scans for motion before starting the pipeline. This automation would allow the pipeline to move advanced fetal image analysis outside of specialist centers and into a more standard workflow. The pipeline worked from 18–40 GA, with only 2 cases (which were <20 weeks) discarded due to motion, and was able to confirm literature trends for all organs except the liver and lungs [2, 15], where the opposite trend was observed. This may be due to the different GA ranges included in the literature. The increasing mean T2* values in the lungs may be due to the increasing vascularity, thereby increasing the amount of fetal haemoglobin present. The long T2* values achieved with the low field MRI allows for excellent contrast in the fetal body, which assists in the reconstruction. This pipeline will provide

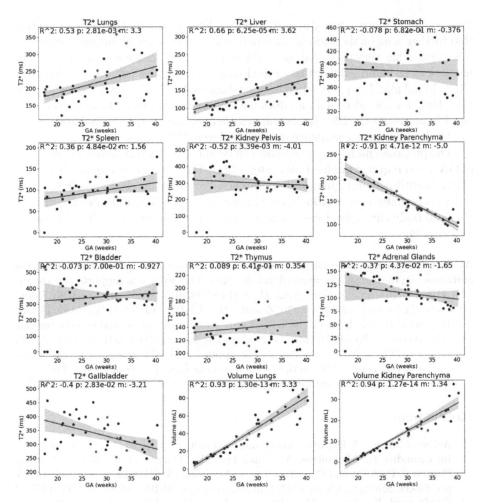

Fig. 4. T2* growth curves for the lungs (R^2: 0.53), liver (R^2: 0.66), stomach (R^2: −0.08), spleen (R^2: 0.36), kidney pelvis (R^2: −0.52), kidney parenchyma (R^2: −0.9), bladder (R^2: −0.07), thymus (R^2: 0.09), adrenal glands (R^2: −0.37), and gallbladder (R^2: −0.40). Two of the graphs in the bottom row (middle, right) display the lung and kidney parenchyma volumes. All organ volumes demonstrated a strong relationship with GA. Blue: Fetuses with normal organs; Red: Fetuses with a pathology potentially impacting body organs.

insight into the development of fetal body organs not yet explored, such as the adrenal glands. A more comprehensive study with more cases at every GA is needed to further validate these normative curves.

While structural T2-weighted images are more suited for volumetry, the fact that all organs follow the expected growths trend further validates the segmentations. It also indicates that for some of the smaller organs (kidney pelvis, adrenal glands), our network has difficulties in the segmentation step for younger fetuses.

Two organs (thymus, adrenal glands) had poor DSC values (below 0.45), indicating that further work on the segmentation network for these difficult organs is required. The thymus has very poor contrast and is difficult to delineate even manually. The network often identified heart tissue as thymus, skewing the T2* values calculated. The adrenal gland is a very small organ, which makes it difficult to segment. The poor DSC for the adrenal gland does not necessarily translate to incorrect average T2* values, as the DSC is a volumetric metric. However, improved DSC scores would allow for more confidence in the calculated T2* values.

The proposed multi-organ pipeline can be successfully run across a wide range of GAs, and requires minimal user interaction. The combination of low field fetal MRI, quantitative imaging, and comprehensive image analysis pipelines could potentially make a substantial impact in our understanding of the development of the fetal body throughout gestation, as well as possibly provide clinical prenatal biomarkers.

Acknowledgments. The authors thank all the participating families as well as the midwives and radiographers involved in this study. This work was supported by the the the NIH (Human Placenta Project-grant 1U01HD087202-01), Wellcome Trust Sir Henry Wellcome Fellowship (201374/Z/16/Z and /B), UKRI FLF (MR/T018119/1), the NIHR Clinical Research Facility (CRF) at Guy's and St Thomas' and by core funding from the Wellcome/EPSRC Centre for Medical Engineering [WT203148/ Z/16/Z]. For the purpose of Open Access, the Author has applied a CC BY public copyright license to any Author Accepted Manuscript version arising from this submission. The views expressed are those of the authors and not necessarily those of the NHS or the NIHR.

References

1. Aviles, J., et al.: A fast anatomical and quantitative MRI fetal exam at low field. In: Licandro, R., Melbourne, A., Abaci Turk, E., Macgowan, C., Hutter, J. (eds.) Perinatal, Preterm and Paediatric Image Analysis. PIPPI 2022. LNCS, vol. 13575, pp. 13–24. Springer, Cham (2022). https://doi.org/10.1007/978-3-031-17117-8_2
2. Baadsgaard, K., Hansen, D.N., Peters, D.A., Frøkjær, J.B., Sinding, M., Sørensen, A.: T2* weighted fetal mri and the correlation with placental dysfunction. Placenta **131**, 90–97 (2023). https://doi.org/10.1016/j.placenta.2022.12.002, https://www.sciencedirect.com/science/article/pii/S0143400422004775
3. Blazejewska, A.I., et al.: 3D in utero quantification of T2* relaxation times in human fetal brain tissues for age optimized structural and functional MRI. Magn. Reson. Med. **78**(3), 909–916 (2017)
4. Cordero-Grande, L., Christiaens, D., Hutter, J., Price, A.N., Hajnal, J.V.: Complex diffusion-weighted image estimation via matrix recovery under general noise models. NeuroImage **200**, 391–404 (2019). https://doi.org/10.1016/j.neuroimage.2019.06.039, https://www.sciencedirect.com/science/article/pii/S1053811919305348
5. Ebner, M., et al.: An automated framework for localization, segmentation and super-resolution reconstruction of fetal brain MRI. NeuroImage **206**, 116324 (2020). https://doi.org/10.1016/j.neuroimage.2019.116324, http://www.sciencedirect.com/science/article/pii/S1053811919309152
6. Hutter, J., Jackson, L., Ho, A., Pietsch, M., Story, L., Chappell, L.C., Hajnal, J.V., Rutherford, M.: T2* relaxometry to characterize normal placental development over gestation in-vivo at 3T. Technical report 4:166, Wellcome Open Research (2019)

7. Hutter, J., et al.: Multi-modal functional MRI to explore placental function over gestation. Magn. Reson. Med. **81**(2), 1191–1204 (2019)

8. Isensee, F., Jaeger, P.F., Kohl, S.A.A., Petersen, J., Maier-Hein, K.H.: nnU-Net: a self-configuring method for deep learning-based biomedical image segmentation. Nat. Methods **18**(2), 203–211 (2021). https://doi.org/10.1038/s41592-020-01008-z

9. Kuklisova-Murgasova, M., Quaghebeur, G., Rutherford, M.A., Hajnal, J.V., Schnabel, J.A.: Reconstruction of fetal brain MRI with intensity matching and complete outlier removal. Med. Image Anal. **16**(8), 1550–1564 (2012). https://doi.org/10.1016/j.media.2012.07.004

10. Lajous, H., et al.: T2 mapping from super-resolution-reconstructed clinical fast spin echo magnetic resonance acquisitions. In: Martel, A.L., et al. (eds.) MICCAI 2020. LNCS, vol. 12262, pp. 114–124. Springer, Cham (2020). https://doi.org/10.1007/978-3-030-59713-9_12

11. MONAI Consortium: MONAI: Medical Open Network for AI, March 2020. https://doi.org/10.5281/zenodo.4323058, https://github.com/Project-MONAI/MONAI

12. Ogawa, S., Lee, T.M., Kay, A.R., Tank, D.W.: Brain magnetic resonance imaging with contrast dependent on blood oxygenation. Proc. Natl. Acad. Sci. **87**(24), 9868–9872 (1990). https://doi.org/10.1073/pnas.87.24.9868, publisher: Proceedings of the National Academy of Sciences

13. Pauling, L., Coryell, C.D.: The magnetic properties and structure of hemoglobin, oxyhemoglobin and carbonmonoxyhemoglobin. Proc. Natl. Acad. Sci. **22**(4), 210–216 (1936). https://doi.org/10.1073/pnas.22.4.210, publisher: Proceedings of the National Academy of Sciences

14. Schmidbauer, V., et al.: Mapping human fetal brain maturation in vivo using quantitative MRI. Am. J. Neuroradiol. (2021). https://doi.org/10.3174/ajnr.A7286, http://www.ajnr.org/content/early/2021/09/23/ajnr.A7286

15. Sethi, S., et al.: Quantification of 1.5 T T1 and T2* relaxation times of fetal tissues in uncomplicated pregnancies. J. Magn. Reson. Imaging **54**(1), 113–121 (2021). https://doi.org/10.1002/jmri.27547, https://onlinelibrary.wiley.com/doi/abs/10.1002/jmri.27547, _eprint: https://onlinelibrary.wiley.com/doi/pdf/10.1002/jmri.27547

16. Story, L., et al.: Foetal lung volumes in pregnant women who deliver very preterm: a pilot study. Pediatr. Res. **87**(6), 1066–1071 (2020). https://doi.org/10.1038/s41390-019-0717-9

17. Tournier, J.D., et al.: Mrtrix3: a fast, flexible and open software framework for medical image processing and visualisation. NeuroImage **202**, 116137 (2019). https://doi.org/10.1016/j.neuroimage.2019.116137, https://www.sciencedirect.com/science/article/pii/S1053811919307281

18. Uus, A., et al.: Deformable slice-to-volume registration for reconstruction of quantitative T2* placental and fetal MRI. In: Hu, Y., et al. (eds.) ASMUS/PIPPI -2020. LNCS, vol. 12437, pp. 222–232. Springer, Cham (2020). https://doi.org/10.1007/978-3-030-60334-2_22

19. Uus, A., et al.: Deformable slice-to-volume registration for motion correction in fetal body MRI. IEEE TMI **39**(9), 2750–2759 (2020). https://doi.org/10.1109/TMI.2020.2974844

20. Uus, A.U., et al.: Automated 3D reconstruction of the fetal thorax in the standard atlas space from motion-corrupted MRI stacks for 21–36 weeks GA range. Med. Image Anal. **80**, 102484 (2022). https://doi.org/10.1016/j.media.2022.102484, https://www.sciencedirect.com/science/article/pii/S1361841522001311

Clinical Applications – Lung

Clinical Applications – Lung

Medical Phrase Grounding with Region-Phrase Context Contrastive Alignment

Zhihao Chen[1], Yang Zhou[2], Anh Tran[3], Junting Zhao[1], Liang Wan[1(✉)],
Gideon Su Kai Ooi[4], Lionel Tim-Ee Cheng[3], Choon Hua Thng[4], Xinxing Xu[2],
Yong Liu[2], and Huazhu Fu[2(✉)]

[1] College of Intelligence and Computing, Tianjin University, Tianjin, China
lwan@tju.edu.cn
[2] Institute of High Performance Computing (IHPC), Agency for Science, Technology
and Research (A*STAR), 1 Fusionopolis Way, #16-16 Connexis, 138632 Singapore,
Republic of Singapore
hzfu@ieee.org
[3] Singapore General Hospital, Singapore, Singapore
[4] National Cancer Center Singapore, Singapore, Singapore

Abstract. Medical phrase grounding (MPG) aims to locate the most
relevant region in a medical image, given a phrase query describing certain medical findings, which is an important task for medical image
analysis and radiological diagnosis. However, existing visual grounding
methods rely on general visual features for identifying objects in natural
images and are not capable of capturing the subtle and specialized features of medical findings, leading to a sub-optimal performance in MPG.
In this paper, we propose **MedRPG**, an end-to-end approach for MPG.
MedRPG is built on a lightweight vision-language transformer encoder
and directly predicts the box coordinates of mentioned medical findings,
which can be trained with limited medical data, making it a valuable tool
in medical image analysis. To enable MedRPG to locate nuanced medical findings with better region-phrase correspondences, we further propose **T**ri-attention **Co**ntext contrastive alignment (**TaCo**). TaCo seeks
context alignment to pull both the features and attention outputs of
relevant region-phrase pairs close together while pushing those of irrelevant regions far away. This ensures that the final box prediction depends
more on its finding-specific regions and phrases. Experimental results on
three MPG datasets demonstrate that our MedRPG outperforms state-
of-the-art visual grounding approaches by a large margin. Additionally,
the proposed TaCo strategy is effective in enhancing finding localization
ability and reducing spurious region-phrase correlations.

Z. Chen and Y. Zhou—Contributed equally to this work.

Supplementary Information The online version contains supplementary material
available at https://doi.org/10.1007/978-3-031-43990-2_35.

372   Z. Chen et al.

Keywords: Medical phrase grounding · vision-language model · contrastive learning

1 Introduction

Medical phrase grounding (MPG) is the task of associating text descriptions with corresponding regions of interest (ROIs) in medical images. It enables machines to understand and interpret medical findings mentioned in medical reports in the context of medical images, which is crucial in medical image analysis and radiological diagnosis. Figure 1 illustrates how an MPG system facilitates the radiological diagnosis process. Radiologists first review the medical images (*e.g.*, X-rays, CT scans, and MRI scans) to find out possible abnormalities and then write a report that summarizes their findings. Then, given the image and report, the MPG system can help doctors to locate and link ROIs to the corresponding phrases in the reports, which reduces the time of the diagnostic process and improves the quality of risk stratification and treatment planning.

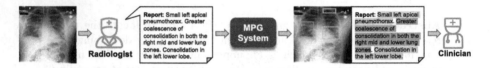

Fig. 1. Illustration on how MPG helps radiological diagnosis.

In this paper, we study the MPG problem and focus on a typical setting to learn the grounding between Chest X-ray images and medical reports. As far as we know, there are only a few related works on the medical phrase grounding problem. (This is probably because medical annotations of grounding data require specialized expertise and are time-consuming and expensive to be collected.) Benedikt *et al.* [1] made use of text semantics to improve biomedical vision-language processing. They first evaluated the grounding performance of self-supervised biomedical vision-language models by proposing an MPG benchmark. However, their focus is on vision-language pre-training rather than addressing the MPG problem. Qin *et al.* [19] proposed to transfer the knowledge of general vision-language models for detection tasks in medical domains. The key idea is to guide the vision-language model through hand-crafted prompting of visual attributes such as color, shape, and location that may be shared between natural and medical domains. This approach fails to consider unique characteristics in radiological images and reports and is inapplicable to MPG for radiological images.

Although visual grounding has been well studied for natural images [3,4,22–24], it is non-trivial to apply these approaches to radiological images. Specifically, MPG requires learning specialized visual-textual features so that the model can identify medical findings with subtle differences in texture and shape and interpret the relative positions mentioned in the medical reports. In contrast, general grounding methods often rely on visual features that are useful for object detection or classification but not specific to medical images, leading to inaccurate region-phrase correlations and thus sub-optimal results. In addition, many grounding models for general domains are too heavy to be trained with limited annotated data, which is common. Such heavy model structures are generally difficult to be trained with limited annotated data.

In this work, we propose **MedRPG**, an end-to-end approach for MPG. MedRPG has a lightweight model architecture and explicitly captures the finding-specific correlations between ROIs and report phrases. Specifically, we propose to stack a few vision-language transformer layers to encode both the medical images and report phrases and directly predict the box coordinates of desired medical findings. Compared to general grounding methods with heavy model architectures, this design is more robust against overfitting for MPG with limited training data. To locate nuanced medical findings with better region-phrase correspondences, we further propose **Tri-attention Context** contrastive alignment (TaCo). TaCo seeks *context alignment* to learn finding-specific representations that jointly align the region, phrase, and box prediction under the same context of a vision-language transformer encoder. It pulls both the *features* and *attention outputs* close together for semantically relevant region-phrase pairs while pushing those of irrelevant pairs far away. This encourages the alignment between regions and phrases at both feature and attention levels, leading to enhanced finding-identification ability and reduced spurious region-phrase correlations. Experimental results on three medical datasets demonstrate that our MedRPG is more effective in localizing medical findings, achieves better region-phrase correspondences, and significantly outperforms general visual grounding approaches on the MPG task.

2   Proposed Method

The MPG problem can be defined as follows: Given a radiological image $\mathbf{I}$ associated with medical phrases $\mathbf{T}$ written by specialist radiologists, MPG aims to locate the described findings and then output a 4-dim bounding box (bbox) coordinates $\mathbf{b} = (x, y, w, h)$, where (x, y) is the box center coordinates and w, h are the box width and height, respectively.

2.1   Model Architecture

Figure 2 illustrates the framework of our method. Given image $\mathbf{I}$ and phrase $\mathbf{T}$, we first leverage the Vision Encoder and Language Encoder to generate the image and text embeddings. Next, we concatenate the multi-modal feature embeddings and append a learnable token (named [REG] token), and then feed them into a lightweight Vision-Language Transformer to encode the intra and inter-modality context in a common semantic space. Finally, the output state of the [REG]

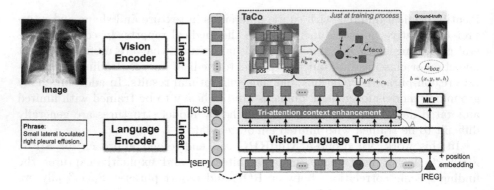

Fig. 2. An overview of the proposed MedRPG method.

token is employed to predict the 4-dim bbox via a grounding head. Additionally, to ensure a consistent representation of medical findings across modalities, we introduce TaCo, which aligns the context of region and phrase embeddings at both the feature and attention levels.

Vision Encoder: Following the common practice [2,4,13], the visual encoder starts with a CNN backbone, followed by the visual transformer. We choose the ResNet-50 [9] as the CNN backbone. The visual transformer includes 6 stacked transformer encoder layers [5]. Given a radiological image $\mathbf{I} \in \mathbb{R}^{3 \times W \times H}$, it is fed into the CNN backbone to obtain the high-level deep features. Next, we apply a 1×1 ConV layer to project the deep features into a C_v-dimensional subspace. Finally, we exploit the visual transformer to mine the long-range visual relations and further output the visual features $\mathbf{F}_v = [\boldsymbol{f}_v^n]_{n=1}^{N_v}$, where N_v is the number of visual tokens and $\boldsymbol{f}_v^n \in \mathbb{R}^{C_v}$ is the n-th token of $\mathbf{F}_v$.

Language Encoder: We leverage pre-trained language models such as BERT [14] as the language encoder, which includes 12 transformer encoder layers. Given a medical phrase $\mathbf{T}$, we first utilize the BERT tokenizer to convert it into a sequence of tokens. Next, we follow the common practice to append a [CLS] token at the beginning as the global representation of the input medical phrases and append a [SEP] token at the end, and then pad the sequence to a fixed length. Finally, we use BERT to encode the tokens into the text embeddings $\mathbf{F}_l = [\boldsymbol{f}^{cls}, \boldsymbol{f}_l^n]_{n=2}^{N_l}$, where N_l is the number of text tokens, C_l is the feature dimensions, and $\boldsymbol{f}_l^n \in \mathbb{R}^{C_l}$ is the n-th token of $\mathbf{F}_l$.

Vision-Language Transformer: After the individual vision and language encoding, we obtain $\mathbf{F}_v$ and $\mathbf{F}_l$. To capture the correspondence between the image and phrase embeddings, we first project them into the common space (channel= C_{vl}) and then fed them into a Vision-Language Transformer (VLT), together with an extra learnable [REG] token, which is further used to predict the bbox:

$$\mathbf{H} = \mathrm{VLT}\left([\varphi_v(\mathbf{F}_v), \varphi_l(\mathbf{F}_l), \boldsymbol{r}]\right), \tag{1}$$

where $\varphi_v(\cdot)$ and $\varphi_l(\cdot)$ denote the project functions for vision and language tokens, respectively. $r \in \mathbb{R}^{C_{vl}}$ is the [REG] token and VLT($\cdot$) denotes the VLT encoder with learnable position embeddings. $\mathbf{H} \in \mathbb{R}^{C_{vl} \times N_{vl}}$ (where $N_{vl} = N_v + N_l + 1$) is the output of VLT that consist of three parts: vision embeddings $H_v = [\mathbf{h}_v^n]_{n=1}^{N_v}$, language embeddings $H_l = [\mathbf{h}^{cls}, \mathbf{h}_l^n]_{n=2}^{N_l}$, and [REG] embedding $\mathbf{h}^{reg}$.

To perform the final grounding results, we further feed $\mathbf{h}^{reg}$ into a 3-layer MLP to predict the final 4-dim box coordinates $\mathbf{b} = \text{MLP}(\mathbf{h}^{reg})$. Given the grounding-truth box $\mathbf{b}_0$, we leverage smooth L1 loss [7] and GIoU [20] loss which are popular in grounding and detection tasks to optimize our model:

$$\mathcal{L}_{box} = \Phi_{l1}(\mathbf{b}, \mathbf{b}_0) + \lambda \cdot \Phi_{giou}(\mathbf{b}, \mathbf{b}_0), \tag{2}$$

where $\mathcal{L}_{box}$ is the box loss. Φ_{l1} and Φ_{giou} are the smooth L1 and GIoU loss functions, respectively. λ is the trade-off parameter.

2.2 Tri-Attention Context Contrastive Alignment

Medical findings often share subtle differences in texture and brightness level due to the low contrast of the medical images, which makes it challenging for the MPG methods to capture accurate region-phrase correlations. To identify nuanced medical findings with better region-phrase correspondences, we propose the Tri-attention Context Contrastive Alignment (TaCo) strategy to learn finding-specific representations with accurate region-phrase correlations by explicitly aligning relevant regions and phrases at both feature and attention levels.

Feature-Level Alignment. The feature-level alignment aims to make visual and textual embeddings with the same semantics meaning to be similar. To this end, given the bbox $\mathbf{b}_0$ related to a given phrase query, we first obtain the positive ROI embeddings $\mathbf{h}_0^{box} \in \mathbb{R}^{C_{vl}} = \text{Pool}(\mathbf{H}_v, \mathbf{b}_0)$ by aggregating visual embeddings $\mathbf{H}_v$ within the bbox $\mathbf{b}_0$. Next, we randomly select K bbox $\{\mathbf{b}_k\}_{k=1}^K$ that have low IoUs with $\mathbf{b}_0$ (i.e., regions that are irrelevant to the given phrase query) and obtain K negative region embeddings $\{\mathbf{h}_k^{box} \in \mathbb{R}^{C_{vl}}\}_{k=1}^K$. Let $\mathbf{h}^{cls}$ be the features of the input phrases. We want to make the positive ROI embedding $\mathbf{h}_0^{box}$ close to the corresponding phrase embedding $\mathbf{h}^{cls}$ whereas negative region embeddings $\{\mathbf{h}_k^{box}\}_{k=1}^K$ far away. This is achieved by exploiting the InfoNCE [8,17] loss as:

$$\mathcal{L}_{fea} = -\log \frac{\exp(\mathbf{h}^{cls} \cdot \mathbf{h}_0^{box}/\tau)}{\sum_{k=0}^K \exp(\mathbf{h}^{cls} \cdot \mathbf{h}_k^{box}/\tau)}, \tag{3}$$

where $\mathcal{L}_{fea}$ denotes the feature-level alignment loss, τ is a temperature hyper-parameter and '$\cdot$' represents the inner (dot) product.

Attention-Level Alignment. In addition to the feature-level alignment, we also consider attention-level alignment, which encourages the attention outputs of VLT for relevant region-phrase pairs to be similar. To realize this, we extract the attention weight $\mathbf{A} \in \mathbb{R}^{N_{vl} \times N_{vl}}$ from the last multi-head attention layer of

VLT. We denote a^{reg}, a^{cls} and $\{a_k^{box}\}_{k=0}^{K}$ as the attention weights for the embeddings of the [REG] token, the [CLS] token, and the $K+1$ bboxes, respectively, where $a_k^{box} = \texttt{Pool}(\mathbf{A}, b_k)$. Given the k-th bbox embedding, we calculate the joint attention weights of bbox, [CLS], and [REG] embeddings and then further product $\mathbf{H}$ to get the triple-attention context pooling c_k as follows:

$$c_k = \sum_{j=0}^{N_{vl}} \left(t_k^{(j)} \cdot \mathbf{H}[:,j] \right), \quad \text{where } t_k = \texttt{Norm}(a^{cls} \cdot a^{reg} \cdot a_k^{box}), \tag{4}$$

where t_k represents the joint attention weights, $t_k^{(j)}$ denotes the j-th element of t_k, $\mathbf{H}[:,j]$ denotes the j-th column of $\mathbf{H}$, and $\texttt{Norm}(\cdot)$ is the L2 normalization operation to constrain the sum of the squared weights to be equal to 1. Such triple-attention context pooling c_k characterizes the contextual dependencies among regions, phrases, and box predictions in the VLT. Intuitively, the box prediction of certain medical findings should be made based on its relevant regions and phrases rather than irrelevant ones. Therefore, the attention outputs c_0 for relevant region-phrase pairs should be similar to their individual embeddings h_0^{box} and h^{cls}, leading to attention-level alignment.

Table 1. Grounding results on MS-CXR [1], ChestX-ray8 [21], and the in-house datasets with respect to Acc and mIoU.

Method	Vision Encoder	Language Encoder	MS-CXR		ChestX-ray8		In-house	
			Acc↑	mIoU↑	Acc↑	mIoU↑	Acc↑	mIoU↑
BioViL[†] [1]	ResNet-50	CXR-BERT	7.78	19.99	6.56	12.78	3.65	13.55
GLoRIA[†] [10]		BioClinicalBERT	28.74	31.17	8.58	16.39	5.74	14.91
RefTR [15]		BERT	53.69	50.11	29.27	29.59	46.03	40.99
VGTR [6]		Bi-LSTM	60.27	53.58	32.65	34.02	47.37	41.92
SeqTR [24]		Bi-GRU	63.20	56.63	32.88	33.09	44.42	39.45
TransVG [4]		BERT	65.87	58.91	34.51	33.98	48.30	43.35
Ours	ResNet-50	BERT	**69.86**	**59.37**	**36.02**	**34.59**	**49.87**	**43.86**

Context Contrastive Alignment. With the above results, we can propose our TaCo strategy by integrating both feature and attention-level alignments. Specifically, we modify the InfoNCE loss (3) to simultaneously perform feature and attention-level alignments by adding the triple-attention context pooling c_k to the respective region and phrase features (i.e., h^{cls}, h_k^{box}). This leads to the TaCo loss as follows:

$$\mathcal{L}_{taco} = -\log \frac{\exp((h^{cls} + c_0) \cdot (h_0^{box} + c_0)/\tau)}{\sum_{k=0}^{K} \exp((h^{cls} + c_k) \cdot (h_k^{box} + c_k)/\tau)}. \tag{5}$$

Finally, we combine the TaCo loss (5) and box loss (2) to get the overall loss functions of MedRPG:

$$\mathcal{L}_{MedRPG} = \mathcal{L}_{box} + \mu \cdot \mathcal{L}_{taco}. \qquad (6)$$

where $\mathcal{L}_{MedRPG}$ denotes total loss for MedRPG and μ is the trade-off parameter.

3   Experiments

Dataset. Our experiments are conducted on two public datasets, i.e., MS-CXR [1], ChestX-ray8 [21], and one in-house datase[1]. MS-CXR is sourced from MIMIC-CXR [11,12] and consists of 1,153 samples of Image-Phrase-BBox triples. We pre-process MS-CXR to make sure a given phrase query corresponds to only one bounding box, which results in 890 samples from 867 patients. ChestX-ray8 is a large-scale dataset for diagnosing 8 common chest diseases, of which 984 images with pathology are provided with hand-labeled bounding boxes. Due to the lack of finding-specific phrases from medical reports, we use category labels as the phrase queries to build the Image-Report-BBox triples for our task. Our in-house dataset comprises 1,824 Image-Phrase-BBox samples from 635 patients, including 23 categories of chest abnormalities with more complex phrases. For a fair comparison, all datasets are split into train-validation-test sets by 7:1:2 based on the patients.

Evaluation Metrics. To evaluate the quality of the MPG task, we follow the standard protocol of nature image grounding [4] to report Acc(%), where a predicted region will be regarded as a positive sample if its intersection over union (IoU) with the ground-truth bounding box is greater than 0.5. Besides, we also report mIoU (%) metric for a more comprehensive comparison.

Table 2. Ablation study for network components on MS-CXR dataset.

Metrics	Only images	Only phrases	Baseline	w/ Feature-level	w/ TaCo
Acc ↑	41.91	40.12	66.86	67.26	**69.86**
mIoU ↑	38.44	39.34	58.93	59.25	**59.37**

Baselines. We compare our MedRPG with SOTA methods for general visual grounding, such as RefTR [15], TransVG [4], VGTR [6], and SeqTR [24]. We choose their official implementations for a fair comparison. Since the medical datasets are too small to train a data-hungry transformer-based model from scratch, we initialize our MedRPG (encoders) from the general grounding models pre-trained on natural images. The compared methods share the same settings. We also compare two self-supervised biomedical vision-language processing methods BioViL [1] and GLoRIA [10] which pre-trained on MIMIC-CXR [11].

---

[1] The ethical approval of this dataset was obtained from the Ethical Committee.

378   Z. Chen et al.

Table 3. Ablation study for hyperparameters μ and K on MS-CXR dataset.

Metrics	$\mu = 0.1$	$\mu = 0.05$	$\mu = 0.025$	$\mu = 0.01$	K=3	K=5	K=7
Acc ↑	66.86	**69.86**	68.86	68.66	67.66	**69.86**	66.67
mIoU ↑	58.92	**59.37**	59.12	59.14	59.19	**59.37**	58.59

Implementation Details. The experiments are conducted on the PyTorch [18] platform with an NVIDIA RTX 3090 GPU. The input image size is 640×640. The channel numbers C_v, C_l, and C_{vl} are 256, 768, and 256. The sample number K is set to 5. The trade-off parameter λ in Eq. 2 and μ in Eq. 6 are set to 1 and 0.05, respectively. The base learning rates for the vision encoder, language encoder, and vision-language transformer are set to 1×10^{-5}, 1×10^{-5}, and 5×10^{-5}, respectively. We train our MedRPG model by the AdamW [16] optimizer for 90 epochs with a learning rate dropped by a factor of 10 after 60 epochs. For all the baselines and MedRPG, we select the best checkpoint for testing based on validation performance and report the average performance metrics computed by repeating each experiment with three different random seeds.

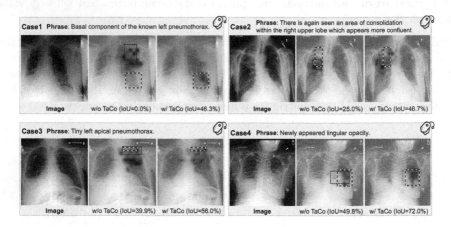

Fig. 3. Visualized grounding results for MedRPG w/ and w/o TaCo. We show the ground-truth box (red box), prediction box (cyan or yellow box), and the [REG] token's attention to visual tokens (a heatmap with high values in red). (Color figure online)

Experimental Results. Table 1 provides the grounding results on the MS-CXR, ChestX-ray8, and in-house datasets. As can be seen, our MedRPG consistently achieves the best performance in all cases. In particular, we note that lightweight models like TransVG and our MedRPG generally perform better, which indicates lightweight models are more applicable for MPG. Despite this, our method still outperforms TransVG by a margin of 6.1% in Acc on MS-CXR.

This can be attributed to the proposed TaCo strategy in learning finding-specific representations and improving region-phrase alignment. On ChestX-ray8, all methods get degraded results due to the lack of position cues in the phrase queries. Nevertheless, our method still outperforms the second-best method by 4.3% in Acc and 1.6% in mIoU. On the in-house dataset, our method is still the best even when there exist much more types of findings to be grounded. Note that the self-supervised methods BioViL and GloRIA achieve very poor results compared to other methods on all three datasets.

Ablation Study. We conduct ablative experiments on the MS-CXR dataset to verify the effectiveness of each component in MedRPG. Table 2 shows the quantitative results of each combination. To verify how the vision and language modalities contribute to the MPG performance, we perform MedRPG with either image or test inputs. As expected, MedRPG can only achieve poor results under the unimodal setting. Next, we consider the inputs with both images and phrases and observe a significant improvement in performance compared to MedRPG trained from a single modality. Then, we equip MedRPG with feature-level alignment and gain the improvement of 0.6% in Acc and 0.5% in mIoU, which suggests it is helpful but still not good enough to learn the accurate region-phrase correspondences. Finally, with the proposed TaCo, MedRPG further gains a significant improvement by 3.8% in Acc. This shows that TaCo is effective in improving the MPG performance with better region-phrase correlations. In addition, we study the impact of hyper-parameters (trade-off parameter μ and number of negative samples K). Table 3 shows two metrics of our MedRPG method with varying hyper-parameters onni the MS-CXR dataset. As can be seen, our method is not very sensitive to hyper-parameter choices. Additional analysis of the model's confidence intervals and its ability to generalize can be found in the supplementary material.

Qualitative Results. In Fig. 3, we show the box predictions and attention maps obtained by MedRPG with and without TaCo to demonstrate the effectiveness of TaCo in identifying abnormal medical findings and capturing region-phrase correlations. For instance, in Case 1, pneumothorax is present in an uncommon location (i.e., lower left lung) and the phrase does not provide an accurate location cue. Without TaCo, MedRPG overfits the upper lung regions where pneumothorax appears more frequently. In contrast, with TaCo, the model can better learn pneumothorax representations and identify the corresponding ROI even without accurate location information. In other cases, although the method without TaCo can also roughly find the location of the medical findings, MedRPG with TaCo can obtain more focused attention maps on the medical findings. It suggests that TaCo is effective in reducing spurious region-phrase correlations, leading to more accurate and interpretable bbox predictions.

4   Conclusion

This study introduces MedRPG, a lightweight and efficient method for medical phrase grounding. A novel tri-attention context contrastive alignment (TaCo) is proposed to learn finding-specific representations and improve region-phrase alignment in feature and attention levels. Experimental results show that MedRPG outperforms existing visual grounding methods and achieves more consistent correlations between phrases and mentioned regions.

Acknowledgements. This work is supported by the grant from Tianjin Natural Science Foundation (Grant No. 21JCYBJC00510), the Agency for Science, Technology and Research (A*STAR) through its AME Programmatic Funding Scheme Under Project A20H4b0141, the A*STAR Central Research Fund "A Secure and Privacy Preserving AI Platform for Digital Health". This research/project is supported by the National Research Foundation, Singapore under its AI Singapore Programme (AISG Award No: AISG2-TC-2021-003).

References

1. Boecking, B., et al.: Making the most of text semantics to improve biomedical vision-language processing. In: Avidan, S., Brostow, G., Cissé M., Farinella, G.M., Hassner, T. (eds.) Computer Vision. ECCV 2022. LNCS, vol. 13696, pp. 1–21. Springer, Cham (2020). https://doi.org/10.1007/978-3-031-20059-5_1
2. Carion, N., Massa, F., Synnaeve, G., Usunier, N., Kirillov, A., Zagoruyko, S.: End-to-end object detection with transformers. In: Vedaldi, A., Bischof, H., Brox, T., Frahm, J.-M. (eds.) ECCV 2020. LNCS, vol. 12346, pp. 213–229. Springer, Cham (2020). https://doi.org/10.1007/978-3-030-58452-8_13
3. Chen, S., Li, B.: Multi-modal dynamic graph transformer for visual grounding. In: proceedings of CVPR (2022)
4. Deng, J., Yang, Z., Chen, T., Zhou, W., Li, H.: TranSVG: end-to-end visual grounding with transformers. In: Proceedings of ICCV (2021)
5. Dosovitskiy, A., et al.: An image is worth 16×16 words: transformers for image recognition at scale. In: Proceedings of ICLR (2021)
6. Du, Y., Fu, Z., Liu, Q., Wang, Y.: Visual grounding with transformers. In: Proceedings of ICME (2022)
7. Girshick, R.: Fast R-CNN. In: Proceedings of ICCV (2015)
8. Gutmann, M., Hyvärinen, A.: Noise-contrastive estimation: a new estimation principle for unnormalized statistical models. In: Proceedings of International Conference on Artificial Intelligence and Statistics (2010)
9. He, K., Zhang, X., Ren, S., Sun, J.: Deep residual learning for image recognition. In: Proceedings of CVPR (2016)
10. Huang, S.C., Shen, L., Lungren, M.P., Yeung, S.: Gloria: a multimodal global-local representation learning framework for label-efficient medical image recognition. In: Proceedings of ICCV (2021)
11. Johnson, A.E., et al.: MIMIC-CXR, a de-identified publicly available database of chest radiographs with free-text reports. Sci. Data **6**(1), 317 (2019)
12. Johnson, A.E., Pollard, T.J., Mark, R.G., Berkowitz, S.J., Horng, S.: MIMIC-CXR database (version 2.0.0). In: PhysioNet (2019)
13. Kamath, A., Singh, M., LeCun, Y., Synnaeve, G., Misra, I., Carion, N.: Mdetr-modulated detection for end-to-end multi-modal understanding. In: Proceedings of ICCV (2021)
14. Kenton, J.D.M.W.C., Toutanova, L.K.: BERT: pre-training of deep bidirectional transformers for language understanding. In: Proceedings of NAACL-HLT (2019)

15. Li, M., Sigal, L.: Referring transformer: a one-step approach to multi-task visual grounding. In: Proceedings of NeurIPS (2021)
16. Loshchilov, I., Hutter, F.: Decoupled weight decay regularization. In: Proceedings of ICLR (2019)
17. Oord, A.v.d., Li, Y., Vinyals, O.: Representation learning with contrastive predictive coding. arXiv preprint arXiv:1807.03748 (2018)
18. Paszke, A., et al.: PyTorch: an imperative style, high-performance deep learning library. In: Proceedings of NeurIPS (2019)
19. Qin, Z., Yi, H., Lao, Q., Li, K.: Medical image understanding with pretrained vision language models: a comprehensive study. In: Proceedings of ICLR (2023)
20. Rezatofighi, H., Tsoi, N., Gwak, J., Sadeghian, A., Reid, I., Savarese, S.: Generalized intersection over union: a metric and a loss for bounding box regression. In: Proceedings of CVPR (2019)
21. Wang, X., Peng, Y., Lu, L., Lu, Z., Bagheri, M., Summers, R.M.: ChestX-ray8: hospital-scale chest x-ray database and benchmarks on weakly-supervised classification and localization of common thorax diseases. In: Proceedings of CVPR (2017)
22. Yang, Z., Gong, B., Wang, L., Huang, W., Yu, D., Luo, J.: A fast and accurate one-stage approach to visual grounding. In: Proceedings of ICCV (2019)
23. Yu, L., et al.: MattNet: modular attention network for referring expression comprehension. In: Proceedings of CVPR (2018)
24. Zhu, C., et al.: SeqTR: a simple yet universal network for visual grounding. In: Avidan, S., Brostow, G., Cissé, M., Farinella, G.M., Hassner, T. (eds.) Computer Vision. ECCV 2022. LNCS, vol. 13695. Springer, Cham (2022). https://doi.org/10.1007/978-3-031-19833-5_35

Topology Repairing of Disconnected Pulmonary Airways and Vessels: Baselines and a Dataset

Ziqiao Weng[1], Jiancheng Yang[2(✉)], Dongnan Liu[1], and Weidong Cai[1]

[1] School of Computer Science, University of Sydney, Sydney, Australia
[2] Computer Vision Laboratory, Swiss Federal Institute of Technology Lausanne (EPFL), Lausanne, Switzerland
jiancheng.yang@epfl.ch

Abstract. Accurate segmentation of pulmonary airways and vessels is crucial for the diagnosis and treatment of pulmonary diseases. However, current deep learning approaches suffer from disconnectivity issues that hinder their clinical usefulness. To address this challenge, we propose a post-processing approach that leverages a data-driven method to repair the topology of disconnected pulmonary tubular structures. Our approach formulates the problem as a keypoint detection task, where a neural network is trained to predict keypoints that can bridge disconnected components. We use a training data synthesis pipeline that generates disconnected data from complete pulmonary structures. Moreover, the new Pulmonary Tree Repairing (PTR) dataset is publicly available, which comprises 800 complete 3D models of pulmonary airways, arteries, and veins, as well as the synthetic disconnected data. Our code and data are available at https://github.com/M3DV/pulmonary-tree-repairing.

Keywords: pulmonary airways · pulmonary vessels · tree structure repairing · geometric deep learning · shape analysis

1 Introduction

Pulmonary diseases pose significant health risks, and computed tomography (CT) analysis of pulmonary airways and vessels has become a valuable clinical tool for revealing tomographic patterns [10,25]. Precise representation of the airway tree is essential for quantifying morphological changes, diagnosing respiratory disorders such as bronchial stenosis, acute respiratory distress syndrome, idiopathic pulmonary fibrosis, chronic obstructive pulmonary disease (COPD), obliterative bronchiolitis, and pulmonary contusion, as well as for virtual bronchoscopy and endobronchial navigation in surgery [3,10]. Furthermore, accurate modeling pulmonary arteries and veins improves computer-aided diagnosis

Supplementary Information The online version contains supplementary material available at https://doi.org/10.1007/978-3-031-43990-2_36.

of pulmonary embolism, chronic pulmonary hypertension [11,17], and lobectomy/segmentectomy [12,26,27].

In recent years, deep learning methods have spawned research on airway and vessel segmentation. Convolutional neural networks (CNNs) have been widely employed in various existing studies to learn robust and discriminative features for automatic airway/artery/vein segmentation [4,8–10,14,15,25]. However, accurately reconstructing complete airway or vessel tree branches remains a major challenge. Current state-of-the-art segmentation models, such as nnU-Net [5], still suffer from inadequate precision due to the minute scale and scattered spatial distribution of peripheral bronchi and vessels, which causes a severe class imbalance between the foreground and background, leading to degraded segmentation accuracy. The implications of such degraded performance can have negative consequences on clinical judgments and diagnoses, as it can lead to disconnections of pulmonary tubular structures of airways or vessels, as depicted in Fig. 1, potentially impeding accurate medical assessments.

In this paper, we formulate the problem of disconnected pulmonary tubular structures as a key point detection task. The primary objective is to repair the topology structures of two disconnected components by accurately identifying the centers of the disconnected parts located at both ends of the components. Endpoints corresponding to the broken centerline of the pulmonary tubular structure are treated as two key points. The identification of these key points is critical in recognizing disconnections in pulmonary tubular structures

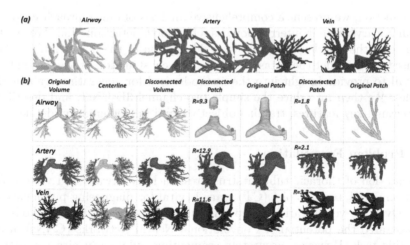

Fig. 1. Visualizations of the PTR dataset. (a): Examples of disconnected predictions from nnU-Net. (b): From left to right: original volumes, centerlines extracted by Vesselvio (with different edge radii indicated by colors), disconnection synthesis, subvolume views of disconnected parts, and corresponding original parts. The volumes have been smoothed for better visualization. The average edge radius, measured in micrometers, is denoted by "R" in the figures. Airways, arteries, and veins are respectively colored in gray, red, and blue.

for diagnosing pulmonary diseases, which has significant research implications. To address this issue, we propose a training data synthesis pipeline that generates disconnected data from complete pulmonary structures. We further explore the training strategy and thus build a strong basline based on 3D-UNet to predict the key points that can bridge disconnected components. Our contributions can be briefly summarized as follows:

- **A novel formulation of a practical research problem**: We have formulated the problem of pseudo disconnection pulmonary tubular structures as a key point detection task, which is a significant contribution to the field as it has not been extensively explored before.
- **An effective yet simple baseline with efficient 3D-UNet**: We propose a two-channel 3D neural network that efficiently identifies key points and bridges disconnected components. Our model demonstrates decent performance, providing a strong baseline for future studies.
- **An open-source benchmark**: To evaluate the proposed model, we have constructed a new pulmonary dataset named Pulmonary Tree Repairing (PTR), and designed proper metrics for performance examination. This dataset will be publicly available soon and will enable reproducibility and comparison of future studies in this field.

2   Method

In this section, we present a comprehensive analysis of our approach for detecting pulmonary tubular interruptions as a keypoint detection task. We start by formulating the problem, followed by a description of the data simulation process used to construct the dataset. The dataset construction process is explained in detail to provide insight into the methods used for generating realistic data samples. We then introduce the simple two-channel 3D-UNet, and describe its architecture, key features, training objective, and implementation details.

2.1   Problem Formulation

Segmentation of thoracic tubular structures, such as airways and vessels, from lung Computed Tomography (CT) scans is vital for diagnosing pulmonary diseases. Over the years, various deep learning-based segmentation methods have demonstrated the potential of Convolutional Neural Networks (CNNs) in handling this task. However, accurately segmenting pulmonary airways, arteries, and veins without interruption remains challenging due to the unique properties of the thoracic tubular structure. The trachea and blood vessels constitute only a small fraction of the whole thoracic CT image, which leads to severe class imbalance between the tubular foreground and background, hindering 3-D CNNs learning from sufficient supervisory signals [15]. Moreover, airways and vessels are complex tree-like structures with numerous bifurcations and branches of various sizes and lengths, making it difficult for CNNs to capture fine-grained patterns without encountering memory/parameter explosion and overfitting [10]. Segmentation networks often produce unsatisfactory predictions, resulting in disconnection or interruption of the estimated tubular structure, which could affect clinicians' judgment in clinical practice. Therefore, identifying the location of disconnections is of great research importance. In this paper, we have formulated the problem as a key point detection task, with the two endpoints of the interrupted centerline of the tubular structure serving as the two key points. We aim to use neural networks to predict the location of the disconnection part of vessels/airways, which has significant research implications.

Keypoint detection is a popular computer vision technique for identifying object parts in images, with applications ranging from face recognition, pose estimation to medical landmark detection [13,16,18,23]. Heatmap regression has emerged as a standard approach for keypoint detection, where ground-truth heatmaps are generated for each keypoint using a Gaussian kernel [7,24]. The network outputs multi-channel heatmaps, with each channel corresponding to a specific keypoint. Our work adopts this approach for detecting two keypoints located at the endpoints of interrupted airway/vessel.

2.2 Training Data Synthesis

We generated synthetic data from lung CT scans with carefully annotated pulmonary airways, arteries, and veins, as no public medical dataset was available for the task at hand. The synthetic data simulates the scenario of vascular/trachea disconnection and serves as a benchmark dataset for the keypoint detection task. To generate the data, binary masks of the tubular structures were extracted from 800 CT scans [6], and VesselVio software [2] was used to identify the centerlines of binarized airway/vessel volumes and create tree-like graphs. Random sampling was performed to select a branch of the vessel or airway, and two keypoints were sampled along the pre-extracted centerline. The keypoints were then subjected to morphological operations (from SimpleITK Python library [1,22]) to create near-true vascular disconnections. The resulting keypoints were labeled KP_1 and KP_2, and the data was visualized in Fig. 1. It is important to note that discontinuities in real-world scenarios are mainly observed in thinner blood vessels. Due to the random sampling process and the prevalence of small branches within the entire tubular structure, the generated discontinuities are predominantly manifested in small blood vessels. Including the subfigures in Fig. 1(b) aims to clearly illustrate the visual appearance of these generated discontinuities.

2.3 The Keypoint Detection Network

The framework and training pipeline of our network are depicted in Fig. 2. In the following sections, we will introduce each component in detail.

Data Sampling. The generated raw data is too large for training the network due to the high-resolution nature of CT scans, which have dimensions of 512×512

for the x-y plane and variable dimensions for the z plane. Directly feeding the entire 3-D volume into the network can cause significant memory overhead and slow down the training convergence, especially with limited computing resources. Therefore, we crop a subvolume with a size of $80 \times 80 \times 80$ around where the disconnection occurs from the original volume. Specifically, since the location of interrupted blood vessels cannot be known in advance and the small connected component where KP_2 is located can be found using morphological operations, we randomly select a point in that small object as the center point of our subvolume. This approach also serves as a new form of data augmentation. For each selected branch in an original volume, we randomly crop one subvolume for training purposes and three subvolumes for validation and testing.

Network Design. We propose an encoder-decoder network that is based on the widely used 3D U-Net architecture. As depicted in Fig. 2, the inputs to the network are obtained by cropping subvolumes of the same size as the original volume. The first input contains only KP_1 and its connected component in the whole volume but is presented in a subvolume view. The second input exclusively comprises the small vessel/airway segment of KP_2. The output heatmaps of the two keypoints correspond to the KP_1 input and KP_2 input, respectively, which avoids learning ambiguity. The 3D U-Net is a neural network architecture that features three encoder and three decoder stages. Each stage includes a convolution block and a downsampling or upsampling layer. The convolution block consists of two convolution layers, each using a kernel size of $3 \times 3 \times 3$, followed by batch normalization and rectified linear unit (ReLU) activation.

The network receives two binarized subvolumes $I \in \mathbb{R}^{2 \times D \times H \times W}$ as inputs, where D, H, W represent the spatial dimensions of the cropped volume. In this study, we decided to set the output heatmaps to the same size as the inputs, without downsampling, in order to avoid the loss of coordinate accuracy.

In the neural network design phase, we prioritized formulating the problem, constructing an open-source dataset, and proposing a comprehensive training and testing pipeline. We refrained from incorporating sophisticated modules,

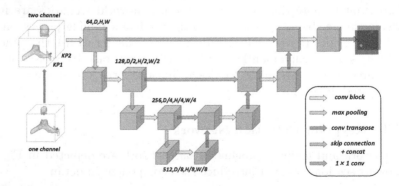

Fig. 2. Keypoint Detection Network. Given two disconnected components, the 3D-UNet outputs two heatmaps corresponding to KP_1 and KP_2.

such as attention mechanisms, transformer blocks, or distillation, and fine-tuning hyper-parameters. Hence, we do not delve into detailed network architecture design in this paper. However, we obtained promising results using a simple two-channel 3D-UNet model and explored various training techniques. Our work lays a solid foundation for future researchers to improve upon our findings by incorporating advanced techniques and innovative modules.

Loss Function. We adopt the state-of-the-art keypoint detection framework to represent the problem as heatmap estimation, where the coordinate with the highest confidence in each heatmap of $H \in \mathbb{R}^{k \times D \times H \times W}$ corresponds to the location of the kth keypoint. The ground-truth heatmaps are generated by placing a 3D Gaussian kernel at the center of each ground-truth keypoint location. For simplicity, we define the Keypoint Mean-Squared Error (KMSE) loss function as follows:

$$\mathcal{L}_{kmse} = \frac{1}{K} \sum_{k=1}^{K} \delta\left(V_k > 0\right) \cdot \left\| H_k - \hat{H}_k \right\|_2^2, \tag{1}$$

where H_k and $\hat{H}_k$ refer to the ground-truth and the predicted heatmaps for the kth keypoint, and K is fixed to 2 in our study. To reduce memory cost, we limit the size of subvolumes to $80 \times 80 \times 80$, which may result in invisible keypoints if the branch is long and the two keypoints are too far apart. Here, V_k indicates the visibility of the ground truth keypoint, where $V_k = 1$ and $\delta(V_k) = 1$ if the keypoint is visible, and vice versa.

Implementation Details. During the training phase, we employ a sampling strategy that randomly crops one volume, which introduces data augmentation, mitigates overfitting, and ensures training convergence. To reduce testing time and enable fair comparisons between models, we generate and save three random crops for each vessel branch during validation and testing. The size of the ground-truth heatmaps is $80 \times 80 \times 80$, and the sigma of the 3D Gaussian kernel used to generate them is set to 2.5. All networks were trained using AdamW optimizer with a learning rate of 0.0001 and beta hyperparameters of 0.5 and 0.999. The training was performed on a single NVIDIA 3090ti GPU with a batch size of 16. PyTorch framework was used for implementation, and early stopping strategy was adopted to prevent overfitting. To speed up training, we initialize the artery and vessel models with the trained airway model. We combined artery and vein training data to increase the training samples and reduce the training time.

2.4   Model Inference

Models trained using our proposed training paradigm may not be directly applicable to real-world data due to several assumptions made during training. Specifically, during training, we assume that the interrupted segmentation mask consists of only two continuous components representing KP_1 and KP_2, and that the location of KP_2 is known a priori, which is used to randomly crop subvolumes. Additionally, we limit the subvolume's size to ensure efficient training.

However, in real-world scenarios, the location of KP_1 and KP_2 components is unknown, and there may be small disconnected objects and noises scattered throughout the volume's entire original size. The only prior knowledge available is that KP_1 is located in the volume's largest connected component (i.e., the main vessel/airway), and KP_2 is in one of the small isolated components. To address this issue, we have developed an algorithm that bridges the gap between model training and inference, and accurately predicts disconnections in real-world situations. The pseudo-code of the inference algorithm is detailed in the supplementary materials.

3   Experiments

3.1   Datasets

A total of 800 CT scans with annotations of pulmonary airways, arteries, and veins are utilized to construct our dataset. The CT scans from multiple medical centers are manually annotated by a junior radiologist and confirmed by a senior radiologist [6]. The data is divided into training, validation, and test subsets with a ratio of 7:1:2. Each CT scan is pre-processed into three binarized volumes of airways, arteries, and veins. Subsequently, 30 distinct branches per volume were randomly selected for each binarized volume under specific criteria to create 30 volumes with vascular interruptions. The Pulmonary Tree Repairing (PTR) dataset includes 3D models represented by binarized ground-truth segmentation masks, centerlines, disconnected volumes, and a corresponding json file for each subject. The json file contains comprehensive information, such as the coordinates of bifurcations, endpoints, and all points along each branch, capturing diverse characteristics specific to each blood vessel. Note that the keypoint detection of airways, arteries, and veins disconnection is treated as three independent tasks, with each task having a dataset size of 800×30. The results are optimized on the validation set and reported on the test set.

3.2   Evaluation Metrics

Based on the definition of Object Keypoint Similarity (OKS) in pose estimation tasks, we have adapted this metric to align with the features of our dataset. Our modifications to the OKS are reflected in the following metric formulations:

$$OKS_k = \exp\left\{-d_k^2/2S\lambda^2\right\}, E_{d_k} = \exp\left\{-d_k^2\right\}, \tag{2}$$

Here d_k is the Euclidean distance between the predicted keypoint and the ground-truth keypoint, along with the vessel volume S of the corresponding branch. To maintain a consistent scale for OKS, we have introduced λ, a constant which we set to 0.2. OKS_k refers to the OKS of kth keypoint ($k = 2$ in our study).

$$OKS_i = \frac{\sum_k OKS_k \cdot \delta\left(V_k > 0\right)}{\sum_k \delta\left(V_k > 0\right)}, \tag{3}$$

where OKS_i denotes the OKS of ith sample and V_k is the visibility flag.

$$AP^\tau = \frac{\sum_i \delta\left(OKS_i > \tau\right)}{\sum_i 1}, \tag{4}$$

where we use standard AP_τ, which measures prediction precision of a model given a specific threshold τ. In order to provide a comprehensive and nuanced evaluation of the model's performance, we report average precision across various thresholds. Specifically, we report AP^{50} (AP at $\tau = 0.5$), AP^{75} (AP at $\tau = 0.75$), AP (the mean AP across 10 τ positions, where $\tau = \{0.5, 0.55, ..., 0.95\}$), AP^S (for small vessels with edge radius within the range of $(0, 2]$), AP^M (for medium vessels with edge radius within the range of $(2, 3]$), AP^L (for large vessels with edge radius greater than 3), AP^{k1} (AP for KP_1), AP^{k2} (AP for KP_2), E_d (mean E_{d_k}), E_d^{k1}, E_d^{k2} (E_d for KP_1, KP_2).

Table 1. Keypoint detection performance on the PTR dataset. $UNet^1$: The input is one subvolume with both KP_1 and KP_2 components; $UNet^2$: The inputs are two concatenated subvolumes of KP_1 and KP_2 components; Metrics are expressed in percentage (%) format.

Task	Method	AP	AP^{k1}	AP^{k2}	AP^{50}	AP^{75}	AP^S	AP^M	AP^L	E_d	E_d^{k1}	E_d^{k2}
Airway	$UNet^1$	80.89	79.23	86.32	94.21	90.47	75.15	85.32	80.02	18.81	15.39	22.07
	$UNet^2$	**87.18**	83.80	90.88	98.48	94.89	79.56	90.29	93.17	**28.54**	23.47	33.37
Artery	$UNet^1$	71.32	70.19	81.90	85.99	78.89	62.98	81.11	68.97	16.71	12.55	20.65
	$UNet^2$	**80.58**	77.46	87.07	94.09	86.85	70.45	88.31	84.25	**25.49**	20.83	29.90
Vein	$UNet^1$	69.10	69.09	79.79	82.80	76.97	59.17	78.25	69.38	15.49	12.26	18.54
	$UNet^2$	**78.78**	76.27	85.45	93.23	84.95	67.26	85.71	85.22	**24.40**	20.81	27.79

3.3 Results

To analyze the performance of our methods on topology repairing of disconnected pulmonary airways and vessels, we report several methods on the proposed PTR dataset, as shown in Table 1. The keypoint heatmap visualization is provided in the supplementary materials.

The study demonstrates that the two-channel 3D-UNet model surpasses the performance of the one-channel counterpart on airway and vessel segmentation tasks. Specifically, the two-channel model yields significant improvements in AP of approximately 7%, 9%, and 15% for airway, artery, and vein tasks, respectively. Additionally, the two-channel model achieves the highest performance on all evaluation metrics for all three tasks. These results suggest that the separation of KP_1 and KP_2 components as two-channel input can effectively improve their interaction in multiple feature levels, leading to improved performances. This is likely due to the high correlation between these two keypoints throughout the topological structure. However, detecting KP_1 was significantly challenging due to the random selection of cropping center points during data sampling, leading to a weaker performance for E_d and AP metrics. Additionally, the sparse distribution of keypoints on small pulmonary vessels posed a considerable challenge for capturing subtle features. Notably, the two-channel networks exhibited superior performance over one-channel methods by a substantial margin, which emphasizes the advantages of separating the two components. In the future work, it will be beneficial to design models that capture this characteristic.

4   Conclusion

In this study, we introduce a data-driven post-processing approach that addresses the challenge of disconnected pulmonary tubular structures, which is crucial for the diagnosis and treatment of pulmonary diseases. The proposed approach utilizes the newly created Pulmonary Tree Repairing (PTR) dataset, comprising 800 complete 3D models of pulmonary structures and synthetic disconnected data. A two-channel simple yet effective neural network is trained to detect keypoints that bridge disconnected components, utilizing a training data synthesis pipeline that generates disconnected data from complete pulmonary structures. Our approach yields promising results and holds great potential for clinical applications. While our study primarily focuses on addressing the disconnection issue, we recognize that more complex scenarios, such as handling multiple disconnected components, distinguishing between arteries and veins, and implementing our method in real-world settings, require further investigation in future work. Point or implicit representations [19–21] learning the geometric structures have high potentials in this application.

Acknowledgment. This research was supported by Australian Government Research Training Program (RTP) scholarship, and supported in part by a Swiss National Science Foundation grant.

References

1. Beare, R., Lowekamp, B., Yaniv, Z.: Image segmentation, registration and characterization in R with simpleitk. J. Stat. Softw. **86** (2018). https://doi.org/10.18637/jss.v086.i08
2. Bumgarner, J.R., Nelson, R.J.: Open-source analysis and visualization of segmented vasculature datasets with vesselvio. Cell Rep. Methods **2**(4), 100189 (2022). https://doi.org/10.1016/j.crmeth.2022.100189. https://www.sciencedirect.com/science/article/pii/S2667237522000443
3. Fetita, C.I., Prêteux, F., Beigelman-Aubry, C., Grenier, P.: Pulmonary airways: 3-D reconstruction from multislice CT and clinical investigation. IEEE Trans. Med. Imaging **23**(11), 1353–1364 (2004)
4. Garcia-Uceda, A., Selvan, R., Saghir, Z., Tiddens, H.A., de Bruijne, M.: Automatic airway segmentation from computed tomography using robust and efficient 3-D convolutional neural networks. Sci. Rep. **11**(1), 1–15 (2021)

5. Isensee, F., Jaeger, P.F., Kohl, S.A., Petersen, J., Maier-Hein, K.H.: nnU-Net: a self-configuring method for deep learning-based biomedical image segmentation. Nat. Methods **18**(2), 203–211 (2021)
6. Kuang, K., et al.: What makes for automatic reconstruction of pulmonary segments. In: Wang, L., Dou, Q., Fletcher, P.T., Speidel, S., Li, S. (eds.) MICCAI 2022. LNCS, vol. 13431, pp. 495–505. Springer, Cham (2022). https://doi.org/10.1007/978-3-031-16431-6_47
7. Luo, Z., Wang, Z., Huang, Y., Wang, L., Tan, T., Zhou, E.: Rethinking the heatmap regression for bottom-up human pose estimation. In: Conference on Computer Vision and Pattern Recognition (2021)
8. Pan, C., et al.: Deep 3D vessel segmentation based on cross transformer network. In: 2022 IEEE International Conference on Bioinformatics and Biomedicine, pp. 1115–1120 (2022)
9. Park, J., et al.: Deep learning based airway segmentation using key point prediction. Appl. Sci. **11**(8), 3501 (2021)
10. Qin, Y., et al.: Learning tubule-sensitive CNNs for pulmonary airway and artery-vein segmentation in CT. IEEE Trans. Med. Imaging **40**(6), 1603–1617 (2021)
11. Rahaghi, F., et al.: Pulmonary vascular morphology as an imaging biomarker in chronic thromboembolic pulmonary hypertension. Pulm. Circ. **6**(1), 70–81 (2016)
12. Saji, H., et al.: Segmentectomy versus lobectomy in small-sized peripheral non-small-cell lung cancer (JCOG0802/WJOG4607L): a multicentre, open-label, phase 3, randomised, controlled, non-inferiority trial. Lancet **399**(10335), 1607–1617 (2022)
13. Sun, K., Xiao, B., Liu, D., Wang, J.: Deep high-resolution representation learning for human pose estimation. In: Conference on Computer Vision and Pattern Recognition (2019)
14. Tetteh, G., et al.: Deepvesselnet: Vessel segmentation, centerline prediction, and bifurcation detection in 3-D angiographic volumes. Front. Neurosci. 1285 (2020)
15. Wang, A., Tam, T., Poon, H., Yu, K.C., Lee, W.N.: Naviairway: a bronchiole-sensitive deep learning-based airway segmentation pipeline for planning of navigation bronchoscopy. arXiv Preprint (2022). https://doi.org/10.36227/techrxiv.19228296
16. Wang, J., et al.: Deep high-resolution representation learning for visual recognition. IEEE Trans. Pattern Anal. Mach. Intell. **43**(10), 3349–3364 (2019)
17. Wittenberg, R., et al.: Acute pulmonary embolism: effect of a computer-assisted detection prototype on diagnosis-an observer study. Radiology **262**, 305–13 (2012). https://doi.org/10.1148/radiol.11110372
18. Xiao, B., Wu, H., Wei, Y.: Simple baselines for human pose estimation and tracking. In: European Conference on Computer Vision (2018)
19. Yang, J., Gu, S., Wei, D., Pfister, H., Ni, B.: RibSeg dataset and strong point cloud baselines for rib segmentation from CT scans. In: de Bruijne, M., et al. (eds.) MICCAI 2021. LNCS, vol. 12901, pp. 611–621. Springer, Cham (2021). https://doi.org/10.1007/978-3-030-87193-2_58
20. Yang, J., Shi, R., Wickramasinghe, U., Zhu, Q., Ni, B., Fua, P.: Neural annotation refinement: development of a new 3D dataset for adrenal gland analysis. In: Wang, L., Dou, Q., Fletcher, P.T., Speidel, S., Li, S. (eds.) MICCAI 2022. LNCS, vol. 13434, pp. 503–513. Springer, Cham (2022). https://doi.org/10.1007/978-3-031-16440-8_48
21. Yang, J., Wickramasinghe, U., Ni, B., Fua, P.: Implicitatlas: learning deformable shape templates in medical imaging. In: Conference on Computer Vision and Pattern Recognition, pp. 15861–15871 (2022)

392   Z. Weng et al.

22. Yaniv, Z., Lowekamp, B.C., Johnson, H.J., Beare, R.: Simpleitk image-analysis notebooks: a collaborative environment for education and reproducible research. J. Digit. Imaging **31**(3), 290–303 (2018)
23. Yao, Q., Quan, Q., Xiao, L., Kevin Zhou, S.: One-shot medical landmark detection. In: de Bruijne, M., et al. (eds.) MICCAI 2021. LNCS, vol. 12902, pp. 177–188. Springer, Cham (2021). https://doi.org/10.1007/978-3-030-87196-3_17
24. Yu, B., Tao, D.: Heatmap regression via randomized rounding. IEEE Trans. Pattern Anal. Mach. Intell. **44**(11), 8276–8289 (2021)
25. Zhang, M., et al.: Multi-site, multi-domain airway tree modeling (ATM'22): a public benchmark for pulmonary airway segmentation. arXiv Preprint (2023)
26. Zhao, W., et al.: 3D deep learning from CT scans predicts tumor invasiveness of subcentimeter pulmonary adenocarcinomas. Can. Res. **78**(24), 6881–6889 (2018)
27. Zhao, Z.R., et al.: Invasiveness assessment by artificial intelligence against intraoperative frozen section for pulmonary nodules ≤ 3 CM. J. Cancer Res. Clin. Oncol. 1–7 (2023)

AirwayFormer: Structure-Aware Boundary-Adaptive Transformers for Airway Anatomical Labeling

Weihao Yu[1], Hao Zheng[1,3], Yun Gu[1(✉)], Fangfang Xie[2], Jiayuan Sun[2], and Jie Yang[1(✉)]

[1] Institute of Medical Robotics, Shanghai Jiao Tong University, Shanghai, China
{geron762,jieyang}@sjtu.edu.cn
[2] Shanghai Chest Hospital, Shanghai, China
[3] Tencent Jarvis Lab, Shenzhen, China

Abstract. Pulmonary airway labeling identifies anatomical names for branches in bronchial trees. These fine-grained labels are critical for disease diagnosis and intra-operative navigation. Recently, various methods have been proposed for this task. However, accurate labeling of each bronchus is challenging due to the fine-grained categories and inter-individual variations. On the one hand, training a network with limited data to recognize multitudinous classes sets an obstacle to the design of algorithms. We propose to maximize the use of latent relationships by a transformer-based network. Neighborhood information is properly integrated to capture the priors in the tree structure, while a U-shape layout is introduced to exploit the correspondence between different nomenclature levels. On the other hand, individual variations cause the distribution overlapping of adjacent classes in feature space. To resolve the confusion between sibling categories, we present a novel generator that predicts the weight matrix of the classifier to produce dynamic decision boundaries between subsegmental classes. Extensive experiments performed on publicly available datasets demonstrate that our method can perform better than state-of-the-art methods. The code is publicly available at https://github.com/EndoluminalSurgicalVision-IMR/AirwayFormer.

Keywords: Airway anatomical labeling · Structural prior · Dynamic decision boundary

This work was partly supported by National Key R&D Program of China (2019YFB1311503, 2017YFC0112700), Committee of Science and Technology, Shanghai, China (19510711200), Shanghai Sailing Program (20YF1420800), NSFC (61661010, 61977046, 62003208), SJTU Trans-med Awards Research (20210101), Science and Technology Commission of Shanghai Municipality (20DZ2220400).

W. Yu and H. Zheng—Equal contribution.

Supplementary Information The online version contains supplementary material available at https://doi.org/10.1007/978-3-031-43990-2_37.

1   Introduction

Automatic airway labeling aims to assign the corresponding anatomical names to the branches in airway trees. The identification of peripheral branches plays an essential role in bronchoscopic navigation. Figure 1 illustrates the hierarchical nomenclature [11] from lobar to subsegmental levels in the bronchial tree. In the first level, the naming is based on the five lung lobes, and similarly, the subtrees are further divided according to the 18 lung segments. The nomenclature of subsegmental bronchi is more complex, which contains six classes (a, b, c, a+b, a+c, and b+c) for the subtrees and their common branches.

Various methods have been proposed for airway labeling [2,6,7,9,10,14,16, 21,22] in recent years. Several works [2,6,16] adopted graph-matching-based algorithms to clarify the candidate trees based on a reference tree. However, the performance is limited at the subsegmental level due to individual variation. Deep learning methods [14,21,22] are also developed for this task. Tan et al. [14] proposed a multi-task U-Net [12] with a structure-aware graph convolutional network (GCN) [5] to segment the airway tree semantically. Yu et al. [21] converted the airway tree from the image to the graph space and designed a multi-stage framework with hypergraph neural networks (HGNN) for node classification. There are two main challenges for these learning-based methods. First, 127 classes are included in the nomenclature up to subsegmental level. The annotation is quite time-consuming and labor-intensive. Classification for over one hundred categories with limited training data is an extremely hard problem. Second, the inter-individual differences regarding the location, direction, length, and diameter of branches become increasingly significant from the lobar order to the subsegmental level. Figure 1(d) demonstrates the t-distributed stochastic neighbor embedding (t-sne) results of the learned features of RB10 branches in two examples. The distribution of RB10b in the first case is overlapped with the distribution of RB10c in the second example. Such individual variation dramatically affects the generalization ability of models.

To resolve the first problem, in this work, we propose to fully use the latent relationships within the tree structure and airway nomenclature by a novel network named AirwayFormer. The tree structure provides inherent message transmission roads, while both global and local information is critical for anatomical labeling. To this end, we first adopt a neighborhood information encoding module based on the transformer block to aggregate both global and local features within the tree structure. Another structural relationship appears in the nomenclature of bronchial trees. As shown in Fig. 1(a), (b), and (c), the subsegmental bronchi RB10a belong to RB10 in the segmental level and the right lower lobe bronchi (RLL) in the lobar level, respectively. The consistency between classes at different levels is an important cue in this task. In previous works, Yu et al. [21] introduced the classification results of the current stage to the next stage as extra constraints. However, the inference is only single-directional, where the prediction is not used to refine the classification in the upper level. To achieve a bi-directional refinement, a U-shape layout is designed to explore the correspondence between different nomenclature levels.

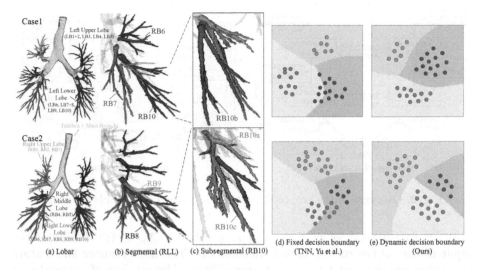

Fig. 1. Three-level nomenclature of the bronchial tree, and visualization of branch features and decision boundaries of different methods. The text color is consistent with the classes. Colors in t-sne figures are consistent with those in the airway tree. RLL is short for the right lower lobe.

For the second challenge, an ideal solution is that the model can adaptively adjust itself to deal with individual variation. Especially in subsegmental level, the same category in different cases may have distinct features. Actually, the nomenclature of subsegmental bronchi depends on their relative positions to the sibling branches within a segment. Based on this prior knowledge, we design a novel generator that learns to capture the relative relationships by predicting the weights of the last fully connected layer. More specifically, the weight for a subsegmental class is generated based on the feature representation of the corresponding segmental class. As illustrated in Fig. 1, compared with using fixed weights in the classifier, the predicted weights dynamically adjust the decision boundaries according to the characteristics of the segmental bronchi in each case, alleviating the overfitting problem caused by individual variation.

Our main contributions can be summarized as follows: (1) AirwayFormer is proposed for accurate airway labeling up to subsegmental level by exploiting the latent structural relationships. (2) A weight generator is designed to mitigate the overfitting caused by individual variation via adaptive decision boundary adjustment. (3) With extensive experiments on the public dataset, our method achieves state-of-the-art results in lobar, segmental, and subsegmental levels.

2   Method

An overview of our proposed method is illustrated in Fig. 2. Following previous work [21], each branch is regarded as a node in a graph space with well-designed 20-dimensional features. Five transformer blocks are used to classify these nodes hierarchically. The weight matrix of the subsegmental classifier is dynamically updated according to the segmental features. We denote the function of the

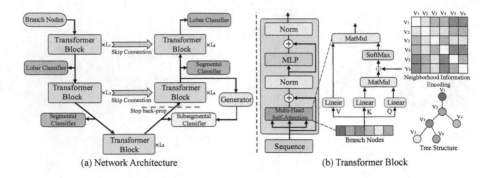

Fig. 2. Overview of the proposed method for bronchi anatomical labeling.

m-th transformer blocks as $\mathcal{F}_m(\cdot), 1 \leq m \leq M = 5$. $X_m \in \mathbb{R}^{n \times d}$ denotes the input feature of the corresponding transformers. The node number and feature dimension are denoted by n and d, respectively. The number of categories of the m-th classifier is c_m. More details about our method are introduced in the following sections.

2.1 Structure-Aware Transformers

We propose a transformer-based model named AirwayFormer to exploit the structural relationships in airway labeling. The tree structure is introduced into the self-attention calculation, while the class consistency in the hierarchical nomenclature is encoded in the U-shape layout.

Self-attention module in transformers enables nodes to aggregate messages from a global scale while neglecting the local structure prior. Recent studies [1,8,20] showed that incorporating graph information into vanilla transformers can achieve competitive performance. As the labeling of peripheral branches needs both global and local information, we adopt a neighborhood information encoding (NIE) module, which integrates the structural prior into the self-attention mechanism. Given an airway tree with n branches, each branch can be seen as a node v. The parent bronchus and children bronchus is adjacent. Then we adopt a distance function $\psi(v_i, v_j)$ to measure the spatial relation between nodes v_i and v_j. Here, $\psi(v_i, v_j)$ is defined as the shortest path distance (SPD) between v_i and v_j. Finally, a codebook $\mathcal{C}_m$ is used to convert the SPD matrix $D \in \mathbb{R}^{n \times n}$ to learnable scale parameters, which can serve as a graph bias term of the attention map $A \in \mathbb{R}^{n \times n}$:

$$A = \frac{X_m Q (X_m K)^T}{\sqrt{d}} + \mathcal{C}_m(D). \tag{1}$$

Here, $Q \in \mathbb{R}^{d \times d}$ and $K \in \mathbb{R}^{d \times d}$ are trainable parameters. Matrix D encodes the neighborhood information of the tree structure, and the network can adaptively concentrate on the local structure by optimizing $\mathcal{C}_m$.

To simultaneously utilize the bi-directional correspondence between different levels, AirwayFormer adopts a U-shape layout. Concretely, the network performs nodes classification hierarchically from lobar level to subsegmental level and then back to lobar level:

$$G_m = \mathcal{F}_m(X_m). \tag{2}$$

$$X_{m+1} = \begin{cases} G_m, & 1 \le m < \frac{M}{2}, \\ \mathcal{CAT}(G_m, G_{M-m}), & \frac{M}{2} \le m < M. \end{cases} \tag{3}$$

$$P_m = \mathcal{Z}_m(G_m). \tag{4}$$

Here, $G_m \in \mathbb{R}^{n \times d}$ is the output feature of the m-th transformer blocks, and $\mathcal{CAT}(\cdot)$ denotes the concatenate operation. $\mathcal{Z}_m(\cdot)$ is the m-th classifier consisting of a simple linear layer while $P_m \in \mathbb{R}^{n \times c_m}$ is the prediction result. The advantages of this design are two-fold. First, the output of coarser level is used as the input of finer level, which is helpful for the network to learn the corresponding relationship of the categories from coarse to fine. Second, the results of coarser level can be directly deduced from the predictions of finer level. Fine-grained features feedback to coarser level reduces the classification difficulty of coarser level and promotes the labeling performance. Besides, transformers conducting labeling of the same level are connected directly using highways. The same-level features can prevent performance degradation and ameliorate training stability. We further stop gradient back-propagation from segmental level to subsegmental level. The reason is that the severe distribution overlap in subsegments makes the feature learning in this level markedly different from others. This is further discussed in the supplementary materials.

2.2   Boundary-Adaptive Classifier

Individual variations cause the distribution of adjacent classes to intersect, especially in subsegmental level. Although most subsegmental bronchi in the same tree are separable according to their relative positions, the inter-individual differences seriously affect the generalization performance of the model. To introduce case-specific information, we propose a boundary-adaptive classifier that adopts a novel generator to predict case-specific weights for each subsegmental category. Let $G_k \in \mathbb{R}^{n \times d}$ denote the subsegmental output feature. We first use the next segmental level feature $G_{k+1} \in \mathbb{R}^{n \times d}$ and prediction $P_{k+1} \in \mathbb{R}^{n \times c_{k+1}}$ to obtain coarser representation for each segmental class:

$$S_{k+1} = Softmax(P_{k+1}). \tag{5}$$

$$H = (S_{k+1}{}^{\alpha})^T G_{k+1}. \tag{6}$$

Here, $S_{k+1} \in \mathbb{R}^{n \times c_{k+1}}$ denotes the probability matrix of the segmental level while α is a learnable cluster parameter. $H = (h_1, h_2, ..., h_{c_{k+1}})^T \in \mathbb{R}^{c_{k+1} \times d}$ are representations for c_{k+1} segmental categories. However, the segmental prediction P_{k+1} is not always perfect and may contain some classification errors. To avoid potential error propagation and obtain better class representations, these cluster centers are refined using a vanilla transformer. Specifically, we take H as query vectors and G_{k+1} as key and value vectors:

$$\tilde{H} = transformer(q = H, k = v = G_{k+1}).\qquad(7)$$

After getting the enhanced segmental class representations, a generator network $\mathcal{T}(\cdot)$ is used to produce classifier weights for subsegmental categories. Since each segment contains six subsegmental bronchi, $\mathcal{T}(\cdot)$ transforms each segmental center into the seven (including itself) classifier weights:

$$w_i = \mathcal{T}(\tilde{h}_i) = U(Gelu(V(\tilde{h}_i))), \quad 1 \le i \le c_{k+1}.\qquad(8)$$

Here, $Gelu(\cdot)$ denotes the GELU activation function [3]. A bottleneck architecture is used to limit the number of parameters: $V \in \mathbb{R}^{d \times b}$ and $U \in \mathbb{R}^{b \times 7 \times d}$ are down-projection and up-projection learnable matrices respectively where $b \ll d$. Finally, the weight matrix of the subsegmental classifier can be obtained by stacking $w_i \in \mathbb{R}^{7 \times d}$.

We use cross-entropy loss with labeling smoothing [13] as the loss function for each task. Then the total loss function can be formulated as

$$\mathcal{L}_{total} = \sum_{m=1}^{M} \gamma_m \mathcal{L}_{ce},\qquad(9)$$

where γ_m is the weight of the m-th loss function.

3   Experiments and Results

3.1   Dataset and Implementation Details

We evaluated our method on the public Airway Tree Labeling (ATL) Dataset[1] [21]. The dataset contains 104 labeled bronchial trees from CT scans whose slice thickness ≤ 0.67 mm and spatial resolution ranges from 0.78 mm to 0.82 mm. The annotation is three-level, including six lobar bronchi, 19 segmental bronchi, and 127 subsegmental bronchi. We conducted 4-fold cross-validation on the dataset.

We set the label smoothing hyperparameter σ of the loss function to 0.02. The weight of loss function γ_m was set to 1. More experiments about hyperparameters can be found in the supplementary materials. We stacked two transformer blocks for airway labeling of each level. The model was trained using Adam optimizer ($\beta_1 = 0.9, \beta_2 = 0.999$) with a learning rate of $5e^{-4}$ for 800 epoches. Predictions of the last lobar and segmental classifiers were used to be the final results of these levels. Four evaluation metrics were used: (a) accuracy (ACC), (b) precision(PR), (c) recall (RC), and (d) F1 score (F1). All the networks were implemented in PyTorch framework with a GeForce GTX TITAN XP GPU.

---

[1] https://github.com/yuyouxixi/airway-labeling/tree/main/dataset.

Table 1. Comparison with different methods in airway anatomical labeling (%).

Method	Lobar				Segmental				Subsegmental			
	ACC	PR	RC	F1	ACC	PR	RC	F1	ACC	PR	RC	F1
(A) GNNs												
GCN [5]	98.0	96.8	96.1	96.5	85.4	86.3	84.0	85.2	60.2	59.5	62.2	60.8
GAT [18]	98.6	98.0	97.0	97.5	88.6	89.6	88.7	89.2	70.7	63.7	66.6	65.1
DAGNN [15]	98.5	97.7	97.2	97.4	89.1	88.6	88.1	88.3	74.0	67.8	71.1	69.4
(B) HGNNs												
HyperGCN [19]	98.4	98.1	96.5	97.3	87.6	83.9	85.1	84.5	61.8	48.2	52.5	50.3
UniSAGE [4]	98.6	98.0	97.0	97.5	89.0	89.6	88.6	89.1	70.9	66.5	68.7	67.6
TNN [21]	98.6	98.4	96.4	97.4	93.6	93.0	93.6	93.3	82.0	76.9	79.7	78.2
(C) CNNs												
SGNet [14]	96.8	96.5	95.8	96.1	83.2	76.6	75.1	74.8	–	–	–	–
(D) Conventional												
Kitaoka et al. [6]	92.5	92.2	91.3	91.8	78.7	77.3	77.4	77.3	55.9	54.2	53.6	53.9
Tschirren et al. [16]	92.8	92.6	92.2	92.4	79.5	78.6	77.5	78.0	57.1	54.5	56.1	55.4
Feragen et al. [2]	94.1	94.1	93.6	93.8	81.8	80.1	83.2	81.7	62.8	60.2	56.2	58.1
(E) Transformers												
Transformer [17]	98.8	98.5	97.6	98.0	93.1	93.0	92.8	92.9	82.5	77.8	80.8	79.3
Ours	**99.2**	**98.8**	**98.2**	**98.5**	**95.7**	**95.9**	**95.9**	**95.9**	**86.0**	**83.5**	**84.5**	**84.0**

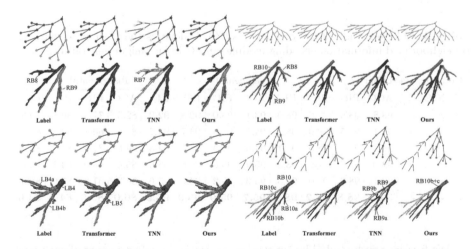

Fig. 3. The results of airway labeling are displayed from the perspective of graph and 3D. The first (last) two rows are the results of segmental (subsegmental) bronchi.

3.2   Evaluations

Table 1 compares the quantitative results of our proposed method with other methods. GNNs methods such as GCN [5] and GAT [18] only aggregate information from neighborhoods, while DAGNN [15] gradually performs message passing and updating from root nodes to the end. These methods merely encode the local structure of airway trees and are seriously disturbed by distribution overlapping in subsegments. Thus, they fail to achieve satisfactory performance, especially in subsegmental level. In HGNN methods, HyperGCN [19] and UniSAGE [4] use hyperedges to represent subtrees. TNN [21] further proposes a sub-network for subtrees to exchange information and outperforms other HGNNs methods. SGNet [14] uses CNNs to conduct semantic segmentation directly on CT images, which can be easily affected by class imbalance. Conventional methods [2,6,16] usually adopt graph-matching-based algorithms, but the reference tree cannot meet all situations due to individual variations. Thus, their performance drops a lot in the subsegmental level. Vanilla Transformer [17] achieves the same effect as TNN owing to the global-scale self-attention module while lacking the utilization of structural prior. Compared to all the above methods, our approach improves the accuracy by more than 2% and 4% respectively in segmental and subsegmental levels, demonstrating the effectiveness of the proposed method.

Qualitative results are displayed in Fig. 3 to demonstrate the superior performance of our model at different levels. In segmental labeling, transformer lacks the local structural prior, in which way the sibling categories cannot be classified well. The problem is more severe in subsegments. TNN adopts a subtree interaction module to learn relative information, but the correspondence between different nomenclature level is not fully utilized. By contrast, our method can perform accurate airway labeling both in segmental and subsegmental levels.

Table 2. Ablation study of key components (%). U denotes the U-shape layout, NIE is neighborhood information encoding module, and G is the weight generator network.

Components			Lobar				Segmental				Subsegmental			
U	NIE	G	ACC	PR	RC	F1	ACC	PR	RC	F1	ACC	PR	RC	F1
✗	✗	✗	98.8	98.5	97.6	98.0	93.1	93.0	92.8	92.9	82.5	77.8	80.8	79.3
✓	✗	✗	99.1	98.5	98.0	98.2	94.7	94.8	94.7	94.7	83.6	80.7	82.2	81.4
✗	✓	✗	99.0	98.2	98.0	98.1	94.6	94.5	94.1	94.3	84.8	81.6	82.7	82.2
✓	✓	✗	99.0	98.6	97.9	98.3	95.4	95.6	95.6	95.6	85.4	83.0	84.1	83.6
✓	✗	✓	99.1	98.6	98.0	98.3	95.0	95.2	94.9	95.0	84.3	81.8	83.0	82.4
✓	✓	✓	**99.2**	**98.8**	**98.2**	**98.5**	**95.7**	**95.9**	**95.9**	**95.9**	**86.0**	**83.5**	**84.5**	**84.0**

We further conducted ablation studies to verify the effectiveness of each component of AirwayFormer. Table 2 shows the results. Vanilla transformer performing bronchi labeling for each level independently is used as the baseline. The U-shape layout encodes the correspondence of nomenclature and improves the results in all three levels. NIE module introduces the local structure prior to self-attention calculation and also promotes the labeling performance, especially in the subsegmental level. Combining the two modules, the performance is further ameliorated. The weight generator network dynamically adjusts the classification weights and improves the subsegmental results. When applying the three modules simultaneously, the network demonstrates the most powerful ability for airway labeling. The experimental results indicate that these proposed modules do contribute to the satisfactory performance of our method.

4   Conclusion

This paper presented a transformer-based method named AirwayFormer for airway anatomical labeling. A U-shape layout integrating graph information is used to exploit the latent relationships fully. Meanwhile, a weight generator that produces the dynamic decision boundaries is designed to capture the relative relationships between sibling categories. Extensive experiments showed that our proposed method achieved superior performance in the bronchi labeling of all three levels, leading to an efficient clinical tool for intra-operative navigation.

References

1. Dwivedi, V.P., Bresson, X.: A generalization of transformer networks to graphs. arXiv preprint arXiv:2012.09699 (2020)
2. Feragen, A., et al.: Geodesic atlas-based labeling of anatomical trees: application and evaluation on airways extracted from CT. IEEE Trans. Med. Imaging **34**(6), 1212–1226 (2014)
3. Hendrycks, D., Gimpel, K.: Gaussian error linear units (GELUs). arXiv preprint arXiv:1606.08415 (2016)
4. Huang, J., Yang, J.: Unignn: a unified framework for graph and hypergraph neural networks. arXiv preprint arXiv:2105.00956 (2021)
5. Kipf, T.N., Welling, M.: Semi-supervised classification with graph convolutional networks. arXiv preprint arXiv:1609.02907 (2016)
6. Kitaoka, H., et al.: Automated nomenclature labeling of the bronchial tree in 3D-CT lung images. In: Dohi, T., Kikinis, R. (eds.) MICCAI 2002. LNCS, vol. 2489, pp. 1–11. Springer, Heidelberg (2002). https://doi.org/10.1007/3-540-45787-9_1
7. Lo, P., van Rikxoort, E.M., Goldin, J., Abtin, F., de Bruijne, M., Brown, M.: A bottom-up approach for labeling of human airway trees. MICCAI Int. WS. Pulm. Im. Anal. (2011)
8. Min, E., et al.: Transformer for graphs: an overview from architecture perspective. arXiv preprint arXiv:2202.08455 (2022)
9. Mori, K., Hasegawa, J.I., Suenaga, Y., Toriwaki, J.I.: Automated anatomical labeling of the bronchial branch and its application to the virtual bronchoscopy system. IEEE Trans. Med. Imaging **19**(2), 103–114 (2000)
10. Mori, K., et al.: Automated anatomical labeling of bronchial branches extracted from CT datasets based on machine learning and combination optimization and its application to bronchoscope guidance. In: Yang, G.-Z., Hawkes, D., Rueckert, D., Noble, A., Taylor, C. (eds.) MICCAI 2009. LNCS, vol. 5762, pp. 707–714. Springer, Heidelberg (2009). https://doi.org/10.1007/978-3-642-04271-3_86

11. Netter, F.H.: Atlas of human anatomy, Professional Edition E-Book: including NetterReference. com Access with full downloadable image Bank. Elsevier health sciences (2014)
12. Ronneberger, O., Fischer, P., Brox, T.: U-Net: convolutional networks for biomedical image segmentation. In: Navab, N., Hornegger, J., Wells, W.M., Frangi, A.F. (eds.) MICCAI 2015. LNCS, vol. 9351, pp. 234–241. Springer, Cham (2015). https://doi.org/10.1007/978-3-319-24574-4_28
13. Szegedy, C., Vanhoucke, V., Ioffe, S., Shlens, J., Wojna, Z.: Rethinking the inception architecture for computer vision. In: Proceedings of the IEEE Conference on Computer Vision and Pattern Recognition, pp. 2818–2826 (2016)
14. Tan, Z., Feng, J., Zhou, J.: SGNet: structure-aware graph-based network for airway semantic segmentation. In: de Bruijne, M., et al. (eds.) MICCAI 2021. LNCS, vol. 12901, pp. 153–163. Springer, Cham (2021). https://doi.org/10.1007/978-3-030-87193-2_15
15. Thost, V., Chen, J.: Directed acyclic graph neural networks. arXiv preprint arXiv:2101.07965 (2021)
16. Tschirren, J., McLennan, G., Palágyi, K., Hoffman, E.A., Sonka, M.: Matching and anatomical labeling of human airway tree. IEEE Trans. Med. Imaging 24(12), 1540–1547 (2005)
17. Vaswani, A., et al.: Attention is all you need. In: Advances in Neural Information Processing Systems, vol. 30 (2017)
18. Veličković, P., Cucurull, G., Casanova, A., Romero, A., Lio, P., Bengio, Y.: Graph attention networks. arXiv preprint arXiv:1710.10903 (2017)
19. Yadati, N., Nimishakavi, M., Yadav, P., Nitin, V., Louis, A., Talukdar, P.: Hypergcn: a new method for training graph convolutional networks on hypergraphs. In: Advances in Neural Information Processing Systems, vol. 32 (2019)
20. Ying, C., et al.: Do transformers really perform badly for graph representation? Adv. Neural. Inf. Process. Syst. 34, 28877–28888 (2021)
21. Yu, W., et al.: TNN: tree neural network for airway anatomical labeling. IEEE Trans. Med. Imaging 42(1), 103–118 (2022)
22. Zhao, T., Yin, Z., Wang, J., Gao, D., Chen, Y., Mao, Y.: Bronchus segmentation and classification by neural networks and linear programming. In: Shen, D., et al. (eds.) MICCAI 2019. LNCS, vol. 11769, pp. 230–239. Springer, Cham (2019). https://doi.org/10.1007/978-3-030-32226-7_26

CLIP-Lung: Textual Knowledge-Guided Lung Nodule Malignancy Prediction

Yiming Lei[1], Zilong Li[1], Yan Shen[2], Junping Zhang[1],
and Hongming Shan[3,4,5](✉) (iD)

[1] Shanghai Key Lab of Intelligent Information Processing, School of Computer
Science, Fudan University, Shanghai, China
[2] School of Pharmacy, China Pharmaceutical University, Nanjing, China
[3] Institute of Science and Technology for Brain-Inspired Intelligence and MOE
Frontiers Center for Brain Science, Fudan University, Shanghai, China
hmshan@fudan.edu.cn
[4] Key Laboratory of Computational Neuroscience and Brain-Inspired Intelligence
(Fudan University), Ministry of Education, Shanghai, China
[5] Shanghai Center for Brain Science and Brain-Inspired Technology, Shanghai, China

Abstract. Lung nodule malignancy prediction has been enhanced by
advanced deep-learning techniques and effective tricks. Nevertheless, current methods are mainly trained with cross-entropy loss using one-hot
categorical labels, which results in difficulty in distinguishing those nodules with closer progression labels. Interestingly, we observe that clinical
text information annotated by radiologists provides us with discriminative knowledge to identify challenging samples. Drawing on the capability of the contrastive language-image pre-training (CLIP) model to learn
generalized visual representations from text annotations, in this paper,
we propose CLIP-Lung, a textual knowledge-guided framework for lung
nodule malignancy prediction. First, CLIP-Lung introduces both class
and attribute annotations into the training of the lung nodule classifier without any additional overheads in inference. Second, we design
a channel-wise conditional prompt (CCP) module to establish consistent relationships between learnable context prompts and specific feature maps. Third, we align image features with both class and attribute
features via contrastive learning, rectifying false positives and false negatives in latent space. Experimental results on the benchmark LIDC-IDRI
dataset demonstrate the superiority of CLIP-Lung, in both classification
performance and interpretability of attention maps. Source code is available at https://github.com/ymLeiFDU/CLIP-Lung.

Keywords: Lung nodule classification · vision-language model ·
prompt learning

1 Introduction

Lung cancer is one of the most fatal diseases worldwide, and early diagnosis
of the pulmonary nodule has been identified as an effective measure to prevent lung cancer. Deep learning-based methods for lung nodule classification

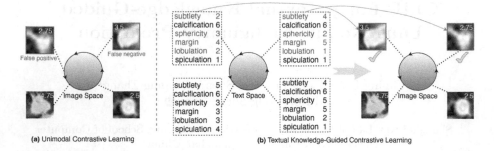

Fig. 1. Motivation of CLIP-Lung. (a) Unimodal contrastive learning. (b) Proposed textual knowledge-guided contrastive learning. Yellow values are the annotated malignancy scores. Dashed boxes contain pairs of textual attributes and annotated values. (Color figure online)

have been widely studied in recent years [9, 12]. Usually, the malignancy prediction is often formulated as benign-malignant binary classification [9, 10, 19], and the higher classification performance and explainable attention maps are impressive. Most previous works employ a learning paradigm that utilizes cross-entropy loss between predicted probability distributions and ground-truth one-hot labels. Furthermore, inspired by ordered labels of nodule progression, researchers have turned their attention to ordinal regression methods to evaluate the benign-unsure-malignant classification task [2, 11, 13, 18, 21], where the training set additionally includes nodules with uncertain labels. Indeed, the ordinal regression-based methods are able to learn ordered manifolds and to further enhance the prediction accuracy.

However, the aforementioned methods still face challenges in distinguishing visually similar samples with adjacent rank labels. For example, in Fig. 1(a), since we conduct unimodal contrastive learning and map the samples onto a spherical space, the false positive nodule with a malignancy score of 2.75 has a closer distance to that with a score of 4.75, and the false negative one should not be closer to that of score 2.5. To address this issue, we found that the text attributes, such as "subtlety", "sphericity", "margin", and "lobulation", annotated by radiologists, can exhibit the differences between these hard samples. Therefore, we propose leveraging text annotations to guide the learning of visual features. In practice, this also aligns with the fact that the annotated text information represents the direct justification for identifying lesion regions in the clinic. As shown in Fig. 1, this text information is beneficial for distinguishing visually similar pairs, while we conduct this behavior by applying contrastive learning that pulls semantic-closer samples and pushes away semantic-farther ones.

To integrate text annotations into the image-domain learning process, an effective text encoder providing accurate textual features is required. Fortunately, recent advances in vision-language models, such as contrastive language-image pre-training (CLIP) [16], provide us with a powerful text encoder pre-trained with text-based supervisions and have shown impressive results in downstream vision tasks. Nevertheless, it is ineffective to directly transfer CLIP to medical tasks due to the data covariate shift. Therefore, in this paper, we pro-

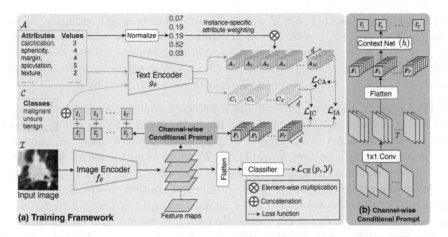

Fig. 2. Illustration of the proposed CLIP-Lung.

pose CLIP-Lung, a framework to classify lung nodules using image-text pairs. Specifically, CLIP-Lung constructs learnable text descriptions for each nodule from both class and attribute perspectives. Inspired by CoCoOp [20], we propose a channel-wise conditional prompt (CCP) module to allow nodule descriptions to guide the generation of informative feature maps. Different from CoCoOp, CCP constructs specific learnable prompts conditioned on grouped feature maps and triggers more explainable attention maps such as Grad-CAM [17], whereas CoCoOp provides only the common condition for all the prompt tokens. Then, we design a textual knowledge-guided contrastive learning based on obtained image features and textual features involving classes and attributes. Experimental results on LIDC-IDRI [1] dataset demonstrate the effectiveness of learning with textual knowledge for improving lung nodule malignancy prediction.

The contributions of this paper are summarized as follows.

1) We propose CLIP-Lung for lung nodule malignancy prediction, which leverages clinical textual knowledge to enhance the image encoder and classifier.
2) We design a channel-wise conditional prompt module to establish consistent relationships among the correlated text tokens and feature maps.
3) We simultaneously align the image features with class and attribute features through contrastive learning while generating more explainable attention maps.

2 Methodology

2.1 Overview

Problem Formulation. In this paper, we arrange the lung nodule classification dataset as $\{\mathcal{I}, \mathcal{Y}, \mathcal{C}, \mathcal{A}\}$, where $\mathcal{I} = \{\boldsymbol{I}_i\}_{i=1}^N$ is an image set containing N lung nodule images. $\mathcal{Y} = \{y_i\}_{i=1}^N$ is the corresponding class label set and $y_i \in \{1, 2, \ldots, K\}$, and K is the number of classes. $\mathcal{C} = \{c_k\}_{k=1}^K$ is a set of text embeddings of classes. Finally, $\mathcal{A} = \{a_m\}_{m=1}^M$ is the set of attribute embeddings, where each element $a_m \in \mathbb{R}^{d \times 1}$ is a vector representing the embedding of an attribute word such as "spiculation". Then, for a given sample $\{I_i, y_i\}$, our aim is to learn a mapping $f_\theta : I_i \mapsto y_i$, where f is a deep neural network parameterized by θ.

CLIP-Lung. In Fig. 2(a), the training framework contains an image encoder f_θ and a text encoder g_ϕ. First, the input image I_i is fed into f_θ and then generates the feature maps. According to Fig. 2(b), the feature maps are converted to channel-wise feature vectors $f_\theta(I_i) = F_{t,:}$ and then to learnable tokens l'_t. Second, we initialize the context tokens l_t and add them with l'_t to construct the learnable prompts, where T is the number of context words. Next, the concatenation of the class token and $l_t + l'_t$ is used as input of text encoder yielding the class features $g_\phi(c_k) = C_{k,:}$, note that $C_{k,:}$ is conditioned on channel-wise feature vectors $F_{t,:}$. Finally, the attribute tokens a_m are also fed into the text encoder to yield corresponding attribute features $g_\phi(a_m) = A_{m,:}$. Note that the vectors $F_{t,:}$, $l_{t,:}$, $l'_{t,:}$, and $C_{k,:}$ are with the same dimension $d = 512$ in this paper. Consequently, we have image feature $F \in \mathbb{R}^{T \times d}$, class feature $C \in \mathbb{R}^{K \times d}$, and attribute feature $A \in \mathbb{R}^{M \times d}$ to conduct the textual knowledge-guided contrastive learning.

2.2   Instance-Specific Attribute Weighting

For the attribute annotations, all the lung nodules in the LIDC-IDRI dataset are annotated with the *same* eight attributes: "subtlety", "internal structure", "calcification", "sphericity", "margin", "lobulation", "spiculation", and "texture" [4,8], and the annotated value for each attribute ranges from 1 to 5 except for "calcification" that is ranged from 1 to 6. In this paper, we fix the parameters of a pre-trained text encoder so that the generated eight text feature vectors are the same for all the nodules. Therefore, we propose an instance-specific attribute weighting scheme to distinguish different nodules. For the i-th sample, the weight for each a_m is calculated through normalizing the annotated values:

$$w_m = \frac{\exp(v_m)}{\sum_{m=1}^M \exp(v_m)}, \tag{1}$$

where v_m denotes the annotated value for a_m. Then the weight vectors of the i-th sample is represented as $w_i = [w_1, w_2, \ldots, w_M]^\top \in \mathbb{R}^{M \times 1}$. Hence, the element-wise multiplication $w_i \cdot A_i$ is unique to I_i.

2.3   Channel-Wise Conditional Prompt

CoCoOp [20] firstly proposed to learn language contexts for vision-language models conditioned on visual features. However, it is inferior to align context words with partial regions of the lesion. Therefore, we propose a channel-wise conditional prompt (CCP) module, in Fig. 2(b), to split latent feature maps into T groups and then flatten them into vectors $F_{t,:}$. Next, we denote $h(\cdot)$ as a context network that is composed of a multi-layer perceptron (MLP) with one hidden layer, and each learnable context token is now obtained by $l_t' = h(F_{t,:})$. Hence, the conditional prompt for the t-th token is $l_t + l_t'$. In addition, CCP also outputs the $F_{t,:}$ for image-class and image-attribute contrastive learning.

2.4 Textual Knowledge-Guided Contrastive Learning

Recall that our aim is to enable the visual features to be similar to the textual features of the annotated classes or attributes and be dissimilar to those of irrelevant text annotations. Consequently, we accomplish this goal through contrastive learning [3,5,7]. In this paper, we conduct such image-text contrastive learning by utilizing pre-trained CLIP text encoder [16]. In Fig. 2, we align $F \in \mathbb{R}^{T \times d}$ with $C \in \mathbb{R}^{K \times d}$ and $A \in \mathbb{R}^{M \times d}$, i.e., using class and attribute knowledge to regularize the feature maps.

Image-Class Alignment. First, the same to CLIP, we align the image and class information by minimizing the cross-entropy (CE) loss for the sample $\{I_i, y_i\}$:

$$\mathcal{L}_{IC} = -\sum_{t=1}^{T}\sum_{k=1}^{K} y_i \log \frac{\exp(\sigma(F_{t,:}, C_{k,:})/\tau)}{\sum_{k'=1}^{K} \exp(\sigma(F_{t,:}, C_{k',:})/\tau)}, \tag{2}$$

where $C_{k,:} = g_\phi(c_k \bigoplus (l_1 + l_1', l_2 + l_2', \ldots, l_T + l_T')) \in \mathbb{R}^{d \times 1}$ and "$\bigoplus$" denotes concatenation, i.e., $C_{k,:}$ is conditioned on learnable prompts $l_t + l_t'$. $\sigma(\cdot, \cdot)$ calculates the cosine similarity and τ is the temperature term. Therefore, $\mathcal{L}_{IC}$ implements the contrastive learning between channel-wise features and corresponding class features, i.e., the ensemble of grouped image-class alignment results.

Image-Attribute Alignment. In addition to image-class alignment, we further expect the image features to correlate with specific attributes. So we conduct image-attribute alignment by minimizing the InfoNCE loss [5,16]:

$$\mathcal{L}_{IA} = -\sum_{t=1}^{T}\sum_{m=1}^{M} \log \frac{\exp(\sigma(F_{t,:}, w_{m,:} \cdot A_{m,:})/\tau)}{\sum_{m'=1}^{M} \exp(\sigma(F_{t,:}, w_{m',:} \cdot A_{m',:})/\tau)}. \tag{3}$$

Hence, $\mathcal{L}_{IA}$ indicates which attribute the $F_{t,:}$ is closest to since each vector $F_{t,:}$ is mapped from the t-th group of feature maps through the context network $h(\cdot)$. Therefore, certain feature maps can be guided by specific annotated attributes.

Class-Attribute Alignment. Although the image features have been aligned with classes and attributes, the class embeddings obtained by the pre-trained CLIP encoder may shift in the latent space, which may result in inconsistent class space and attribute space, i.e., annotated attributes do not match the corresponding classes, which is contradictory to the actual clinical diagnosis. To avoid this weakness, we further align the class and attribute features:

$$\mathcal{L}_{CA} = -\sum_{k=1}^{K}\sum_{m=1}^{M} \log \frac{\exp(\sigma(C_{k,:}, w_{m,:} \cdot A_{m,:})/\tau)}{\sum_{m'=1}^{M} \exp(\sigma(C_{k,:}, w_{m',:} \cdot A_{m',:})/\tau)}, \tag{4}$$

and this loss implies semantic consistency between classes and attributes.

Finally, the total loss function is defined as follows:

$$\mathcal{L} = \mathbb{E}_{I_i \in \mathcal{I}}[\mathcal{L}_{CE} + \mathcal{L}_{IC} + \alpha \cdot \mathcal{L}_{IA} + \beta \cdot \mathcal{L}_{CA}], \tag{5}$$

where α and β are hyperparameters for adjusting the losses and are set as 1 and 0.5, respectively. $\mathcal{L}_{CE}$ denotes the cross-entropy loss between predicted probabilities obtained by the classifier and the ground-truth labels. Note that during the inference phase, test images are only fed into the trained image encoder and classifier. As a result, CLIP-Lung does not introduce any additional computational overhead in inference.

3   Experiments

3.1   Dataset and Implementation Details

Dataset. LIDC-IDRI [1] is a dataset for pulmonary nodule classification or detection based on low-dose CT, which involves 1,010 patients. According to the annotations, we extracted 2,026 nodules, and all of them were labeled with scores from 1 to 5, indicating the malignancy progression. We cropped all the nodules with a square shape of a doubled equivalent diameter at the annotated center, then resized them to the volume of $32 \times 32 \times 32$. Following [9,11], we modified the first layer of the image encoder to be with 32 channels. According to existing works [11,18], we regard a nodule with an average score between 2.5 and 3.5 as *unsure* nodules, *benign* and *malignant* categories are those with scores lower than 2.5 and larger than 3.5, respectively. In this paper, we construct three sub-datasets: LIDC-A contains three classes of nodules both in training and test sets; according to [11], we construct the LIDC-B, which contains three classes of nodules *only* in the training set, and the test set contains benign and malignant nodules; LIDC-C includes benign and malignant nodules both in training and test sets.

Experimental Settings. In this paper, we apply the CLIP pre-trained ViT-B/16 as the text encoder for CLIP-Lung, and the image encoder we used is ResNet-18 [6] due to the relatively smaller scale of training data. The image encoder is initialized randomly. Note that for the text branch, we froze the parameters of the text encoder and updated the learnable tokens l and l' during training. The learning rate is 0.001 following the cosine decay, while the optimizer is stochastic gradient descent with momentum 0.9 and weight decay 0.00005. The temperature τ is initialized as 0.07 and updated during training. All of our experiments are implemented with PyTorch [15] and trained with NVIDIA A100 GPUs. The experimental results are reported with average values through five-fold cross-validation. We report the recall and F1-score values for different classes and use "$\pm$" to indicate standard deviation.

3.2   Experimental Results and Analysis

Performance Comparisons. In Table 1, we compare the classification performances on the LIDC-A dataset, where we regard the benign-unsure-malignant

Table 1. Classification results on the test set of LIDC-A.

Method	Accuracy	Benign		Malignant		Unsure	
		Recall	F1	Recall	F1	Recall	F1
CE Loss	54.2 ± 0.6	72.2	62.0	64.4	61.3	29.0	36.6
Poisson [2]	52.7 ± 0.7	60.5	56.8	58.4	58.7	41.0	44.1
NSB [13]	53.4 ± 0.7	**80.7**	63.0	**67.3**	63.8	16.0	24.2
UDM [18]	54.6 ± 0.4	76.7	64.3	49.5	53.5	32.5	39.5
CORF [21]	56.8 ± 0.4	71.3	63.3	61.3	62.3	38.5	44.3
CLIP [16]	56.6 ± 0.3	59.5	59.2	55.2	60.0	53.9	52.2
CoCoOp [20]	56.8 ± 0.6	59.0	59.2	55.2	60.0	**55.1**	52.8
CLIP-Lung	**60.9 ± 0.4**	67.5	**64.4**	60.9	**66.3**	53.4	**54.1**

Table 2. Classification results on test sets of LIDC-B and LIDC-C.

Method	LIDC-B					LIDC-C				
	Accuracy	Benign		Malignant		Accuracy	Benign		Malignant	
		Recall	F1	Recall	F1		Recall	F1	Recall	F1
CE Loss	83.3 ± 0.6	92.4	88.4	63.4	70.3	85.5 ± 0.5	91.5	89.7	72.3	75.6
Poisson [2]	81.8 ± 0.4	94.2	87.7	54.5	65.1	84.0 ± 0.3	87.9	88.3	75.2	74.5
NSB [13]	78.1 ± 0.5	90.6	85.8	50.5	60.7	84.9 ± 0.7	91.0	89.2	71.3	74.6
UDM [18]	79.3 ± 0.4	87.0	86.2	62.4	67.7	84.6 ± 0.5	88.8	88.8	75.2	75.2
CORF [21]	81.5 ± 0.3	**95.9**	87.8	49.5	62.8	83.0 ± 0.2	87.9	87.7	72.3	72.6
CLIP [16]	83.6 ± 0.6	92.0	88.7	64.4	70.4	87.5 ± 0.3	92.0	91.0	77.0	78.8
CoCoOp [20]	86.8 ± 0.7	94.5	90.9	69.0	75.9	88.2 ± 0.6	**95.0**	91.8	72.4	78.8
CLIP-Lung	**87.5 ± 0.3**	94.5	**91.7**	**72.3**	**79.0**	**89.5 ± 0.4**	94.0	**92.8**	**80.5**	**82.8**

as an ordinal relationship. Compared with ordinal classification methods such as Poisson, NSB, UDM, and CORF, CLIP-Lung achieves the highest accuracy and F1-scores for the three classes, demonstrating the effectiveness of textual knowledge-guided learning. CLIP and CoCoOp also outperform ordinal classification methods and show the superiority of large-scale pre-trained text encoders. Furthermore, CLIP-Lung obtained higher recalls than CLIP and CoCoOp $w.r.t.$ benign and malignant classes, however, the recall of unsure is lower than theirs. We argue that this is due to the indistinguishable textual annotations, such as similar attributes of different nodules. In addition, we verify the effect of textual branch of CLIP-Lung using MV-DAR [12] on LIDC-A dataset. The obtained accuracy values with and without the textual branch are 58.9% and 57.3%, respectively, demonstrating the effectiveness of integrating textual knowledge.

Table 2 presents a performance comparison of CLIP-Lung on the LIDC-B and LIDC-C datasets. Notably, CLIP-Lung obtains higher evaluation values other than recalls of benign class. This disparity is likely attributed to the similarity in appearances and subtle variations in text attributes among the benign nodules. Consequently, aligning these distinct feature types becomes challenging, resulting in a bias towards the text features associated with malignant nodules.

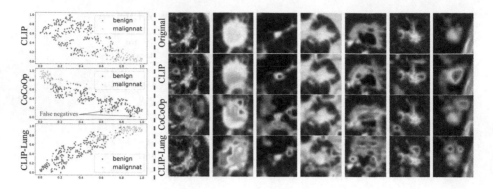

Fig. 3. The t-SNE (**Left**) and Grad-CAM (**Right**) results.

Table 3. Ablation study on different losses. We report classification accuracies.

$\mathcal{L}_{IC}$	$\mathcal{L}_{IA}$	$\mathcal{L}_{CA}$	LIDC-A	LIDC-B	LIDC-C
✓			56.8 ± 0.6	86.8 ± 0.7	88.2 ± 0.6
✓	✓		59.4 ± 0.4	86.8 ± 0.6	86.7 ± 0.4
	✓	✓	58.1 ± 0.2	85.7 ± 0.6	87.5 ± 0.5
✓		✓	56.9 ± 0.3	84.7 ± 0.4	84.0 ± 0.7
✓	✓	✓	$\mathbf{60.9} \pm 0.4$	$\mathbf{87.5} \pm 0.5$	$\mathbf{89.5} \pm 0.4$

Visual Features and Attention Maps. To illustrate the influence of incorporating class and attribute knowledge, we provide the t-SNE [14] and Grad-CAM [17] results obtained by CLIP, CoCoOp, and CLIP-Lung. In Fig. 3, we can see that CLIP yields a non-compact latent space for two kinds of nodules. CoCoOp and CLIP-Lung alleviate this phenomenon, which demonstrates that the learnable prompts guided by nodule classes are more effective than fixed prompt engineering. Unlike CLIP-Lung, CoCoOp does not incorporate attribute information in prompt learning, leading to increased false negatives in the latent space. From the attention maps, we can observe that CLIP cannot precisely capture spiculation and lobulation regions that are highly correlated with malignancy. Simultaneously, our CLIP-Lung performs better than CoCoOp, which demonstrates the guidance from textual descriptions such as "spiculation".

Ablation Studies. In Table 3, we verify the effectiveness of different loss components on the three constructed datasets. Based on $\mathcal{L}_{IC}$, $\mathcal{L}_{IA}$ and $\mathcal{L}_{CA}$ improve the performances on LIDC-A, indicating the effectiveness of capturing fine-grained features of ordinal ranks using class and attribute texts. However, they perform relatively worse on LIDC-B and LIDC-C, especially the $\mathcal{L}_{IC} + \mathcal{L}_{CA}$. That is to say, $\mathcal{L}_{IA}$ is more important in latent space rectification, *i.e.*, image-attribute consistency. In addition, we observe that $\mathcal{L}_{IC} + \mathcal{L}_{IA}$ performs better than $\mathcal{L}_{IA} + \mathcal{L}_{CA}$, which is attributed to that $\mathcal{L}_{CA}$ regularizes the image features indirectly.

4   Conclusion

In this paper, we proposed a textual knowledge-guided framework for pulmonary nodule classification, named CLIP-Lung. We explored the utilization of clinical textual annotations based on large-scale pre-trained text encoders. CLIP-Lung aligned the different modalities of features generated from nodule classes, attributes, and images through contrastive learning. Most importantly, CLIP-Lung establishes correlations between learnable prompt tokens and feature maps using the proposed CCP module, and this guarantees explainable attention maps localizing fine-grained clinical features. Finally, CLIP-Lung outperforms compared methods, including CLIP on LIDC-IDRI benchmark. Future work will focus on extending CLIP-Lung with more diverse textual knowledge.

Acknowledgements. This work was supported in part by Natural Science Foundation of Shanghai (No. 21ZR1403600), National Natural Science Foundation of China (Nos. 62101136 and 62176059), China Postdoctoral Science Foundation (No. 2022TQ0069), Shanghai Municipal of Science and Technology Project (No. 20JC1419500), and Shanghai Center for Brain Science and Brain-inspired Technology.

References

1. Armato, S.G., III., McLennan, G., Bidaut, L., et al.: The lung image database consortium (LIDC) and image database resource initiative (IDRI): a completed reference database of lung nodules on CT scans. Med. Phys. **38**(2), 915–931 (2011)
2. Beckham, C., Pal, C.: Unimodal probability distributions for deep ordinal classification. In: Proceedings of the International Conference on Machine Learning, pp. 411–419 (2017)
3. Chen, X., He, K.: Exploring simple Siamese representation learning. In: Proceedings of the IEEE/CVF Conference on Computer Vision and Pattern Recognition, pp. 15750–15758 (2021)
4. Hancock, M.C., Magnan, J.F.: Predictive capabilities of statistical learning methods for lung nodule malignancy classification using diagnostic image features: an investigation using the lung image database consortium dataset. In: Medical Imaging 2017: Computer-Aided Diagnosis, vol. 10134, pp. 558–569 (2017)
5. He, K., Fan, H., Wu, Y., Xie, S., Girshick, R.: Momentum contrast for unsupervised visual representation learning. In: Proceedings of the IEEE/CVF Conference on Computer Vision and Pattern Recognition, pp. 9729–9738 (2020)
6. He, K., Zhang, X., Ren, S., Sun, J.: Deep residual learning for image recognition. In: Proceedings of the IEEE Conference on Computer Vision and Pattern Recognition, pp. 770–778 (2016)
7. Henaff, O.: Data-efficient image recognition with contrastive predictive coding. In: International Conference on Machine Learning, pp. 4182–4192 (2020)
8. Joshi, A., Sivaswamy, J., Joshi, G.D.: Lung nodule malignancy classification with weakly supervised explanation generation. J. Med. Imaging **8**(4), 044502–044502 (2021)
9. Lei, Y., Tian, Y., Shan, H., Zhang, J., Wang, G., Kalra, M.K.: Shape and margin-aware lung nodule classification in low-dose CT images via soft activation mapping. Med. Image Anal. **60**, 101628 (2020)

10. Lei, Y., Zhang, J., Shan, H.: Strided self-supervised low-dose CT denoising for lung nodule classification. Phenomics **1**, 257–268 (2021)
11. Lei, Y., Zhu, H., Zhang, J., Shan, H.: Meta ordinal regression forest for medical image classification with ordinal labels. IEEE/CAA J. Autom. Sin. **9**(7), 1233–1247 (2022)
12. Liao, Z., Xie, Y., Hu, S., Xia, Y.: Learning from ambiguous labels for lung nodule malignancy prediction. IEEE Trans. Med. Imaging **41**(7), 1874–1884 (2022)
13. Liu, X., Zou, Y., Song, Y., Yang, C., You, J., K Vijaya Kumar, B.: Ordinal regression with neuron stick-breaking for medical diagnosis. In: Proceedings of the European Conference on Computer Vision, pp. 335–344 (2018)
14. Van der Maaten, L., Hinton, G.: Visualizing data using t-SNE. J. Mach. Learn. Res. **9**(11) (2008)
15. Paszke, A., et al.: Automatic differentiation in PyTorch (2017)
16. Radford, A., et al.: Learning transferable visual models from natural language supervision. In: International Conference on Machine Learning, pp. 8748–8763 (2021)
17. Selvaraju, R.R., Cogswell, M., Das, A., Vedantam, R., Parikh, D., Batra, D.: Grad-CAM: visual explanations from deep networks via gradient-based localization. In: Proceedings of the IEEE International Conference on Computer Vision, pp. 618–626 (2017)
18. Wu, B., Sun, X., Hu, L., Wang, Y.: Learning with unsure data for medical image diagnosis. In: Proceedings of the IEEE International Conference on Computer Vision, pp. 10590–10599 (2019)
19. Xie, Y., Zhang, J., Xia, Y.: Semi-supervised adversarial model for benign-malignant lung nodule classification on chest CT. Med. Image Anal. **57**, 237–248 (2019)
20. Zhou, K., Yang, J., Loy, C.C., Liu, Z.: Conditional prompt learning for vision-language models. In: Proceedings of the IEEE/CVF Conference on Computer Vision and Pattern Recognition, pp. 16816–16825 (2022)
21. Zhu, H., et al.: Convolutional ordinal regression forest for image ordinal estimation. IEEE Trans. Neural Netw. Learn. Syst. **33**(8), 4084–4095 (2022)

CTFlow: Mitigating Effects of Computed Tomography Acquisition and Reconstruction with Normalizing Flows

Leihao Wei[1,2], Anil Yadav[2], and William Hsu[2(✉)]

[1] Department of Electrical and Computer Engineering, Samueli School of Engineering, University of California, Los Angeles, CA 90095, USA
[2] Medical and Imaging Informatics, Department of Radiological Sciences, David Geffen School of Medicine at UCLA, Los Angeles, CA 90024, USA
whsu@mednet.ucla.edu

Abstract. Mitigating the effects of image appearance due to variations in computed tomography (CT) acquisition and reconstruction parameters is a challenging inverse problem. We present CTFlow, a normalizing flows-based method for harmonizing CT scans acquired and reconstructed using different doses and kernels to a target scan. Unlike existing state-of-the-art image harmonization approaches that only generate a single output, flow-based methods learn the explicit conditional density and output the entire spectrum of plausible reconstruction, reflecting the underlying uncertainty of the problem. We demonstrate how normalizing flows reduces variability in image quality and the performance of a machine learning algorithm for lung nodule detection. We evaluate the performance of CTFlow by 1) comparing it with other techniques on a denoising task using the AAPM-Mayo Clinical Low-Dose CT Grand Challenge dataset, and 2) demonstrating consistency in nodule detection performance across 186 real-world low-dose CT chest scans acquired at our institution. CTFlow performs better in the denoising task for both peak signal-to-noise ratio and perceptual quality metrics. Moreover, CTFlow produces more consistent predictions across all dose and kernel conditions than generative adversarial network (GAN)-based image harmonization on a lung nodule detection task. The code is available at https://github.com/hsu-lab/ctflow.

Keywords: Image harmonization · computed tomography · normalizing flows

1 Introduction

The increased availability of radiological data and rapid advances in medical image analysis has led to an exponential growth in prediction models that utilize features extracted from clinical imaging scans to detect and diagnose diseases and predict response to treatment [1–4]. However, variations in the acquisition and reconstruction of CT scans

---

Supplementary Information The online version contains supplementary material available at https://doi.org/10.1007/978-3-031-43990-2_39.

H. Greenspan et al. (Eds.): MICCAI 2023, LNCS 14226, pp. 413–422, 2023.
https://doi.org/10.1007/978-3-031-43990-2_39

result in quantitative image features with poor reproducibility [5, 6]. Several studies have demonstrated that differences in dose, slice thickness, reconstruction method, and reconstruction kernel negatively impact radiomic feature reproducibility. Predicting prediction model performance is confounded by how medical images are acquired and reconstructed [7–10]. Many studies have developed techniques to address sources of CT parameter variability [5, 11]. However, inverse problems such as recovering full radiation dose scans from lower dose scans are inherently ill-posed. A range of outputs may be possible when using image restoration algorithms, including potential artifacts that impact the performance of downstream algorithms. Like GANs or variational autoencoders, normalizing flows is a method for learning complex data representations but with an explicit ability to infer the output as a probability distribution and with the added benefit of more stable training [12]. While normalizing flows has shown success in image synthesis tasks for natural images [13], few studies have examined them in medical image harmonization tasks. Denker et al. employed a normalizing flow model conditioned on LDCT reconstruction by filtered backprojection to improve reconstruction quality from the raw sinogram data [14].

This paper presents CTFlow, which aims to utilize normalizing flows to harmonize variations in image appearance of CT scans by maximizing the explicit likelihood of a target condition (e.g., 100% dose, medium kernel, 1 mm slice thickness) given a CT scan that was acquired using different parameters (50% dose, sharp kernel, 1 mm slice thickness). Normalizing flow has two important advantages: 1) the translated low-dose CTs have minimal artifacts because the output is a maximum likelihood estimate that closely matches the target reference distribution, and 2) unlike GANs, which are susceptible to mode collapse, CTFlow can generate multiple solutions to reduce inference uncertainty. We demonstrate how CTFlow compares with current state-of-the-art methods for mitigating dose and reconstruction kernel differences. We evaluated using image quality metrics and a lung nodule detection task.

2   Methods

2.1   Datasets

This study used two unique datasets: (1) the UCLA low-dose chest CT dataset, a collection of 186 exams acquired using Siemens CT scanners at an equivalent dose of 2 mGy following an institutional review board-approved protocol. The raw projection data of scans were exported, and Poisson noise was introduced, as described in Zabic *et al.* [15], at levels equivalent to 10% of the original dose. Projection data were then reconstructed into an image size of 512×512 using three reconstruction kernels (smooth, medium, sharp) at 1.0 mm slice thickness. The dataset was split into 80 scans for training, 20 for validation, and 86 for testing. (2) AAPM-Mayo Clinic Low-Dose CT "Grand Challenge" dataset, a publicly available Grand Challenge dataset consisting of 5,936 abdominal CT images from 10 patient cases reconstructed at 1.0 mm slice thickness. Each case consists of a paired 100% "normal dose" scan and a simulated 25% "low dose" scan. Images from eight patient cases were used for training, and two cases were reserved for validation. All images were randomly cropped into patches of 128×128 pixels, generated from regions containing the body. This dataset was only used for evaluating image quality against other harmonization techniques.

2.2 Normalizing Flows

In this section, we describe the normalizing flows and modifications that were made to improve computational efficiency. Deterministic approaches to image translation (e.g., using a convolutional neural network) attempt to find a mapping function $y = g_\theta(x)$ that takes an input image x and outputs an image y that mimics the appearance of a target condition. For example, x could be an image acquired using a low dose protocol (e.g., 25% dose, smooth kernel), and y represents the image acquired at the target acquisition and reconstruction parameter (e.g., 100% dose, medium kernel). Flow-based image translation aims to approximate the density function $\prod_{y|x}(y|x, \theta)$ using maximum likelihood estimation.

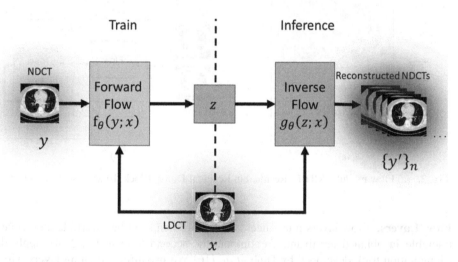

Fig. 1. Relationship between forward flow and inverse flow. LDCT: low (25%) dose computed tomography scan; NDCT: normal (100%) dose computed tomography scan.

Figure 1 summarizes the normalizing flow approach. Normalizing flow gradually transforms an initial (Gaussian) density function $p_z(z)$ to a target distribution $\prod(y|x)$ using an invertible neural network $y = g_\theta(z; x) \leftrightarrow z = g_\theta^{-1}(y; x) = f_\theta(y; x)$, where g and f are the decoding (inverse flow) and encoding (forward flow) functions, respectively. By the change of variables theorem, we can calculate the density function

$$\prod_{y|x}(y|x, \theta) = p_z(f_\theta(y; x))\left|det\frac{df_\theta(y; x)}{dy}\right| \tag{1}$$

which can be trained by maximizing the log-likelihood. In practice, a multilayer flow operation is preferred because a single-layer flow cannot represent complex non-linear relationships within the data. f is decomposed into a series of invertible neural network layers h^n where $h^n = f_\theta^n(h^{n-1}; e(x))$, n represents the number of layers, and $e(x)$ represents a deep convolutional neural network that extracts salient feature maps of x upon which the flow layers are conditioned. For an N-layer flow model, the objective is to maximize

$$\hat{\theta} = \underset{\theta}{\mathrm{argmax}}\, logp_z(z) + \sum_{n=1}^{N} log \left| det \frac{df_\theta^n(h^{n-1}; e(x))}{dh^{n-1}} \right| \qquad (2)$$

Once the training is complete, the decoding function $g_\theta(z; x)$ is applied using random latent variable z, which is drawn from the independent and identically distributed Gaussian density function. The use of z allows us to generate a range of possible restored images $y\prime$, conditioned on the same input image x.

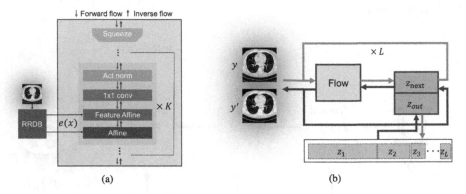

Fig. 2. (a) Flow module. RRDB: Residual in Residual Dense Block (b) Multiscale architecture

Flow Layers. Flow layers must meet two requirements: 1) be invertible and 2) be a tractable Jacobian determinant. To compute the second term in Eq. 2, we apply the triangulation trick developed by Dinh *et al.* [16] We use affine coupling layers with a conditional variable. We first equally split the channels into (h_1^n, h_2^n) and apply an affine transformation on h_2^n while keeping an identity transform on h_1^n. We apply scale and shift factor computed by a shallow convolutional neural network (CNN) given h_1^n in spatial coordinates i, j to compute the n + 1 layer flow of h_2^{n+1}. Finally, we concatenated the splitting components back to obtain the next layer flow $h^{n+1} = concat\left(h_1^{n+1}, h_2^{n+1}\right)$ Thus, by definition, Jacobian of h^{n+1} is a lower triangular matrix. Figure 2a depicts the components of the flow module, which are described below:

- **Activation normalization:** A channel-wise batch normalization [17] was applied, yielding an output with zero mean and unit variance.
- **Invertible 1 x 1 conv:** Following the approach in [18], we utilized a learnable 1x1 convolution $h_{i,j}^n = W h_{i,j}^{n-1}$ where W is a square matrix with dimension $c \times c$ (c is the number of channels). Each spatial element i, j in h is multiplied by this 1x1 convolution matrix. The log determinant is computed using PLU factorization.

- **Feature conditional affine.** We compute the scale and shift factor from $e(x)$ again using a shallow CNN to apply the n-th layer flow transformation h. The motivation is to impose a relationship between feature maps extracted $e(x)$ and activation maps h.

The deep convolutional neural network extractor $e(x)$ is based on Residual-in-Residual Dense Blocks (RRDB) [19]. This network contains 14 RRDB blocks and is our feature extractor for low-dose images. The RRDB network was trained using L_1 loss for 60k iterations. The batch size was 16 and the learning rate was set to 2e−4. The Adam optimizer was used with $\beta_1 = 0.9$, $\beta_2 = 0.99$. After training, all layers of RRDB were frozen and used only for feature extraction. Feature maps were derived from 2, 6, 10, 14 block outputs. Afterward, the outputs of each block were concatenated into $e(x)$.

Multiscale Architecture. Since the flow approach is invertible, input x and latent space vector z must have the same dimensions. However, in most cases, $\prod_{y|x}(y|x, \theta)$ is a low-dimensional manifold in high-dimensional input space. Computation is inefficient when a flow model is imposed with a higher dimensionality than the dimension of true latent space. Given the multiscale architecture in RealNVP, we can simplify the model and improve the density estimation at multiple levels. The overall multiscale architecture is depicted in Fig. 2b, where we equally divide each output z into (z_{out}, z_{next}), while recursively feeding z_{next} to the next level. Once all levels have been reached, z_{out} is outputted, representing the maximum log-likelihood estimation.

Network Training. We trained CTFlow using a batch size of 16 and 50k iterations. The learning rate was set to 1e−4 and halved at 50%, 75%, 90%, and 95% of the total training steps. A negative log-likelihood loss was used.

2.3 Experiments

We conducted two experiments to evaluate CTFlow: image quality metrics and impact on the performance of a lung nodule computer-aided detection (CADe) algorithm.

Image Quality. Using the Grand Challenge dataset, we assessed image quality and compared it with other previously published low-dose CT denoising techniques. We computed image quality metrics using the peak signal-to-noise ratio (PSNR), structural similarity (SSIM), and Learned Perceptual Image Patch Similarity (LPIPS) [20]. Our comparison was conducted using adversarial-based approaches (WGAN using mean squared error loss, WGAN using perceptual loss, and a 3D spectral-norm GAN called SNGAN [21–23], previously developed in our group), a convolutional neural network-based approach (SRResNet) [24], and a denoising algorithm based on collaborative filtering Block-matching and 3D filtering (BM3D) [25].

Nodule Detection. We evaluated the ability of CTFlow to harmonize differences in reconstruction kernels and their effect on the performance of a lung nodule detection algorithm. Our CADe system was based on the RetinaNet model, a composite model comprised of a backbone network called feature pyramid net and two subnetworks responsible for object classification with bounding box regression. The model was trained and validated on the LIDC-IDRI dataset, a public de-identified dataset of diagnostic and

418   L. Wei et al.

low-dose CT scans with annotations from four experienced thoracic radiologists. As part of the training process, we only considered nodules annotated by at least three readers in the LIDC dataset. A total of 7,607 slices (with 4,234 nodule annotations) were used for training and 2,323 slices (with 1,454 nodule annotations) for testing in a single train-test split. A bounding box was then created around the union of all the annotator contours to serve as the reference for the detection model. After training for 200 epochs with Focal loss and Adam optimizer, the model achieved an average precision (AP@0.5) of 0.62 on the validation set.

We hypothesized that the CTFlow models should yield better consistency in lung nodule detection performance compared with not normalizing or other state-of-the-art methods. As a comparison, we trained a 3D SNGAN model using the same training and validation set as CTFlow to perform the same task. We trained three separate CTFlow and SNGAN models to map scans reconstructed using smooth, medium, or sharp kernels to a reference condition. We computed the F1 score (the harmonic mean of the CADe algorithm's precision and recall) when executing the model on the CTFlow and SNGAN normalized scans. We then determined the Concordance Correlation Coefficient [26] on the F1 scores, comparing the F1 score of the model when executed on the normalized scan to when executed on the reference scan.

3   Results

3.1   Network Training

On the Grand Challenge dataset, CTFlow took 3 days to train on an NVIDIA RTX 8000 GPU. The peak GPU memory usage was 39 GB. Unlike GANs that required two loss functions, our network was optimized with only one loss function. The negative log-likelihood loss was stable and decreased monotonically.

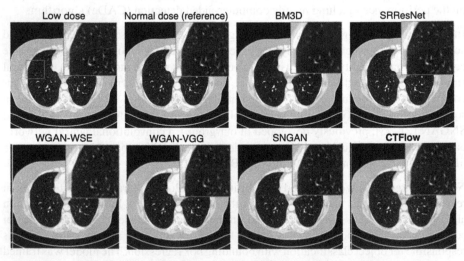

Fig. 3. Visual comparison of Grand Challenge dataset results.

3.2 Image Quality

Table 1 summarizes the results for image quality metrics, while Fig. 3 depicts the same representative slice outputted by each method. While BM3D and SRResNet generated the highest PSNR and SSIM, the images were overly smooth and lacked high-frequency components. Important texture details were lost in the restoration, which may negatively impact downstream tasks (e.g., radiologist interpretation, CAD algorithm performance) that rely on maintaining texture features to characterize lesions. CTFlow achieved 6% better perceptual quality than SNGAN.

Table 1. Image quality metrics generated from the Grand Challenge dataset validation images.

	PSNR	SSIM	LPIPS
BM3D	**32.81**	0.847	0.175
SRResNet	31.89	**0.891**	0.087
WGAN-MSE	31.32	0.864	0.036
WGAN-VGG	31.37	0.869	0.035
SNGAN	31.28	0.865	0.035
CTFlow	31.50	0.863	**0.032**

3.3 Nodule Detection

Table 2 summarizes the CCC values for each kernel pair. McBride [27] suggested the following guidelines for interpreting Lin's concordance correlation coefficient. Poor: $<0:9$; moderate: 0.90 to 0.95; substantial: 0.95 to 0.99; perfect: >0.99 and above. CTFlow achieved CCC scores within the "perfect" range when assessing the agreement in F1 scores when given images reconstructed using varying kernels.

Table 2. Consistency in lung nodule detection performance measured by CCC scores for three pairwise kernel combinations.

	Smooth to Medium	Medium to Sharp	Smooth to Sharp	Mean
SNGAN	0.853	0.844	0.941	0.879
CTFlow	**0.991**	**0.997**	**0.991**	**0.993**

4 Conclusion

We developed CTFlow, a normalizing flows approach to mitigating variations in CT scans. We demonstrated that CTFlow achieved consistent performance across image quality metrics, yielding the best perceptual quality score. Moreover, CTFlow was better than a GAN-based method in maintaining consistent lung nodule detection performance. Compared to generative models, the normalizing flows approach offers exact and efficient likelihood computation and generates diverse outputs that are closer to the target distribution.

We note several limitations of this work. In our evaluations, we trained separate CTFlow and comparison models for each mapping (e.g., transforming a 'smooth' kernel to a 'medium' kernel scan), allowing us to troubleshoot models more easily. A single model conditioned on different doses and kernels would be more practical. Also, CTFlow depends on tuning a variance parameter; better PSNR and SSIM may have been achieved with the optimization of this parameter. Finally, this study focused on mitigating the effect of a single CT parameter, either dose (in image quality) or kernel (in nodule detection). In the real world, multiple CT parameters interact (dose and kernel); these more complex interactions are being investigated as part of future work.

One underexplored area of normalizing flow is its ability to generate the full distribution of possible outputs. Using this information, we can estimate where high uncertainty exists in the model output, providing information to downstream image processing steps, such as segmentation, object detection, and classification. For example, Chan *et al.* [28] applied an approximate Bayesian inference scheme based on posterior regularization to improve uncertainty quantification on covariate-shifted data sets, resulting in improved prognostic models for prostate cancer. Investigating how this uncertainty can be incorporated into downstream tasks, such as our lung nodule CADe algorithm, is also part of future work.

Acknowledgments. This work was supported by the National Institute of Biomedical Imaging and Bioengineering of the National Institutes of Health under awards R56 EB031993 and R01 EB031993. The authors thank John M. Hoffman, Nastaran Emaminejad, and Michael McNitt-Gray for providing access to the UCLA low-dose CT dataset. The content is solely the responsibility of the authors and does not necessarily represent the official views of the National Institutes of Health.

References

1. Sala, E., et al.: Unravelling tumour heterogeneity using next-generation imaging: radiomics, radiogenomics, and habitat imaging. Clin. Radiol. **72**(1), 3 (2017). https://doi.org/10.1016/j.crad.2016.09.013
2. Fave, X., et al.: Delta-radiomics features for the prediction of patient outcomes in non-small cell lung cancer. Sci. Rep. **7**(1), 588 (2017). https://doi.org/10.1038/s41598-017-00665-z
3. Aerts, H.J.: The potential of radiomic-based phenotyping in precision medicine: a review. JAMA Oncol. **2**(12), 1636–1642 (2016). https://doi.org/10.1001/jamaoncol.2016.2631
4. Aerts, H.J., et al.: Decoding tumour phenotype by noninvasive imaging using a quantitative radiomics approach. Nat. Commun. **5**(1), 4006 (2014). https://doi.org/10.1038/ncomms5006
5. Traverso, A., Wee, L., Dekker, A., Gillies, R.: repeatability and reproducibility of radiomic features: a systematic review. Int. J. Radiat. Oncol. Biol. Phys. **102**(4), 1143–1158 (2018). https://doi.org/10.1016/j.ijrobp.2018.05.053
6. Mackin, D., et al.: Measuring computed tomography scanner variability of radiomics features. Invest. Radiol. **50**(11), 757–765 (2015). https://doi.org/10.1097/RLI.0000000000000180

7. Lu, L., Ehmke, R.C., Schwartz, L.H., Zhao, B.: Assessing agreement between radiomic features computed for multiple CT imaging settings. PLoS ONE **11**(12), e0166550 (2016). https://doi.org/10.1371/journal.pone.0166550

8. Zhao, B., et al.: Reproducibility of radiomics for deciphering tumor phenotype with imaging. Sci. Rep. **6**, 23428 (2016). https://doi.org/10.1038/srep23428

9. Kalpathy-Cramer, J., et al.: Radiomics of lung nodules: a multi-institutional study of robustness and agreement of quantitative imaging features. Tomography. **2**(4), 430–437 (2016). https://doi.org/10.18383/j.tom.2016.00235

10. Lo, P., Young, S., Kim, H.J., Brown, M.S., McNitt-Gray, M.F.: Variability in CT lung-nodule quantification: effects of dose reduction and reconstruction methods on density and texture based features. Med. Phys. **43**(8), 4854 (2016). https://doi.org/10.1118/1.4954845

11. Nan, Y., et al.: Data harmonisation for information fusion in digital healthcare: a state-of-the-art systematic review, meta-analysis and future research directions. Inf. Fusion **82**, 99–122 (2022). https://doi.org/10.1016/j.inffus.2022.01.001

12. Dinh, L., Sohl-Dickstein, J., Bengio, S.: Density estimation using Real NVP (2016)

13. Lugmayr, A., Danelljan, M., Van Gool, L., Timofte, R.: SRFlow: learning the super-resolution space with normalizing flow. In: Vedaldi, A., Bischof, H., Brox, T., Frahm, J.-M. (eds.) ECCV 2020. LNCS, vol. 12350, pp. 715–732. Springer, Cham (2020). https://doi.org/10.1007/978-3-030-58558-7_42

14. Denker, A., Schmidt, M., Leuschner, J., Maass, P., Behrmann, J.: Conditional normalizing flows for low-dose computed tomography image reconstruction (2020)

15. Zabic, S., Wang, Q., Morton, T., Brown, K.M.: A low dose simulation tool for CT systems with energy integrating detectors. Med. Phys. **40**(3), 031102 (2013). https://doi.org/10.1118/1.4789628

16. Dinh, L., Krueger, D., Bengio, Y.: Nice: non-linear independent components estimation. arXiv preprint arXiv:14108516 (2014)

17. Ioffe, S., Szegedy, C.: Batch normalization: accelerating deep network training by reducing internal covariate shift. In: International Conference on Machine Learning, pp. 448–56. PMLR (2015)

18. Kingma, D.P., Dhariwal, P.: Glow: generative flow with invertible 1x1 convolutions. In: Advances in Neural Information Processing Systems, vol. 31 (2018)

19. Wang, X., et al.: EsrGAN: enhanced super-resolution generative adversarial networks. In: Leal-Taixé, L., Roth, S. (eds.) ECCV 2018. LNCS, vol. 11133, pp. 63–79. Springer, Cham (2019). https://doi.org/10.1007/978-3-030-11021-5_5

20. Zhang, R., Isola, P., Efros, A.A., Shechtman, E., Wang, O.: The unreasonable effectiveness of deep features as a perceptual metric. In: Proceedings of the IEEE Conference on Computer Vision and Pattern Recognition, pp. 586–595 (2018)

21. Wolterink, J.M., Leiner, T., Viergever, M.A., Isgum, I.: Generative adversarial networks for noise reduction in low-dose CT. IEEE Trans. Med. Imaging **36**(12), 2536–2545 (2017). https://doi.org/10.1109/TMI.2017.2708987

22. Yang, Q., et al.: Low-dose CT image denoising using a generative adversarial network with Wasserstein distance and perceptual loss. IEEE Trans. Med. Imaging. **37**(6), 1348–1357 (2018). https://doi.org/10.1109/TMI.2018.2827462

23. Wei, L., Lin, Y., Hsu, W.: Using a generative adversarial network for CT normalization and its impact on radiomic features. In: IEEE International Symposium on Biomedical Imaging. Iowa City, IA (2020)

24. Lim, B., Son, S., Kim, H., Nah, S., Lee, K.M.: Enhanced deep residual networks for single image super-resolution (2017)

25. Dabov, K., Foi, A., Katkovnik, V., Egiazarian, K.: Image denoising by sparse 3-D transform-domain collaborative filtering. IEEE Trans Image Process. **16**(8), 2080–2095 (2007). https://doi.org/10.1109/tip.2007.901238

26. Lawrence, I., Lin, K.: A concordance correlation coefficient to evaluate reproducibility. Biometrics, 255–268 (1989)
27. McBride, G.: A proposal for strength-of-agreement criteria for Lin's concordance correlation coefficient. NIWA client report: HAM2005-062, p. 62 (2005)
28. Chan, A., Alaa, A., Qian, Z., Van Der Schaar, M.: Unlabelled data improves Bayesian uncertainty calibration under covariate shift. In: International Conference on Machine Learning, pp. 1392–402. PMLR (2020)

Automated CT Lung Cancer Screening Workflow Using 3D Camera

Brian Teixeira[1]([⊠]), Vivek Singh[1], Birgi Tamersoy[2], Andreas Prokein[3], and Ankur Kapoor[1]

[1] Digital Technology and Innovation, Siemens Healthineers, Princeton, NJ, USA
BRIAN.TEIXEIRA@SIEMENS-HEALTHINEERS.COM
[2] Digital Technology and Innovation, Siemens Healthineers, Erlangen, Germany
[3] Computed Tomography, Siemens Healthineers, Forchheim, Germany

Abstract. Despite recent developments in CT planning that enabled automation in patient positioning, time-consuming scout scans are still needed to compute dose profile and ensure the patient is properly positioned. In this paper, we present a novel method which eliminates the need for scout scans in CT lung cancer screening by estimating patient scan range, isocenter, and Water Equivalent Diameter (WED) from 3D camera images. We achieve this task by training an implicit generative model on over 60,000 CT scans and introduce a novel approach for updating the prediction using real-time scan data. We demonstrate the effectiveness of our method on a testing set of 110 pairs of depth data and CT scan, resulting in an average error of 5 mm in estimating the isocenter, 13 mm in determining the scan range, 10 mm and 16 mm in estimating the AP and lateral WED respectively. The relative WED error of our method is 4%, which is well within the International Electrotechnical Commission (IEC) acceptance criteria of 10%.

Keywords: CT · Lung Screening · Dose · WED · 3D Camera

1 Introduction

Lung cancer is the leading cause of cancer death in the United States, and early detection is key to improving survival rates. CT lung cancer screening is a low-dose CT (LDCT) scan of the chest that can detect lung cancer at an early stage, when it is most treatable. However, the current workflow for performing CT lung scans still requires an experienced technician to manually perform pre-scanning steps, which greatly decreases the throughput of this high volume procedure. While recent advances in human body modeling [4,5,12,13,15] have allowed for automation of patient positioning, scout scans are still required as they are used by automatic exposure control system in the CT scanners to compute the dose to be delivered in order to maintain constant image quality [3].

Since LDCT scans are obtained in a single breath-hold and do not require any contrast medium to be injected, the scout scan consumes a significant portion of the scanning workflow time. It is further increased by the fact that tube rotation has to be adjusted between the scout and actual CT scan. Furthermore,

H. Greenspan et al. (Eds.): MICCAI 2023, LNCS 14226, pp. 423–431, 2023.
https://doi.org/10.1007/978-3-031-43990-2_40

any patient movement during the time between the two scans may cause misalignment and incorrect dose profile, which could ultimately result in a repeat of the entire process. Finally, while minimal, the radiation dose administered to the patient is further increased by a scout scan.

We introduce a novel method for estimating patient scanning parameters from non-ionizing 3D camera images to eliminate the need for scout scans during pre-scanning. For LDCT lung cancer screening, our framework automatically estimates the patient's lung position (which serves as a reference point to start the scan), the patient's isocenter (which is used to determine the table height for scanning), and an estimate of patient's Water Equivalent Diameter (WED) profiles along the craniocaudal direction which is a well established method for defining Size Specific Dose Estimate (SSDE) in CT imaging [8,9,11,18]. Additionally, we introduce a novel approach for updating the estimated WED in real-time, which allows for refinement of the scan parameters during acquisition, thus increasing accuracy. We present a method for automatically aborting the scan if the predicted WED deviates from real-time acquired data beyond the clinical limit. We trained our models on a large collection of CT scans acquired from over 60,000 patients from over 15 sites across North America, Europe and Asia. The contributions of this work can be summarized as follows:

- A novel workflow for automated CT Lung Cancer Screening without the need for scout scan
- A clinically relevant method meeting IEC 62985:2019 requirements on WED estimation.
- A generative model of patient WED trained on over 60,000 patients.
- A novel method for real-time refinement of WED, which can be used for dose modulation

2   Method

Water Equivalent Diameter (WED) is a robust patient-size descriptor [17] used for CT dose planning. It represents the diameter of a cylinder of water having the same averaged absorbed dose as the material contained in an axial plane at a given craniocaudal position z [2]. The WED of a patient is thus a function taking as input a craniocaudal coordinate and outputting the WED of the patient at that given position. As WED is defined in an axial plane, the diameter needs to be known on both the Anterior-Posterior (AP) and lateral (Left- Right) axes noted respectively $WED_{AP}(z)$ and $WED_L(z)$. As our focus here is on lung cancer screening, we define 'WED profile' to be the 1D curve obtained by uniformly sampling the WED function along the craniocaudal axis within the lung region. Our method jointly predicts the AP and lateral WED profiles.

While WED can be derived from CT images, paired CT scans and camera images are rarely available, making direct regression through supervised learning challenging. We propose a semi-supervised approach to estimate WED from depth images. First, we train a WED generative model on a large collection of

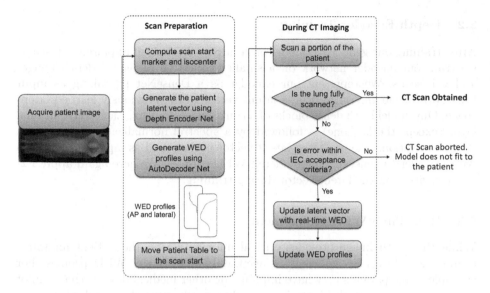

Fig. 1. Overview of the proposed workflow.

CT scans. We then train an encoder network to map the patient depth image to the WED manifold. Finally, we propose a novel method to refine the prediction using real-time scan data.

2.1 WED Latent Space Training

We use an AutoDecoder [10] to learn the WED latent space. Our model is a fully connected network with 8 layers of 128 neurons each. We used layer normalization and ReLU activation after each layer except the last one. Our network takes as input a latent vector together with a craniocaudal coordinate z and outputs $WED_{AP}(z)$ and $WED_L(z)$, the values of the AP and lateral WED at the given coordinate. In this approach, our latent vector represents the encoding of a patient in the latent space. This way, a single AutoDecoder can learn patient-specific continuous WED functions. Since our network only takes the craniocaudal coordinate and the latent vector as input, it can be trained on partial scans of different sizes. The training consists of a joint optimization of the AutoDecoder and the latent vector: the AutoDecoder is learning a realistic representation of the WED function while the latent vector is updated to fit the data.

During training, we initialize our latent space to a unit Gaussian distribution as we want it to be compact and continuous. We then randomly sample points along the craniocaudal axis and minimize the L1 loss between the prediction and the ground truth WED. We also apply L2-regularization on the latent vector as part of the optimization process.

2.2   Depth Encoder Training

After training our generative model on a large collection of unpaired CT scans, we train our encoder network on a smaller collection of paired depth images and CT scans. We represent our encoder as a DenseNet [1] taking as input the depth image and outputting a latent vector in the previously learned latent space. Our model has 3 dense blocks of 3 convolutional layers. Each convolutional layer (except the last one) is followed by a spectral normalization layer and a ReLU activation. The predicted latent vector is then used as input to the frozen AutoDecoder to generate the predicted WED profiles. We here again apply L2-regularization on the latent vector during training.

2.3   Real-Time WED Refinement

While the depth image provides critical information on the patient anatomy, it may not always be sufficient to accurately predict the WED profiles. For example, some patients may have implants or other medical devices that cannot be guessed solely from the depth image. Additionally, since the encoder is trained on a smaller data collection, it may not be able to perfectly project the depth image to the WED manifold. To meet the strict safety criteria defined by the IEC, we propose to dynamically update the predicted WED profiles at inference time using real-time scan data. First, we use our encoder network to initialize the latent vector to a point in the manifold that is close to the current patient. Then, we use our AutoDecoder to generate initial WED profiles. As the table moves and the patient gets scanned, CT data is being acquired and ground truth WED can be computed for portion of the body that has been scanned, along with the corresponding craniocaudal coordinate. We can then use this data to optimize the latent vector by freezing the AutoDecoder and minimizing the L1 loss between the predicted and ground truth WED profiles through gradient descent. We can then feed the updated latent vector to our AutoDecoder to estimate the WED for the remaining portions of the body that have not yet been scanned and repeat the process.

In addition to improving the accuracy of the WED profiles prediction, this approach can also help detect deviation from real data. After the latent vector has been optimized to fit the previously scanned data, a large deviation between the optimized prediction and the ground truth profiles may indicate that our approach is not able to find a point in the manifold that is close to the data. In this case, we may abort the scan, which further reduces safety risks. Overall flowchart of the proposed approach is shown in Fig. 1.

3   Results

3.1   Data

Our CT scan dataset consists of 62, 420 patients from 16 different sites across North America, Asia and Europe. Our 3D Camera dataset consists of 2, 742 pairs

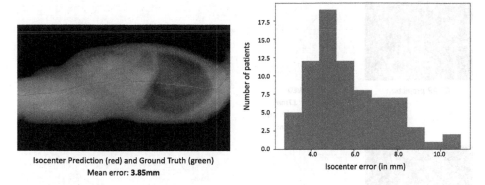

Isocenter Prediction (red) and Ground Truth (green)
Mean error: **3.85mm**

Isocenter error (in mm)

Fig. 2. Isocenter results on our evaluation set. Left column presents a qualitative result from our evaluation set. The red line corresponds to our model prediction and the green line is the ground truth computed from the CT. The right column presents a histogram of the errors in mm. (Color figure online)

of depth image and CT scan from $2,742$ patients from 6 different sites across North America and Europe acquired using a ceiling-mounted Kinect 2 camera. Our evaluation set consists of 110 pairs of depth image and CT scan from 110 patients from a separate site in Europe.

3.2 Patient Preparation

Patient positioning is the first step in lung cancer screening workflow. We first need to estimate the table position and the starting point of the scan. We propose to estimate the table position by regressing the patient isocenter and the starting point of the scan by estimating the location of the patient's lung top.

Starting Position. We define the starting position of the scan as the location of the patient's lung top. We trained a DenseUNet [7] taking the camera depth image as input and outputting a Gaussian heatmap centered at the patient's lung top location. We used 4 dense blocks of 4 convolutional layers for the encoder and 4 dense blocks of 4 convolutional layers for the decoder. Each convolutional layer (except the last one) is followed by a batch normalization layer and a ReLU activation. We trained our model on $2,742$ patients using Adaloss [14] and the Adam [6] optimizer with a learning rate of 0.001 and a batch size of 32 for 400 epochs. Our model achieves a mean error of **12.74 mm** and a 95th percentile error of **28.32 mm**. To ensure the lung is fully visible in the CT image, we added a 2 cm offset on our prediction towards the outside of the lung. We then defined the accuracy as whether the lung is fully visible in the CT image when using the offset prediction. We report an accuracy of **100%** on our evaluation set of 110 patients.

428   B. Teixeira et al.

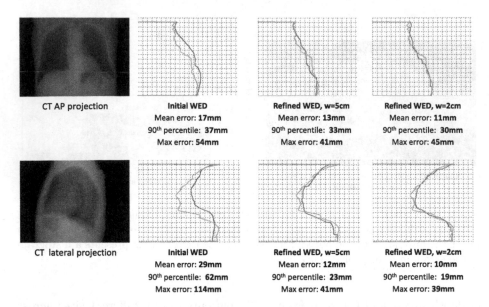

Fig. 3. AP (top) and lateral (bottom) WED profile regression with and without real-time refinement. w corresponds to the portion size of the body that gets scanned before updating the prediction (in cm). First column shows a lateral projection view of the CT. Second column shows the performance of our model without real-time refinement. Third and fourth columns show the performance of our model with real-time refinement every 5 cm and 2 cm respectively. Ground truth is depicted in green and our prediction is depicted in red. While the original prediction was off towards the center of the lung, the real-time refinement was able to correct the error.

Isocenter. The patient isocenter is defined as the centerline of the patient's body. We trained a DenseNet [1] taking the camera depth image as input and outputting the patient isocenter. Our model is made of 4 dense blocks of 3 convolutional layers. Each convolutional layer (except the last one) is followed by a batch normalization layer and a ReLU activation. We trained our model on 2,742 patients using Adadelta [16] with a batch size of 64 for 300 epochs. On our evaluation set, our model outperforms the technician's estimates with a mean error of **5.42 mm** and a 95th percentile error of **8.56 mm** compared to 6.75 mm and 27.17 mm respectively. Results can be seen in Fig. 2.

3.3 Water Equivalent Diameter

We trained our AutoDecoder model on our unpaired CT scan dataset of 62,420 patients with a latent vector of size 32. The encoder was trained on our paired CT scan and depth image dataset of 2,742 patients. We first compared our method against a simple direct regression model. We trained a DenseUNet [7] taking the camera depth image as input and outputting the Water Equivalent Diameter profile. We trained this baseline model on 2,742 patients using the Adadelta [6] optimizer with a learning rate of 0.001 and a batch size of 32. We

Table 1. WED profile errors on our testing set (in *mm*). 'w' corresponds to the portion size of the body that gets scanned before updating the prediction (in cm). Top of the table corresponds to lateral WED profile, bottom corresponds to AP WED profile. Updating the prediction every 20 mm produces the best results.

Method (lateral)	Mean error	90th perc error	Max error
Direct Regression	45.07	76.70	101.50
Proposed (initial)	27.06	52.88	79.27
Proposed (refined, w = 5)	19.18	42.44	73.69
Proposed (refined, w = 2)	**15.93**	**35.93**	**61.68**
Method (AP)			
Direct Regression	45.71	71.85	82.84
Proposed (initial)	16.52	31.00	40.89
Proposed (refined, w = 5)	12.19	25.73	37.36
Proposed (refined, w = 2)	**10.40**	**22.44**	**33.85**

then measured the performance of our model before and after different degrees of real-time refinement, using the same optimizer and learning rate. We report the comparative results in Table 1.

We observed that our method largely outperforms the direct regression baseline with a mean lateral error **40%** lower and a 90th percentile lateral error over **30%** lower. Bringing in real-time refinement greatly improves the results with a mean lateral error over **40%** and a 90th percentile lateral error over **20%** lower than before refinement. AP profiles show similar results with a mean AP error improvement of nearly **40%** and a 90th percentile AP error improvement close to **30%**. When using our proposed method with a 20 mm window refinement, our proposed approach outperforms the direct regression baseline by over **60%** for lateral profile and nearly **80%** for AP.

Figures 3 highlights the benefits of using real-time refinement. Overall, our approach shows best results with an update frequency of 20 mm, with a mean lateral error of **15.93 mm** and a mean AP error of **10.40 mm**. Figure 4 presents a qualitative evaluation on patients with different body morphology.

Finally, we evaluated the clinical relevancy of our approach by computing the relative error as described in the International Electrotechnical Commission (IEC) standard IEC 62985:2019 on *Methods for calculating size specific dose estimates (SSDE) for computed tomography* [2]. The Δ_{REL} metric is defined as:

$$\Delta_{REL}(z) = \left| \frac{W\hat{E}D(z) - WED(z)}{WED(z)} \right| \qquad (1)$$

where:

- $W\hat{E}D(z)$ is the predicted water equivalent diameter
- $WED(z)$ is the ground truth water equivalent diameter
- z is the position along the craniocaudal axis of the patient.

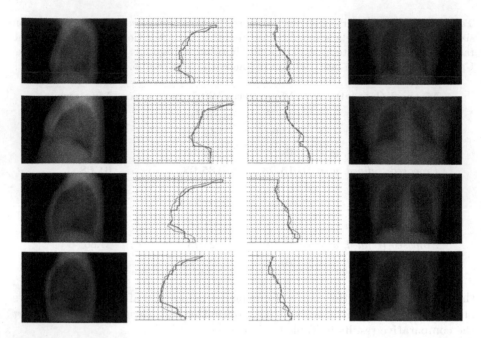

Fig. 4. Qualitative analysis of the proposed method with 2 cm refinement on patient with different morphology. From left to right: Lateral CT projection, Lateral WED profile, AP WED profile, AP CT projection.

IEC standard states the median value of the set of $\Delta_{REL}(z)$ along the craniocaudal axis (noted Δ_{REL}) should be below **0.1**. Our method achieved a mean lateral Δ_{REL} error of **0.0426** and a mean AP Δ_{REL} error of **0.0428**, falling well within the acceptance criteria.

4   Conclusion

We presented a novel 3D camera based approach for automating CT lung cancer screening workflow without the need for a scout scan. Our approach effectively estimates start of scan, isocenter and Water Equivalent Diameter from depth images and meets the IEC acceptance criteria of relative WED error. While this approach can be used for other thorax scan protocols, it may not be applicable to trauma (e.g. with large lung resections) and inpatient settings, as the deviation in predicted and actual WED would likely be much higher. In future, we plan to establish the feasibility as well as the utility of this approach for other scan protocols and body regions.[1]

---

[1] Disclaimer: The concepts and information presented in this paper are based on research results that are not commercially available. Future commercial availability cannot be guaranteed.

References

1. Huang, G., Liu, Z., Weinberger, K.Q.: Densely connected convolutional networks. CoRR (2016)
2. International.Electrotechnical.Commission: IEC 62985:2019 (2019)
3. Kalra, M.K., et al.: Techniques and applications of automatic tube current modulation for CT. Radiology 233(3), 649–657 (2004)
4. Karanam, S., Li, R., Yang, F., Hu, W., Chen, T., Wu, Z.: Towards contactless patient positioning. IEEE Trans, Medical Imaging (2020)
5. Keller, M., Zuffi, S., Black, M.J., Pujades, S.: Osso: Obtaining skeletal shape from outside (2022)
6. Kingma, D.P., Ba, J.: Adam: A method for stochastic optimization. In: ICLR (2015)
7. Li, X., Chen, H., Qi, X., Dou, Q., Fu, C., Heng, P.: H-denseunet: hybrid densely connected UNet for liver and liver tumor segmentation from CT volumes. CoRR (2017)
8. American Association of Physicists in Medicine: Use of water equivalent diameter for calculating patient size and size-specific dose estimates (SSDE) in CT (2014)
9. Mihailidis, D., Tsapaki, V., Tomara, P.: A simple manual method to estimate water-equivalent diameter for calculating size-specific dose estimate in chest computed tomography. British J. Radiol. 94(1117), 20200473 (2021)
10. Park, J.J., Florence, P., Straub, J., Newcombe, R.A., Lovegrove, S.: Deepsdf: Learning continuous signed distance functions for shape representation. CoRR (2019)
11. Rajaraman, V., Ponnusamy, M., Halanaik, D.: Size specific dose estimate (SSDE) for estimating patient dose from CT used in myocardial perfusion spect/ct. Asia Oceania J. Nuclear Med. Biol. 8(1), 58(2020)
12. Singh, V., et al.: Darwin: Deformable patient avatar representation with deep image network (2017)
13. Teixeira, B., et al.: Generating synthetic x-ray images of a person from the surface geometry. In: IEEE Conference on Computer Vision and Pattern Recognition (CVPR) (2018)
14. Teixeira, B., Tamersoy, B., Singh, V.K., Kapoor, A.: Adaloss: Adaptive loss function for landmark localization. CoRR (2019)
15. Wu, Y., et al.: Towards generating personalized volumetric phantom from patient's surface geometry (2018)
16. Zeiler, M.D.: Adadelta: An adaptive learning rate method (2012)
17. Zhang, D., Liu, X., Duan, X., Bankier, A.A., Rong, J., Palmer, M.R.: Estimating patient water equivalent diameter from CT localizer images - a longitudinal and multi-institutional study of the stability of calibration parameters. Med. Phys. 47(5), 2139–2149 (2020)
18. Zhang, D., Mihai, G., Barbaras, L.G., Brook, O.R., Palmer, M.R.: A new method for CT dose estimation by determining patient water equivalent diameter from localizer radiographs: geometric transformation and calibration methods using readily available phantoms. Med. Phys. 45(7), 3371–3378 (2018)

Clinical Applications – Musculoskeletal

Speech Audio Synthesis from Tagged MRI and Non-negative Matrix Factorization via Plastic Transformer

Xiaofeng Liu[1(✉)], Fangxu Xing[1], Maureen Stone[2], Jiachen Zhuo[2], Sidney Fels[3], Jerry L. Prince[4], Georges El Fakhri[1], and Jonghye Woo[1]

[1] Massachusetts General Hospital and Harvard Medical School, Boston, MA, USA
xliu61@mgh.harvard.edu
[2] University of Maryland, Baltimore, MD, USA
[3] University of British Columbia, Vancouver, BC, Canada
[4] Johns Hopkins University, Baltimore, MD, USA

Abstract. The tongue's intricate 3D structure, comprising localized functional units, plays a crucial role in the production of speech. When measured using tagged MRI, these functional units exhibit cohesive displacements and derived quantities that facilitate the complex process of speech production. Non-negative matrix factorization-based approaches have been shown to estimate the functional units through motion features, yielding a set of building blocks and a corresponding weighting map. Investigating the link between weighting maps and speech acoustics can offer significant insights into the intricate process of speech production. To this end, in this work, we utilize two-dimensional spectrograms as a proxy representation, and develop an end-to-end deep learning framework for translating weighting maps to their corresponding audio waveforms. Our proposed plastic light transformer (PLT) framework is based on directional product relative position bias and single-level spatial pyramid pooling, thus enabling flexible processing of weighting maps with variable size to fixed-size spectrograms, without input information loss or dimension expansion. Additionally, our PLT framework efficiently models the global correlation of wide matrix input. To improve the realism of our generated spectrograms with relatively limited training samples, we apply pair-wise utterance consistency with Maximum Mean Discrepancy constraint and adversarial training. Experimental results on a dataset of 29 subjects speaking two utterances demonstrated that our framework is able to synthesize speech audio waveforms from weighting maps, outperforming conventional convolution and transformer models.

1 Introduction

Intelligible speech is produced by the intricate three-dimensional structure of the tongue, composed of localized functional units [26]. These functional units, when measured using tagged magnetic resonance imaging (MRI), exhibit cohesive displacements and derived quantities that serve as intermediate structures linking

H. Greenspan et al. (Eds.): MICCAI 2023, LNCS 14226, pp. 435–445, 2023.
https://doi.org/10.1007/978-3-031-43990-2_41

tongue muscle activity to tongue surface motion, which in turn facilitates the production of speech. A framework based on sparse non-negative matrix factorization (NMF) with manifold regularization can be used to estimate the functional units given input motion features, which yields a set of building blocks (or basis vectors) and a corresponding sparse weighting map (or encoding) [27]. The building blocks can form and dissolve with remarkable speed and agility, yielding highly coordinated patterns that vary depending on the specific speech task at hand. The corresponding weighting map can then be used to identify the cohesive regions and reveal the underlying functional units [25]. As such, by elucidating the relationship between the weighting map and intelligible speech, we can gain valuable insights for the development of speech motor control theories and the treatment of speech-related disorders.

Despite recent advances in cross-modal speech processing, translating between varied-size of wide 2D weighting maps and high-frequency 1D audio waveforms remains a challenge. The first obstacle is the inherent heterogeneity of their respective data representations, compounded by the tendency of losing pitch information in audio [1,6]. By contrast, transforming a 1D audio waveform into a 2D spectrogram provides a rich representation of the audio signal's energy distribution over the frequency domain, capturing both pitch and resonance information along the time axis [9,12]. Second, the input sizes of the weighting maps vary between $20 \times 5,745$ and $20 \times 11,938$, while the output spectrogram has a fixed size for each audio section. Notably, fully connected layers used in [1] require fixed size input, while the possible fully convolution neural networks (CNN) can have varied output sizes and unstable performance [23]. Third, modeling global correlations for the long column dimension of the weighting map and the lack of spatial local neighboring relationships in the row dimension presents further difficulties for conventional CNNs that rely on deep hierarchy structure for expanding the reception field [2,21]. Furthermore, the limited number of training pairs available hinders the large model learning process.

To address the aforementioned challenges, in this work, we propose an end-to-end translator that generates 2D spectrograms from 2D weighting maps via a heterogeneous plastic light transformer (PLT) encoder and a 2D CNN decoder. The lightweight backbone of PLT can efficiently capture the global dependencies with a wide matrix input in every layer [14]. Our PLT module is designed with directional product relative position bias and single-level spatial pyramid pooling to enable flexible global modeling of weighting maps with variable sizes, producing fixed-size spectrograms without information loss or dimension expansion due to cropping, padding, or interpolation for size normalization. To deal with a limited number of training samples, we explore pair-wise utterance consistency as prior knowledge with Maximum Mean Discrepancy (MMD) [8] in a disentangled latent space as an additional optimization objective. Additionally, a generative adversarial network (GAN) [10] can be incorporated to enhance the realism of the generated spectrograms.

The main contributions of this work are three-fold:

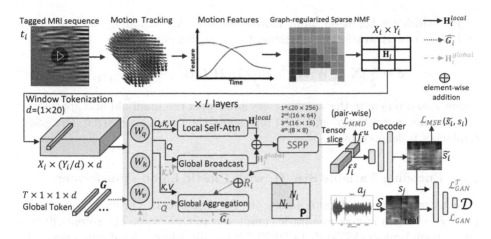

Fig. 1. Illustration of our translation framework. Only the NMF and translator with heterogeneous PLT encoder and 2D CNN decoder are used for testing.

- To our knowledge, this is the first attempt at relating functional units with audio waveforms by means of intermediate representations, including weighting maps and spectrograms.
- We developed a plastic light-transformer to achieve efficient global modeling of position sensitive weighting maps with variable sizes and long dimensions.
- We further explored the pair-wise utterance consistency constraint with MMD minimization and adversarial training as additional supervision signals to deal with relatively limited training samples.

Both quantitative and qualitative evaluation results demonstrate superior synthesis performance over comparison methods. Our framework has the potential to support clinicians and researchers in deepening their understanding of the interplay between tongue movements and speech waveforms, thereby improving treatment strategies for patients with speech-related disorders.

2 Methods

2.1 Preprocessing

During the training phase, we are given M pairs of synchronized tagged MRI sequences t_i and audio waveforms a_i, i.e., $\{t_i, a_i\}_{i=1}^{M}$. First, we apply a non-linear transformation using librosa to convert a_i into mel-spectrograms, denoted as s_i with the function $\mathcal{S} : a_i \rightarrow s_i$. This transformation uses a Hz-scale to emphasize human voice frequencies ranging from 40 to 1000 Hz Hz, while suppressing high-frequency instrument noise. Second, for each tagged MRI sequence t_i, we use a phase-based diffeomorphic registration method [29] to track the internal motion of the tongue. This allows us to generate corresponding weighting maps denoted as $\mathbf{H}_i$, which are based on input motion features $\mathbf{X}_i$, including the magnitude and angle of each track, by optimizing the following equation.

$$\mathcal{E} = \frac{1}{2}\|\mathbf{X}_i - \mathbf{W}_i\mathbf{H}_i\|_F^2 + \frac{1}{2}\lambda\mathrm{Tr}(\mathbf{H}_i\mathbf{L}_i\mathbf{H}_i^\top) + \eta\,\|\mathbf{H}_i\|_{1/2}, \tag{1}$$

where λ and η denote the weights associated with the manifold and sparse regularizations, respectively, and $\mathrm{Tr}(\cdot)$ represents the trace of a matrix. The graph Laplacian is denoted by $\mathbf{L}$.

2.2 Encoding Variable Size $\mathbf{H}_i$ with Plastic Light-Transformer

Directly modeling correlations among any two elements in a given weighting map $\mathbf{H}_i \in \mathbb{R}^{X_i \times Y_i}$ can impose quadratic complexity of $\mathcal{O}(X_i^2 Y_i^2)$. The recent efficient vision transformers (ViTs) [5,14,20,20,32] usually adopt a local patch design to compute local self-attention and correlate patches with CNNs. Specifically, the input is divided into $N_i = \frac{X_i}{P_x} \times \frac{Y_i}{P_y}$ patches[1], each of which is flattened to a token vector with a length of $d = P_x \times P_y$ [7]. The local self-attention is then formulated with a complexity of $\mathcal{O}(N_i d^2 = X_i Y_i d)$ as follows:

$$\mathbf{H}_i^{\text{local}} = \mathrm{Attn}(\mathbf{H}_i^q, \mathbf{H}_i^k, \mathbf{H}_i^v) = \mathrm{SoftMax}(\frac{\mathbf{H}_i^q \mathbf{H}_i^{k\top}}{\sqrt{d}})\mathbf{H}_i^v, \in \mathbb{R}^{X_i \times Y_i}, \tag{2}$$

where vectors $\mathbf{H}_i^q$, $\mathbf{H}_i^k$, $\mathbf{H}_i^v \in \mathbb{R}^{N_i \times d}$ are produced by the linear projections of query (W_q), key (W_k), and value (W_v) branches, respectively [5,7,32]. The global correlation of ViTs with CNN [5,31,32] or window shifting [20], however, may not be efficient for our wide matrix $\mathbf{H}_i$, which lacks explicit row-wise neighboring features and may have a width that is too long for hierarchical convolution modeling. To address these challenges, we follow the lightweight ViT design [14], which uses a global embedding $\mathbf{G} \in \mathbb{R}^{T \times d}$ with $T \ll N_i$ randomly generated global tokens as the anchor for global information aggregation $\hat{\mathbf{G}}_i$. The aggregation is performed with attention of $\mathbf{G}^q, \mathbf{H}_i^k, \mathbf{H}_i^v$, which is then broadcasted with attention of $\mathbf{H}_i^q, \hat{\mathbf{G}}_i^k, \hat{\mathbf{G}}_i^v$ to leverage global contextual information [14].

While LightViT backbones have been shown to achieve wide global modeling within each layer [14], they are not well-suited for our variable size input and fixed size output translation. Although the self-attention scheme used in ViTs does not constrain the number of tokens, the absolute patch-position encoding in conventional ViTs [7] can only be applied to a fixed N_i [32], and the attention module will keep the same size of input and output. Notably, the number of tokens N_i will change depending on the size of $X_i \times Y_i$. As such, in this work, we resort to the directional product relative position bias [28] to add $\mathbf{R}_i \in \mathbb{R}^{N_i \times N_i}$, where element $r_{a,b} = \mathbf{p}_{\delta_{a,b}^x, \delta_{a,b}^y}$ is a trainable scalar, indicating the relative position weight between the patches a and b[2]. We set the offset of patch position in

---

[1] The bottom-right boundary is padded with 0 to ensure $X_i \% P_x = 0$ and $Y_i \% P_y = 0$.
[2] a learnable matrix $\mathbf{p} \in \mathbb{R}^{(2P_x-1) \times (2P_y-1)}$ is initialized with trunc_normal_, where $P_x = 20$ and $P_y = \frac{12000}{20} = 600$ are the maximum patch dimensions in our task.

x and y directions $\delta_{a,b}^x = x_a - x_b + P_x, \delta_{a,b}^y = y_a - y_b + M_y$ as the index in $\mathbf{p}$. Furthermore, the product relative position bias utilized in this work can distinguish between vertical or horizontal offsets, whereas the popular cross relative position bias [28] in computer vision tasks does not need to differentiate between time and spatial neighboring relationships in two dimensions.

Therefore, for global attention, we can aggregate the information of local tokens by modeling their global dependencies with

$$\hat{\mathbf{G}}_i = \text{Attn}(\mathbf{G}^q, \mathbf{H}_i^k, \mathbf{H}_i^v) = \text{SoftMax}(\frac{\mathbf{G}^q\mathbf{H}_i^{k\top} + \mathbf{R}_i}{\sqrt{d}})\mathbf{H}_i^v, \in \mathbb{R}^{X_i \times Y_i}. \quad (3)$$

Then, these global dependencies are broadcasted to every local token:

$$\mathbf{H}_i^{\text{global}} = \text{Attn}(\mathbf{H}_i^q, \hat{\mathbf{G}}_i^k, \hat{\mathbf{G}}_i^v) = \text{SoftMax}(\frac{\mathbf{H}_i^q\hat{\mathbf{G}}_i^{k\top} + \mathbf{R}_i}{\sqrt{d}})\hat{\mathbf{G}}_i^v, \in \mathbb{R}^{X_i \times Y_i}. \quad (4)$$

By adding $\mathbf{H}_i^{\text{local}}$ and $\mathbf{H}_i^{\text{global}}$, each token can benefit from both local and global features, while maintaining linear complexity with respect to the input size. This brings noticeable improvements with negligible FLOPs increment. However, the sequentially proportional patch merging used in [5,14,32] still generates output sizes that vary with input sizes. Therefore, we utilize the single-level Spatial Pyramid Pooling (SSPP) [13] to extract a fixed-size feature for arbitrary input sizes. As illustrated in Fig. 1, the output of our channel-wise SSPP module with 20×256 bins has the size of $20 \times 256 \times d$, which can be a token merging scheme that adapts to the input size. Therefore, the final output of a layer is given by

$$\mathbf{H}_i' = \text{SSPP}(\mathbf{H}_i^{\text{local}} + \mathbf{H}_i^{\text{global}}) \in \mathbb{R}^{X_i \times Y_i}. \quad (5)$$

We cascade four PLT layers with SSPP as our encoder to extract the feature representation $f_i \in \mathbb{R}^{8 \times 8 \times d}$. For the decoder, we adopt a simple 2D CNN with three deconvolutional layers to synthesize the spectrogram $\tilde{s}_i$.

2.3   Overall Training Protocol

We utilize the intermediate pairs of $\{\mathbf{H}_i, s_i\}_{i=1}^M$ to train our translator $\mathcal{T}$, which consists of a PLT encoder and a 2D CNN decoder. The quality of the generated spectrograms $\tilde{s}_i$ is evaluated using the mean square error (MSE) with respect to the ground truth spectrograms s_i:

$$\mathcal{L}_{\text{MSE}} = ||\tilde{s}_i - s_i||_2^2 = ||\mathcal{T}(\mathbf{H}_i) - \mathcal{S}(a_i)||_2^2. \quad (6)$$

Additionally, we utilize the utterance consistency in the latent feature space as an additional optimization constraint. Specifically, we propose to disentangle f_i into two parts, i.e., utterance-related f_i^u and subject-related f_i^s. In practice, we split the utterance/subject-related parts channel-wise using tensor slicing method. Following the idea of deep metric learning [19], we aim to minimize the discrepancy between the latent features f_i^u and f_j^u of two samples

t_i and t_j that belong to the same utterance. Therefore, we use MMD [8] as an efficient discrepancy loss $\mathcal{L}_{\text{MMD}} = \gamma \text{MMD}(f_i^u, f_j^u)$, where $\gamma = 1$ or 0 for same or different utterance pairs, respectively.

Of note, the f_i^s is implicitly encouraged to incorporate the subject-related style of the articulation other than f_i^u with a complementary constraint [16,18] for reconstruction. Therefore, the decoder, which takes f_i^s conditioned on f_i^u can be considered as the utterance-conditioned spectrogram distribution modeling. This approach follows a divide-and-conquer strategy [3,17] for each utterance and can be particularly efficient for relatively few utterance tasks.

A GAN model can be further utilized to boost the realism of $\tilde{s}_i$. A discriminator $\mathcal{D}$ is employed to differentiate whether the mel-spectrogram is real $s_i = \mathcal{S}(a_i)$ or generated $\tilde{s}_i = \mathcal{T}(\mathbf{H}_i)$ with the following binary cross-entropy loss:

$$\mathcal{L}_{\text{GAN}} = \mathbb{E}_{s_i}\{\log(\mathcal{D}(s_i))\} + \mathbb{E}_{\tilde{s}_i}\{\log(1 - \mathcal{D}(\tilde{s}_i))\}. \tag{7}$$

In adversarial training, the translator $\mathcal{T}$ attempts to confuse $\mathcal{D}$ by optimizing $\mathcal{L}_{GAN}^{\mathcal{T}} = \mathbb{E}_{\tilde{s}_i}\{-\log(1 - \mathcal{D}(\tilde{s}_i))\}$. Of note, $\mathcal{T}$ does not involve real spectrograms in $\log(\mathcal{D}(s_i'))$ [24]. Therefore, the overall optimization objectives of our translator $\mathcal{T}$ and discriminator $\mathcal{D}$ are expressed as:

$$\min_{\mathcal{T}} \mathcal{L}_{\text{MSE}} + \beta\mathcal{L}_{\text{MMD}} + \lambda\mathcal{L}_{\text{GAN}}^{\mathcal{T}}; \quad \min_{\mathcal{D}} \mathcal{L}_{\text{GAN}}, \tag{8}$$

where β and λ represent the weighting parameters. Notably, only $\mathcal{T}$ is utilized in testing, and we do not need pairwise inputs for utterance consistency. Recovering audio waveform from mel-spectrogram can be achieved by the well-established Griffin-Lim algorithm [11] in the Librosa toolbox.

3 Experiments and Results

For evaluation, we collected paired 3D tagged MRI sequences and audio waveforms from a total of 29 subjects, while performing the speech words "a souk" or "a geese," with a periodic metronome-like sound as guidance [15,30]. The tagged-MRI sequences consisted of 26 frames, which were resized to 128×128. The resulting $\mathbf{H}$ matrix varied in size from $20 \times 5{,}745$ to $20 \times 11{,}938$ (we set one dimension to a constant value of 20.) The audio waveforms had varying lengths between 21,832 to 24,175. To augment the dataset, we employed a sliding window technique on each audio, allowing us to crop sections with 21,000 time points, resulting in 100 audio waveforms. Then, we utilized the Librosa library to convert all audio waveforms into mel-spectrograms with a size of 64×64. For our evaluation, we utilized a subject-independent leave-one-out approach. For the data augmentation of the $\mathbf{H}$ matrix, we randomly drop the column to round Y_i to the nearest hundred, e.g., 9,882 to 9,800, generating 100 versions of $\mathbf{H}$. We utilized the leave-one-out evaluation, following a subject-independent manner.

In our implementation, we set $P_x = 1$ and $P_y = 20$, i.e., $d = 20$. Our encoder consisted of four PLT encoder layers with SSPP, to extract a feature f_i with the size of $8 \times 8 \times 20$. Specifically, the first $8 \times 8 \times 4$ component was

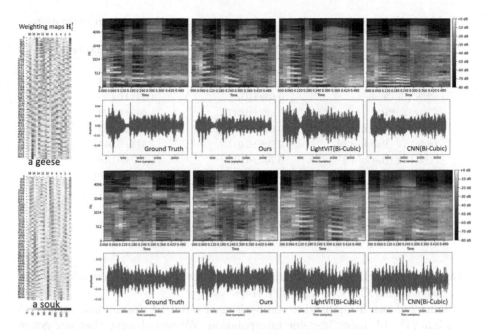

Fig. 2. Comparisons of our PLT with CNN and LightViT using bi-cubic interpolation. We show $\mathbf{H}_i^T$ for compact layout. Audios are attached in supplementary.

set as the utterance-related factors, and the remaining 16 channels were for the subject-specific factors. Then, the three 2D de-convolutional layers were applied as our decoder to generate the 64×64 mel-spectrogram. The activation units in our model were rectified linear units (ReLU), and we normalized the final output of each pixel using the sigmoid function. The discriminator in our model consisted of three convolutional layers and two fully connected layers, and had a sigmoid output. A detailed description of the network structure is provided in the supplementary material, due to space limitations.

Our model was implemented using PyTorch and trained 200 epochs for approximately 6 h on a server equipped with an NVIDIA V100 GPU. Notably, the inference from a $\mathbf{H}$ matrix to audio took less than 1 s, depending on the size of $\mathbf{H}$. Also, the pairwise utterance consistency and GAN training were only applied during the training phase and did not affect inference. For our method and its ablation studies, we consistently set the learning rates of our heterogeneous translator and discriminator to $lr^{\mathcal{T}} = 10^{-3}$ and $lr^{\mathcal{D}} = 10^{-4}$, respectively, with a momentum of 0.5. The loss trade-off hyperparameters were set as $\beta = 0.75$, and we set $\lambda = 1$.

It is important to note that without NMF, generating intelligible audio with a small number of subjects using video-based audio translation models, such as Lip2AudSpect [1], is not feasible. As an alternative, we pre-processed the input by cropping, padding with zeros, or using bi-cubic interpolation to obtain

Table 1. Numerical comparisons during testing using leave-one-out evaluation

Encoder Models	Corr2D for spectrogram ↑	PESQ for waveform ↑
CNN (Crop)	0.614±0.013	1.126±0.021
CNN (Padding 0)	0.684±0.010	1.437±0.018
CNN (Bi-Cubic)	0.689±0.012	1.451±0.020
CNN+SSPP	0.692±0.017	1.455±0.022
LightViT (Crop)	0.635±0.015	1.208±0.022
LightViT (Padding 0)	0.708±0.011	1.475±0.015
LightViT (Bi-Cubic)	0.702±0.012	1.492±0.018
Ours	**0.742**±0.012	**1.581**±0.020
Ours with cross embedding	0.720±0.013	1.550±0.021
Ours w/o Pair-wise Disentangle	0.724±0.010	1.548±0.019
Ours w/o GAN	0.729±0.011	1.546±0.020

a fixed-size input **H**. We then compared the performance of our encoder module with conventional CNN or LightViT [14].

Figure 2 shows a qualitative comparison of our PLT framework with CNN and LightViT [14] using bi-cubic interpolation. We can observe that our generated spectrogram and the corresponding audio waveforms demonstrate superior alignment with the ground truth. It is worth noting that the CNN model or the CNN-based global modeling ViTs [5,31] require deep models to achieve large receptive fields [2,21]. Moreover, the interpolation process adds significant computational complexity for both CNN and LightViT, making it difficult to train on a limited dataset. In Fig. 3(a), we show that our proposed PLT framework achieves a stable performance gain along with the training and outperforms CNN with the crop, which lost the information of some functional units.

Following [1], we used 2D Pearson's correlation coefficient (Corr2D) [4], and Perceptual Evaluation of Speech Quality (PESQ) [22] as our evaluation metrics to measure the synthesis quality of spectrograms in the frequency domain, and waveforms in the time domain, respectively. The numerical comparisons of different encoder structures with conventional CNN or LightViT with different crop or padding strategies and our PLT framework are provided in Table 1. The standard deviation was obtained from three independent random trials. Our framework outperformed CNN and lightViT consistently. In addition, the synthesis performance was improved by pair-wise disentangled utterance consistency MMD loss and GAN loss, as demonstrated in our ablation studies. Furthermore, it outperformed the in-directional cross relative position bias [28], since two dimensions in the weighting map indicate time and spatial relationship, respectively. Notably, even though CNN with SSPP can process varied size inputs, it suffers from limited long-term modeling capacity [2,21] and unstable performance [23]. The sensitivity analysis of our loss weights are given in Fig. 3(b) and (c), where the performance was relatively stable for $\beta \in [0.75, 1.5]$ and $\lambda \in [1, 2]$.

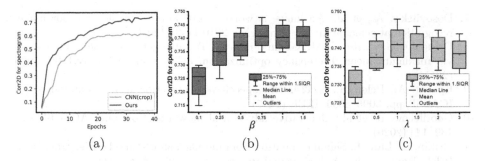

Fig. 3. (a) Comparison of Corr2D using our plastic light transformer and CNN with crop. Sensitivity analysis of β (b) and λ (c).

4 Conclusion

This work aimed to explore the relationship between tongue movements and speech acoustics by translating weighting maps, which represent the functional units of the tongue, to their corresponding audio waveforms. To achieve this, we proposed a deep PLT framework that can handle variable-sized weighting maps and generated fixed-sized spectrograms, without information loss or dimension expansion. Our framework efficiently modeled global correlations in wide matrix input. To improve the realism of the generated spectrograms, we applied pairwise utterance consistency with MMD constraint and adversarial training. Our experimental results demonstrated the potential of our framework to synthesize audio waveforms from weighting maps, which can aid clinicians and researchers in better understanding the relationship between the two modalities.

Acknowledgements. This work is supported by NIH R01DC014717, R01DC018511, R01CA133015, and P41EB022544.

References

1. Akbari, H., Arora, H., Cao, L., Mesgarani, N.: Lip2audspec: speech reconstruction from silent lip movements video. In: ICASSP, pp. 2516–2520. IEEE (2018)
2. Araujo, A., Norris, W., Sim, J.: Computing receptive fields of convolutional neural networks. Distill 4(11), e21 (2019)
3. Che, T., et al.: Deep verifier networks: verification of deep discriminative models with deep generative models. In: AAAI (2021)
4. Chi, T., Ru, P., Shamma, S.A.: Multiresolution spectrotemporal analysis of complex sounds. J. Acoust. Soc. Am. **118**(2), 887–906 (2005)
5. Chu, X., et al.: Twins: revisiting the design of spatial attention in vision transformers. Adv. Neural. Inf. Process. Syst. **34**, 9355–9366 (2021)
6. Chung, J.S., Zisserman, A.: Lip reading in the wild. In: Lai, S.-H., Lepetit, V., Nishino, K., Sato, Y. (eds.) ACCV 2016. LNCS, vol. 10112, pp. 87–103. Springer, Cham (2017). https://doi.org/10.1007/978-3-319-54184-6_6

7. Dosovitskiy, A., et al.: An image is worth 16x16 words: transformers for image recognition at scale. arXiv preprint arXiv:2010.11929 (2020)
8. Dziugaite, G.K., Roy, D.M., Ghahramani, Z.: Training generative neural networks via maximum mean discrepancy optimization. arXiv preprint arXiv:1505.03906 (2015)
9. Ephrat, A., Peleg, S.: Vid2speech: speech reconstruction from silent video. In: ICASSP, pp. 5095–5099. IEEE (2017)
10. Goodfellow, I., et al.: Generative adversarial networks. Commun. ACM **63**(11), 139–144 (2020)
11. Griffin, D., Lim, J.: Signal estimation from modified short-time Fourier transform. IEEE Trans. Acoust. Speech Signal Process. **32**(2), 236–243 (1984)
12. He, G., Liu, X., Fan, F., You, J.: Image2audio: facilitating semi-supervised audio emotion recognition with facial expression image. In: Proceedings of the IEEE/CVF Conference on Computer Vision and Pattern Recognition Workshops, pp. 912–913 (2020)
13. He, K., Zhang, X., Ren, S., Sun, J.: Spatial pyramid pooling in deep convolutional networks for visual recognition. IEEE Trans. Pattern Anal. Mach. Intell. **37**(9), 1904–1916 (2015)
14. Huang, T., Huang, L., You, S., Wang, F., Qian, C., Xu, C.: Lightvit: towards light-weight convolution-free vision transformers. arXiv preprint arXiv:2207.05557 (2022)
15. Lee, J., Woo, J., Xing, F., Murano, E.Z., Stone, M., Prince, J.L.: Semi-automatic segmentation of the tongue for 3D motion analysis with dynamic MRI. In: ISBI, pp. 1465–1468. IEEE (2013)
16. Liu, X., Chao, Y., You, J.J., Kuo, C.C.J., Vijayakumar, B.: Mutual information regularized feature-level Frankenstein for discriminative recognition. IEEE TPAMI **44**, 5243–5260 (2021)
17. Liu, X., et al.: Domain generalization under conditional and label shifts via variational Bayesian inference. IJCAI (2021)
18. Liu, X., et al.: Feature-level Frankenstein: eliminating variations for discriminative recognition. In: CVPR, pp. 637–646 (2019)
19. Liu, X., Vijaya Kumar, B., You, J., Jia, P.: Adaptive deep metric learning for identity-aware facial expression recognition. In: CVPR, pp. 20–29 (2017)
20. Liu, Z., et al.: Swin transformer: hierarchical vision transformer using shifted windows. In: Proceedings of the IEEE/CVF International Conference on Computer Vision, pp. 10012–10022 (2021)
21. Luo, W., Li, Y., Urtasun, R., Zemel, R.: Understanding the effective receptive field in deep convolutional neural networks. In: Advances in Neural Information Processing Systems, vol. 29 (2016)
22. Recommendation, I.T.: Perceptual evaluation of speech quality PESQ): an objective method for end-to-end speech quality assessment of narrow-band telephone networks and speech codecs. Rec. ITU-T P. 862 (2001)
23. Richter, M.L., Byttner, W., Krumnack, U., Wiedenroth, A., Schallner, L., Shenk, J.: (Input) size matters for CNN classifiers. In: Farkaš, I., Masulli, P., Otte, S., Wermter, S. (eds.) ICANN 2021. LNCS, vol. 12892, pp. 133–144. Springer, Cham (2021). https://doi.org/10.1007/978-3-030-86340-1_11
24. Salimans, T., Goodfellow, I., Zaremba, W., Cheung, V., Radford, A., Chen, X.: Improved techniques for training GANs. NIPS **29**, 2234–2242 (2016)
25. Woo, J., et al.: A sparse non-negative matrix factorization framework for identifying functional units of tongue behavior from MRI. IEEE Trans. Med. Imaging **38**(3), 730–740 (2018)

26. Woo, J., et al.: A deep joint sparse non-negative matrix factorization framework for identifying the common and subject-specific functional units of tongue motion during speech. Med. Image Anal. **72**, 102131 (2021)
27. Woo, J., et al.: Identifying the common and subject-specific functional units of speech movements via a joint sparse non-negative matrix factorization framework. In: Medical Imaging 2020: Image Processing, vol. 11313, pp. 446–451. SPIE (2020)
28. Wu, K., Peng, H., Chen, M., Fu, J., Chao, H.: Rethinking and improving relative position encoding for vision transformer. In: Proceedings of the IEEE/CVF International Conference on Computer Vision, pp. 10033–10041 (2021)
29. Xing, F., et al.: Phase vector incompressible registration algorithm for motion estimation from tagged magnetic resonance images. IEEE TMI **36**(10), 2116–2128 (2017)
30. Xing, F., Woo, J., Murano, E.Z., Lee, J., Stone, M., Prince, J.L.: 3D tongue motion from tagged and cine MR images. In: Mori, K., Sakuma, I., Sato, Y., Barillot, C., Navab, N. (eds.) MICCAI 2013. LNCS, vol. 8151, pp. 41–48. Springer, Heidelberg (2013). https://doi.org/10.1007/978-3-642-40760-4_6
31. Yang, J., et al.: Focal self-attention for local-global interactions in vision transformers. arXiv preprint arXiv:2107.00641 (2021)
32. Zhang, Q., Yang, Y.B.: Rest: an efficient transformer for visual recognition. Adv. Neural. Inf. Process. Syst. **34**, 15475–15485 (2021)

Robust Hough and Spatial-To-Angular Transform Based Rotation Estimation for Orthopedic X-Ray Images

Magdalena Bachmaier[1,2], Maximilian Rohleder[1,2], Benedict Swartman[3],
Maxim Privalov[3], Andreas Maier[1], and Holger Kunze[1,2(✉)]

[1] Pattern Recognition Lab, Department of Computer Science, Friedrich-Alexander
University Erlangen-Nürnberg, Erlangen, Germany
Holger.hk.kunze@siemens-healthineers.com
[2] Siemens Healthcare GmbH, Forchheim, Germany
[3] Department for Trauma and Orthopaedic Surgery, BG Trauma Center
Ludwigshafen, Ludwigshafen, Germany

Abstract. Standardized image rotation is essential to improve reading
performance in interventional X-ray imaging. To minimize user interaction and streamline the 2D imaging workflow, we present a new automated image rotation method. Image rotation can follow two steps:
First, an anatomy specific centerline image is predicted which depicts the
desired anatomical axis to be aligned vertically after rotation. In a second step, the necessary rotation angle is calculated from the orientation
of the predicted line image. We propose an end-to-end trainable model
with the Hough transform (HT) and a differentiable spatial-to-angular
transform (DSAT) embedded as known operators. This model allows
to robustly regress a rotation angle while maintaining an explainable
inner structure and allows to be trained with both a centerline segmentation and angle regression loss. The proposed method is compared to a
Hu moments-based method on anterior-posterior X-ray images of spine,
knee, and wrist. For the wrist images, the HT based method reduces the
mean absolute angular error (MAE) from $9.28°$ using the Hu moments-
based method to $3.54°$. Similar results for the spinal and knee images
can be reported. Furthermore, a large improvement of the 90th percentile
of absolute angular error by a factor of 3 indicates a better robustness
and reduction of outliers for the proposed method.

Keywords: Machine Learning · Image Rotation · Hough Transform

1 Introduction

Intra-operative X-ray imaging supports the assessment of fracture reduction
and implant positioning, which significantly increases surgical precision [7] and
the overall outcome [1]. Therefore, standardized image rotation is essential to
improve reading performance and interpretation. While in diagnostic imaging,
this can be reached by careful alignment of the imaging system with the patient,
in intra-operative setups, this alignment is hard to achieve due to mechanical

constraints of the widely used C-arm devices and limited space in the operating room. The rotation of the images must then be corrected in a post-processing step [11]. Therefore, a common task in the radiography workflow is to manually rotate digital X-ray images to a preferred orientation suitable for diagnostic reading. However, during a surgery, user interaction of the surgeon with the system must be minimized due to sterility considerations. Thus, even improved user interaction with the system does not improve the situation for the surgeon much. Supporting staff is often not well trained in operating the vendor-specific systems and changes job positions frequently. Therefore, the speed and quality of the image alignment depends on the experience of the operator and leaves room for mistakes. An automatic X-ray alignment system can help ensure proper alignment resulting in improved reading performance and interpretation also [2]. Previous attempts [3,14,17,18] to automate this rotation typically only allow the identification of a limited set of orientations. In addition, they are often specially designed for certain examinations. More recent publications [4] work on the regression of the correct angular offset. Conventional vision algorithms based on feature extraction with subsequent feature registration to an atlas, typically fail to automate the image rotation of intra-operative images due to the enormous variety of images with different collimation settings, fracture types, and instruments/casts that obscure many image features. Additionally, they hardly generalize to a large number of body regions due to the nature of the hand crafted image features. In this paper, we extend the indirect regression approach of Kunze et al. [11]. It bases on the insight that for many anatomical structures a line can be defined that strongly corresponds with the upright position of the X-ray image. After this line has been determined as a heatmap by a Convolutional neural network (CNN) according to the idea of Kordon et al. [8] this heatmap can be post-processed to calculate the orientation of the line. Kunze et al. [11] have demonstrated that this indirect approach is more robust than the direct regression approaches for angle regression [2,6] based on ResNet or PoseNet. In applications with different crops of anatomic regions, Kordon et al. [10] have shown that the quality of the heatmap can be improved if a companion objective function is integrated into the training that optimizes the regression target directly. Therefore, the derivable calculation of the position and direction based on the Hu moments (HM) was added to the cost function of the segmentation task, enabling end-to-end gradient flow. This approach can be also adapted for the angle regression. However, the HM are an indirect measure leaving much room for the algorithm to shape the heatmap in an uncontrolled manner. Therefore, to limit the space of solutions, we propose to substitue the HM based angle regression with a Hough transform (HT) based one. By doing so, the HM based cost term is replaced by a term, that favors straight, narrow lines. Thereto, we present an algorithm to implement the computation of the line angle based on the HT differentiable to keep end-to-end training based on angular values.

The main contribution of the paper can be summarized as follows:

- proposition of a derivable calculation of the image rotation angle derived from the HT heatmap in the manner of a differentiable spatial to numerical transform (DSNT) [16]
- ablation study which examines the properties of the proposed method and shows that the proposed method increases the robustness of the image rotation compared to state-of-the-art indicated by the 90th percentile of the absolute angular error.
- verification of the approach on a the body regions spine, wrist, and knee.

2   Materials and Methods

2.1   Determination of Rotation Angle

In Kunze et al. [11], two classes of algorithms are compared with which the rotation angle of images can be derived: The direct regression method, uses an encoder structure determining festures of the image followed by a regression network converting them into an angular valued, represented by its sine and cosine value for stability reasons. The indirect method computes a heatmap of a line which represents the upwards direction of the image. Analysing the direction of the line returns the rotation angle of the image. While the first approach consisting of a modified version of ResNet-34 with the fully connected (FC) layer being substituted by two FC layers serves as baseline algorithm [11], in the following the angle determination of the indirect approch is revised.

Angle Regression Exploiting the Hough Transform. As pointed out in the introduction, the training with a cost function based on the HM does not enforce a narrow destinct line, but any structure with a defined main axis contributes to a low cost. A cost function which enforces this behaviour would potentially more distinctive heatmap.

A common approach to detect lines in pattern recognition applications is the HT: Given a line $\mathbb{L}_{\rho,\theta}$, with ρ the angle between line and the x-axis and θ its distance from the origin, its corresponding pixels $\{x_{\rho,\theta,i}, y_{\rho,\theta,i}\}$ vote in the Hough space for the closest bin (ρ, θ). In contrast to the HM based direction calculation, the maximum of the HT is only obtained for a straight line of given extent. The extent can be controlled by the bin width of the HT. This feature of the HT enables the training of a narrow structure with a defined width punishing wide structures. However, the HT does not return an angle itself but returns a new 2-dimensional representation of the heatmap, with the line to be sought represented as its maximum. When this representation shall be integrated in an end-to-end trainable network, a differentiable calculation of the position of this maximum is needed. Thereto, we follow the idea of the DSNT layer introduced by Nibali et al. [16] that calculates the channel-wise centroid of the input. For the current scenario, merely the centroid $\bar{\theta}$ of the θ coordinate indicating the direction of the line is of interest. The $\bar{\theta}$ can be written as

$$\bar{\theta} = \frac{\sum_\theta \sum_\rho \theta g(\theta, \rho)}{\sum_\theta \sum_\rho g(\theta, \rho)} = \frac{\sum_\theta \theta \sum_\rho g(\theta, \rho)}{\sum_\theta \sum_\rho g(\theta, \rho)}. \tag{1}$$

Equation (1) involves a summation over ρ coordinate. Since the zeroth-order Helgason-Ludwig condition of the Radon transform [13] states that the sum over each projection is constant, also the Hough histogram yields a constant first-order moment. This renders the direct implementation incapable for the angle computation.

To overcome this shortcoming, A non-linear function needs to be applied to the Hough histogram to assign high pixel values a greater weight than small ones. Keeping pixel values between 0 and 1 while weighting maximum values and suppressing low values, a steep sigmoid function is employed, which maps input values to a range from 0 to 1. At best, only values from roughly 0.8 to 1 should be used. The steeper the sigmoid function is, the more it approximates the step function, which ideally assigns a weight of 1 to values of 0.8 and higher.

For the end-to-end training, the same issue holds as for the direct regression: the angular value θ has a discontinuity for $\theta = -90°$ and $\theta = 90°$. Both θ values represent the same line. To pass this boundy problem, the cosine- and sine-value are computed instead of the angular value itself. This can be incorporated in Eq. (1) by using a sine- and cosine-weighting instead of linear one. So the total angular value calculation given the HT can be written as

$$(cos(\hat{\theta}), sin(\hat{\theta})) = \left(\frac{\sum_\theta cos\theta \sum_\rho \hat{g}(\theta, \rho)}{\sum_\theta \sum_\rho \hat{g}(\theta, \rho)}, \frac{\sum_\theta sin\theta \sum_\rho \hat{g}(\theta, \rho)}{\sum_\theta \sum_\rho \hat{g}(\theta, \rho)} \right) \quad (2)$$

with

$$\hat{g}(\theta, \rho) = sigmoid\left(\alpha \cdot g(\theta, \rho) - \beta\right), \quad (3)$$

where $\alpha = 20$ and $\beta = 17$ were determined heuristically. $\hat{\theta}$ can be calculated by the atan2 function from its sine and cosine value. We call the layer, that implements Eq. (2) in combination with the atan2 method differential spatial to angular transform (DSAT).

Neither the HT layer parameters nor the DSAT layer have parameters, that need to be trained: The HT can be written as a multiplication of a vector representing the values of the image with a sparse matrix mapping the input image into a $[N_\rho \times N_\theta]$ Hough histogram Y, where N_ρ and N_θ are the numbers of discrete offsets and angles [12]. For the DSAT layer, Eq. 1 can be implemented. So both parts of the algorithms can be treated as known operators [15].

2.2   Data

For the experiments, three different datasets containing the spine, the knee and the wrist, were used. The spinal dataset consists of 958 images derived from 148 patients, where 289 images represent the cervical, 150 the lumbar, and 519 the thoracic region. The wrist dataset contains 257 and the knee dataset 113 images. The 16-bit, gray-scale X-ray images were selected randomly retrospectively from anonymized databases which were acquired using the Cios Spin mobile C-arm system of Siemens Healthcare GmbH during orthopedic surgeries. The images are depicted in the AP view and may contain screws, plates, and other surgical tools.

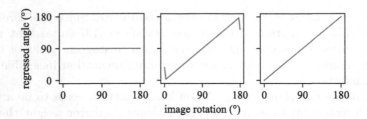

Fig. 1. Regression results for different calculations of the image rotation based on the Hough transform (HT). Left: Linear central moment on HT, Middle: Linear central moment on sigmoid activated HT, Right: Sine-Cosine weighted central moment on sigmoid activated HT

In these images, the centerline was marked using the labelme tool [19] by a trained medical engineer. During annotation, special care has been taken to ensure that the rotation angle of the image can be calculated as the angle between the line and the y-axis by consistently drawing the orientation of the line.

2.3 Training Protocol, Experiments, and Evaluation

To identify the effect of the post processing method on the angle regression error, an ablation study was set up. Following Kunze et al. [11], for the indirect (segmentation-based) method, a D-LinkNet [20] was chosen as underlying network architecture. The output of the D-LinkNet, which is the heatmap of the line, is fed into the HT layer and its result is processed by the DSAT layer. As loss function for the training, the weighted sum of the segmentation loss and the regression loss was chosen. The segmentation loss consists of the sum of Binary Cross Entropy and Mean Squared Error(MSE). For the regression loss, the MSE of the regressed rotation's sine- and cosine-values was selected. Thus, the total loss can be defined as

$$\mathcal{L} = \lambda \cdot \mathcal{L}_{seg} + (1 - \lambda) \cdot \mathcal{L}_{reg}, \tag{4}$$

where $\lambda \in \mathbb{R}$ is a multiplicative weighting term. Using this we examined the following 3 cases:

1. For $\lambda = 1$, the training corresponds to one without the regression loss. In this case, the performance of the HM based and the HT based angle regression methods can be compared on the heatmap.
2. With a constant weighting term $\lambda = 0.5$, the regression loss is considered along with the segmentation loss during the training.
3. When decreasing *lambda* over the epochs from 1 to 0, we obtain a training which is kick-started using the segmentation loss only. But in the end, only the regression loss is used for the training.

For the parameter selection of the HT layer, a grid search was performed on synthetic line heatmaps, resulting in discretization values $d\theta = 0.4°$ and $d\rho = 1\,8px$.

For comparison reasons, the direct regression method based on a ResNet-34 was trained using the MSE based on the sine- and cosine-vales as cost function.

During all trainings, online augmentation for rotation ($\alpha \in [-45°, 45°], p = 0.5$), inverting ($p = 0.5$), shifting ($s \in [-40px, 40px], p = 0.5$) and contrast enhancement was applied. After the augmentation, the images were scaled to the dimension [H:256 × W:256] px, using zero-padding the image to preserve the ratio. Thereafter, sample-wise data normalization using z-scoring was used and combined with batch normalization [9]. The heatmaps for segmentation training were created by placing a Gaussian function of width $7px$ in a neighborhood of $21px$ on the line connecting the augmented start and end point of the center line. The Gaussian was scaled such that the center point obtains a value of 1 [11].

For the evaluation, different matrices based on the angular error were used. The mean error (ME) reveals a directional offset of the algorithm. Further on, the mean absolute error (MSE) and the standard deviation of the absolute error (Std.) are reported. Practically more relevant are the percentiles P_{50}, P_{90} and P_{95} based on the absolute error as they reveal to what extent manual rotation corrections are required after the automatic correction was performed. Based on the MAE a Wilcoxon signed-rank test was performed to verify the significance of the observed differences.

To evaluate the performance of the presented methods, a 5-fold cross-validation scheme is employed. The mean performance score across all folds served as final result of the evaluated algorithm. Training was performed using the Adam optimizer for 500 epochs with a learning rate of $l = 0.001$ divided by two every 50 epochs. The batch size was selected to be 8. The weights of the segmentation network as well as of the direct regression model were initialized by He's method [5]. Implementation was done in PyTorch v1.11 (Python v3.9.12, CUDA v11.0), training was performed on a Windows 10 system with 64 GB RAM and 24 GB NVIDIA Titan RTX. Reproducibility was confirmed by repeated trainings.

3   Results

The results of the experiments are shown in Table 1. A comparison of the segmentation results is made in Fig. 1.

The results show for all three anatomies, that the HT based methods are superior to the HM based method. So already as post-processing method, it significantly outperforms the HM based method (spine: $p = 1.0 \times 10^{-95}$, wrist: $p = 5.6 \times 10^{-21}$, knee: $p = 6.7 \times 10^{-8}$). Further improvements of the angle regression results can be realized by incorporating the regression loss in the training (spine: $p = 9.5 \times 10^{-91}$, wrist: $p = 5.7 \times 10^{-21}$, knee: $p = 4.5 \times 10^{-9}$). Best results except for wrist are obtained when the final training is performed on the regression loss only (spine: $p = 1.3 \times 10^{-2}$, wrist: $p = 7.7 \times 10^{-1}$, knee: $p = 2.1 \times 10^{-2}$). For the wrist adding the regression loss to the overall loss does not improve the training. The comparable P_{50} value for most of the anatomies shows, that HT based and HM based angle regression both work well for standard images.

Table 1. Angular error in degree for the different network topologies and angle calculation methods. The direct regression uses the values obtained by the fully connected layers, while the Hu moments and Hough transform-based methods analyze the detected line at the heatmap output and are integrated as post-processing step as well as additional layers.

Anatomy	Model	ME	MAE	Std.	P_{50}	P_{90}	P_{95}
spine	ResNet-34	0.07	6.95	6.32	5.37	14.54	18.95
	Seg loss, HM post-proc	−0.17	2.14	3.23	1.19	4.87	7.22
	Seg loss, HT post-proc	0.13	1.80	2.36	1.08	3.92	5.69
	Seg loss & HM loss	−0.01	2.05	2.91	1.28	4.33	6.08
	Seg loss & HT loss	0.10	1.79	2.50	1.13	3.61	5.21
	adapt. Seg loss & HM loss	−0.08	1.98	3.50	1.24	4.31	5.97
	adapt. Seg loss & HT loss	0.15	1.61	2.19	1.05	3.40	4.72
wrist	ResNet-34	0.32	7.31	5.12	6.25	14.43	17.28
	Seg loss, HM post-proc	2.26	9.28	16.02	3.13	26.51	43.80
	Seg loss, HT post-proc	0.39	3.73	7.72	1.97	7.63	10.01
	Seg loss & HM loss	4.40	11.24	16.93	3.80	31.13	48.33
	Seg loss & HT loss	0.23	3.54	6.60	2.07	6.75	11.46
	adapt. Seg loss & HM loss	3.24	9.93	16.75	3.66	24.81	49.09
	adapt. Seg loss & HT loss	0.63	4.80	11.67	2.05	6.97	11.47
knee	ResNet-34	0.22	5.63	4.97	4.01	12.48	14.27
	Seg loss, HM post-proc	4.99	8.77	9.18	5.71	19.53	29.97
	Seg loss, HT post-proc	0.15	4.43	11.83	1.68	7.16	10.73
	Seg loss & HM loss	6.00	10.00	15.27	4.77	24.57	40.45
	Seg loss & HT loss	0.44	3.71	4.39	2.23	7.76	11.13
	adapt. Seg loss & HM loss	0.49	8.31	13.14	3.68	16.72	32.62
	adapt. Seg loss & HT loss	1.05	3.48	8.07	1.71	6.29	8.34

For all anatomies, an improvement in the 90th percentile of the absolute angular error can be observed. While for the spine the improvement is about 17%, for the wrist and knee, this value an reduction by a factor of 4.6 and 3.1, respectively, can be observed.

A case-by-case analysis shows that on images for which the centerline is well determined, the HT analysis returns roughly the same result as the HM analysis. However, since the dataset also contains many challenging cases, e.g., images with cropped anatomical structures of interest or images containing metal implants, the centerline cannot always be well determined by the segmentation network. Then occasionally, the heatmap generates short and fragmented estimates of the centerline. For these images, the error of the regressed angle by the HT based method typically is thinner compared to that by the HM computed one (Fig. 2).

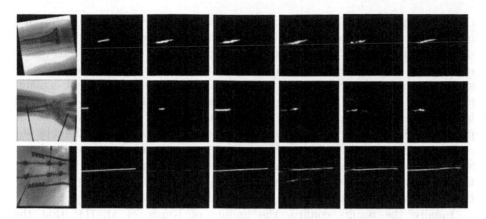

Fig. 2. Comparison of the segmentation results for upper ankle, wrist and spine (top to bottom) trained with segmentation loss only, Hu moments, Hough loss, Hu moments with loss adaption, and Hough loss with loss adaption (column 3–6) with corresponding input images (1st column) and ground truth heatmaps (2nd column).

4 Discussion

We introduced a new HT based method to calculate the rotation of straight heatmaps. For the implementation, special care needs to be taken to cope with constant first-order moments of the HT and boundary problems at $\theta = \pm 90°$.

Doing so, the proposed method has similar performance to HM for X-ray images of good quality. In more difficult cases, when the segmentation algorithm returns multiple and only short segments, like for the wrist or knee, the proposed algorithm is more robust than the HM based one.

The largest benefit of the HT based image rotation calculation can be achieved with its embedding into the cost function of a segmentation network. Then it helps to improve the accuracy compared to a segmentation-only approach. Especially outliers can be reduced resulting in a lower MAE and standard deviation of the error. The presented results confirm the work of Kordon et al. [10] that direct optimization of the target measure in addition to a proxy objective is beneficial to the underlying heatmap-generating problem. That is especially true for body regions like the wrist or knee. For these regions, the lengths of the structures and the annotated lines defining the orientation of the image vary. Then, the algorithm can benefit from a more generic description. For the spine, a line from the top to the bottom of the image can be drawn. Therefore, the regression-related term provides less additional information during the training.

The inspection of the generated heatmaps reveals that the HT based approach can enforce more pronounced, thinner heatmaps compared to the HM - as desired. This observation can be explained by the property of the HT based cost function: it penalizes structures that are not oriented along the desired direction. Only thin straight heatmaps along the direction have a positive effect on the loss term. Structures that are parallel, like the centerlines of the radius and ulna for the wrist images, are penalized. In contrast to that, the HM derived cost term does not punish broad heatmaps as long as the largest main axis points toward the correct direction. Thus, this regression term does not enforce one single, thin line. Also, clusters are tolerated as the results depict. That explains the reduced number of outliers for the HT bases method compared to the HM based, indicated by P_{90} and P_{95} values. The fact that the adaptation of the parameter λ during the training has only a small effect can be explained by the regression loss being larger compared to the segmentation loss by magnitudes as soon as a certain segmentation level is reached. So the segmentation loss only at the beginning of the training has a distinct influence on the training. Thus, the training process is adapting the influence of the segmentation loss by itself.

Finally, the results confirm the finding of Kunze et al. [11] that the segmentation-based rotation regression is more robust compared to the direct regression by a ResNet-34, at the cost of the uncertainty of an up-down flip. For the wrist, slightly worse results are reported compared to [11]. That can be attributed to the dataset for this study containing only intra-operative X-ray images which are acquired not as standardized as the diagnostic images used in [11].

Data Use Declaration: The data was obtained retrospectively from anonymized databases and not generated intentionally for the study. The acquisition of data from patients had a medical indication. Informed consent was not required.

References

1. Atesok, K.: The use of intraoperative three-dimensional imaging (ISO-C-3D) in fixation of intraarticular fractures. Injury **38**(10), 1163–1169 (2007). https://doi.org/10.1016/j.injury.2007.06.014
2. Baltruschat, I.M., Saalbach, A., Heinrich, M.P., Nickisch, H., Jockel, S.: Orientation regression in hand radiographs: a transfer learning approach. In: Proceedings of SPIE Medical Imaging, vol. 10574, pp. 473–480 (2018). https://doi.org/10.1117/12.2291620
3. Boone, J.M., Seshagiri, S., Steiner, R.M.: Recognition of chest radiograph orientation for picture archiving and communications systems display using neural networks. J. Digit. Imaging **5**(3), 190–193 (1992). https://doi.org/10.1007/BF03167769
4. Fonseca, A., Vieira, G.S., Felix, J., Freire Sobrinho, P., Silva, A.V.P., Soares, F.: Automatic orientation identification of pediatric chest x-rays. In: 2020 IEEE 44th Annual Computers, Software, and Applications Conference (COMPSAC), pp. 1449–1454 (2020). https://doi.org/10.1109/COMPSAC48688.2020.00-51
5. He, K., Zhang, X., Ren, S., Sun, J.: Delving deep into rectifiers: surpassing human-level performance on imagenet classification. In: Proceedings of the IEEE International Conference on Computer Vision, pp. 1026–1034 (2015). https://doi.org/10.1109/ICCV.2015.123
6. Kausch, L., et al.: Toward automatic c-arm positioning for standard projections in orthopedic surgery. Int. J. Comput. Assist. Radiol. Surg. **15**(7), 1095–1105 (2020). https://doi.org/10.1007/s11548-020-02204-0

7. Keil, H., Beisemann, N., Swartman, B., Vetter, S.Y., Grützner, P.A., Franke, J.: Intra-operative imaging in trauma surgery. EFORT Open Rev. **3**(10), 541–549 (2018). https://doi.org/10.1302/2058-5241.3.170074

8. Kordon, F., et al.: Multi-task localization and segmentation for x-ray guided planning in knee surgery. In: Shen, D., et al. (eds.) MICCAI 2019. LNCS, vol. 11769, pp. 622–630. Springer, Cham (2019). https://doi.org/10.1007/978-3-030-32226-7_69

9. Kordon, F., et al.: Improved x-ray bone segmentation by normalization and augmentation strategies. In: Bildverarbeitung für die Medizin 2019. I, pp. 104–109. Springer, Wiesbaden (2019). https://doi.org/10.1007/978-3-658-25326-4_24

10. Kordon, F., Maier, A., Swartman, B., Privalov, M., El Barbari, J.S., Kunze, H.: Deep geometric supervision improves spatial generalization in orthopedic surgery planning. In: Wang, L., Dou, Q., Fletcher, P.T., Speidel, S., Li, S. (eds.) Medical Image Computing and Computer Assisted Intervention - MICCAI 2022. LNCS, pp. 615–625. Springer Nature Switzerland, Cham (2022). https://doi.org/10.1007/978-3-031-16449-1_59

11. Kunze, H., Kordon, F., Maier, A., Breininger, K.: Direct and indirect image rotation estimation methods of orthopedic X-ray images. In: Proceedings SPIE Medical Imaging, pp. 640–647 (2022). https://doi.org/10.1117/12.2606045

12. Lin, Y., Pintea, S., Gemert, J.: Semi-supervised lane detection with deep hough transform. pp. 1514–1518 (2021). https://doi.org/10.1109/ICIP42928.2021.9506299

13. Ludwig, D.: The radon transform on Euclidean space. Commun. Pure Appl. Math. **19**(1), 49–81 (1966). https://doi.org/10.1002/cpa.3160190105

14. Luo, H., Luo, J.: Robust online orientation correction for radiographs in PACs environments. IEEE Trans. Med. Imaging **25**(10), 1370–1379 (2006). https://doi.org/10.1109/tmi.2006.880677

15. Maier, A.K., et al.: Learning with known operators reduces maximum error bounds. Nat. Mach. Intell. **1**(8), 373–380 (2019). https://doi.org/10.1038/s42256-019-0077-5

16. Nibali, A., He, Z., Morgan, S., Prendergast, L.: Numerical coordinate regression with convolutional neural networks. arXiv preprint arXiv:1801.07372 (2018). https://doi.org/10.48550/arXiv.1801.07372

17. Nose, H., Unno, Y., Koike, M., Shiraishi, J.: A simple method for identifying image orientation of chest radiographs by use of the center of gravity of the image. Radiol. Phys. Technol. **5**(2), 207–212 (2012). https://doi.org/10.1007/s12194-012-0155-4

18. Pietka, E., Huang, H.: Orientation correction for chest images. J. Digit. Imaging **5**(3), 185–189 (1992). https://doi.org/10.1007/BF03167768

19. Wada, K.: Labelme: image polygonal annotation with python (2016)

20. Zhou, L., Zhang, C., Wu, M.: D-LinkNet: LinkNet with pretrained encoder and dilated convolution for high resolution satellite imagery road extraction. In: 2018 IEEE/CVF Conference on Computer Vision and Pattern Recognition Workshops (CVPRW), pp. 192–1924 (2018). https://doi.org/10.1109/CVPRW.2018.00034

3D Dental Mesh Segmentation Using Semantics-Based Feature Learning with Graph-Transformer

Fan Duan and Li Chen[✉]

School of Software, BNRist, Tsinghua University, Beijing, China
chenlee@tsinghua.edu.cn

Abstract. Accurate segmentation of digital 3D dental mesh plays a crucial role in various specialized applications within oral medicine. While certain deep learning-based methods have been explored for dental mesh segmentation, the current quality of segmentation fails to meet clinical requirements. This limitation can be attributed to the complexity of tooth morphology and the ambiguity of gingival line. Further more, the semantic information of mesh cells which can provide valuable insights into their categories and enhance local geometric attributes is usually disregarded. Therefore, the segmentation of dental mesh presents a significant challenge in digital oral medicine. To better handle the issue, we propose a novel semantics-based feature learning for dental mesh segmentation that can fully leverage the semantic information to grasp the local and non-local dependencies more accurately through a well-designed graph-transformer. Moreover, we perform adaptive feature aggregation of cross-domain features to obtain high-quality cell-wise 3D dental mesh segmentation results. We validate our method using real 3D dental mesh, and the results demonstrate that our method outperforms the state-of-the-art one-stage methods on 3D dental mesh segmentation. Our Codes are available at https://github.com/df-boy/SGTNet.

Keywords: Dental mesh segmentation · Graph-Transformer · Adaptive feature aggregation

1 Introduction

The 3D dental mesh segmentation is aimed to accurately separate the dental mesh into distinct components, namely individual teeth and gums. Therefore stable and accurate 3D dental mesh segmentation plays an essential role in various areas of oral medicine, including orthodontics and denture design, where precise tooth segmentation is of great importance for subsequent procedures and treatments. However, this task is accompanied by notable challenges arising from the inherent limitations in scan accuracy and the presence of considerable noise within the reconstructed 3D dental mesh, consequently leading to the blurring of tooth boundary.

© The Author(s), under exclusive license to Springer Nature Switzerland AG 2023
H. Greenspan et al. (Eds.): MICCAI 2023, LNCS 14226, pp. 456–465, 2023.
https://doi.org/10.1007/978-3-031-43990-2_43

To solve these challenges, some methods have been widely explored. Some conventional methods usually utilize specific geometric properties [15,16] like coordinates and normal vectors to perform threshold segmentation. These methods mainly use the pre-defined attributes, making it hard to achieve high-quality results through automated segmentation.

In recent years, many deep learning-based methods have been proposed to perform more accurate automated dental mesh segmentation. Some methods [10,12] pack the point-wise features into 2D image-like inputs which are fed into a multi-layer CNN-like network and some other methods [4,5,9,13] extend the point cloud feature learning frameworks to perform the segmentation. These methods tend to ignore the inherent topology of mesh and thus the quality of the segmentation is not good enough. Some methods [2,8] design a two-stage network consisting of tooth centroid extraction and tooth segmentation. But these methods are not equally effective for models with crowding or missing teeth which are common in real scenes. And they need centroid labels which will incur additional computational cost. Some recent methods like TSGCNet [14] design a two-stream graph convolution-based feature extraction network to extract the features of the C-stream (coordinate) and the N-stream (normal vector) separately and predict the cell/vertex-wise segmentation results according to the concatenated features of two streams. TSGCNet pioneered a new dental mesh feature process paradigm of decoupling the initial feature into C-stream (coordinate) and N-stream (normal vector), and used the graph convolution [11] to consider the topological continuity when updating the features. However, these methods still suffer from the following deficiencies.

First of all, these methods still have some difficulty on the tooth boundary especially for the crowded teeth. Although there always exist individual differences in the shape of teeth from person to person, the teeth at different positions do have distinctive shape priors that distinguish them from other teeth, such as canine teeth and molar teeth. If we can make full use of the shape priors attached to the semantic information of the teeth, these segmentation errors on the boundaries can be decreased greatly. Thus it provides a more promising way to perform the segmentation guided by the semantic information. Secondly, these methods are often confused at the molars since they only use graph convolution to model the dependencies. Therefore a better alternative would be utilizing the local and non-local semantic information at the same time to further enhance the features, that means the long distance dependencies also need to be considered. Lastly, the existing methods always directly perform concatenation on the features from different angles, resulting in the incorrect segmentation on the misaligned teeth. This is due to the fact that the importance of features from different perspectives can vary a lot in different regions. Hence regressing a specific weight and fusing the features with these weight parameters can effectively eliminate the feature imbalance and pay more emphasis on salient features.

To address these issues, we propose a novel semantics-based feature learning network to fully utilize the semantic information and grasp the local and non-local dependencies. We first follow the TSGCNet [14] to decouple the features into coordinate domain (C-domain) and normal domain (N-domain) which indicate the spatial and geometric features respectively. And then we design a multi-scale encoder network and at each scale we utilize a coarse classifier which accepts the adaptively fused features from the C-domain and the embedded N-domain to predict a semantic pseudo label. Then with the semantic label we use the graph-transformer module to model the long distance dependencies in the neighbourhood of cells with the minimal semantic distance and perform feature aggregation according to the dependencies. Last but not least, we use the global graph-transformer module on the cross-domain features to further learn the semantic information and fuse them by adaptive feature fusion module before fed into the decoder.

To conclude, our contributions are three-fold: (1) We propose a novel semantics-based feature learning which can fully utilize semantic information to enhance the local and global mesh features. (2) We design a new feature fusion module that obtains global dependencies in C-domain and N-domain to further utilize the semantic information and adaptively fuses cross-domain features. (3) We compare with several recent methods on the real 3D dental mesh collected by the hospital. The OA (Overall Accuracy) and mIoU (mean Intersection over Union) both indicates that we perform superior performance on the 3D dental mesh segmentation task. Extensive evaluations prove that our method significantly outperforms the state-of-the-art one-stage methods.

2   Method

2.1   Overview

Our network mainly consists of a **semantics-based graph-transformer** module and an **adaptive cross-domain feature fusion** module, as shown in Fig. 1. We follow the TSGCNet [14] to fetch the initial cross-domain features as two $N \times 12$ matrices, but our N-domain features serve as the embedding domain instead. The semantics-based graph-transformer is a multi-scale encoder and at each scale we aggregate the features in the neighbourhood of cells with minimal semantic distance provided by the coarse semantics prediction module. The N-domain features are embedded through the adaptive feature fusion module to extract more accurate semantic information. And for the concatenated features from the different scales, we perform the global graph-transformer block respectively in the two domains and fuse them through the same adaptive fusion strategy for the subsequent cell-wise segmentation.

2.2   Semantics-Based Graph-Transformer

The semantics-based graph-transformer module aims to generate the multi-scale cell-wise feature vectors in C-domain embedded with N-domain which can represent the geometric features of dental mesh at different positions accurately and discriminatively. We denote the initial feature vectors extracted from the dental

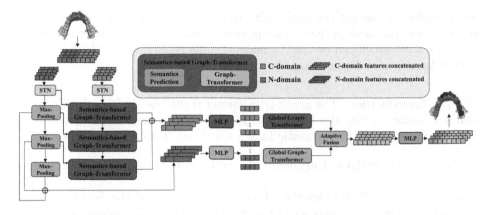

Fig. 1. An overview of the proposed framework.

mesh as a $N \times 24$ matrix, N is the number of the cells and the 24-dimensional vector consists of 12-dimensional relative coordinates and 12-dimensional normal vectors. Then through a normal STN module [3], we make the C-domain and N-domain space invariant due to the fact that the position and orientation of dental mesh can be various. Formed as two domains of $N \times 12$ matrices which represent the spatial position and the geometric features respectively, we perform semantic prediction on the C-domain with the N-domain features embedded to generate a pseudo semantic label $L = \{l_1, l_2, ..., l_n\}$ for each cell. And then according to the semantic information, we use the graph-transformer for each cell in its local neighbourhood where the cells have the minimal semantic differences and update their features to make the difference between the cells with different labels greater. For N-domain, in a geometric sense, they can help classify the semantic label and enhance the local features, hence we mainly upsample it to adapt to the different scales of C-domain without any other modifications.

Semantics Prediction. The semantic prediction is mainly used to generate a pseudo cell-wise label for each cell which can effectively extract the semantic information. Denote the C-domain feature as C that has a shape of $N \times k$, and the N-domain feature as N that has a shape of $N \times k$, and we regress a C-domain weight and a N-domain weight which indicates the weights of the domain fusion. So we have the cross-domain features F adaptively fused as:

$$F_j = c_j \cdot C_j \oplus n_j \cdot N_j \tag{1}$$

where $\oplus$ is the channel-wise concatenation, and j indicates the layer of the semantics-based graph-transformer modules, and c_j, n_j is the adaptive weights of C-domain and N-domain in layer j, and C_j, N_j is the output from the previous layer. And after that we perform a simple MLP to generate the final pseudo semantic cell-wise label formed as:

$$L_j = \mathbf{max}(\mathbf{softmax}(\mathbf{MLP}(F_j))) \tag{2}$$

where **softmax** can get the probability that the cells belong to each class and **max** outputs the index of the maximum value. Thus we have the cell-wise pseudo semantic label which can be used to make the difference between the features of cells belonging to different categories greater.

Graph-Transformer. The graph-transformer is composed of a semantic KNN and a Transformer Encoder Block. For the input C-domain, we first construct a KNN graph based on the semantic biased Euclidean distance formed as:

$$Distance(cell_i, cell_j) = Euclidean(cell_i, cell_j) + Semantic\_Dist(cell_i, cell_j)$$

(3)

where $cell_i$ and $cell_j$ indicate the ith cell and jth cell and the $Semantic\_Dist$ function measures the difference of the two cells which is formulated as:

$$Semantic\_Dist(cell_i, cell_j) = \begin{cases} 0, \ l_i = l_j \\ \lambda, \ l_i \neq l_j \end{cases}$$

(4)

where λ is a positive parameter that can be set according to the specific task and l_i indicates the pseudo label of the ith cell. Then we perform a transformer encoder block on each cell to get the local dependencies which can enhance the local features belong to each class. The attention we used is a standard multi-head attention, and we set the query and value as the matrices of neighbour features of the cells and the key as the distance matrix between cells and their neighbours.

2.3   Adaptive Cross-Domain Feature Fusion

The adaptive cross-domain feature fusion module aims to fuse the C-domain and N-domain features for the cell-wise segmentation. Through the above semantics-based graph-transformer module we have obtained the accurate multi-scale C-domain features embedded by the N-domain features. Then we need to fuse the features to integrate the spatial information and the geometric information of the cells. Therefore, we first perform the concatenation on the features of different scales and use MLP to fuse the multi-scale features together which can balance the features at different scales. This process is formulated as:

$$\begin{cases} F_c = \mathbf{MLP}(F_c^1 \oplus F_c^2 \oplus F_c^3) \\ F_n = \mathbf{MLP}(F_n^1 \oplus F_n^2 \oplus F_n^3) \end{cases}$$

(5)

where $\oplus$ is the channel-wise concatenation while c and n represent C-domain and N-domain features. Then through a global graph-transformer block, we just use the standard multi-head attention on the C-domain and N-domain respectively, so as to capture long distance dependencies and have global knowledge of the semantic information.

Since the learnable weights can fuse the cross-domain features adaptively, we use the same cross-domain feature fusion strategy similar to that we used in the semantics prediction module. Further more, we use a single MLP to generate a feature mask and perform the dot product on the feature and the mask which is formulated as:

$$\hat{F} = \mathbf{MLP}(F) \odot F \tag{6}$$

where $\odot$ is the element-wise multiplication and F is the fused cross-domain features for segmentation.

3   Experiments and Results

3.1   Implementation Details

Our network is implemented with PyTorch 1.11.0 on four NVIDIA GeForce RTX 3090 GPUs. The input meshes are all downsampled to 12000 cells. And in the training process, we optimize the network through minimizing the cross-entropy loss which is very commonly used in the segmentation task. The learning rate was empirically set as 10^{-3}, and reduced by 0.2 decay every 20 epochs.

3.2   Dataset

Our dataset consists of 200 3D dental meshes obtained by an intraoral scanner on real orthodontic patients from hospital and each raw mesh contains even more than 150,000 cells, so we down sample the raw mesh to 12000 cells while preserving the surface topology. We randomly split the whole dataset as a training set with 160 meshes and a testing set with 40 meshes. And we segment the raw mesh into 14 teeth and gums, following the FDI World Dental Federation notation [1]. This means that each input mesh has 15 labels. For convenience, we do not distinguish between the maxillary teeth and mandibular teeth, and treat the teeth on the opposite side (from maxillary and mandibular teeth respectively) as the same class. And we augment the training set by the random translation between $[-10, 10]$, the random rotation between $[-\pi, \pi]$ and the random scaling between $[0.8, 1.2]$.

For evaluation, overall accuracy (OA) and mean intersection over union (mIoU) are adopted for quantitative comparison.

3.3   Comparing with SOTA Methods

We compare our network against five recent methods, including PointNet++ [7], DGCNN [11], PVCNN [6], MeshSegNet [5] and TSGCNet [14]. For a fair comparison, we utilize the public implementations of compared methods to fine-tune their network for generating their best segmentation results. All the methods are trained for 200 epochs.

The segmentation results are shown in Table 1. All the metrics show that our method outperforms the other methods a lot. Specifically, the TSGCNet [14] is the state-of-the-art one-stage method of the 3D dental mesh segmentation task which pioneered a two-stream graph convolution network to extract the features more accurately and TSGCNet [14] improve the segmentation performance on

Table 1. The segmentation results comparing with five methods on OA and mIoU. Gum-mIoU and teeth-mIoU indicates the mIoU computed on the gums and teeth respectively.

	Method	OA	mIoU	gum-mIoU	teeth-mIoU
Point-based	PointNet++ [7]	87.18	0.708	0.867	0.697
	DGCNN [11]	91.56	0.793	0.927	0.784
	MeshSegNet [5]	93.15	0.825	0.931	0.816
Voxel-based	PVCNN [6]	94.54	0.849	0.934	0.843
Graph-based	TSGCNet [14]	95.63	0.890	0.942	0.886
	Ours	**96.97**	**0.921**	**0.956**	**0.918**

the teeth greatly compared to MeshSegNet [5]. Compared with TSGCNet [14], we still increase the OA and m-IoU on the 3D dental mesh segmentation task by 1.34% and 3.1% respectively.

We also perform the qualitative experiments to further evaluate the segmentation results in Fig. 2. From the visualization results, we can find out that our method also outperforms other methods. In particular, we present some complicated cases where there may exist wisdom teeth, the crowded arrangement or the misplaced teeth. In the first row, the raw mesh has a total of 16 teeth which is different from our setting, and in this case, other methods tend to merge some of the small teeth which will cause confusion, while with semantic information, our method successfully labels all the teeth except the two wisdom teeth. In the second and third row where there exist misplaced teeth and worn teeth, we can see that the previous methods cannot perform segmentation correctly. But with adaptive feature fusion which balances the coordinate features and normal vector features, our method can segment the worn teeth accurately as well as the misplaced teeth guided by the semantic information. And the fourth row demonstrates that in terms of extracting the local geometric features, we can also achieve the superior performance. This qualitative experiment further suggests that our design is more effective and accurate on this segmentation task.

3.4   Ablation Study

We evaluate the effectiveness of semantics prediction, graph-transformer and adaptive feature fusion as three critical components of our method. We perform the evaluation by excluding one of these critical components each time. Specifically, when we remove the semantics prediction module, we will use a naive KNN to construct the KNN graph. When we remove the graph-transformer module, we replace it with max-pooling layers. In the absence of the adaptive feature fusion module, we directly concatenate the features from C-domain and N-domain for cross-domain feature fusion. The results of the ablation study are presented in Table 2. It turns out that semantics prediction, graph-transformer and adaptive feature fusion all bring performance improvement on the 3D den-

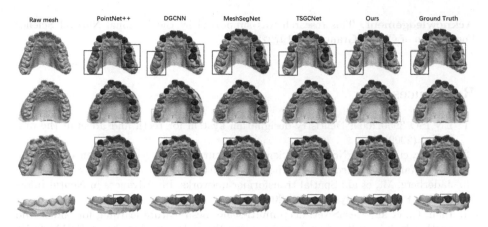

Fig. 2. Visualization of segmentation results comparing with four methods along with the raw dental meshes and ground-truth labels.

tal mesh segmentation task. And we can see that although semantics prediction module improves a little in metrics, it is primarily attributed to the fact that the segmentation errors focus on the boundary cells and the number of such errors is relatively small. But the impact of these boundary cells on the overall segmentation quality is significant.

Table 2. Ablation study of critical components of our method.

Method	OA	mIoU
Ours	**96.97**	**0.921**
Ours w/o semantics prediction	96.52	0.916
Ours w/o graph-transformer	96.18	0.911
Ours w/o adaptive feature fusion	96.05	0.909

4   Conclusion

We propose a novel semantics-based feature learning to make full use of local and unlocal semantic information to enhance the features extracted. This architecture can decouple the spatial and geometric features into C-domain and N-domain and embed the N-domain into C-domain to further utilize the semantic pseudo label to perform the local graph-transformer module. Lastly we fuse the features from spatial and geometric domain adaptively by using the global graph-transformer module and adaptive feature fusion module. The effectiveness of our proposed method is evaluated on the real dental mesh from real orthodontic patients. In the future work, we will try to utilize the feature decoupling and fusing strategy in other segmentation tasks.

Acknowledgement. This research was supported by the National Natural Science Foundation of China (Grant No. 61972221).

References

1. EN ISO 3950:2009: Dentistry-designation system for teeth and areas of the oral cavity (2009)
2. Cui, Z., et al.: TSegNet: an efficient and accurate tooth segmentation network on 3D dental model. Med. Image Anal. **69**, 101949 (2021)
3. Jaderberg, M., et al.: Spatial transformer networks. In: Advances in Neural Information Processing Systems 28 (2015)
4. Lian, C., et al.: MeshSNet: deep multi-scale mesh feature learning for end-to-end tooth labeling on 3D dental surfaces. In: Shen, D., et al. (eds.) MICCAI 2019. LNCS, vol. 11769, pp. 837–845. Springer, Cham (2019). https://doi.org/10.1007/978-3-030-32226-7_93
5. Lian, C., et al.: Deep multi-scale mesh feature learning for automated labeling of raw dental surfaces from 3D intraoral scanners. IEEE Trans. Med. Imaging **39**(7), 2440–2450 (2020)
6. Liu, Z., Tang, H., Lin, Y., Han, S.: Point-voxel CNN for efficient 3D deep learning.In: Advances in Neural Information Processing Systems 32 (2019)
7. Qi, C.R., Yi, L., Su, H., Guibas, L.J.: PointNet++: deep hierarchical feature learning on point sets in a metric space. In: Advances in Neural Information Processing Systems 30 (2017)
8. Qiu, L., Ye, C., Chen, P., Liu, Y., Han, X., Cui, S.: DArch: dental arch prior-assisted 3D tooth instance segmentation with weak annotations. In: Proceedings of the IEEE/CVF Conference on Computer Vision and Pattern Recognition, pp. 20752–20761 (2022)
9. Sun, D., et al.: Automatic tooth segmentation and dense correspondence of 3D dental model. In: Martel, A.L., et al. (eds.) MICCAI 2020. LNCS, vol. 12264, pp. 703–712. Springer, Cham (2020). https://doi.org/10.1007/978-3-030-59719-1_68
10. Tian, S., Dai, N., Zhang, B., Yuan, F., Yu, Q., Cheng, X.: Automatic classification and segmentation of teeth on 3D dental model using hierarchical deep learning networks. IEEE Access **7**, 84817–84828 (2019)
11. Wang, Y., Sun, Y., Liu, Z., Sarma, S.E., Bronstein, M.M., Solomon, J.M.: Dynamic graph CNN for learning on point clouds. ACM Trans. Graph. (TOG) **38**(5), 1–12 (2019)
12. Xu, X., Liu, C., Zheng, Y.: 3D tooth segmentation and labeling using deep convolutional neural networks. IEEE Trans. Vis. Comput. Graph. **25**(7), 2336–2348 (2018)
13. Zanjani, F.G., et al.: Deep learning approach to semantic segmentation in 3D point cloud intra-oral scans of teeth. In: International Conference on Medical Imaging with Deep Learning, pp. 557–571. PMLR (2019)

14. Zhang, L., et al.: TSGCNet: discriminative geometric feature learning with two-stream graph convolutional network for 3D dental model segmentation. In: Proceedings of the IEEE/CVF Conference on Computer Vision and Pattern Recognition, pp. 6699–6708 (2021)

15. Zhao, M., Ma, L., Tan, W., Nie, D.: Interactive tooth segmentation of dental models. In: 2005 IEEE Engineering in Medicine and Biology 27th Annual Conference, pp. 654–657. IEEE (2006)

16. Zou, B.J., Liu, S.J., Liao, S.H., Ding, X., Liang, Y.: Interactive tooth partition of dental mesh base on tooth-target harmonic field. Comput. Biol. Med. **56**, 132–144 (2015)

Full Image-Index Remainder Based Single Low-Dose DR/CT Self-supervised Denoising

Yifei Long[1], Jiayi Pan[1], Yan Xi[2], Jianjia Zhang[1(✉)], and Weiwen Wu[1(✉)]

[1] School of Biomedical Engineering, Shenzhen Campus of Sun Yat-sen University,
Shenzhen 518107, China
{zhangjj225,wuweiw7}@mail.sysu.edu.cn

[2] Shanghai First-Imaging Information Technology Co., LTD., Shanghai, China

Abstract. Low-dose digital radiography (DR) and computed tomography (CT) play a crucial role in minimizing health risks during clinical examinations and diagnoses. However, reducing the radiation dose often leads to lower signal-to-noise ratio measurements, resulting in degraded image quality. Existing supervised and self-supervised reconstruction techniques have been developed with noisy and clean image pairs or noisy and noisy image pairs, implying they cannot be adapted to single DR and CT image denoising. In this study, we introduce the Full Image-Index Remainder (FIRE) method. Our method begins by dividing the entire high-dimensional image space into multiple low-dimensional sub-image spaces using a full image-index remainder technique. By leveraging the data redundancy present within these sub-image spaces, we identify similar groups of noisy sub-images for training a self-supervised denoising network. Additionally, we establish a sub-space sampling theory specifically designed for self-supervised denoising networks. Finally, we propose a novel regularization optimization function that effectively reduces the disparity between self-supervised and supervised denoising networks, thereby enhancing denoising training. Through comprehensive quantitative and qualitative experiments conducted on both clinical low-dose CT and DR datasets, we demonstrate the remarkable effectiveness and advantages of our FIRE method compared to other state-of-the-art approaches.

Keywords: Digital radiography · Computed tomography ·
Self-supervised · Remainder · Image denoising

1 Introduction

Digital Radiography (DR) and Computed Tomography (CT) techniques are extensively utilized in the diagnosis of various clinical conditions [15,24]. However, one major concern associated with these imaging methods is the exposure of

Supplementary Information The online version contains supplementary material available at https://doi.org/10.1007/978-3-031-43990-2_44.

patients to X-ray radiation [12]. Reducing the X-ray radiation dose unavoidably leads to a decline in the number of photons detected, resulting in measurements with a low signal-to-noise ratio and consequent regression in image quality. Consequently, accurately diagnosing clinical conditions based on degraded DR/CT images poses significant challenges. Hence, the development of sophisticated and efficient image-denoising techniques that can effectively address both DR and CT modalities becomes imperative and urgent in clinical applications.

Traditional image denoising methods have relied on exploring spatial pixel features and properties in the transform domain [2,5,8]. These methods include approaches like non-local mean (NLM) [23] and BM3D [5]. However, the need for complex parameter adjustment and their relatively slow speed have limited the practical applications of these traditional denoising methods.

With the advancement of neural networks, deep learning-based denoising techniques have shown superior performance compared to traditional methods [1,3,9]. Supervised denoising methods, such as U-Net [6], DnCNN [25], and RED-CNN [4], have demonstrated promising results. However, when it comes to digital radiography (DR) and computed tomography (CT), these supervised methods face challenges since they rely on paired noisy and clean image data, which are not readily available in DR and CT imaging [7].

To overcome the limitations of supervised techniques, self-supervised or unsupervised denoising methods have been proposed and developed, leveraging the similarity between noisy images [18,26]. Several unsupervised learning-based methods have been introduced for image denoising, including Noise2Noise [10,13,16], Noise2Sim [20], Noise2Void [14], and Neighbor2Neighbor [11] among others [17]. However, these methods have defects in the DR and CT denoising tasks. They either require noisy and noisy image pairs in training, or sampling strategies resulting in information missing and the same noise level of the sub-image pairs.

Therefore, we designed a new self-supervised image denoising technique, i.e., the full image-index remainder (FIRE), adapting to low-dose DR and CT to address the challenge. Specifically, our FIRE method first divided the whole high-dimensional image space into a series of low-dimensional sub-image spaces with the image-index remainder technique. Based on the remainder of the full image index, a specific image sampler is designed to sample the sub-images. With this strategy, a set of sampled noisy sub-images from a single noisy image is obtained for self-supervised training of the denoising neural network without clean images. We further proved that our proposed sampling strategy is effective in supplementary materials. In addition, we proposed a new loss function to train the unsupervised image-denoising network with dedicated parameter tuning. For further optimizing the feasible domain, a regularization strategy is introduced to reduce the gap between the self-supervised and the supervised denoising network. We evaluated the FIRE on several large-scale clinical DR and CT datasets without clean data. The experimental results show that our FIRE method achieves the best results in noise suppression and visual perception. We also compared our FIRE and other methods using public clinical data with ground truth, and the quantitative and qualitative results demonstrate the out-performance of our method.

The remaining sections are organized as follows. Sections 2 and 3, introduce our proposed FIRE method in terms of the network architectures, sampler design, and loss functions. In Sect. 4, we will test our model on large-scale low-dose DR and CT image datasets and compare it with other well-performing methods. Finally, a summary will be given.

2   Methodology

2.1   Related Theories

Given a clean image x and its corresponding noisy image y, the supervised image denoising methods try to train the denoising network f parameterized by θ minimizing the loss function below:

$$E_{\{x,y\}} = \{f_\theta(y), x\}. \tag{1}$$

Implementing the supervised denoising network on noisy and clean image pairs, the usual loss function can be formulated as:

$$\theta = \arg\min_\theta E_{x,y}\|f_\theta(y) - x\|_2^2. \tag{2}$$

However, in actual situations, it is difficult to obtain the corresponding clean and noisy image pairs, especially in medical imaging scans. To address this issue, Noise2Noise proposed a self-denoising method that does not require real clean images, but it depends on pairs of independent noisy images of the same scene. The method proves that the results obtained by minimizing the following equation are the same as the results obtained by the supervised case above:

$$\theta' = \arg\min_{\theta'} E_{x,y,z}\|f'_\theta(y) - z\|_2^2 \tag{3}$$

where y and z are two independent noisy images conditioned on x. In previous studies [16], it has been proved that the results of training using only noisy and noisy image pairs can be approximately equal to supervised cases. Neighbor2Neighbor [11] focuses on sampling a single noisy image with independent noisy and noisy image pairs from the same scene for network training. However, there are still challenges in applying this method to DR and CT image denoising tasks. Specifically, DR and CT images have a larger range of pixel values and more details in the image, while the current method uses only the part image information in each iteration. In the following section, we propose our FIRE method for DR and CT image denoising tasks.

2.2   Framework

In this section, we propose a self-supervised framework for training a single noisy image-denoising network. First, we design a new subspace sampling technique for generating subspace image groups to train the network. Next, for recovering finer image details and features, a specialized loss function consisting of reconstruction terms and regularization terms is proposed and used for training the network. The overall of our FIRE framework is shown in Fig. 1(a) and (b).

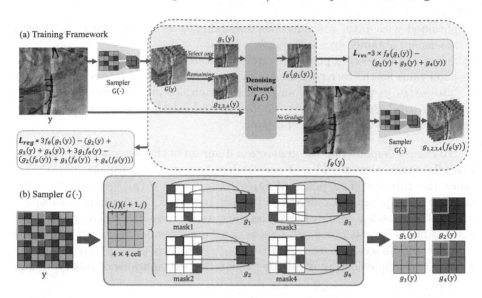

Fig. 1. (a) An overview of our proposed FIRE framework. The regularization loss L is calculated as following two stages. First, the subsampled image is calculated as the image reconstruction term L_{rec} between the network output and the target. Second, we added the regularization term L_{reg} to correct the difference of implying ground truth between the subsampled noisy image. (b) Details of the remainder sampler based on pixel coordinates. The different color boxes, including blue and pink, respectively represent the transformation from one part of the image to this part after sampling.

Subspace Sampling: The remainder with respect to a positive integer (i.e., N) of image pixel coordination is excellent to be used to develop a sampler. With it, a raw noisy image can be divided into a series of sub-image spaces, and such design can fully use the advantages of the image pixels index resulting in satisfying the condition of independence. Let $y \in R^{(W*H)}$ be one noisy image sample, where W and H represent the width and height of the raw image y. Considering the size of the medical image is usually a multiple of 256, the N is chosen as 4 in this study. In fact, the N can be adjusted depending on the image size. In this case, the sampler G consists of four sub-sampler $G = (g_1, g_2, g_3, g_4)$. The details of the remainder sampler G can be developed as follows:

a) Making the sampled pictures contain all the pixels within the original noisy image, it is necessary to ensure that the length and width of the image y can be divisible by N, which is chosen as 4;

b) Encoding pixel index of the image y from the left-upper corner, all image pixels can be accessed by the abscissa i and ordinate j. The image indexes can range from $(0,0)$ to $(W-1, H-1)$.

c) Calculating the remainder of $(2 \times i + j)\%4$ and defined it as k. k is an integer in $[0,3]$. All pixel coordinations satisfying $k = 0$ are retained, and the remaining pixel values are set to 0 to generate the first mask. The second mask is obtained with pixel coordinations satisfying $k = 1$. The third and fourth masks are obtained by satisfying $k = 2$ and $k = 3$.

d) Selecting the reserved pixel value from the four pixels in each 2×2 area as the pixel value for the corresponding position of the subsampled image, and put it into $(g_1, g_2, g_3$ and $g_4)$. By doing this, four sampled pictures are obtained, i.e., $(g_1(y), g_2(y), g_3(y)$ and $g_4(y))$. The width and length of all sampled subspace images are $W/2$ and $H/2$ respectively.

To clearly demonstrate the generation diagram of the subspace image group, Fig. 1(b) summarizes the idea of our proposed remainder subspace subsampler. Since the corresponding pixels of $(g_1(y), g_2(y), g_3(y), g_4(y))$ generated by the sampler are adjacent but different in the original image. Each sampled subspace image is independent of the others, and the completeness and difference of image content in the four sampled sub-images can be guaranteed. Here we can divide the four sub-images into two groups of $g_1(y)$ and $(g_2(y), g_3(y), g_4(y))$. Please refer to the supplementary materials for detailed explanations of this grouping.

Regularization Optimization: For self-supervised training, the following minimization optimization problem is formulated by taking advantage of the constraint

$$min_\theta E_{y|x} \|3f_\theta(g_1(y)) - (g_2(y) + g_3(y) + g_4(y))\|_2^2 + \lambda E_{y|x} \|3f_\theta(g_1(y)) - (g_2(y)$$
$$+ g_3(y) + g_4(y)) + (3g_1(f_\theta(y)) - (g_2(f_\theta(y)) + g_3(f_\theta(y)) + g_4(f_\theta(y)))\|_2^2. \quad (4)$$

According to $E_{x,y} = E_x E_{y|x}$, it can be converted into the following regularization optimization problem

$$min_\theta E_{x,y} \|3f_\theta(g_1(y)) - (g_2(y) + g_3(y) + g_4(y))\|_2^2 + \lambda E_{x,y} \|3f_\theta(g_1(y)) - (g_2(y)$$
$$+ g_3(y) + g_4(y)) + (3g_1(f_\theta(y)) - (g_2(f_\theta(y)) + g_3(f_\theta(y)) + g_4(f_\theta(y)))\|_2^2. \quad (5)$$

Finally, the loss function incorporating the regularization term is proposed to train the denoising network with

$$L = L_{rec} + \lambda \times L_{reg}$$
$$= \|3f_\theta(g_1(y)) - (g_2(y) + g_3(y) + g_4(y))\|_2^2 + \lambda \|3f_\theta(g_1(y)) - (g_2(y) + g_3(y)$$
$$+ g_4(y)) + (3g_1(f_\theta(y)) - (g_2(f_\theta(y)) + g_3(f_\theta(y)) + g_4(f_\theta(y)))\|_2^2 \quad (6)$$

where f_θ is the denoising network and λ is a hyperparameter to balance the regularization term.

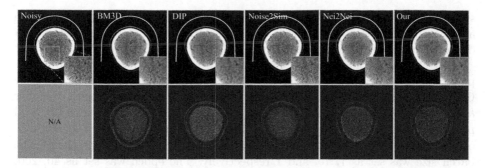

Fig. 2. Visualization comparison of different methods in the clinical brain CT results. The visualization of the results after denoising by different methods is shown above, and the noise gap level between the output image and the original noisy image is shown below.

3   Experiments

In this section, we first introduce the details and configuration of clinical experiments. To evaluate the effectiveness of this method, several advanced image-denoising methods for CT and DR were involved in the comparison. The experiments were conducted on several large-scale CT and DR data sets.

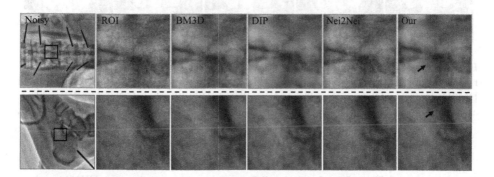

Fig. 3. Visual comparison of different methods in real DR denoising results. The boxes of different colors represent images of different parts. The green image is represented as the spine and the blue as the hand. (Color figure online)

Experimental Configuration: The network architecture of our proposed FIRE framework was a modified U-Net [21]. All experiments were conducted on a PC server equipped with Python3.9.7, PyTorch1.8, and NVIDIA TITAN RTX graphics processors.

In terms of the comparisons, we include BM3D [5], DIP [17], Noise2Sim [20], Noise2Void [14], Blind2Unblind [22], Neighbor2Neighbor(Nei2Nei) [11] and two supervised denoising algorithms DnCNN [25] and DD-net [27]. According to the applicability of denoising methods, we chose different comparison methods in different experiments. More details can be found in Figs. 2 to 4.

The clinical DR and brain CT image data are collected from the hospital. The public CT image data came from the 2016 NIH-AAPM Mayo Clinical low-dose CT competition [19] (Moen, Chen, Holmes III, Duan, Yu, Yu, Leng, Fletcher & McCollough 2021). In the three experiments of brain CT, clinical DR, and public clinical CT, the image size was 512×512 pixels. The initial learning rate of the ADAM optimizer was set as 0.0003, 0.0003, and 0.0004 respectively. The batch size was set as 1, 16, and 1 respectively. The epoch was set as 40, 200, and 10 respectively to ensure the model was convergent. The regularization term parameter λ was set to $\lambda = 2$, 10, and 1 respectively.

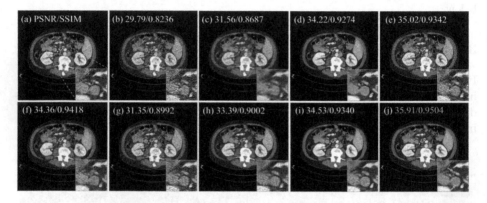

Fig. 4. Visual comparison of different methods on the public CT dataset. (a) is the reference obtained with the high dose, (b) is the noisy image, (c)-(j) are results of BM3D, DnCNN, DDnet, Noise2Sim, Noise2Void, Blind2Unblind, Neighbor2Neighbor, and our proposed FIRE. PSNR and SSIM values are marked at the top of the picture.

Experimental Results: In three different denoising tasks, our FIRE method showed the best results. For the clinical brain CT image denoising, the excellent denoising results of FIRE can be seen in Fig. 2. Due to a certain gap in ground truth between adjacent frames images used in network training, Noise2Sim cannot achieve good enough results. DIP produces a lot of artifacts after removing the noise and misses a lot of details and features. Besides, BM3D and DIP consume long time, and the parameters need to be manually adjusted on different denoising conditions. For the real DR image dataset, Fig. 3 demonstrates the DR image denoising results of different body parts for different patients. It can be seen that the denoising performance of Neighbor2Neighbor is limited. It may be because the Neighbor2Neighbor only uses information of half pixels for training, and the entire structure and pixel information within the DR image is partially missed. On the public CT image data set, we can intuitively compare the advantages of our method over other methods from the PSNR and SSIM measures. Figure 4 shows the typical slice denoising results. Figure 5(a)-(j) presents the noise power spectrum(NPS) of all methods, where blue indicates it is closer to the reference. As seen, our FIRE obtains the most blue and the least red results. Figure 5(k) reflects the comparison of pixel values on the profile line of a slice of the test set. Among them, our FIRE is closest to the reference.

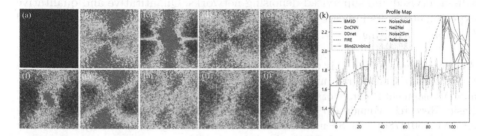

Fig. 5. (a)-(j) correspond to the noise power spectrum of Fig. 4. (k) is the profile map of different methods.

Influence of Regularization Term: We proposed a regularization term in Sect. 3, and the hyperparameter λ is used to adjust the regularization term. Figure 6 shows the visual results of FIRE with different λ values. When $\lambda = 0$, the regularization term is removed. When $\lambda = 1$, the denoising effect of the network is the best. As λ increases, the residual degree of noise increases. This shows that the regularization term can adjust the smoothness and noisiness of denoising. Therefore, choosing an appropriate λ value helps to obtain better denoising results.

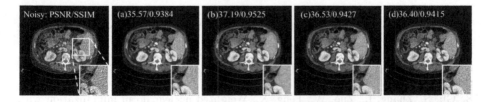

Fig. 6. The ablation study of the influence of regularization terms. The following results are the results of different λ values. (a) $\lambda = 0$, (b) $\lambda = 1$, (c) $\lambda = 8$, and (d) $\lambda = 16$.

4   Conclusion

We proposed a full image-index remainder method (FIRE) using only a single noisy image. The proposed FIRE first divided the whole high-dimensional image space into a series of low-dimensional sub-image spaces with the full image-index remainder technique. Our FIRE retains the complete information within the original noisy image. In addition, we proposed a new regularization optimization function to regularize sub-space image training by reducing the gap between the self-supervised and supervised denoising networks. Quantitative and qualitative experiment results indicate that the proposed FIRE is effective in both DR and CT image denoising.

Acknowledgements. This work was supported in part by National Natural Science Foundation of China (grant numbers 62101611 and 62201628), National Key Research and Development Program of China (2022YFA1204200), Guangdong Basic and Applied Basic Research Foundation (grant number 2022A1515011375, 2023A1515012278, 2023A1515011780) and Shenzhen Science and Technology Program (grant number JCYJ20220530145411027, JCYJ20220818102414031).

References

1. Anwar, S., Barnes, N.: Real image denoising with feature attention. In: 2019 IEEE/CVF International Conference on Computer Vision (ICCV), pp. 3155–3164 (2019)
2. Buades, A., Coll, B., Morel, J.M.: A non-local algorithm for image denoising. In: 2005 IEEE Computer Society Conference on Computer Vision and Pattern Recognition (CVPR'05), vol. 2, pp. 60–65 (2005)
3. Chang, M., Li, Q., Feng, H., Xu, Z.: Spatial-adaptive network for single image denoising (2020)
4. Chen, H., et al.: Low-dose CT with a residual encoder-decoder convolutional neural network. IEEE Trans. Med. Imaging **36**(12), 2524–2535 (2017)
5. Dabov, K., Foi, A., Katkovnik, V., Egiazarian, K.: Image denoising by sparse 3-D transform-domain collaborative filtering. IEEE Trans. Image Process. **16**(8), 2080–2095 (2007)
6. Falk, T., et al.: U-Net: deep learning for cell counting, detection, and morphometry. Nat. Methods **16**, 67–70 (2019)
7. Freedman, M.T., Artz, D.S.: Image processing in digital radiography. Semin. Roentgenol. **32**(1), 25–37 (1997), digital Radiography Using Storage Phosphor Technology
8. Gu, S., Zhang, L., Zuo, W., Feng, X.: Weighted nuclear norm minimization with application to image denoising, pp. 2862–2869 (2014)
9. Guo, S., Yan, Z., Zhang, K., Zuo, W., Zhang, L.: Toward convolutional blind denoising of real photographs. In: 2019 IEEE/CVF Conference on Computer Vision and Pattern Recognition (CVPR), pp. 1712–1722 (2019)
10. Hasan, A.M., Mohebbian, M.R., Wahid, K.A., Babyn, P.: Hybrid-collaborative Noise2Noise denoiser for low-dose CT images. IEEE Trans. Radiat. Plasma Med. Sci. **5**(2), 235–244 (2021)

11. Huang, T., Li, S., Jia, X., Lu, H., Liu, J.: Neighbor2Neighbor: a self-supervised framework for deep image denoising. IEEE Trans. Image Process. **31**, 4023–4038 (2022)

12. Immonen, E., et al.: The use of deep learning towards dose optimization in low-dose computed tomography: a scoping review. Radiography **28**(1), 208–214 (2022)

13. Kashyap, M., Tambwekar, A., Manohara, K., Subramanyam, N.: Speech denoising without clean training data: a Noise2Noise approach (2021)

14. Krull, A., Buchholz, T.O., Jug, F.: Noise2Void - learning denoising from single noisy images. In: 2019 IEEE/CVF Conference on Computer Vision and Pattern Recognition (CVPR), pp. 2124–2132 (2019)

15. Krupinski, E.A., et al.: Digital radiography image quality: image processing and display. J. Am. Coll. Radiol. **4**(6), 389–400 (2007)

16. Lehtinen, J., et al.: Noise2Noise: learning image restoration without clean data (2018)

17. Lempitsky, V., Vedaldi, A., Ulyanov, D.: Deep image prior. In: 2018 IEEE/CVF Conference on Computer Vision and Pattern Recognition, pp. 9446–9454 (2018)

18. Li, X., Fan, C., Zhao, C., Zou, L., Tian, S.: NIRN: self-supervised noisy image reconstruction network for real-world image denoising. Appl. Intell. **52**, 1–18 (2022)

19. McCollough, C., et al.: Low dose CT image and projection data, LDCT and projection data, version 5, data set, the cancer imaging archive (2020). https://doi.org/10.7937/9NPB-2637

20. Niu, C., et al.: Suppression of correlated noise with similarity-based unsupervised deep learning (2020)

21. Ronneberger, O., Fischer, P., Brox, T.: U-Net: convolutional networks for biomedical image segmentation (2015)

22. Wang, Z., Liu, J., Li, G., Han, H.: Blind2Unblind: self-supervised image denoising with visible blind spots. In: 2022 IEEE/CVF Conference on Computer Vision and Pattern Recognition (CVPR), pp. 2017–2026 (2022)

23. Wiest-Daesslé, N., Prima, S., Coupé, P., Morrissey, S.P., Barillot, C.: Non-local means variants for denoising of diffusion-weighted and diffusion tensor MRI. In: Ayache, N., Ourselin, S., Maeder, A. (eds.) Medical Image Computing and Computer-Assisted Intervention - MICCAI 2007, pp. 344–351. Springer, Berlin, Heidelberg (2007). https://doi.org/10.1007/978-3-540-75759-7_42

24. Williams, M.B., et al.: Digital radiography image quality: image acquisition. J. Am. Coll. Radiol. **4**(6), 371–388 (2007)

25. Yin, Z., Wan, B., Yuan, F., Xia, X., Shi, J.: A deep normalization and convolutional neural network for image smoke detection. IEEE Access **5**, 18429–18438 (2017)

26. Zainulina, E., Chernyavskiy, A., Dylov, D.V.: Self-supervised physics-based denoising for computed tomography. ArXiv abs/2211.00745 (2022)

27. Zhang, Z., Liang, X., Dong, X., Xie, Y., Cao, G.: A sparse-view CT reconstruction method based on combination of DenseNet and deconvolution. IEEE Trans. Med. Imaging **37**(6), 1407–1417 (2018)

Shape-Based Pose Estimation
for Automatic Standard Views
of the Knee

Lisa Kausch[1,2,3]([✉]), Sarina Thomas[4], Holger Kunze[5], Jan Siad El Barbari[6],
and Klaus H. Maier-Hein[1,2,3]

[1] Division of Medical Image Computing, German Cancer Research Center (DKFZ),
Heidelberg, Germany
`l.kausch@dkfz-heidelberg.de`
[2] National Center for Tumor Diseases (NCT), Heidelberg, Germany
[3] Pattern Analysis and Learning Group, Department of Radiation Oncology,
Heidelberg University Hospital, Heidelberg, Germany
[4] Department of Informatics, University of Oslo, Oslo, Norway
[5] Advanced Therapy Systems Division, Siemens Healthineers, Erlangen, Germany
[6] MINTOS Research Group, Trauma Surgery Clinic Ludwigshafen, Heidelberg,
Germany

Abstract. Surgical treatment of complicated knee fractures is guided by
real-time imaging using a mobile C-arm. Immediate and continuous control is achieved via 2D anatomy-specific standard views that correspond to
a specific C-arm pose relative to the patient positioning, which is currently
determined manually, following a trial-and-error approach at the cost of
time and radiation dose. The characteristics of the standard views of the
knee suggests that the shape information of individual bones could guide
an automatic positioning procedure, reducing time and the amount of
unnecessary radiation during C-arm positioning. To fully automate the C-
arm positioning task during knee surgeries, we propose a complete framework that enables (1) automatic laterality and standard view classification
and (2) automatic shape-based pose regression toward the desired standard view based on a single initial X-ray. A suitable shape representation
is proposed to incorporate semantic information into the pose regression
pipeline. The pipeline is designed to handle two distinct standard views
with one architecture. Experiments were conducted to assess the performance of the proposed system on 3528 synthetic and 1386 real X-rays for
the a.-p. and lateral standard. The view/laterality classificator resulted in
an accuracy of 100%/98% on the simulated and 99%/98% on the real X-
rays. The pose regression performance was $d\theta_{a.-p} = 5.8 \pm 3.3°$, $d\theta_{lateral} =
3.7 \pm 2.0°$ on the simulated data and $d\theta_{a.-p} = 7.4 \pm 5.0°$, $d\theta_{lateral} =
8.4 \pm 5.4°$ on the real data outperforming intensity-based pose regression.

Keywords: Shape-based pose estimation · Standard projections ·
Knee

Supplementary Information The online version contains supplementary material
available at https://doi.org/10.1007/978-3-031-43990-2_45.

1   Introduction

Intraoperative imaging employing a mobile C-arm enables immediate and continuous control during orthopedic and trauma interventions. For optimal fracture reduction and implant placement, correct acquisition of standard views that correspond to a specific C-arm pose relative to the patient is essential [18]. Incorrect standard views can exhibit superimposed anatomical structures, leading to overlooked errors that can result in malunion of fractures, functional impairment, or require revision surgeries. To enable deducing all three dimensions of the trauma case, at least two 2D fluoroscopic views are acquired in two distinct planes usually at right angles to each other. The current manual C-arm positioning procedure results in only 20% surgically relevant acquisitions while the remaining 80% are caused by the iterative positioning process, exposing patients and clinical staff to unnecessary radiation [16].

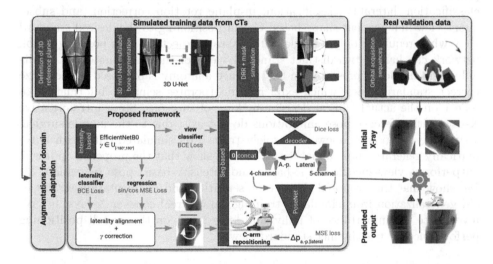

Fig. 1. Proposed shape-based pose estimation framework for automatic acquisition of standard views. A single architecture is trained for the representation of 2 distinct standard views simultaneously. The 2-step pipeline consists of a direct intensity-based combined view classification and in-plane rotation regression, followed by a segmentation-based pose regression focusing on the out-of-plane rotation. The pipeline is solely trained on synthetic data with automatically generated ground truth annotations and evaluated on real X-rays.

Recent developments towards robotic C-arms ask for automatic positioning methods. Many state of the art approaches require a patient-specific CT for intraoperative real-time simulation [4,5,8] or 2D-3D registration [2,7,17], external tracking equipment [6,16], manual landmark annotation [1] or do not estimate an optimal pose but reproduce intraoperatively recorded C-arm views employing augmented reality [6,20]. The inherent prior assumptions and severe inference with the clinical workflow hinder broad clinical applicability until today. In contrast to the majority of anatomical regions, the standard planes of the knee are not orthogonal to each other. The a.-p. standard view is characterized by symmetric projection of the joint gap, femoral and tibial condyles. The tibia surface projects line-shaped and the medial half of the fibula head is superimposed by the tibia. In the lateral standard view, both femoral condyles are aligned and the joint gap is maximized. Automatic deep learning-based positioning for standard views involves specific challenges for image understanding due to overlapping anatomical structures, the presence of surgical implants, and changing viewing directions and showed to benefit from extracting semantic information [10]. Inspired by that and considering that standard views of the knee anatomy are characterized by the shape information of the individual projected bones, we propose a complete framework to fully automate the C-arm positioning tasks during knee surgeries (Fig. 1). Our contribution is 4-fold: (1) We propose a novel framework that enables simultaneous automatic standard view classification, laterality classification, in-plane rotation correction, and subsequent view-independent shape-based C-arm positioning to the desired standard view while requiring only a single initial X-ray projection. One pose regression network can handle two distinct standard views of the knee anatomy. A suitable segmentation representation for the knee anatomy is proposed to recognize correct standard views, which explicitly incorporates semantic information to reflect on the actual clinical decision-making of surgeons. Since intraoperative X-rays with reference pose annotations do not exist due to the mobile surgical setup, the proposed framework is solely trained on simulated data with automatically generated pose annotations. (2) We show that the proposed approach outperforms view-specific shape-based and intensity-based pose regression. (3) We show that the proposed shape representation and augmentation strategies aid generalization from simulated training data to real cadaver X-rays. (4) We investigate the importance of individual knee bones on the overall positioning performance for two distinct standard views.

2   Materials and Methods

An overview of the complete framework for fully automated C-arm positioning towards desired standard views during knee surgeries is given in Fig. 1. The anterior-posterior (a.-p.) and the lateral standard view showed to be sufficient for various diagnostic entities [3].

2.1   Training Data Simulation

To address the interventional data scarcity problem, simulated training data was generated from a collection of CT and C-arm volumes using a realistic DRR simulation framework [21] complemented with corresponding 2D segmentations. Preprocessing involved the following steps: **(1) Field-of-view cropping**: Prevents superposition of the other laterality in the projection domain.

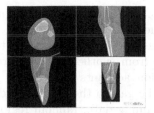
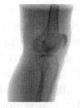

(a) Separation of femur condyles in segmentation   (b) Exemplary simulation

Fig. 2. Suitable segmentation representation to recognize the lateral standard view of the knee: In an ideal standard view the condyles' overlap each other.

| (a) In-plane rotation γ | (b) Detector translation | (c) Random region dropout | (d) Transparent edges | (e) Border overlays |

Fig. 3. Training data augmentations.

(2) Laterality alignment: Based on the laterality classification left knees are horizontally flipped to mirror the right anatomy, simplify the pose estimation task, and prohibit ambiguities during pose estimation. **(3) Definition of 3D reference planes**: Two independent raters defined the 3D reference planes in the CT volumes utilizing a DRR preview integrated into the open-source Medical Imaging Toolkit [22] with interactive plane positioning. They serve as ground truth pose reference during simulation. **(4) 3D automatic bone segmentation**: To compute automatic 3D segmentations, a 3D nnU-Net [9] is trained on a subset of 10 manually annotated CTs for the task of multilabel bone segmentation, segmenting the femur, tibia, fibula, and patella. **(5) Suitable segmentation representation**: The two femur condyles are not distinguishable in the shape-based representation, however, this is relevant for optimal lateral positioning. In an optimal lateral view, femur condyles would overlap. Annotating the condyles as line features would result in an increased manual labeling effort in the projection domain. Alternatively, we propose to incorporate this information in the segmentation by separating the femur annotation symmetrically along the femoral shaft (Fig. 2a). This results in one additional segmentation label for the lateral standard to recognize condyles' congruence and derive the directional pose offset. **(6) DRR and mask simulation**: DRRs are simulated for varying angulations of orbital and angular rotation α, $\beta \in [-40°, 40°]$ around the defined reference standard. The system parameters are defined according to a Siemens Cios Spin® with $300\,mm$ detector and $1164\,mm$ source-detector distance. Simulation was performed for an increased field-of-view to create in-plane

γ augmentations without introducing cut-off regions at the image borders. The DeepDRR simulation framework was extended to allow the forward projection of corresponding masks (Fig. 2b). A set of augmentations is applied to the simulated dataset, to bridge the domain gap from synthetic data to real X-rays (Fig. 3). The in-plane rotation augmentation accounts for variable patient to C-arm alignment, the translation bridges the gap to the validation data where the joint gap is not centered like in the training data, random region dropout reflects superposition artifacts, transparent edge overlays reflect projection artifacts resulting from, e.g., the operating table, and border overlays account for the gamma correction interpolation artifacts.

For simulation, a set of 24 CTs was considered, 15 CTs without metal and 9 C-arm volumes with metal. The data was divided $60 - 20 - 20\,\%$ for training (15 CTs), validation (5 CTs), test (4 CTs).

2.2   Shape-Based Positioning Framework:

The proposed shape-based positioning framework was trained jointly for both standard views (Fig. 1). It consists of two modules: The first is responsible for a view classification, in-plane rotation and laterality alignment, directly estimated from the image intensities. Thereby, pose ambiguities and data variation are addressed, simplifying the task for the subsequent module to estimate the optimal C-arm positioning for the desired standard view, employing shape features.

(1) Intensity-based multi-task classification and regression module: For simultaneous in-plane rotation regression, view recognition, and laterality classification, an EfficientNet-B0 feature extractor [14,19] was extended with two binary classification heads with one output neuron, followed by sigmoid activation, and one regression head, with the same architecture, but 2 outputs, omitting the activation. The in-plane rotation γ is mapped to sin/cos-space to ensure a continuous Loss function during optimization. All training examples were aligned with the same laterality during data simulation which would otherwise result in pose ambiguities. Thus, to train the laterality classifier, the training examples were randomly flipped horizontally with $p = 0.5$ and the corresponding γ label was adapted accordingly. The weights were optimized using Binary Cross Entropy Loss for the classification heads, and Mean Squared Error for the regression head.

(2) Shape-based pose regression: Following surgical characteristics for recognizing correct standard views of the knee, a view-independent shape-based pose regression framework was developed. The architecture is based on a 2D U-Net [12] with two view-specific segmentation heads, because the segmentation labels differ for both views. Considering the results of the view classification, the extracted segmentation map of the corresponding segmentation head are used as input for the pose regression network that outputs the necessary C-arm pose update $(\alpha,\ \beta,\ \gamma,\ \mathbf{t}) \in \mathbf{R}^6$ to acquire the desired standard view [11]. To ensure equal number of input channels to the PoseNet for both standard views, a zero channel is appended to the a.-p. multi-label segmentation head output.

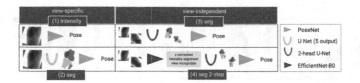

Fig. 4. Variants for performance comparison.

2.3 Real X-Ray Test Data:

Real X-rays for validation were sampled from single Siemens Cios Spin®
sequences generated during 3D acquisition of 6 knee cadavers. Preprocessing
consisted of (1) definition of 3D standard reference planes, (2) laterality check,
(3) sampling of X-rays around the defined reference standards in the interval
α, $\beta \in [-30°, 30°]$, and generation of ground truth pose labels. Since the Spin
sequences are orbital acquisition sequences, only the orbital rotation is equidis-
tantly covered in the test set, while the angular rotation is constant for all X-rays
sampled from the same sequence. The number of sampled X-rays per standard
and view may differ, if the reference standard is located close to the edge of the
orbital sequence (range: 102–124).

3   Experiments and Results

The proposed pipeline was evaluated considering the following research questions:

(RQ1) Does the proposed shape-based pipeline outperform view-specific
intensity-based and shape-based pose regression? How does it influence
the generalization from synthetic to real data? (Sect. 3.1)
(RQ2) How do individual bones influence the overall positioning performance?
(Sect. 3.2)
(RQ3) How accurate is the performance of the view and laterality classification?
(Sect. 3.3)

Positioning performance was evaluated based on the angle $\theta = \arccos\left(\langle v_{pred}, v_{gt}\rangle\right)$ between the principal rays of the ground truth v_{gt} and predicted pose v_{pred}
and the mean absolute error (AE) of in-plane rotation γ. The interrater variation
of the reference standard planes defined by two independent raters serves as an
upper bound for the reachable accuracy of a C-arm positioning approach trained
on the reference annotations. It was assessed in terms of orientation differences
θ ($\theta_{a.-p.} = 4.1 \pm 2.6°$, $\theta_{lateral} = 1.8 \pm 1.3°$). The models were implemented using
PyTorch 1.6.0, trained with an 11 GB GeForce RTX 2080 Ti, and optimized
with the Adam optimizer with a base learning rate of $\eta = 10^{-4}$ and batchsize 8,
pre-trained independently, and jointly fine-tuned until convergence.

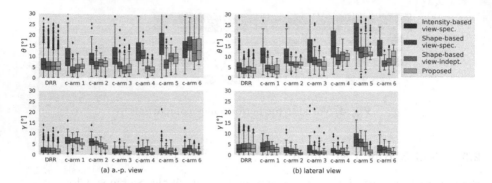

Fig. 5. Pose regression performance on simulated and real X-rays compared for different pipeline variants.

3.1 Importance of Pipeline Design Choices (RQ1)

In an ablation study, the proposed shape-based view-independent pose regression was compared to view-specific direct intensity-based pose regression [11]. Further, the complete pipeline (2-step) is compared to a 1-step segmentation-based approach trained view-specific and view-independent (Fig. 4). Evaluation was performed on the simulated test DRRs and cadaveric X-rays (Fig. 5). Significant performance differences were confirmed by a paired t-test.

View-independent vs. view-specific networks: While the view-specific networks perform significantly better (lateral) or comparable (a.-p.) on the simulated data, the view-independent networks perform significantly better (a.-p.) or comparable (lateral) on the real data.

1-step vs. 2-step: The proposed 2-step approach performs significantly better or comparable than a 1-step shape-based pose regression approach on most validation cases (8/12) in viewing direction θ. Regarding the γ rotation, the 2-step approach improves performance across all validation cases.

Generalization from DRR to X-ray: The shape-based pose regression network combined with joint view-independent training clearly boosts the performance compared to direct intensity-based pose regression from $d\theta_{a.-p}^{X-ray} = 12.2 \pm 6.8°$, $d\theta_{lateral}^{X-ray} = 14.4 \pm 7.6°$ to $d\theta_{a.-p}^{X-ray} = 7.4 \pm 5.0°$, $d\theta_{lateral}^{X-ray} = 8.4 \pm 5.4°$. Figure 6 shows qualitative results on exemplary real test X-rays.

3.2 Importance of Individual Bones on Overall Performance (RQ2)

Figure 7 shows the importance of individual segmented bone classes on the overall positioning performance evaluated on the test DRRs (3528 DRRs). The fibula has very little influence on the positioning for both views. The patella is only important for the a.-p. view, while tibia and femur are relevant for both views. The condyle assignment for the lateral view determines the rotation direction for the orbital and angular rotation (α, β). Inverting the assignment of left and right femur condyle results in a sign flip in α, β.

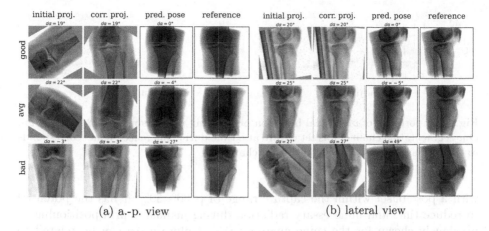

(a) a.-p. view   (b) lateral view

Fig. 6. Exemplary visual results on the real X-ray test dataset for the a.-p. and lateral standard positioning with good, average, and bad performance measured with respect to the orbital rotation offset α to the reference standard pose. The initial X-ray projection is visualized along with the in-plane corrected projection with the overlaid segmentation mask, and the predicted standard view side-by-side with the desired reference standard.

3.3 Accuracy of View and Laterality Classification (RQ3)

The classificator performances were assessed on the synthetic (3528 DRRs, 4 CTs) and real data (1386 X-rays, 6 C-arm scans). The view classificator (a.-p. / lateral) achieved an accuracy of 100% on the test DRRs and 99% on the X-rays. The laterality classificator (left / right) resulted in an accuracy of 98% on the test DRRs and 98% on the X-rays.

4 Discussion and Conclusion

A complete framework for automatic acquisition of standard views of the knee is proposed that can handle several standard views with a single architecture. The complete pipeline is trained on simulated data with automatically generated annotations and evaluated on real intraoperative X-rays. To bridge the domain gap, different augmentation strategies are suggested that address intraoperative confounding factors, e.g., the OR table. View-independent training and multi-label shape features improve the generalization from simulated training to real X-rays and outperform direct intensity-based approaches. View-independent networks result in more training data which showed to improve the generalization from simulated training to real X-rays. The 2-step approach increases robustness and simultaneously automates necessary preprocessing tasks like laterality and standard view recognition, which can be performed with very high accuracy on simulated (100%, 98%) and real data (99%, 98%). The approach is fast and easy to translate into the operating room as it does not require any additional technical equipment. Assuming that the surgeon acquires the initial X-ray

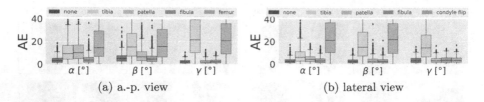

(a) a.-p. view   (b) lateral view

Fig. 7. Importance of individual bones on positioning performance (3528 test DRRs). One segmentation channel was set to zero at a time during inference, 'none' corresponds to the reference performance utilizing all channels.

with a pose offset within the capture range of $[-30°, 30°]$, it has the potential to reduce time and unnecessary radiation during manual C-arm positioning as previously shown for the spine anatomy [15]. Failure cases can be related to inaccurate intermediate segmentations which may result from patella baja not represented in the training dataset. The segmentation features can serve as a sanity check and indicate the reliability of the pose regression result. Further experiments with a larger training set covering more anatomical variation, e.g., patella baja, different flexion angles, and fractures [13] can potentially address observed failure cases.

Data use declaration: The data was obtained retrospectively from anonymized databases and not generated intentionally for the study. The acquisition of data from living patients had a medical indication and informed consent was not required. The corresponding consent for body donation for these purposes has been obtained.

References

1. Binder, N., Bodensteiner, C., Matthäus, L., Burgkart, R., Schweikard, A.: Image guided positioning for an interactive c-arm fluoroscope. Comput. Assist. Radiol. Surg., 5–7 (2006)
2. Bott, O.J., Dresing, K., Wagner, M., Raab, B.W., Teistler, M.: Use of a C-arm fluoroscopy simulator to support training in intraoperative radiography. Radiographics **31**(3), E31–E41 (2011)
3. Cockshott, W.P., Racoveanu, N., Burrows, D., Ferrier, M.: Use of radiographic projections of knee. Skeletal Radiol. **13**(2), 131–133 (1985)
4. De Silva, et al.: C-arm positioning using virtual fluoroscopy for image-guided surgery. In: Medical Imaging: Image-guided Procedures, Robotic Interventions, and Modeling, vol. 10135, p. 101352K. International Society for Optics and Photonics (2017)
5. Fallavollita, P., et al.: *Desired-View* controlled positioning of angiographic C-arms. In: Golland, P., Hata, N., Barillot, C., Hornegger, J., Howe, R. (eds.) MICCAI 2014. LNCS, vol. 8674, pp. 659–666. Springer, Cham (2014). https://doi.org/10. 1007/978-3-319-10470-6_82

6. Fotouhi, J., et al.: Interactive flying frustums (IFFs): spatially aware surgical data visualization. Int. J. Comput. Assist. Radiol. Surg. **14**(6), 913–922 (2019). https://doi.org/10.1007/s11548-019-01943-z

7. Gong, R.H., Jenkins, B., Sze, R.W., Yaniv, Z.: A cost effective and high fidelity fluoroscopy simulator using the image-guided surgery toolkit (IGSTK). In: Medical Imaging: Image-Guided Procedures, Robotic Interventions, and Modeling, vol. 9036, p. 903618. International Society for Optics and Photonics (2014). https://doi.org/10.1117/12.2044112

8. Haiderbhai, M., Turrubiates, J.G., Gutta, V., Fallavollita, P.: Automatic C-arm positioning using multi-functional user interface. In: Canadian Medical and Biological Engineering Society Proceedings, vol. 42 (2019)

9. Isensee, F., Jaeger, P.F., Kohl, S.A., Petersen, J., Maier-Hein, K.H.: nnU-Net: a self-configuring method for deep learning-based biomedical image segmentation. Nat. Methods **18**(2), 203–211 (2021). https://doi.org/10.1038/s41592-020-01008-z

10. Kausch, L., et al.: C-arm positioning for spinal standard projections in different intra-operative setting. In: International Conference on Medical Image Computing and Computer-Assisted Intervention, pp. 352–362 (2021). https://doi.org/10.1007/978-3-030-87202-1_34

11. Kausch, L., et al.: Toward automatic C-arm positioning for standard projections in orthopedic surgery. Int. J. Comput. Assist. Radiol. Surg. **15**(7), 1095–1105 (2020). https://doi.org/10.1007/s11548-020-02204-0

12. Klein, A., Wasserthal, J., Greiner, M., Zimmerer, D., Maier-Hein, K.H.: basic_unet_example (v2019.01) (2019). https://doi.org/10.5281/zenodo.2552439

13. Krönke, S., et al.: CNN-based pose estimation for assessing quality of ankle-joint X-ray images. In: Medical Imaging 2022: Image Processing, vol. 12032, pp. 344–352. SPIE (2022)

14. Mairhöfer, D., et al.: An AI-based framework for diagnostic quality assessment of ankle radiographs. In: Medical Imaging with Deep Learning (2021)

15. Mandelka, E., et al.: Intraoperative adjustment of radiographic standard projections of the spine: interrater-and intrarater variance and consequences of 'fluoro-hunting'considering time and radiation exposure-a cadaveric study. medRxiv (2022)

16. Matthews, F., et al.: Navigating the fluoroscope's C-arm back into position: an accurate and practicable solution to cut radiation and optimize intraoperative workflow. J. Orthopaedic Trauma **21**(10), 687–692 (2007). https://doi.org/10.1097/BOT.0b013e318158fd42

17. Miao, S., et al.: Dilated FCN for multi-agent 2D/3D medical image registration. In: Proceedings of the AAAI Conference on Artificial Intelligence, vol. 32 (2018)

18. Norris, B.L., Hahn, D.H., Bosse, M.J., Kellam, J.F., Sims, S.H.: Intraoperative fluoroscopy to evaluate fracture reduction and hardware placement during acetabular surgery. J. Orthopaedic Trauma **13**(6), 414–417 (1999). https://doi.org/10.1097/00005131-199908000-00004

19. Tan, M., Le, Q.: EfficientNet: rethinking model scaling for convolutional neural networks. In: International Conference on Machine Learning, pp. 6105–6114. PMLR (2019)

20. Unberath, M., et al.: Augmented reality-based feedback for technician-in-the-loop C-arm repositioning. Healthc. Technol. Lett. **5**(5), 143–147 (2018). https://doi.org/10.1049/htl.2018.5066

21. Unberath, M., et al.: DeepDRR – a catalyst for machine learning in fluoroscopy-guided procedures. In: Frangi, A.F., Schnabel, J.A., Davatzikos, C., Alberola-López, C., Fichtinger, G. (eds.) MICCAI 2018. LNCS, vol. 11073, pp. 98–106. Springer, Cham (2018). https://doi.org/10.1007/978-3-030-00937-3_12
22. Wolf, I., et al.: The medical imaging interaction toolkit. Med. Image Anal. 9(6), 594–604 (2005)

Optimizing the 3D Plate Shape
for Proximal Humerus Fractures

Marilyn Keller[1(✉)], Marcell Krall[2], James Smith[2], Hans Clement[2],
Alexander M. Kerner[2], Andreas Gradischar[3], Ute Schäfer[2], Michael J. Black[1],
Annelie Weinberg[2], and Sergi Pujades[4]

[1] Max Planck Institute for Intelligent Systems, Tuebingen, Germany
marilyn.keller@tuebingen.mpg.de
[2] Medical University of Graz, Graz, Austria
[3] CAE Simulation & Solutions GmbH, Vienna, Austria
[4] Univ. Grenoble Alpes, Inria, CNRS, Grenoble INP, LJK, Grenoble, France

Abstract. To treat bone fractures, implant manufacturers produce 2D
anatomically contoured plates. Unfortunately, existing plates only fit a
limited segment of the population and/or require manual bending during
surgery. Patient-specific implants would provide major benefits such as
reducing surgery time and improving treatment outcomes but they are
still rare in clinical practice. In this work, we propose a patient-specific
design for the long helical 2D PHILOS (Proximal Humeral Internal Locking System) plate, used to treat humerus shaft fractures. Our method
automatically creates a custom plate from a CT scan of a patient's bone.
We start by designing an optimal plate on a template bone and, with
an anatomy-aware registration method, we transfer this optimal design
to any bone. In addition, for an arbitrary bone, our method assesses if a
given plate is fit for surgery by automatically positioning it on the bone.
We use this process to generate a compact set of plate shapes capable of
fitting the bones within a given population. This plate set can be preprinted in advance and readily available, removing the fabrication time
between the fracture occurrence and the surgery. Extensive experiments
on ex-vivo arms and 3D-printed bones show that the generated plate
shapes (personalized and plate-set) faithfully match the individual bone
anatomy and are suitable for clinical practice.

Keywords: Personalized plates · 3D printing · ex-vivo evaluation

1 Introduction

Bone shapes vary significantly across the world's population [1], making the
design of personalized medical plates for repairing fractures challenging. Yet

Supplementary Information The online version contains supplementary material
available at https://doi.org/10.1007/978-3-031-43990-2_46.

patient-specific implants would provide major benefits for surgeons and patients: surgery time would be reduced and treatment outcomes improved. While some plates are malleable enough to be contoured during surgery to improve overall fitness, these are more prone to fatigue failure when compared to rigid pre-contoured plates. But anatomically pre-contoured plates only fit a subset of the population [2–4], causing malalignment at the fracture site [5]. Even with the extra contouring step during the surgery, the plates do not fit all bone shapes [6].

Our work contributes several key steps to create personalized plates: (i) we first design an optimal PHILOS 3D plate shape for the humerus bone, fulfilling the clinical constraints given by a senior surgeon; (ii) we propose an anatomy-aware transfer of the plate shape into any new bone, resulting in a personalized plate shape; (iii) we propose a method to position an arbitrary plate on an arbitrary bone according to clinical constraints, allowing surgical compatibility to be evaluated; (iv) we leverage the compatibility assessment to obtain a compact set of plates that accommodates a given population of bones; (v) we validate our methodology with extensive experiments on ex-vivo and 3D-printed bones, demonstrating the relevance of the designed personalized plate shapes for the clinical setting; (vi) we make available for research purposes the humerus and plate 3D models, as well as the plate extraction and fitting code[1].

The creation of surgical plates consists of several tasks, which have been partially addressed by the existing state of the art. In Table 1 of Sup. Mat. we provide a schematic comparison of how our approach goes beyond existing methods.

Most methods start with a base plate or template that is designed for a specific bone surgery (tibia, humerus, clavicle, etc.) [2,4,6–13]. One important step is how to position a given plate on a given bone. Existing approaches are either manual, semi-automatic [2,3,7,11], or fully automatic [4,9,10,12,14]. Our fully automatic positioning strategy differs from the state of the art as it considers the surrounding anatomy; i.e. to minimize the risk of radial nerve damage, the plate is twisted around the bone so that the proximal end is fixed on the lateral side of the humeral head, and the distal end of the plate to the ventral surface of the humerus [15,16].

Once the plate is positioned, one needs to evaluate if the plate can be used in surgery. Some works compute plate-to-bone metrics [2,6,13,17], but do not propose a binary decision criterion stating whether the plate is valid for surgery or not. We argue that this binary decision, also provided by existing works [3,4,7–9,11,12,14], is important, as aggregate distance numbers can be misleading; e.g. one part of the plate could have a perfect fit but another part could make the plate not suitable for surgery. In our work, we define a three section-based criterion that takes into account the two plate-to-bone fixation regions and the middle plate section that transitions from one fixation region to the other. In addition, we perform a case study evaluation with an expert surgeon, in which the validity of the numerical criterion is confirmed.

---

[1] https://humerusplate.is.tue.mpg.de.

Once the plate is positioned on a bone, and the fit evaluated, one can further consider the question of *how to deform the plate to fulfill the fit criterion*. For instance, [6] aims to develop a plate shape to reduce in-situ plate manipulation. They approximate the plate and target bone surfaces by planar sections and measure the bending necessary for each section of the plate to fit the bone surface. In contrast, our approach is similar to the one of [3], in which the plate shape is not deformed after evaluating the fit, but rather directly extracted from the shape of an individual bone. We propose a plate transfer strategy that takes into account the matching of anatomic regions. One advantage over existing methods is that our approach simultaneously ensures both: (i) a personalized plate shape matching the bone's shape and (ii) its proper placement.

Prior work proposes a set of plate shapes that, together, can accommodate a range of patients [8,9]. This set is created by manually modifying an original plate to improve its plate-to-bone distance. We propose an automatic approach in which we first create many plates, and then, with a greedy algorithm, select the plate set that accommodates the most bones.

2   Method

Dataset. To create and evaluate our plate designs, we use a dataset consisting of 97 3D meshes of humerus bones, divided into two groups (A and B). Group A has 54 bones scanned with a FARO laser scanner (25 females and 29 males, 50% Black and 50% White, age range 17 and 45). Group B has 43 Computed Tomography (CT) scans of bones from autotomized body donors. The CT volumes were segmented and cleaned to reconstruct a mesh of the bone. Left-side humerus scans were mirrored along the z-axis to work with right-side humeri only.

Fig. 1. Left: surgeon annotations showing where the plate should be positioned and areas to be avoided. Middle: the designed plate on the template bone. Right: Plate-on-bone fit criteria. Colors show the vertices defined as the fixation area on the plate (Blue: $\mathcal{P}_h$; Red: $\mathcal{P}_s$) and on the bone (Pale blue: $\mathcal{B}_h$; Pale red: $\mathcal{B}_s$). (Color figure online)

Plate Design. Given a bone, our goal is to automatically generate an optimal plate shape and determine its position on the bone. The optimality of our plate design is defined by an experienced surgeon, who established fixation points, areas to avoid, and plate-to-bone distance tolerances. These choices were validated by a second expert surgeon. A plate fulfilling these constraints on a bone is considered optimal for surgery. To obtain the plate design we 3D printed 7 bone meshes from our dataset with diverse shapes and asked the surgeon to annotate each bone with a marker. The surgeon indicated the bone areas that should be in contact with an ideal plate and the areas to be avoided (Fig. 1 left). From these annotations, we designed an ideal plate mesh $\mathbf{P_T}$ contoured to a humerus template bone mesh $\mathbf{T}$ (Fig. 1 middle). This plate has similar dimensions as the actual INTEOS PROXIMAL HUMERAL PLATE 3.5. Following the surgeon's advise, the plate should not be in contact with the bone at the bone neck level, hence, an offset was added in this region. The fit criteria between a plate and a bone was defined such that the distance between them should not exceed $2mm$ at the fixation points, located at the head $\mathcal{P}_h$ and the shaft $\mathcal{P}_s$, and should be less than $5mm$ in the other regions, with the exception of the plate neck, which must not be in direct contact with the bone (Fig. 1 right).

Bone Registration for Optimal Plate Transfer. Once the optimal plate is defined on the template bone, the goal is to register any new bone to the template, by preserving the relevant anatomic regions for the plate fixation. To do so, we adapt the registration technique designed for vertebrae [18] to the humerus.

We first register each bone of set A by optimizing their Eqs. 1, 2 and 3 from [18]. Given these initial registrations we build a statistical shape model using principal component analysis (PCA). This model is parameterized with the shape vector β, and constructs a bone mesh $\mathbf{T}(\beta)$: an array of size $N_b \times 3$ containing the mesh vertex positions.

Second, we register the whole dataset using the learned bone shape model by minimizing

$$\mathrm{E}(\beta, \mathbf{t}, \mathbf{r}, \mathbf{F}; \mathcal{S}) = E_{p2m}(\mathcal{S}, \mathbf{r} \cdot (\mathbf{T}(\beta) + \mathbf{F}) + \mathbf{t}) + \lambda_\beta ||\beta||_2 + \lambda_L \Delta(\mathbf{T}(\beta)) \cdot \mathbf{F} \quad (1)$$

where $\mathcal{S}$ denotes the scanned bone mesh, $\mathbf{t}$ and $\mathbf{r}$ are, respectively, the 3D rigid translation and rotation applied to the bone mesh vertices, and $\mathbf{F}$ is a 3D per-vertex offset applied to each mesh vertex. $E_{p2m}(\mathcal{S}, .)$ is the point-to-triangle distance between the scan vertices and the mesh triangles. The function $\Delta(\mathbf{T})$ is the mesh Laplacian operator, which is used to regularize the offsets $\mathbf{F}$. We minimize Eq. (1) in three successive steps: with respect to $(\mathbf{t}, \mathbf{r})$, β, and $\mathbf{F}$. For each scan $\mathcal{S}_i$ we obtain its corresponding bone registration $\mathbf{B}'_i = \mathbf{T}(\beta) + \mathbf{F}$.

Due to the elongated shape of the humeri and their axial rotation similarity, the registrations $\mathbf{B}'_i$ can contain sliding, i.e. the same vertex does not correspond precisely to the same anatomic location on two different bones. To correct this, we perform a second pass of registration using smooth shells [19] to deform the initial template $\mathbf{T}$ to the surface of $\mathbf{B}'_i$ and we obtain the final bone registrations $\mathbf{B}_i$ matching the scan $\mathcal{S}_i$ (see Fig. 2 left). Note that smooth shells [19] can not directly register $\mathbf{T}$ to the scans $\mathcal{S}_i$, and the intermediate meshes $\mathbf{B}'_i$ are required.

To validate the bone correspondences we define anatomic regions on the bone template $\mathbf{T}$, and transfer them to all registrations $\mathbf{B}_i$ on a per-vertex index basis. Figure 2 shows that the annotated anatomic parts are faithfully preserved.

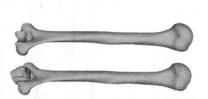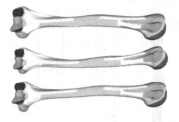

Fig. 2. Left: Registered bone $\mathbf{B}_i$ (green) and superposition to the input scan $\mathcal{S}_i$ (pink). Right: Registrations $\mathbf{B}_i$ to multiple scans. Despite the different individual bends and twists of each bone, the colored anatomic regions are consistent. (Color figure online)

Plate Extraction. As the registration process preserves anatomic regions, given a bone registration $\mathbf{B}_i$, we can extract a new optimal plate $\mathbf{P}_i$. For each vertex of the designed plate $\mathbf{P_T}$ we compute its offset to the template bone $\mathbf{T}$. By applying these offsets to the new bone $\mathbf{B}_i$, we obtain the personalized plate $\mathbf{P}_i$.

Positioning a Plate on a Bone. Now that we can create personalized plates, we want to study if a created plate can be used for surgery on another bone. For that, given an arbitrary plate $\mathbf{P}_i$ and bone registration $\mathbf{B}_j$, we need to position the plate and determine if it fits. We automate the plate positioning by minimizing the distance between the plate and bone fixation points. We start by computing an initial 3D rigid transformation $\mathbf{r}_0, \mathbf{t}_0$ by optimizing Eq. 2 in Sup. Mat. and then obtaining the final positioning $\mathbf{r}_f, \mathbf{t}_f$ by optimizing Eq. 3 in Sup. Mat. These equations enforce the fixation areas of the plate $\mathcal{P}_h^i$ to match $\mathcal{B}_h^j$ and $\mathcal{P}_s^i$ to match $\mathcal{B}_s^j$ (see Fig. 1 right), while avoiding plate-to-bone inter-penetrations. Once the plate is positioned on the bone, we evaluate the binary fit criteria (see Fig. 1 right) to conclude whether the plate shape is fit for surgery or not.

Plate Shapes Set. A single plate design can not accommodate the whole population due to its morphological variance [1]. Thus we propose to build a set of plate shapes with a greedy algorithm. The algorithm is summarized in pseudo-code in Alg. 1 of Sup. Mat. Given a bone dataset, we start by creating as many personalized plates as bones. Then, for each plate we use the previous optimization to determine how many bones it accommodates, i.e how well a plate shape generalizes to different bones. We then select the plate that fits most bones and remove the plate and the fitted bones from the current sets. We iterate until no bones are left. The result is an ordered plate-set that accommodates the input bone dataset.

3 Experiments

3.1 Numerical Evaluations

Bone Registration Accuracy. We validate that our registrations $\mathbf{B}_i$ accurately match the scans $\mathcal{S}_i$ by computing the mean distance (MD) between each registration to the closest vertex in the scan. We obtain a MD of 0.08 mm (std=0.04)

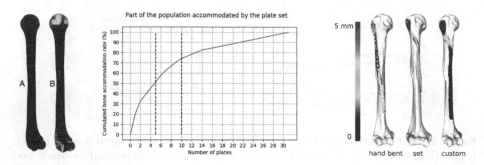

Fig. 3. Left: Per vertex max distance between the registrations $\mathbf{B}_i$ and scans S for bone set A (left) and B (right). Blue (0 mm) to red (1 mm). Max distances are higher for bone set B because of their lower scan quality (segmentation artifacts). Middle: Accommodation percentage as a function of the number of plates. Right: Plate-to-bone distance for three plates: surgeon's hand bent, best from the plate set, and custom plate. Blue (0mm) to red (5mm or more). (Color figure online)

for group A, and a MD of 0.27 mm (std=0.55) for group B. For both sets, the registered bones match the scans with sub-millimeter accuracy. In addition, all distances are less than 1mm on the surface where the plate is positioned (Fig. 3 left).

Plate Set Coverage. One application of our design is that it can generate a plate set that could be preprinted in the hospital and be readily available for immediate use. Using our greedy algorithm on our bone dataset, a set $\mathcal{P}_S^{N=5}$ made of the first 5 plates accommodates already 51.04% of the bones and $\mathcal{P}_S^{N=10}$ accommodates 73.96% (Fig. 3 middle). In Sec. 4 we discuss how this coverage could be improved.

3.2   Ex-Vivo Evaluations

Comparison with State-of-the-Art Plates. We CT-scanned an isolated bone (not included in the original dataset) and asked a surgeon to manually contour a state-of-the-art plate, as done in clinical practice. We compare it to the *custom plate* $\mathbf{P}_C$ and the best *set plate* $\mathbf{P}_S$ by computing the plate-to-bone distances (Fig. 3 right). The proposed plates are clearly closer to the bone than the hand-contoured plate. This proximity is known to be beneficial for bone recovery [20]. While $\mathbf{P}_S$ is less accurate than the custom plate - and shorter, as it was generated from a shorter bone - the surgeon considered the plate and its placement as suitable. This experiment reveals the difficulty of closely fitting the bone by manually bending a plate and the benefit of the proposed design.

Ex-vivo Surgery. We performed an ex-vivo experiment on a cadaveric arm, mimicking an actual minimally-invasive surgical operation setting. One goal was to test whether the designed 3D plates can be properly inserted along the bone under the muscles. We CT-scanned 3 cadaveric arms (not included in the original dataset), reconstructed the bone scans and registered them to obtain $\mathbf{B}_1$, $\mathbf{B}_2$

and $\mathbf{B}_3$. For each bone we generated and 3D printed their custom plate $\mathbf{P}_C^{1,2,3}$ and the best fitting set plate $\mathbf{P}_S^{1,2,3}$. The plates were drilled and coated with metallic paint to be visible on the CT scan. During the experiment, the plates were inserted in the cadaver arm, as it would be done in a standard operation, and fixed to the humerus with screws. For the 3 bones, the surgeon estimated that the fit of both the custom and set plates was satisfactory. He noted that the surface roughness of the 3D print and metallic coating was too high, making the plate hard to slide across the humerus into place.

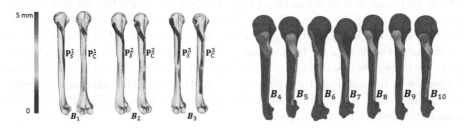

Fig. 4. Left: Color-coded plate-to-bone distances (blue: $0\,\mathrm{mm}$; red: $\geq 5\,\mathrm{mm}$) for the set plate and custom plate on bones $\mathbf{B}_1$, $\mathbf{B}_2$, and $\mathbf{B}_3$. Right: Placement of the 3D printed plates $\mathbf{P}_S$ on the corresponding 3D printed bone: green: algorithm's placement; orange: the surgeon's placement. (Color figure online)

We CT-scanned each plate screwed to the bone and used the scans to asses the fit quality. Figure 4 left shows the plate-to-bone distances for each bone. The plate $\mathbf{P}_C^1$ fits closely on the proximal side near the sulcus bicipitalis. For $\mathbf{P}_C^2$ and $\mathbf{P}_C^3$ the fits are not as good as $\mathbf{P}_C^1$. This was credited to a handling/alignment issue raised by the surgeon. Consequently, greater deviation in alignment is seen on the distal end of the bone on the lateral side. There is also a greater distance between the ventral part of the plate and the sulcus bicipitalis. The plate is too lateral on the humerus. In each case, the surgeon found the custom plate better. The limit in terms of fit was mainly attributed to the operative process rather than to the plate shape itself. The surgeon was satisfied with the 6 fits and noted that the 3D-printed plates were less stiff than the commercial plates he usually uses, making them harder to manipulate.

3.3   Evaluations with 3D-Printed Bones

To further evaluate our design, we generated 7 bones $\{\mathbf{B}_i\}_{i\in[4;10]}$. $\mathbf{B}_7$ is the mean bone shape and the 6 others are generated by uniformly sampling the bone PCA space $\mathcal{B}$ ($\beta \in \{-2, 2\}$) in the first 3 PCA dimensions. For each bone, we 3D printed its custom plate, $\mathbf{P}_C$, and the best set plate, $\mathbf{P}_S$, from the plate-set.

First, we presented pairs of bones and plates to the surgeon and asked whether they were fit for surgical use. Then, we asked them to position the plates on the 3D bones to evaluate our positioning strategy. Last, to evaluate the adequacy of a plate set, we presented one bone to the surgeon and asked them to choose the best plate from the set.

Ready-for-Surgery Evaluation. For each bone $\mathbf{B}_i$, the surgeon considered both plates as candidates and evaluated if their shape was suitable for surgery. The surgeon stated that 100% of the plates were fit for surgery and that the custom plates $\mathbf{P}_C$ were always clearly better fitting than the plates $\mathbf{P}_S$ from the set.

Positioning Evaluation. To evaluate the plate positioning computed by our algorithm, we compare it to the surgeon's positioning. We asked the surgeon to position the set plates $\mathbf{P}_S$ on each bone, secured them with blue tack and scanned the whole with a FARO hand scanner. Figure 4 right shows the overlay of both positions of the used plate for each bone. For most bones, the placement of the plate by the surgeon and the algorithm were very similar. The differences can be explained as there are multiple plate positions that fit. Both placements (orange and green), even in the most dissimilar case (first bone in Fig. 4 right) were considered correct by the surgeon.

Choosing from the Plate Set. Our algorithm iteratively positions the plates $\mathbf{P}_S^i$ from $i = 0$ to $i = N$ and picks the best-fitting plate, so we asked the surgeon to perform a similar trial-and-error task to evaluate the relevance of our plate set.

Out of the 7 bones, the surgeon only chose the same plate as our algorithm for two of them ($\mathbf{B}_7$ and $\mathbf{B}_{10}$). For the other cases, the surgeon compared his selection to the algorithms' choice and confirmed that both plates were fit for surgery: different plates with slightly different placements can work for the same bone. Most interestingly, some plates selected by the surgeon did not fit the bone according to the strict numeric fit criteria. We discuss this finding in the next section. The surgeon noted that the plates of our set were always preferred over the state-of-the-art plates.

4   Conclusions, Take-Aways and Future Work

Conclusion. We propose to automatically generate a custom humerus plate that specifically matches the 3D shape of a bone and specifically meets the requirements of a surgeon for minimally invasive surgery. We also generate a set of plate shapes that accommodates a given bone population. We extensively evaluate our approach on cadaverous arms and 3D printed bones.

Experimental Takeaways. All the proposed plates (individual and set plate) match the shape of the bones, while the individual ones are considered the best by the surgeon. Furthermore, our experiments show the importance of evaluating on 3D-printed bones and ex-vivo arms. With the former, the surgeon can easily assess the plate-to-bone fit. With the latter, the insertion procedure can affect the plate placement. Moreover, our results on 3D printed bones argue for a relaxation of the *theoretical* fitting constraints: the plate-to-bone distance could be higher in some areas while still obtaining a proper surgical fit. Loosening the tolerances would allow more bones to be fit with the same plates and potentially the number of plates required in the plate set could be further reduced.

Future Work. To reach actual clinical use, two elements need consideration: (i) the current method takes as input the shape of a healthy bone, whereas patients have a fracture. One could perform a CT scan of the other arm and generate the symmetric bone or leverage methods that reconstruct a full bone from partial observations [18]. (ii) Our work focuses on the geometric design of the plate but does not consider the physical properties of the plate, such as thickness and stiffness. Future work should optimize these physical properties for the stresses that the plate must endure, e.g. using finite element methods.

Acknowledgements. This work was supported by the project CAMed (COMET K- Project 871132) which is funded by the Austrian Federal Ministry of Transport, Innovation and Technology (BMVIT) and the Austrian Federal Ministry for Digital and Economic Affairs (BMDW) and the Styrian Business Promotion Agency (SFG). Michael J. Black (MJB) has received research gift funds from Adobe, Intel, Nvidia, Meta/Facebook, and Amazon. MJB has financial interests in Amazon, Datagen Technologies, and Meshcapade GmbH. MJB's research was performed solely at MPI. We thank Karoline Seibert at Hofer GmbH & Co KG, Fürstenfeld, Austria for the 3D printing of bones and plates.

References

1. White, T.D., Folkens, P.A.: The human bone manual. Elsevier (2005)
2. Goyal, K.S., Skalak, A.S., Marcus, R.E., Vallier, H.A., Cooperman, D.R.: Analysis of anatomic periarticular tibial plate fit on normal adults. Clin. Orthop. Relat. Res. **1976–2007**(461), 245–257 (2007)
3. Schmutz, B., Rathnayaka, K., Albrecht, T.: Anatomical fitting of a plate shape directly derived from a 3D statistical bone model of the tibia. J. Clin. Orthopaed. Trauma **10**, S236–S241 (2019)
4. Kozic, N., et al.: Optimisation of orthopaedic implant design using statistical shape space analysis based on level sets. Med. Image Anal. **14**(3), 265–275 (2010)
5. Hwang, J.H., Oh, J.K., Oh, C.W., Yoon, Y.C., Choi, H.W.: Mismatch of anatomically pre-shaped locking plate on Asian femurs could lead to malalignment in the minimally invasive plating of distal femoral fractures: a cadaveric study. Arch. Orthop. Trauma Surg. **132**(1), 51–56 (2012)
6. Bou-Sleiman, H., Ritacco, L.E., Nolte, L.-P., Reyes, M.: Minimization of intraoperative shaping of Orthopaedic fixation plates: a population-based design. In: Fichtinger, G., Martel, A., Peters, T. (eds.) MICCAI 2011. LNCS, vol. 6892, pp. 409–416. Springer, Heidelberg (2011). https://doi.org/10.1007/978-3-642-23629-7_50
7. Schmutz, B., Wullschleger, M.E., Kim, H., Noser, H., Schütz, M.A.: Fit assessment of anatomic plates for the distal medial tibia. J. Orthop. Trauma **22**(4), 258–263 (2008)
8. Schmutz, B., Wullschleger, M.E., Noser, H., Barry, M., Meek, J., Schütz, M.A.: Fit optimisation of a distal medial tibia plate. Comput. Methods Biomech. Biomed. Engin. **14**(04), 359–364 (2011)

9. Harith, H., Schmutz, B., Malekani, J., Schuetz, M.A., Yarlagadda, P.K.: Can we safely deform a plate to fit every bone? Population-based fit assessment and finite element deformation of a distal tibial plate. Med. Eng. Phys. **38**(3), 280–285 (2016)

10. Carrillo, F., Vlachopoulos, L., Schweizer, A., Nagy, L., Snedeker, J., Fürnstahl, P.: A time saver: optimization approach for the fully automatic 3D planning of forearm osteotomies. In: Descoteaux, M., Maier-Hein, L., Franz, A., Jannin, P., Collins, D.L., Duchesne, S. (eds.) MICCAI 2017. LNCS, vol. 10434, pp. 488–496. Springer, Cham (2017). https://doi.org/10.1007/978-3-319-66185-8_55

11. Wu, X., et al.: Preliminary exploration of a quantitative assessment index for the matching performance of anatomical bone plates using computer. J. Orthop. Surg. Res. **14**(1), 1–8 (2019)

12. Tkany, L., Hofstätter, B., Petersik, A., Miehling, J., Wartzack, S., Sesselmann, S.: New design process for anatomically enhanced osteosynthesis plates. J. Orthop. Res. **37**(7), 1508–1517 (2019)

13. Zenker, M., et al.: Quantifying osteosynthesis plate prominence-mathematical definitions and case study on a clavicle plate. Comput. Methods Biomech. Biomed. Eng. **25**, 1–10 (2022)

14. Schulz, A.P., et al.: Evidence based development of a novel lateral fibula plate (variAx fibula) using a real CT bone data based optimization process during device development. Open Orthop. J. **6**, 1–7 (2012)

15. Gill, D.R., Torchia, M.E.: The spiral compression plate for proximal humeral shaft nonunion: a case report and description of a new technique. J. Orthop. Trauma **13**(2), 141–144 (1999)

16. Da Silva, T., Rummel, F., Knop, C., Merkle, T.: Comparing iatrogenic radial nerve lesions in humeral shaft fractures treated with helical or straight philos plates: a 10-year retrospective cohort study of 62 cases. Arch. Orthop. Trauma Surg. **140**(12), 1931–1937 (2020)

17. Petersik, A., et al.: A numeric approach for anatomic plate design. Injury **49**, S96–S101 (2018)

18. Meng, D., Keller, M., Boyer, E., Black, M., Pujades, S.: Learning a statistical full spine model from partial observations. In: Reuter, M., Wachinger, C., Lombaert, H., Paniagua, B., Goksel, O., Rekik, I. (eds.) ShapeMI 2020. LNCS, vol. 12474, pp. 122–133. Springer, Cham (2020). https://doi.org/10.1007/978-3-030-61056-2_10

19. Eisenberger, M., Lahner, Z., Cremers, D.: Smooth shells: multi-scale shape registration with functional maps. In: Proceedings of the IEEE/CVF Conference on Computer Vision and Pattern Recognition, pp. 12265–12274 (2020)

20. Ahmad, M., Nanda, R., Bajwa, A., Candal-Couto, J., Green, S., Hui, A.: Biomechanical testing of the locking compression plate: when does the distance between bone and implant significantly reduce construct stability? Injury **38**(3), 358–364 (2007)

MSKdeX: Musculoskeletal (MSK) Decomposition from an X-Ray Image for Fine-Grained Estimation of Lean Muscle Mass and Muscle Volume

Yi Gu[1,2](✉), Yoshito Otake[1], Keisuke Uemura[3], Masaki Takao[4], Mazen Soufi[1], Yuta Hiasa[1], Hugues Talbot[2], Seiji Okada[3], Nobuhiko Sugano[5], and Yoshinobu Sato[1]

[1] Division of Information Science, Graduate School of Science and Technology, Nara Institute of Science and Technology, Ikoma, Japan
{gu.yi.gu4,otake,yoshi}@is.naist.jp
[2] CentraleSupélec, Université Paris-Saclay, Paris, France
[3] Department of Orthopaedics, Osaka University Graduate School of Medicine, Suita, Japan
[4] Department of Bone and Joint Surgery, Ehime University Graduate School of Medicine, Toon, Japan
[5] Department of Orthopaedic Medical Engineering, Osaka University Graduate School of Medicine, Suita, Japan

Abstract. Musculoskeletal diseases such as sarcopenia and osteoporosis are major obstacles to health during aging. Although dual-energy X-ray absorptiometry (DXA) and computed tomography (CT) can be used to evaluate musculoskeletal conditions, frequent monitoring is difficult due to the cost and accessibility (as well as high radiation exposure in the case of CT). We propose a method (named MSKdeX) to estimate fine-grained muscle properties from a plain X-ray image, a low-cost, low-radiation, and highly accessible imaging modality, through musculoskeletal decomposition leveraging fine-grained segmentation in CT. We train a multi-channel quantitative image translation model to decompose an X-ray image into projections of CT of individual muscles to infer the lean muscle mass and muscle volume. We propose the object-wise intensity-sum loss, a simple yet surprisingly effective metric invariant to muscle deformation and projection direction, utilizing information in CT and X-ray images collected from the *same* patient. While our method is basically an unpaired image-to-image translation, we also exploit the nature of the bone's rigidity, which provides the paired data through 2D-3D rigid registration, adding strong pixel-wise supervision in unpaired training. Through the evaluation using a 539-patient dataset, we showed that the proposed method significantly outperformed conventional methods. The average Pearson correlation coefficient between the predicted and

---

Supplementary Information The online version contains supplementary material available at https://doi.org/10.1007/978-3-031-43990-2_47.

CT-derived ground truth metrics was increased from 0.460 to 0.863. We believe our method opened up a new musculoskeletal diagnosis method and has the potential to be extended to broader applications in multi-channel quantitative image translation tasks.

Keywords: Muscles · Radiography · Generative adversarial networks (GAN) · Sarcopenia · Image-to-image translation

1   Introduction

Sarcopenia is a prevalent musculoskeletal disease characterized by the inevitable loss of skeletal muscle, causing increased risks of all-cause mortality and disability that result in heavy healthcare costs [1–6]. Measuring body composition, such as lean muscle mass (excluding fat contents), is essential for diagnosing musculoskeletal diseases, where dual-energy X-ray absorptiometry (DXA) [7,8] and computed tomography (CT) [9–11] are often used. However, DXA and CT require special equipment that is much less accessible in a small clinic. Furthermore, CT requires high radiation exposure, and DXA allows the measurement of only overall body composition, which lacks details in individual muscles such as the iliacus muscle, which overlays with the gluteus maximus muscle in DXA images. Although several recent works used X-ray images for bone mineral density (BMD) estimation and osteoporosis diagnosis [12–15], only a few works estimated muscle metrics and sarcopenia diagnosis [16,17], and the deep learning technology used is old. Recently, BMD-GAN [15] was proposed for estimating BMD through X-ray image decomposition using X-ray and CT images aligned by 2D-3D registration. However, they did not target muscles. Nakanishi et al. [17] proposed an X-ray image decomposition for individual muscles. However, they calculated only the affected and unaffected muscle volumes ratio without considering the absolute volume and lean mass, which is more relevant to sarcopenia diagnosis.

In this study, we propose **MSKdeX**: Musculoskeletal (**MSK**) **de**composition from a plain **X**-ray image for the fine-grained estimation of lean muscle mass and volume of each individual muscle, which are useful metrics for evaluating muscle diseases including sarcopenia. Figure 1 illustrates the meaning of our fine-grained muscle analysis and its challenges. The contribution of this paper is three-fold: 1) proposal of the *object-wise intensity-sum* (OWIS) loss, a simple yet effective metric invariant to muscle deformation and projection direction, for quantitative learning of the absolute volume and lean mass of the muscles, 2) proposal of partially aligned training utilizing the aligned (paired) dataset for the rigid object for the pixel-wise supervision in an unpaired image translation task, 3) extensive evaluation of the performance using a 539-patient dataset.

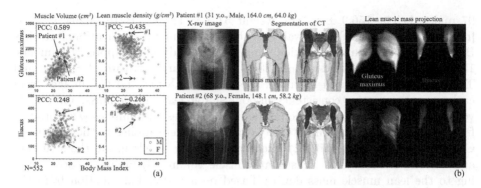

Fig. 1. Variations in the muscle volume and lean muscle density among the 552 patients in our dataset. (a) Relationships of muscle volume and lean muscle density with respect to body mass index (BMI). Moderate and weak correlations of BMI were observed with muscle volume and lean muscle density, respectively, in the gluteus maximus (Glu. max.), while little correlations were observed in the iliacus. (b) Visualization of two representative cases. Patient #1 (young, male) and Patient #2 (old, female) had similar BMI and almost the same gluteus maximus volume, while the lean muscle mass was significantly different, likely due to the fatty degeneration in Patient #2, which was clearly observable in the projections of the lean muscle mass volume.

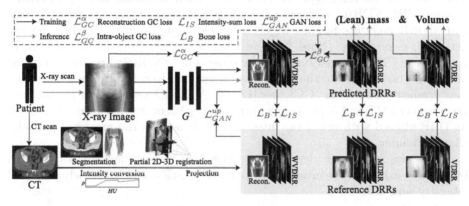

Fig. 2. Overview of the proposed MSKdeX. Three types of object-wise DRRs (of segmented individual muscle/bone regions) were obtained from CT through segmentation [18], intensity conversion [19,20], 2D-3D registration for bones [21], and projection, embedding information of volume and mass. A decomposition model was trained using GAN loss and proposed GC loss chain, OWIS loss, and bone loss to decompose an X-ray image into DRRs whose intensity sum derives the metric of volume and mass.

2   Method

2.1   Dataset Preparation

Figure 2 illustrates the overview of the proposed MSKdeX. We collected a dataset of 552 patients subject to the total hip arthroplasty surgery (455 females and

97 males, height 156.9 ± 8.3 cm, weight 57.5 ± 11.9 kg, BMI 23.294 ± 3.951 [mean ± std]). Ethical approval was obtained from the Institutional Review Boards of the institutions participating in this study (IRB approval numbers: 15056-3 for Osaka University and 2019-M-6 for Nara Institute of Science and Technology). We acquired a pair of pre-operative X-ray and CT images from each patient, assuming consistency in bone shape, lean muscle mass, and muscle volume. Automated segmentation of individual bones and muscles was obtained from CT [18]. Three different intensity conversions were applied to the segmented CT; 1) the original intensity, 2) intensity of 1.0 for voxels inside the structure and 0.0 for voxels outside to estimate muscle volume, 3) intensity corresponding to the lean muscle mass density based on a conversion function from the Hounsfield unit (HU) to the mass density [19,20] to estimate lean muscle mass. Following [19], we assumed the voxels with less than –30 HU consisted of the fat, more than +30 HU consisted of the lean muscle, and the voxels in between –30 to +30 HU contained the fat and lean muscle with the ratio depending on linear interpolation of the HU value. The mass of the lean muscle was calculated by the conversion function proposed in [20]. Then, object-wise DRRs for the three conversions were generated for each segmented individual object (bone/muscle) region. (Note: When we refer to a DRR in this paper, it is object-wise.) We call the three types of the DRRs weighted volume DRR (WVDRR), volume DRR (VDRR), and mass DRR (MDRR). The intensity sum of VDRR and MDRR amounts to each object's muscle volume (cm^3) and lean muscle mass (g), respectively. (Note: "Muscle volume" includes the fat in addition to lean muscle.) The summation of all the objects of WVDRRs becomes an image with a contrast similar to the real X-ray image used to calculate the reconstruction gradient correlation (GC) loss [17,22]. A 2D-3D registration [21] of each bone between CT and X-ray image of the same patient was performed to obtain its DRR aligned with the X-ray image, which is used in the proposed partially aligned training.

Since muscles deform depending on the joint angle, they are not aligned. Instead, we exploited the invariant property of muscles using the newly proposed intensity-sum loss.

2.2   Model Training

We train a decomposition model G to decompose an X-ray image into the $DRRs = \{VDRR, MDRR, WVDRR\}$ to infer the lean muscle mass and muscle volume, adopting CycleGAN [23]. The model backbone is replaced with HRNet [24]. The GAN loss $\mathcal{L}_{GAN}^{up}$ we use is formulated in supplemental materials.

Structural Consistency. We call the summation of a DRR over all the channels (objects) the virtual X-ray image defined as $I^{VX} = V(I^{DRR}) = \sum_i I_i^{DRR}$, where I_i^{DRR} is the i-th object image of a DRR. We applies reconstruction GC loss [17] $\mathcal{L}_{GC}^{\alpha}$ defined as

$$\mathcal{L}_{GC}^{\alpha}(G) = \mathbb{E}_{I^X} - GC(I^X, V(G(I^X)^{WVDRR})) \tag{1}$$

to maintain the structure consistency between an X-ray image and decomposed DRR, where $G(I^X)^{WVDRR}$ is the decomposed WVDRR. However, we do not apply reconstruction GC loss for VDRR and MDRR because of lacking attenuation coefficient information. Instead, we propose inter-DRR/intra-object GC loss $\mathcal{L}^\beta_{GC}$ defined as

$$
\mathcal{L}^\beta_{GC}(G) = \mathbb{E}_{I^X} - \frac{1}{N} \sum_i \Big[GC(st(G(I^X)^{WVDRR}_i), G(I^X)^{VDRR}_i)
$$
$$
+ GC(st(G(I^X)^{WVDRR}_i), G(I^X)^{MDRR}_i) \Big] \tag{2}
$$

to chain the structural constraints from WVDRR to VDRR and MDRR, where the $G(I^X)^{WVDRR}_i$, $G(I^X)^{VDRR}_i$, and $G(I^X)^{MDRR}_i$ are i-th object image of the decomposed WVDRR, VDRR, and MDRR, respectively. The $st(\cdot)$ operator stops the gradient from being back-propagated in which the decomposed VDRR and MDRR are expected to be structurally closer to WVDRR (not vice-versa) to stabilize training. Thus, our structural consistency constant $\mathcal{L}^{up}_{GC}$ is defined as $\mathcal{L}^{up}_{GC} = \lambda_{gca} \mathcal{L}^\alpha_{GC}(G) + \mathcal{L}^\beta_{GC}(G)$ where the λ_{gca} balances the two GC losses.

Intensity Sum Consistency. Unlike general images, our DRRs embedded specific information so that the intensity sum represents physical metrics (mass and volume). Furthermore, the conventional method did not utilize the paired information of an X-ray image and DRR (obtained from the same patient). We took advantage of the paired information, proposing the *object-wise intensity-sum* loss, a simple yet effective metric invariant to patient pose and projection direction, for quantitative learning. The OWIS loss $\mathcal{L}_{IS}$ is defined as:

$$
\mathcal{L}_{IS}(DRR) = \mathbb{E}_{(I^X, I^{DRR})} \frac{1}{NHW} \sum_i \Big| S(G(I^X)^{DRR}_i) - S(I^{DRR}_i) \Big|, \tag{3}
$$

where I^{DRR}_i and $S(\cdot)$ are the i-th object image of DRR and the intensity summation operator (sum over the intensity of an image), respectively. The H and W are the image height and weight, respectively, served as temperatures for numeric stabilizability. The intensity consistency objective $\mathcal{L}^{all}_{IS}$ is defined as $\mathcal{L}^{all}_{IS} = \mathcal{L}_{IS}(WVDRR) + \mathcal{L}_{IS}(VDRR) + \mathcal{L}_{IS}(MDRR)$.

Partially Aligned Training. A previous study [15] suggested that supervision by the aligned (paired) data can improve the quantitative translation. Therefore, we incorporated 2D-3D registration [21] to align the pelvis and femur DRRs with the paired X-ray images for partially aligned training to improve overall performance, including muscle metrics estimation. We applied $L1$ and GC loss to maintain quantitative and structural consistencies, respectively. However, we preclude using GAN loss and *feature matching* loss to avoid the training burden by additional discriminators. The paired bone loss for a DRR is defined as

$$\mathcal{L}_B(DRR) = \mathbb{E}_{(I^X, I^{DRR})} \frac{1}{N_b} \sum_{i \in K} \Big[\lambda_{l1} \left\| G(I^X)_i^{DRR} - I_i^{DRR} \right\|_1$$
$$-GC(G(I^X)_i^{DRR}, I_i^{DRR}) \Big], \tag{4}$$

where the K is a set of indexes containing aligned bone indexes. The N_b is the size of the set K. The λ_{l1} tries to balance structural faithfulness and quantitative accuracy. The objective of partially aligned pixel-wise learning is defined as $\mathcal{L}_B^{all} = \mathcal{L}_B(WVDRR) + \mathcal{L}_B(VDRR) + \mathcal{L}_B(MDRR)$.

Full Objective. The full objective, aiming for realistic decomposition while maintaining structural faithfulness and quantitative accuracy, is defined as

$$\mathcal{L} = \min_{G,F} \max_{D^X, D^{DRRs}} \left(\mathcal{L}_{GAN}^{up} + \mathcal{L}_{GC} + \lambda_{is}\mathcal{L}_{IS}^{all} + \mathcal{L}_B^{all} \right), \tag{5}$$

where the λ_{is} re-weights the penalty on the proposed OWIS loss.

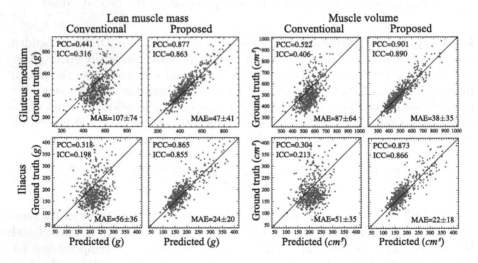

Fig. 3. Lean muscle mass (left) and muscle volume (right) estimation results by the conventional and proposed methods for the gluteus medius (top) and iliacus (bottom).

3   Experiments and Results

The automatic segmentation results of 552 CTs were visually verified, and 13 cases with severe segmentation failures were omitted from our analysis, resulting in 539 CTs. Four-fold cross-validation was performed, i.e., 404 or 405 training data and 134 or 135 test data per fold. The baseline of our experiment was the vanilla CycleGAN with the reconstruction GC loss proposed in [17]. We evaluated the predicted lean muscle mass and volume using the ground truth derived

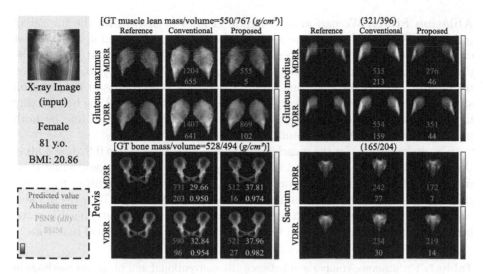

Fig. 4. Visualization of decomposition results of the MDRR and VDRR for gluteus maximus, gluteus medius, pelvis, and sacrum on a representative case.

from 3D CT images with three metrics, Pearson correlation coefficient (PCC), intra-class correlation coefficient (ICC), and mean absolute error (MAE). Additionally, we evaluated the image quality of predicted DRRs of the bones by comparing them with the aligned DRRs using peak-signal-noise-ratio (PSNR) and structural similarity index measure (SSIM). Implementation details are described in supplemental materials.

Figure 3 shows the prediction results on the gluteus medius and iliacus muscles. The conventional method (without using the intensity constraints) resulted in low PCCs of 0.441 and 0.522 for the lean muscle mass and muscle volume estimations, respectively, for the gluteus medius, and 0.318 and 0.304, respectively, for the iliacus. Significant improvements by the proposed method were observed, achieving high PCCs of 0.877 and 0.901 of the lean muscle mass and muscle volume estimations, respectively, for the gluteus medius, and 0.865 and 0.873, respectively, for the iliacus. Figure 4 visualized the decomposed VDRR and MDRR of four objects of a representative case. Our method (MSKdeX) reduced the hallucinating features in the decomposed DRRs by the proposed losses. The overall intensity of the conventional method was clearly different from the reference, while the proposed method decomposed the X-ray image considering the structural faithfulness and quantitative accuracy, outperforming the conventional method significantly. Table 1 shows evaluation for other objects. Statistical test (one-way ANOVA) was performed on the conventional and proposed methods using prediction absolute error, where the differences are significant ($p < 0.001$) for all the objects. More detailed results and a visualization video can be found in supplemental materials.

Ablation Study. We performed ablation studies to investigate the impact of proposed OWIS loss and the use of aligned bones using 404 training and 135 test data. The re-weighting parameter λ_{is} of 0, 10, and 1000 with and without the partially aligned training $\mathcal{L}_B$ was tested. The λ_{is} of 0 without partially aligned training [$\lambda_{is} = 0$ (False)] is considered our baseline. The results of the ablation study were summarized in Table 2, where the bold font indicated the best setting in a column. We observed significant improvements from the baseline by both proposed features, OWIS loss and partially aligned training $\mathcal{L}_B$.

The average PCC for the muscles was improved from 0.457 to 0.826 by adding the OWIS loss ($\lambda_{is} = 100$) and to 0.796 by adding the bone loss, while their combination achieved the best average PCC of 0.855, demonstrating the superior ability of quantitative learning of the proposed MSKdeX. The results also suggested that the weight balance for loss terms needs to be made to achieve the best performance. More detailed results are shown in supplemental materials.

Table 1. Performance comparison between the conventional and proposed methods in a cross-validation study using 539 data.

Lean muscle mass and Bone mass estimation accuracy (PCC)								
Method	Glu. max.	Glu. med.	Glu. min.	Ilia.	Obt. ext.	Pec.	Pelv.	Sac.
Conv.	.417	.441	.415	.318	.416	.457	.612	.478
Prop.	**.831**	**.877**	**.864**	**.865**	**.825**	**.832**	**.950**	**.878**
Muscle volume and bone volume estimation accuracy (PCC)								
Method	Glu. max.	Glu. med.	Glu. min.	Ilia.	Obt. ext.	Pec.	Pelv.	Sac.
Conv.	.490	.522	.463	.304	.402	.460	.586	.538
Prop.	**.882**	**.901**	**.890**	**.873**	**.864**	**.849**	**.956**	**.873**
Bone decomposition accuracy [mean PSNR(mean SSIM)]								
Method	MDRR Pelv., Fem.		VDRR Pelv., Fem.			MVDRR Pelv., Fem.		
Conv.	31.5(.940), 31.7(.962)		31.7(.940), 32.4(.963)			34.5(.956), 31.6(.971)		
Prop.	**37.1(.978), 37.8(.987)**		**37.7(.982), 39.5(.991)**			**39.0(.980), 37.1(.987)**		

Table 2. Summary of the ablation study for 135 test data.

λ_{is}	With $\mathcal{L}_B$	Lean muscle mass and bone mass estimation accuracy (PCC)							
		Glu. max.	Glu. med.	Glu. min.	Ilia.	Obt. ext.	Pec.	Pelv.	Sac.
0	(False)	.415	.419	.469	.368	.473	.600	.542	.265
0	(True)	.734	.799	.788	.855	.784	.813	**.954**	.798
100	(False)	.799	.815	.829	.854	.815	.842	.925	.774
100	(True)	**.854**	**.857**	**.837**	**.883**	**.854**	**.846**	.947	**.898**
1000	(False)	.798	.765	.770	.839	.704	.840	.870	.767
1000	(True)	.795	.776	.767	.828	.748	.820	.929	.745

4   Summary

We proposed MSKdeX, a method for fine-grained estimation of the lean muscle mass and volume from a plain X-ray image (2D) through the musculoskeletal decomposition, which, in fact, recovers CT (3D) information. Our method decomposes an X-ray image into DRRs of objects to infer the lean muscle mass and volume considering the structural faithfulness (by the gradient correlation loss chain) and quantitative accuracy (by the object-wise intensity-sum loss and aligned bones training), outperforming the conventional method by a large margin as shown in Sect. 3. The results suggested a high potential of MSKdeX for opportunistic screening of musculoskeletal diseases in routine clinical practice, providing a new approach to accurately monitoring musculoskeletal health. The aligned bone DRRs positively affected the quantification of the density and volume of the muscles as shown in the ablation study in Sect. 3, implying the deep connection between muscles and bones. The prediction of muscles overlapped with the pelvis in the X-ray image can leverage the strong pixel-wise supervision by the aligned pelvis's DRR, which can be considered as a type of *calibration*. Our future works are the validation with a large-scale dataset and extension to the decomposition into a larger number of objects.

Acknowledgement. The research in this paper was funded by MEXT/JSPS KAKENHI (19H01176, 20H04550, 21K16655).

Code Availability. The code is available from the authors ({gu.yi.gu4,otake, yoshi}@is.naist.jp) upon reasonable request for research activity.

References

1. Kitamura, A., Seino, S., Takumi, A., et al.: Sarcopenia: prevalence, associated factors, and the risk of mortality and disability in Japanese older adults. J. Cachexia Sarcopenia Muscle **12**(1), 30–38 (2021). https://doi.org/10.1002/jcsm.12651
2. Chen, L.-K., Lee, W.-J., Peng, L.-N., Liu, L.-K., et al.: Recent advances in sarcopenia research in Asia: 2016 update from the Asian Working Group for Sarcopenia. JAMDA **17**(8), 767.e1-767.e7 (2016). https://doi.org/10.1016/j.jamda.2016.05.016
3. Marzetti, E., Calvani, R., Tosato, M., Cesari, M., et al.: Sacopenia: an overview. Aging Clin. Exp. Res. **29**, 11–17 (2017). https://doi.org/10.1007/s40520-016-0704-5
4. Petermann-Rocha, F., Balntze, V., Gray, S.R., et al.: Global prevalence of sarcopenia and severe sarcopenia: a systematic review and meta-analysis. J. Cachexia Sarcopenia Muscle **13**(1), 86–99 (2022). https://doi.org/10.1002/jcsm.12783
5. Shu, X., Lin, T., Wang, H., Zhao, Y., et al.: Diagnosis, prevalence, and mortality of sarcopenia in dialysis patients: a systematic review and meta-analysis. J. Cachexia Sarcopenia Muscle **13**(1), 145–158 (2022). https://doi.org/10.1002/jcsm.12890
6. Edwards, M.H., Dennision, E.M., Sayer, A.A., et al.: Osteoporosis and sarcopenia in older age. Bone **80**, 126–130 (2015). https://doi.org/10.1016/j.bone.2015.04.016
7. Shepherd, J., Ng, B., Sommer, M., Heymsfield, S.B.: Body composition by DXA. Bone **104**, 101–105 (2017). https://doi.org/10.1016/j.bone.2017.06.010

8. Nana, A., Slater, G.J., Stewart, A.D., Burke, L.M.: Methodology review: using dual-energy X-ray absorptiometry (DXA) for the assessment of body composition in athletes and active people. Int. J. Sport Nutr. Exerc. Metab. **25**(2), 198–215 (2015). https://doi.org/10.1123/ijsnem.2013-0228

9. Feliciano, E.M.C., et al.: Evaluation of automated computed tomography segmentation to assess body composition and mortality associations in cancer patients. J. Cachexia Sarcopenia Muscle **11**(5), 1258–1269 (2020). https://doi.org/10.1002/jcsm.12573

10. Paris, M.T., Tandon, P., Heyland, D.K., et al.: Automated body composition analysis of clinically acquired computed tomography scans using neural networks. Clin. Nutr. **39**(10), 3049–3055 (2020). https://doi.org/10.1016/j.clnu.2020.01.008

11. Ogawa, T., Takao, M., et al.: Validation study of the CT-based cross-sectional evaluation of muscular atrophy and fatty degeneration around the pelvis and the femur. J. Orthop. **25**(1), 139–144 (2020). https://doi.org/10.1016/j.jos.2019.02.004

12. Hsieh, C.-I., Zheng, K., Lin, C., Mei, L., et al.: Automated bone mineral density prediction and fracture risk assessment using plain radiographs via deep learning. Nat. Commun. **12**(1), 5472 (2021). https://doi.org/10.1038/s41467-021-25779-x

13. Wang, F., Zheng, K., Lu, Le, et al.: Lumbar bone mineral density estimation from chest X-ray images: anatomy-aware attentive multi-ROI modeling. IEEE Trans. Med. Imaging. **42**(1), 257–267 (2023). https://doi.org/10.1109/TMI.2022.3209648

14. Ho, C.-S., Chen, Y.-P., Fan, T.-Y., Kuo, C.-F., et al.: Application of deep learning neural network in predicting bone mineral density from plain X-ray radiography. Arch. Osteoporosis **16**(1), 153 (2021). https://doi.org/10.1007/s11657-021-00985-8

15. Gu, Y., et al.: BMD-GAN: bone mineral density estimation using X-Ray image decomposition into projections of bone-segmented quantitative computed tomography using hierarchical learning. In: Wang, L., Dou, Q., Fletcher, P.T., Speidel, S., Li, S. (eds.) Medical Image Computing and Computer Assisted Intervention – MICCAI 2022. MICCAI 2022. LNCS, vol. 13436, pp. 644–654. Springer, Cham (2022). https://doi.org/10.1007/978-3-031-16446-0_61

16. Ryu, J., et al.: Chest X-ray-based opportunistic screening of sarcopenia using deep learning. J. Cachexia Sarcopenia Muscle **14**(1), 418–428 (2023). https://doi.org/10.1002/jcsm.13144

17. Nakanishi, N., Otake, Y., Hiasa, Y., Gu, Y., Uemura, K., et al.: Decomposition of musculoskeletal structures from radiography using an improved CycleGAN framework. Sci. Rep. **13**, 8482 (2023). https://doi.org/10.1038/s41598-023-35075-x

18. Hiasa, Y., Otake, Y., et al.: Automated muscle segmentation from clinical CT using Bayesian U-Net for personalized musculoskeletal modeling. IEEE Trans. Med. Imaging **39**(4), 1030–1040 (2019). https://doi.org/10.1109/TMI.2019.2940555

19. Aubrey, J., Esfandiari, N., Baracos, V.E., Buteau, F.A., et al.: Measurement of skeletal muscle radiation attenuation and basis of its biological variation. Acta Physiol. (Oxf) **210**(3), 489–497 (2014). https://doi.org/10.1111/apha.12224

20. Wilfried, S., Thomas, B., Wolfgang, S.: Correlation between CT numbers and tissue parameters needed for Monte Carlo simulations of clinical dose distributions. Phys. Med. Biol. **45**(2), 459 (2000). https://doi.org/10.1088/0031-9155/45/2/314

21. Otake, Y., et al.: Intraoperative image-based multiview 2D/3D registration for image-guided orthopaedic surgery: incorporation of fiducial-based C-arm tracking and GPU-acceleration. IEEE Trans. Med. Imaging **31**(4), 948–962 (2012)

22. Hiasa, Y., et al.: Cross-modality image synthesis from unpaired data using Cycle-GAN. In: Gooya, A., Goksel, O., Oguz, I., Burgos, N. (eds.) SASHIMI 2018. LNCS, vol. 11037, pp. 31–41. Springer, Cham (2018). https://doi.org/10.1007/978-3-030-00536-8_4
23. Zhu, J.-Y., Park, T., Isola, P., Efros, A.A.: Unpaired image-to-image translation using cycle-consistent adversarial networks. In: ICCV 2017, pp. 2242–2251 (2017)
24. Wang, J., et al.: Deep high-resolution representation learning for visual recognition. IEEE TPAMI **43**(10), 3349–3364 (2021). https://doi.org/10.1109/TPAMI.2020.2983686

Clinical Applications – Oncology

Clinical Applications – Oncology

Second-Course Esophageal Gross Tumor Volume Segmentation in CT with Prior Anatomical and Radiotherapy Information

Yihua Sun[1], Hee Guan Khor[1], Sijuan Huang[2], Qi Chen[3], Shaobin Wang[1,3], Xin Yang[2(✉)], and Hongen Liao[1(✉)]

[1] Department of Biomedical Engineering, School of Medicine, Tsinghua University, Beijing, China
wsb20@mails.tsinghua.edu.cn, liao@tsinghua.edu.cn
[2] Sun Yat-Sen University Cancer Center, State Key Laboratory of Oncology in South China, Collaborative Innovation Center for Cancer Medicine, Guangdong Key Laboratory of Nasopharyngeal Carcinoma Diagnosis and Therapy, Guangzhou, China
yangxin@sysucc.org.cn
[3] MedMind Technology Co., Ltd., Beijing, China

Abstract. Esophageal cancer is a significant global health concern, and radiotherapy (RT) is a common treatment option. Accurate delineation of the gross tumor volume (GTV) is essential for optimal treatment outcomes. In clinical practice, patients may undergo a second round of RT to achieve complete tumor control when the first course of treatment fails to eradicate cancer completely. However, manual delineation is labor-intensive, and automatic segmentation of esophageal GTV is difficult due to the ambiguous boundary of the tumor. Detailed tumor information naturally exists in the previous stage, however the correlation between the first and second course RT is rarely explored. In this study, we first reveal the domain gap between the first and second course RT, and aim to improve the accuracy of GTV delineation in the second course RT by incorporating prior information from the first course. We propose a novel prior **A**natomy and **RT** information enhanced **S**econd-course **E**sophageal **GTV** segmentation network (**ARTSEG**). A region-preserving attention module (RAM) is designed to understand the long-range prior knowledge of the esophageal structure, while preserving the regional patterns. Sparsely labeled medical images for various isolated tasks necessitate efficient utilization of knowledge from relevant datasets and tasks. To achieve this, we train our network in an information-querying manner. ARTSEG incorporates various prior knowledge, including: 1) Tumor volume variation between first and second RT courses, 2) Cancer cell proliferation, and 3) Reliance of GTV on esophageal anatomy. Extensive quantitative and qualitative experiments validate our designs.

H. Liao and X. Yang are the co-corresponding authors.

Supplementary Information The online version contains supplementary material available at https://doi.org/10.1007/978-3-031-43990-2_48.

Keywords: Second course radiotherapy · Esophageal gross tumor volume · Data efficient learning · Prior anatomical information · Attention

1   Introduction

Esophageal cancer is a significant contributor to cancer-related deaths globally [3,15]. One effective treatment option is radiotherapy (RT), which utilizes high-energy radiation to target cancerous cells [4]. To ensure optimal treatment outcomes, both the cancerous region and the adjacent organ-at-risk (OAR) must be accurately delineated, to focus the high-energy radiation solely on the cancerous area while protecting the OARs from any harm. Gross tumor volume (GTV) represents the area of the tumor that can be identified with a high degree of certainty and is of paramount importance in clinical practice.

In the clinical setting, patients may undergo a second round of RT treatment to achieve complete tumor control when initial treatment fails to completely eradicate cancer [16]. However, the precise delineation of the GTV is labor-intensive, and is restricted to specialized hospitals with highly skilled RT experts. The automatic identification of the esophagus presents inherent challenges due to its elongated soft structure and ambiguous boundaries between it and adjacent organs [12]. Moreover, the automatic delineation of the GTV in the esophagus poses a significant difficulty, primarily attributable to the low contrast between the esophageal GTV and the neighboring tissue, as well as the limited datasets.

Recently, advances in deep learning [21] have promoted research in automatic esophageal GTV segmentation from computed tomography (CT) [18,19]. Since the task is challenging, Jin et al. [9,10] improve the segmentation accuracy by incorporating additional information from paired positron emission tomography (PET). Nevertheless, such approaches require several imaging modalities, which can be both costly and time-consuming, while disregarding any knowledge from previous treatment or anatomical understanding. Moreover, the correlation between the first and second courses of RT is rarely investigated, where detailed prior tumor information naturally exists in the previous RT planning.

In this paper, we present a comprehensive study on accurate GTV delineation for the second course RT. We proposed a novel prior **A**natomy and **RT** information enhanced **S**econd-course **E**sophageal **G**TV segmentation network (**ARTSEG**). A region-preserving attention module (RAM) is designed to effectively capture the long-range prior knowledge in the esophageal structure, while preserving regional tumor patterns. To the best of our knowledge, we are the first to reveal the domain gap between the first and second courses for GTV segmentation, and explicitly leverage prior information from the first course to improve GTV segmentation performance in the second course.

The medical images are labeled sparsely, which are isolated by different tasks [20]. Meanwhile, an ideal method for automatic esophageal GTV segmentation in the second course of RT should consider three key aspects: 1) Changes in tumor volume after the first course of RT, 2) The proliferation of cancerous cells from a tumor to neighboring healthy cells, and 3) The anatomical-dependent

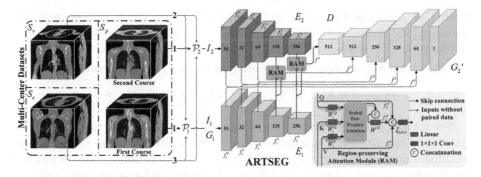

Fig. 1. Our training approach leverages multi-center datasets containing relevant annotations, that challenges the network to retrieve information from E_1 using the features from E_2. The decoder D utilizes the prior knowledge obtained from I_1 and G_1 to generate the mask prediction. Our training strategy leverages three datasets that introduce prior knowledge to the network of the following three key aspects: 1) Tumor volume variation, 2) Cancer cell proliferation, and 3) Reliance of GTV on esophageal anatomy.

nature of GTV on esophageal locations. To achieve this, we efficiently exploit knowledge from multi-center datasets that are not tailored for second-course GTV segmentation. Our training strategy does not specific to any tasks but challenges the network to retrieve information from another encoder with augmented inputs, which enables the network to learn from the above three aspects. Extensive quantitative and qualitative experiments validate our designs.

2 Network Architecture

In the first course of RT, a CT image denoted as I_1 is utilized to manually delineate the esophageal GTV, G_1. During the second course of RT, a CT image I_2 of the same patient is acquired. However, I_2 is not aligned with I_1 due to soft tissue movement and changes in tumor volume that occurred during the first course of treatment. Both images $I_{1/2}$ have the spatial shape of $H \times W \times D$.

Our objective is to predict the esophageal GTV G_2 of the second course. It would be advantageous to leverage insights from the first course, as it comprises comprehensive information pertaining to the tumor in its preceding phase. Therefore, the input to encoder E_1 consists of the concatenation of I_1 and G_1 to encode the prior information (features f_1^d) from the first course, while encoder E_2 embeds both low- and high-level features f_2^d of the local pattern of I_2 (Fig. 1),

$$f_1^d = E_1(I_1, G_1), \ f_2^d = E_2(I_2), \ d = 0, 1, 2, 3, 4 \tag{1}$$

where the spatial shape of $f_{1/2}^d$ is $\frac{H}{2^d} \times \frac{W}{2^d} \times \frac{D}{2^d}$, with 2^{d+4} channels.

Region-Preserving Attention Module. To effectively learn the prior knowledge in the elongated esophagus, we design a region-preserving attention module

(RAM), as shown in Fig. 1. The multi-head attention (MHA) [17] is employed to gather long-range informative values in f_1^d with f_2^d as queries and f_1^d as keys. The features $f_{1/2}^d$ are reshaped to $\frac{HWD}{2^{3d}} \times C$ before passed to the MHA, where C is the channel dimension. The attentive features f_A^d can be formulated as:

$$f_A^d = MHA(Q, K, V) = MHA(f_2^d, f_1^d, f_1^d), d = 3, 4. \tag{2}$$

Since MHA perturbs the positional information, we preserve the tumor local patterns by concatenating original features to the attentive features at the channel dimension, followed by a $1 \times 1 \times 1$ bottleneck convolution $\xi_{1 \times 1 \times 1}$ to squeeze the channel features (named as RAM), as shown in the following equations,

$$f^d = \begin{cases} Concat(f_1^d, f_2^d), & d = 0, 1, 2, \\ \xi_{1 \times 1 \times 1}\left(Concat(f_1^d, f_2^d, f_A^d)\right), & d = 3, 4, \end{cases} \tag{3}$$

where the lower-level features from both encoders are fused by concatenation. The decoder D generates a probabilistic prediction $G_2' = D(f^0, \cdots, f^4)$ with skip connections (Fig. 1). We utilize the 3D Dice [14] loss function, $\mathcal{L}_{DICE}(G_2', G_2)$.

3   Training Strategy

The network should learn from three aspects: 1) **Tumor volume variation**: the structural changes of the tumor from the first to the second course; 2) **Cancer cell proliferation**: The tumor in esophageal cancer tends to infiltrate into the adjacent tissue; 3) **Reliance of GTV on esophageal anatomy**: The anatomical dependency between esophageal GTV and the position of the esophagus.

Medical images are sparsely labeled which are isolated by different tasks [20], and are often inadequate. In this study, we use a paired first-second course GTV dataset S_p, an unpaired GTV dataset S_v, and a public esophagus dataset S_e.

In order to fully leverage both public and private datasets, the training objective should not be specific to any tasks. Here, we denote G_1/G_2 as prior/target annotations respectively, which are not limited only to the GTV areas. As shown in Fig. 1, our strategy is to challenge the network to retrieve information from augmented inputs in E_1 using the features from E_2, which can incorporate a wide range of datasets that are not tailored for second-course GTV segmentation.

3.1   Tumor Volume Variation

The differences in tumor volume between the first and second courses following an RT treatment can have a negative impact on the state-of-the-art (SOTA) learning-based techniques, which will be discussed in Sect. 4.2. To adequately monitor changes in tumor volume and integrate information from the initial course into the subsequent course, a paired first-second courses dataset $S_p = \{i_p^1, i_p^2, g_p^1; g_p^2\}$ is necessary for training. In S_p, i_p^1 and i_p^2 are the first and second course CT images, while g_p^1 and g_p^2 are the corresponding GTV annotations.

3.2   Cancer Cell Proliferation

The paired dataset S_p for the first and second courses is limited, whereas an unpaired GTV dataset $S_v = \{i_v; g_v\}$ can be easily obtained in a standard clinical workflow with a substantial amount. S_v lacks its counterpart for the second course, in which i_v/g_v are the CT image and the corresponding annotation for GTV. To address this, we apply two distinct randomized augmentations, $\mathcal{P}_1, \mathcal{P}_2$, to mimic the unregistered issue of the first and second course CT. The transformed data is feed into the encoders $E_{1/2}$ as shown in the following equations:

$$I_1, G_1, I_2, G_2 = \begin{cases} \mathcal{P}_1(i_p^1), \mathcal{P}_1(g_p^1), \mathcal{P}_2(i_p^2), \mathcal{P}_2(g_p^2), & \text{when } i_p^1, i_p^2, g_p^1, g_p^2 \in S_p, \\ \mathcal{P}_1(i_v), \mathcal{P}_1(g_v), \mathcal{P}_2(i_v), \mathcal{P}_2(g_v), & \text{when } i_v, g_v \in S_v, \\ \mathcal{P}_1(i_e), \mathcal{P}_1(g_e), \mathcal{P}_2(i_e), \mathcal{P}_2(g_e), & \text{when } i_e, g_e \in S_e. \end{cases} \quad (4)$$

The esophageal tumor can proliferate with varying morphologies into the surrounding tissues. Although not paired, S_v contains valuable information about the tumor. Challenging the network to query information within GTV will enhance the capacity to retrieve pertinent information for the tumor positions.

3.3   Reliance of GTV on Esophageal Anatomy

To make full use of the datasets of relevant tasks, we incorporate a public esophagus segmentation dataset, denoted as $S_e = \{i_e; g_e\}$, where i_e/g_e represent the CT images and corresponding annotations of the esophagus structure. By augmenting the data as described in Eq. (4), S_e challenges the network to extract information from the entire esophagus, which enhances the network's embedding space with anatomical prior knowledge of the esophagus. Similarly, data from the paired S_p is also augmented by $\mathcal{P}_{1/2}$ to increase the network's robustness.

In summary, our training strategy is not dataset-specific or target-specific, thus allowing the integration of prior knowledge from multi-center esophageal GTV-related datasets, which effectively improves the network's ability to retrieve information for the second course from the three key aspects stated in Sect. 3.

4   Experiments

4.1   Experimental Setup

Datasets. The paired first-second course dataset, S_p, is collected from Sun Yat-Sen University Cancer Center (Ethics Approval Number: B2023-107-01), comprising paired CT scans of 69 distinct patients from South China. We collected the GTV dataset S_v from MedMind Technology Co., Ltd., which has CT scans from 179 patients. For both S_p and S_v, physicians annotated the esophageal cancer GTV in each CT. The GTV volume statistics (cm^3, mean $\pm$ std.) in S_v is 40.60 ± 29.75, and is $83.70 \pm 55.97/71.66 \pm 49.36$ for the first/second course RT in S_p respectively. Additionally, we collect S_e from SegTHOR [12], consisting of CT scans and esophagus annotations from 40 patients who did not

Table 1. The results suggest a domain gap between the first and second courses, which indicates increased difficulty in GTV segmentation for the second course. Asterisks indicate p-value < 0.05 for the performance gap between the first and second course.

Methods	First Course				Second Course			
	DSC (%) ↑		ASD (mm) ↓		DSC (%) ↑		ASD (mm) ↓	
	mean ± std.	med.	mean ± std.	med.	mean ± std.	med.	mean ± std.	med.
UNETR [7]	59.77 ± 20.24	62.90	10.57 ± 14.66	7.06	53.03 ± 17.62*	55.17	11.29 ± 11.44	8.42
Swin UNETR [6]	60.84 ± 19.74	64.07	10.29 ± 17.78	6.67	57.04 ± 20.16*	60.73	9.76 ± 15.43	6.21
DenseUnet [19]	63.95 ± 18.23	68.11	8.94 ± 13.82	6.04	55.35 ± 18.59*	58.54	9.84 ± 6.91*	8.54
3D U-Net [5]	66.73 ± 17.21	69.86	8.04 ± 16.83	4.19	57.50 ± 19.49*	62.62	9.14 ± 12.03*	6.09

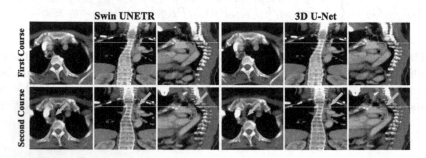

Fig. 2. The domain gap observed between the first and second courses RT. The blue arrows indicate that the methods tend to exhibit false delineation in the second course, suggesting a lack of consideration for tumor changes after the first course. (Yellow: Locations of the transverse planes; Green: GTV ground truth contours; Red: predictions.) (Color figure online)

have esophageal cancer. We randomly split S_p into training and test datasets at the patient-level. The training dataset includes S_v, S_e, and 41 patients from S_p (denoted as S_p^{train}), while the test dataset comprises 28 patients from S_p (denoted as S_p^{test}).

Implementation Details. The CT volumes from the first and second course in S_p are aligned based on the center of the lung mask [8]. The CT volumes are applied with a windowing of $[-100, 300]$ HU, and resampled to 128^3, with a voxel size of $1.2 \times 1.2 \times 3$ mm^3. The augmentations $\mathcal{P}_{1/2}$ involve a combination of random 3D resized cropping, flipping, rotation in the transverse plane, and Gaussian noise. We employ the Adam [11] optimizer with $(\beta_1, \beta_2, lr) = (0.9, 0.999, 0.001)$ for training for 500 epoches. The network is implemented using PyTorch [2] and MONAI [1], and detailed configurations are in the supplementary material. Experiments are performed on an NVIDIA RTX 3090 GPU with 24GB memory.

Performance Metrics. Dice score (DSC), averaged surface distance (ASD) and Hausdorff distance (HSD) are used as metrics for evaluation. The Wilcoxon signed-rank test is used to compare the performance of different methods.

4.2   Domain Gap Between the First and Second Course

As previously mentioned, the volume of the tumors changes after the first course of RT. To demonstrate the presence of a domain gap between the first and second courses, we train SOTA methods with datasets S_p^{train} and S_v, by feeding the data sequentially into the network. We then evaluate the models on S_p^{test}. The results presented in Table 1 indicate a performance gap between GTV segmentation in the first and second courses, with the latter being more challenging. Notably, the paired first-second course dataset S_p^{test} pertains to the same group of patients, thereby ensuring that any performance drop can be attributed solely to differences in courses of RT, rather than variations across different patients.

Figure 2 illustrates the reduction in the GTV area after the initial course of RT, where the transverse plane is taken from the same location relative to the vertebrae (yellow lines). The blue arrows indicate that the networks failed to track these changes and produced false predictions in the second course of RT. This suggests that deep learning-based approaches may not rely solely on the identification of malignant tissue patterns, as doctors do, but rather predict high-risk areas statistically. Therefore, for accurate second-course GTV segmentation, we need to explicitly propagate prior information from the first course using dual encoders in ARTSEG, and incorporate learning about tumor changes.

4.3   Evaluations of Second-Course GTV Segmentation Performance

Combination of Various Datasets. Table 2 presents the information gain derived from multi-center datasets using quantified metrics for segmentation performance. We first utilize a standard ARTSEG (w/o RAM) as an ablation network. When prior information from the first course is explicitly introduced using S_p, ARTSEG outperforms other baselines for GTV segmentation in the second course, which reaches a DSC of 66.73%. However, in Fig. 3, it can be observed that the model failed to accurately track the GTV area along the esophagus (orange arrows) due to the soft and elongated nature of the esophageal tissue, which deforms easily during CT scans performed at different times.

By subsequently incorporating S_e for structural esophagus prior knowledge, the DSC improved to 69.42%. Meanwhile, the esophageal tumor comprises two primary regions, the original part located in the esophagus and the extended part that has invaded the surrounding tissue. As shown in Fig. 3, identifying the tumor proliferation into the surrounding tissue without comprehensive knowledge of tumor morphology can be challenging (blue arrows). To address this, incorporating S_v to comprehensively learn the tumor morphology is required.

When S_v is incorporated for learning tumor proliferation, the DSC improved to 72.64%. We can observe from Case 2 in Fig. 3 that the network has a better understanding of the tumor proliferation with S_v, while it still fails to track the GTV area along the esophagus as pointed by the orange arrow. Therefore, S_v and S_e improve the network from two distinct aspects and are both valuable. Our proposed training strategy fully exploits the datasets S_p, S_v, and S_e, and

Table 2. Quantitative comparison of GTV segmentation performance in the second course. Our proposed ARTSEG+RAM achieved better overall performance, where asterisks indicate ARTSEG+RAM outperforms other methods with p-value < 0.05.

Methods	S_p^{train}	S_v	S_e	DSC (%) ↑ mean ± std	med.	ASD(mm) ↓ mean ± std	med.	HSD(mm) ↓ mean ± std	med.	Inference speed(ms)
Swin UNETR§ [6]	✓	✓		57.04 ± 20.16*	60.73	9.76 ± 15.43*	6.21	58.73 ± 46.67*	51.00	59.25
3D U-Net§ [5]	✓	✓		57.50 ± 19.49*	62.62	9.14 ± 12.03*	6.09	63.54 ± 49.59*	48.29	4.01
ARTSEG w/o RAM	✓			66.73 ± 16.20*	71.85	4.45 ± 4.10*	3.10	38.22 ± 27.51*	31.22	9.05
	✓		✓	69.42 ± 11.05*	71.12	3.98 ± 3.03*	2.99	47.89 ± 50.82*	33.04	
	✓	✓		72.64 ± 13.53*	74.64	2.69 ± 1.69	2.23	21.71 ± 11.95	19.69	
ARTSEG w/o RAM	✓	✓	✓	74.54 ± 13.33	**76.84**	2.51 ± 1.85	1.83	27.00 ± 41.79	16.35	
ARTSEG+MHA [17]	✓	✓	✓	74.34 ± 13.27	76.25	2.49 ± 1.62	1.97	27.33 ± 33.25	15.99	12.24
ARTSEG+RAM	✓	✓	✓	**75.26 ± 12.24**	76.40	**2.39 ± 1.57**	**1.73**	**19.75 ± 11.83**	**15.54**	12.60

§Without explicit information from the first-course. (Best methods in Table 1)

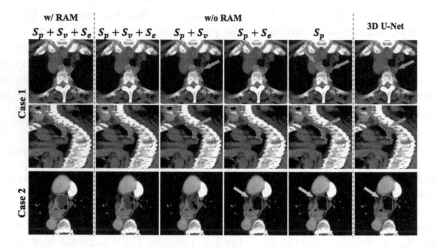

Fig. 3. The impact of different prior knowledge on esophageal tumor detection. Networks with inadequate knowledge of the esophagus may fail in identifying the tumor within the esophagus (orange arrows), whereas a limited understanding of tumor morphology can deteriorate the ability to detect the tumor in the adjacent area (blue arrows). Our proposed approach, encompassing comprehensive prior knowledge, shows superior performance. (Green: GTV ground truth contours; Red: predictions.) (Color figure online)

further improve the DSC to 74.54% by utilizing comprehensive knowledge of both the tumor morphology and esophageal structures.

Region-Preserving Attention Module. Although introducing the esophageal structural prior knowledge using S_e can improve the performance in DSC and ASD (Table 2), the increase in HSD (38.22 to 47.89 mm; 21.71 to 27.00 mm) indicates that there are outliers far from the ground truth boundaries. This may be attributed to the convolution that cannot effectively handle the long-range knowledge of the esophagus structure. The attention mechanism can effectively capture the long-range relationship as shown recently in [13].

However, there is no performance gain with MHA as shown in Table 2, and the HSD further increased to 27.33 mm. We attribute the drawback is due to the location-agnostic nature of the operations in MHA, where the local regional correlations are perturbed.

To tackle the aforementioned problem, we propose RAM which involves the concatenation of the original features with attention outputs, allowing for the preservation of convolution-generated regional tumor patterns while effectively comprehending long-range prior knowledge specific to the esophagus. Finally, our proposed ARTSEG with RAM achieves the best DSC/HSD of 75.26%/19.75 mm, and outperforms its ablations as well as other baselines, as shown in Table 2.

Limitations. For the method's generalizability, analysis of diverse imaging protocols and segmentation backbones are inadequate. Besides, ARTSEG requires more computational resources due to its dual-encoder and attention design.

5   Conclusion

In this paper, we reveal the domain gap between the first and second courses of RT for esophageal GTV segmentation. To improve the accuracy of GTV declination in the second course, we explicitly incorporated the naturally existing prior information from the first course. Besides, to efficiently leverage prior knowledge contained in various medical CT datasets, we train the network in an information-querying manner. We proposed RAM to capture long-range prior knowledge in the esophageal structure, while preserving the regional tumor patterns. Our proposed ARTSEG incorporates prior knowledge of the tumor volume variation, cancer cell proliferation, and reliance of GTV on esophageal anatomy, which enhances the GTV segmentation accuracy in the second course RT. Our future research includes accurate delineation for multiple targets in the second course and knowledge transferring through the time series of multiple courses.

Acknowledgments. Thanks to National Key Research and Development Program of China (2022YFC2405200), National Natural Science Foundation of China (82027807, U22A2051), Beijing Municipal Natural Science Foundation (7212202), Institute for Intelligent Healthcare, Tsinghua University (2022ZLB001), Tsinghua-Foshan Innovation Special Fund (2021THFS0104), Guangdong Esophageal Cancer Institute Science and Technology Program (Q202221, Q202214, M-202016).

References

1. Medical Open Network for Artificial Intelligence (MONAI). https://monai.io/
2. PyTorch. https://pytorch.org/
3. Bray, F., Ferlay, J., Soerjomataram, I., Siegel, R.L., Torre, L.A., Jemal, A.: Global cancer statistics 2018: GLOBOCAN estimates of incidence and mortality worldwide for 36 cancers in 185 countries. CA: Cancer J. Clin. **68**(6), 394–424 (2018)
4. Burnet, N.G., Thomas, S.J., Burton, K.E., Jefferies, S.J.: Defining the tumour and target volumes for radiotherapy. Cancer Imaging **4**(2), 153–161 (2004)

5. Falk, T., et al.: U-net: deep learning for cell counting, detection, and morphometry. Nat. Methods **16**(1), 67–70 (2019)
6. Hatamizadeh, A., Nath, V., Tang, Y., Yang, D., Roth, H.R., Xu, D.: Swin UNETR: swin transformers for semantic segmentation of brain tumors in MRI images. In: Crimi, A., Bakas, S. (eds.) BrainLes 2021. LNCS, vol. 12962, pp. 272–284. Springer, Cham (2022). https://doi.org/10.1007/978-3-031-08999-2_22
7. Hatamizadeh, A., et al.: UNETR: transformers for 3D medical image segmentation. In: Proceedings of the IEEE/CVF Winter Conference on Applications of Computer Vision (WACV), pp. 574–584 (2022)
8. Hofmanninger, J., Prayer, F., Pan, J., Röhrich, S., Prosch, H., Langs, G.: Automatic lung segmentation in routine imaging is primarily a data diversity problem, not a methodology problem. Eur. Radiol. Exp. **4**(1), 1–13 (2020). https://doi.org/10.1186/s41747-020-00173-2
9. Jin, D., et al.: DeepTarget: gross tumor and clinical target volume segmentation in esophageal cancer radiotherapy. Med. Image Anal. **68**, 101909 (2021)
10. Jin, D., et al.: Accurate esophageal gross tumor volume segmentation in PET/CT using two-stream chained 3D deep network fusion. In: Shen, D., et al. (eds.) MICCAI 2019. LNCS, vol. 11765, pp. 182–191. Springer, Cham (2019). https://doi.org/10.1007/978-3-030-32245-8_21
11. Kingma, D.P., Ba, J.L.: Adam: a method for stochastic optimization. In: International Conference on Learning Representations (2015)
12. Lambert, Z., Petitjean, C., Dubray, B., Kuan, S.: Segthor: segmentation of thoracic organs at risk in CT images. In: 2020 Tenth International Conference on Image Processing Theory, Tools and Applications (IPTA), pp. 1–6 (2020)
13. Li, J., Chen, J., Tang, Y., Wang, C., Landman, B.A., Zhou, S.K.: Transforming medical imaging with transformers? a comparative review of key properties, current progresses, and future perspectives. Med. Image Anal. **85**, 102762 (2023)
14. Milletari, F., Navab, N., Ahmadi, S.A.: V-net: fully convolutional neural networks for volumetric medical image segmentation. In: 2016 Fourth International Conference on 3D Vision (3DV), pp. 565–571 (2016)
15. Pennathur, A., Gibson, M.K., Jobe, B.A., Luketich, J.D.: Oesophageal carcinoma. Lancet **381**(9864), 400–412 (2013)
16. Van Andel, J.G., et al.: Carcinoma of the esophagus: results of treatment. Ann. Surg. **190**(6), 684–689 (1979)
17. Vaswani, A., et al.: Attention is all you need. In: Advances in Neural Information Processing Systems, vol. 30 (2017)
18. Yousefi, S., et al.: Esophageal tumor segmentation in CT images using a dilated dense attention Unet (DDAUnet). IEEE Access **9**, 99235–99248 (2021)
19. Yousefi, S., et al.: Esophageal gross tumor volume segmentation using a 3D convolutional neural network. In: Frangi, A.F., Schnabel, J.A., Davatzikos, C., Alberola-López, C., Fichtinger, G. (eds.) MICCAI 2018. LNCS, vol. 11073, pp. 343–351. Springer, Cham (2018). https://doi.org/10.1007/978-3-030-00937-3_40
20. Zhou, S.K., et al.: A review of deep learning in medical imaging: imaging traits, technology trends, case studies with progress highlights, and future promises. Proc. IEEE **109**(5), 820–838 (2021)
21. Zhou, S.K., Rueckert, D., Fichtinger, G.: Handbook of Medical Image Computing and Computer Assisted Intervention. Academic Press (2019)

Self-feedback Transformer: A Multi-label Diagnostic Model for Real-World Pancreatic Neuroendocrine Neoplasms Data

Mingyu Wang[1,2], Yi Li[1,2], Bin Huang[1,2], Chenglang Yuan[1,2], Yangdi Wang[3], Yanji Luo[3(✉)], and Bingsheng Huang[1,2(✉)]

[1] Medical AI Lab, School of Biomedical Engineering, Shenzhen University Medical School, Shenzhen University, Shenzhen, China
huangb@szu.edu.cn
[2] Marshall Laboratory of Biomedical Engineering, Shenzhen University, Shenzhen, China
[3] Department of Radiology, The First Affiliated Hospital, Sun Yat-sen University, Guangzhou, China
luoyj26@mail.sysu.edu.cn

Abstract. CAD is an emerging field, but most models are not equipped to handle missing and noisy data in real-world medical scenarios, particularly in the case of rare tumors like pancreatic neuroendocrine neoplasms (pNENs). Multi-label models meet the needs of real-world study, but current methods do not consider the issue of missing and noisy labels. This study introduces a multi-label model called Self-feedback Transformer (SFT) that utilizes a transformer to model the relationships between labels and images, and uses a ingenious self-feedback strategy to improve label utilization. We evaluated SFT on 11 clinical tasks using a real-world dataset of pNENs and achieved higher performance than other state-of-the-art multi-label models with mAUCs of 0.68 and 0.76 on internal and external datasets, respectively. Our model has four inference modes that utilize self-feedback and expert assistance to further increase mAUCs to 0.72 and 0.82 on internal and external datasets, respectively, while maintaining good performance even with input label noise ratios up to 40% in expert-assisted mode.

Keywords: Computer Aided Diagnosis · Real-world · Multi-label · Self-feedback Transformer

1 Introduction

Computer-aided Diagnosis (CAD) systems have achieved success in many clinical tasks [5,6,12,17]. Most CAD studies were developed on regular and selected

---

M. Wang, Y. Li and B. Huang—Contributed equally to this work.

Supplementary Information The online version contains supplementary material available at https://doi.org/10.1007/978-3-031-43990-2_49.

H. Greenspan et al. (Eds.): MICCAI 2023, LNCS 14226, pp. 521–530, 2023.
https://doi.org/10.1007/978-3-031-43990-2_49

datasets in the laboratory environment, which avoided the problems (data noise, missing data, etc.) in the clinical scenarios [3,6,9,13,18]. In a real clinical scenario, the clinicians generally synthesize all aspects of information, and conduct consultations with Multidisciplinary Team (MDT), to accurately diagnose and plan the treatment [9,10,13]. Real-world studies have received increasing attention [11,16], and it is challenging for the CAD in the real-world scenarios as: 1) Consistent with the clinical workflow, CAD needs to consider multidisciplinary information to obtain multidimensional diagnosis; 2) Due to information collection, storage and manual evaluation, there are missing and noisy medical data. This phenomenon is especially common in rare tumors like pancreatic neuroendocrine neoplasms (pNENs).

In order to overcome above challenges, some studies [3,9,13,18] used multi-label method because of the following advantages: 1) The input of the model is only a single modality such as images, which is easy to apply clinically; 2) The model learns multi-label and multi-disciplinary knowledge, which is consistent with clinical logic; 3) Multi-label simultaneous prediction, which meets the need of clinical multi-dimensional description of patients. For the above advantages, multi-label technology is suitable for real-world CAD. The previous multi-label CAD studies were designed based on simple parameter sharing methods [9,15,20] or Graph Neural Network (GNN) method [2]. The former implicitly interacts with multi-label information, making it difficult to fully utilize the correlation among labels; And the latter requires the use of word embeddings pre-trained on public databases, which is not friendly to many medical domain proper nouns. The generalizability of previous multi-label CAD studies is poor due to these disadvantages. In addition, none of the current multi-label CAD studies have considered the problem of missing labels and noisy labels.

Considering these real-world challenges, we propose a multi-label model named Self-feedback Transformer (SFT), and validate our method on a real-world pNENs dataset. The main contributions of this work are listed: 1) A transformer multi-label model based on self-feedback mechanism was proposed, which provided a novel method for multi-label tasks in real-world medical application; 2) The structure is flexibility and interactivity to meet the needs of real-world clinical application by using four inference modes, such as expert-machine combination mode, etc.; 3) SFT has good noise resistance, and can maintain good performance under noisy label input in expert-assisted mode.

2   Method

Transformer has achieved success in many fields [4,19]. Inspired by DETR [1] and C-Tran [8], we propose a multi-label model based on transformer and self-feedback mechanism. As shown in Fig. 1, 1) image is embedded by Convolutional Neural Network (CNN) firstly; 2) then all labels are embedded and combined with their state embeddings; 3) finally, all embeddings are fed into a transformer, and the output label tokens are fed into Fully Connection (FC) layers for final predictions. Based on this network, we further introduce a self-feedback strategy,

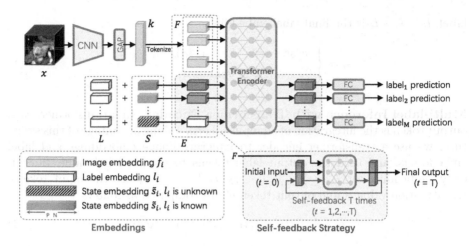

Fig. 1. SFT architecture and illustration of self-feedback strategy.

which allows the label information (including the missing labels) to be reused iteratively for enhancing the utilization of labels.

2.1   Transformer-Based Multi-label Model

Image Embeddings F. Given input image $x \in \mathbb{R}^{L \times W \times H}$, the feature vector $k \in \mathbb{R}^C$ is extracted by a CNN after Global Average Pooling (GAP), where the output channel $C = 256$. Then k is split along the channel dimension into N ($N = 8$) sub-feature vectors $F = \{f_1, f_2, \ldots, f_N\}$, $f_i \in \mathbb{R}^d$, $d = C/N$ for tokenization. We choose 3D VGG8, a simple CNN with 8 convolution layers.

Label Embeddings L. In order to realize the information interaction among labels, and between labels and image features, we embed labels by an embedding layer. Each image x has M labels, and all labels are embedded into a vector set $L = \{l_1, l_2, \ldots, l_M\}$, $l_i \in \mathbb{R}^d$ by the learnable embedding layer of size $d \times M$.

Soft State Embeddings S. There is a correlation between labels, e.g. the lesions with indistinct borders tend to be malignant. Therefore, we hypothesize that the states (GT values) of some labels can be a context for helping predict the remaining labels. We use a soft state embedding method. Specifically, we first embed the positive and negative states into s^p and s^n, both $\in \mathbb{R}^d$, and then the final state embedding $\tilde{s}_i$ is the weighted sum of s^p and s^n as shown in Equation (1). The state weight w_i^p and w_i^n is the true label value (eg. $w_i^p = 1.0$ when label is positive), where $w_i^p + w_i^n = 1$. For labels with continuous values such as age, the value normalized to $0 \sim 1$ is w_i^p. The $\tilde{s}_i$ is set as a zero vector for unknown label. $\widetilde{l}_i = \widetilde{s}_i + l_i$ is the final label embedding.

$$\widetilde{s}_i = \begin{cases} w_i^p s^p + w_i^n s^n, & \text{if } l_i \text{ is known} \\ 0, & \text{if } l_i \text{ is unknown} \end{cases} . \tag{1}$$

Multi-label Inference with Transformer Encoder. In a transformer, each output token is the integration of all input tokens. Taking advantage of this structure, we use a transformer encoder to integrate image embeddings and label embeddings, and used the output label tokens to predict label value. Specifically, embedding set $E = \{f_1, f_2, \cdots, f_N, \widetilde{l}_1, \widetilde{l}_2, \cdots, \widetilde{l}_M\}$ are the input tokens, the attention value α and output token e' are computed as follows:

$$\alpha_{ij} = softmax((W^q e_i)^\mathsf{T}(W^k e_j)/\sqrt{d}), \tag{2}$$

$$\bar{e}_i = \sum_{j=1}^{M} \alpha_{ij} W^v e_j, \tag{3}$$

$$e'_i = ReLU\left(\bar{e}_i W^r + b_1\right) W^o + b_2, \tag{4}$$

where e_i is from E, W^q, W^k and W^v are weight matrices of query, key and value, respectively, W^r and W^o are transformation matrices, and b_1 and b_2 are bias vectors. This update procedure is repeated for L layers, where the e'_i are fed to the successive transformer layer. Finally, all e'_i which are label output tokens are fed into M independent FC layers for predicting value of each label.

2.2   Self-feedback Strategy

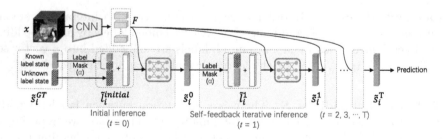

Fig. 2. Illustration of the self-feedback strategy.

The states of unknown labels cannot provide context, thus, the information interaction between known labels and unknown labels may be weaken. To overcome this problem, we propose a Self-feedback Strategy (SFS) inspired by Recurrent Neural Networks (RNN) to enhance the interaction of labels.

Training Progress and Loss Function. As shown in Fig. 2, at time point $t = 0$, the state embedding is initialized to $\tilde{s}_i^{GT}$ by Ground Truth (GT) value, and the initial label embedding $\tilde{l}_i^{initial}$ is computed by $\tilde{l}_i^{initial} = \tilde{s}_i^{GT} + l_i$. The $\tilde{l}_i^{initial}$ is combined with f_i as the initial input, and then the output predicted value is converted into state embedding $\tilde{s}_i^0$ by Equation (1). When t>0, the label embedding $\tilde{l}_i^t$ is updated iteratively by $\tilde{l}_i^t = \tilde{s}_i^{t-1} + l_i$, and then fed into the transformer T times. For classification and regression labels, we use focal cross-entropy loss and L2 loss respectively, and use the method in [7] to auto-weight the loss value of each label. The backpropagation of gradients and parameter updates are performed immediately after calculating the loss at each time point t. In the regular inference phase, the state of all labels is initialized as unknown.

Label Mask Strategy. To avoid predicting with labels' own input state, we use a Label Mask Strategy (LMS) during training phase to randomly mask a certain proportion a of known labels, which causes the labels' states to be embedded as zero vectors. Meanwhile, only the loss of the masked known label is calculated.

3 Experiments and Results

3.1 Dataset and Evaluation

Real-World pNENs Dataset. We validated our method on a real-world pNENs dataset from two centers. All patients with arterial phase Computed Tomography (CT) images were included. The dataset contained 264 and 28 patients in center 1 and center 2, and a senior radiologist annotated the bounding boxes for all 408 and 28 lesions. We extracted 37 labels from clinical reports, including survival, immunohistochemical (IHC), CT findings, etc. Among them, 1)RECIST drug response (RS), 2)tumor shrink (TS), 3)durable clinical benefit (DCB), 4)progression-free survival (PFS), 5)overall survival (OS), 6)grade (GD), 7)somatostatin receptor subtype 2(SSTR2), 8)Vascular Endothelial Growth Factor Receptor 2 (VEFGR2), 9)O6-methylguanine methyltransferase (MGMT), 10)metastatic foci (MTF), and 11)surgical recurrence (RT) are main tasks, and the remaining are auxiliary tasks. 143 and 28 lesions were segmented by radiologists, and the radiomics features of them were extracted, of which 162 features were selected and binarized as auxiliary tasks because of its statistically significant correlation with the main labels. The label distribution and the overlap ratio (Jaccard index) of lesions between pairs of labels are shown in Fig. 3. It is obvious that the real-world dataset has a large number of labels with randomly missing data, thus, we used an adjusted 5-fold cross-validation. Taking a patient as a sample, we chose the dataset from center 1 as the internal dataset, of which the samples with most of the main labels were used as Dataset 1 (219 lesions) and was split into 5 folds, and the remaining samples are randomly divided into the training set Dataset 2 (138 lesions) and the validation set Dataset 3 (51 lesions), the training set and the validation set of the corresponding folds were added during cross-validation, respectively. All samples in Center 2 left as external test set. Details of each dataset are in the Supplementary Material.

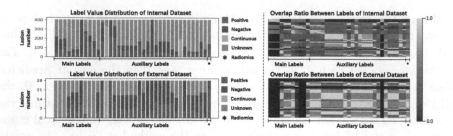

Fig. 3. Label value distribution and overlap ratio of lesions between pairs of labels.

Dataset Evaluation Metrics. We evaluate the performance of our method on the 10 main tasks for internal dataset, and due to missing labels and too few SSTR2 labels, only the performance of predicting RT, PFS, OS, GD, MTF are evaluated for external dataset. We employ accuracy (ACC), sensitivity (SEN), specificity (SPC), F1-score (F1) and area under the receiver operating characteristic (AUC) for each task, and compute the mean value of them (e.g. mAUC).

3.2   Implementation Details

The CT window width and window level were adjusted to 310 and 130 HU refer to [17], and the image inside the bounding box was cropped and scaled to $128 \times 128 \times 64$ pixels. The numbers of convolutional kernels of VGG8 are [3, 32, 32, 64, 64, 128, 128, 256]. Transformer encoder contained 2 layers and 8 heads. Layer normalization was used in transformer. The LMS a was set as 0.5, and the training feedback times T was 4. We used Adam optimiser, and used cosine annealing to reduce the learning rate from 1e–4 to 1e–12 over all 200 epochs. Our method was implemented in Pytorch, using an NVIDIA RTX TITAN GPU.

3.3   Comparison and Ablation Experiments

We compared our method with Single-task(ST), Parameters Sharing (PS), ML-Decoder [14] and C-tran [8]. Specifically, a ST model is trained using single label, and PS model uses a FC layer followed by the CNN to predict all labels. It should be noted that the CNN backbone of each method was replaced as 3D VGG8 to ensure fair comparison. In the ablation experiment, we removed the LMS and SFS to analyze their impact. The AUC of each main label is shown in Fig. 4, and the average performance is shown in Table 1. It can be seen that multi-label models is better than that of ST due to using the relationship among labels, and SFT outperform other methods on most tasks and the average performance. The ablation experiments results showed that removing the LMS and SFS components causes performance degradation, indicating the necessity of them.

Table 1. Results of the comparison experiments and the ablation experiments (%). SFT** means SFT w/o SFS and LMS, and SFT* means SFT w/o SFS.

Method	Internal Dataset (219+51 lesions)					External Dataset (28 lesions)				
	mAUC	mACC	mSEN	mSPC	mF1	mAUC	mACC	mSEN	mSPC	mF1
ST	62.56	63.76	64.36	70.09	54.77	64.47	66.98	70.00	70.68	59.57
PS	62.04	64.27	68.45	65.97	60.71	69.85	71.00	**86.50**	64.09	68.40
ML-Decoder	59.36	59.13	**73.81**	60.23	57.26	67.66	65.02	79.50	61.94	60.86
C-tran	62.53	63.05	60.23	73.11	56.14	70.96	**80.05**	63.50	**86.59**	66.54
SFT**	61.01	64.90	60.98	72.77	54.46	62.27	74.35	67.50	76.83	65.46
SFT*	64.39	64.41	71.75	64.48	60.82	74.19	76.12	73.50	79.21	68.53
SFT	**67.76**	**66.53**	69.51	**73.62**	**61.51**	75.50	78.08	73.00	80.71	**71.33**

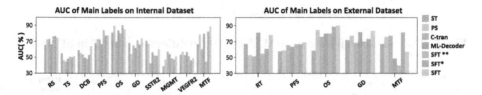

Fig. 4. Predictive performance of different methods on the main tasks.

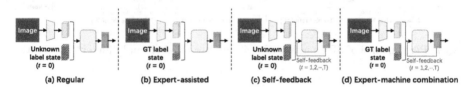

Fig. 5. Different inference modes of SFT.

3.4 Performance of Different Inference Modes

As shown in the Fig. 5, we designed 4 inference modes: 1) Regular, only input images; 2) Expert-assisted (EA), certain information is provided by clinicians; 3) Self-feedback (SF), iteratively inference T times by using prediction; 4) Expert-machine Combination (EMC), expert-assisted and self-feedback inference are both performed. Only the auxiliary labels states were input in EA mode. The results is shown in Table 2. Both SF and EA perform better than regular mode, and EMC outperforms other modes with a mAUC of 0.72 (0.82 on external dataset). We tested the SFT with and without SFS training under different feedback times T in SF mode, and results (Fig. 6.) showed that the performance of SFT with SFS training increases gradually with the increase of T, while the SFT without SFS training has a general trend of decreasing performance after continuous iteration.

Table 2. Performance of different inference modes by using SFT (%).

Mode	Internal Dataset (219+51 lesions)					External Dataset (28 lesions)				
	mAUC	mACC	mSEN	mSPC	mF1	mAUC	mACC	mSEN	mSPC	mF1
Regular	67.76	66.53	69.51	73.62	61.51	75.50	78.08	73.00	80.71	71.33
SF ($T = 1$)	69.94	67.24	**72.24**	72.24	62.29	77.33	81.95	**84.00**	74.13	**78.35**
EA	71.26	68.40	68.42	77.58	62.88	81.30	82.29	66.00	**94.44**	74.44
EMC	**72.22**	**71.26**	66.87	**81.77**	**63.65**	82.45	**83.01**	70.00	93.81	75.24

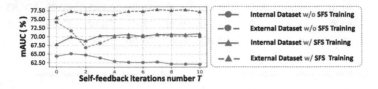

Fig. 6. The relationship between mAUC and the iterations number T in SF mode.

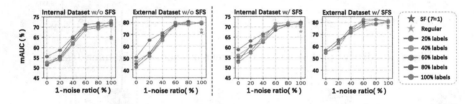

Fig. 7. Performance of SFT under different number of labels and different noise ratios.

3.5   Analysis of Noise Resistance

To explore the noise resistance of SFT in EA mode, we selected 20, 40, 60, 80, and 100 percent of the known labels respectively, and further negated 0, 20, 40, 60, 80, 100 percent of the selected labels to simulate noisy labels. The way to negate a label is to change the label value x to $1 - x$. As shown in Fig. 7, as the noise ratio increases, the performance shows a decreasing trend, and the performance decreases slightly when the noise ratio ≤ 40 %. Finally, it can be observed that the SFT using the SFS training strategy is relatively less affected by noise. When using 100 percent labels, the mAUC of the internal dataset decreased from 0.71 (noise ratio $= 0.0$) to 0.53 (noise ratio $= 1.0$), a decrease of 0.19 (0.26 on external dataset); the corresponding internal mAUC of the model without SFS training decreased by 0.21 (0.38 on external dataset). So SFS training can improve a certain anti-noise ability.

4   Conclusion

We proposed a novel model SFT for multi-label prediction on real-world pNENs data. The model integrates label and image informations based on a transformer encoder, and iteratively uses its own prediction based on a self-feedback mechanism to improve the utilization of missing labels and correlation among labels. Experiment results demonstrated our proposed model outperformed other multi-label models, showed flexibility by multiple inference modes, and had a certain ability to maintain performance when the input context noise was less than 40%.

Acknowledgements. This work was supported by National Natural Science Foundation of China (81971684), Marshall Lab of Biomedical Engineering open fund: Medical-Engineering Project.

References

1. Carion, N., Massa, F., Synnaeve, G., Usunier, N., Kirillov, A., Zagoruyko, S.: End-to-end object detection with transformers. In: Vedaldi, A., Bischof, H., Brox, T., Frahm, J.-M. (eds.) ECCV 2020. LNCS, vol. 12346, pp. 213–229. Springer, Cham (2020). https://doi.org/10.1007/978-3-030-58452-8_13
2. Chen, B., Li, J., Lu, G., Yu, H., Zhang, D.: Label co-occurrence learning with graph convolutional networks for multi-label chest x-ray image classification. IEEE J. Biomed. Health Inform. **24**(8), 2292–2302 (2020)
3. Choi, H., Ha, S., Kang, H., Lee, H., Lee, D.S., Initiative, A.D.N., et al.: Deep learning only by normal brain pet identify unheralded brain anomalies. EBioMedicine **43**, 447–453 (2019)
4. Dosovitskiy, A., et al.: An image is worth 16x16 words: transformers for image recognition at scale. arXiv preprint arXiv:2010.11929 (2020)
5. Eweje, F.R., et al.: Deep learning for classification of bone lesions on routine MRI. EBioMedicine **68**, 103402 (2021)
6. Jiang, Y., et al.: Development and validation of a deep learning CT signature to predict survival and chemotherapy benefit in gastric cancer: a multicenter, retrospective study. Ann. Surg. **274**(6), e1153–e1161 (2021)
7. Kendall, A., Gal, Y., Cipolla, R.: Multi-task learning using uncertainty to weigh losses for scene geometry and semantics. In: Proceedings of the IEEE Conference on Computer Vision and Pattern Recognition, pp. 7482–7491 (2018)
8. Lanchantin, J., Wang, T., Ordonez, V., Qi, Y.: General multi-label image classification with transformers. In: Proceedings of the IEEE/CVF Conference on Computer Vision and Pattern Recognition, pp. 16478–16488 (2021)
9. Lin, D., et al.: Application of comprehensive artificial intelligence retinal expert (care) system: a national real-world evidence study. Lancet Digit. Health **3**(8), e486–e495 (2021)
10. Partelli, S., et al.: European cancer organisation essential requirements for quality cancer care (erqcc): pancreatic cancer. Cancer Treat. Rev. **99**, 102208 (2021)
11. Penberthy, L.T., Rivera, D.R., Lund, J.L., Bruno, M.A., Meyer, A.M.: An overview of real-world data sources for oncology and considerations for research. CA: Cancer J. Clin. **72**, 287–300 (2021)
12. Peng, S., et al.: Deep learning-based artificial intelligence model to assist thyroid nodule diagnosis and management: a multicentre diagnostic study. Lancet Digit. Health **3**(4), e250–e259 (2021)
13. Ravizza, S., et al.: Predicting the early risk of chronic kidney disease in patients with diabetes using real-world data. Nat. Med. **25**(1), 57–59 (2019)

14. Ridnik, T., Sharir, G., Ben-Cohen, A., Ben-Baruch, E., Noy, A.: ML-Decoder: scalable and versatile classification head. arXiv preprint arXiv:2111.12933 (2021)
15. Shen, S., Han, S.X., Aberle, D.R., Bui, A.A., Hsu, W.: An interpretable deep hierarchical semantic convolutional neural network for lung nodule malignancy classification. Expert Syst. Appl. **128**, 84–95 (2019)
16. Sherman, R.E., et al.: Real-world evidence—what is it and what can it tell us? (2016)
17. Song, C., et al.: Predicting the recurrence risk of pancreatic neuroendocrine neoplasms after radical resection using deep learning radiomics with preoperative computed tomography images. Ann. Transl. Med. **9**(10), 833 (2021)
18. Tai, Y., Gao, B., Li, Q., Yu, Z., Zhu, C., Chang, V.: Trustworthy and intelligent COVID-19 diagnostic IoMT through XR and deep-learning-based clinic data access. IEEE Internet Things J. **8**(21), 15965–15976 (2021)
19. Vaswani, A., et al.: Attention is all you need. In: Advances in Neural Information Processing Systems 30 (2017)
20. Zhang, S., et al.: A novel interpretable computer-aided diagnosis system of thyroid nodules on ultrasound based on clinical experience. IEEE Access **8**, 53223–53231 (2020)

Anatomy-Informed Data Augmentation for Enhanced Prostate Cancer Detection

Balint Kovacs[1,2,3](✉)(iD), Nils Netzer[2,3](iD), Michael Baumgartner[1,4,5](iD),
Carolin Eith[2,3], Dimitrios Bounias[1,3], Clara Meinzer[2], Paul F. Jäger[5,6](iD),
Kevin S. Zhang[2], Ralf Floca[1], Adrian Schrader[2,3](iD), Fabian Isensee[1,5](iD),
Regula Gnirs[2], Magdalena Görtz[7,8], Viktoria Schütz[7], Albrecht Stenzinger[9],
Markus Hohenfellner[7], Heinz-Peter Schlemmer[2], Ivo Wolf[10](iD),
David Bonekamp[2](iD), and Klaus H. Maier-Hein[1,5,11](iD)

[1] Division of Medical Image Computing, German Cancer Research Center (DKFZ),
Heidelberg, Germany
balint.kovacs@dkfz-heidelberg.de
[2] Division of Radiology, German Cancer Research Center (DKFZ),
Heidelberg, Germany
[3] Medical Faculty Heidelberg, Heidelberg University, Heidelberg, Germany
[4] Faculty of Mathematics and Computer Science, Heidelberg University,
Heidelberg, Germany
[5] Helmholtz Imaging, German Cancer Research Center (DKFZ),
Heidelberg, Germany
[6] Interactive Machine Learning Group, German Cancer Research Center (DKFZ),
Heidelberg, Germany
[7] Department of Urology, University of Heidelberg Medical Center,
Heidelberg, Germany
[8] Junior Clinical Cooperation Unit 'Multiparametric Methods for Early Detection
of Prostate Cancer', German Cancer Research Center (DKFZ), Heidelberg, Germany
[9] Institute of Pathology, University of Heidelberg Medical Center,
Heidelberg, Germany
[10] Mannheim University of Applied Sciences, Mannheim, Germany
[11] Pattern Analysis and Learning Group, Department of Radiation Oncology,
Heidelberg University Hospital, Heidelberg, Germany

Abstract. Data augmentation (DA) is a key factor in medical image
analysis, such as in prostate cancer (PCa) detection on magnetic resonance images. State-of-the-art computer-aided diagnosis systems still rely
on simplistic spatial transformations to preserve the pathological label
post transformation. However, such augmentations do not substantially
increase the organ as well as tumor shape variability in the training set,
limiting the model's ability to generalize to unseen cases with more diverse
localized soft-tissue deformations. We propose a new anatomy-informed
transformation that leverages information from adjacent organs to simulate typical physiological deformations of the prostate and generates
unique lesion shapes without altering their label. Due to its lightweight

---

D. Bonekamp and K. H. Maier-Hein—Equal contribution.

Supplementary Information The online version contains supplementary material
available at https://doi.org/10.1007/978-3-031-43990-2_50.

H. Greenspan et al. (Eds.): MICCAI 2023, LNCS 14226, pp. 531–540, 2023.
https://doi.org/10.1007/978-3-031-43990-2_50

computational requirements, it can be easily integrated into common DA frameworks. We demonstrate the effectiveness of our augmentation on a dataset of 774 biopsy-confirmed examinations, by evaluating a state-of-the-art method for PCa detection with different augmentation settings.

Keywords: Data augmentation · Soft-tissue deformation · Prostate cancer detection

1 Introduction

Data augmentation (DA) is a key factor in the success of deep neural networks (DNN) as it artificially enlarges the training set to increase their generalization ability as well as robustness [22]. It plays a crucial role in medical image analysis [8] where annotated datasets are only available with limited size. DNNs have already successfully supported radiologists in the interpretation of magnetic resonance images (MRI) for prostate cancer (PCa) diagnosis [3]. However, the DA scheme received less attention, despite its potential to leverage the data characteristic and address overfitting as the root of generalization problems.

State-of-the-art approaches still rely on simplistic spatial transformations, like translation, rotation, cropping, and scaling by globally augmenting the MRI sequences [12,20]. They exclude random elastic deformations, which can change the lesion outline but might alter the underlying label and thus produce counterproductive examples for training [22]. However, soft tissue deformations, which are currently missing from the DA schemes, are known to significantly affect the image morphology and therefore play a critical role in accurate diagnosis [6].

Both lesion and prostate shape geometrical appearance influence the clinical assessment of Prostate Imaging-Reporting and Data System (PI-RADS) [24]. The prostate constantly undergoes soft tissue deformation dependent on muscle contractions, respiration, and more importantly variable filling of the adjacent organs, namely the bladder and the rectum. Among these sources, the rectum has the largest influence on the prostate and lesion shape variability due to its large motion [4] and the fact that the majority of the lesions are located in the adjacent peripheral prostate zone [1]. However, only one snapshot of all these functional states is captured within each MRI examination, and almost never will be exactly the same on any repeat or subsequent examination. Ignoring these deformations in the DA scheme can potentially limit model performance.

Model-driven transformations attempting to simulate organ functions - like respiration, urinary excretion, cardiovascular- and digestion mechanics - offer a high degree of diversity while also providing realistic transformations. Currently, the finite element method (FEM) is the standard for modeling biomechanics [13]. However, their computation is overly complex [10] and therefore does not scale to on-the-fly DA [7]. Recent motion models rely on DNNs using either a FEM model [15] or complex training with population-based models [18]. Motion models have not been integrated into any deep learning framework as an online data augmentation yet, thereby leaving the high potential of inducing application-specific knowledge into the training procedure unexploited.

In this work we propose an anatomy-informed spatial augmentation, which leverages information from adjacent organs to mimic typical deformations of the prostate. Due to its lightweight computational requirements, it can be easily integrated into common DA frameworks. This technique allows us to simulate different physiological states during the training and enrich our dataset with a wider range of organ and lesion shapes. Inducing this kind of soft tissue deformation ultimately led to improved model performance in patient- and lesion-level PCa detection on an independent test set.

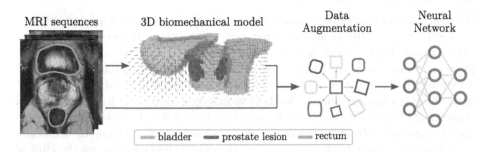

Fig. 1. The proposed anatomy-informed prostate augmentation. Simulating typical physiologic changes in adjacent organs enlarges the training set with realistic soft tissue deformations of the prostate thereby increasing the generalization ability as well as the robustness of the network. Due to its lightweight computational requirements, it can be easily integrated into online network training.

2 Methods

2.1 Mathematical Model of the Anatomy-Informed Deformation

Model-driven spatial transformations simulate realistic soft-tissue deformations, which are part of the physiology, and can highly affect the shape of the prostate as well as the lesions in it. As the computation of state-of-the-art FEM models does not scale to on-the-fly DA, we introduce simplifications to be able to integrate such a biomechanical model as an online DA into the model training:

- soft tissue deformation of the prostate is mostly the result of morphological changes in the bladder and rectal space [4,6],
- due to the isotropic mechanical behavior of the rectum and the bladder [19], we apply isotropic deformation to them,
- we assume similar elastic modulus between the prostate and surrounding muscles [16], allowing us to approximate these tissue classes as homogeneous,
- we introduce a non-linear component into the model by transforming the surrounding tissue proportionally to the distance from the rectum and bladder in order to generate realistic deformations [23].

Based on them, we define the vector field V for the transformation as the gradient of the convolution between the Gaussian kernel G_σ and the indicator function S_{organ}, multiplied by a scalar C to control deformation amplitude and direction:

$$V = \nabla(G_\sigma * S_{organ}(x, y, z)) \cdot C. \tag{1}$$

The resulting V serves as the deformation field for an MRI sequence $I(x, y, z)$:

$$I_{deformed}(x, y, z) = I(x + V_x(x, y, z), y + V_y(x, y, z), z + V_z(x, y, z)). \quad (2)$$

It allows us to simulate the distension or evacuation of the bladder or rectal space. We refer to this transformation as anatomy-informed deformation. We make it publicly available in Batchgenerators [9] and integrate it into a nnU-Net trainer https://github.com/MIC-DKFZ/anatomy_informed_DA.

2.2 Experimental Setting

We evaluate our anatomy-informed DA qualitatively as well as quantitatively.

First, we visually inspect whether our assumptions in Sect. 2.1 regarding pelvic biomechanics resulted in realistic transformations. We apply either our proposed transformation to the rectum or the bladder, random deformable or no transformation in randomly selected exams and conduct a strict Turing test with clinicians having different levels of radiology expertise (a freshly graduated clinician (C.E.) and resident radiologists (C.M., K.S.Z.), 1.5 - 3 years of experience in prostate MRI) to determine if they can notice the artificial deformation.

Finally, we quantify the effect of our proposed transformation on the clinical task of patient-level PCa diagnosis and lesion-level PCa detection. We derive the diagnosis through semantic segmentation of the malignant lesions following previous studies [5,11,12,20,21]. Semantic segmentation provides interpretable predictions that are sensitive to spatial transformations, making it appropriate for testing spatial DAs. To compare the performance of the trained models to radiologists, we calculate their performance using the clinical PI-RADS scores and histopathological ground truths. To consider clinically informative results, we use the partial area under the Receiver Operating Characteristic (pAUROC) for patient-level evaluation with the sensitivity threshold of 78.75%, which is 90% of the sensitivity of radiologists for PI-RADS $\geq$ 4. Additionally, we calculate the F_1-score at the sensitivity of PI-RADS $\geq$ 4. Afterward, we evaluate model performances on object-level using the Free-Response Receiver Operating Characteristic (FROC) and the number of detections at the radiologists' lesion level performance for PI-RADS $\geq$ 4, at 0.32 average number of False Positives per scan. Objects were derived by applying a threshold of 0.5 to the softmax outputs followed by connected component analysis to identify connected regions in the segmentation maps. Predictions with an Intersection over Union of 0.1 with a ground truth object were considered True Positives. To systematically compare the effect of our proposed anatomy-informed DA with the commonly used settings, we create three main DA schemes:

1. **Basic** DA setting of nnU-Net [8], which is an extensive augmentation pipeline containing simple spatial transformations, namely translation, rotation and scaling. This setting is our reference DA scheme.
2. **Random deformable** transformations as implemented in the nnU-Net [8] DA pipeline extending the basic DA scheme (1) to test its presence in the medical domain. Our hypothesis is that it will produce counterproductive examples, resulting in inferior performance compared to our proposed DA.

3. Proposed **anatomy-informed** transformation in addition to the simple DA scheme (1). We define two variants of it:
   (a) Deforming only the rectum, as rectal distension has the highest influence among the organs on the shapes of the prostate lesions [4].
   (b) Deforming the bladder in addition to the rectum, as bladder deformations also have an influence on lesions, although smaller.

2.3   Prostate MRI Data

774 consecutive bi-parametric prostate MRI examinations are included in this study, which were acquired in-house during the clinical routine. The ethics committee of the Medical Faculty Heidelberg approved the study (S-164/2019) and waived informed consent to enable analysis of a consecutive cohort. All experiments were performed in accordance with the declaration of Helsinki [2] and relevant data privacy regulations. For every exam, PI-RADS v2 [24] interpretation was performed by a board-certified radiologist. Every patient underwent extended systematic and targeted MRI trans-rectal ultrasound-fusion transperineal biopsy. Malignancy of the segmented lesions was determined from a systematic-enhanced lesion ground-truth histopathological assessment, which has demonstrated reliable ground-truth assessment with sensitivity comparable to radical prostatectomy [17]. The samples were evaluated according to the International Society of Urological Pathology (ISUP) standards under the supervision of a dedicated uropathologist. Clinically significant prostate cancer (csPCa) was defined as ISUP grade 2 or higher. Based on the biopsy results, every csPCa lesion was segmented on the T2-weighted sequences retrospectively by multiple in-house investigators under the supervision of a board-certified radiologist. In addition to the lesions, the rectum and the bladder segmentations were automatically predicted by a model built upon nnU-Net [8] trained iteratively on an in-house cohort initially containing a small portion of our cohort. Multiple radiologists confirmed the quality of the predicted segmentations.

2.4   Training Protocol

774 exams were split into 80% training set (619 exams) and 20% test set (155 exams) by stratifying them based on the prevalence of csPCa (36.3%). The MRI sequences were registered using B-spline transformation based on mutual information to match the ground-truth segmentations across all modalities [12,14]. As the limited number of exams with csPCa and the small lesion size compared to the whole image can cause instability during training, we adapted the cropping strategy from [21] by keeping the organ segmentations to use the anatomy-informed DA (offsets of ± 9 mm axial to the prostate and ± 11.25 mm in the axial plane to the rectum and the bladder). The images are preprocessed by the automated algorithm of nnU-Net [8]. We trained 3D nnU-Net models in 5-fold cross-validation with different spatial DA schemes, see Sect. 2.2. The hyperparameter C of the anatomy-informed DA was optimized using validation

results, sampled during training with uniform distribution constrained by amplitude values in positive and negative directions of $C = \{300, 600, 900, 1200, 1500\}$. $C_{rectum} = 1200$ and $C_{bladder} = 600$ were selected for the final models. Compared to the standard nnU-Net settings, we implemented balanced sampling regarding the prevalence of csPCa and reduced the number of epochs to 350 to avoid overfitting. We used Mish activation function, Ranger optimizer, cosine anneal learning rate scheduler, and initial learning rate of 0.001 following [12]. The final models are ensembled and evaluated on the independent test set using bootstrapping with 1000 replications to provide standard deviation and to calculate p-values for the F_1-score and for the number of detected lesions using two-sided t-test to determine statistical significance.

3 Results

The anatomy-informed transformation produced highly realistic soft tissue deformations. Figure 2 shows an example of the transformation simulating rectum distensions with prostate lesions at different distances from the rectum. 92% of the rectum and 93% of the bladder deformation from the randomly picked exams became so realistic that our freshly graduated clinician did not detect them, but our residents noticed 87.5% of the rectum and 25% of the bladder deformations based on small transformation artifacts and their expert intuition. Irregularities resulted from the random elastic deformations can be easily detected, in contrast to our method being challenging to detect its artificial nature.

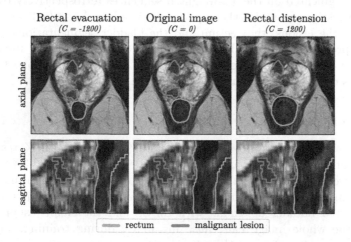

Fig. 2. Results of the proposed anatomy-informed deformation on the rectum, showing the outlines of the rectum and malignant lesions. The middle images show the original MRI sequence, the left images simulate rectal space evacuation, while the right images rectal distension. The transformation induces localized soft tissue deformations, resulting in changes in the shape of lesions only in the adjacent peripheral prostate zone.

In Table 1 we summarize the patient-level pAUROC and F_1-scores; and lesion-level FROC results on the independent test set showing the advantage of using anatomy-informed DA. To further highlight the practical advantage of the proposed augmentation, we compare the performance of the trained models to the radiologists' diagnostic performance for PI-RADS $\geq$ 4, which locate the most informative performance point clinically on the ROC diagram, see Fig. 3.

Table 1. Prostate cancer detection results on our independent test set

DA scheme	pAUROC	F_1-score	FROC
1. basic (reference)	$44.33 \pm 11.65\%$	$57.31 \pm 3.14\%$	$58.14 \pm 5.79\%$
2. random elastic	$38.94 \pm 14.38\%$	$56.98 \pm 3.08\%$	$58.63 \pm 5.42\%$
3.a) proposed (rectum)	$59.92 \pm 13.27\%$	$61.64 \pm 3.61\%$	$59.55 \pm 5.97\%$
3.b) proposed (rectum + bladder)	$53.27 \pm 13.42\%$	$62.42 \pm 3.84\%$	$59.93 \pm 5.53\%$

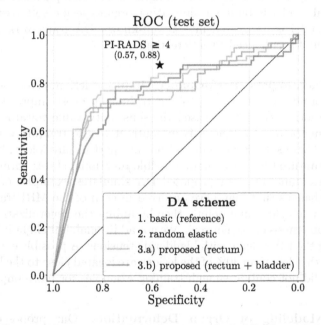

Fig. 3. The effect of different spatial DA schemes on the ROC. The radiologists' performance with PI-RADS $\geq$ 4 is marked to locate the most informative performance point clinically. Both variants of the proposed anatomy-informed DA (3.a and 3.b) increased the sensitivity value around the clinical PI-RADS $\geq$ 4 performance point compared to the simple (1) and random elastic (3) DA schemes, approaching it closely.

Extending the basic DA scheme with the proposed anatomy-informed deformation not only increased the sensitivity closely matching the radiologists' patient-level diagnostic performance but also improved the detection of PCa on a lesion level. Interestingly, while the use of random deformable transformation also improved lesion-level performance, it did not approach the diagnostic performance of the radiologists, unlike the anatomy-informed DA.

At the selected patient- and object-level working points, the model with the proposed rectum- and bladder-informed DA scheme reached the best results with significant improvements ($p < 0.05$) compared to the model with the basic DA setting by increasing the F_1-score with 5.11% and identifying 4 more lesions (5.3%) from the 76 lesions in our test set.

The time overhead introduced by anatomy-informed augmentation caused no increase in the training time, the GPU remained the main bottleneck.

4   Discussion

This paper addresses the utilization of anatomy-informed spatial transformations in the training procedure to increase lesion, prostate, and adjacent organ shape variability for the task of PCa diagnosis. For this purpose, a lightweight mathematical model is built for simulating organ-specific soft tissue deformations. The model is integrated into a well-known DA framework and used in model training for enhanced PCa detection.

Towards Radiologists' Performance. Inducing lesion shape variability via anatomy-informed augmentation to the training process improved the lesion detection performance and increased the sensitivity value towards radiologist-level performance in PCa diagnosis in contrast to the training with the basic DA setting. These soft tissue deformations are part of physiology, but only one snapshot is captured from the many possible functional states within each individual MR examination. Our proposed DA simulates examples of physiologic anatomical changes that may have occurred in each of the MRI training examples at the same exam time points, thereby aiding the generalization ability as well as the robustness of the network. We got additional, but slight improvements by extending the DA scheme with bladder distensions. A possible explanation for this result is that less than 30% of the lesions are located close to the bladder, and our dataset did not contain enough training examples for more improvements.

Realistic Modeling of Organ Deformation. Our proposed anatomy-informed transformation was designed to mimic real-world deformations in order to preserve essential image features. Most of the transformed sequences successfully passed the Turing test against a freshly graduated clinician with prostate MRI expertise, and some were even able to pass against radiology residents with more expertise. To support the importance of realism in DA quantitatively, we compared the performance of the basic and our anatomy-informed DA scheme with that of the random deformable transformation. The random deformable DA scheme generated high lesion shape variability, but it resulted in lower performance values. This could be due to the fact that it can also cause implausible or even harmful image warping, distorting important features, and producing counterproductive training examples. In comparison, our proposed anatomy-informed DA outperformed the basic and random deformable DA, demonstrating the significance of realistic transformations for achieving superior model performance.

High Applicability with Limitations. The easy integration into DA frameworks and no increase in the training time make our proposed anatomy-informed DA highly applicable. Its limitation is the need for additional organ segmentations, which requires additional effort from the annotator. However, pre-trained networks for segmenting anatomical structures like nnU-Net [8] have been introduced recently, which can help to overcome this limitation. Additionally, our transformation computation allows certain errors in the organ segmentations compared to applications where fully accurate segmentations are needed. The success of anatomy-informed DA opens the research question of whether it enhances performance across diverse datasets and model backbones.

5   Conclusion

In this work, we presented a realistic anatomy-informed augmentation, which mimics typical organ deformations in the pelvis. Inducing realistic soft-tissue deformations in the model training via this kind of organ-dependent transformation increased the diagnostic accuracy for PCa, closely approaching radiologist-level performance. Due to its simple and fast calculation, it can be easily integrated into DA frameworks and can be applied to any organ with similar distension properties. Due to these advantages, the shown improvements in the downstream task strongly motivate to utilize this model as a blueprint for other applications.

References

1. Ali, A., Du Feu, A., Oliveira, P., Choudhury, A., Bristow, R.G., Baena, E.: Prostate zones and cancer: lost in transition? Nat. Rev. Urol. **19**(2), 101–115 (2022)
2. Association, W.M., et al.: 64th WMA general assembly Fortaleza Brazil (2013). WMA Declaration of Helsinki-Ethical Principles for Medical Research Involving Human Subjects (2018)
3. Bhattacharya, I., et al.: A review of artificial intelligence in prostate cancer detection on imaging. Ther. Adv. Urol. **14**, 17562872221128792 (2022)
4. Boubaker, M.B., Ganghoffer, J.F.: Bladder/prostate/rectum: biomechanical models of the mobility of pelvic organs in the context of prostate radiotherapy. In: Biomechanics of Living Organs, pp. 307–324. Elsevier (2017)
5. Duran, A., Dussert, G., Rouvière, O., Jaouen, T., Jodoin, P.M., Lartizien, C.: ProstAttention-Net: a deep attention model for prostate cancer segmentation by aggressiveness in MRI scans. Med. Image Anal. **77**, 102347 (2022)
6. Engels, R.R., Israël, B., Padhani, A.R., Barentsz, J.O.: Multiparametric magnetic resonance imaging for the detection of clinically significant prostate cancer: what urologists need to know. Part 1: acquisition. Eur. Urol. **77**(4), 457–468 (2020)

7. Hu, Y., et al.: Adversarial deformation regularization for training image registration neural networks. In: Frangi, A.F., Schnabel, J.A., Davatzikos, C., Alberola-López, C., Fichtinger, G. (eds.) MICCAI 2018. LNCS, vol. 11070, pp. 774–782. Springer, Cham (2018). https://doi.org/10.1007/978-3-030-00928-1_87

8. Isensee, F., Jaeger, P.F., Kohl, S.A., Petersen, J., Maier-Hein, K.H.: nnU-Net: a self-configuring method for deep learning-based biomedical image segmentation. Nat. Methods 18(2), 203–211 (2021)

9. Isensee, F., et al.: Batchgenerators - a python framework for data augmentation (2020). https://github.com/MIC-DKFZ/batchgenerators

10. Khallaghi, S., et al.: Statistical biomechanical surface registration: application to MR-TRUS fusion for prostate interventions. IEEE Trans. Med. Imaging 34(12), 2535–2549 (2015)

11. Kohl, S., et al.: Adversarial networks for the detection of aggressive prostate cancer. In: Workshop on Machine Learning for Health (NIPS ML4H 2017) (2017)

12. Netzer, N., et al.: Fully automatic deep learning in bi-institutional prostate magnetic resonance imaging: effects of cohort size and heterogeneity. Invest. Radiol. 56(12), 799–808 (2021)

13. Payan, Y., Ohayon, J.: Biomechanics of living organs: hyperelastic constitutive laws for finite element modeling. World Bank Publications (2017)

14. Pellicer-Valero, O.J., et al.: Deep learning for fully automatic detection, segmentation, and Gleason grade estimation of prostate cancer in multiparametric magnetic resonance images. Sci. Rep. 12(1), 1–13 (2022)

15. Pfeiffer, M., Riediger, C., Weitz, J., Speidel, S.: Learning soft tissue behavior of organs for surgical navigation with convolutional neural networks. Int. J. Comput. Assist. Radiol. Surg. 14(7), 1147–1155 (2019). https://doi.org/10.1007/s11548-019-01965-7

16. Qasim, M., et al.: Biomechanical modelling of the pelvic system: improving the accuracy of the location of neoplasms in MRI-TRUS fusion prostate biopsy. BMC Cancer 22(1), 1–10 (2022)

17. Radtke, J.P., et al.: Multiparametric magnetic resonance imaging (MRI) and MRI-transrectal ultrasound fusion biopsy for index tumor detection: correlation with radical prostatectomy specimen. Eur. Urol. 70(5), 846–853 (2016)

18. Romaguera, L.V., Mezheritsky, T., Mansour, R., Carrier, J.F., Kadoury, S.: Probabilistic 4D predictive model from in-room surrogates using conditional generative networks for image-guided radiotherapy. Med. Image Anal. 74, 102250 (2021)

19. Rubod, C., et al.: Biomechanical properties of human pelvic organs. Urology 79(4), e17–e22 (2012)

20. Saha, A., Hosseinzadeh, M., Huisman, H.: End-to-end prostate cancer detection in bpMRI via 3d CNNs: effects of attention mechanisms, clinical priori and decoupled false positive reduction. Med. Image Anal. 73, 102155 (2021)

21. Sanyal, J., Banerjee, I., Hahn, L., Rubin, D.: An automated two-step pipeline for aggressive prostate lesion detection from multi-parametric MR sequence. AMIA Summits Transl. Sci. Proceed. 2020, 552 (2020)

22. Shorten, C., Khoshgoftaar, T.M.: A survey on image data augmentation for deep learning. J. Big Data 6(1), 1–48 (2019)

23. Wang, Y., Ni, D., Qin, J., Xu, M., Xie, X., Heng, P.A.: Patient-specific deformation modelling via elastography: application to image-guided prostate interventions. Sci. Rep. 6(1), 1–10 (2016)

24. Weinreb, J.C., et al.: PI-RADS prostate imaging-reporting and data system: 2015, version 2. Eur. Urol. 69(1), 16–40 (2016)

Clinical Evaluation of AI-Assisted Virtual Contrast Enhanced MRI in Primary Gross Tumor Volume Delineation for Radiotherapy of Nasopharyngeal Carcinoma

Wen Li[1], Dan Zhao[2], Zhi Chen[1], Zhou Huang[2], Saikit Lam[1], Yaoqin Xie[3], Wenjian Qin[3], Andy Lai-Yin Cheung[4], Haonan Xiao[1], Chenyang Liu[1], Francis Kar-Ho Lee[5], Kwok-Hung Au[5], Victor Ho-Fun Lee[4], Jing Cai[1,6(✉)], and Tian Li[1(✉)]

[1] The Hong Kong Polytechnic University, Hong Kong SAR, China
Jing.cai@polyu.eud.hk, litian.li@polyu.edu.hk
[2] Department of Radiation Oncology, Beijing Cancer Hospital & Institute, Beijing, China
[3] Shenzhen Institute of Advanced Technology, Chinese Academy of Science, Shenzhen, China
[4] Department of Clinical Oncology, The University of Hong Kong, Hong Kong SAR, China
[5] Department of Clinical Oncology, Queen Elizabeth Hospital, Hong Kong SAR, China
[6] Research Institute for Smart Ageing, The Hong Kong Polytechnic University, Hong Kong SAR, China

Abstract. This study aims to investigate the clinical efficacy of AI generated virtual contrast-enhanced MRI (VCE-MRI) in primary gross-tumor-volume (GTV) delineation for patients with nasopharyngeal carcinoma (NPC). We retrospectively retrieved 303 biopsy-proven NPC patients from three oncology centers. 288 patients were used for model training and 15 patients were used to synthesize VCE-MRI for clinical evaluation. Two board-certified oncologists were invited for evaluating the VCE-MRI in two aspects: image quality and effectiveness in primary tumor delineation. Image quality of VCE-MRI evaluation includes distinguishability between real contrast-enhanced MRI (CE-MRI) and VCE-MRI, clarity of tumor-to-normal tissue interface, veracity of contrast enhancement in tumor invasion risk areas, and efficacy in primary tumor staging. For primary tumor delineation, the GTV was manually delineated by oncologists. Results showed the mean accuracy to distinguish VCE-MRI from CE-MRI was 53.33%; no significant difference was observed in clarity of tumor-to-normal tissue interface between VCE-MRI and CE-MRI; for the veracity of contrast enhancement in tumor invasion risk areas and efficacy in primary tumor staging, a Jaccard Index of 76.04% and accuracy of 86.67% were obtained, respectively. The image quality evaluation suggests that the quality of VCE-MRI is approximated to real CE-MRI. In tumor delineation evaluation, the Dice Similarity Coefficient and Hausdorff Distance of the GTVs that delineated from VCE-MRI and CE-MRI were 0.762 (0.673–0.859) and 1.932 mm (0.763 mm–2.974 mm) respectively, which were clinically acceptable according to the experience of the radiation oncologists. This study demonstrated the VCE-MRI is highly promising in replacing the use of gadolinium-based CE-MRI for NPC delineation.

Keywords: MRI · Nasopharyngeal Carcinoma · Tumor Delineation

© The Author(s), under exclusive license to Springer Nature Switzerland AG 2023
H. Greenspan et al. (Eds.): MICCAI 2023, LNCS 14226, pp. 541–550, 2023.
https://doi.org/10.1007/978-3-031-43990-2_51

1 Introduction

Nasopharyngeal carcinoma (NPC), also known as lymphoepithelioma, is a highly aggressive malignancy that originates in nasopharynx [1]. NPC is characterized by a distinct geographical distribution in Southeast Asia, North Africa, and Arctic [2]. In China, NPC accounts for up to 50% of all head and neck cancers, while in Southeast Asia, NPC accounts for more than 70% of all head and neck cancers [3]. Radiotherapy (RT) is currently the main treatment remedy, which needs precise tumor delineation to ensure a satisfactory RT outcome. However, accurately delineating the NPC tumor is challenging due to the highly infiltrative nature of NPC and its complex location, which is surrounded by critical organs such as brainstem, spinal cord, temporal lobes, etc. To improve the visibility of NPC tumor for precise gross-tumor-volume (GTV) delineation, contrast-enhanced MRI (CE-MRI) is administrated through injection of gadolinium-based contrast agents (GBCAs) during MRI scanning. Despite the superior tumor-to-normal tissue contrast of CE-MRI, the use of GBCAs during MRI scanning can result in a fatal systemic disease known as nephrogenic systemic fibrosis (NSF) in patients with renal insufficiency [4]. NSF can cause severe physical impairment, such as joint contractures of fingers, elbows, and knees, and can progress to involve critical organs such as the heart, diaphragm, pleura, pericardium, kidney, liver, and lung [5]. It was reported that the incidence rate of NSF is around 4% after GBCA administration in patients with severe renal insufficiency, and the mortality rate can reach 31% [6]. Currently, there is no effective treatment for NSF, making it crucial to find a CE-MRI alternative for patients at risk of NSF.

In recent years, artificial intelligence (AI), especially deep learning, plays a game-changing role in medical imaging [7, 8], which showed great potential to eliminate the use of the toxic GBCAs through synthesizing virtual contrast-enhanced MRI (VCE-MRI) from gadolinium-free sequences, such as T1-weighted (T1w) and T2-weighted (T2w) MRI [9–12]. In 2018, Gong et al. [11] utilized pre-contrast and 10% low-dose T1w MRI to synthesize the VCE-MRI for brain disease diagnosis using a U-shape model, they found that gadolinium dose is able to be reduced by 10-fold by deep learning while the contrast information could be preserved. Followed by their work, Kleesiek et al. [10] proposed a Bayesian model to explore the feasibility of synthesizing VCE-MRI from contrast-free sequences, their study demonstrated that deep learning is highly feasible to totally eliminate the use of GBCAs. In the area of RT, Li et al. [9] developed a multi-input model to synthesize VCE-MRI for NPC RT. In addition to the advantage of eliminating the use of GBCA, VCE-MRI synthesis can also speed up the clinical workflow by eliminating the need for acquiring CE-MRI scan, which saves time for both clinical staff and patients. However, current studies mostly focus on algorithms development while lack comprehensive clinical evaluations to demonstrate the efficacy of the synthetic VCE-MRI in clinical settings.

The clinical evaluation of AI-based techniques is of paramount importance in healthcare. Rigorous clinical evaluations can establish the safety and efficacy of AI-based techniques, identify potential biases and limitations, and facilitate the integration of clinical expertise to ensure accurate and meaningful results [13]. Furthermore, the clinical evaluation of AI-based techniques can help identify areas for improvement and optimization, leading to development of more effective algorithms.

To bridge this bench-to-bedside research gap, in this study, we conducted a series of clinical evaluations to assess the effectiveness of synthetic VCE-MRI in NPC delineation, with a particular focus on assessment in VCE-MRI image quality and primary GTV delineation. This study has two main novelties: (i) To the best of our knowledge, this is the first clinical evaluation study of the VCE-MRI technique in RT; and (ii) multi-institutional MRI data were included in this study to obtain more reliable results. The success of this study would fill the current knowledge gap and provide the medical community with a clinical reference prior to clinical application of the novel VCE-MRI technique in NPC RT.

2 Materials and Methods

2.1 Data Description

Patient data was retrospectively collected from three oncology centers in Hong Kong. This dataset included 303 biopsy-proven (stage I-IVb) NPC patients who received radiation treatment during 2012–2016. The three hospitals were labelled as Institution-1 (110 patients), Institution-2 (58 patients), and Institution-3 (135 patients), respectively. For each patient, T1w MRI, T2w MRI, gadolinium-based CE-MRI, and planning CT were retrieved. MRI images were automatically registered as MRI images for each patient were scanned in the same position. The use of this dataset was approved by the Institutional Review Board of the University of Hong Kong/Hospital Authority Hong Kong West Cluster (HKU/HA HKW IRB) with reference number UW21-412, and the Research Ethics Committee (Kowloon Central/Kowloon East) with reference number KC/KE-18-0085/ER-1. Due to the retrospective nature of this study, patient consent was waived. For model development, 288 patients were used for model development and 15 patients were used to synthesize VCE-MRI for clinical evaluation. The details of patient characteristics and the number split for training and testing of each dataset were illustrated in Table 1. Prior to model training, MRI images were resampled to 256*224 by bilinear interpolation [14] due to the inconsistent matrix sizes of the three datasets.

2.2 VCE-MRI Synthesis Network

The multimodality-guided synergistic neural network (MMgSN-Net) was applied to learn the mapping from T1w MRI and T2w MRI to CE-MRI. The MMgSN-Net was a 2D network. The effectiveness of this network in VCE-MRI synthesis for NPC patients has been demonstrated by Li et al. in [9]. T1w MRI and T2w MRI were used as input and corresponding CE-MRI was used as learning target. In this work, we obtained 12806 image pairs for model training. Different from the original study, which used single institutional data for model development and utilized min-max value of the whole dataset for data normalization, in this work, we used mean and standard deviation of each individual patient to normalize MRI intensities due to the heterogeneity of the MRI intensities across institutions [15].

544   W. Li et al.

Table 1. Details of the multi-institutional patient characteristics. FS: field strength; TR: repetition time; TE: echo time; No.: Number; Avg: average.

Institution (Vendor-FS)	Patient No. (train/test)	Avg. age	Modality	TR (ms)	TE (ms)
Institution-1 (Siemens-1.5T)	110 (105/5)	56 ± 11	T1w	562–739	13–17
			T2w	7640	97
			CE-MRI	562–739	13–17
Institution-2 (Philips-3T)	58 (53/5)	49 ± 15	T1w	4.8–9.4	2.4–8.0
			T2w	3500–4900	50–80
			CE-MRI	4.8–9.4	2.4–8.0
Institution-3 (Siemens-3T)	135 (130/5)	57 ± 12	T1w	620	9.8
			T2w	2500	74
			CE-MRI	3.42	1.11

2.3 Clinical Evaluations

The evaluation methods used in this study included image quality assessment of VCE-MRI and primary GTV delineation. Two board-certified radiation oncologists (with 8 years' and 6 years' clinical experience, respectively) were invited to perform the VCE-MRI quality assessment and GTV delineation according to their clinical experience. Considering the clinical burden of oncologists, 15 patients were included for clinical evaluations. All clinical evaluations were performed on an Eclipse workstation (V5.0.10411.00, Varian Medical Systems, USA) with the same monitor, and the window/level can be adjusted freely by the oncologists. The results were obtained under the consensus of the two oncologists.

Image Quality Assessment of VCE-MRI. To evaluate the image quality of synthetic VCE-MRI against the real CE-MRI, we conducted four RT-related evaluations: (i) distinguishability between CE-MRI and VCE-MRI; (ii) clarity of tumor-to-normal tissue interface; (iii) veracity of contrast enhancement in tumor invasion risk areas; and (iv) efficacy in primary tumor staging. The VCE-MRI and CE-MRI volumes were imported as individual patients to Eclipse system and randomly and blindly shown to oncologists for evaluation. The MRI volumes were shown in axial view, sagittal view and coronal view, and the oncologists can scroll through the slices to view adjacent image slices.

(i) *Distinguishability between CE-MRI and VCE-MRI.* To evaluate the reality of VCE-MRI, oncologists were invited to differentiate the synthetic patients (i.e., image volumes that generated from synthetic VCE-MRI) from real patients (i.e., image volumes that generated from real CE-MRI). Different from the previous studies that utilized limited number (20-50 slices, axial view) of 2D image slices for reality evaluation [9, 10], we used 3D volumes in this study to help oncologists visualize the inter-slice adjacent information. The judgement results were recorded, and the accuracy of each institution and the overall accuracy were calculated.

(ii) *Clarity of tumor-to-normal tissue interface.* The clarity of tumor-normal tissue interface is critical for tumor delineation, which directly affects the delineation outcomes. Oncologists were asked to use a 5-point Likert scale ranging from 1 (poor) to 5 (excellent) to evaluate the clarity of tumor-to-normal tissue interface. Paired two-tailed t-test (with a significance level of $p = 0.05$) was applied to analyses if the scores obtained from real patients and synthetic patients are significantly different.

(iii) *Veracity of contrast enhancement in tumor invasion risk areas.* In addition to the critical tumor-normal tissue interface, the areas surrounding the NPC tumor will also be considered during delineation. To better evaluate the veracity of contrast enhancement in VCE-MRI, we selected 25 tumor invasion risk areas according to [16], including 13 high-risk areas and 12 medium-risk areas, and asked oncologists to determine whether these areas were at risk of being invaded according to the contrast-enhanced tumor regions. The 13 high-risk areas include: retropharyngeal space, parapharyngeal space, levator veli palatine muscle, prestyloid compartment, Tensor veli palatine muscle, poststyloid compartment, nasal cavity, pterygoid process, basis of sphenoid bone, petrous apex, prevertebral muscle, clivus, and foramen lacerum. The 12 medium-risk areas include foramen ovale, great wing of sphenoid bone, medial pterygoid muscle, oropharynx, cavernous sinus, sphenoidal sinus, pterygopalatine fossa, lateral pterygoid muscle, hypoglossal canal, foramen rotundum, ethmoid sinus, and jugular foramen. The areas considered at risk of tumor invasion were recorded.

The Jaccard index (JI) [17] was utilized to quantitatively evaluate the results of recorded risk areas from CE-MRI and VCE-MRI. The JI could be calculated by:

$$JI = |R_{CE} \cap R_{VCE}| / |R_{CE} \cup R_{VCE}| \tag{1}$$

where R_{CE} and R_{VCE} represents the set of risk areas that recorded from CE-MRI and corresponding VCE-MRI, respectively. JI measures similarity of two datasets, which ranges from 0% to 100%. Higher JI indicates more similar of the two sets.

(iv) *Efficacy in primary tumor staging.* A critical RT-related application of CE-MRI is tumor staging, which plays a critical role in treatment planning and prognosis prediction [18]. To assess the efficacy of VCE-MRI in NPC tumor staging, oncologists were asked to determine the stage of the primary tumor shown in CE-MRI and VCE-MRI. The staging results from CE-MRI were taken as the ground truth and the staging accuracy of VCE-MRI was calculated.

Primary GTV Delineation. GTV delineation is the foremost prerequisite for a successful RT treatment of NPC tumor, which demands excellent precision [19]. An accurate tumor delineation improves local control and reduce toxicity to surrounding normal tissues, thus potentially improving patient survival [20]. To evaluate the feasibility of eliminating the use of GBCA by replacing CE-MRI with VCE-MRI in tumor delineation, oncologists were asked to contour the primary GTV under assistance of VCE-MRI. For comparison, CE-MRI was also imported to Eclipse for tumor delineation but assigned as a different patient, which were shown to oncologists in a random and blind manner. To mimic the real clinical setting, contrast-free T1w, T2w MRI and corresponding CT of each patient were imported into the Eclipse system since sometimes T1w and T2w MRI will also be referenced during tumor delineation. Due to both real patients and synthetic patients were involved in delineation, to erase the delineation memory of the same patient, we separated the patients to two datasets, each with the same number of patients, both two datasets with mixed real patients and synthetic patients without overlaps (i.e., the CE-MRI and VCE-MRI from the same patient are not in the same dataset).When finished the first dataset delineation, there was a one-month interval before the delineation of the second dataset. After the delineation of all patients, the Dice similarity coefficient (*DSC*) [21] and Hausdorff distance (*HD*) [22] of the GTVs delineated from real patients and corresponding synthetic patients were calculated to evaluate the accuracy of delineated contours.

Dice Similarity Coefficient (DSC). DSC is a broadly used metric to compare the agreement between two segmentations [23]. It measures the spatial overlap between two segmentations, which ranges from 0 (no spatial overlap) to 1 (complete overlap). The *DSC* can be expressed as:

$$DSC = 2 * |C_{CE} \cap C_{VCE}| / (|C_{CE}| + |C_{VCE}|) \tag{2}$$

where C_{CE} and C_{VCE} represent the contours delineated from real patients and synthetic patients, respectively.

Hausdorff Distance (HD). Even though DSC is a well-accepted segmentation comparison metric, it is easily influenced by the size of contours. Small contours typically receive lower *DSC* than larger contours [24].Therefore, *HD* was applied as a supplementary to make a more thorough comparison. *HD* is a metric to measure the maximum distance between two contours. Given two contours C_{CE} and C_{VCE}, the HD could be calculated as:

$$HD = max(max_{x \in C_{CE}} d(x, C_{VCE}), max_{y \in C_{VCE}} d(y, C_{CE})) \tag{3}$$

where $d(x, C_{VCE})$ and $d(y, C_{CE})$ represent the distance from point x in contour C_{CE} to contour C_{VCE} and the distance from point y in contour C_{VCE} to contour C_{CE}.

3 Results and Discussion

3.1 Image Quality of VCE-MRI

Table 2 summarizes the results of the four VCE-MRI quality evaluation metrics, including: (i) distinguishability between CE-MRI and VCE-MRI; (ii) clarity of tumor-to-normal tissue interface; (iii) veracity of contrast enhancement in tumor invasion risk areas; and (iv) efficacy in primary tumor staging.

(i) **Distinguishability between CE-MRI and VCE-MRI.** The overall judgement accuracy for the MRI volumes was 53.33%, which is close to a random guess accuracy (i.e., 50%). For Institution-1, 2 real patients were judged as synthetic and 1 synthetic patient was considered as real. For Institution-2, 2 real patients were determined as synthetic and 4 synthetic patients were determined as real. For Institution-3, 2 real patients were judged as synthetic and 3 synthetic patients were considered as real. In total, 6 real patients were judged as synthetic and 8 synthetic patients were judged as real.

(ii) **Clarity of tumor-to-normal tissue interface.** The overall clarity scores of tumor-to-normal tissue interface for real and synthetic patients were 3.67 with a median of 4 and 3.47 with a median of 4, respectively. No significant difference was observed between these two scores ($p = 0.38$). The average scores for real and synthetic patients were 3.6 and 3, 3.6 and 3.8, 3.8 and 3.6 for Institution-1, Institution-2, and Institution-3, respectively. 5 real patients got a higher score than synthetic patients and 3 synthetic patients obtained a higher score than real patients. The scores of the other 7 patient pairs were the same.

(iii) **Veracity of contrast enhancement in tumor invasion risk areas.** The overall *JI* score between the recorded tumor invasion risk areas from CE-MRI and VCE-MRI was 74.06%. The average *JI* obtained from Institution-1, Institution-2, and Institution-3 dataset were similar with a result of 71.54%, 74.78% and 75.85%, respectively. In total, 126 risk areas were recorded from the CE-MRI for all of the evaluation patients, while 10 (7.94%) false positive high risk invasion areas and 9 (7.14%) false negative high risk invasion areas were recorded from VCE-MRI.

(iv) **Efficacy in primary tumor staging.** A T-staging accuracy of 86.67% was obtained using VCE-MRI. 13 patient pairs obtained the same staging results. For the Institution-2 data, all synthetic patients observed the same stages as real patients. For the two T-stage disagreement patients, one synthetic patient was staged as phase IV while the corresponding real patient was staged as phase III, the other synthetic patient was staged as I while corresponding real patient was staged as phase III.

Table 2. Image quality evaluation results of VCE-MRI: (A) Distinguishability between CE-MRI and VCE-MRI; (B) Clarity of tumor-to-normal tissue interface; (C) Veracity of contrast enhancement in risk areas; and (D) T-staging. Abbreviations: Inst: Institution; C.A.: Center-based average; O.A.: Overall average; Syn: Synthetic.

	(A)			(B)					
	Inst-1	Inst-2	Inst-3	Inst-1		Inst-2		Inst-3	
	/	/	/	Real	Syn	Real	Syn	Real	Syn
C.A.	70%	40%	50%	3.6	3	3.6	3.8	3.8	3.6
O.A.	53.33%			Real: 3.67		Syn: 3.47			
	(C)			(D)					
	Inst-1	Inst-2	Inst-3	Inst-1		Inst-2		Inst-3	
C.A.	71.54%	74.78%	75.85%	80%		100%		80%	
O.A.	74.06%			86.67%					

Figure 1 illustrates an example of the synthetic VCE-MRI. Compared to CE-MRI, T1w MRI has the similar tumor shape but with indistinguishable tumor-to-normal tissue interface, while T2w MRI shows superior lesion contrast but with smaller tumor volume (yellow boxes). The deep learning model integrated the complementary information of T1w MRI and T2w MRI, and successfully synthesized VCE-MRI with similar contrast and tumor volume as CE-MRI, with no obvious contrast differences in tumor regions, as shown in difference map between CE-MRI and VCE-MRI.

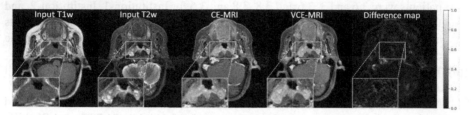

Fig. 1. Illustration of the synthetic VCE-MRI.

3.2 Primary GTV delineation

The average DSC and HD between the C_{CE} and C_{VCE} was 0.762 (0.673–0.859) with a median of 0.774, and 1.932 mm (0.763 mm–2.974 mm) with a median of 1.913 mm, respectively. For Institution-1, Institution-2, and Institution-3, the average DSC were 0.741, 0.794 and 0.751 respectively, while the average HD were 2.303 mm, 1.456 mm, and 2.037 mm respectively. Figure 2 illustrated the delineated primary GTV contours from an average patient with the DSC of 0.765 and HD of 1.938 mm. The green contour shows the primary GTV that delineated form the synthetic patient, while the red contour was delineated from corresponding real GBCA-based patient.

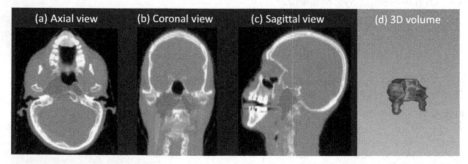

Fig. 2. Illustration of the primary GTVs from a typical patient with an average DSC and HD.

4 Conclusion

In this study, we conducted a series of clinical evaluations to validate the clinical efficacy of VCE-MRI in RT of NPC patients. Results showed the VCE-MRI has great potential to provide an alternative to GBCA-based CE-MRI for NPC delineation.

Acknowledgement. This research was partly supported by research grants of General Research Fund (GRF 15102219, GRF 15103520), the University Grants Committee, and Project of Strategic Importance Fund (P0035421), Projects of RISA (P0043001), One-line Budget (P0039824, P0044474), The Hong Kong Polytechnic University, and Shenzhen-Hong Kong-Macau S&T Program (Category C) (SGDX20201103095002019), Shenzhen Basic Research Program (R2021A067), Shenzhen Science and Technology Innovation Committee (SZSTI).

Data use Declaration:. The use of this dataset was approved by the Institutional Review Board of University of Hong Kong/Hospital Authority Hong Kong West Cluster (HKU/HA HKW IRB) with reference number UW21-412, and the Research Ethics Committee (Kowloon Central/Kowloon East) with reference number KC/KE-18-0085/ER-1. Due to the retrospective nature of this study, patient consent was waived.

References

1. Chen, Y.-P., et al.: Nasopharyngeal carcinoma. The Lancet **394**(10192), 64–80 (2019)
2. Chang, E.T., et al.: The evolving epidemiology of nasopharyngeal carcinoma. Cancer Epidemiol. Biomarkers Prev. **30**(6), 1035–1047 (2021)
3. Sturgis, E.M., Wei, Q., Spitz, M.R.: Descriptive epidemiology and risk factors for head and neck cancer. In: Seminars in Oncology. Elsevier (2004)
4. Sadowski, E.A., et al.: Nephrogenic systemic fibrosis: risk factors and incidence estimation. Radiology **243**(1), 148–157 (2007)
5. Thomsen, H.S.: Nephrogenic systemic fibrosis: a serious late adverse reaction to gadodiamide. Eur. Radiol. **16**(12), 2619–2621 (2006)
6. Schlaudecker, J.D., Bernheisel, C.R.: Gadolinium-associated nephrogenic systemic fibrosis. Am. Fam. Phys. **80**(7), 711–714 (2009)
7. Qin, W., et al.: Superpixel-based and boundary-sensitive convolutional neural network for automated liver segmentation. Phys. Med. Biol. **63**(9), 095017 (2018)
8. Qin, W., et al.: Automated segmentation of the left ventricle from MR cine imaging based on deep learning architecture. Biomed. Phys. Eng. Express **6**(2), 025009 (2020)
9. Li, W., et al.: Virtual contrast-enhanced magnetic resonance images synthesis for patients with nasopharyngeal carcinoma using multimodality-guided synergistic neural network. Int. J. Radiat. Oncol. Biol. Phys. **112**(4), 1033–1044 (2022)
10. Kleesiek, J., et al.: Can virtual contrast enhancement in brain MRI replace gadolinium? A feasibility study. Invest. Radiol. **54**(10), 653–660 (2019)
11. Gong, E., et al.: Deep learning enables reduced gadolinium dose for contrast-enhanced brain MRI. J. Magn. Reson. Imaging **48**(2), 330–340 (2018)
12. Li, W., et al. Multi-institutional investigation of model generalizability for virtual contrast-enhanced MRI synthesis. In: Wang, L., Dou, Q., Fletcher, P.T., Speidel, S., Li, S. (eds.) Medical Image Computing and Computer Assisted Intervention–MICCAI 2022. MICCAI 2022. LNCS. Vol. 13437, pp. 765–773. Springer, Cham (2022). https://doi.org/10.1007/978-3-031-16449-1_73

13. Topol, E.J.: High-performance medicine: the convergence of human and artificial intelligence. Nat. Med. **25**(1), 44–56 (2019)
14. Gribbon, K.T., Bailey, D.G.: A novel approach to real-time bilinear interpolation. In: Proceedings. DELTA 2004. Second IEEE International Workshop on Electronic Design, Test and Applications. IEEE (2004)
15. Li, W., et al.: Model generalizability investigation for GFCE-MRI synthesis in NPC radiotherapy using multi-institutional patient-based data normalization (2022). TechRxiv.Preprint
16. Liang, S.-B., et al.: Extension of local disease in nasopharyngeal carcinoma detected by magnetic resonance imaging: improvement of clinical target volume delineation. Int. J. Radiat. Oncol. Biol. Phys. **75**(3), 742–750 (2009)
17. Fletcher, S., Islam, M.Z.: Comparing sets of patterns with the Jaccard index. Australas. J. Inf. Syst. **22** (2018)
18. Lee, A.W., et al.: International guideline for the delineation of the clinical target volumes (CTV) for nasopharyngeal carcinoma. Radiother. Oncol. **126**(1), 25–36 (2018)
19. Jager, E.A., et al.: GTV delineation in supraglottic laryngeal carcinoma: interobserver agreement of CT versus CT-MR delineation. Radiat. Oncol. **10**(1), 1–9 (2015)
20. Jameson, M.G., et al.: Correlation of contouring variation with modeled outcome for conformal non-small cell lung cancer radiotherapy. Radiother. Oncol. **112**(3), 332–336 (2014)
21. Balagopal, A., et al.: A deep learning-based framework for segmenting invisible clinical target volumes with estimated uncertainties for post-operative prostate cancer radiotherapy. Med. Image Anal. **72**, 102101 (2021)
22. Yang, J., et al.: A multimodality segmentation framework for automatic target delineation in head and neck radiotherapy. Med. Phys. **42**(9), 5310–5320 (2015)
23. Chang, H.-H., et al.: Performance measure characterization for evaluating neuroimage segmentation algorithms. Neuroimage **47**(1), 122–135 (2009)
24. Schreier, J., et al.: Clinical evaluation of a full-image deep segmentation algorithm for the male pelvis on cone-beam CT and CT. Radiother. Oncol. **145**, 1–6 (2020)

Multi-task Learning of Histology and Molecular Markers for Classifying Diffuse Glioma

Xiaofei Wang[1], Stephen Price[1], and Chao Li[1,2,3,4(✉)]

[1] Department of Clinical Neurosciences, University of Cambridge, Cambridge, UK
cl647@cam.ac.uk
[2] Department of Applied Mathematics and Theoretical Physics,
University of Cambridge, Cambridge, UK
[3] School of Science and Engineering, University of Dundee, Dundee, UK
[4] School of Medicine, University of Dundee, Dundee, UK

Abstract. Most recently, the pathology diagnosis of cancer is shifting to integrating molecular makers with histology features. It is a urgent need for digital pathology methods to effectively integrate molecular markers with histology, which could lead to more accurate diagnosis in the real world scenarios. This paper presents a first attempt to jointly predict molecular markers and histology features and model their interactions for classifying diffuse glioma bases on whole slide images. Specifically, we propose a hierarchical multi-task multi-instance learning framework to jointly predict histology and molecular markers. Moreover, we propose a co-occurrence probability-based label correction graph network to model the co-occurrence of molecular markers. Lastly, we design an inter-omic interaction strategy with the dynamical confidence constraint loss to model the interactions of histology and molecular markers. Our experiments show that our method outperforms other state-of-the-art methods in classifying diffuse glioma, as well as related histology and molecular markers on a multi-institutional dataset.

Keywords: Diffuse Glioma · Digital Pathology · Multi-task learning · Muti-label Classification

1 Introduction

Diffuse glioma is the most common and aggressive primary brain tumors in adults, accounting for more deaths than any other type [7]. Pathology diagnosis is the gold standard for diffuse glioma but is usually time-consuming and highly depends on the expertise of senior pathologists [13]. Hence, automatic algorithms based on histology whole slide images (WSIs) [15], namely digital pathology, promise to offer rapid diagnosis and aid precise treatment.

Supplementary Information The online version contains supplementary material available at https://doi.org/10.1007/978-3-031-43990-2_52.

Recently, deep learning has achieved success in diagnosing various tumors [2,21]. Most methods are mainly predicting histology based on WSI, less concerning molecular markers. However, the paradigm of pathological diagnosis of glioma has shifted to molecular pathology, reflected by the 2021 WHO Classification of Tumors of the Central Nervous System [14]. The role of key molecular markers, i.e., isocitrate dehydrogenas (IDH) mutations, co-deletion of chromosome 1p/19q and homozygous deletion (HOMDEL) of cyclin-dependent kinase inhibitor 2A/B (CDKN), have been highlighted as major diagnostic markers for glioma, while histology features that are traditionally emphasized are now considered as reference, although still relevant in many cases. For instance, in the new pathology scheme, glioblastoma is increasingly diagnosed according to IDH mutations, while previously its diagnosis mostly relies on histology features, including necrosis and microvascular proliferation (NMP).[1]

However, the primary approaches to assess molecular markers include gene sequencing and immuno-staining, which are time-consuming and expensive than histology assessment. As histology features are closely associated with molecular alterations, algorithm predicting molecular markers based on histology WSIs is feasible and have clinical significance. Moreover, under the new paradigm of integrating molecular markers with histological features into tumor classification, it is helpful to model the interaction of histology and molecular makers for a more accurate diagnosis. Therefore, there is an urgent need for developing novel digital pathology methods based on WSI to predict molecular markers and histology jointly and modeling their interactions for final tumor classification, which could be valuable for the clinically relevant diagnosis of diffuse glioma.

This paper proposes a deep learning model (DeepMO-Glioma) for glioma classification based on WSIs, aiming to reflect the molecular pathology paradigm. Previous methods are proposed to integrate histology and genomics for tumour diagnosis [3,10,20]. For instance, Chen et al. [3] proposed a multimodal fusion strategy to integrate WSIs and genomics for survival prediction. Xing et al. [20] devised a self-normalizing network to encode genomics. Nevertheless, most existing approaches of tumor classification only treat molecular markers as additional input, incapable to simultaneously predict the status of molecular markers, thus clinically less relevant under the current clinical diagnosis scheme. To jointly predict histology and molecular markers following clinical diagnostic pathway, we propose a novel hierarchical multi-task multi-instance learning (HMT-MIL) framework based on vision transformer [4], with two partially weight-sharing parts to jointly predict molecular markers and histology.

Moreover, multiple molecular markers are needed for classifying cancers, due to complex tumor biology. To reflect real-world clinical scenarios, we formulate predicting multiple molecular markers as a multi-label classification (MLC) task. Previous MLC methods have successfully modeled the correlation among labels [12,22]. For example, Yazici et al. [22] proposed an orderless recurrent method, while Li et al. designed a label attention transformer network with graph

---

[1] Similar changes of the diagnostic protocol can also be found in endometrial cancer [9], renal neoplasia [18], thyroid carcinomas [19], etc.

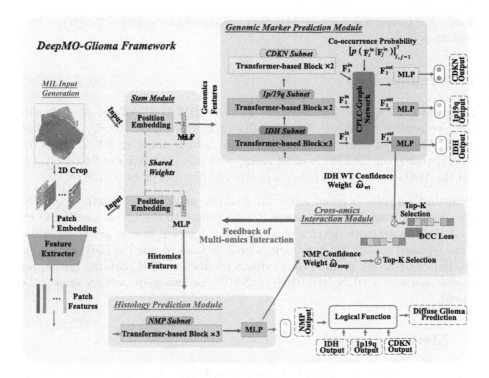

Fig. 1. Architecture of DeepMO-Glioma.

embedding. In medical domain, Zhang *et al.* [25] devised a dual-pool contrastive learning for classifying fundus and X-ray images. Despite success, when applied to predicting multiple molecular markers, most existing methods may ignore the co-occurrence of molecular markers, which have intrinsic associations [23]. Hence, we propose a co-occurrence probability-based, label-correlation graph (CPLC-Graph) network to model the co-occurrence of molecular markers, i.e., intra-omic relationship.

Lastly, we focus on modeling the interaction between molecular markers and histology. Specifically, we devise a novel inter-omic interaction strategy to model the interaction between the predictions of molecular markers and histology, e.g., IDH mutation and NMP, both of which are relevant in diagnosing glioblastoma. Particularly, we design a dynamical confidence constraint (DCC) loss that constrains the model to focus on similar areas of WSIs for both tasks. To the best of our knowledge, this is the first attempt to classify diffuse gliomas via modeling the interaction of histology and molecular markers.

Our main contributions are: (1) We propose a multi-task multi-instance learning framework to jointly predict molecular markers and histology and finally classify diffuse glioma, reflecting the new paradigm of pathology diagnosis. (2) We design a CPLC-Graph network to model the intra-omic relationship of multiple molecular markers. (3) We design a DCC learning strategy to model the inter-omic interaction between histology and molecular markers for glioma classification.

2   Preliminaries

Database: We use publicly available TCGA GBM-LGG dataset [6]. Following [15], we remove the WSIs of low quality or lack of labels. Totally, we include 2,633 WSIs from 940 cases, randomly split into training (2,087 WSIs of 752 cases), validation (282 WSIs of 94 cases) and test (264 WSIs of 94 cases) sets. All the WSIs are crop into patches of size $224\,\mathrm{px} \times 224\,\mathrm{px}$ at $0.5\,\mu\mathrm{m}\,\mathrm{px}^{-1}$.

Training Labels: Original lables for genomic markers and histology of WSIs are obtained from TCGA database [6]. According to the up-to-date WHO criteria [14], we generate the classification labels for each case as grade 4 glioblastoma (defined as IDH widetype), oligodendroglioma (defined as IDH mutant and 1p/19q co-deletion), grade 4 astrocytoma (defined as IDH mutant, 1p/19q non co-deletion with CDKN HOMDEL or NMP), or low-grade astrocytoma (other cases).

3   Methodology

Figure 1 illustrates the proposed DeepMO-Glioma. As shown above, the up-to-date WHO criteria incorporates molecular markers and histology features. Therefore, our model is designed to jointly learn the tasks of predicting molecular markers and histology features in a unified framework. DeepMO-Glioma consists four modules, i.e., stem, genomic marker prediction, histology prediction and cross-omics interaction. Given the cropped patches $\{\mathbf{X}_i\}_1^N$ as the input, DeepMO-Glioma outputs 1) the status of molecular markers, including IDH mutation $\hat{l}_{idh} \in \mathbb{R}^2$, 1p/19q co-deletion $\hat{l}_{1p/19q} \in \mathbb{R}^2$ and CDKN HOMDEL $\hat{l}_{cdkn} \in \mathbb{R}^2$, 2) existence of NMP $\hat{l}_{nmp} \in \mathbb{R}^2$ and 3) final diagnosis of diffuse gliomas $\hat{l}_{glio} \in \mathbb{R}^4$. Of note, the final diagnosis of diffuse gliomas is generated via a decision tree-based logical function with the input of predicted molecular markers and histology, consistent with the up-to-date WHO criteria.

3.1   Hierarchical Multi-task Multi-instance Learning

To extract global information from input $\{\mathbf{X}_i\}_1^N$, we propose a hierarchical multi-task multi-instance learning (HMT-MIL) framework for both histology and molecular marker predictions. Different from methods using one [24] or several [3,20] representative patches per slide, HMT-MIL framework can extract information from $N = 2,500$ patches per WSI via utilizing the MIL learning paradigm with transformer blocks [4] embedded. Note for WSIs with patch number< N, we adopt a biological repeat strategy for dimension alignment.

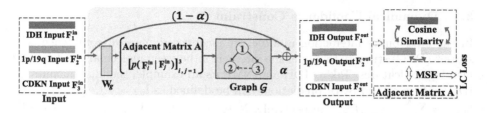

Fig. 2. Pipelines of CPLC-Graph network and LC loss.

3.2 Co-occurrence Probability-Based, Label-Correlation Graph

In predicting molecular markers, i.e., IDH, 1p/19q and CDKN, existing MLC methods based on label correlation may ignore the co-occurrence of the labels. In the genomic marker prediction module, we proposed a co-occurrence probability-based, label-correlation graph (CPLC-Graph) network and a label correlation (LC) loss for intra-omic modeling of the co-occurrence probability of the three markers.

1) **CPLC-Graph network:** CPLC-Graph (Fig. 2) is defined as $\mathcal{G} = (\mathbf{V}, \mathbf{E})$, where $\mathbf{V}$ indicates the nodes, while $\mathbf{E}$ represents the edges. Given the intermediate features in predicting the three molecular markers subnets $\mathbf{F}^{in} = [\mathbf{F}_i^{in}]_{i=1}^3 \in \mathbb{R}^{3 \times C}$ as input nodes, we construct a co-occurrence probability based correlation matrix $\mathbf{A} \in \mathbb{R}^{3 \times 3}$ to reflect the relationships among each node feature, with a weight matrix $\mathbf{W}_g \in \mathbb{R}^{C \times C}$ to update the value of $\mathbf{F}^{in}$. Formally, the output nodes $\mathbf{F}^{mid} \in \mathbb{R}^{3 \times C}$ are formulated by a single graph convolutional network layer as

$$\mathbf{F}^{mid} = \delta(\mathbf{A}\mathbf{F}^{in}\mathbf{W}_g), \text{where } \mathbf{A} = [A_i^j]_{i,j=1}^3, A_i^j = \frac{1}{2}\big(p(\mathbf{F}_i^{in}|\mathbf{F}_j^{in}) + p(\mathbf{F}_j^{in}|\mathbf{F}_i^{in})\big). \quad (1)$$

In (1), $\delta(\cdot)$ is an activation function and $p(\mathbf{F}_i^{in}|\mathbf{F}_j^{in})$ denote the probability of the status of i-th marker given the status of j-th marker. Besides, residual structure is utilized to generate the final output $\mathbf{F}^{out}$ of CPLC-Graph network, defined as $\mathbf{F}^{out} = \alpha\mathbf{F}^{mid} + (1 - \alpha)\mathbf{F}^{in}$, where α is a graph balancing hyper-parameter.

2) **LC loss:** In order to fully exploit the co-occurrence probability of different molecular markers, we further devise the LC loss that constrains the similarity between any two output molecular markers $\mathbf{F}_i^{out}$ and $\mathbf{F}_j^{out}$ to approach their correspondent co-occurrence probability A_i^j. Formally, the LC loss is defined as

$$\mathcal{L}_{LC} = \mathcal{MSE}(\mathbf{A}, \mathbf{D}_{cos}), \text{where } \mathbf{D}_{cos} = [D_{cos}^{i,j}]_{i,j=1}^3, D_{cos}^{i,j} = \frac{(\mathbf{F}_i^{out})^\top \mathbf{F}_j^{out}}{\|\mathbf{F}_i^{out}\| \|\mathbf{F}_j^{out}\|}. \quad (2)$$

In (2), $\mathcal{MSE}$ denotes the function of mean square error, while $D_{cos}^{i,j}$ is the cosine similarity of features $\mathbf{F}_i^{out}$ and $\mathbf{F}_j^{out}$.

3.3   Dynamical Confidence Constraint

In the cross-omics interaction module, we design a dynamical confidence constraint (DCC) strategy to model the interaction between molecular markers and histological features. Taking IDH and NMP as an example, the final outputs for IDH widetype[2] and NMP predictions can be defined as $\hat{l}_{wt} = \sum_{n=1}^{N} \omega_{wt}^n f_{wt}^n$ and $\hat{l}_{nmp} = \sum_{n=1}^{N} \omega_{nmp}^n f_{nmp}^n$, respectively. Note that f_{wt}^n and ω_{wt}^n are values of the extracted feature and the corresponding decision weight of n-th patch, respectively. We then reorder $[\omega_{wt}^n]_{n=1}^N$ to $[\hat{\omega}_{wt}^n]_{n=1}^N$ based on their values. Similarly, we obtain $[\hat{\omega}_{nmp}^n]_{n=1}^N$ for NMP confidence weights.

Based on ordered confidence weights, we constrain the prediction networks of histology and molecular markers to focus on the WSI areas important for both predictions, thus modeling inter-omic interactions. Specifically, we achieve the confidence constraint through a novel DCC loss focusing on top K important patches for both prediction. Formally, the DCC loss in m-th training epoch is defined as:

$$\mathcal{L}_{\text{DCC}} = \frac{1}{2K_m} \sum_{k=1}^{K_m} \left(\mathcal{S}(\hat{\omega}_{wt}^k, \hat{\omega}_{nmp}) + \mathcal{S}(\hat{\omega}_{nmp}^k, \hat{\omega}_{wt}) \right), \tag{3}$$

where $\mathcal{S}(\hat{\omega}_{wt}^k, \hat{\omega}_{nmp})$ is the indicator function taking the value 1 when the k-th important patch of IDH widetype is in the set of top K_m important patches for NMP, and vice versa. In addition, to facilitate the learning process with DCC loss, we adopt a curriculum-learning based training strategy dynamically focusing on hard-to-learn patches, regarded as the patches with higher decision importance weight, as patches with lower confidence weight, e.g., patches with fewer nuclei, are usually easier to learn in both tasks. Hence, K_m is further defined as

$$K_m = K_0 \beta^{\lfloor \frac{m}{m_0} \rfloor}. \tag{4}$$

In (4), K_0 and m_0 are hyper-parameters to adjust $\mathcal{L}_{\text{DCC}}$ in training process.

4   Experiments and Results

4.1   Implementation Details

The proposed DeepMO-Glioma is trained on the training set for 70 epochs, with batch size of 8 and initial learning rate of 0.003 with Adam optimizer [11] together with weight decay. Key hyper-parameters are listed in Table 1 of supplementary material. All hyper-parameters are tuned to achieve the best performance over the validation set. All experiments are conducted on a computer with an Intel(R) Xeon(R) E5-2698 CPU @2.20 GHz, 256 GB RAM and 4 Nvidia Tesla V100 GPUs. Additionally, our method is implemented on PyTorch with Python environment.

---

[2] Note that, IDH widetype is incorporated in diagnosing glioblasoma in current clinical paradigm; while previously, diagnosis of glioblasoma is puley based on NMP.

Table 1. Performance of classifying glioma based on WHO 2021 criteria [1].

Method	Diffuse glioma classification, %						Ablation (w/o), %		
	Ours	CLAM	TransMIL	ResNet*	DenseNet*	VGG-16*	Graph	LC loss	DCC
Acc.	**77.3**	71.2	68.2	59.1	62.9	60.6	65.5	71.2	68.2
Sen.	**76.0**	62.9	60.2	51.6	52.5	49.8	47.7	61.0	59.8
Spec.	**86.6**	82.9	79.9	71.4	83.5	74.5	83.0	82.3	84.7
F_1-score	**71.0**	60.0	59.2	51.6	50.9	49.6	49.4	61.2	53.5

* We modified baselines to the MIL setting, since they are not originally designed for MIL.

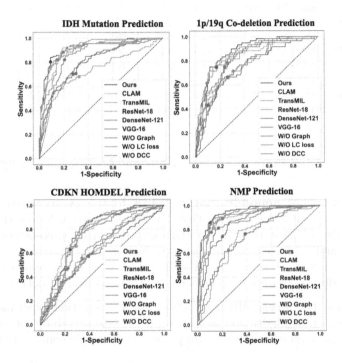

Fig. 3. ROCs of our model, comparison and ablation models for predicting 4 molecular markers. Refer to supplementary Fig. 2 for the zoom-in windows of ROCs.

4.2   Performance Evaluation

1) Glioma classification. We compare our model with five other state-of-the-art methods: CLAM[15], TransMIL [16], ResNet-18 [5], DenseNet-121 [8] and VGG-16 [17]. Note CLAM [15] and TransMIL [16] are MIL framework, while others are commonly-used image classification methods, set as our baselines. The left panel of Table 1 shows that DeepMO-Glioma performs the best, achieving at least 6.1%, 13.1%, 3.1% and 11.0% improvement over other models in accuracy, sensitivity, specificity and AUC, respectively, indicating that our model

Table 2. Performance in predicting genomic markers, histology and ablation studies.

		Ours	CLAM	TransMIL	ResNet	DenseNet	VGG-16	No Graph	No LC loss	No DCC
IDH	Acc.	**86.4**	81.4	83.7	67.8	72.0	70.5	80.7	84.1	84.1
	Sen	80.5	93.8	82.3	68.1	70.8	70.8	82.3	82.3	88.5
	Spec	90.7	72.2	84.8	67.5	72.8	70.2	79.5	85.4	80.8
	AUC.	**92.0**	91.1	90.7	72.7	80.3	79.6	86.1	90.8	89.1
1p19q	Acc.	81.4	81.8	80.3	70.1	72.3	79.2	76.1	**84.1**	75.0
	Sen.	75.0	48.3	43.3	71.7	66.7	43.3	75.0	61.7	78.3
	Spec.	83.3	91.7	91.2	69.6	74.0	89.7	76.5	90.7	74.0
	AUC.	**88.1**	82.0	82.9	76.8	77.1	75.5	83.0	86.7	85.2
CDKN	Acc.	**68.6**	67.8	60.2	58.7	59.1	58.7	60.2	58.3	60.2
	Sen.	63.2	65.8	55.9	59.9	57.9	57.2	47.4	38.2	46.1
	Spec.	75.9	70.5	66.1	57.1	60.7	60.7	77.7	85.7	79.5
	AUC.	**77.2**	77.0	65.5	62.8	62.9	59.9	72.6	76.7	76.7
NMP	Acc.	**87.5**	83.7	85.6	68.6	69.3	76.9	82.2	84.8	83.0
	Sen.	85.7	81.4	81.4	62.9	76.4	74.3	81.4	89.3	77.9
	Spec.	89.5	86.3	90.3	75.0	61.3	79.8	83.1	79.8	88.7
	AUC.	**94.5**	90.7	92.7	74.0	74.7	86.1	86.7	93.4	91.7

could effectively integrate molecular markers and histology in classifying diffuse gliomas.

2) Predictions of genomic markers and histology features. From the left panel of Table 2, we observe that DeepMO-Glioma achieves the AUC of 92.0%, 88.1%, 77.2% and 94.5% for IDH mutation, 1p/19q co-deletion, CDKN HOMDEL and NMP prediction, respectively, considerably better than all the comparison models. Figure 3 plots the ROCs of all models, demonstrating the superior performance of our model over other comparison models.

3) Network interpretability. An additional visualization experiment is conducted based on patch decision scores to test the interpretability of our method. Due to the page limit, the results are presented in supplementary Fig. 1.

4.3   Results of Ablation Experiments

1) CPLC-Graph network. The right panels of Table 1 shows that, by setting graph balancing weight α to 0 for the proposed CPLC-Graph, the accuracy, sensitivity, specificity and F_1-score decreases by 7.8%, 29.0%, 3.6% and 21.6%, respectively. Similar results are observed for the prediction tasks of molecular markers and histology (Table 2). Also, the ROC of removing the CPLC-Graph network is shown in Fig. 3. These results indicate the utility of the proposed CPLC-Graph network.

2) LC loss. The right panels of Table 1 shows that the performance after removing LC loss decreases in all metrics, causing a reduction of 6.1%, 15.0%, 4.3% and 9.8%, in accuracy, sensitivity, specificity and F_1-score, respectively. Similar results for the tasks of molecular marker and histology prediction are observed in the right panel of Table 2 with ROCs in Fig. 3, indicating the effectiveness of the proposed LC loss.

3) DCC loss. From Table 1, we observe that the proposed DCC loss improves the performance in terms of accuracy by 9.1%. Similar results can be found for sensitivity, specificity and F_1-score. From Table 2, we observe that the AUC decreases 2.9%, 2.9%, 0.5% and 2.8% for the prediction of IDH, 1p/19q, CDKN and NMP, respectively, when removing the DCC loss. Such performance is also found in comparing the ROCs in Fig. 3, suggesting the importance of the DCC loss for all the tasks.

5   Summary

The paradigm of pathology diagnosis has shifted to integrating molecular makers with histology features. In this paper, we aim to classify diffuse gliomas under up-to-date diagnosis criteria, via jointly learning the tasks of molecular marker prediction and histology classification. Inputting histology WSIs, our model incorporates a novel HMT-MIL framework to extract global information for both predicting both molecular markers and histology. We also design a CPLC-Graph network and a DCC loss to model both intra-omic and inter-omic interactions. Our experiments demonstrate that our model has achieved superior performance over other state-of-the-art methods, serving as a potentially useful tool for digital pathology based on WSIs in the era of molecular pathology.

Acknowledgments. This work was supported by the NIHR Brain Injury MedTech Co-operative and the NIHR Cambridge Biomedical Research Centre (NIHR203312). This publication presents independent research funded by the National Institute for Health and Care Research (NIHR). The views expressed are those of the author(s) and not necessarily those of the NHS, the NIHR or the Department of Health and Social Care. XW is grateful for the China Scholarship Council (CSC) Cambridge scholarship. SJP is funded by National Institute for Health and Care Research (NIHR) Clinician Scientist Fellowship (NIHR/CS/009/011). CL is funded by Guarantors of Brain Fellowship.

References

1. Bale, T.A., Rosenblum, M.K.: The 2021 who classification of tumors of the central nervous system: an update on pediatric low-grade gliomas and glioneuronal tumors. Brain Pathol. **32**(4), e13060 (2022)
2. Campanella, G., et al.: Clinical-grade computational pathology using weakly supervised deep learning on whole slide images. Nat. Med. **25**(8), 1301–1309 (2019)
3. Chen, R.J., et al.: Pathomic fusion: an integrated framework for fusing histopathology and genomic features for cancer diagnosis and prognosis. IEEE Trans. Med. Imaging **41**(4), 757–770 (2020)
4. Dosovitskiy, A., et al.: An image is worth 16x16 words: transformers for image recognition at scale. arXiv preprint arXiv:2010.11929 (2020)

5. He, K., Zhang, X., Ren, S., Sun, J.: Deep residual learning for image recognition. In: Proceedings of the IEEE Conference on Computer Vision and Pattern Recognition, pp. 770–778 (2016)
6. https://portal.gdc.cancer.gov/
7. https://www.cancer.net/cancer-types/brain-tumor/statistics
8. Huang, G., Liu, Z., Van Der Maaten, L., Weinberger, K.Q.: Densely connected convolutional networks. In: Proceedings of the IEEE Conference on Computer Vision and Pattern Recognition, pp. 4700–4708 (2017)
9. Imboden, S., et al.: Implementation of the 2021 molecular ESGO/ESTRO/ESP risk groups in endometrial cancer. Gynecol. Oncol. **162**(2), 394–400 (2021)
10. Jiang, S., Zanazzi, G.J., Hassanpour, S.: Predicting prognosis and IDH mutation status for patients with lower-grade gliomas using whole slide images. Sci. Rep. **11**(1), 16849 (2021)
11. Kingma, D.P., Ba, J.: Adam: a method for stochastic optimization. arXiv preprint arXiv:1412.6980 (2014)
12. Li, X., Wu, H., Li, M., Liu, H.: Multi-label video classification via coupling attentional multiple instance learning with label relation graph. Pattern Recogn. Lett. **156**, 53–59 (2022)
13. Liang, S., et al.: Clinical practice guidelines for the diagnosis and treatment of adult diffuse glioma-related epilepsy. Cancer Med. **8**(10), 4527–4535 (2019)
14. Louis, D.N., et al.: The 2021 who classification of tumors of the central nervous system: a summary. Neuro Oncol. **23**(8), 1231–1251 (2021)
15. Lu, M.Y., Williamson, D.F., Chen, T.Y., Chen, R.J., Barbieri, M., Mahmood, F.: Data-efficient and weakly supervised computational pathology on whole-slide images. Nat. Biomed. Eng. **5**(6), 555–570 (2021)
16. Shao, Z., Bian, H., Chen, Y., Wang, Y., Zhang, J., Ji, X., et al.: TransMil: transformer based correlated multiple instance learning for whole slide image classification. Adv. Neural. Inf. Process. Syst. **34**, 2136–2147 (2021)
17. Simonyan, K., Zisserman, A.: Very deep convolutional networks for large-scale image recognition. arXiv preprint arXiv:1409.1556 (2014)
18. Trpkov, K., et al.: New developments in existing who entities and evolving molecular concepts: The genitourinary pathology society (gups) update on renal neoplasia. Mod. Pathol. **34**(7), 1392–1424 (2021)
19. Volante, M., Lam, A.K., Papotti, M., Tallini, G.: Molecular pathology of poorly differentiated and anaplastic thyroid cancer: what do pathologists need to know? Endocr. Pathol. **32**, 63–76 (2021)
20. Xing, X., Chen, Z., Zhu, M., Hou, Y., Gao, Z., Yuan, Y.: Discrepancy and gradient-guided multi-modal knowledge distillation for pathological glioma grading. In: Wang, L., Dou, Q., Fletcher, P.T., Speidel, S., Li, S. (eds.) MICCAI 2022. LNCS, vol. 13435, pp. 636–646. Springer, Cham (2022). https://doi.org/10.1007/978-3-031-16443-9_61
21. Yang, H., et al.: Deep learning-based six-type classifier for lung cancer and mimics from histopathological whole slide images: a retrospective study. BMC Med. **19**, 1–14 (2021)
22. Yazici, V.O., Gonzalez-Garcia, A., Ramisa, A., Twardowski, B., Weijer, J.v.d.: Orderless recurrent models for multi-label classification. In: Proceedings of the IEEE/CVF Conference on Computer Vision and Pattern Recognition, pp. 13440–13449 (2020)
23. Yip, S., et al.: Concurrent CIC mutations, IDH mutations, and 1p/19q loss distinguish oligodendrogliomas from other cancers. J. Pathol. **226**(1), 7–16 (2012)

24. Zhang, L., Wei, Y., Fu, Y., Price, S., Schönlieb, C.B., Li, C.: Mutual contrastive low-rank learning to disentangle whole slide image representations for glioma grading. arXiv preprint arXiv:2203.04013 (2022)
25. Zhang, Y., Luo, L., Dou, Q., Heng, P.A.: Triplet attention and dual-pool contrastive learning for clinic-driven multi-label medical image classification. Med. Image Anal. 102772 (2023)

Bridging Ex-Vivo Training and Intra-operative Deployment for Surgical Margin Assessment with Evidential Graph Transformer

Amoon Jamzad[1]([✉]), Fahimeh Fooladgar[4], Laura Connolly[1],
Dilakshan Srikanthan[1], Ayesha Syeda[1], Martin Kaufmann[2], Kevin Y.M. Ren[3],
Shaila Merchant[2], Jay Engel[2], Sonal Varma[3], Gabor Fichtinger[1],
John F. Rudan[2], and Parvin Mousavi[1]

[1] School of Computing, Queen's University, Kingston, Canada
a.jamzad@queensu.ca
[2] Department of Surgery, Queen's University, Kingston, Canada
[3] Department of Pathology and Molecular Medicine,
Queen's University, Kingston, Canada
[4] Department of Electrical and Computer Engineering, University of British
Columbia, Vancouver, Canada

Abstract. PURPOSE: The use of intra-operative mass spectrometry along with Graph Transformer models showed promising results for margin detection on ex-vivo data. Although highly interpretable, these methods lack the ability to handle the uncertainty associated with intra-operative decision making. In this paper for the first time, we propose Evidential Graph Transformer network, a combination of attention mapping and uncertainty estimation to increase the performance and interpretability of surgical margin assessment. METHODS: The Evidential Graph Transformer was formulated to output the uncertainty estimation along with intermediate attentions. The performance of the model was compared with different baselines in an ex-vivo cross-validation scheme, with extensive ablation study. The association of the model with clinical features were explored. The model was further validated for a prospective ex-vivo data, as well as a breast conserving surgery intra-operative data. RESULTS: The purposed model outperformed all baselines, statistically significantly, with average balanced accuracy of 91.6%. When applied to intra-operative data, the purposed model improved the false positive rate of the baselines. The estimated attention distribution for status of different hormone receptors agreed with reported metabolic findings in the literature. CONCLUSION: Deployment of ex-vivo models is challenging due to the tissue heterogeneity of intra-operative data. The proposed Evidential Graph Transformer is a powerful tool that while providing the attention distribution of biochemical subbands, improve the surgical deployment power by providing decision confidence.

Keywords: Intra-operative deployment · Uncertainty estimation · Interpretation · Graph transformer network · Breast cancer margin

© The Author(s), under exclusive license to Springer Nature Switzerland AG 2023
H. Greenspan et al. (Eds.): MICCAI 2023, LNCS 14226, pp. 562–571, 2023.
https://doi.org/10.1007/978-3-031-43990-2_53

1   Introduction

Achieving complete tumor resection in surgical oncology like breast conserving surgery (BCS) is challenging as boundaries of tumors are not always visible/palpable [10]. In BCS the surgeon removes breast cancer while attempting to preserve as much healthy tissue as possible to prevent permanent deformation and to enhance cosmesis. The current standard of care for evaluating surgical success is to investigate the resection margins, which refers to the area surrounding the excised tumor. Up to 30% of surgeries result in incomplete tumor resection and require a revision operation [10]. The intelligent knife (iKnife) is a mass spectrometry device that can address this challenge by analyzing the biochemical signatures of resected tissue using the smoke that is released during tissue incineration [3]. Each spectrum contains the distribution of sampled ions with respect to their mass to charge ratio (m/z). Previously, learning models have been used in combination with iKnife data for ex-vivo tissue characterization and real-time margin detection [16,17].

The success of clinical deployment of learning models heavily relies on approaches that are not only accurate but also interpretable. Therefore, it should be clear how models reach their decisions and the confidence they have in such decision. Studies suggest that one way to improve these factors is through data centric approaches i.e. to focus on appropriate representation of data. Specifically, representation of data as graphs has been shown to be effective for medical diagnosis and analysis [1]. It has also been shown that graph neural networks can accurately capture the biochemical signatures of iKnife and determine the tissue type. Particularly, Graph Transformer Networks (GTN) has have shown to further enhance the transparency of underlying relation between the graph nodes and decision making via attention mechanism [11].

Biological data, specially those acquired intra-opertively, are heterogeneous by nature. While the use of ex-vivo data collected under specific protocols are beneficial to develop baseline models, intra-operative deployment of these models is challenging. For iKnife, the ex-vivo data is usually collected from homogeneous regions of resected specimens under the guidance of a trained pathologist, versus the intra-operative data is recorded continuously while the surgeon cutting through tissues with different heterogeneity and pathology. Therefore, beyond predictive power and explainable decision making, intra-operative models must be able to handle mixed and unseen pathology labels.

Uncertainty-aware models in computer-assisted interventions can provide clinicians with feedback on prediction confidence to increase their reliability during deployment. Deep ensembles [15] and Bayesian networks [9] incur high runtime and computational cost both at training and inference time and thus, less practical for real-time computer-assisted interventions. Evidential Deep Learning [18] is another approach that has been proposed based on the evidence framework of Dempster-Shafer Theory [12]. Since the evidential approach jointly generates the network prediction and uncertainty estimation, it seems more suitable for computationally efficient intra-operative deployment.

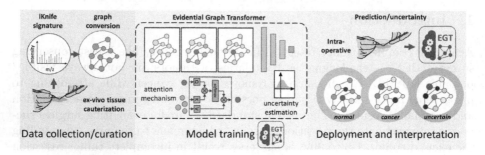

Fig. 1. An overview of the proposed approach including data collection and preprocessing, graph conversion, and interpretation of uncertainty and attentions.

In this paper, we propose Evidential Graph Transformer (EGT), a combination of graph-based feature-level attention mechanism with sample-level uncertainty estimation, to increase the performance and interpretability of surgical margin assessment. This is done by implementing the evidential loss and prediction functions within a graph transformer model to output the uncertainty, intermediate attention, and model prediction. To demonstrate the state-of-the-art performance of the proposed approach on mass spectrometry data, the model is compared with different baselines in both cross-validation and prospective schemes on ex-vivo data. Furthermore, the performance of model is also investigated intraoperatively. In addition to the proposed model, we present a new visualization approach to better correlate the graph nodes with the spectral content of the data, which improves interpretability. In addition to the ablation study on the network and graph strictures, we also investigate the metabolic association of breast cancer hormone receptor status.

2   Materials and Methods

Figure 1 presents the overview of the proposed approach. Following data collection and curation, each burn (spectrum) is converted to a single graph structure. The proposed graph model learns from the biochemical signatures of the tissue to classify cancer versus normal tissue. The uncertainty and intermediate attentions generated by the model are visualized and explored for their association with the biochemical mechanisms of cancer.

2.1   Data Curation

Ex-vivo: Data is collected from fresh breast tissue samples from the patients referred to BCS at Kingston Health Sciences Center over two years. The study is approved by the institutional research ethics board and patients consent to be included. Peri-operatively, a pathologist guides and annotates the ex-vivo point-burns, referred to as spectra, from normal or cancerous breast tissue immediately after excision. In addition to spectral data, clinicopathological details such as the

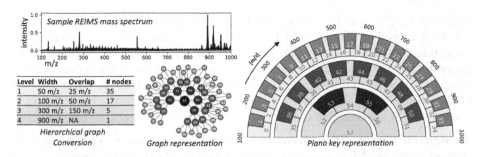

Fig. 2. Graph conversion: each spectrum is converted to a multi-level hierarchical graph. For intuitive interpretation of the process, Piano-key visualization is introduced. Each key represent a node with m/z range equal to the angular extent of the key.

status of hormone receptors is also provided post-surgically. In total 51 cancer and 149 normal spectra are collected and stratified into five folds (4 for cross validation and 1 prospectively) with each patient restricted to one fold only.

Intra-operative: A stream of iKnife data is collected during a BCS case (27 min) at Kingston Health Sciences Center. At the sampling rate of 1 Hz, a total of 1616 spectra are recorded. Each spectrum is then labeled based both on surgeons comments during the operation and post-operative pathology report.

Preprocessing: Each spectrum is converted to a hierarchical graph as illustrated in Fig. 2. The nodes are generated from a specific subband in each spectrum. Different subband widths (50, 100, 300, and 900 m/z) are used to create different levels of hierarchy (Fig. 2). The edges connect nodes with overlapping subbands within and between levels. As a result, each graph (spectrum) consists of 58 nodes and 135 edges. For details on graph conversion please refer to [2] and [11]. For easier interpretation of nodes with respect to their corresponding subbands, we visualize the graph as a Piano-key plot in Fig. 2, where each key represents a node with m/z range equal to the angular extent of the key. The dark keys show the subband overlaps between adjacent nodes.

2.2   Network Architecture and Training

Graph Transformer Network: The GTN consists of a node embedding layer, L Graph Transformer Layers (GTL), a node aggregation layer, multiple dense layers, and a prediction layer [8]. Assume a graph G with N nodes and $h_i \in R^{d \times 1}$ as node features of node i. In each GTL, the H headed attention mechanism updates the features of node i based on all neighboring node features h_j that are directly connected to node i via e_{ij} edges. The attention mechanism for node update at layer $l + 1$ is formulated as:

$$w_{ij}^{k,l} = softmax_j \left(\frac{Q^{k,l} h_i^l \cdot K^{k,l} h_j^l}{\sqrt{d}} \right), \quad \hat{h}_i^{l+1} = O^l \overset{H}{\underset{k=1}{\|}} \left(\sum_{j \in N_i} w_{ij}^{k,l} V^{k,l} h_j^l \right) \quad (1)$$

where $Q^{k,l}$, $K^{k,l}$, and $V^{k,l}$ are trainable linear weights. The weights w_{ij}^{kl} defines the k-th attention that is paid by node j to update node i at layer l. The concatenation of all H attention heads multiplied by trainable parameters O^l generates final attention $\hat{h}_i^{l+1}$, which is passed through batch normalization and residual layers to update the node features for the next layer. After the last GTL, features from all nodes are aggregated, then passed to the dense layers to construct a final prediction output.

Evidential Graph Transformer: Evidential deep learning provides a well-defined theoretical framework to jointly quantify classification prediction and uncertainty modeling by assuming the class probability follows a Dirichlet distribution [18]. We propose to modify the loss and prediction layer of GTN, considering the same assumption, to formulate the Evidential Graph Transformer model. Therefore, there are two mechanisms embedded in EGT: i) node-level attention calculation - via aggregation of neighboring nodes according to their relevance to the predictions, and ii) graph-level uncertainty estimation - via fitting the Dirichlet distribution to the predictions.

In the context of surgical margin assessment, the attentions reveal the relevant metabolic ranges to cancerous tissue, while uncertainty helps identify and filter data with unseen pathology. Specifically, the attentions affect the predictions by selectively emphasizing the contributions of relevant nodes, enabling the model to make more accurate predictions. On the other hand, the spread of the outcome probabilities as modeled by the Dirichlet distribution represents the confidence in the final predictions. Combining the two provides interpretable predictions along with the uncertainty estimation.

Mathematically, the Dirichlet distribution is characterized by $\boldsymbol{\alpha} = [\alpha_1, ..., \alpha_C]$ where C is the number of classes in the classification task. The parameters can be estimates as $\boldsymbol{\alpha} = f(x_i|\Theta) + 1$ where $f(x_i|\Theta)$ is the output of the Evidential Graph Transformer parameterized by Θ for each sample(x_i). Then, the expected probability for the c-th class p_c and the total uncertainty u for each sample (x_i) can be calculated as $p_c = \frac{\alpha_c}{S}$, and $u = \frac{C}{S}$, respectively, where $S = \sum_{c=1}^{C} \alpha_c$. To fit the Dirichlet distribution to the output layer of our network, we use a loss function consisting of the prediction error $\mathcal{L}_i^p$ and the evidence adjustment $\mathcal{L}_i^e$

$$\mathcal{L}_i(\Theta) = \mathcal{L}_i^p(\Theta) + \lambda \mathcal{L}_i^e(\Theta) \tag{2}$$

where λ is the annealing coefficient to balance the two terms. $\mathcal{L}_i^p$ can be cross-entropy, negative log-likelihood, or mean square error , while $\mathcal{L}_i^e(\Theta)$ is KL divergence to the uniform Dirichlet distribution [18].

2.3   Experiments

Network/Graph Ablation: We explore the hyper-parameters of the proposed model in an extensive ablation study. The attention parameters include the number of attention heads (1–15 with step size of 2) and the number of hidden features (7–14). For the evidential loss, we evaluate the choice of loss function (the 3 previously mentioned), and the annealing coefficient (5–50 with step size of 5). The number of GTLs and dense layers are both fixed at 3. Additionally, we run ablation studies on the graph structure themselves to show the importance of presenting the data as graphs. We try randomizing the edge connections and dropping the nodes with overlapping m/z subbands.

Ex-vivo Evaluation: The performance of the proposed network is compared with 3 baseline models including GTN, graph convolution network [14], and non-graph convolution network. Four-fold cross validation is used for comparison of the different approaches, to increase the generalizability (3 folds for train/validation, test on remaining unseen fold, report average test performance). Separate ablation studies are performed for the baseline models to fine tune their structural parameters. All experiments are implemented using PyTorch with Adam optimizer, learning rate of 10^{-4}, batch size of 32, and early stopping based on validation loss. To demonstrate the robustness of the model and ensure it is not overfitting, we also report the performance of the ensemble model from the 4-fold cross validation study on the 5th unseen prospective test fold.

Clinical Relevance: Hormone receptor status plays an important role in determining breast cancer prognosis and tailoring treatment plans for patients [6]. Here, we explore the correlation of the attention maps generated by EGT with the status of HER2 and PR hormones associated with each spectrum. These hormones are involved in different types of signaling that the cell depends on [5].

Intra-operative Deployment: To explore the intra-operative capability of the models, we deploy the ensemble models of the proposed method as well as the baselines from the cross-validation study to the BCS iKnife stream.

3   Results and Discussion

Ablation Study and Ex-vivo Evaluation: According to our ablation study, hyper parameters of 11 attention heads, 11 hidden features per attention head, the cross entropy loss function, and annealing coefficient of 30, result in higher performances when compared to other configurations (370k learnable parameters). The performance of EGT in comparison with the mentioned baselines are summarized in Table 1. As can be seen, the proposed EGT model with average accuracy of 94.1% outperformed all the baselines statistically significantly (maximum p-values of 0.02 in one-tail paired Wilcoxon Signed-Rank test). The lower standard deviation of parameters shows the robustness of EGT compared to other baselines. The regularization term in EGT loss prevents overconfident estimation of incorrect predictions [18] that could lead to superior results, compared to GTN, without overfitting. Lastly, when compared to other state-of-the-art baselines with uncertainty estimation mechanisms, the proposed Evidential Graph Transformer network (average balanced accuracy of $91.6 \pm 4.3\%$ in Table 1) outperforms MC Dropout [9], Deep Ensembles [15], and Masksembles [7] ($86.1 \pm 5.7\%$, $88.5 \pm 6.8\%$, and $89.2 \pm 5.4\%$ respectively [19]).

The estimated probabilities in evidence based models are directly correlated with model confidence and therefore more interpretable. To demonstrate this,

Table 1. Average(standard deviation) of accuracy (ACC), balanced accuracy (BAC) Sensitivity (SEN), Specificity (SPC), and the area under the curve (AUC) for the proposed Evidential Graph Transformer in comparison with graph transformer (GTN), graph convolution (GCN), and non-graph convolution (CNN) baselines.

Model		ACC%	BAC%	SEN%	SPC%	AUC
Evidential Graph Transformer		**94.1(3.2)**	**91.6(4.3)**	**97.2(3.2)**	**85.9(7.9)**	**0.96(0.03)**
Graph	GT	90.4(6.3)	88.7(8.3)	92.6(6.4)	84.8(15.1)	0.96(0.05)
	GCN	91.6(4.0)	88.1(5.9)	96.3(4.0)	80.0(11.7)	0.92(0.07)
Non-graph	CNN	91.5(4.3)	87.6(7.5)	96.2(4.4)	79.0(16.1)	0.89(0.08)

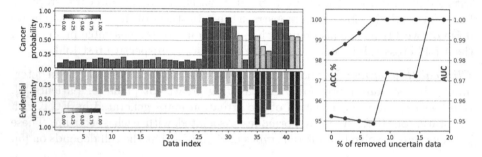

Fig. 3. *Left* Estimated probabilities and uncertainty scores for data samples in test set. *Right* Effect of uncertain data exclusion on accuracy and AUC during model deployment.

the probability of cancer predictions and uncertainty scores for all test samples are visualized in the left plot of Fig. 3. As seen, the higher the uncertainty score (bottom bar plot), the closer the estimated cancer probability is to 0.5 (top bar plot). This information can be provided during deployment to further augment surgical decision making for uncertain data instances. This is demonstrated in the right plot of Fig. 3, where the samples with high uncertainties are gradually disregarded. It can be seen that by not using the network prediction for up to 10% of most uncertain test data, the AUC increases to 1. Providing surgeons with not only the model decision but also a measure of model confidence will improve their intervention decisions. For example, if the model has low confidence in a prediction they can reinforce their decision by other means.

The result of our graph structure ablation shows the drop of average ACC to 85.6% by randomizing the edges in the graph (p-value 0.004). Dropping overlapping nodes further decreased the ACC to 82.3% (p-value 0.001). Although the model still trained due to node aggregation, random graph structure acts as noise and affects the performance. Multi-level graphs were shown to outperform other structures for masspect data [Akbarifar 2021] as they preserve the receptive field in the neighborhood of subbands (metabolites).

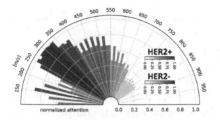 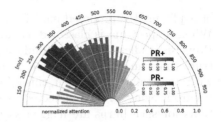

Fig. 4. Visualization of attention distribution for HER2 (*left*) and PR (*right*) hormone receptors in cancerous spectra.

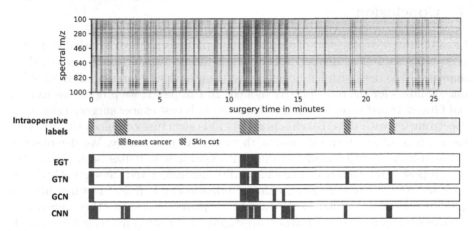

Fig. 5. Intra-operative data and label from a BCS case (top) and the temporal prediction of different ex-vivo models (bottom).

Clinical Relevance: An appropriate visualization of the attention map for samples can be used to help with this exploration. Accumulating the attentions maps from the cancerous burns based on their hormone receptor status results in the representative maps demonstrated in Fig. 4. The polar bars in this figure show the attention level paid to the nodes in the associated m/z subband. It can be seen that more attention is paid to the amino acids range (100–350 m/z) in HER2 positive breast cancer in comparison to HER2 negative breast cancer, which is in accordance with previous literature that has found evidence for higher glutamine metabolism activity in HER2+ [13]. we have also found that there's more attention in this range for PR negative breast cancer in comparison PR positive, which is in concordance with previous literature demonstrating that these subtypes have higher glutamine metabolic activity [4,5].

Intra-operative Deployment: The raw intra-operative iKnife data (y-axis is m/z spectral range and x-axis is the surgery timeline) along with the temporal reference labels extracted from surgeon's call-outs and pathology report are shown in Fig. 5, top. As seen, the iKnife stream contains spectra from skin cuts, which is considered as an unseen label for the ex-vivo models. The results of deploying the proposed models and baselines are presented in Fig. 5, bottom. When a spectrum is classified as cancer, a red line is overlaid on the timeline. Previous studies showed the similarity between skin and breast cancer mass spectrum that can confuse the binary models. Since our proposed EGT is equipped with uncertainty estimation, this information can be used to eliminate skin spectra from being wrongly detected as cancer. By integrating uncertainty, predictions for such burns are flagged as uncertain so clinicians can compensate for surgical decision making with other sources of information.

4   Conclusion

Intra-operative deployment of deep learning solutions requires a measure of interpretability as well as predictive confidence. These two factors are particularly importance to deal with heterogeneity of tissues which represented as mixed or unseen labels for the retrospective models. In this paper, we propose an Evidential Graph Transformer for margin detection in breast cancer surgery using mass spectrometry with these benefits in mind. This structure combines the attention mechanisms of graph transformer with predictive uncertainty. We demonstrate the significance of this model in different experiments. It has been shown that the proposed architecture can provide additional insight and consequently clearer interpretation of surgical margin characterization and clinical features like status of hormone receptors. In the future, we plan to work on other uncertainty estimation approaches and further investigate the graph conversion technique to be more targeted on the metabolic pathways, rather than regular conversion.

References

1. Ahmedt-Aristizabal, D., Armin, M.A., Denman, S., Fookes, C., Petersson, L.: Graph-based deep learning for medical diagnosis and analysis: past, present and future. Sensors **21**(14), 4758 (2021). https://doi.org/10.3390/S21144758
2. Akbarifar, F., et al.: Graph-based analysis of mass spectrometry data for tissue characterization with application in basal cell carcinoma surgery. In: SPIE Medical Imaging: Image-Guided Procedures, Robotic Interventions, and Modeling, vol. 11598 (2021)
3. Balog, J., et al.: In vivo endoscopic tissue identification by rapid evaporative ionization mass spectrometry (REIMS). Angewandte Chemie Int. Ed. **54**(38), 11059–11062 (2015). https://doi.org/10.1002/anie.201502770
4. Budczies, J., et al.: Glutamate enrichment as new diagnostic opportunity in breast cancer. Int. J. Cancer **136**(7), 1619–1628 (2015). https://doi.org/10.1002/ijc.29152
5. Demas, D.M., et al.: Glutamine metabolism drives growth in advanced hormone receptor positive breast cancer. Front. Oncol. **9** (2019). https://doi.org/10.3389/fonc.2019.00686
6. Dunnwald, L.K., Rossing, M.A., Li, C.I.: Hormone receptor status, tumor characteristics, and prognosis: a prospective cohort of breast cancer patients. Breast Cancer Res. **9**(1), R6 (2007). https://doi.org/10.1186/bcr1639

7. Durasov, N., Bagautdinov, T., Baque, P., Fua, P.: Masksembles for uncertainty estimation. In: Proceedings of the IEEE/CVF Conference on Computer Vision and Pattern Recognition (CVPR), pp. 13539–13548 (2021)
8. Dwivedi, V.P., Bresson, X.: A generalization of transformer networks to graphs. In: Methods and Applications, AAAI Workshop on Deep Learning on Graphs (2021)
9. Gal, Y., Ghahramani, Z.: Dropout as a Bayesian approximation: Representing model uncertainty in deep learning. In: Balcan, M.F., Weinberger, K.Q. (eds.) Proceedings of The 33rd International Conference on Machine Learning. Proceedings of Machine Learning Research, vol. 48, pp. 1050–1059. PMLR, New York, New York, USA, 20–22 June 2016
10. Hargreaves, A.C., Mohamed, M., Audisio, R.A.: Intra-operative guidance: methods for achieving negative margins in breast conserving surgery. J. Surg. Oncol. **110**(1), 21–25 (2014). https://doi.org/10.1002/JSO.23645
11. Jamzad, A., et al.: Graph transformers for characterization and interpretation of surgical margins. In: de Bruijne, M., et al. (eds.) MICCAI 2021. LNCS, vol. 12907, pp. 88–97. Springer, Cham (2021). https://doi.org/10.1007/978-3-030-87234-2_9
12. Jsang, A.: Subjective Logic: A Formalism for Reasoning Under Uncertainty. Springer, Cham Verlag (2016). https://doi.org/10.1007/978-3-319-42337-1
13. Kim, S., Kim, D.H., Jung, W.H., Koo, J.S.: Expression of glutamine metabolism-related proteins according to molecular subtype of breast cancer. Endocrine-Related Cancer **20**(3), 339–348 (2013). https://doi.org/10.1530/ERC-12-0398
14. Kipf, T.N., Welling, M.: Semi-supervised classification with graph convolutional networks. In: 5th International Conference on Learning Representations. ICLR (2017)
15. Lakshminarayanan, B., Pritzel, A., Blundell, C.: Simple and scalable predictive uncertainty estimation using deep ensembles. In: Advances in Neural Information Processing Systems, vol. 30 (2017)
16. Santilli, A., et al.: Domain adaptation and self-supervised learning for surgical margin detection. Int. J. Comput. Assist. Radiol. Surg. 1–9 (2021). https://doi.org/10.1007/s11548-021-02381-6
17. Santilli, A., et al.: Self-supervised learning for detection of breast cancer in surgical margins with limited data. In: Proceedings - International Symposium on Biomedical Imaging, April 2021, pp. 980–984, April 2021. https://doi.org/10.1109/ISBI48211.2021.9433829
18. Sensoy, M., Kaplan, L., Kandemir, M.: Evidential deep learning to quantify classification uncertainty. In: Advances in Neural Information Processing Systems, vol. 31 (2018)
19. Syeda, A.: Self-supervision and uncertainty estimation in surgical margin detection (2023)

Probabilistic Modeling Ensemble Vision Transformer Improves Complex Polyp Segmentation

Tianyi Ling[1,2], Chengyi Wu[2,3], Huan Yu[2,4], Tian Cai[5], Da Wang[1], Yincong Zhou[2], Ming Chen[2(✉)], and Kefeng Ding[1(✉)]

[1] Department of Colorectal Surgery and Oncology (Key Laboratory of Cancer Prevention and Intervention, China National Ministry of Education, Key Laboratory of Molecular Biology in Medical Sciences, Zhejiang Province, China), The Second Affiliated Hospital, Zhejiang University School of Medicine, Hangzhou, Zhejiang, China
dingkefeng@zju.edu.cn

[2] Department of Bioinformatics, College of Life Sciences, Zhejiang University, Hangzhou, Zhejiang, China
mchen@zju.edu.cn

[3] Department of Hepatobiliary and Pancreatic Surgery, The First Affiliated Hospital, School of Medicine, Zhejiang University, Hangzhou, Zhejiang, China

[4] Department of Thoracic Surgery, The Second Affiliated Hospital, School of Medicine, Zhejiang University, Hangzhou, Zhejiang, China

[5] Department of Hepatobiliary and Pancreatic Surgery, The Second Affiliated Hospital, Zhejiang University School of Medicine, Hangzhou, Zhejiang, China

Abstract. Colorectal polyps detected during colonoscopy are strongly associated with colorectal cancer, making polyp segmentation a critical clinical decision-making tool for diagnosis and treatment planning. However, accurate polyp segmentation remains a challenging task, particularly in cases involving diminutive polyps and other intestinal substances that produce a high false-positive rate. Previous polyp segmentation networks based on supervised binary masks may have lacked global semantic perception of polyps, resulting in a loss of capture and discrimination capability for polyps in complex scenarios. To address this issue, we propose a novel Gaussian-Probabilistic guided semantic fusion method that progressively fuses the probability information of polyp positions with the decoder supervised by binary masks. Our **P**robabilistic Modeling **E**nsemble Vision **T**ransformer **Net**work(***PETNet***) effectively suppresses noise in features and significantly improves expressive capabilities at both pixel and instance levels, using just simple types of convolutional decoders. Extensive experiments on five widely adopted datasets show that *PETNet* **outperforms** existing methods in identifying polyp

T. Ling and C. Wu—Equal contributions.

Supplementary Information The online version contains supplementary material available at https://doi.org/10.1007/978-3-031-43990-2_54.

camouflage, appearance changes, and small polyp scenes, and achieves a speed about **27FPS** in edge computing devices. Codes are available at: https://github.com/Seasonsling/PETNet.

Keywords: Colonoscopy · Polyp Segmentation · Vision Transformer

1 Introduction

Colorectal cancer (CRC) remains a major health burden with elevated mortality worldwide [1]. Most cases of CRC arise from adenomatous polyps or sessile serrated lesions in 5 to 10 years [9]. Colonoscopy is considered the gold standard for the detection of colorectal polyps. Polyp segmentation is a fundamental task in the computer-aided detection (CADe) of polyps during colonoscopy, which is of great significance in the clinical prevention of CRC.

Traditional machine learning approaches in polyp segmentation primarily focus on learning low-level features, such as texture, shape, or color distribution [14]. In recent years, encoder-decoder based deep learning models such as U-Net [12], UNet++ [22], ResUNet++ [6], and PraNet [4] have dominated the field. Furthermore, Transformer [3,17,19,20] models have also been proposed for polyp segmentation, and achieve the state-of-the-art(SOTA) performance.

Despite significant progress made by these binary mask supervised models, challenges remain in accurately locating polyps, particularly in complex clinical scenarios, due to their insensitivity to complex lesions and high false-positive rates. More specifically, most polyps have an elliptical shape with well-defined boundaries. However, supervised segmentation learning solely based on binary masks may not be effective in discriminating polyps in complex clinical scenarios. Endoscopic images often contain pseudo-polyp objects with strong boundaries, such as colon folds, blood vessels, and air bubbles, which can result in false positives. In addition, sessile and flat polyps have ambiguous and challenging boundaries to delineate. To address these limitations, Qadir et al. [11] proposed using Gaussian masks for supervised model training. This approach reduces false positives significantly by assigning less attention to outer edges and prioritizing surface patterns. However, this method has limitations in accurately segmenting polyp boundaries, which are crucial for clinical decision-making.

Therefore, the primary challenge lies in enhancing polyp segmentation performance in complex scenarios by precisely preserving the polyp segmentation boundaries, while simultaneously maximizing the decoder's attention on the overall pattern of the polyps.

In this paper, we propose a novel transformer-based polyp segmentation framework, *PETNet*, which addresses the aforementioned challenges and achieves SOTA performance in locating polyps with high precision. Our contributions are threefold:

- We propose a novel Gaussian-Probabilistic guided semantic fusion method for polyp segmentation, which improves the decoder's global perception of polyp locations and discrimination capability for polyps in **complex scenarios**.

- We evaluate the performance of *PETNet* on five widely adopted datasets, demonstrating its superior ability to identify polyp camouflage and small polyp scenes, achieving **state-of-the-art** performance in locating polyps with high precision. Furthermore, we show that *PETNet* can achieve a speed of about **27FPS** in edge computing devices (Nvidia Jetson Orin).
- We design several polyp instance-level evaluation metrics, considering that conventional pixel-level calculation methods cannot explicitly and comprehensively evaluate the overall performance of polyp segmentation algorithms.

2   Methods

2.1   Architecture Overview

As shown in Fig. 1, *PETNet* is an end-to-end polyp segmentation framework consists of three core module groups. (1) The **Encoder Group** employs a vision transformer backbone [18] cascaded with a mixed transformer attention layer to encode long-range dependent features at four scales. (2) The **Gaussian-Probabilistic Modeling Group** consists of a Gaussian Probabilistic Guided UNet-like decoder branch(GUDB) and Gaussian Probabilistic-Induced transition(GIT) modules. (3) The **Ensemble Binary Decoders Group** includes a UNet-like structure branch(UDB) [12], a fusion module(Fus), and a cascaded fusion module(CFM) [3].

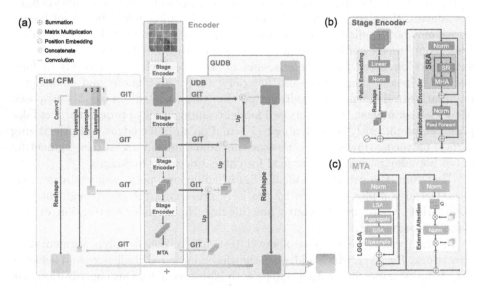

Fig. 1. Proposed PETNet Framework: (a) Comprises three critical module groups. (b) Depicts the Stage Encoder. (c) Illustrates the Mixed Transformer Attention layer (MTA).

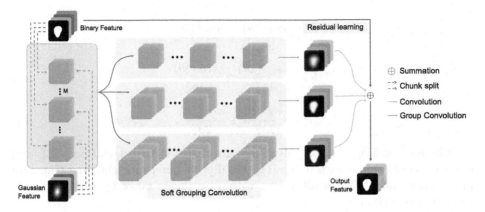

Fig. 2. Illustration of Gaussian Probabilistic-induced transition (GIT).

2.2 Encoder Group

To balance the trade-off between computational speed and feature representation capability, we utilize the pre-trained PVTv2-B2 model [18] as the backbone. Mixed transformer attention(MTA) layer is composed of Local-Global Gaussian-Weighted Self-Attention (LGG-SA) and External Attention (EA). We add a MTA layer to encode the last level features, enhancing the model's semantic representation and accelerating the training process [16]. Moreover, the encoder output features are presented as $\{\mathbf{X}_i^E\}_{i=1}^4$ with channels of $[2C, 4C, 8C, 16C]$.

2.3 Gaussian-Probabilistic Modeling Group

To incorporate both polyp location probability and surface pattern information in a progressive manner, we propose the Gaussian Probabilistic-induced Transition (GIT) method. This method involves the interaction between a Gaussian auxiliary decoder and multiple binary decoders in a layer-wise fashion, as shown in Fig. 2.

Gaussian Probabilistic Mask. Inspired by [11] and [21], in addition to utilizing binary representation, polyps can also be represented as probability heatmaps with blurred borders. We present a method of converting the binary polyp mask $f(x,y) \in \{0,1\}^{W \times H \times 1}$ into Gaussian masks $Y(x,y) \in [0,1]^{W \times H \times 1}$ by utilizing elliptical Gaussian kernels. Specifically, for every polyp in a binary mask, after masking other polyp pixels as background, we calculate

$$Y = \exp\left(-\left(a(x-x_o)^2 + 2b(x-x_o)(y-y_o) + (y-y_o)^2\right)\right) \tag{1}$$

where (x_o, y_o) is the mass of each polyp in the binary image $f(x,y)$. To rotate the output 2D Gaussian masks according to the orientation, we set a, b, c as followings,

$$a = \frac{cos^2(\theta)}{2\sigma_x^2} + \frac{sin^2(\theta)}{2\sigma_y^2}, \tag{2}$$

$$b = \frac{-sin(2\theta)}{4\sigma_x^2} + \frac{sin(2\theta)}{4\sigma_y^2}, \tag{3}$$

$$c = \frac{cos^2(\theta)}{2\sigma_x^2} + \frac{sin^2(\theta)}{2\sigma_y^2}, \tag{4}$$

where σ_x^2 and σ_y^2 are the polyp size-adaptive standard deviations [21], and θ is the orientation of each polyp [11]. Finally, we determine the final Gaussian probabilistic mask $\mathcal{P_G}$ for all polyps within an image mask by computing the element-wise maximum.

Gaussian Guided UNet-Like Decoder Branch. The Gaussian Guided UNet-like decoder branch(GUDB) module is a simple UNet-like decoding branch supervised by Gaussian Probabilistic masks. We employ four levels of encoder output features, and adjust encoder features $\{\mathbf{X}_i^E\}_{i=1}^4$ to the features $\{\mathbf{X}_i^G\}_{i=1}^4$ with channels of $[C, 2C, 2C, 2C]$ in each level. At the final layer, a 1×1 convolution is used to convert the feature vector to one channel, producing a size of $\mathcal{H} \times \mathcal{W} \times 1$ Gaussian mask.

Gaussian Probabilistic-Induced Transition Module. We use the Gaussian probabilistic-induced transition module(GIT) to achieve transition between binary features and gaussian features. Given the features originally sent to the Decoder as binary features $\{\mathbf{X}_i^B\}_{i=1}^4$, and the transformed encoder features sent to GUDB as $\mathbf{X}^G$. We first splits 4 levels of $\mathbf{X}^B$ and $\mathbf{X}^G$ into fixed groups as:

$$\{\mathbf{X}_{i,m}^D\}_{m=1}^M \in \mathbb{R}^{C_d/M \times H_i \times W_i} \leftarrow \mathbf{X}_i^D \in \mathbb{R}^{C_d \times H_i \times W_i}, \tag{5}$$

where M is the corresponding number of groups. Then, we periodically arrange groups of $\mathbf{X}_{i,m}^B$ and $\mathbf{X}_{i,m}^G$ for each level, and generate the regrouped feature $\mathbf{Q}_i \in \mathbb{R}^{(C_i+C_g) \times H_i \times W_i}$ in an Multi-layer sandwiches manner. Soft grouping convolution [7] is then applied to provide parallel nonlinear projections at multiple fine-grained sub-spaces (Fig. 2). We further introduce residual learning in a parallel manner at different group-aware scales. The final output $\{\mathbf{Z}_i^T\}_{i=1}^4 \in \mathbb{R}^{C_i \times H_i \times W_i}$ is obtained for the UDB decoder. Considering the computation cost, The binary features $\mathbf{X}^B \leftarrow \mathbf{X}^E$ for the Fus decoder have channel numbers of $[4C, 4C, 4C, 4C]$. The Fus decoder and CFM share identical transited output features, while CFM exclusively utilizes the last three levels of features.

2.4   Ensemble Binary Decoders Group

During colonoscopy, endoscopists often use the two-physician observation approach to improve the detection rate of polyps. Building on this manner, we propose the ensemble method that integrates multiple simple decoders to enhance the detection and discrimination of difficult polyp samples. We demonstrate the effectiveness of our approach using three commonly used convolutional decoders. After GIT process, diverse level of Gaussian probabilistic-induced binary features were sent to these decoders. The output mask $\mathcal{P}$ is obtained by element-wise summation of $\mathcal{P}_i$, where i represents the binary decoder index.

Fusion Module. As shown in Fig. 1, Set $\mathcal{X}i$, $i \in (1, 2, 3, 4)$ represent multi-scale mixed features. Twice convolution following with bilinear interpolation are applied to transform these feature with same 4C channels as $\mathcal{X}_1{'}, \mathcal{X}_2{'}, \mathcal{X}_3{'}, \mathcal{X}_4{'}$. Afterward, we get $\mathcal{X}_{out}$ with the resolution of $\mathcal{H}/4 \times \mathcal{W}/4 \times C1$ through following formula, where $\mathcal{F}$ represents twice 3×3 convolution:

$$\mathcal{X}_{out} = \mathcal{F}(Concat(\mathcal{X}_1{'}, \mathcal{X}_2{'}, \mathcal{X}_3{'}, \mathcal{X}_4{'}) \tag{6}$$

UNet Decoder Branch and CFM Module. The structure of the UDB is similar to that of the GUDB, except for the absence of channel reduction prior to decoding. In our evaluation, we also examine the decoder CFM utilized in [3], which shares the same input features (excluding the first level) as the Fus.

3   Experiments

3.1   Datasets Settings

To evaluate models fairly, we completely follow *PraNet* [4] and use five public datasets, including 548 and 900 images from ClinicDB [2] and Kvasir-SEG [5] as training sets, and the remaining images as validation sets. We also test the generalization capability of all models on three unseen datasets (ETIS [13] with 196 images, CVC-ColonDB [8] with 380 images, and EndoScene [15] with 60 images). Training settings are the same as [3].

3.2   Loss Setting

Our loss function formulates as $\mathcal{L} = \sum_{i=1}^{N} \mathcal{L}_i + \lambda \mathcal{L}_g$, and

$$\mathcal{L}_i = \sum_{i=1}^{n}(\mathcal{L}_{IOU}(\mathcal{P}_i, \mathcal{G}_B) + \mathcal{L}_{BCE}(\mathcal{P}_i, \mathcal{G}_B)) \tag{7}$$

where N is the total number of binary decoders, $\mathcal{L}_g$ represents the L1 loss between the ground truth Gaussian mask $\mathcal{G}_G$ and GUDB prediction mask $\mathcal{P}_G$. λ is a hyperparameter used to balance the binary and Gaussian losses. Furthermore, we employ intermediate decoder outputs to calculate auxiliary losses for convergence acceleration.

3.3   Evaluation Metrics

Conventional evaluation metrics for polyp segmentation are typically limited to pixel-level calculations. However, metrics that consider the entire polyp are also crucial. Here we assess our model from both **pixel-level** and **instance-level** perspectives.

Pixel-level evaluation is based on mean intersection over union ($mIoU$), mean Dice coefficient ($mDic$), and weighted F_1 score (wF_m). For polyp instance evaluation, a true positive (TP) is defined when the detection centroid is located within the polyp mask. False positives (FPs) occur when a wrong detection output is provided for a negative region, and false negatives (FNs) occur when a polyp is missed in a positive image. Finally, we compute sensitivity $n\mathcal{S}en = TP/(TP + FN) \times 100$, precision $n\mathcal{P}re = TP/(TP + FP) \times 100$, and $n\mathcal{F}_1 = 2 \times (\mathcal{S}en \times \mathcal{P}re)/(\mathcal{S}en + \mathcal{P}re) \times 100$ based on the number count for instance evaluation.

3.4   Results

Training and Learning Ability. Table S1 displays the results of our model's training and learning performance. Our model achieves comparable performance to the SOTA model on the Kvasir-SEG and ClinicDB datasets. Notably, our model yields superior results in false-positive instance evaluation.

Generalization Ability. The generalization results are shown in Table 1. We conduct three unseen datasets to test models' generalizability. Results show that $PETNet$ achieves excellent generalization performance compared with previous models. Most importantly, our false-positive instance counts(45 in ETIS and 55 in CVC-ColonDB) reduce significantly of other models. We also observe a performance mismatch phenomenon in pixel-level evaluation and instance-level evaluation.

Small Polyp Detection Ability. The detection capability results of small polyps are shown in Table 2. Diminutive polyps are hard to precisely detect, while they are the major targets of optical biopsies performed by endoscopists. We selected images from two unseen datasets with 0∼2% polyp labeled area to perform the test. As shown, $PETNet$ demonstrates great strength in both datasets, which indicates that one of the major advantages of our model lies in detecting small polyps with lower false-positive rates.

Ablation Analysis. Table 3 presents the results of our ablation study, where we investigate the contribution of the two key components of our model, namely the Gaussian-Probabilistic Guided Semantic Fusion method and ensemble decoders. We observe that while the impact of each binary decoder varies, all sub binary decoders contribute to the overall performance. Furthermore, the GIT method significantly enhances instance-level evaluation without incurring performance penalty in pixel-level evaluation, especially in unseen datasets.

Table 1. Quantitative results of the test datasets EndoScene, CVC-ColonDB and ETIS-LaribPolypDB.

	Methods	$mDic$	$mIoU$	$nSen$	$nPre$	nF_1	TP	FP	FN
EndoScene	PraNet (MICCAI'20) [4]	0.748	0.634	**1.000**	0.526	0.690	60	54	0
	EU-Net (CRV'21) [10]	0.882	0.809	**1.000**	0.938	0.968	60	4	0
	LDNet (MICCAI'22) [19]	0.888	0.814	**1.000**	0.984	0.992	60	1	0
	SSFormer-S (MICCAI'22) [17]	0.881	0.812	**1.000**	0.938	0.968	60	4	0
	Polyp-PVT(CAAI AIR'23) [3]	0.894	0.829	**1.000**	0.968	0.984	60	2	0
	PETNet **(Ours)**	**0.899**	**0.834**	**1.000**	**1.000**	**1.000**	60	0	0
ColonDB	PraNet (MICCAI'20) [4]	0.643	0.542	0.882	0.499	0.638	351	352	47
	EU-Net (CRV'21) [10]	0.757	0.672	0.889	0.760	0.819	354	112	44
	LDNet (MICCAI'22) [19]	0.751	0.667	0.905	0.726	0.805	360	136	38
	SSFormer-S (MICCAI'22) [17]	0.787	0.708	0.900	0.840	0.869	358	68	40
	Polyp-PVT(CAAI AIR'23) [3]	0.808	0.724	**0.937**	0.772	0.847	**373**	110	**25**
	PETNet **(Ours)**	**0.817**	**0.740**	0.935	**0.871**	**0.902**	372	55	26
ETIS	PraNet (MICCAI'20) [4]	0.584	0.482	0.904	0.390	0.545	188	294	20
	EU-Net (CRV'21) [10]	0.687	0.610	0.870	0.557	0.679	181	144	27
	LDNet (MICCAI'22) [19]	0.702	0.611	0.909	0.659	0.764	189	98	19
	SSFormer-S (MICCAI'22) [17]	0.744	0.672	0.856	0.674	0.754	178	86	30
	Polyp-PVT(CAAI AIR'23) [3]	0.765	0.687	**0.923**	0.667	0.774	**192**	96	**16**
	PETNet **(Ours)**	**0.782**	**0.703**	0.904	**0.807**	**0.853**	188	45	20

Table 2. Quantitative results of the **small polyp detection** in ETIS and CVC-ColonDB dataset. Small polyps are defined as the polyp area accounts for 0~2% of the entire image.

	Methods	$mDic$	$mIoU$	$nSen$	$nPre$	nF_1	TP	FP	FN
ETIS	PraNet [4]	0.43	0.34	0.87	0.34	0.49	87	167	13
	Polyp-PVT [3]	0.68	0.60	**0.93**	0.61	0.74	**93**	60	**7**
	PETNet **(Ours)**	**0.69**	**0.60**	0.88	**0.73**	**0.80**	88	33	12
ColonDB	PraNet [4]	0.45	0.34	0.82	0.40	0.54	80	120	17
	Polyp-PVT [3]	**0.68**	**0.58**	**0.93**	0.71	0.80	**90**	37	**7**
	PETNet **(Ours)**	0.67	0.58	**0.93**	**0.76**	**0.84**	**90**	28	**7**

Table 3. Ablation study for *PETNet* on five datasets. wF_m: pixel-based weighted F1 score, nF_1: instance-based weighted F1 score. w/o: without.

Dataset	Metric	Baseline	w/o Fus	w/o UDB	w/o CFM	w/o GUDB	PETNet
ClinicDB	wF_m	0.819	0.929	0.929	**0.941**	0.933	0.932
	nF_1	0.883	**0.978**	0.964	0.971	0.971	0.964
Kvasir	wF_m	0.804	0.904	**0.914**	0.899	0.911	0.912
	nF_1	0.853	0.882	0.879	**0.895**	0.881	0.891
EndoScene	wF_m	0.705	0.859	0.877	0.876	0.874	**0.884**
	nF_1	0.851	0.992	0.952	0.976	0.976	**1.000**
ColonDB	wF_m	0.635	0.800	0.783	0.775	**0.812**	0.802
	nF_1	0.761	0.898	0.894	0.881	0.882	**0.902**
ETIS	wF_m	0.467	0.738	0.744	0.725	0.736	**0.747**
	nF_1	0.577	0.840	0.810	0.805	0.850	**0.853**

3.5   Comparative Analysis

Fig.S1 shows that our proposed model, *PETNet*, outperforms SOTA models in accurately identifying polyps under complex scenarios, including lighting disturbances, water reflections, and motion blur. Faluire cases are shown in Fig.S2.

3.6   Running in the Real World

Furthermore, we deployed *PETNet* on the edge computing device Nvidia Jetson Orin and optimized its performance using TensorRT. Our results demonstrate that *PETNet* achieves real-time denoising and segmentation of polyps with high accuracy, achieving a speed of 27 frames per second on the device(Video S1).

4   Conclusion

Based on intrinsic characteristics of the endoscopic polyp image, we specifically propose a novel segmentation framework named **PETNet** consisting of three key module groups. Experiments show that *PETNet* consistently outperforms most current cutting-edge models on five challenging datasets, demonstrating its solid robustness in distinguishing other intestinal analogs. Most importantly, *PETNet* shows better sensitivity to complex lesions and diminutive polyps.

Acknowledgement. This work was supported by the National Natural Sciences Foundation of China (Nos. 31771477, 32070677), the Fundamental Research Funds for the Central Universities (No. 226-2022-00009), and the Key R&D Program of Zhejiang (No. 2023C03049).

References

1. Ahmed, A.M.A.A.: Generative adversarial networks for automatic polyp segmentation. arXiv:2012.06771 [cs, eess] (2020)
2. Bernal, J., Sánchez, F.J., Fernández-Esparrach, G., Gil, D., Rodríguez, C., Vilariño, F.: WM-DOVA maps for accurate polyp highlighting in colonoscopy: Validation vs. saliency maps from physicians. Comput. Med. Imaging Graph.: Official J. Comput. Med. Imaging Soc. **43**, 99–111 (2015). https://doi.org/10.1016/j.compmedimag.2015.02.007
3. Dong, B., Wang, W., Fan, D.P., Li, J., Fu, H., Shao, L.: Polyp-PVT: polyp segmentation with pyramid vision transformers. arXiv:2108.06932 [cs] (2021)
4. Fan, D.P., et al.: PraNet: parallel reverse attention network for polyp segmentation. arXiv:2006.11392 [cs, eess] (2020)
5. Jha, D., et al.: Kvasir-SEG: a segmented polyp dataset. arXiv:1911.07069 [cs, eess] (2019)
6. Jha, D., et al.: ResUNet++: an advanced architecture for medical image segmentation. arXiv:1911.07067 [cs, eess] (2019)
7. Ji, G.P., Fan, D.P., Chou, Y.C., Dai, D., Liniger, A., Van Gool, L.: Deep gradient learning for efficient camouflaged object detection. Tech. Rep. arXiv:2205.12853, arXiv (2022)

8. Mamonov, A.V., Figueiredo, I.N., Figueiredo, P.N., Tsai, Y.H.R.: Automated polyp detection in colon capsule endoscopy. IEEE Trans. Med. Imaging **33**(7), 1488–1502 (2014). https://doi.org/10.1109/TMI.2014.2314959, http://arxiv.org/abs/1305.1912

9. National Health Commission of the People's Republic of China: [Chinese Protocol of Diagnosis and Treatment of Colorectal Cancer (2020 edition)]. Zhonghua Wai Ke Za Zhi [Chinese Journal of Surgery] **58**(8), 561–585 (2020). https://doi.org/10.3760/cma.j.cn112139-20200518-00390

10. Patel, K., Bur, A.M., Wang, G.: Enhanced U-Net: a feature enhancement network for polyp segmentation. In Proceedings of the International Robots & Vision Conference. International Robots & Vision Conference **2021**, 181–188 (2021). https://doi.org/10.1109/crv52889.2021.00032, https://www.ncbi.nlm.nih.gov/pmc/articles/PMC8341462/

11. Qadir, H.A., Shin, Y., Solhusvik, J., Bergsland, J., Aabakken, L., Balasingham, I.: Toward real-time polyp detection using fully CNNs for 2D Gaussian shapes prediction. Med. Image Anal. **68**, 101897 (2021). https://doi.org/10.1016/j.media.2020.101897, https://linkinghub.elsevier.com/retrieve/pii/S1361841520302619

12. Ronneberger, O., Fischer, P., Brox, T.: U-Net: convolutional networks for biomedical image segmentation. arXiv:1505.04597 [cs] (2015), http://arxiv.org/abs/1505.04597

13. Silva, J., Histace, A., Romain, O., Dray, X., Granado, B.: Toward embedded detection of polyps in WCE images for early diagnosis of colorectal cancer. Int. J. Comput. Assist. Radiol. Surg. **9**(2), 283–293 (2013). https://doi.org/10.1007/s11548-013-0926-3

14. Tajbakhsh, N., Gurudu, S.R., Liang, J.: A comprehensive computer-aided polyp detection system for colonoscopy videos. Inf. Process. Med. Imaging **24**, 327–38 (2015). https://doi.org/10.1007/978-3-319-19992-4_25

15. Vázquez, D., et al.: A benchmark for endoluminal scene segmentation of colonoscopy images. J. Healthc. Eng. **2017**, 4037190 (2017). https://doi.org/10.1155/2017/4037190

16. Wang, H., et al.: Mixed transformer U-Net for medical image segmentation. arXiv:2111.04734 [cs, eess] (2021)

17. Wang, J., Huang, Q., Tang, F., Meng, J., Su, J., Song, S.: Stepwise feature fusion: local guides global. In: Wang, L., Dou, Q., Fletcher, P.T., Speidel, S., Li, S. (eds.) Medical Image Computing and Computer Assisted Intervention - MICCAI 2022, pp. 110–120. Lecture Notes in Computer Science, Springer Nature Switzerland, Cham (2022). https://doi.org/10.1007/978-3-031-16437-8_11

18. Wang, W., et al.: PVTv2: improved baselines with pyramid vision transformer. arXiv:2106.13797 [cs] (2022)

19. Zhang, R., Lai, P., Wan, X., Fan, D.J., Gao, F., Wu, X.J., Li, G.: Lesion-aware dynamic kernel for polyp segmentation. In: Wang, L., Dou, Q., Fletcher, P.T., Speidel, S., Li, S. (eds.) Medical Image Computing and Computer Assisted Intervention - MICCAI 2022, pp. 99–109. Lecture Notes in Computer Science, Springer Nature Switzerland, Cham (2022). https://doi.org/10.1007/978-3-031-16437-8_10

20. Zhang, Y., Liu, H., Hu, Q.: TransFuse: fusing transformers and CNNs for medical image segmentation. arXiv:2102.08005 [cs] (2021)

21. Zhou, X., Wang, D., Krähenbühl, P.: Objects as points (2019). https://doi.org/10.48550/arXiv.1904.07850

22. Zhou, Z., Siddiquee, M.M.R., Tajbakhsh, N., Liang, J.: UNet++: a nested U-Net architecture for medical image segmentation. arXiv:1807.10165 [cs, eess, stat] (2018)

Clinical Applications – Ophthalmology

Retinal Thickness Prediction from Multi-modal Fundus Photography

Yihua Sun[1], Dawei Li[2], Seongho Kim[3], Ya Xing Wang[4], Jinyuan Wang[4], Tien Yin Wong[5,6], Hongen Liao[1(✉)], and Su Jeong Song[3(✉)]

[1] Department of Biomedical Engineering, School of Medicine, Tsinghua University, Beijing, China
liao@tsinghua.edu.cn
[2] College of Future Technology, Peking University, Beijing, China
[3] Department of Ophthalmology, Kangbuk Samsung Hospital, Sungkyunkwan University School of Medicine, Seoul, Republic of Korea
sjsong7@gmail.com
[4] Beijing Institute of Ophthalmology, Beijing Tongren Hospital, Capital University of Medical Science, Beijing Ophthalmology and Visual Sciences Key Laboratory, Beijing, China
[5] Tsinghua Medicine, Tsinghua University, Beijing, China
[6] Singapore Eye Research Institute, Singapore National Eye Centre, Singapore, Singapore

Abstract. Retinal thickness map (RTM), generated from OCT volumes, provides a quantitative representation of the retina, which is then averaged into the ETDRS grid. The RTM and ETDRS grid are often used to diagnose and monitor retinal-related diseases that cause vision loss worldwide. However, OCT examinations can be available to limited patients because it is costly and time-consuming. Fundus photography (FP) is a 2D imaging technique for the retina that captures the reflection of a flash of light. However, current researches often focus on 2D patterns in FP, while its capacity of carrying thickness information is rarely explored. In this paper, we explore the capability of infrared fundus photography (IR-FP) and color fundus photography (C-FP) to provide accurate retinal thickness information. We propose a Multi-Modal Fundus photography enabled Retinal Thickness prediction network (M²FRT). We predict RTM from IR-FP to overcome the limitation of acquiring RTM with OCT, which boosts mass screening with a cost-effective and efficient solution. We first introduce C-FP to provide IR-FP with complementary thickness information for more precise RTM prediction. The misalignment of images from the two modalities is tackled by the Transformer-CNN hybrid design in M²FRT. Furthermore, we obtain the ETDRS grid prediction solely from C-FP using a lightweight decoder,

S. J. Song and H. Liao are the co-corresponding authors.

Supplementary Information The online version contains supplementary material available at https://doi.org/10.1007/978-3-031-43990-2_55.

which is optimized with the guidance of the RTM prediction task during the training phase. Our methodology utilizes the easily acquired C-FP, making it a valuable resource for providing retinal thickness quantification in clinical practice and telemedicine, thereby holding immense clinical significance.

Keywords: Retinal thickness prediction · Multi-modality · Transformer · Color fundus photography · Infrared fundus photography

1   Introduction

Retinal thickness map (RTM), generated from optical coherence tomography (OCT) volumes, provides a quantitative representation of various retina pathologic conditions [3]. The ETDRS grid is an array comprising nine values representing the averaged thickness in nine regions in RTM [5]. The RTM and ETDRS grid, are widely employed diagnostic and monitoring techniques for retinal disorders including age-related macular degeneration, glaucoma, and diabetic retinopathy [14], which are prevalent causes of visual impairment worldwide [6]. On the other hand, OCT has been a critical diagnostic tool in ophthalmology due to its exceptional sensitivity and precision in identifying major eye diseases.

However, OCT exams are only available to limited patients as it is both costly and time-consuming, which impedes the acquisition of RTM and ETDRS grid. The recent advances in deep learning [20,21] have prompted research efforts aimed at addressing this limitation. There have been attempts to predict center-involving macular edema from color fundus photographs (C-FP) [17]. Although these studies showed high sensitivity and specificity, they only provided a binary classification for the presence of macular edema. The lack of quantitative retina thickness prediction results mandated further study.

Fundus photography (FP) is widely used to image the retina, which captures the reflected signal of emitted signal from the retinal surface with a flash of light [13]. As the retina is partially-transparent, a minority of light would pass through the surface [19] and reflect back, which might carry information about the retinal thickness. This hypothesis motivates us to explore the connection between the RTM/ETDRS grid and the IR-FP/C-FP, which is rarely explored. Nonetheless, the FPs hold substantial clinical value in facilitating large-scale screening by acquiring RTM and ETDRS grid much faster and more affordable.

Recently, Holmberg et al. [10] presented DeepRT, a convolutional neural network (CNN) designed for predicting retinal thickness using only infrared fundus photographs (IR-FP), disregarding C-FP. Exploring the capacity of **C-FP** to provide depth information has two major advantages: 1) **More precise RTM prediction**: Different from IR-FP, C-FP is acquired using light of multiple wavelengths that penetrate different depths in the retina [19]. We assume that this can provide richer thickness information, which can lead to more precise RTM prediction when combined with IR-FP; 2) **Clinical significance**: C-FP is the most commonly used diagnostic tool in ophthalmology, and can be obtained even using a smartphone [7]. The ability to derive thickness information from C-FP alone, without OCT scans, will make C-FP a potential tool for high functioning telemedicine platform which has the ability to diagnose, monitor treatment response, and even screen high-risk patients for diabetic macular edema (DME).

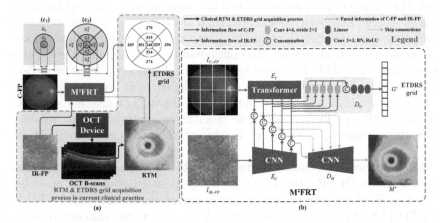

Fig. 1. (a) M^2FRT predicts the RTM with enhanced thickness information from muti-modal FPs without OCT scans, and the ETDRS grid can be predicted with C-FP only. (b) M^2FRT utilizes Transformer/CNN hybrid encoders E_T/E_C for C-FP/IR-FP, to tackle the unregistered issue and extract 2D aligned thickness information in an end-to-end learning manner. A CNN decoder D_M is employed to predict the RTM. Besides, a lightweight decoder D_G is designed to predict the ETDRS grid base on the information from C-FP only, which is guided by the RTM prediction task during training. (c_1) Areas for evaluations in the RTM prediction task, and (c_2) in the ETDRS grid (left eye) prediction task. In (c_1), G_3 is the remaining area of $G_{1,2}$ in RTM.

In this paper, we explore the capability of IR-FP and C-FP to provide accurate retinal thickness information, with a cohort of patients with DME of different grades. We propose a **M**ulti-**M**odal **F**undus photography enabled **R**etinal **T**hickness prediction network (**M^2FRT**). It is comprised of two separate encoders, a CNN E_C and a Transformer E_T, that encode localized information and rich depth information form IR-FP and C-FP respectively. We utilize the features extracted from E_C to facilitate the learning process of E_T in gathering 2D aligned thickness information via its attention mechanism. The enriched features are subsequently fed into a decoder to predict the RTM.

Furthermore, we obtain the ETDRS grid prediction, i.e. nine values representing averaged thickness in the predefined areas in Fig. 1 (c_2), solely from the C-FP by processing the features extracted from E_T through another lightweight decoder, which has significant clinical implications. To the best of our knowledge, we are the first to demonstrate the benefit of C-FP for RTM prediction and derive the ETDRS grid prediction solely from C-FP.

2   Methodology

In this study, we exclusively concentrate on DME to explore the predictive capacity of FPs regarding the retinal thickness. The rationale behind this is that, apart from DME, predicting retinal thickness itself has relatively less clinical value. For example, for age-related macular degeneration, the ophthalmologist needs to look for subtle changes in abnormal OCT features (e.g. subretinal fluid, pigmentary epithelial detachments [16]), rather than just the retinal thickness.

In standard clinical settings, the ophthalmologist will acquire the C-FP upon patients' arrival. If RTM is deemed necessary for diagnosis, a separate device will capture IR-FP and conduct OCT scanning. Figure 1 (a) illustrates the acquisition process of RTM using OCT, where each B-scan is registered with the 2D positions in IR-FP. The ETDRS grid is an array comprising nine values indicating the average thickness (μm) in nine predefined regions in RTM (Fig. 1 (c_2)).

Dovetailed with the clinical settings, M^2FRT aims to predict the RTM corresponding to the IR-FP, utilizing enriched depth information from pre-collected C-FP. The RTM requires precise pixel-wise correspondences to the IR-FP, while the ETDRS grid is a regional concept. Therefore, we can manage to derive an ETDRS grid prediction using only easier acquired C-FP, even in the absence of IR-FP, which holds importance within clinical scenarios and telemedicine.

As mentioned above, the FPs from the two modalities are captured by different machines. So, the FPs are not registered and have a distinct field of view (FoV). The recent advances in vision Transformers [4,12,18] have inspired us to address this challenge, because the multi-head attention mechanism is location-agnostic, but rather leverages patch embedding and position encoding to introduce positional information.

2.1   Encoder

The overall pipeline of M^2FRT is presented in Fig. 1 (b). The notations used for the images in the modality of IR-FP and C-FP are I_{IR-FP} and I_{C-FP}, respectively. The objective is to predict the thickness map M in the FoV of I_{IR-FP} and ETDRS grid G, which represents the central area of the retina and is the major concern in clinical practices.

The convolution and concatenation pose "hard" operations on the spatial dimensions. Thus, whether we concatenate I_{IR-FP} and I_{C-FP} as input or in the feature space under a CNN backbone, the misalignment of I_{IR-FP} and I_{C-FP} will deteriorate the performance for M prediction. In contrast, the spatial information is "softly" incorporated into the Transformer architecture, where the subsequent operations in the feature space are location-agnostic.

Therefore, we utilize a CNN encoder E_C from U-Net [15] to extract features from I_{IR-FP}, and a Transformer encoder E_T from 2D ViT/UNETR [4,8] to extract features from I_{C-FP}. Notably, the deep features extracted by E_T are spatially perturbed. M^2FRT leverages attention mechanisms in E_T to gather 2D aligned thickness information from I_{C-FP}, guided by the features extracted from

$I_{IR\text{-}FP}$ by E_C. The extracted multi-level features from $I_{IR\text{-}FP}$ and $I_{C\text{-}FP}$ are denoted as $f_{IR\text{-}FP}$ and $f_{C\text{-}FP}$ respectively, as shown in the following equations:

$$f_{IR\text{-}FP} = E_C(I_{IR\text{-}FP}), \quad f_{C\text{-}FP} = E_T(I_{C\text{-}FP}). \tag{1}$$

2.2 Decoder

M²FRT extracts 2D aligned depth information from C-FP, which enrich the depth representations acquired from IR-FP in an end-to-end learning manner. The extracted features are fused by concatenation and passed to the decoder D_M to generate the thickness map prediction M', where $M' = D_M(f_{IR\text{-}FP}, f_{C\text{-}FP})$.

With fine-grained thickness information extracted for the RTM prediction task, the encoded features obtained from E_T are ready to be decoded to predict the ETDRS grid using a lightweight decoder D_G. In D_G, the features from multiple levels are combined using a series of convolutions and concatenations. Then the final prediction for G is generated by a linear projection. The predicted ETDRS grid is denoted as G', where $G' = D_G(f_{C\text{-}FP})$.

2.3 Loss Functions

The loss functions $\mathcal{L}_1^M$ and $\mathcal{L}_1^G$ are employed in the prediction of the RTM and ETDRS grid using L_1 criteria, respectively, as shown in the following equations,

$$\mathcal{L}_1^M = \|M - M'\|_1, \ \mathcal{L}_1^G = \frac{1}{9}\sum_{i=1}^{9}\left|G^{(i)} - G'^{(i)}\right|, \tag{2}$$

where $G^{(i)}$ and $G'^{(i)}$ are the i-th number in the ETDRS grid ground truth G and prediction G'. The final loss function is $\mathcal{L} = \mathcal{L}_1^M + \mathcal{L}_1^G$.

3 Experiments

3.1 Experimental Setup

Dataset. A total of 967 retinal images were gathered from 361 distinct patients diagnosed with DME of different grades who underwent intravitreal injections. The dataset is collected in Kangbuk Samsung Hospital (IRB Approval Number: KBSMC 2022-12-016-001) between 2020 and 2021. The averaged retinal thickness (μm) in the dataset is 275.92 ± 20.91 (mean $\pm$ std.). For each patient, 31 B-scans are obtained by a Heidelberg OCT device, which are used to calculate the retinal thickness between the internal limiting membrane and the Bruch's membrane. The segmentations of the membrane layers are directly exported from the OCT machine. Images with poor fixation or OCTs with major segmentation errors are excluded by an experienced ophthalmologist.

Data Pre-processing. For IR-FP, we center-crop the area corresponding to the OCT scanning area with a resolution of 544×544, and then calculate RTM

Table 1. Quantitative comparison of different methods for RTM prediction, with MAE (μm) and PSNR (dB). The top-2 methods are highlighted in bold and underlined. By incorporating multi-modal FP as input, networks can access more comprehensive thickness information, resulting in improved performance. The most efficient way to tackle the unregistered problem is to utilize encoders E_C and E_T for IR-FP and C-FP respectively. Asterisks indicate M^2FRT outperforms the baselines with p-values<0.01.

Inputs	Methods			RTM		G_1		G_2		G_3	
				MAE↓	PSNR↑	MAE↓	PSNR↑	MAE↓	PSNR↑	MAE↓	PSNR↑
IR-FP	UNet++ [22]			29.28*	25.93*	64.52*	22.09*	31.71*	27.68*	27.92*	26.11*
	DeepRT [10]			25.49*	28.04*	66.57*	21.88*	28.96*	28.93*	23.78*	28.81*
	E_T, D_M			27.21*	27.87*	40.42*	27.64*	29.23	29.46	26.48*	28.05*
	U-Net [15] (E_C, D_M)			27.84*	27.38*	60.98*	22.63*	30.90*	28.43*	26.40*	27.86*
IR-FP & C-FP	U-Net [15]			25.16*	28.36*	44.07*	26.32*	28.92*	29.19*	23.95*	28.72*
	IR-FP Enc.	C-FP Enc.	Dec.	–	–	–	–	–	–	–	–
	E_T	E_T	D_M	26.44*	27.99*	40.01	27.88	28.54	29.54	25.68*	28.19*
	E_T	E_C	D_M	25.33*	28.49*	39.52	28.05	28.58	<u>29.76</u>	24.33*	28.76*
	E_C	E_C	D_M	24.80*	28.54*	43.19*	26.87*	28.93*	29.32*	23.54*	28.92*
	E_C	E_T	D_M	<u>23.82</u>	<u>28.91</u>	38.92	**28.16**	**27.80**	**29.78**	<u>22.65</u>	<u>29.28</u>
	E_C	E_T	D_M, D_G	**23.80**	**28.92**	<u>38.60</u>	<u>28.12</u>	<u>28.29</u>	29.64	**22.54**	**29.33**

Enc.: Encoder; Dec.: Decoder. M^2FRT is comprised of E_C, E_T, D_M, D_G.

ground truth within. With respect to the B-scans, the retinal thickness is calculated for 31 lines in the 2D IR-FP, and then linearly interpolated to match the resolution of IR-FP. For C-FP, we resize it to 544×544 from an original resolution of 3608×3608. The dataset is randomly split into training and test datasets at the patient level. The training/test dataset consisted of 657/310 images from 252/109 patients, respectively.

Implementation Details. The M^2FRT is implemented with PyTorch [2] and MONAI [1], and detailed configurations are in the supplementary material. Random flipping and rotation are utilized for data augmentation. We use the Adam [11] optimizer with $(\beta_1, \beta_2) = (0.9, 0.999)$ for training for 300 epochs. The initial learning rate is 0.001 and exponentially decayed with $\gamma = 0.999$.

Performance Metrics. For the RTM predictions, we use mean absolute error (MAE) and peak signal-to-noise ratio (PSNR) for evaluation in the areas $G_{1,2,3}$ as shown in Fig. 1 (c_1), where the peak signal is set to $800\mu m$. For the ETDRS grid predictions, we calculate the MAE of the predictions of the nine grids, as shown in Fig. 1 (c_2). For the right eye, the grid must be mirrored horizontally, i.e., $G_3^2 \leftrightarrow G_3^4$ and $G_2^2 \leftrightarrow G_2^4$. The Wilcoxon signed-rank test is employed to compare the performance of M^2FRT with the baselines.

3.2 Quantitative and Qualitative Evaluations on RTM Predictions

To better illustrate the problem and our solution, we begin with the most concise design, U-Net [15]. In Table 1, the MAE/PSNR for U-Net with IR-FP as input are $27.84\,\mu m / 27.38\,dB$. By concatenating multi-modal IR-FP and C-FP as input to the U-Net, the performance improved to $25.16\,\mu m / 28.36\,dB$, indicating that C-FP has the potential of containing additional thickness information.

However, the multi-modal FPs are unregistered and have a distinct FoV, in which case a mere concatenation of these inputs would diminish the network's

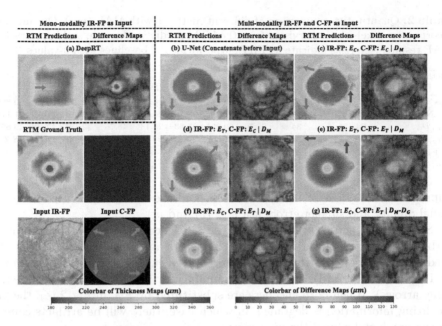

Fig. 2. The use of IR-FP alone to predict RTM will cause larger errors, particularly in the central area (pink arrows). Artifacts can be generated by the annular boundary (grey arrows) and vessels (purple arrows) from C-FP when misaligned information is roughly fused. Additionally, the dual E_T design weakens the localized 2D correspondence, where patch embedding can generate artifacts (brown arrows). Our proposed methods (f) and (g), with E_C and E_T extracting features for IR-FP and C-FP respectively, better leverage 2D aligned thickness information and lead to lower errors. (Color figure online)

capacity to effectively exploit the thickness information from paired 2D positions. A simple solution is to encode the multi-modal FPs with two separated convolutional encoders E_C, where the features are deeply fused along the downsampling path. The unregistered problem is eased by the higher-level features with a larger receptive field, and the MAE/PSNR are improved to $24.80\,\mu m/28.54\,dB$.

After all, 2D convolution and concatenation pose a "hard" operation to the spatial dimensions, which is still interfered with by the unregistered problem. As shown in Fig. 2, in (b) and (c), there are artifacts in the RTM predictions caused by the annular boundary (grey arrows) and misaligned vessels from C-FP (purple arrows). On the contrary, the attention operations in Transformer are location-agnostic, where the spatial information is more "softly" introduced into the network by patch embedding and position encoding [4].

Therefore, we employ distinct encoders of a CNN E_C and a Transformer E_T to IR-FP and C-FP respectively. The attention mechanism in E_T is encouraged to gather 2D aligned thickness information from the perturbed patch embeddings, with the guidance from the decoder D_M and the $\mathcal{L}_1^M$ loss function. With this CNN-Transformer hybrid design, the MAE/PSNR performance are

Table 2. Quantitative comparison of different methods with MAE (μm). Our method incorporates pixel-wise supervision from the RTM prediction branch, and improves the MAE results. Asterisks indicate M²FRT outperforms the baselines with p-values<0.05.

Methods	Mean Absolute Error (μm)									
	ETDRS Grid	G_1	G_2^1	G_2^2	G_2^3	G_2^4	G_3^1	G_3^2	G_3^3	G_3^4
ResNet-50 [9]	25.12*	37.46*	27.22	29.73*	25.89	26.43*	19.12*	22.27	17.65	20.36*
ResNet-101 [9]	25.06*	36.41*	27.05	29.29	25.76	26.18*	19.32	22.60*	17.86	21.11*
E_T, D_G	24.42*	34.88	26.70	29.29*	25.63	25.13	18.76	22.18*	**17.38**	19.85*
E_T, D_G, E_C, D_M	**23.84**	**34.36**	**26.15**	**28.24**	**25.25**	**24.53**	**18.50**	**21.42**	17.44	**18.71**

M²FRT is comprised of E_T, D_G, E_C, D_M.

improved to $23.82\,\mu m/28.91\,dB$ in Table 1, and the network produced the best visual quality and smaller errors in Fig. 2 (f) and (g).

Since IR-FP acts as a localizer for the OCT scan and RTM, spatially perturbing the features from IR-FP with E_T is not appropriate for the accurate prediction of RTM, and thus not yielding better quantitative results, as shown in Table 1. In Fig. 2 (d), the annular boundary artifacts from C-FP still exist (grey arrows). When both encoders are substituted by E_T, in Fig. 2 (e), the 2D localizing information is degraded, in which case, there will be artifacts caused by the patch embedding (brown arrows).

Our proposed M²FRT utilizes a combination of multi-modal IR-FP and C-FP to predict the RTM. M²FRT outperforms the state-of-the-art (SOTA) RTM prediction technique, DeepRT [10], which uses mono IR-FP as input. Besides, methods with multi-modal FPs surpass methods with mono IR-FP as input, especially in the central G_1 area, as shown in Table 1 and pink arrows in Fig. 2. The results demonstrate that C-FP has the ability to provide complementary depth information with IR-FP. The effectiveness of our methodology is validated through the ablation study on the encoders and decoders, as presented in Table 1.

Additionally, when E_T is guided to gather aligned features for RTM using the attention mechanism, the deep features from E_T are ready to be decoded by D_G for ETDRS grid predictions, which involves computing the averaged thickness in nine predefined regions. Notably, the ETDRS grid prediction task does not have a significant impact on the performance of the RTM prediction (the last two rows in Table 1), while the ETDRS grid prediction task can benefit from the supervision provided by the RTM prediction task, which will be discussed in Sect. 3.3.

3.3   Quantitative Evaluations on ETDRS Grid Predictions

Following the clinical settings, we predict the full RTM based on the IR-FP localizer in place of the OCT scanning procedure, which can boost mass screening. We gather enriched thickness information from C-FP and improve the performance with a hybrid CNN-Transformer design, as elaborated in Sect. 3.2.

In addition to identifying 2D disease patterns in C-FP, predicting the ETDRS grid solely from C-FP can exploit additional information in the C-FP and hold significant clinical value for rapid diagnosis, especially in the field of telemedicine. To achieve this, we can adopt a conventional learning-based method to predict the nine numbers in the ETDRS grid, i.e. ResNet [9], as shown in Table 2.

However, simply approximating the nine numbers will neglect the fine-grained thickness information. To address this issue, following the design in Sect. 3.2, the encoder E_T for C-FP is guided by the encoder E_C from the IR-FP part for detailed RTM predictions. Therefore, E_T has been trained to extract fine-grained depth information from C-FP, which can be decoded for the averaged thickness for ETDRS grid predictions with D_G. The fine-grained thickness supervision from the RTM prediction task can benefit the ETDRS grid prediction task, as shown in the last two rows of Table 2. Besides, our proposed M^2FRT outperforms its ablation and other baselines, as shown in Table 2. We can also observe from Table 1 and 2 that the central thickness in G_1 area is more challenging to predict than the surrounding area for the RTM and ETDRS grid prediction task.

4   Conclusion

In this paper, we demonstrate the advantages of leveraging multi-modal information from C-FP for RTM prediction with respect to IR-FP, which overcomes the limitations of OCT and has the potential to enhance mass screening. Additionally, we propose a novel method for predicting the ETDRS grids solely from C-FP, which has significant clinical importance for fast diagnosis, telemedicine, etc. Our results indicate that additional fine-grained supervision from the RTM prediction task is beneficial for ETDRS grid prediction, where the ETDRS grid is decoded from the encoder of C-FP by a lightweight decoder during the training procedure of the RTM prediction task. Further research could be conducted for: 1) Predicting RTM of multiple retinal layers simultaneously, and 2) Improving RTM prediction's resolution and detail by acquiring finer OCT as ground truth.

Acknowledgments. The authors acknowledge supports from National Key Research and Development Program of China (2022YFC2405200), National Natural Science Foundation of China (82027807, U22A2051), Beijing Municipal Natural Science Foundation (7212202), Institute for Intelligent Healthcare, Tsinghua University (2022ZLB001), and Tsinghua-Foshan Innovation Special Fund (2021THFS0104). We would like to thank Hee Guan Khor for discussions on experiments and writing, and Zhuxin Xiong for discussions on data pre-processing.

References

1. Medical open network for artificial intelligence (MONAI). https://monai.io/
2. PyTorch. https://pytorch.org/
3. Bhende, M., Shetty, S., Parthasarathy, M.K., Ramya, S.: Optical coherence tomography: a guide to interpretation of common macular diseases. Indian J. Ophthalmol. **66**(1), 20–35 (2018)

594   Y. Sun et al.

4. Dosovitskiy, A., et al.: An image is worth 16×16 words: transformers for image recognition at scale. In: International Conference on Learning Representations (2021)
5. Early Treatment Diabetic Retinopathy Study Research Group: grading diabetic retinopathy from stereoscopic color fundus photographs-an extension of the modified airlie house classification: ETDRS report number 10. Ophthalmology **98**(5, Supplement), pp. 786–806 (1991)
6. Flaxman, S.R., et al.: Global causes of blindness and distance vision impairment 1990–2020: a systematic review and meta-analysis. Lancet Glob. Health **5**(12), e1221–e1234 (2017)
7. Haddock, L.J., Kim, D.Y., Mukai, S.: Simple, inexpensive technique for high-quality smartphone fundus photography in human and animal eyes. J. Ophthalmol. **2013**, 518479 (2013)
8. Hatamizadeh, A., et al.: UNETR: transformers for 3D medical image segmentation. In: Proceedings of the IEEE/CVF Winter Conference on Applications of Computer Vision (WACV), pp. 574–584 (2022)
9. He, K., Zhang, X., Ren, S., Sun, J.: Deep residual learning for image recognition. In: 2016 IEEE Conference on Computer Vision and Pattern Recognition (CVPR), pp. 770–778 (2016)
10. Holmberg, O.G., et al.: Self-supervised retinal thickness prediction enables deep learning from unlabelled data to boost classification of diabetic retinopathy. Nat. Mach. Intell. **2**(11), 719–726 (2020)
11. Kingma, D.P., Ba, J.L.: Adam: a method for stochastic optimization. In: International Conference on Learning Representations (2015)
12. Li, J., Chen, J., Tang, Y., Wang, C., Landman, B.A., Zhou, S.K.: Transforming medical imaging with transformers? a comparative review of key properties, current progresses, and future perspectives. Med. Image Anal. **85**, 102762 (2023)
13. Panwar, N., Huang, P., Lee, J., Keane, P.A., Chuan, T.S., Richhariya, A., Teoh, S., Lim, T.H., Agrawal, R.: Fundus photography in the 21st century-a review of recent technological advances and their implications for worldwide healthcare. Telemedicine and e-Health **22**(3), 198–208 (2016)
14. Röhlig, M., Prakasam, R.K., Stüwe, J., Schmidt, C., Stachs, O., Schumann, H.: Enhanced grid-based visual analysis of retinal layer thickness with optical coherence tomography. Information **10**(9) (2019)
15. Ronneberger, O., Fischer, P., Brox, T.: U-NET: convolutional networks for biomedical image segmentation. In: Medical Image Computing and Computer-Assisted Intervention – MICCAI 2015, pp. 234–241 (2015)
16. Schmidt-Erfurth, U., Waldstein, S.M., Deak, G.G., Kundi, M., Simader, C.: Pigment epithelial detachment followed by retinal cystoid degeneration leads to vision loss in treatment of neovascular age-related macular degeneration. Ophthalmology **122**(4), 822–832 (2015)
17. Varadarajan, A.V., et al.: Predicting optical coherence tomography-derived diabetic macular edema grades from fundus photographs using deep learning. Nat. Commun. **11**(1), 130 (2020)
18. Vaswani, A., et al.: Attention is all you need. In: Advances in Neural Information Processing Systems, vol. 30 (2017)
19. Wang, L.V., Wu, H.I.: Biomedical Optics: Principles and Imaging. John Wiley & Sons (2012)
20. Zhou, S.K., et al.: A review of deep learning in medical imaging: imaging traits, technology trends, case studies with progress highlights, and future promises. Proc. IEEE **109**(5), 820–838 (2021)

21. Zhou, S.K., Rueckert, D., Fichtinger, G.: Handbook of Medical Image Computing and Computer Assisted Intervention. Academic Press (2019)
22. Zhou, Z., Rahman Siddiquee, M.M., Tajbakhsh, N., Liang, J.: UNet++: a nested U-Net architecture for medical image segmentation. In: Stoyanov, D., et al. (eds.) DLMIA/ML-CDS -2018. LNCS, vol. 11045, pp. 3–11. Springer, Cham (2018). https://doi.org/10.1007/978-3-030-00889-5_1

Reliable Multimodality Eye Disease Screening via Mixture of Student's t Distributions

Ke Zou[1,2], Tian Lin[3,4], Xuedong Yuan[1,2(✉)], Haoyu Chen[3,4(✉)], Xiaojing Shen[1,5], Meng Wang[6], and Huazhu Fu[6]

[1] National Key Laboratory of Fundamental Science on Synthetic Vision, Sichuan University, Sichuan, China
yxd@scu.edu.cn
[2] College of Computer Science, Sichuan University, Sichuan, China
[3] Joint Shantou International Eye Center, Shantou University and the Chinese University of Hong Kong, Guangdong, China
drchenhaoyu@gmail.com
[4] Medical College, Shantou University, Guangdong, China
[5] College of Mathematics, Sichuan University, Sichuan, China
[6] Institute of High Performance Computing, A*STAR, Singapore, Singapore

Abstract. Multimodality eye disease screening is crucial in ophthalmology as it integrates information from diverse sources to complement their respective performances. However, the existing methods are weak in assessing the reliability of each unimodality, and directly fusing an unreliable modality may cause screening errors. To address this issue, we introduce a novel multimodality evidential fusion pipeline for eye disease screening, EyeMoSt, which provides a measure of confidence for unimodality and elegantly integrates the multimodality information from a multi-distribution fusion perspective. Specifically, our model estimates both local uncertainty for unimodality and global uncertainty for the fusion modality to produce reliable classification results. More importantly, the proposed mixture of Student's t distributions adaptively integrates different modalities to endow the model with heavy-tailed properties, increasing robustness and reliability. Our experimental findings on both public and in-house datasets show that our model is more reliable than current methods. Additionally, EyeMos$t$ has the potential ability to serve as a data quality discriminator, enabling reliable decision-making for multimodality eye disease screening.

Keywords: Multimodality · uncertainty estimation · eye disease

---

K. Zou and T. Lin—Equal contribution.

Supplementary Information The online version contains supplementary material available at https://doi.org/10.1007/978-3-031-43990-2_56.

1   Introduction

Retinal fundus images and Optical Coherence Tomography (OCT) are common 2D and 3D imaging techniques used for eye disease screening. Multimodality learning usually provides more complementary information than unimodality learning [3,4,31]. This motivates researchers to integrate multiple modalities to improve the performance of eye disease screening. Current multimodality learning methods can be roughly classified into early, intermediate, and late fusion, depending on the fusion stage [2]. For multimodality ophthalmic image learning, recent works have mainly focused on the early fusion [10,15,23] and intermediate fusion stages [3,4,16,27]. Early fusion-based approaches integrate multiple modalities directly at the data level, usually by concatenating the raw or preprocessed multimodality data. Hua *et al.* [10] combined preprocessed fundus images and wide-field swept-source optical coherence tomography angiography at the early stage and then extracted representational features for diabetic retinopathy recognition. Intermediate fusion strategies allow multiple modalities to be fused at different intermediate layers of the neural networks. He *et al.* [9] extracted different modality features with convolutional block attention module [28] and modality-specific attention mechanisms, then concatenated them to realize the multimodality fusion for retinal image classification. However, few studies have explored multimodality eye disease screening at the late fusion stage. Furthermore, the above methods do not adequately assess the reliability of each unimodality, and may directly fuse an unreliable modality with others. This could lead to screening errors and be challenging for real-world clinical safety deployment. To achieve this goal, we propose a reliable framework for the multimodality eye disease screening, which provides a confidence (uncertainty) measure for each unimodality and adaptively fuses multimodality predictions in principle.

Uncertainty estimation is an effective way to provide a measure of reliability for ambiguous network predictions. The current uncertainty estimation methods mainly include Bayesian neural networks, deep ensemble methods, and deterministic-based methods. Bayesian neural networks [18,21,22] learn the distribution of network weights by treating them as random variables. However, these methods are affected by the challenge of convergence and have a large number of computations. The dropout method has alleviated this issue to a certain extent [12]. Another uncertainty estimation way is to learn an ensemble of deep networks [14]. Recently, to alleviate computational complexity and overconfidence [25], deterministic-based methods [17,19,25,26,32] have been proposed to directly output uncertainty in a single forward pass through the network. For multimodal uncertainty estimation, the Trusted Multi-view Classification (TMC) [8] is a representative method that proposes a new paradigm of multi-view learning by dynamically integrating different views at the evidence level. However, TMC has a limited ability to detect Out-Of-Distribution (OOD) samples [11]. This attributes to TMC is particularly weak in modeling epistemic uncertainty for each single view [12]. Additionally, the fusion rule in TMC fails to account for conflicting views, making it unsuitable for safety-critical deployment [30]. To address these limitations, we propose EyeMoSt, a novel evidential

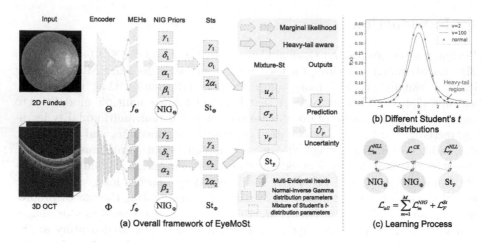

Fig. 1. Reliable multimodality Eye Disease Screening pipeline. (a) Overall framework of EyeMoSt. (b) Student's t Distributions with different degrees of freedom. (c) The overall learning process of EyeMoSt.

fusion method that models both aleatoric and epistemic uncertainty in unimodality, while efficiently integrating different modalities from a multi-distribution fusion perspective.

In this work, we propose a novel multimodality eye disease screening method, called EyeMoSt, that conducts Fundus and OCT modality fusion in a reliable manner. Our EyeMoSt places Normal-inverse Gamma (NIG) prior distributions over the pre-trained neural networks to directly learn both aleatoric and epistemic uncertainty for unimodality. Moreover, Our EyeMoSt introduces the Mixture of Student's t (MoSt) distributions, which provide robust classification results with global uncertainty. More importantly, MoSt endows the model with robustness under heavy-tailed property awareness. We conduct sufficient experiments on two datasets for different eye diseases (*e.g.*, glaucoma grading, age-related macular degeneration, and polypoid choroidal vasculopathy) to verify the reliability and robustness of the proposed method. In summary, the key contributions are as follows:

1) We propose a novel multimodality eye disease screening method, EyeMoSt, which conducts reliable fusion of Fundus and OCT modalities.

2) Our EyeMoSt introduces the MoSt distributions, which provide robust classification results with local and global uncertainty.

3) We conduct extensive experiments on two datasets for different eye diseases.[1]

2   Method

In this section, we introduce the overall framework of our EyeMoSt, which efficiently estimates the aleatoric and epistemic uncertainty for unimodality and

---

[1] Our code has been released in https://github.com/Cocofeat/EyeMoSt.

adaptively integrates Fundus and OCT modalities in principle. As shown in Fig. 1 (a), we first employ the 2D/3D neural network encoders to capture different modality features. Then, we place multi-evidential heads after the trained networks to model the parameters of higher-order NIG distributions for unimodality. To merge these predicted distributions, We derive the posterior predictive of the NIG distributions as Student's t (St) distributions. Particularly, the Mixture of Student's t (MoSt) distributions is introduced to integrate the distributions of different modalities in principle. Finally, we elaborate on the training pipeline for the model evidence acquisition.

Given a multimodality eye dataset $\mathcal{D} = \left\{ \{ \mathbf{x}_m^i \}_{m=1}^{M} \right\}$ and the corresponding label y^i, the intuitive goal is to learn a function that can classify different categories. Fundus and OCT are common imaging modalities for eye disease screening. Therefore, here M = 2, $\mathbf{x}_1^i$ and $\mathbf{x}_2^i$ represent Fundus and OCT input modality data, respectively. We first train 2D encoder Θ of Res2Net [7] and 3D encoder Φ of MedicalNet [5] to identify the feature-level informativeness, which can be defined as $\Theta \left(\mathbf{x}_1^i \right)$ and $\Phi \left(\mathbf{x}_2^i \right)$, respectively.

2.1 Uncertainty Estimation for Unimodality

We extend the deep evidential regression model [1] to multimodality evidential classification for eye disease screening. To this end, to model the uncertainty for Fundus or OCT modality, we assume that the observe label y^i is drawn from a Gaussian $\mathcal{N} \left(y^i | \mu, \sigma^2 \right)$, whose mean and variance are governed by an evidential prior named the NIG distribution:

$$\text{NIG} \left(\mu, \sigma^2 | \mathbf{p}_m \right) = \mathcal{N} \left(\mu | \gamma_m, \frac{\sigma^2}{\delta_m} \right) \Gamma^{-1} \left(\sigma^2 | \alpha_m, \beta_m \right), \tag{1}$$

where Γ^{-1} is an inverse-gamma distribution, $\gamma_m \in \mathbb{R}, \delta_m > 0, \alpha_m > 1, \beta_m > 0$ are the learning parameters. Specifically, the multi-evidential heads will be placed after the encoders Θ and Φ (as shown in Fig. 1 (a)), which outputs the prior NIG parameters $\mathbf{p}_m = (\gamma_m, \delta_m, \alpha_m, \beta_m)$. As a result, the aleatoric (AL) and epistemic (EP) uncertainty can be estimated by the $\mathbb{E} \left[\sigma^2 \right]$ and the $\text{Var} \left[\mu \right]$, respectively, as:

$$\text{AL} = \mathbb{E} \left[\sigma^2 \right] = \frac{\alpha_m}{\beta_m - 1}, \qquad \text{EP} = \text{Var} \left[\mu \right] = \frac{\beta_m}{\delta_m \left(\alpha_m - 1 \right)}. \tag{2}$$

Then, given the evidence distribution parameter $\mathbf{p}_m$, the marginal likelihood is calculated by marginalizing the likelihood parameter:

$$p \left(y^i | x_m^i, \mathbf{p}_m \right) = \int_\mu \int_{\sigma^2} p \left(y^i | x_m^i, \mu, \sigma^2 \right) \text{NIG} \left(\mu, \sigma^2 | \mathbf{p}_m \right) d\mu d\sigma^2. \tag{3}$$

Interacted by the prior and the Gaussian likelihood of each unimodality [1], its analytical solution does exist and yields an St prediction distribution as:

$$p\left(y^i | x_m^i, \mathbf{p}_m\right) = \frac{\Gamma\left(\alpha_m + \frac{1}{2}\right)}{\Gamma\left(\alpha_m\right)} \sqrt{\frac{\delta_m}{2\pi\beta_m\left(1 + \delta_m\right)}} \left(1 + \frac{\delta_m\left(y^i - \gamma_m\right)^2}{2\beta_m\left(1 + \delta_m\right)}\right)^{-\left(\alpha_m + \frac{1}{2}\right)}$$

$$= St\left(y^i; \gamma_m, o_m, 2\alpha_m\right), \tag{4}$$

with $o_m = \dfrac{\beta_m\left(1 + \delta_m\right)}{\delta_m\alpha_m}$. The complete derivations of Eq. 4 are available in Supplementary S1.1. Thus, the two modalities distributions are transformed into the student's t Distributions $St\left(y^i; u_m, \Sigma_m, v_m\right) = St\left(y^i; \gamma_m, o_m, 2\alpha_m\right)$, with $u_m \in \mathbb{R}, \Sigma_m > 0, v_m > 2$.

2.2   Mixture of Student's t Distributions (MoSt)

Then, we focus on fusing multiple St Distributions from different modalities. How to rationally integrate multiple Sts into a unified St is the key issue. To this end, the joint modality of distribution can be denoted as:

$$St\left(y^i; u_F, \Sigma_F, v_F\right) = St\left(y^i; \begin{bmatrix} u_1^i \\ u_2^i \end{bmatrix}, \begin{bmatrix} \Sigma_{11} & \Sigma_{12} \\ \Sigma_{21} & \Sigma_{22} \end{bmatrix}, \begin{bmatrix} v_1^i \\ v_2^i \end{bmatrix}\right). \tag{5}$$

In order to preserve the closed St distribution form and the heavy-tailed properties of the fusion modality, the updated parameters are given by [24]. In simple terms, we first adjust the degrees of freedom of the two distributions to be consistent. As shown in Fig. 1 (b), the smaller values of degrees of freedom (DOF) v has heavier tails. Therefore, we construct the decision value $\tau_m = v_m$ to approximate the parameters of the fused distribution. We assume that multiple St distributions are still an approximate St distribution after fusion. Assuming that the degrees of freedom of τ_1 are smaller than τ_2, then, the fused St distribution $St\left(y^i; u_F, \Sigma_F, v_F\right)$ will be updated as:

$$v_F = v_1, \qquad u_F = u_1, \qquad \Sigma_F = \frac{1}{2}\left(\Sigma_1 + \frac{v_2\left(v_1 - 2\right)}{v_1\left(v_2 - 2\right)}\Sigma_2\right). \tag{6}$$

More intuitively, the above formula determines the modality with a stronger heavy-tailed attribute. That is, according to the perceived heavy-tailed attribute of each modality, the most robust modality is selected as the fusion modality. Finally, the prediction and uncertainty of the fusion modality is given by:

$$\hat{y}^i = \mathbb{E}_{p\left(x_F^i, \mathbf{p}_F\right)}\left[y^i\right] = u_F, \qquad \hat{U}_F = \mathbb{E}\left[\sigma_F^2\right] = \Sigma_F \frac{v_F}{v_F - 2}. \tag{7}$$

2.3   Learning the Evidential Distributions

Under the evidential learning framework, we expect more evidence to be collected for each modality, thus, the proposed model is expected to maximize the likelihood function of the model evidence. Equivalently, the model is expected to minimize the negative log-likelihood function, which can be expressed as:

$$\mathcal{L}_m^{NLL} = \log \frac{\Gamma(\alpha_m)\sqrt{\frac{\pi}{\delta_m}}}{\Gamma\left(\alpha_m + \frac{1}{2}\right)} - \alpha_m \log\left(2\beta_m\left(1 + \delta_m\right)\right)$$
$$+ \left(\alpha_m + \frac{1}{2}\right) \log\left(\left(y^i - \gamma_m\right)^2 \delta_m + 2\beta_m\left(1 + \delta_m\right)\right). \tag{8}$$

Then, to fit the classification tasks, we introduce the cross entropy term $\mathcal{L}_m^{CE}$:

$$\mathcal{L}_m^{NIG} = \mathcal{L}_m^{NLL} + \lambda \mathcal{L}_m^{CE}, \tag{9}$$

where λ is the balance factor set to 0.5. For further information on the selection of hyperparameter λ, please refer to Supplementary S2. Similarly, for the fusion modality, we first maximize the likelihood function of the model evidence as follows:

$$\mathcal{L}_F^{NLL} = \log \Sigma_F + \log \frac{\Gamma\left(\frac{v_F}{2}\right)}{\Gamma\left(\frac{v_F+1}{2}\right)} + \log \sqrt{v_F\pi} + \frac{(v_F+1)}{2} \log\left(1 + \frac{(y^i - u_F)^2}{v_F \Sigma_F}\right),$$
$$\tag{10}$$

Complete derivations of Eq. 8 are available in Supplementary S1.2. Then, to achieve better classification performance, the cross entropy term $\mathcal{L}_m^{CE}$ is also introduced into Eq. 8 as below:

$$\mathcal{L}_F^{St} = \mathcal{L}_F^{NLL} + \lambda \mathcal{L}_F^{CE}, \tag{11}$$

Totally, the evidential learning process for multimodality screening can be denoted as:

$$\mathcal{L}_{all} = \sum_{m=1}^{M} \mathcal{L}_m^{NIG} + \mathcal{L}_F^{St}. \tag{12}$$

In this paper, we mainly consider the fusion of two modalities, $M = 2$.

3   Experiments

Datasets: In this paper, we verify the effectiveness of EyeMoSt on the two datasets. For the glaucoma recognition, We validate the proposed method on the GAMMA [29] dataset. It contains 100 paired cases with a three-level glaucoma grading. They are divided into the training set and test set with 80 and 20 respectively. We conduct the five-fold cross-validation on it to prevent performance improvement caused by accidental factors. Then, we test our method on the in-house collected dataset, which includes Age-related macular degeneration (AMD) and polypoid choroidal vasculopathy (PCV) diseases. They are divided into training, validation, and test sets with 465, 69, and 70 cases respectively. More details of the dataset can be found in Supplementary S2. Both of these datasets are including the paired cases of Fundus (2D) and OCT (3D).[2]

---

[2] The ethical approval of this dataset was obtained from the Ethical Committee.

Table 1. Comparisons with different algorithms on the GAMMA and in-house dataset. F and O denote the Fundus and OCT modalities, respectively. The top-2 results are highlighted in Red and Blue. Higher ACC and Kappa, and Lower ECE mean better.

Methods	GAMMA dataset						In-house dataset					
	Original		Gaussian noise				Original		Gaussian noise			
			$\sigma = 0.1$ (F)		$\sigma = 0.3$ (O)				$\sigma = 0.1$ (F)		$\sigma = 0.3$ (O)	
	ACC	Kappa	ACC	Kappa	ACC	Kappa	ACC	ECE	ACC	ECE	ACC	ECE
B-IF	0.700	0.515	0.623	0.400	0.530	0.000	0.800	0.250	0.593	0.450	0.443	0.850
B-EF [10]	0.660	0.456	0.660	0.452	0.500	0.000	0.800	0.200	0.777	0.223	0.443	0.557
M^2LC [28]	0.710	0.527	0.660	0.510	0.500	0.000	0.814	0.186	0.786	0.214	0.443	0.557
MCDO [6]	0.758	0.636	0.601	0.341	0.530	0.000	0.786	0.214	0.771	0.304	0.429	0.571
DE [14]	0.710	0.539	0.666	0.441	0.530	0.000	0.800	0.200	0.800	0.200	0.609	0.391
TMC [8]	0.810	0.658	0.430	0.124	0.550	0.045	0.829	0.171	0.814	0.186	0.443	0.557
Our	0.850	0.754	0.663	0.458	0.830	0.716	0.829	0.171	0.800	0.200	0.829	0.171

Training Details: Our proposed method is implemented in PyTorch and trained on NVIDIA GeForce RTX 3090. Adam optimization [13] is employed to optimize the overall parameters with an initial learning rate of 0.0001. The maximum of epoch is 100. The data augmentation techniques for GAMMA dataset are similar to [3], including random grayscaling, random color jitter, and random horizontal flipping. All inputs are uniformly adjusted to 256×256 and $128 \times 256 \times 128$ for Fundus and OCT modalities. The batch size is 16.

Compared Methods and Metrics: We compare the following six methods: For different fusion stage strategies, **a) B-EF** Baseline of the early fusion [10] strategy, **b) B-IF** Baseline of the intermediate typical fusion method, **c)** M^2**LC** [28] of the intermediate fusion method and the later fusion method **d) TMC** [8] are used as comparisons. B-EF is first integrated at the data level, and then passed through the same MedicalNet [5]. B-IF first extracts features by the encoders (same with us), and then concatenates their output features as the final prediction. For the uncertainty quantification methods, **e) MCDO** (Monte Carlo Dropout) employs the test time dropout as an approximation of a Bayesian neural network [6]. **f) DE** (Deep ensemble) quantifies the uncertainties by ensembling multiple models [14]. We adopt the accuracy (ACC) and Kappa metrics for intuitive comparison with different methods. Particularly, expected calibration error (ECE) [20] is used to compare the calibration of the uncertainty algorithms.

Comparison and Analysis: We reported our algorithm with different methods on the GAMMA and in-house datasets in Table 1. First, we compare these methods under the clean multimodality eye data. Our method obtained competitive results in terms of ACC and Kappa. Then, to verify the robustness of our model, we added Gaussian noise to Fundus or OCT modality ($\sigma = 0.1/0.3$) on

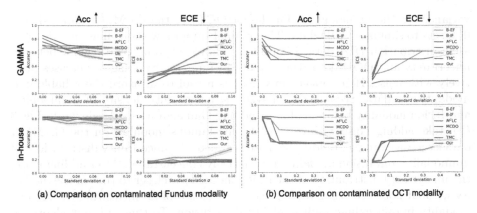

(a) Comparison on contaminated Fundus modality   (b) Comparison on contaminated OCT modality

Fig. 2. Accuracy and ECE performance of different algorithms in contaminated single modality with different levels of noise on GAMMA and in-house datasets.

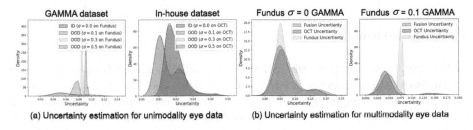

(a) Uncertainty estimation for unimodality eye data   (b) Uncertainty estimation for multimodality eye data

Fig. 3. Uncertainty density of unimodality and multimodality eye data.

the two datasets. Compared with other methods, our EyeMoSt maintains classification accuracy in noisy OCT modality, while comparable in noisy Fundus modality. More generally, we added different Gaussian noises to Fundus or OCT modality, as shown in Fig. 2. The same conclusion can be drawn from Fig. 2. This is attributed to the perceived long tail in the data when fused. The visual comparisons of different noises to the Fundus/OCT modality on the in-house dataset can be found in Supplementary S2. To further quantify the reliability of uncertainty estimation, we compared different algorithms using the ECE indicator. As shown in Table 1 and Fig. 2, our proposed algorithm performs better in both clean and single pollution modalities. The inference times of the uncertainty-based methods on two modalities on the in-house dataset are 5.01 s (MCDO), 8.28 s (DE), 3.98 s (TMC), and 3.22 s (Ours). It can be concluded that the running time of EyeMos$t$ is lower than other methods. In brief, we conclude that our proposed model is more robust and reliable than the above methods.

Understanding Uncertainty for Unimodality/Multimodality Eye Data: To make progress towards the multimodality ophthalmic clinical application of uncertainty estimation, we conducted unimodality and multimodality uncertainty analysis for eye data. First, we add more Gaussian noise with varying variances to the unimodality (Fundus or OCT) in the GAMMA and in-house datasets to simulate OOD data. The original samples without noise are denoted as in-distribution (ID) data. Figure 3 (a) shows a strong relationship between uncertainty and OOD data. Uncertainty in unimodality images increases positively with noise. Here, uncertainty acts as a tool to measure the reliable unimodality eye data. Second, we analyze the uncertainty density of unimodality and fusion modality before and after adding Gaussian noise. As shown in Fig. 3 (b), take adding noise with $\sigma = 0.1$ to the Fundus modality on the GAMMA dataset as an example. Before the noise is added, the uncertainty distributions of unimodality and fusion modality are relatively concentrated. After adding noise, the uncertainty distribution of the fusion modality is closer to that of the modality without noise. Hence, EyeMoSt can serve as a tool for measuring the reliable modality in ophthalmic multimodality data fusion. To this end, our algorithm can be used as an out-of-distribution detector and data quality discriminator to inform reliable and robust decisions for multimodality eye disease screening.

4   Conclusion

In this paper, we propose the EyeMoSt for reliable and robust screening of eye diseases using evidential multimodality fusion. Our EyeMoSt produces the uncertainty for unimodality and then adaptively fuses different modalities in a distribution perspective. The different NIG evidence priors are employed to model the distribution of encoder observations, which supports the backbones to directly learn aleatoric and epistemic uncertainty. We then derive an analytical solution to the Student's t distributions of the NIG evidence priors on the Gaussian likelihood function. Furthermore, we propose the MoSt distributions in principle adaptively integrates different modalities, which endows the model with heavy-tailed properties and is more robust and reliable for eye disease screening. Extensive experiments show that the robustness and reliability of our method in classification and uncertainty estimation on GAMMA and in-house datasets are competitive with previous methods. Overall, our approach has the potential to multimodality eye data discriminator for trustworthy medical AI decision-making. In future work, our focus will be on incorporating uncertainty into the training process and exploring the application of reliable multimodality screening for eye diseases in a clinical setting.

Acknowledgements. This work was supported by the National Research Foundation, Singapore under its AI Singapore Programme (AISG Award No: AISG2-TC-2021-003), A*STAR AME Programmatic Funding Scheme Under Project A20H4b0141, A*STAR Central Research Fund, the Science and Technology Department of Sichuan Province (Grant No. 2022YFS0071 & 2023YFG0273), and the China Scholarship Council (No. 202206240082).

References

1. Amini, A., Schwarting, W., Soleimany, A., Rus, D.: Deep evidential regression. Adv. Neural. Inf. Process. Syst. **33**, 14927–14937 (2020)
2. Baltrušaitis, T., Ahuja, C., Morency, L.P.: Multimodal machine learning: a survey and taxonomy. IEEE Trans. Pattern Anal. Mach. Intell. **41**(2), 423–443 (2018)
3. Cai, Z., Lin, L., He, H., Tang, X.: COROLLA: an efficient multi-modality fusion framework with supervised contrastive learning for glaucoma grading. In: IEEE International Symposium on Biomedical Imaging (ISBI), pp. 1–4. IEEE (2022)
4. Cai, Z., Lin, L., He, H., Tang, X.: Uni4Eye: unified 2D and 3D self-supervised pre-training via masked image modeling transformer for ophthalmic image classification. In: Wang, L., Dou, Q., Fletcher, P.T., Speidel, S., Li, S. (eds.) Medical Image Computing and Computer Assisted Intervention – MICCAI 2022. MICCAI 2022. LNCS, vol. 13438. Springer, Cham (2022). https://doi.org/10.1007/978-3-031-16452-1_9
5. Chen, S., Ma, K., Zheng, Y.: Med3D: transfer learning for 3D medical image analysis. arXiv preprint arXiv:1904.00625 (2019)
6. Gal, Y., Ghahramani, Z.: Dropout as a Bayesian approximation: representing model uncertainty in deep learning. In: International Conference on Machine Learning, pp. 1050–1059. PMLR (2016)
7. Gao, S.H., Cheng, M.M., Zhao, K., Zhang, X.Y., Yang, M.H., Torr, P.: Res2net: a new multi-scale backbone architecture. IEEE Trans. Pattern Anal. Mach. Intell. **43**(2), 652–662 (2021)
8. Han, Z., Zhang, C., Fu, H., Zhou, J.T.: Trusted multi-view classification. In: International Conference on Learning Representations (2020)
9. He, X., Deng, Y., Fang, L., Peng, Q.: Multi-modal retinal image classification with modality-specific attention network. IEEE Trans. Med. Imaging **40**(6), 1591–1602 (2021)
10. Hua, C.H., et al.: Convolutional network with twofold feature augmentation for diabetic retinopathy recognition from multi-modal images. IEEE J. Biomed. Health Inform. **25**(7), 2686–2697 (2020)
11. Jung, M.C., Zhao, H., Dipnall, J., Gabbe, B., Du, L.: Uncertainty estimation for multi-view data: the power of seeing the whole picture. In: Advances in Neural Information Processing Systems (2022)
12. Kendall, A., Gal, Y.: What uncertainties do we need in Bayesian deep learning for computer vision? In: Advances in Neural Information Processing Systems (2017)
13. Kingma, D., Ba, J.: Adam: a method for stochastic optimization. Computer Science (2014)
14. Lakshminarayanan, B., Pritzel, A., Blundell, C.: Simple and scalable predictive uncertainty estimation using deep ensembles. In: Advances in Neural Information Processing Systems 30 (2017)
15. Li, X., Jia, M., Islam, M.T., Yu, L., Xing, L.: Self-supervised feature learning via exploiting multi-modal data for retinal disease diagnosis. IEEE Trans. Med. Imaging **39**(12), 4023–4033 (2020)
16. Li, Y., El Habib Daho, M., Conze, P.H., et al.: Multimodal information fusion for glaucoma and diabetic retinopathy classification. In: Antony, B., Fu, H., Lee, C.S., MacGillivray, T., Xu, Y., Zheng, Y. (eds.) Ophthalmic Medical Image Analysis. OMIA 2022. LNCS, vol. 13576. Springer, Cham (2022). https://doi.org/10.1007/978-3-031-16525-2_6

17. Liu, J.Z., Lin, Z., Padhy, S., Tran, D., Bedrax-Weiss, T., Lakshminarayanan, B.: Simple and principled uncertainty estimation with deterministic deep learning via distance awareness. In: Proceedings of the 34th International Conference on Neural Information Processing Systems (2020)
18. MacKay, D.J.: A practical Bayesian framework for backpropagation networks. Neural Comput. **4**(3), 448–472 (1992)
19. Malinin, A., Gales, M.: Predictive uncertainty estimation via prior networks. In: Advances in Neural Information Processing Systems 31 (2018)
20. Maronas, J., Paredes, R., Ramos, D.: Calibration of deep probabilistic models with decoupled Bayesian neural networks. Neurocomputing **407**, 194–205 (2020)
21. Neal, R.M.: Bayesian learning for neural networks, vol. 118. Springer New York, NY (2012). https://doi.org/10.1007/978-1-4612-0745-0
22. Ranganath, R., Gerrish, S., Blei, D.: Black box variational inference. In: Artificial Intelligence and Statistics, pp. 814–822. PMLR (2014)
23. Rodrigues, E.O., Conci, A., Liatsis, P.: Element: multi-modal retinal vessel segmentation based on a coupled region growing and machine learning approach. IEEE J. Biomed. Health Inform. **24**(12), 3507–3519 (2020)
24. Roth, M., Özkan, E., Gustafsson, F.: A student's t filter for heavy tailed process and measurement noise. In: 2013 IEEE International Conference on Acoustics, Speech and Signal Processing, pp. 5770–5774. IEEE (2013)
25. Sensoy, M., Kaplan, L., Kandemir, M.: Evidential deep learning to quantify classification uncertainty. In: Proceedings of the 32nd International Conference on Neural Information Processing Systems, pp. 3183–3193 (2018)
26. Van Amersfoort, J., Smith, L., Teh, Y.W., Gal, Y.: Uncertainty estimation using a single deep deterministic neural network. In: International Conference on Machine Learning, pp. 9690–9700. PMLR (2020)
27. Wang, W., et al.: Learning two-stream CNN for multi-modal age-related macular degeneration categorization. IEEE J. Biomed. Health Inform. **26**(8), 4111–4122 (2022)
28. Woo, S., Park, J., Lee, J.-Y., Kweon, I.S.: CBAM: convolutional block attention module. In: Ferrari, V., Hebert, M., Sminchisescu, C., Weiss, Y. (eds.) ECCV 2018. LNCS, vol. 11211, pp. 3–19. Springer, Cham (2018). https://doi.org/10.1007/978-3-030-01234-2_1
29. Wu, J., Fang, H., Li, F., et al.: Gamma challenge: glaucoma grading from multi-modality images. arXiv preprint arXiv:2202.06511 (2022)
30. Zadeh, L.A.: Review of a mathematical theory of evidence. AI Mag. **5**(3), 81 (1984)
31. Zhou, T., Ruan, S., Canu, S.: A review: deep learning for medical image segmentation using multi-modality fusion. Array **3**, 100004 (2019)
32. Zou, K., et al.: EvidenceCap: towards trustworthy medical image segmentation via evidential identity cap. arXiv preprint arXiv:2301.00349 (2023)

Polar-Net: A Clinical-Friendly Model for Alzheimer's Disease Detection in OCTA Images

Shouyue Liu[1,2], Jinkui Hao[1], Yanwu Xu[3,4(✉)], Huazhu Fu[5], Xinyu Guo[1], Jiang Liu[6], Yalin Zheng[7], Yonghuai Liu[8], Jiong Zhang[1], and Yitian Zhao[1(✉)]

[1] Cixi Institute of Biomedical Engineering, Ningbo Institute of Materials Technology and Engineering, Chinese Academy of Sciences, Ningbo, China
yitian.zhao@nimte.ac.cn
[2] Cixi Biomedical Research Institute, Wenzhou Medical University, Ningbo, China
[3] School of Future Technology, South China University of Technology, Guangzhou, China
ywxu@ieee.org
[4] Pazhou Lab, Guangzhou, China
[5] Institute of High Performance Computing, A*STAR, Singapore, Singapore
[6] Department of Computer Science, Southern University of Science and Technology, Shenzhen, China
[7] Department of Eye and Vision Science, University of Liverpool, Liverpool, England
[8] Department of Computer Science, Edge Hill University, Ormskirk, England

Abstract. Optical Coherence Tomography Angiography (OCTA) is a promising tool for detecting Alzheimer's disease (AD) by imaging the retinal microvasculature. Ophthalmologists commonly use region-based analysis, such as the ETDRS grid, to study OCTA image biomarkers and understand the correlation with AD. In this work, we propose a novel deep-learning framework called Polar-Net. Our approach involves mapping OCTA images from Cartesian coordinates to polar coordinates, which allows for the use of approximate sector convolution and enables the implementation of the ETDRS grid-based regional analysis method commonly used in clinical practice. Furthermore, Polar-Net incorporates clinical prior information of each sector region into the training process, which further enhances its performance. Additionally, our framework adapts to acquire the importance of the corresponding retinal region, which helps researchers and clinicians understand the model's decision-making process in detecting AD and assess its conformity to clinical observations. Through evaluations on private and public datasets, we have demonstrated that Polar-Net outperforms existing state-of-the-art methods and provides more valuable pathological evidence for the association between retinal vascular changes and AD. In addition, we also show that the two innovative modules introduced in our framework have a significant impact on improving overall performance.

Supplementary Information The online version contains supplementary material available at https://doi.org/10.1007/978-3-031-43990-2_57.

Keywords: OCTA · Alzheimer's Disease · Polar Transformation

1 Introduction

Alzheimer's disease (AD) is a progressive and debilitating neurological disorder that affects millions of people worldwide. Although primary detection of AD can be achieved through a combination of cognitive function tests and neuroimaging techniques, such as magnetic resonance imaging (MRI) and cerebrospinal fluid (CSF) analysis [14]. However, these approaches suffer from being invasive, time-consuming, or expensive, hindering their use in routine clinical practice. The convergence of tissue origin, structural characteristics, and functional mechanisms between the eyes and the brain has been previously reported [18]. For example, patients with AD have significantly decreased blood vessel density in superficial parafoveal and choriocapillaris (CC) [25]. To this end, the automated AD detection using fundus image has emerged as an active research field in the last two years [1,13,20]. Color fundus photography (CFP) has commonly used for AD studies, but the CFP has limitations in capturing the information of deep layer vessels. Optical coherence tomography angiography (OCTA) is an innovative non-invasive technology that generates high-resolution images of depth-resolved retinal microvasculature projections [8], including SVC, DVC, and CC.

Studies on clinical biomarkers of OCTA images are mainly based on regional analysis, e.g., the early treatment of diabetic retinopathy study (ETDRS) grid, which divides a target area into 9 regions with three concentric circles and two orthogonal lines, as shown in the right three sub-figures in Fig. 1. The region-based analysis allows a more specific evaluation of retinal changes and their correlation with AD, which can provide a more nuanced understanding of the disease. Research following ETDRS and IE grid demonstrated the significance of many regions, e.g., in the three sub-regions of nasal-outer, superior-inner, and inferior-inner in inner vascular complexes, which present a substantial decrease in vascular area density and vascular length density for the AD participants [23].

Over the past few years, deep-learning-based algorithms have achieved remarkable success in the analysis of medical images. As for AD detection, several methods use an integration of multiple modalities [20,21]. However, these methods rarely follow the clinical region-based analysis routine, which limits their ability to incorporate valuable clinical statistical findings and generate easily interpretable results. To address the above issues, we proposed a novel deep-learning framework to take full advantage of clinical region-based analysis, for AD detection in OCTA images. To obtain a more accurate and interpretable result, we specifically designed an approximate sector convolution, based on the polar transformation and a multi-kernel feature extraction module. The main contributions of the paper can be summarised as follows: **(1)** Based on the well-known clinically used ETDRS grids for retinal image analysis, we incorporate the regional importance prior in the training process through a weight matrix, so as to better understand the correlations between retinal structure alterna-

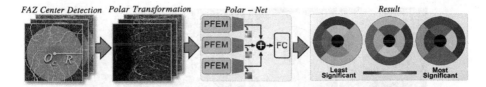

Fig. 1. The proposed AD detection model utilizes the polar transformation (left two) inspired by ETDRS grids commonly used in ophthalmic analysis. This model allows for easy understanding by ophthalmologists, as the output of the Polar-Net can be interpreted through an intuitive color map illustration (right) that indicates different levels of significance.

tions and AD. **(2)** We introduce an approximate sector convolution through polar transformation, to mimic the clinical region-based analysis, by mapping the OCTA image from the Cartesian system to the polar system, as shown in the left two sub-figures in Fig. 1. **(3)** We further performed the explainability analysis on the well-trained model. The interpretable results showed consistency with the conclusions of the previous clinical studies, indicating that the proposed method can be a potential tool, to investigate the pathological evidence of the relationship between the fundus and AD.

2   Methodology

Figure 1 shows the flowchart of our AD detection method using SVC, DVC, and CC projections of OCTA as input. First, we utilize VAFF-Net [6] to locate the center of the FAZ on SVC. We then transform the original images into the polar coordinates with the FAZ center as the origin. The transformed images are then fed into our Polar-Net, which produces the final detection result and the corresponding region importance matrix.

2.1   Polar Coordinate Transformation for OCTA Image

We introduce a method called polar transformation, to realize region-based analysis. As shown in Fig. 2, the polar transformation converts the region of interest (blue circle) into a polar coordinate system Fig. 2(c), with the center of the FAZ (according to the definition [4] of ETDRS), $O_c(u_o, v_o)$ as the origin. The original image is represented as points in the Cartesian system $p(u, v)$, and the corresponding points in the polar system are represented by $p\prime(\theta, r)$. The relationship between these two coordinate systems is given by the following equations:

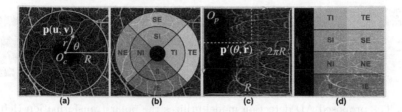

Fig. 2. Illustrations of the Cartesian coordinate system (a), an ETDRS grid on an OCTA projection (b), the polar coordinate system (c), and the mapping relationship after the transformation (d). Definition: temporal-inner (TI), temporal-external (TE), superior-inner (SI), superior-external (SE), nasal-inner (NI), nasal-external (NE), inferior-inner (II), and inferior-external (IE).

$$\begin{cases} u = r\cos\theta \\ v = r\sin\theta \end{cases} \Leftrightarrow \begin{cases} r = \sqrt{u^2 + v^2} \\ \theta = \tan^{-1} v/u \end{cases}. \tag{1}$$

The width of the transformed image is equal to the distance R, the minimal length from the center $O_c(u_o, v_o)$ to the edge in the original image, and the height is $2\pi R$. Since the corners are cropped, the outermost pixels of the region of interest are kept in order to preserve the original information as much as possible, and the part near O_c is filled by nearest neighbor interpolation. The polar transformation represents the original image in the polar coordinate system by pixel-wise mapping [3], and has the following properties:

1) Approximate Sector-Shaped Convolution. Convolution is widely used in convolutional neural networks (CNNs), where the shape of the convolution kernel is always rectangular. However, in the real world, many semantics are non-rectangular, such as circle and sector, which makes the adaptability of the receptive field in CNNs suboptimal. For the polar transformation, the mapping relationship is fixed, enabling us to approximate the sector convolution with a rectangular convolution kernel at a lower computational cost. The mapping relationship shown in Fig. 2(b)(d) explains this, and for the sake of simplicity and clarity, we use ETDRS girds as an example. When we perform convolution with a rectangular kernel along the $TI \rightarrow SI$ direction on the transformed image, it is equivalent to performing convolution with a sector-shaped kernel counterclockwise around the FAZ center in the original image.

2) Equivalent Augmentation. Applying data augmentation to the original image is the same as applying data augmentation in the polar system since the transformation is a pixel-wise mapping [3]. For instance, by changing the start angle and the transformation center $O_c(u_o, v_o)$, we can realize the drift cropping operation in the polar system. It is analogous to applying various cropping factors for data augmentation by changing the transformation radius R.

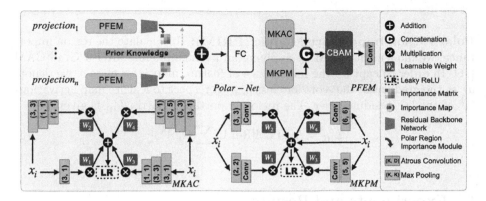

Fig. 3. The details of Polar-Net. It contains multiple input branches, and each branch starts with a polar feature extractor module (PFEM) and ends with a residual network. PFEM consists of a multi-kernel atrous convolution module (MKAC), a multi-kernel pooling module (MKPM), and a convolutional block attention module (CBAM). Parameters K and D denote the kernel size and dilation respectively.

2.2 Network Architecture

In the transformed image, we can extract features around the FAZ. Rectangular features at different scales correspond to different sectoral features in the retina. Therefore, it is critical to extract information across different sizes of the visual field. To this end, we design the Polar-Net. As shown in Fig. 3, it contains several branches, the number of which varies according to the number of projections. Each branch starts with a polar feature extractor module (PFEM) and ends with a residual network. To take full advantage of all the branches, middle fusion is used. To generate the region importance matrix, a polar region importance module (PRIM) is proposed, which follows the residual network. Furthermore, Polar-Net can receive a prior knowledge matrix to utilize clinical knowledge.

Polar Feature Extractor Module (PFEM): To extract shallow features in different views, we propose PFEM, which consists of a multi-kernel atrous convolution module (MKAC), a multi-kernel pooling module (MKPM), and a convolutional block attention module (CBAM) [22]. For each projection x_i, MKAC $H(\cdot)$ applies multiple scale atrous convolutions to enlarge the field of view [24], and MKPM $G(\cdot)$ applies a series of max-pooling operations with different pooling kernels to discover microscopic changes, by extracting the most salient feature [9]. Finally, CBAM $T(\cdot)$ is applied to exploit the inter-channel hidden information of features. During feature extraction, W_n is used to adaptively adjust the weights of the above processes. The mathematical notation of the above is:

$$\begin{cases} F_{\text{MKAC}} = \text{LeakyReLU}\left[\sum_n H(x_i) W_n\right], \\ F_{\text{MKPM}} = \text{LeakyReLU}\left[x_i + \sum_n G(x_i) W_n\right], \\ F_{\text{PFEM}} = T\left(\text{Concat}\left(F_{\text{MKAC}}, F_{\text{MKPM}}\right)\right). \end{cases} \tag{2}$$

Polar Region Importance Module (PRIM): To calculate the region importance, we implement PRIM by applying an average pooling after a Grad-CAM [16]. In order to capture the importance of feature map k for class c, we denote a_k^c as the gradient of the score for class c, with respect to feature map activations A^k of the last residual layer. The region importance matrix L_{RI}^c is given by:

$$L_{RI}^c = \text{AvgPool}[\text{ReLU}(\sum_k \alpha_k^c A^k)]. \tag{3}$$

3   Experiments and Results

Data Description: An in-house dataset was conducted for this study. It includes 199 images from 114 AD patients and 566 images from 291 healthy subjects. All data were collected with the approval of the relevant authorities and the consent of the patients, following the Declaration of Helsinki. All the patients conform to the standards of the National Institute on Aging and Alzheimer's Association (NIA-AA). The images were captured with a swept-source OCTA (VG200S, SVision Imaging). The images were captured in a $3 \times 3\,\text{mm}^2$ area centered on the fovea. We make sure that the images from a single patient will only be used as training or testing sets once. In the cross-validation experiment subset, we sample the categories from each dataset at the same ratio.

Table 1. The detection results (mean ± std) over the in-house dataset.

Model	ACC	AUROC	Kappa
ResNet-34 [7]	0.8125 ± 0.0267	0.7960 ± 0.0479	0.4909 ± 0.0717
EfficientNet-B3 [17]	0.7942 ± 0.0104	0.7908 ± 0.0157	0.4177 ± 0.0267
ConvNeXt-S [12]	0.7562 ± 0.0138	0.5903 ± 0.0313	0.1660 ± 0.0437
HorNet-S$_{GF}$ [15]	0.7602 ± 0.0113	0.5921 ± 0.0286	0.1738 ± 0.0458
VAN-B6 [5]	0.7707 ± 0.0124	0.6911 ± 0.0298	0.2939 ± 0.0485
ViT-Base [2]	0.7904 ± 0.0183	0.7726 ± 0.0286	0.3641 ± 0.0715
SwinV2-T [11]	0.7601 ± 0.0117	0.7528 ± 0.0343	0.3242 ± 0.0448
MUCO-Net [20]	0.7968 ± 0.0369	0.7773 ± 0.0414	0.3985 ± 0.0789
Polar-Net w/o PFEM	**0.8191 ± 0.0140**	**0.8315 ± 0.0245**	**0.5279 ± 0.0224**
Polar-Net w/o trans	**0.8388 ± 0.0142**	**0.8401 ± 0.0168**	**0.5279 ± 0.0264**
Polar-Net	**0.8518 ± 0.0169**	**0.8484 ± 0.0295**	**0.5766 ± 0.0685**
Polar-Net w prior	**0.8532 ± 0.0174**	**0.8523 ± 0.0320**	**0.5817 ± 0.0600**

Implementation Details: We implemented our proposed method with Pytorch. The model was trained on an Ubuntu 20.04 server equipped with two

Nvidia RTX 3090 GPUs. We employed Adam as the optimizer with an initial learning rate of 2e-5 and a batch size of 28. We also applied data augmentation by randomly rotating the images by $\pm 20°$ around their centers. The model was trained for 200 epochs. During the transformation, we considered the difference between the left and right eyes and used nearest-neighbor interpolation. The width of the transformed images was resized to 224 pixels. Five-fold crossvalidation was employed to fully utilize the data and make the results more reliable. Since there is no standard way to convert existing prior knowledge into matrices, for prior knowledge, we manually generated a 4×2 weight matrix according to the study [23]. The weights were 1 by default. The regions with p-values less than 0.05 had a weight of 1.5, and regions with p-values less than 0.01 had a weight of 2. For the entire DVC, the weight was set to 2.

Evaluation and Interpretability Assessment: We evaluate the performance of the model on the test set using the accuracy score (ACC), area under the receiver operating characteristic (AUROC), and kappa. To evaluate the performance, we compared our method with several state-of-the-art methods in the computer vision field and one in the AD detection field. Table 1 shows that our method outperforms the others in ACC, AUROC, and Kappa, with an improvement of up to 4.07%, 5.63%, and 9.08%, respectively. Prior knowledge did not bring much performance improvement, partly due to the crude method of generating the prior matrices, and partly probably due to the fact that the network's adaptive algorithm may have already learned similar prior knowledge. During the testing phase, we activated PRIM and generated a 4×2 importance matrix for the entire testing set. An inverse operation of the polar transformation was applied to generate the importance map.

For the sake of simplicity, here we take the EDTRS grid for analysis. As shown in Fig. 4 (a), it can be seen that globally, the importance of CC is highest and DVC is the second. This matches the findings that AD patients have a considerably lower density in choriocapillaris flow [25]. Meanwhile, the significance of DVC coincides with research findings that there is a considerable reduction in vascular area density and other factors in DVC [23]. In the DVC and CC, the parafovea is more important. This may relate to the loss of ganglion cells in the parafoveal retina [19]. For single projection (Fig. 4 (b)), different regions have different importance and the contributions of *NI*, *SI* and *II* (illustrated in Fig. 2) are higher. In summary, we found a pattern that high importance always occurs where there are more micro-vessels, such as the CC, DVC, and the parafovea. This finding coincides with the conclusion that the microvasculature of the brain and retina is significantly decreased in AD patients [25]. Our interpretable results roughly match the clinical study results, because we have made the network follow the clinical analysis method. This also proves the clinical relevance of ours. The minor difference is perhaps because the network unearthed the high-dimensional features that have not yet been discovered clinically.

Ablation Study: To evaluate the effectiveness of the polar transformation and Polar-Net, we performed an ablation study. To validate the proposed Polar-Net, we removed the PFEM. The results are shown at the bottom of Table 1. To

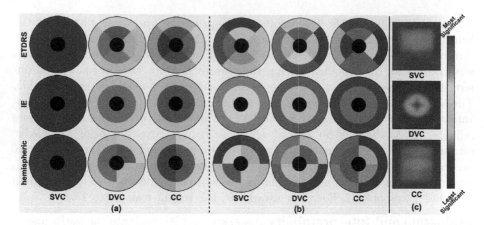

Fig. 4. Importance maps according to ETDRS, IE, and hemispheric grids. Group (a) takes into account the differences in importance between the projections, while (b) reflects the relative importance of regions within a single projection. The redder the color, the greater the importance. (c) shows a set of examples using the Grad-CAM visualization method, based on ResNet-34 in AD detection.

Table 2. The ablation results (mean ± std) of the polar transformation over the in-house dataset.

Model	Input	ACC	AUROC	Kappa
ResNet-34 [7]	$w/o\ trans$	0.8125 ± 0.0267	0.7960 ± 0.0479	0.4909 ± 0.0717
	$w\ trans$	**0.8244 ± 0.0226**	**0.8471 ± 0.0279**	**0.5137 ± 0.0417**
EfficientNet-B3 [17]	$w/o\ trans$	0.7942 ± 0.0104	0.7908 ± 0.0157	0.4177 ± 0.0267
	$w\ trans$	**0.8335 ± 0.0157**	**0.8295 ± 0.0279**	**0.5287 ± 0.0600**
ViT-Base [2]	$w/o\ trans$	0.7904 ± 0.0183	0.7726 ± 0.0286	0.3641 ± 0.0715
	$w\ trans$	**0.7982 ± 0.0177**	**0.8025 ± 0.0509**	**0.4235 ± 0.0748**
MUCO-Net [20]	$w/o\ trans$	0.7968 ± 0.0369	0.7773 ± 0.0414	0.3985 ± 0.0789
	$w\ trans$	**0.7916 ± 0.0255**	**0.7956 ± 0.0500**	**0.4271 ± 0.0659**

validate the transformation, we used the transformed images and the original images respectively. The results are shown in Table 2. All the results showed the effectiveness of the proposed components and modules.

Extended Experiment: To further verify our detection method's stability and generalisability, we conducted an additional experiment on a public dataset OCTA-500 [10]. It contains 189 images from 29 subjects with diabetic retinopathy and 160 healthy control. The details of the implementation are the same as the experiments on the in-house dataset. As shown in Table 3, our method achieved the best performances compared to the competitors.

Table 3. The classification results (mean ± std) of different methods over the OCTA-500 dataset.

Model	ACC	AUROC	Kappa
ResNet-34 [7]	0.9641 ± 0.0389	0.9818 ± 0.0306	0.8412 ± 0.1757
EfficientNet-B3 [17]	0.9632 ± 0.0225	0.9741 ± 0.0245	0.8375 ± 0.1006
VAN-B6 [5]	0.9478 ± 0.0245	0.9517 ± 0.0300	0.7691 ± 0.1334
ViT-Base [2]	0.9692 ± 0.0281	0.9768 ± 0.0247	0.8694 ± 0.1199
MUCO-Net [20]	0.9529 ± 0.0204	0.9717 ± 0.0222	0.8086 ± 0.0885
Polar-Net	$\mathbf{0.9898 \pm 0.0140}$	$\mathbf{0.9949 \pm 0.0072}$	$\mathbf{0.9604 \pm 0.0544}$

4 Conclusion

In this paper, we propose a novel framework for AD detection using retinal OCTA images, leveraging clinical prior knowledge and providing interpretable results. Our approach involves polar transformation, allowing for the use of approximate sector convolution and enabling the implementation of the region-based analysis. Additionally, our framework, called Polar-Net, is designed to acquire the importance of the corresponding retinal region, facilitating the understanding of the model's decision-making process in detecting AD and assessing its conformity to clinical observations. We evaluate the performance of our method on both private and public datasets, and the results demonstrate that Polar-Net outperforms state-of-the-art methods. Importantly, our approach produces clinically interpretable results, providing a potential tool for disease research to investigate the underlying pathological mechanisms. Our work presents a promising approach to using OCTA imaging for AD detection. Furthermore, we highlight the importance of incorporating clinical knowledge into AI models to improve interpretability and clinical applicability.

Acknowledgment. This work was supported in part by the National Science Foundation Program of China (62272444, 62103398), Zhejiang Provincial Natural Science Foundation of China (LR22F020008), the Youth Innovation Promotion Association CAS (2021298), the A*STAR AME Programmatic Funding Scheme Under Project A20H4b0141, and A*STAR Central Research Fund.

References

1. Cheung, C.Y., et al.: A deep learning model for detection of Alzheimer's disease based on retinal photographs: a retrospective, multicentre case-control study. Lancet Digit. Health 4(11), e806–e815 (2022)
2. Dosovitskiy, A., et al.: An image is worth 16x16 words: transformers for image recognition at scale. arXiv preprint arXiv:2010.11929 (2020)
3. Fu, H., Cheng, J., Xu, Y., Wong, D.W.K., Liu, J., Cao, X.: Joint optic disc and cup segmentation based on multi-label deep network and polar transformation. IEEE Trans. Med. Imaging 37(7), 1597–1605 (2018)

4. Early photocoagulation for diabetic retinopathy: ETDRS report number 9. Ophthalmology **98**(5), 766–785 (1991)
5. Guo, M.H., Lu, C.Z., Liu, Z.N., Cheng, M.M., Hu, S.M.: Visual attention network. arXiv preprint arXiv:2202.09741 (2022)
6. Hao, J., et al.: Retinal structure detection in octa image via voting-based multitask learning. IEEE Trans. Med. Imaging **41**(12), 3969–3980 (2022)
7. He, K., Zhang, X., Ren, S., Sun, J.: Deep residual learning for image recognition. In: Proceedings of the IEEE Conference on Computer Vision and Pattern Recognition, pp. 770–778 (2016)
8. Jeong, K.: Determining degenerative from vascular dementia using optical coherence tomography biomarkers for tomography and angiography. J. Multiple Sclerosis **09**(11), 001–002 (2022)
9. Ju, Y., Shi, B., Jian, M., Qi, L., Dong, J., Lam, K.M.: NormAttention-PSN: a high-frequency region enhanced photometric stereo network with normalized attention. Int. J. Comput. Vision **130**(12), 3014–3034 (2022)
10. Li, M., et al.: IPN-V2 and OCTA-500: methodology and dataset for retinal image segmentation. arXiv preprint arXiv:2012.07261 (2020)
11. Liu, Z., et al.: Swin transformer v2: scaling up capacity and resolution. In: Proceedings of the IEEE/CVF Conference on Computer Vision and Pattern Recognition, pp. 12009–12019 (2022)
12. Liu, Z., Mao, H., Wu, C.Y., Feichtenhofer, C., Darrell, T., Xie, S.: A convnet for the 2020s. In: Proceedings of the IEEE/CVF Conference on Computer Vision and Pattern Recognition, pp. 11976–11986 (2022)
13. Ma, Y., et al.: ROSE: a retinal oct-angiography vessel segmentation dataset and new model. IEEE Trans. Med. Imaging **40**(3), 928–939 (2021)
14. Palmer, N.P., Ortega, B.T., Joshi, P.: Cognitive impairment in older adults: epidemiology, diagnosis, and treatment. Psychiatr. Clin. (2022)
15. Rao, Y., Zhao, W., Tang, Y., Zhou, J., Lim, S.L., Lu, J.: HorNet: efficient high-order spatial interactions with recursive gated convolutions. In: Advances in Neural Information Processing Systems (2022)
16. Selvaraju, R.R., Cogswell, M., Das, A., Vedantam, R., Parikh, D., Batra, D.: Grad-CAM: visual explanations from deep networks via gradient-based localization. In: Proceedings of the IEEE International Conference on Computer Vision, pp. 618–626 (2017)
17. Tan, M., Le, Q.: EfficientNet: rethinking model scaling for convolutional neural networks. In: International Conference on Machine Learning, pp. 6105–6114. PMLR (2019)
18. Teja, K.V.R., Berendschot, T.T., Steinbusch, H., Webers, A.C., Murthy, R.P., Mathuranath, P.: Cerebral and retinal neurovascular changes: a biomarker for Alzheimer's disease. J. Gerontol. Geriatr. Res. **6**(4) (2017)
19. Un, Y., Alpaslan, F., Dikmen, N.T., Sonmez, M.: Posterior pole analysis and ganglion cell layer measurements in Alzheimer's disease. Hosp. Pract. **50**(4), 282–288 (2022)
20. Wang, X., et al.: Screening of dementia on octa images via multi-projection consistency and complementarity. In: Wang, L., Dou, Q., Fletcher, P.T., Speidel, S., Li, S. (eds.) Medical Image Computing and Computer Assisted Intervention - MICCAI 2022, pp. 688–698. Springer Nature Switzerland, Cham (2022). https://doi.org/10.1007/978-3-031-16434-7_66
21. Wisely, C.E., et al.: Convolutional neural network to identify symptomatic Alzheimer's disease using multimodal retinal imaging. Br. J. Ophthalmol. **106**(3), 388–395 (2022)

22. Woo, S., Park, J., Lee, J.-Y., Kweon, I.S.: CBAM: convolutional block attention module. In: Ferrari, V., Hebert, M., Sminchisescu, C., Weiss, Y. (eds.) ECCV 2018. LNCS, vol. 11211, pp. 3–19. Springer, Cham (2018). https://doi.org/10.1007/978-3-030-01234-2_1

23. Xie, J., et al.: Deep segmentation of octa for evaluation and association of changes of retinal microvasculature with Alzheimer's disease and mild cognitive impairment. Br. J. Ophthalmol. (2023). https://doi.org/10.1136/bjo-2022-321399

24. Yang, K., et al.: RecepNet: network with large receptive field for real-time semantic segmentation and application for blue-green algae. Remote Sens. **14**(21), 5315 (2022)

25. Zhang, S., et al.: Choriocapillaris changes are correlated with disease duration and MoCA score in early-onset dementia. Frontiers Aging Neurosci. **13**, 656750 (2021)

Context-Aware Pseudo-label Refinement for Source-Free Domain Adaptive Fundus Image Segmentation

Zheang Huai, Xinpeng Ding, Yi Li, and Xiaomeng Li[✉]

The Hong Kong University of Science and Technology, Kowloon, Hong Kong
eexmli@ust.hk

Abstract. In the domain adaptation problem, source data may be unavailable to the target client side due to privacy or intellectual property issues. Source-free unsupervised domain adaptation (SF-UDA) aims at adapting a model trained on the source side to align the target distribution with only the source model and unlabeled target data. The source model usually produces noisy and context-inconsistent pseudo-labels on the target domain, i.e., neighbouring regions that have a similar visual appearance are annotated with different pseudo-labels. This observation motivates us to refine pseudo-labels with context relations. Another observation is that features of the same class tend to form a cluster despite the domain gap, which implies context relations can be readily calculated from feature distances. To this end, we propose a context-aware pseudo-label refinement method for SF-UDA. Specifically, a context-similarity learning module is developed to learn context relations. Next, pseudo-label revision is designed utilizing the learned context relations. Further, we propose calibrating the revised pseudo-labels to compensate for wrong revision caused by inaccurate context relations. Additionally, we adopt a pixel-level and class-level denoising scheme to select reliable pseudo-labels for domain adaptation. Experiments on cross-domain fundus images indicate that our approach yields the state-of-the-art results. Code is available at https://github.com/xmed-lab/CPR.

Keywords: Source-free domain adaptation · Context similarity · Pseudo-label refinement · Fundus image

1 Introduction

Accurate segmentation of the optic cup and optic disc in fundus images is essential for the cup-to-disc ratio measurement that is critical for glaucoma screening and detection [6]. Although deep neural networks have achieved great advances in medical image segmentation, they are susceptible to data with

Supplementary Information The online version contains supplementary material available at https://doi.org/10.1007/978-3-031-43990-2_58.

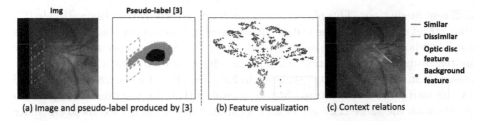

(a) Image and pseudo-label produced by [3]   (b) Feature visualization   (c) Context relations

Fig. 1. (a) Example of context-inconsistent pseudo-labels. Due to domain gap, the pseudo-label of optic disc has irregular protuberance which is inconsistent with adjacent predictions. (b) t-SNE visualization of target pixel features produced by source model. Under domain shift, despite not aligning with source segmentor, target features of the same class still form a cluster. (c) Inspired by (b), context relations can be computed from feature distances.

domain shifts, such as those caused by using different scanning devices or different hospitals [24]. Unsupervised domain adaptation [11] is proposed to transfer knowledge to the target domain with access to the source and target data while not requiring any annotation in the target domain. Recently, source-free unsupervised domain adaptation (SF-UDA) has become a significant area of research [5,10,14,15,19,20], where source data is inaccessible due to privacy or intellectual property concerns.

Existing SF-UDA solutions can be categorized into four main groups: batch normalization (BN) statistics adaptation [16,17,23], approximating source images [9,26], entropy minimization [2], and pseudo-labeling [3,25]. BN statistics adaptation methods aim to address the discrepancy of statistics between different domains. For example, [16,17] update low-order and high-order BN statistics with distinct training objectives, while [23] adapts BN statistics to minimize the entropy of the model's prediction. Approximating source images aims to generate source-like images. For example, [26] first attains a coarse source image by freezing the source model and training a learnable image, then refines the image via mutual Fourier Transform. The refined source-like image provides a representation of the source data distribution and facilitates domain alignment during the adaptation process. For another instance, [9] learns a domain prompt to add to a target domain image so that the sum simulates the source image. Entropy minimization methods aim to produce more confident model predictions. For example, [2] minimizes output entropy with a regularizer of class-ratio. The class-ratio is estimated by an auxiliary network that is pre-trained on the source domain. For pseudo-labeling [12,29], erroneous pseudo-labels are either discarded or corrected. For example, [3] identifies low-confidence pseudo-labels at both the pixel-level and the class-level. On the other hand, [25] performs uncertainty-weighted soft label correction by estimating the class-conditional label error probability. However, all of these methods overlook context relations, which can enhance adaptation performance without the need to access the source data.

We observe in our experiments (see Fig. 1 (a)) that domain gaps can result in the source model making context-inconsistent predictions. For neighboring patches of an image with similar visual appearance, the source model can yield vastly different predictions. This phenomenon can be explained by the observation in [15] that target data shifts in the feature space, causing some data points to shift across the boundary of the source domain segmentor. The issue of context inconsistency motivates us to utilize context relations in refining pseudo-labels. Moreover, it is observed in our experiments (as shown in Fig. 1(b)) that target features produced by the source model still form clusters, meaning that the features of target data points with the same class are closely located. This discovery led us to calculate context relations from feature distances; see Fig. 1(c).

In this paper, we present a novel context-aware pseudo-label refinement (**CPR**) framework for source-free unsupervised domain adaptation. Firstly, we develop a context-similarity learning module, where context relations are computed from distances of features via a context-similarity head. This takes advantage of the intrinsic clustered feature distribution under domain shift [27,28], where target features generated by the source encoder are close for the same class and faraway for different classes (see Fig. 1 (b)). Secondly, context-aware revision is designed to leverage adjacent pseudo-labels for revising bad pseudo-labels, with aid of the learned context relations. Moreover, a calibration strategy is proposed, aiming to mitigate the negative effect brought about by the inaccurate learned context relations. Finally, the refined pseudo-labels are denoised with consideration of model knowledge and feature distribution [3,13] to select reliable pseudo-labels for domain adaptation. Experiments on cross-domain fundus image segmentation demonstrate our proposed framework outperforms the state-of-the-art source-free methods [3,25,26].

2   Method

Figure 2 illustrates our SF-UDA framework via context-aware pseudo-label refinement. In this section, we first introduce the context-similarity learning scheme. Next, we propose the pseudo-label refinement strategy. Finally, we present the model training with the denoised refined pseudo-labels.

2.1   Context-Similarity Learning

In the SF-UDA problem, a source model $f^s : \mathcal{X}_s \to \mathcal{Y}_s$ is trained using the data $\{x_s^i, y_s^i\}_{i=1}^{n_s}$ from the source domain $\mathcal{D}_s = (\mathcal{X}_s, \mathcal{Y}_s)$, where $(x_s^i, y_s^i) \in (\mathcal{X}_s, \mathcal{Y}_s)$. f^s is typically trained with a supervision loss of cross-entropy. Also an unlabeled dataset $\{x_t^i\}_{i=1}^{n_t}$ from the target domain $\mathcal{D}_t$ is given, where $x_t^i \in \mathcal{D}_t$. SF-UDA aims to learn a target model $f^t : \mathcal{X}_t \to \mathcal{Y}_t$ with only the source model f^s and the target dataset $\{x_t^i\}_{i=1}^{n_t}$. In our fundus segmentation problem, $y^i \in \{0,1\}^{H \times W \times C}$, where C is the number of classes and $C = 2$ because there are two segmentation targets, namely optic cup and optic disc.

Architecture of Context-Similarity Head. Although the target features generated by source encoder do not align with the source segmentor, features of the same classes tend to be in the same cluster while those of different classes are faraway, as shown in Fig. 1 (b). This indicates the source feature encoder is useful for

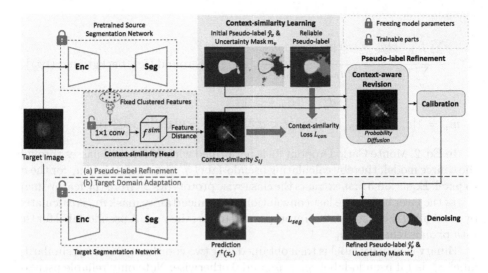

Fig. 2. Overview of the proposed context-aware pseudo-label refinement (CPR) framework for SF-UDA. It consists of two stages: (a) The context-similarity head for computing context relations is trained by reliable pseudo-labels. The learned context similarities are then used to refine the pseudo-labels; (b) Only the refined pseudo-labels with high confidence supervise the training of the segmentation network. The network consists of a feature encoder (Enc) and a segmentor (Seg).

computing context relations. Therefore, we freeze the source encoder and add an additional head to the encoder for learning context semantic relations, motivated by [1]. A side benefit of freezing the source encoder is the training time and required memory can be reduced, as backward propagation is not needed on the encoder. Specifically, the feature map f^{sim} is first obtained, where a 1×1 convolution is applied for adaptation to the target task. Then the semantic similarity between coordinate i and coordinate j on the feature map is defined as

$$S_{ij} = \exp\left\{-\|f^{sim}(x_i, y_i) - f^{sim}(x_j, y_j)\|_1\right\}. \tag{1}$$

Computing similarities between every pair of coordinates in a feature map is computationally costly. Thus, for each coordinate i, only similarities with coordinates j lying within the circle of radius r are considered in our implementation.

Training of Context-Similarity Head. Given a target image x_t, initial pseudo-labels and uncertainty mask can be obtained from the source model f^s and x_t, following previous work [3] as:

$$
\begin{aligned}
p_{v,k} &= f^s(x_t)_v, k = 1, \ldots, K, \\
p_v &= \mathrm{avg}(p_{v,1}, \ldots, p_{v,K}), u_v = \mathrm{std}(p_{v,1}, \ldots, p_{v,K}), \\
\hat{y}_v &= \mathbb{1}[p_v \geq \gamma],
\end{aligned}
\tag{2}
$$

$$z^\omega = \frac{\sum\limits_v f_{l,v} \cdot \mathbb{1}[\hat{y}_v = \omega]\mathbb{1}[u_v < \eta] \cdot p_{v,\omega}}{\sum\limits_v \mathbb{1}[\hat{y}_v = \omega]\mathbb{1}[u_v < \eta] \cdot p_{v,\omega}}, \quad \omega \in \{\text{foreground (fg), background (bg)}\},$$

$$d_v^\omega = \|f_{l,v} - z^\omega\|_2, \tag{3}$$

$$m_v = \mathbb{1}[u_v < \eta](\mathbb{1}[\hat{y}_v = 1]\mathbb{1}[d_v^{fg} < d_v^{bg}] + \mathbb{1}[\hat{y}_v = 0]\mathbb{1}[d_v^{fg} > d_v^{bg}]).$$

In Eq. 2, Monte Carlo Dropout [8] is performed with K forward passes through the source model, thereby calculating pseudo-label $\hat{y}_v$ and uncertainty u_v for the v-th pixel. Equation 3 first extracts the class-wise prototypes z^ω from the feature map $f_{l,v}$ of the layer before the last convolution, then uncertainty mask m_v is calculated by combining the distance to prototypes and uncertainty u_v. A pseudo-label for the v-th pixel is reliable if $m_v = 1$.

Binary similarity label is then obtained. For two coordinates i and j, similarity label S_{ij}^* is 1 if pseudo-labels $\hat{y}_i = \hat{y}_j$, and 0 otherwise. Note only reliable pseudo-labels are considered to provide less noisy supervision.

The context-similarity head is trained with S^*. To address the class imbalance issue, the loss of each type of similarity (fg-fg, bg-bg, fg-bg) is calculated and aggregated [1] as

$$\mathcal{L}_{con} = -\frac{1}{4} \mathop{avg}_{\substack{\hat{y}_i = \hat{y}_j = 1 \\ m_i = m_j = 1}} (\log S_{ij}) - \frac{1}{4} \mathop{avg}_{\substack{\hat{y}_i = \hat{y}_j = 0 \\ m_i = m_j = 1}} (\log S_{ij}) - \frac{1}{2} \mathop{avg}_{\substack{\hat{y}_i \neq \hat{y}_j \\ m_i = m_j = 1}} (\log(1 - S_{ij})). \tag{4}$$

2.2 Context-Similarity-Based Pseudo-Label Refinement

Context-Aware Revision. The trained context-similarity head is utilized to refine the initial coarse pseudo-labels. Specifically, context-similarities S_{ij} are computed by passing the target image through the source encoder and the trained head. Then the refined probability for the i-th coordinate is updated as the weighted average of the probabilities in a local circle around the i-th coordinate as

$$p_i^{re} = \sum_{d(i,j) \leq r} \frac{S_{ij}^\beta}{\sum_{d(i,j) \leq r} S_{ij}^\beta} \cdot p_j \tag{5}$$

where p_i^{re} is the revised probability and $d(\cdot)$ is the Euclidean distance. $\beta \geq 1$, in order to highlight the prominent similarities and ignore the smaller ones. By combining neighboring predictions based on context relations, revised probabilities are more robust. Equation 5 is performed iteratively for t rounds, since revised probabilities can be used for further revision.

Calibration. The probability update by Eq. 5 might be hurt by inaccurate context relations. We observe that for some classes (optic cup for fundus segmentation) with worse pseudo-labels, the context-similarity for "fg-bg" is not learned well. Consequently, the probability of background incorrectly propagates to that of foreground, making the probability of foreground lower. To tackle this issue, the revised probability is calibrated as

$$p'_i = \frac{p_i^{re}}{max_j(p_j^{re})}. \tag{6}$$

The decreased probability is rectified by the maximum value in the image, considering the maximum probability (e.g., in the center of a region) after calibration of a class should be close to 1.

2.3   Model Adaptation with Denoised Pseudo Labels

The refined pseudo-labels can be obtained by $\hat{y}'_v = \mathbb{1}[p'_v \geq \gamma]$. However, noisy pseudo-labels inevitably exist. The combination of model knowledge and target feature distribution shows the best estimation of sample confidence [13]. To this end, reliable pseudo-labels are selected at pixel-level and class-level [3] as

$$m'_{v,p} = \mathbb{1}(p'_v < \gamma_{low} \text{ or } p'_v > \gamma_{high})$$
$$m'_{v,c} = \mathbb{1}(\hat{y}'_v = 1)\mathbb{1}(d_v^{fg} < d_v^{bg}) + \mathbb{1}(\hat{y}'_v = 0)\mathbb{1}(d_v^{fg} > d_v^{bg}), \tag{7}$$

in which γ_{low} and γ_{high} are two thresholds for filtering out pseudo-labels without confident probabilities. d_v^{fg} and d_v^{bg} are the distances to feature prototypes as computed in Eq. 3. The final label selection mask is the intersection of $m'_{v,p}$ and $m'_{v,c}$, i.e., $m'_v = m'_{v,p} \cdot m'_{v,c}$. The target model is trained under the supervision of pseudo-labels selected by m'_v, with cross-entropy loss:

$$\mathcal{L}_{seg} = -\sum_v m'_v \cdot \left[\hat{y}'_v \cdot \log(f^t(x_t)_v) + (1 - \hat{y}'_v) \cdot \log(1 - f^t(x_t)_v)\right]. \tag{8}$$

3   Experiments

Datasets. For a fair comparison, we follow prior work [3] to select three mainstream datasets for fundus image segmentation, i.e., Drishti-GS [22], RIM-ONE-r3 [7], and the validation set of REFUGE challenge [18]. These datasets are split into 50/51, 99/60, and 320/80 for training/testing, respectively.

Implementation Details and Evaluation Metrics. Following prior works [3, 24, 25], our segmentation network is MobileNetV2-adapted [21] DeepLabv3+ [4]. The context-similarity head comprises two branches for optic cup and optic disc, respectively. Each branch includes a 1×1 convolution and a similarity feature map. The threshold γ for determining pseudo-labels is set to 0.75, referring to [24]. The radius r in Eq. 5, the β in Eq. 5 and the iteration number t are set to 4, 2 and 4 respectively. The two thresholds for filtering out unconfident refined pseudo-labels are empirically set as $\gamma_{low} = 0.4$ and $\gamma_{high} = 0.85$, respectively. Each image is pre-processed by clipping a 512×512 optic disc region [24]. The same augmentations as in [3,25] are applied, including Gaussian noise, contrast adjustment, and

Table 1. Comparison with state-of-the-arts on two settings. "W/o adaptation" refers to directly evaluating the source model on the target dataset. "Upper bound" refers to training the model on the target dataset with labels.

Methods	Dice [%]↑			ASD [pixel]↓		
	Optic cup	Optic disc	Avg	Optic cup	Optic disc	Avg
Source: Drishti-GS; Target: RIM-ONE-r3						
W/o adaptation	70.84	89.94	80.39	13.44	10.76	12.10
Upper bound	83.81	96.61	90.21	6.92	2.96	4.94
DPL [3]	71.70	92.52	82.11	12.49	7.34	9.92
FSM [26]	74.34	91.41	82.88	14.52	10.30	12.41
U-D4R [25]	73.48	93.18	83.33	10.18	6.15	8.16
CPR (ours)	**75.02**	**95.03**	**85.03**	**9.84**	**4.32**	**7.08**
Source: REFUGE; Target: Drishti-GS						
W/o adaptation	79.80	93.89	86.84	13.25	6.70	9.97
Upper bound	89.63	96.80	93.22	6.65	3.55	5.10
DPL [3]	82.04	95.27	88.65	12.14	5.32	8.73
FSM [26]	79.30	94.34	86.82	13.79	5.95	9.87
U-D4R [25]	81.82	95.98	88.90	12.21	4.45	8.33
CPR (ours)	**84.49**	**96.16**	**90.32**	**10.19**	**4.23**	**7.21**

random erasing. The Adam optimizer is adopted with learning rates of $3e-2$ and $3e-4$ in the context-similarity learning stage and the target domain adaptation stage respectively. The momentum of the Adam optimizer is set to 0.9 and 0.99. The batch size is set to 8. The context-similarity head is trained for 16 epochs and the target model is trained for 10 epochs. The implementation is carried out via PyTorch on a single NVIDIA GeForce RTX 3090 GPU. For evaluation, we adopt the widely used Dice coefficient and Average Surface Distance (ASD).

Comparison with State-of-the-Arts. Table 1 shows the comparison of our method with the state-of-the-art SF-UDA methods. Besides three SOTA methods, i.e., DPL [3], FSM [26], and U-D4R [25], we also report the adaptation result without adaptation and the result with fully supervised learning (denoted as "upper bound"). The results show that our approach achieves clear improvements over the previous methods, owing to the proposed pseudo-label refinement scheme which takes advantage of the feature distribution property under domain shift to learn context relations and utilizes valuable context information to rectify pseudo-labels. Figure 3 (a) shows a qualitative comparison.

Ablation Study on Different Modules. Table 2 provides a quantitative analysis to investigate the function of each module. Each component shows its importance in improving the adaptation performance. Particularly, without our pseudo-label refinement, an obvious decrease of segmentation performance can be witnessed, revealing its necessity. Without calibration, the segmentation performance

Fig. 3. On the Drishti-GS to RIM-ONE-r3 adaptation: (a) Qualitative comparison of the optic cup and disc segmentation results with different methods. (b) An example of pseudo-label change with the proposed refinement scheme.

Table 2. Quantitative ablation study on the Drishti-GS to RIM-ONE-r3 adaptation.

Pseudo-label refinement		Denoising	Dice [%]		
Context-aware revision	Calibration		Optic cup	Optic disc	Avg
✗	✗	✗	67.25	93.73	80.49
✓	✗	✗	53.73	92.67	73.20
✓	✗	✓	69.80	94.95	82.38
✓	✓	✗	74.68	93.10	83.89
✗	✗	✓	72.34	94.13	83.23
✓	✓	✓	**75.02**	**95.03**	**85.03**

degrades significantly, which is because the probabilities without calibration do not have correct absolute values. This demonstrates calibration is a necessary step after the revision. Denoising filters out unreliable pseudo-labels by taking into account individual probabilities and feature distribution, thus providing more correct guidance. Integrating all the components completes our framework and yields the best result.

Ablation Study on Pseudo-label Refinement. Ablation study is conducted to verify the effectiveness of the pseudo-label refinement strategy. As shown in Table 3, after refinement, the quality of the pseudo-label is clearly promoted, leading to more accurate supervision for target domain adaptation. For the pseudo-label of optic disc which originally has high accuracy, our refinement scheme encouragingly achieves a boost of 3.5%, showing the robustness of our refinement scheme for different quality of initial pseudo-labels. Without calibration, the accu-

Table 3. Comparison of pseudo-label quality of the training set with different methods on the Drishti-GS to RIM-ONE-r3 adaptation.

Methods	Dice [%]	
	Optic cup	Optic disc
Initial pseudo-label [3]	67.66	90.01
Refined pseudo-label	**72.01**	**93.51**
Refined pseudo-label (w/o calibration)	58.34	93.40

racy of the pseudo-label of optic cup is substantially dropped, indicating it is an indispensable part of the overall scheme. Figure 3 (b) visualizes an example of the evolution of the pseudo-label. As can be seen, the context-inconsistent region is clearly improved.

4   Conclusion

This work presents a novel SF-UDA method for the fundus image segmentation problem. We propose to explicitly learn context semantic relations to refine pseudo-labels. Calibration is performed to compensate for the wrong revision caused by inaccurate context relations. The performance is further boosted via the denoising scheme, which provides reliable guidance for adaptation. Our experiments on cross-domain fundus image segmentation show that our method outperforms the state-of-the-art SF-UDA approaches.

Acknowledgement. This work was partially supported by the Hong Kong Innovation and Technology Fund under Project ITS/030/21, as well as by the HKUST-BICI Exploratory Fund (HCIC-004) and Foshan HKUST Projects under Grants FSUST21-HKUST10E and FSUST21-HKUST11E.

References

1. Ahn, J., Kwak, S.: Learning pixel-level semantic affinity with image-level supervision for weakly supervised semantic segmentation. In: Proceedings of the IEEE Conference on Computer Vision and Pattern Recognition, pp. 4981–4990 (2018)
2. Bateson, M., Kervadec, H., Dolz, J., Lombaert, H., Ben Ayed, I.: Source-relaxed domain adaptation for image segmentation. In: Martel, A.L., et al. (eds.) MICCAI 2020. LNCS, vol. 12261, pp. 490–499. Springer, Cham (2020). https://doi.org/10.1007/978-3-030-59710-8_48
3. Chen, C., Liu, Q., Jin, Y., Dou, Q., Heng, P.-A.: Source-free domain adaptive fundus image segmentation with denoised pseudo-labeling. In: de Bruijne, M., et al. (eds.) MICCAI 2021. LNCS, vol. 12905, pp. 225–235. Springer, Cham (2021). https://doi.org/10.1007/978-3-030-87240-3_22
4. Chen, L.-C., Zhu, Y., Papandreou, G., Schroff, F., Adam, H.: Encoder-decoder with atrous separable convolution for semantic image segmentation. In: Ferrari, V., Hebert, M., Sminchisescu, C., Weiss, Y. (eds.) ECCV 2018. LNCS, vol. 11211, pp. 833–851. Springer, Cham (2018). https://doi.org/10.1007/978-3-030-01234-2_49
5. Ding, N., Xu, Y., Tang, Y., Xu, C., Wang, Y., Tao, D.: Source-free domain adaptation via distribution estimation. In: Proceedings of the IEEE/CVF Conference on Computer Vision and Pattern Recognition, pp. 7212–7222 (2022)
6. Fu, H., Cheng, J., Xu, Y., Wong, D.W.K., Liu, J., Cao, X.: Joint optic disc and cup segmentation based on multi-label deep network and polar transformation. IEEE Trans. Med. Imaging **37**(7), 1597–1605 (2018)
7. Fumero, F., Alayón, S., Sanchez, J.L., Sigut, J., Gonzalez-Hernandez, M.: Rim-one: an open retinal image database for optic nerve evaluation. In: 2011 24th International Symposium on Computer-Based Medical Systems (CBMS), pp. 1–6. IEEE (2011)

8. Gal, Y., Ghahramani, Z.: Dropout as a Bayesian approximation: representing model uncertainty in deep learning. In: International Conference on Machine Learning, pp. 1050–1059. PMLR (2016)
9. Hu, S., Liao, Z., Xia, Y.: ProsFDA: prompt learning based source-free domain adaptation for medical image segmentation. arXiv preprint arXiv:2211.11514 (2022)
10. Jing, M., Zhen, X., Li, J., Snoek, C.G.M.: Variational model perturbation for source-free domain adaptation. In: Oh, A.H., Agarwal, A., Belgrave, D., Cho, K. (eds.) Advances in Neural Information Processing Systems (2022). https://openreview.net/forum?id=yTJze_xm-u6
11. Kamnitsas, K., et al.: Unsupervised domain adaptation in brain lesion segmentation with adversarial networks. In: Niethammer, M., et al. (eds.) IPMI 2017. LNCS, vol. 10265, pp. 597–609. Springer, Cham (2017). https://doi.org/10.1007/978-3-319-59050-9_47
12. Lee, D.H., et al.: Pseudo-label: the simple and efficient semi-supervised learning method for deep neural networks. In: Workshop on Challenges in Representation Learning, ICML, vol. 3, p. 896. Atlanta (2013)
13. Lee, J., Jung, D., Yim, J., Yoon, S.: Confidence score for source-free unsupervised domain adaptation. In: International Conference on Machine Learning, pp. 12365–12377. PMLR (2022)
14. Li, R., Jiao, Q., Cao, W., Wong, H.S., Wu, S.: Model adaptation: unsupervised domain adaptation without source data. In: Proceedings of the IEEE/CVF Conference on Computer Vision and Pattern Recognition, pp. 9641–9650 (2020)
15. Liang, J., Hu, D., Feng, J.: Do we really need to access the source data? Source hypothesis transfer for unsupervised domain adaptation. In: International Conference on Machine Learning, pp. 6028–6039. PMLR (2020)
16. Liu, X., Xing, F., El Fakhri, G., Woo, J.: Memory consistent unsupervised off-the-shelf model adaptation for source-relaxed medical image segmentation. Med. Image Anal. **83**, 102641 (2022)
17. Liu, X., Xing, F., Yang, C., El Fakhri, G., Woo, J.: Adapting off-the-shelf source segmenter for target medical image segmentation. In: de Bruijne, Marleen (ed.) MICCAI 2021. LNCS, vol. 12902, pp. 549–559. Springer, Cham (2021). https://doi.org/10.1007/978-3-030-87196-3_51
18. Orlando, J.I., et al.: Refuge challenge: a unified framework for evaluating automated methods for glaucoma assessment from fundus photographs. Med. Image Anal. **59**, 101570 (2020)
19. Roy, S., et al.: Uncertainty-guided source-free domain adaptation. In: Avidan, S., Brostow, G., Cissé, M., Farinella, G.M., Hassner, T. (eds.) ECCV 2022. LNCS, vol. 13685, pp. 537–555. Springer, Cham (2022). https://doi.org/10.1007/978-3-031-19806-9_31
20. Teja, S.P., Fleuret, F.: Uncertainty reduction for model adaptation in semantic segmentation. In: Proceedings of the IEEE/CVF Conference on Computer Vision and Pattern Recognition (CVPR), pp. 9613–9623, June 2021
21. Sandler, M., Howard, A., Zhu, M., Zhmoginov, A., Chen, L.C.: Mobilenetv 2: inverted residuals and linear bottlenecks. In: Proceedings of the IEEE Conference on Computer Vision and Pattern Recognition, pp. 4510–4520 (2018)
22. Sivaswamy, J., Krishnadas, S., Chakravarty, A., Joshi, G., Tabish, A.S., et al.: A comprehensive retinal image dataset for the assessment of glaucoma from the optic nerve head analysis. JSM Biomed. Imaging Data Papers **2**(1), 1004 (2015)
23. Wang, D., Shelhamer, E., Liu, S., Olshausen, B., Darrell, T.: Tent: fully test-time adaptation by entropy minimization. In: International Conference on Learning Representations (2021). https://openreview.net/forum?id=uXl3bZLkr3c

24. Wang, S., Yu, L., Li, K., Yang, X., Fu, C.-W., Heng, P.-A.: Boundary and entropy-driven adversarial learning for fundus image segmentation. In: Shen, D., et al. (eds.) MICCAI 2019. LNCS, vol. 11764, pp. 102–110. Springer, Cham (2019). https://doi.org/10.1007/978-3-030-32239-7_12
25. Xu, Z., et al.: Denoising for relaxing: unsupervised domain adaptive fundus image segmentation without source data. In: Wang, L., Dou, Q., Fletcher, P.T., Speidel, S., Li, S. (eds.) MICCAI 2022. LNCS, vol. 13435, pp. 214–224. Springer, Cham (2022). https://doi.org/10.1007/978-3-031-16443-9_21
26. Yang, C., Guo, X., Chen, Z., Yuan, Y.: Source free domain adaptation for medical image segmentation with Fourier style mining. Med. Image Anal. **79**, 102457 (2022)
27. Yang, S., Wang, Y., Wang, K., Jui, S., van de weijer, J.: Attracting and dispersing: a simple approach for source-free domain adaptation. In: Oh, A.H., Agarwal, A., Belgrave, D., Cho, K. (eds.) Advances in Neural Information Processing Systems (2022). https://openreview.net/forum?id=ZlCpRiZN7n
28. Yang, S., van de Weijer, J., Herranz, L., Jui, S., et al.: Exploiting the intrinsic neighborhood structure for source-free domain adaptation. In: Advances in Neural Information Processing Systems, vol. 34, pp. 29393–29405 (2021)
29. Yao, H., Hu, X., Li, X.: Enhancing pseudo label quality for semi-supervised domain-generalized medical image segmentation. In: Proceedings of the AAAI Conference on Artificial Intelligence, vol. 36, pp. 3099–3107 (2022)

Retinal Age Estimation with Temporal Fundus Images Enhanced Progressive Label Distribution Learning

Zhen Yu[1,2,3], Ruiye Chen[5,6], Peng Gui[3,4], Lie Ju[2,3,7], Xianwen Shang[5,6], Zhuoting Zhu[5,6], Mingguang He[5,6], and Zongyuan Ge[2,3,8(✉)]

[1] Central Clinical School, Faculty of Medicine, Nursing and Health Sciences, Monash University, Melbourne, Australia
[2] AIM for Health Lab, Monash University, Victoria, Australia
zongyuan.ge@monash.edu
[3] Monash Medical AI, Monash University, Victoria, Australia
[4] School of Computer Science, Wuhan University, Hubei, China
[5] Centre for Eye Research Australia, University of Melbourne, Melbourne, Australia
[6] Ophthalmology, Department of Surgery, University of Melbourne, Melbourne, Australia
[7] Faculty of Engineering, Monash University, Melbourne, Australia
[8] Faculty of IT, Monash University, Melbourne, Australia
https://mmai.group

Abstract. Retinal age has recently emerged as a reliable ageing biomarker for assessing risks of ageing-related diseases. Several studies propose to train deep learning models to estimate retinal age from fundus images. However, the limitation of these studies lies in 1) both of them only train models on snapshot images from single cohorts; 2) they ignore label ambiguity and individual variance in the modeling part. In this study, we propose a progressive label distribution learning (LDL) method with temporal fundus images to improve the retinal age estimation on snapshot fundus images from multiple cohorts. First, we design a two-stage LDL regression head to estimate adaptive age distribution for individual images. Then, we eliminate cohort variance by introducing ordinal constraints to align image features from different data sources. Finally, we add a temporal branch to model sequential fundus images and use the captured temporal evolution as auxiliary knowledge to enhance the model's predictive performance on snapshot fundus images. We use a large retinal fundus image dataset which consists of ~130k images from multiple cohorts to verify our method. Extensive experiments provide evidence that our model can achieve lower age prediction errors than existing methods.

Keywords: Retinal age estimation · label distribution learning · temporal fundus image

1 Introduction

Population ageing is a huge health burden worldwide as the risk of morbidity and mortality increases exponentially with age [2]. However, great heterogeneity

H. Greenspan et al. (Eds.): MICCAI 2023, LNCS 14226, pp. 629–638, 2023.
https://doi.org/10.1007/978-3-031-43990-2_59

exists across individuals with the same chronological age, indicating chronological age poorly reflects intra-individual variation [10]. A quest for biomarkers that can accurately determine individual-specific, age-related risk of adverse outcomes has been embarked upon. Among the countless potential candidate ageing biomarkers [4,6,12], retinal age has been verified to be one of the most reliable indicators with the advantages of being rapid, non-invasive, and cost-effective [5,16,17].

With the advent of technology, deep learning (DL) algorithms have found great applications in retinal age prediction. For example, Liu et al. [9] developed a convolutional neural network (CNN) to estimate retinal age with label distribution learning (LDL) on 12k fundus images from healthy Chinese populations. Zhu et al. [17] trained a CNN regression model on the UK Biobank cohort consisting of ∼70k fundus images. The limitations of these studies include: 1) the use of a single source of data in these studies has underestimated the complexity of data variance in real-world scenarios, which limits the generalizability of retinal age prediction. 2) only snapshot databases are used in these studies and failure track a detailed trail of age-specific changes. 3) outputting a single value with direct regressions [17] ignores the ambiguity of age labels, and using fixed label distribution [9] as ground truth did not consider individual variations. Tackling these shortcomings will improve the generalizability as well as reduce technical errors in the age prediction algorithm, providing a more reliable retinal age estimate.

Therefore, in this study, we present an attempt to provide a novel accurate estimate of retinal age by learning adaptive age distribution from multiple cohorts with temporal fundus images available. Instead of learning a model using fixed label distribution as ground truth, we formulate the age estimation as a two-stage LDL task and give an adaptive distribution estimate for individual fundus images. As learning the LDL model with images from different data sources can harm the consistency and ordinality of embedding space, we introduce ordinal constraints to align the image features from different domains. Moreover, to leverage the temporal knowledge from the fundus image sequence, we add a temporal branch to capture the temporal evolution and use this auxiliary information to enhance the predictive performance of our model on snapshot images. We verify our method on a large retinal fundus dataset which consists of approximately 130k images of healthy subjects from the UKB cohort and Chinese cohorts. Extensive experiments prove that our model can achieve lower age prediction errors on multiple cohorts.

2   Method

2.1   Progressive Label Distribution Learning

As shown in Fig. 1, we formulate the retinal age estimation as a two-stage label distribution learning process. In the first stage, the model uses global features to predict a coarse age distribution on roughly discretized age labels. Each coarse age prediction is associated with a query vector corresponding to an age group.

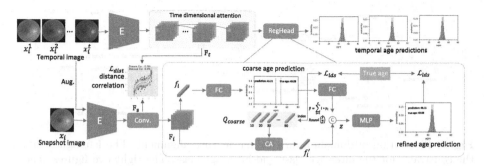

Fig. 1. Overview of the proposed method. The regression is formulated as a two-stage label distribution learning problem and the model further uses temporal knowledge to improve the snapshot learning.

Then, the model performs class attention [13] between the age group query and spatial features to generate fine-level features which are further combined with the coarse age prediction to give refined age predictions.

Formally, given a dataset with N images $\mathcal{X} = \{x_i\}_{i=1}^N$, the corresponding age labels $\mathcal{Y} = \{y_i\}_{i=1}^N$ range in $[a, b]$. The image encoder first transforms an input image x_i into a spatial feature $\mathbf{F}_i' \in \mathbb{R}^{H \times W \times C}$, then a convolutional projection layer maps $\mathbf{F}_i'$ into the base representation $\mathbf{F}_i \in \mathbb{R}^{H \times W \times D}$ and the averaged feature $f_i \in \mathbb{R}^{1 \times D}$ for the following age distribution learning. We discretize the age classes as $\hat{y}_i = \mathrm{R}\,(y_i/\delta_d) * \delta_d$, where $\mathrm{R}\,(\cdot)$ denotes the round operator and δ_d is the age bin for tuning the discretization degree. Therefore, the total discretized age class number is $C_{\delta_d} = \mathrm{R}\left(\frac{|b-a|}{\delta_d}\right)$. For the coarse-level age estimation, we set a large $\delta_d = 10$ which determines the age group queries as $\mathbf{Q}_{coarse} \in \mathbb{R}^{C_{\delta_d} \times D}$. Then, we use an FC layer with softmax applied on the f_i to calculate the coarse age distribution $p_i \in \mathbb{R}^{1 \times C_{\delta_d}}$. Different from the previous study [9] using fixed label distribution as ground truth, we directly learn the distribution from training data with discretized age labels:

$$\mathcal{L}_{lds} = \frac{1}{N} \sum_{i=1}^N -log\,(p_{i,\hat{y}_i}) + \frac{\alpha}{2N} \sum_{i=1}^N (y_i - m_i)^2 + \frac{\beta}{N} \sum_{i=1}^N \sum_{c=1}^{C_{\delta_d}} p_{i,c} * (y_i - m_i)^2 \quad (1)$$

where $m_i = \sum_{c=1}^{C_{\delta_d}} p_{i,c} * \hat{y}_c$ is the expected value of the learned distribution p_i, The first term is the cross-entropy loss which helps the model converges in an early training stage, the last two terms encourage the learned distribution to be centered and concentrated at the true age labels.

In the refining stage, the mean value of coarse age distribution m_i is used to select the age group query from Q_{coarse} to involve the computation of fine-level feature:

$$f' = GAP\left(A\left(Q_{coarse}[R\left(\frac{m_i}{\delta_d}\right)], \mathbf{F}_i; \theta_a\right)\right) \quad (2)$$

where $GAP\,(\cdot)$ is the global averaged pooling and $A\,(\cdot)$ denote the attention function with θ_a as the parameters. The key and value vectors in the atten-

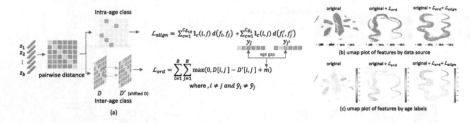

Fig. 2. Illustration of domain-aware ordinal feature alignment. The left figure shows the constraint for the inter-age class and intra-age class; The right two figures denote the feature visualization result.

tion function come from $\mathbf{F}_i$. Finally, we concatenate the f' with the mapped coarse age distribution as the final feature embedding to predict the fine-level age distribution on a small age bin of $\delta_d = 1$:

$$z = \text{concat}\left(f', \text{f}\left(p_i \odot \hat{y}_{1,\ldots,C_{\delta_d}}; \theta_f\right)\right) \tag{3}$$

$$p_i' = \text{softmax}\left(\text{mlp}\left(z; \theta_m\right)\right) \tag{4}$$

where $\text{f}(\cdot)$ denotes an FC layer with parameters θ_f, $\text{mlp}(\cdot)$ represents a multi-layer perceptron with one hidden layer and the parameter is θ_m. The training loss is the same with Eq. 1.

2.2   Cross-Domain Ordinal Feature Alignment

Although existing studies [7,14] show that formulating regression as a classification task to learn the label distribution yields better performance, the ordinal information of age relations is lost in feature space. Moreover, when the training data comes from distinct data sources, the domain variance further damages the coherence of the learned features. In Fig. 2 (b) and Fig. 2 (c), we visualize the intermediate feature learned on our fundus image dataset. As can be seen, in Fig. 2 (c) the original model produces scattered and inconsistent features for ordinal age labels, while in 2 (b) the features exhibit a clear gap for different data sources.

To address the above issues, we propose to introduce ordinal constraints in the label distribution learning and perform feature alignment to eliminate the domain variance. The key idea of imposing ordinal constraints in embedding space is to construct a set of triplets and enforce the feature distance to be consistent with the relative age gap. Specifically, for each batch of input data $\{x_1, \ldots, x_B\}$, we first compute their pairwise feature distance which outputs a distance matrix $D \in \mathbb{R}^{B \times B}$. Then, we construct feature triplets and calculate the distance gap by subtracting shifted distance matrix D' from the original D. In this case, each sample will have a chance to serve as the anchor to be compared with other samples. We formulate the ordinal constraint as following margin loss:

$$\mathcal{L}_{ord} = \sum_{i=1}^{B} \sum_{j=1}^{B} \max\left(0, D\left[i,j\right] - D'\left[i,j\right] + m\right), \text{s.t.} i \neq j, \text{and} \hat{y}_i \neq \hat{y}_j \tag{5}$$

where $D\left[\cdot\right]$ denotes metric of Euclidean distance, m is a dynamic margin depends on the relative age difference gap between $|\hat{y}_i - \hat{y}_j|$ and $|\hat{y}_i - \hat{y}_{j'}|$. To align features from different data domains, we directly select samples from same class and push them closer in the embedding space by minimizing the intra-class distance on both coarse-level features and fine-level features:

$$\mathcal{L}_{align} = \sum_{c=1}^{C_{\delta_{10}}} \mathbb{I}_c\left(i,j\right) d\left(f_i, f_j\right) + \sum_{c=1}^{C_{\delta_1}} \mathbb{I}_c\left(i,j\right) d\left(f_i', f_j'\right) \tag{6}$$

where the $\mathbb{I}_c\left(i,j\right)$ is an indicator function.

2.3   Co-Learning with Temporal Fundus Images

Compared to merely learning from single snapshot images, temporal data capturing more aging information can further boost retinal age prediction. However, in practice, temporal fundus data can be limited because the individuals are often lost to follow-up. Directly learning a temporal model on these small data usually cause poor generalization. Therefore, we propose to co-train our model on limited temporal imaging data and large-scale snapshot imaging data. Our aim is to use the auxiliary knowledge from temporal data to enhance the performance of our model tested on snapshot images.

As illustrated in Fig. 1, the temporal branch consists of an image encoder, a time dimensional attention module (TDA), and a regression head. The temporal image encoder and the regression head are the same as that of the snapshot branch. We first input a fundus image sequence into the temporal branch and extract temporal features $\mathbf{F}_t \in \mathbb{R}^{T \times D}$ from the TDA module which performs time dimensional attention to capturing the correlation of local regions across temporal images. At the same time, we input augmented sequential fundus images into the snapshot branch which outputs feature $\mathbf{F}_s$[1]. Inspired by [15], we encourage the snapshot features to preserve similar relations in temporal features by optimizing the distance correlation loss:

$$\mathcal{L}_{dist} = 1 - \mathcal{R}^2\left(\mathbf{F}_s, \mathbf{F}_t\right) \tag{7}$$

where $\mathcal{R}^2\left(\cdot\right)$ denotes the distance correlation and the detailed definition refers to [15]. We simple sum $\mathcal{L}_{lds}$, $\mathcal{L}_{ord}$, $\mathcal{L}_{align}$ and $\mathcal{L}_{dist}$ as the final loss.

---

[1] We omit the i in feature notions for simplicity.

3 Experiment and Results

3.1 Dataset and Implementation

Dataset : We include four datasets in our experiment: the UK Biobank cohort, CN-A, CN-B and CN-C. The UK Biobank is a publicly available prospective cohort with over 50,000 UK residents recruited in 2006[2] 45-degree fundus images were introduced in 2009 for the study subjects. CN-A was a cross-sectional study recruiting participants in eye hospitals. Another database was from the historical data collected in general hospitals, named CN-B. CN-C is an ongoing prospective cohort study that enrolled a total of 4,939 participants in 2009–2010. Participants were invited to take part in annual follow-up assessments including fundus images.

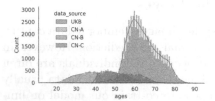

Dataset		$N_{patient}$	N_{image}	Age range
UKB cohort		10891	18909	40-70
Chinese cohort	CN-A	673	4005	31-79
	CN-B	12796	26628	15-91
	CN-C	4663	85298	27-91
Combined dataset		29003	133895	15-91

(a) Age distribution of different cohorts.

(b) Demographics of UKB cohort and Chinese cohort.

Fig. 3. Summary of retinal datasets used in this study.

Training details : As the biological age is normally developed and assessed in healthy populations where biological age is considered equal to chronological age, the model here is trained on 133895 selected snapshot images of healthy subjects without any report of systemic diseases from the four datasets (shown in Fig. 3). The temporal data is a subset of the snapshot data and consists of 2937 sequences with an average length of 5. We split the dataset into training, validation, and testing set with a ratio of 7:1:2. The standard data augmentation techniques such as random resized cropping, color transformation, and flipping are equally used in all experiments. Each image is resized to a fixed input size of 320 × 320. We use ReseNet-50 [3] as the image encoder for all models and train them using ADAM optimizer with a batch size of 100 and a training epoch of 45 with early stopping. The initial learning rates are set to 1×10^{-5} and 3×10^{-4} for the backbone layers and newly added layers, respectively. We divide the learning rate by 10 every 15 epochs.

---

[2] https://biobank.ndph.ox.ac.uk/showcase/browse.cgi.

Table 1. Comparison of the proposed method with existing stuides.

Method	UKB cohort		Chinese cohort		All data	
	MAE	Pearson's R	MAE	Pearson's R	MAE	Pearson's R
Direct regression	3.64	0.820	3.44	0.921	3.47	0.918
Classification	3.51	0.831	3.37	0.923	3.39	0.921
Mean-Variance [11]	3.44	0.829	3.13	0.941	3.28	0.936
Ranking-coral [1]	3.41	0.831	3.21	0.939	3.24	0.935
POE-Reg [8]	3.56	0.836	3.14	0.941	3.20	0.936
POE-CLS	3.23	0.818	3.13	0.942	3.18	0.937
PLDL	3.15	0.854	3.07	0.945	3.08	0.942
PLDL (with temp)	**3.14**	**0.859**	**3.01**	**0.946**	**3.03**	**0.943**

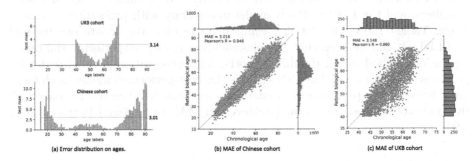

Fig. 4. Error distribution and MAE result on different cohorts.

Evaluation metrics : Consistent with previous studies, we consider the mean absolute error (MAE) and the Pearson correlation as measures for assessing the performance of models.

3.2 Quantitative Results

Comparative Study: We then compare our model with existing popular regression methods which include both direct regression method, classification-

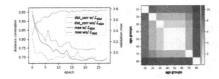

(a) Training curves and similarity matrix of age group queries.

Component	Pearson's R	MAE
coarse stage	0.939	3.23
+ refined stage	0.941	3.11
+ $\mathcal{L}_{ord}$	0.941	3.09
+ $\mathcal{L}_{align}$	0.942	3.08
+ temporal	0.943	3.03

(b) Ablation results on the proposed model.

Fig. 5. Result of ablation study on the proposed method.

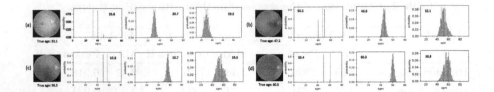

Fig. 6. Show cases of estimated age distributions. For each sample, the left two distributions are coarse prediction and refined prediction from our model, respectively. The rightmost figure denotes the result from the baseline classification model.

based methods [8,11], and ranking-based method [1]. The Mean-var improves the classification model by adding concentration regularization, the ranking-based method explicitly introduces ordinal information by combing a set of binary classifiers, and the POE methods model uncertainty with probabilistic embeddings. Table 1 shows the detailed comparison results. We denote our method as PLDL. It can be seen that the classification model outperforms the direct regression method on all the data cohorts. This observation is consistent with previous studies [11,14]. The POE-CLS is the best-performed model in all baselines, however, the performance is inferior to our model. When trained only with the snapshot images, our method achieves an MAE of 3.08 and Pearson's R of 0.942.

Ablation Study: Here, we give the ablation results of our model to illustrate how different components affect the final performance. Figure 5b shows how the performance changed when adding different components in the proposed method. As can be seen, with only the coarse stage prediction, the model produces an MAE of 3.23 and Pearson's R of 0.939. Performing the refined age stage improves the MAE by ~ 0.1. Introducing ordinal feature alignment gives a margin improvement in age prediction performance, but the feature space shows a clear improvement (see Fig. 2). At last, modelling the temporal fundus images improves the MAE from 3.08 to 3.03. In Fig. 5a, we give the learning curve of distance correlation and the MAE results on the validation set. As we can see, the snapshot features show a high correlation with the feature from the temporal branch and the MAE also becomes lower when optimizing the $\mathcal{L}_{dist}$. In Fig. 4, we give the detailed MAE distribution over different age labels on each cohort. The results indicate that the model produces high MAE on the tail ages in each cohort. Therefore, a future step to improve our model would be to consider the imbalanced learning techniques or group-wise analysis to reduce the MAE bias.

3.3   Visualization Results

In Fig. 6, we illustrate the estimated age distribution for some fundus image samples by our model and the baseline classification model. It can be seen that the refined age distributions are more accurate than the coarse prediction due to more precise discretization. Compared to the estimated age distributions from the baseline model, our method shows a more concentrated age distribution. In Fig. 5a, we visualize the similarity matrix by computing the pair-wise cosine distance between the age group queries. It can be seen that the query vectors for age groups 40~50, 50~60, and 60~70 exhibit a very high similarity which implies that these groups may share more common ageing features.

4   Conclusion

In this study, we present a novel accurate modeling of retinal age prediction. Our model is capable of learning adaptive age distribution from multiple cohorts and leveraging temporal knowledge learned from sequencing images to improve age prediction on snapshot image modeling. Our model demonstrated improved performance in four independent datasets, with an overall MAE much lower than previously proposed algorithms.

References

1. Chen, S., Zhang, C., Dong, M., Le, J., Rao, M.: Using ranking-CNN for age estimation. In: Proceedings of the IEEE Conference on Computer Vision and Pattern Recognition, pp. 5183–5192 (2017)
2. Cheng, X., et al.: Population ageing and mortality during 1990–2017: a global decomposition analysis. PLoS Med. **17**(6), e1003138 (2020)
3. He, K., Zhang, X., Ren, S., Sun, J.: Deep residual learning for image recognition. In: Proceedings of the IEEE Conference on Computer Vision and Pattern Recognition, pp. 770–778 (2016)
4. Horvath, S., Raj, K.: DNA methylation-based biomarkers and the epigenetic clock theory of ageing. Nat. Rev. Genet. **19**(6), 371–384 (2018)
5. Hu, W., et al.: Retinal age gap as a predictive biomarker of future risk of Parkinson's disease. Age and Ageing **51**(3), afac062 (2022)
6. Lee, J., et al.: Deep learning-based brain age prediction in normal aging and dementia. Nature Aging **2**(5), 412–424 (2022)
7. Li, Q., et al.: Unimodal-concentrated loss: Fully adaptive label distribution learning for ordinal regression. In: Proceedings of the IEEE/CVF Conference on Computer Vision and Pattern Recognition, pp. 20513–20522 (2022)
8. Li, W., Huang, X., Lu, J., Feng, J., Zhou, J.: Learning probabilistic ordinal embeddings for uncertainty-aware regression. In: Proceedings of the IEEE/CVF Conference on Computer Vision and Pattern Recognition, pp. 13896–13905 (2021)
9. Liu, C., et al.: Biological age estimated from retinal imaging: a novel biomarker of aging. In: Shen, D., et al. (eds.) MICCAI 2019. LNCS, vol. 11764, pp. 138–146. Springer, Cham (2019). https://doi.org/10.1007/978-3-030-32239-7_16
10. Lowsky, D.J., Olshansky, S.J., Bhattacharya, J., Goldman, D.P.: Heterogeneity in healthy aging. J. Gerontol. Series A: Biomed. Sci. Med. Sci. **69**(6), 640–649 (2014)
11. Pan, H., Han, H., Shan, S., Chen, X.: Mean-variance loss for deep age estimation from a face. In: Proceedings of the IEEE Conference on Computer Vision and Pattern Recognition, pp. 5285–5294 (2018)

12. Peretz, L., Rappoport, N.: Deviation of physiological from chronological age is associated with health. In: Challenges of Trustable AI and Added-Value on Health, pp. 224–228. IOS Press (2022)

13. Touvron, H., Cord, M., Sablayrolles, A., Synnaeve, G., Jégou, H.: Going deeper with image transformers. In: Proceedings of the IEEE/CVF International Conference on Computer Vision, pp. 32–42 (2021)

14. Zhang, S., Yang, L., Mi, M.B., Zheng, X., Yao, A.: Improving deep regression with ordinal entropy. arXiv preprint arXiv:2301.08915 (2023)

15. Zhen, X., Meng, Z., Chakraborty, R., Singh, V.: On the versatile uses of partial distance correlation in deep learning. In: Computer Vision-ECCV 2022: 17th European Conference, Tel Aviv, Israel, October 23–27, 2022, Proceedings, Part XXVI. pp. 327–346. Springer (2022). https://doi.org/10.1007/978-3-031-19809-0_19

16. Zhu, Z., et al.: Association of retinal age gap with arterial stiffness and incident cardiovascular disease. Stroke **53**(11), 3320–3328 (2022)

17. Zhu, Z., et al.: Retinal age gap as a predictive biomarker for mortality risk. British J. Ophthalmol. 107(4), 547–554 (2022)

Fundus-Enhanced Disease-Aware Distillation Model for Retinal Disease Classification from OCT Images

Lehan Wang[1], Weihang Dai[1], Mei Jin[2], Chubin Ou[3], and Xiaomeng Li[1(✉)]

[1] The Hong Kong University of Science and Technology, Hong Kong, China
eexmli@ust.hk
[2] Department of Ophthalmology, Guangdong Provincial Hospital of Integrated
Traditional Chinese and Western Medicine, Guangdong, China
[3] Guangdong Weiren Meditech Co., Ltd, Foshan, China

Abstract. Optical Coherence Tomography (OCT) is a novel and effective screening tool for ophthalmic examination. Since collecting OCT images is relatively more expensive than fundus photographs, existing methods use multi-modal learning to complement limited OCT data with additional context from fundus images. However, the multi-modal framework requires eye-paired datasets of both modalities, which is impractical for clinical use. To address this problem, we propose a novel fundus-enhanced disease-aware distillation model (**FDDM**), for retinal disease classification from OCT images. Our framework enhances the OCT model during training by utilizing unpaired fundus images and does not require the use of fundus images during testing, which greatly improves the practicality and efficiency of our method for clinical use. Specifically, we propose a novel class prototype matching to distill disease-related information from the fundus model to the OCT model and a novel class similarity alignment to enforce consistency between disease distribution of both modalities. Experimental results show that our proposed approach outperforms single-modal, multi-modal, and state-of-the-art distillation methods for retinal disease classification. Code is available at https://github.com/xmed-lab/FDDM.

Keywords: Retinal Disease Classification · Knowledge Distillation · OCT Images

1 Introduction

Retinal diseases are one of the most common eye disorders, which can lead to vision impairment and blindness if left untreated. Computer-aided diagnosis has been increasingly used as a tool to detect ophthalmic diseases at the earliest possible time and to ensure rapid treatment. Optical Coherence Tomography (OCT) [6] is an innovative imaging technique with the ability to capture micrometer-resolution

---

Supplementary Information The online version contains supplementary material available at https://doi.org/10.1007/978-3-031-43990-2_60.

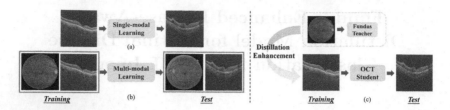

Fig. 1. Comparison between (a) single-modal, (b) multi-modal learning, and (c) our proposed distillation enhancement method. Under our setting, images from an additional modalitiy are only used for model training and are not required for inference.

images of retina layers, which provides a deeper view compared to alternative methods, such as fundus photographs [17], thereby allowing diseases to be detected earlier and more accurately. Because of this, OCT imaging has become the primary diagnostic test for many diseases, such as age-related macular degeneration, central serous chorioretinopathy, and retinal vascular occlusion [2].

Traditional methods manually design OCT features and adopt machine learning classifiers for prediction [11,16,23]. In recent years, deep learning methods have achieved outstanding performance on various medical imaging analysis tasks and have also been successfully applied to retinal disease classification with OCT images [8–10]. However, diagnosing disease with a single OCT modality, as shown in Fig. 1 (a), is still challenging since OCT scans are inadequate compared with fundus photos due to their more expensive cost in data collection. Some methods attempt to use extra layer-related knowledge from the segmentation task to improve prediction despite limited OCT data [3,7,13,15], but this leads to increased training costs since an additional segmentation model is required.

Recent works have attempted to include additional modalities for classification through multi-modal learning shown in Fig. 1 (b), where fundus and OCT images are jointly used to detect various retinal diseases and achieve promising results [4,12,14,18,24–26]. Wang *et al.* [24,25] used a two-stream structure to extract fundus and OCT features, which are then concatenated for prediction. He *et al.* [4] designed modality-specific attention networks to tackle differences in modal characteristics. Nevertheless, there are still limitations in these existing approaches. Firstly, existing multi-modal learning approaches require strictly paired images from both modalities for training and testing. This necessitates the collection of multi-modal images for the same patients, which can be laborious, costly, and not easily achievable in real-world clinical practice. Secondly, previous works mostly focused on a limited set of diseases, such as age-related macular degeneration (AMD), diabetic retinopathy, and glaucoma, which cannot reflect the complexity and diversity of real-world clinical settings.

To this end, we propose **F**undus-enhanced **D**isease-aware **D**istillation **M**odel (**FDDM**) for retinal disease classification from OCT images, as shown in Fig. 1 (c). FDDM is motivated by the observation that fundus images and OCT images provide complementary information for disease classification. For instance, in the case of AMD detection, fundus images can provide information on the number and area of drusen or atrophic lesions of AMD, while OCT can reveal the aggressiveness of subretinal and intraretinal fluid lesions [26]. Utilizing this complementary information from both modalities can enhance AMD detection accuracy.

Our main goal is to extract disease-related information from a fundus teacher model and transfer it to an OCT student model, all without relying on paired training data. To achieve this, we propose a class prototype matching method to align the general disease characteristics between the two modalities while also eliminating the adverse effects of a single unreliable fundus instance. Moreover, we introduce a novel class similarity alignment method to encourage the student to learn similar inter-class relationships with the teacher, thereby obtaining additional label co-occurrence information. Unlike existing works, our method is capable of extracting valuable knowledge from any accessible fundus dataset without additional costs or requirements. Moreover, our approach only needs one modality during the inference process, which can help *greatly reduce the prerequisites for clinical application*.

To summarize, our main contributions include 1) We propose a novel fundus-enhanced disease-aware distillation model for retinal disease classification via class prototype matching and class similarity alignment; 2) Our proposed method offers flexible knowledge transfer from any publicly available fundus dataset, which can significantly reduce the cost of collecting expensive multi-modal data. This makes our approach more accessible and cost-effective for retinal disease diagnosis; 3) We validated our proposed method using a clinical dataset and other publicly available datasets. The results demonstrate superior performance when compared to state-of-the-art alternatives, confirming the effectiveness of our approach for retinal disease classification.

2   Methodology

Our approach is based on two ideas: class prototype matching, which distills generalized disease-specific knowledge unaffected by individual sample noise, and class similarity alignment, which transfers additional label co-occurrence information from the teacher to the student. Details of both components are discussed in the sections below. An overview of our framework is shown in Fig. 2.

We denote the fundus dataset as $D_f = \{x_{f,i}, y_{f,i}\}_{i=1}^N$, and the OCT dataset as $D_o = \{x_{o,j}, y_{o,j}\}_{j=1}^M$. To utilize knowledge from the fundus modality during training, we build a teacher model, denoted F_t, trained on D_f. Similarly, an OCT model F_s is built to learn from OCT images D_o using the same backbone architecture as the fundus model. We use binary cross-entropy loss as the classification loss $\mathcal{L}_{CLS}$ for optimization, to allow the same input to be associated with multiple classes. During inference time, only OCT data is fed into the OCT model to compute the probabilities $p = \{p^c\}_{c=1}^C$ for each disease, c.

2.1   Class Prototype Matching

To distill features from the teacher model into the student model, we aim to ensure that features belonging to the same class are similar. However, we note that individual sample features can be noisy since they contain variations specific

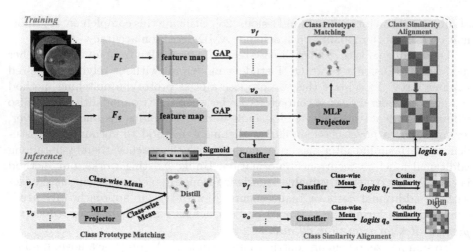

Fig. 2. Overview of our proposed FDDM. Our method is based on class prototype matching, which distills disease-specific features, and class similarity alignment, which distills inter-class relationships.

to the sample instance instead of the class. In order to reduce noise and ensure disease-specific features are learnt, we compress features of each class into a class prototype vector to represent the general characteristics of the disease. During the training per batch, the class prototype vector is the average of all the feature vectors belonging to each category, which is formulated as:

$$e_f^c = \frac{\sum_{i=1}^{B} v_{f,i} * y_{f,i}^c}{\sum_{i=1}^{B} y_{f,i}^c}, \quad e_o^c = \frac{\sum_{j=1}^{B} \mathbf{P}(v_{o,j}) * y_{o,j}^c}{\sum_{j=1}^{B} y_{o,j}^c}, \tag{1}$$

where e_f^c and e_o^c denote the prototype vector for class c of the fundus and OCT modality respectively, $v_{f,i}$ ($v_{o,j}$) represents the feature vector of the input image, and $y_{f,i}^c$ ($y_{o,j}^c$) is a binary number which indicates whether the instance belongs to class c or not. $\mathbf{P}$ demotes an MLP projector that projects OCT features into the same space as fundus features.

In the class prototype matching stage, we apply softmax loss to the prototype vectors of fundus modality to formulate soft targets $\mathcal{E}_f^c = \sigma(e_f^c/\tau)$, where τ is the temperature scale that controls the strength to soften the distribution. Student class prototypes $\mathcal{E}_o^c$ are obtained in the same way. KL divergence is then used to encourage OCT student to learn matched class prototypes with fundus teacher:

$$\mathcal{L}_{CPM} = \sum_{c=1}^{C} \mathcal{E}_f^c \log\left(\frac{\mathcal{E}_f^c}{\mathcal{E}_o^c}\right), \tag{2}$$

By doing so, the OCT model is able to use the global information from fundus modality for additional supervision. Overall, our approach adopts class prototypes from fundus modality instead of noisy features from individual samples, which provides more specific knowledge for OCT student model.

2.2   Class Similarity Alignment

We also note that for multi-label classification tasks, relationships among different classes also contain important information, especially since label co-occurrence is common for eye diseases. Based on this observation, we additionally propose a class similarity alignment scheme to distill knowledge concerning inter-class relationships from fundus model to OCT model.

First, we estimate the disease distribution by averaging the obtained logits of fundus and OCT model in a class-wise manner to get $q_f = \{q_f^c\}_{c=1}^C, q_o = \{q_o^c\}_{c=1}^C$. Then, to transfer information on inter-class relationships, we enforce cosine similarity matrices of the averaged logits to be consistent between teacher and student model. The similarity matrix for teacher model is calculated as $\mathcal{Q}_f^c = \sigma(sim(q_f^c, q_f)/\tau)$ and is obtained similarly for student model, $\mathcal{Q}_o^c$. KL divergence loss is used to encourage alignment between the two similarity matrices:

$$\mathcal{L}_{CSA} = \sum_{c=1}^C \mathcal{Q}_f^c \log\left(\frac{\mathcal{Q}_f^c}{\mathcal{Q}_o^c}\right), \tag{3}$$

In this way, disease distribution knowledge is distilled from fundus teacher model, forcing OCT student model to learn additional knowledge concerning inter-class relationships, which is highly important in multi-label scenarios.

2.3   Overall Framework

The overall loss is the combination of classification loss and distillation enhancement loss:

$$\mathcal{L}_{OCT} = \mathcal{L}_{CLS} + \alpha\mathcal{L}_{CPM} + \beta\mathcal{L}_{CSA}, \tag{4}$$

where α and β are loss weights that control the contribution of each distillation loss. Admittedly, knowledge distillation strategies in computer vision [1,5,20–22,27] can be applied to share multi-modal information as well. Unlike classical distillation methods, our two novel distillation losses allow knowledge about *disease-specific features* and *inter-class relationships* to be transferred, thereby allowing knowledge distillation to be conducted with unpaired data.

3   Experiments

3.1   Experimental Setup

Dataset. To evaluate the effectiveness of our approach, we collect a new dataset TOPCON-MM with paired fundus and OCT images from 369 eyes of 203 patients in Guangdong Provincial Hospital of Integrated Traditional Chinese and Western Medicine using a Topcon Triton swept-source OCT featuring multimodal fundus imaging. For fundus images, they are acquired at a resolution of 2576×1934. For OCT scans, the resolution ranges from 320×992 to 1024×992. Specifically, multiple fundus and OCT images are obtained for each eye,

Table 1. Statistics of our collected TOPCON-MM dataset.

Category	Normal	dAMD	wAMD	DR	CSC	PED	MEM	FLD	EXU	CNV	RVO	Total
Eyes	153	52	30	72	15	23	38	93	90	14	10	369
Fundus Images	299	178	171	502	95	134	200	638	576	143	34	1520
OCT Images	278	160	145	502	95	133	196	613	573	138	34	1435

and each image may reveal multiple diseases with consistent labels specific to that eye. All cases were examined by two ophthalmologists independently to determine the diagnosis label. If the diagnosis from two ophthalmologists disagreed with each other, a third senior ophthalmologist with more than 15 years of experience was consulted to determine the final diagnosis. As shown in Table 1, there are eleven classes, including normal, dry age-related macular degeneration (dAMD), wet age-related macular degeneration (wAMD), diabetic retinopathy (DR), central serous chorioretinopathy (CSC), pigment epithelial detachment (PED), macular epiretinal membrane (MEM), fluid (FLD), exudation (EXU), choroid neovascularization (CNV) and retinal vascular occlusion (RVO).

Implementation Details. Following prior work [24,25], we use contrast-limited adaptive histogram equalization for fundus images and median filter for OCT images as data preprocessing. We adopt data augmentation including random crop, flip, rotation, and changes in contrast, saturation, and brightness. All the images are resized to 448×448 before feeding into the network. For a fair comparison, we apply identical data processing steps, data augmentation operations, model backbones and running epochs in all the experiments. We use SGD to optimize parameters with a learning rate of 1e-3, a momentum of 0.9, and a weight decay of 1e-4. The batch size is set to 8. For weight parameters, τ is set to 4, α is set to 2 and β is set to 1. All the models are implemented on an NVIDIA RTX 3090 GPU. We split the dataset into training and test subsets according to the patient's identity and maintained a training-to-test set ratio of approximately 8:2. To ensure the robustness of the model, the result was reported by five-fold cross-validation.

Evaluation Metrics. We follow previous work [12] to evaluate image-level performance. As each eye in our dataset was scanned multiple times, we use the ensemble results from all the images of the same eye to determine the final prediction. More specifically, if any image indicates an abnormality, the eye is predicted to have the disease.

3.2 Compare with State-of-the-Arts

To prove the effectiveness of our proposed method, we compare our approach with single-modal, multi-modal, and knowledge distillation methods. From Table 2, it is apparent that the model trained with OCT alone performs better than the fundus models. It is noteworthy that current multi-modality

Table 2. Results on our collected TOPCON-MM dataset. "Training" and "Inference" indicate which modalities are required for both phases. "Paired" indicates whether paired fundus-OCT images are required in training. All the experiments use ResNet50 as the backbone. For multi-modal methods, two ResNet50 without shared weights are applied separately for each modality. "Late Fusion" refers to the direct ensemble of the results from models trained with two single modalities. †: we implement multi-modal methods in retinal disease classification. ⋆: we run KD methods in computer vision.

Method	Training	Inference	Paired	MAP	Sensitivity	Specificity	F1 Score	AUC
Single-Modal Methods								
ResNet50	Fundus	Fundus	-	50.56 ± 3.05	43.68 ± 6.58	92.24 ± 0.77	54.95 ± 7.09	79.97 ± 1.63
ResNet50	OCT	OCT	-	66.44 ± 3.81	53.14 ± 6.60	95.28 ± 0.85	64.16 ± 6.24	87.73 ± 1.44
Multi-Modal Methods								
Late Fusion	Both	Both	✓	63.83 ± 1.34	54.45 ± 2.72	94.29 ± 0.74	64.93 ± 3.00	86.92 ± 1.48
Two-Stream CNN† [25]	Both	Both	✓	58.75 ± 2.71	53.47 ± 3.82	92.97 ± 0.91	61.82 ± 4.02	84.79 ± 2.77
MSAN† [4]	Both	Both	✓	59.49 ± 3.43	56.44 ± 3.13	93.37 ± 0.59	63.95 ± 3.77	84.51 ± 1.91
FitNet⋆ [21]	Both	OCT	✓	63.41 ± 3.45	54.44 ± 4.04	94.87 ± 0.58	65.00 ± 4.32	87.17 ± 1.85
KD⋆ [5]	Both	OCT	✓	63.69 ± 2.04	51.70 ± 3.10	95.75 ± 0.62	63.56 ± 2.32	87.90 ± 1.03
RKD⋆ [20]	Both	OCT	✓	63.59 ± 3.04	53.42 ± 1.71	94.42 ± 1.81	63.70 ± 0.72	87.36 ± 2.08
DKD⋆ [27]	Both	OCT	✓	64.40 ± 2.09	53.83 ± 5.23	95.24 ± 0.11	64.00 ± 4.44	87.52 ± 0.58
SimKD⋆ [1]	Both	OCT	✓	65.10 ± 2.63	53.13 ± 5.49	95.09 ± 0.92	63.19 ± 6.53	87.97 ± 1.32
Ours	Both	OCT	✗	$\mathbf{69.06 \pm 3.39}$	$\mathbf{57.15 \pm 5.93}$	$\mathbf{95.93 \pm 0.57}$	$\mathbf{69.17 \pm 6.07}$	$\mathbf{89.06 \pm 0.97}$

methods [4,25] and knowledge distillation methods [1,5,20,21,27] do not yield improved results on our dataset. Table 2 also demonstrates that compared with the single-modal OCT baseline, our method improves MAP from 66.44% to 69.06%, F1 from 64.16% to 69.17%. *This shows that it is still possible to learn valuable information from the fundus modality to assist the OCT model, despite being a weaker modality.* It can be observed that our approach outperforms the state-of-the-art multi-modal retinal image classification method [4] by 9.57% in MAP (69.06% *v.s.* 59.49%). Notably, our method excels the best-performing knowledge distillation method [1] by 3.96% in MAP (69.06% *v.s.* 65.10%). We also note that the alternative methods are limited to training with eye-paired fundus and OCT images only, whilst our approach does not face such restrictions.

To further demonstrate the efficiency of our proposed distillation enhancement approach, we validate our method on a publicly available multi-modal dataset with fundus and OCT images, MMC-AMD [24]. MMC-AMD dataset contains four classes: normal, dry AMD, PCV, and wet AMD. We reproduce single-modal ResNet, Two-Stream CNN [25], and KD methods [5,21] as baselines and show results in Fig. 3 (a). It can be seen that our method improves MAP to 92.29%, largely surpassing existing methods.

3.3 Results Trained with Other Fundus Datasets

Since we implement distillation in a disease-aware manner, multi-modal fundus and OCT training data do not need to be paired. Theoretically, any publicly available fundus dataset could be applied as long as it shares a label space that overlaps with our OCT data. To verify this hypothesis, we separately reproduce our methods with fundus images from two datasets, MMC-AMD [24] and

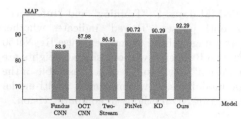
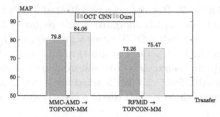

(a) MAP of different models on patient-split   (b) Transfer from fundus datasets.
MMC-AMD dataset.

Fig. 3. Results on other publicly available datasets.

Table 3. Ablation study of our method. "Majority" and "Minority" refers to the average score of classes that represent more than 10% or less than 10% of the total number of images, respectively. "Overall" indicates overall performance on all the classes.

Method		MAP			F1 Score	AUC
CPM	CSA	Overall	Majority	Minority		
✗	✗	66.44 ± 3.81	71.12 ± 4.04	58.26 ± 7.42	64.16 ± 6.24	87.73 ± 1.44
✗	✓	67.50 ± 3.00	70.60 ± 4.91	**62.08** ± 8.72	65.40 ± 5.21	88.09 ± 1.28
✓	✗	67.76 ± 2.34	72.26 ± 3.20	59.90 ± 8.32	65.58 ± 2.58	88.73 ± 1.32
✓	✓	**69.06** ± 3.39	**73.34** ± 3.48	61.47 ± 8.17	**69.17** ± 6.07	**89.06** ± 0.97

RFMiD [19]. To ensure label overlap, we only select fundus and OCT images from common classes for training and validation, namely, 3 classes for MMC-AMD and 6 classes for RFMiD. The results are reported in Fig. 3 (b). Compared with single-modal OCT model, our distillation enhancement can achieve an increase of 4.26% (84.06% *v.s.* 79.80%) and 2.21% (75.47% *v.s.* 73.26%) in MAP. Our results prove that *our method has the flexibility to use any existing fundus dataset to enhance OCT classification*.

3.4   Ablation Studies

Table 3 shows the ablation study of our method. To provide additional insight, we also show the results on majority classes, which contain over 10% images of the dataset, and minority classes with less than 10%. It can be seen that individually using CPM and CSA can improve the overall result by 1.32% and 1.06% in MAP, respectively. Removing either of the components degrades the performance. Results also show that CPM improves classification performance in majority classes by distilling disease-specific knowledge, while CSA benefits minority classes by attending to inter-disease relationships. By simultaneously adopting CPM and CSA, the overall score of all the classes is improved.

4   Conclusion

Our work proposes a novel fundus-enhanced disease-aware distillation module, FDDM, for retinal disease classification. The module incorporates class prototype matching to distill global disease information from the fundus teacher to the OCT student, while also utilizing class similarity alignment to ensure the consistency of disease relationships between both modalities. Our approach deviates from the existing models that rely on paired instances for multi-modal training and inference, making it possible to extract knowledge from any available fundus data and render predictions with only OCT modality. As a result, our approach significantly reduces the prerequisites for clinical applications. Our extensive experiments demonstrate that our method outperforms existing baselines by a considerable margin.

Acknowledgement. This work is supported by grants from Foshan HKUST Projects under Grants FSUST21-HKUST10E and FSUST21-HKUST11E, as well as by the Hong Kong Innovation and Technology Fund under Projects PRP/041/22FX and ITS/030/21.

References

1. Chen, D., Mei, J.P., Zhang, H., Wang, C., Feng, Y., Chen, C.: Knowledge distillation with the reused teacher classifier. In: CVPR, pp. 11933–11942 (2022)
2. Ehlers, J.: The Retina Illustrated. Thieme Medical Publishers, Incorporated (2019)
3. Fang, L., Wang, C., Li, S., Rabbani, H., Chen, X., Liu, Z.: Attention to lesion: lesion-aware convolutional neural network for retinal optical coherence tomography image classification. IEEE Trans. Med. Imaging **38**(8), 1959–1970 (2019)
4. He, X., Deng, Y., Fang, L., Peng, Q.: Multi-modal retinal image classification with modality-specific attention network. IEEE Trans. Med. Imaging **40**, 1591–1602 (2021)
5. Hinton, G., Vinyals, O., Dean, J., et al.: Distilling the knowledge in a neural network. arXiv preprint arXiv:1503.02531 **2**(7) (2015)
6. Huang, D., et al.: Optical coherence tomography. Science **254**(5035), 1178–1181 (1991)
7. Huang, L., He, X., Fang, L., Rabbani, H., Chen, X.: Automatic classification of retinal optical coherence tomography images with layer guided convolutional neural network. IEEE Signal Process. Lett. **26**(7), 1026–1030 (2019)
8. Karri, S.P.K., Chakraborty, D., Chatterjee, J.: Transfer learning based classification of optical coherence tomography images with diabetic macular edema and dry age-related macular degeneration. Biomed. Opt. Express **8**(2), 579–592 (2017)
9. Kermany, D.S., et al.: Identifying medical diagnoses and treatable diseases by image-based deep learning. Cell **172**(5), 1122–1131 (2018)
10. Lee, C.S., Baughman, D.M., Lee, A.Y.: Deep learning is effective for classifying normal versus age-related macular degeneration oct images. Ophthalmol. Retina **1**(4), 322–327 (2017)
11. Lemaître, G., et al.: Classification of SD-OCT volumes using local binary patterns: experimental validation for DME detection. J. Ophthalmol. **2016**, 1–14 (2016). https://doi.org/10.1155/2016/3298606

12. Li, X., et al.: Multi-modal multi-instance learning for retinal disease recognition. In: ACMMM, pp. 2474–2482 (2021)
13. Li, X., Shen, L., Shen, M., Tan, F., Qiu, C.S.: Deep learning based early stage diabetic retinopathy detection using optical coherence tomography. Neurocomputing **369**, 134–144 (2019)
14. Li, Y., et al.: Multimodal information fusion for glaucoma and diabetic retinopathy classification. In: Antony, B., Fu, H., Lee, C.S., MacGillivray, T., Xu, Y., Zheng, Y. (eds.) Ophthalmic Medical Image Analysis: 9th International Workshop, OMIA 2022, Held in Conjunction with MICCAI 2022, Singapore, Singapore, September 22, 2022, Proceedings, pp. 53–62. Springer International Publishing, Cham (2022). https://doi.org/10.1007/978-3-031-16525-2_6
15. Liu, X., Bai, Y., Jiang, M.: One-stage attention-based network for image classification and segmentation on optical coherence tomography image. In: SMC, pp. 3025–3029. IEEE (2021)
16. Liu, Y.Y., Chen, M., Ishikawa, H., Wollstein, G., Schuman, J.S., Rehg, J.M.: Automated macular pathology diagnosis in retinal oct images using multi-scale spatial pyramid and local binary patterns in texture and shape encoding. Med. Image Anal. **15**(5), 748–759 (2011)
17. Müller, P.L., Wolf, S., Dolz-Marco, R., Tafreshi, A., Schmitz-Valckenberg, S., Holz, F.G.: Ophthalmic Diagnostic Imaging: Retina. In: Bille, J.F. (ed.) High Resolution Imaging in Microscopy and Ophthalmology, pp. 87–106. Springer, Cham (2019). https://doi.org/10.1007/978-3-030-16638-0_4
18. Ou, Z., et al.: M 2 LC-Net: A multi-modal multi-disease long-tailed classification network for real clinical scenes. China Commun.D **18**(9), 210–220 (2021)
19. Pachade, S., et al.: Retinal fundus multi-disease image dataset (RFMid): a dataset for multi-disease detection research. Data **6**(2), 14 (2021)
20. Park, W., Kim, D., Lu, Y., Cho, M.: Relational knowledge distillation. In: Proceedings of the IEEE/CVF Conference on Computer Vision and Pattern Recognition, pp. 3967–3976 (2019)
21. Romero, A., Ballas, N., Kahou, S.E., Chassang, A., Gatta, C., Bengio, Y.: Fitnets: Hints for thin deep nets. arXiv preprint arXiv:1412.6550 (2014)
22. Shu, C., Liu, Y., Gao, J., Yan, Z., Shen, C.: Channel-wise knowledge distillation for dense prediction. In: Proceedings of the IEEE/CVF International Conference on Computer Vision, pp. 5311–5320 (2021)
23. Srinivasan, P.P., et al.: Fully automated detection of diabetic macular edema and dry age-related macular degeneration from optical coherence tomography images. Biomed. Opt. Express **5**(10), 3568–3577 (2014)
24. Wang, W., et al.: Learning two-stream CNN for multi-modal age-related macular degeneration categorization. IEEE J. Biomed. Health Inform. **26**(8), 4111-4122 (2022)
25. Wang, W., et al.: Two-stream CNN with loose pair training for multi-modal AMD categorization. In: Shen, D., et al. (eds.) MICCAI 2019. LNCS, vol. 11764, pp. 156–164. Springer, Cham (2019). https://doi.org/10.1007/978-3-030-32239-7_18
26. Yoo, T.K., Choi, J.Y., Seo, J.G., Ramasubramanian, B., Selvaperumal, S., Kim, D.W.: The possibility of the combination of oct and fundus images for improving the diagnostic accuracy of deep learning for age-related macular degeneration: a preliminary experiment. Med. Biol. Eng. Comput. **57**(3), 677–687 (2019)
27. Zhao, B., Cui, Q., Song, R., Qiu, Y., Liang, J.: Decoupled knowledge distillation. In: CVPR, pp. 11953–11962 (2022)

VF-HM: Vision Loss Estimation Using Fundus Photograph for High Myopia

Zipei Yan[1], Dong Liang[2(✉)], Linchuan Xu[1(✉)], Jiahang Li[1], Zhengji Liu[2], Shuai Wang[3], Jiannong Cao[1], and Chea-su Kee[2]

[1] Department of Computing, The Hong Kong Polytechnic University,
Kowloon, Hong Kong
linch.xu@polyu.edu.hk
[2] School of Optometry, The Hong Kong Polytechnic University,
Kowloon, Hong Kong
dong1.liang@connect.polyu.hk
[3] Department of Biomedical Engineering, Tsinghua University, Beijing, China

Abstract. High myopia (HM) is a leading cause of irreversible vision loss due to its association with various ocular complications including myopic maculopathy (MM). Visual field (VF) sensitivity systematically quantifies visual function, thereby revealing vision loss, and is integral to the evaluation of HM-related complications. However, measuring VF is subjective and time-consuming as it highly relies on patient compliance. Conversely, fundus photographs provide an objective measurement of retinal morphology, which reflects visual function. Therefore, utilizing machine learning models to estimate VF from fundus photographs becomes a feasible alternative. Yet, estimating VF with regression models using fundus photographs fails to predict local vision loss, producing stationary nonsense predictions. To tackle this challenge, we propose a novel method for VF estimation that incorporates VF properties and is additionally regularized by an auxiliary task. Specifically, we first formulate VF estimation as an ordinal classification problem, where each VF point is interpreted as an ordinal variable rather than a continuous one, given that any VF point is a discrete integer with a relative ordering. Besides, we introduce an auxiliary task for MM severity classification to assist the generalization of VF estimation, as MM is strongly associated with vision loss in HM. Our method outperforms conventional regression by 16.61% in MAE metric on a real-world dataset. Moreover, our method is the first work for VF estimation using fundus photographs in HM, allowing for more convenient and accurate detection of vision loss in HM, which could be useful for not only clinics but also large-scale vision screenings.

Keywords: Vision loss estimation · Visual field · Fundus photograph · Ordinal classification · Auxiliary learning

---

Z. Yan and D. Liang—Equal contribution.

Supplementary Information The online version contains supplementary material available at https://doi.org/10.1007/978-3-031-43990-2_61.

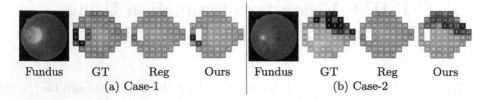

| Fundus | GT | Reg | Ours | Fundus | GT | Reg | Ours |

(a) Case-1   (b) Case-2

Fig. 1. Estimated VF from different methods using fundus. GT denotes the ground truth, Reg denotes the regression baseline, and Ours denotes our method.

1   Introduction

High myopia (HM) has become a global concern for public health, with its markedly growing prevalence [10] and its increased risk of irreversible vision loss and even blindness [14,16,25,27]. In brief, excessive axial elongation in HM eyes will produce mechanical stretching on the posterior segment of eyeballs, leading to various structural changes and HM-related complications, e.g., myopic maculopathy (MM), and consequently, functional changes, resulting in vision loss.

Accurate quantification of vision loss is integral to the early detection and timely treatment for MM and other HM-related complications [16]. Currently, the diagnosis of vision loss is made on the basis of visual field (VF) sensitivity by standard automated perimetry, which is a systematic metric and gold standard to quantify visual function [19]. However, measuring VF is prohibitively time-consuming and subjective as it highly requires patients' concentration and compliance during the test [12].

Conversely, imaging techniques, such as fundus photography (a.k.a., fundus), provide a relatively objective and robust measurement of the retinal morphology, which likely corresponds to the VF with an underlying "structure-function relationship" [27,32]. Actually, fundus is most commonly used for the diagnosis and evaluation of HM and its complications, in particular in rural and developing regions, with its lower cost and convenience of acquisition [17].

Therefore, utilizing machine learning models to estimate VF from fundus becomes a promising and feasible alternative for HM subjects in clinical practice. To the best of our knowledge, there is no existing approach to estimate VF from fundus. Some studies have been proposed to estimate the global indices (e.g., mean deviation) of VF from fundus [3,11], and others estimate VF using retinal thickness [4,18,28,30]. It is worth mentioning that, all these studies were conducted for the glaucoma population [3,4,11,18,28,30], in which most cases of visual abnormality or defect were likely glaucomatous. However, MM and other HM-related complications may lead to non-glaucomatous vision loss.

Actually, estimating VF with conventional regression [18] using fundus fails to predict local vision loss in our HM population, producing stationary nonsense predictions. As shown in Fig. 1, these predictions from regression exhibit a relatively similar and consistent pattern in most HM subjects, failing to capture/learn the

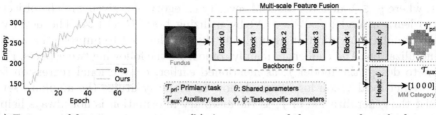

(a) Entropy of feature space   (b) An overview of the proposed method.

Fig. 2. (a) The entropy of feature space on training data during training progress from conventional regression (denoted by Reg) and our VF-HM. (b) An overview of our proposed method: VF-HM.

inter-subject variability and local defects of VF. And these predictions are very close to the mean value of VF in training data (see the supplementary material). The reason for such failure lies in regression's inability to learn high-entropy feature representations [31], which is further confirmed by measuring the entropy of feature representations, as marked in blue in Fig. 2a.

To tackle this challenge, we propose a novel method for estimating VF for HM using fundus, namely VF-HM. In general, VF-HM incorporates VF properties and is additionally regularized by an auxiliary task, thereby learning relatively high-entropy feature representations (see the orange line in Fig. 2a). In detail, we formulate VF estimation as an ordinal classification problem, where each VF point is interpreted as an ordinal variable rather than a continuous one, given that any VF point is a discrete integer with a relative ordering. Besides, we introduce an auxiliary task for MM severity classification to assist the generalization of VF estimation, because MM is strongly associated with vision loss in HM [7,16,21,32] and its symptom can be observed from the fundus directly. As a result, VF-HM significantly outperforms conventional regression and accurately predicts vision loss (see Fig. 1).

Our contributions are summarized as follows:

- We propose a novel method, VF-HM, for estimating VF from fundus for HM. VF-HM more accurately detects the local vision loss and significantly outperforms conventional regression by 16.61% in the MAE metric on a real dataset.
- VF-HM is the first work for VF estimation using fundus for HM, allowing for more convenient and cost-efficient detection of vision loss in HM, which could be useful for not only clinics but also large-scale vision screenings.

2   Problem Formulation

Let $\mathcal{D} = \{(\boldsymbol{x}_i, \boldsymbol{m}_i)\}$ denote the training set, where $\boldsymbol{x}_i \in \mathcal{X}$ denotes the fundus, $\boldsymbol{m}_i \in \mathcal{M}$ denotes its corresponded VF. And $\mathcal{A} = \{(\boldsymbol{x}_i, y_i)\}$ denotes the auxiliary set, where $y_i \in \mathcal{Y}$ denotes the MM severity category of a given $\boldsymbol{x}_i$. The objective is to learn a model $f : \mathcal{X} \rightarrow \mathcal{M}$ by utilizing both $\mathcal{D}$ and $\mathcal{A}$. The novelty of this formulation is additionally utilizing the auxiliary set to improve the model's generalization. And challenges mainly come from the following two aspects. First, how to design the model f, as mentioned earlier, conventional regression fails to predict local vision loss. Second, how to properly utilize the auxiliary set to assist the generalization of f, as the auxiliary information is not always helpful during the learning progress, i.e., sometimes may interfere [5,6,22].

3   Proposed Method: VF-HM

In this section, we first present an overview of the proposed method. Then, we introduce the details of different components.

3.1   Overview

We present an overview of the proposed method in Fig. 2b. Specifically, the primary task (denoted by $\mathcal{T}_{\mathrm{pri}}$) is the VF estimation and the auxiliary task is MM classification (denoted by $\mathcal{T}_{\mathrm{aux}}$). Then, our method aims to solve $\mathcal{T}_{\mathrm{pri}}$ with the assistance of $\mathcal{T}_{\mathrm{aux}}$. We propose to parameterize the solution for $\mathcal{T}_{\mathrm{pri}}$ and $\mathcal{T}_{\mathrm{aux}}$ by two neural networks: $f(\cdot; \theta, \phi)$ and $g(\cdot; \theta, \psi)$, where they share the same backbone θ and have their own task-specific parameters ϕ and ψ. Thereafter, the overall objective function is formulated as follows:

$$\mathcal{L} = \mathcal{L}_{\mathrm{pri}}(\theta, \phi) + \lambda \mathcal{L}_{\mathrm{aux}}(\theta, \psi) \tag{1}$$

where $\mathcal{L}_{\mathrm{pri}}$ and $\mathcal{L}_{\mathrm{aux}}$ denote the loss function for $\mathcal{T}_{\mathrm{pri}}$ and $\mathcal{T}_{\mathrm{aux}}$, respectively. $\lambda \in (0, 1]$ is a hyper-parameter to control the importance of $\mathcal{L}_{\mathrm{aux}}$.

3.2   Primary Task: VF Estimation

The overall interest is *only* the primary task $\mathcal{T}_{\mathrm{pri}}$, which is parameterized by $f(\cdot; \theta, \phi) : \mathcal{X} \rightarrow \mathcal{M}$. Specifically, we formulate $\mathcal{T}_{\mathrm{pri}}$ as an *ordinal classification* (aka, *rank learning*) problem, where each VF point m_i^j represents an *ordinal variable/rank* rather than a continuous one. Such a formulation incorporates the distinct properties of VF, which include: 1) Discretization: $\forall m_i^j \in [0, 40] \cap \mathbb{Z}$, that is, any VF value is a *positive discrete integer*. 2) Ordinalization: $m_i^0 \prec m_i^1 \prec \ldots \prec m_i^j$, there is a *relative order* among VF values. To achieve this goal, we extend the *ordinal variable/rank* into binary labels [2,13], i.e., $\boldsymbol{m}_i^j = [r_i^{j,1}, \ldots, r_i^{j,K-1}]^T$ where $r_i^{j,k} \in \{0, 1\}$ indicates whether $\boldsymbol{m}_i^j$ exceeds k-th rank or not. To ensure rank-monotonic and guarantee prediction consistency, we utilize the *ordinal bias* [2]. In detail, the task-specific parameter ϕ contains independent bias for each ordinal variable. Thereafter, $\mathcal{T}_{\mathrm{pri}}$ can be solved by the binary cross-entropy loss, which is defined as follows:

$$\mathcal{L}_{\mathrm{pri}}(\theta, \phi) = \mathbb{E}_{(\boldsymbol{x}_i, \boldsymbol{m}_i) \in \mathcal{X} \times \mathcal{M}}[L_{\mathrm{BCE}}(f(\boldsymbol{x}_i; \theta, \phi), \boldsymbol{m}_i)] \tag{2}$$

where $L_{\mathrm{BCE}}(\cdot)$ denotes the binary cross-entropy loss

In addition, we propose to reuse the features from different blocks, as they contain distinct spatial information. Specifically, we propose Multi-scale Feature Fusion (MFF) for aggregating features from different blocks. As highlighted in orange in Fig. 2b, MFF aggregates features from all blocks at the last in an addition operation. The detailed implementation is reported in Sect. 4.2.

3.3 Auxiliary Task: MM Classification

The auxiliary task $\mathcal{T}_{\mathrm{aux}}$ is introduced *only* to assist the generalization of $\mathcal{T}_{\mathrm{pri}}$. Specifically, $\mathcal{T}_{\mathrm{aux}}$ is to predict MM severity category y_i from fundus x_i, which is parameterized by $g(\cdot; \theta, \psi) : \mathcal{X} \rightarrow \mathcal{Y}$. MM is highly correlated to vision loss [7, 16, 21, 32], and its symptom can be observed from the fundus directly. According to its increasing severity, MM can be classified into five categories [26], i.e., $C_0 \prec C_1 \ldots \prec C_4$. Therefore, we also interpret the MM category as the *ordinal variable/rank*. Similar to the label extension in $\mathcal{T}_{\mathrm{pri}}$, we extend the MM category into binary labels $y_i = [r_1, r_2, r_3, r_4]^T$. The loss function $\mathcal{L}_{\mathrm{aux}}$ for solving $\mathcal{T}_{\mathrm{aux}}$ is also the binary cross-entropy, which is defined as follows:

$$\mathcal{L}_{\mathrm{aux}}(\theta, \psi) = \mathbb{E}_{(x_i, y_i) \in \mathcal{X} \times \mathcal{Y}}[L_{\mathrm{BCE}}(g(x_i; \theta, \psi), y_i)] \qquad (3)$$

However, the $\mathcal{T}_{\mathrm{aux}}$ is not always helpful for $\mathcal{T}_{\mathrm{pri}}$ because of the negative transfer [5, 6, 22]. The negative transfer refers to a problem that sometimes $\mathcal{T}_{\mathrm{aux}}$ becomes harmful for $\mathcal{T}_{\mathrm{pri}}$. Specifically, let $\nabla_\theta \mathcal{L}$ denote the gradient of Eq. (1) in terms of the shared parameters θ, and it can be decomposed as follows:

$$\nabla_\theta \mathcal{L} = \nabla_\theta \mathcal{L}_{\mathrm{pri}} + \lambda \nabla_\theta \mathcal{L}_{\mathrm{aux}} \qquad (4)$$

$\mathcal{T}_{\mathrm{aux}}$ becomes harmful for $\mathcal{T}_{\mathrm{pri}}$, when the *cosine similarity* between $\nabla_\theta \mathcal{L}_{\mathrm{pri}}$ and $\nabla_\theta \mathcal{L}_{\mathrm{aux}}$ becomes negative [6], i.e., $\cos(\nabla_\theta \mathcal{L}_{\mathrm{aux}}, \nabla_\theta \mathcal{L}_{\mathrm{pri}}) < 0$. Negative transfer is observed in our setting when optimizing Eq. (1) directly, as illustrated in Fig. 3a.

Following [6], we mitigate negative transfer by refining $\nabla_\theta \mathcal{L}_{\mathrm{aux}}$. Specifically, we adapt the weighted cosine similarity to refine $\nabla_\theta \mathcal{L}_{\mathrm{aux}}$, which is defined as follows:

$$\nabla_\theta \mathcal{L}_{\mathrm{aux}} = \max\left(0, \cos(\nabla_\theta \mathcal{L}_{\mathrm{aux}}, \nabla_\theta \mathcal{L}_{\mathrm{pri}})\right) \cdot \nabla_\theta \mathcal{L}_{\mathrm{aux}} \qquad (5)$$

4   Experiments

In this section, we conduct experiments on a clinic-collected real-world dataset to evaluate the performance of our proposed method[1].

---

[1] Our code is available at https://github.com/yanzipei/VF-HM.

Table 1. Main results. 'K-fold' denotes performance from K-fold cross-validation on training data. 'Test' denotes performance on test data (pre-trained on training data). ($\downarrow$) denotes the lower value indicates better performance. RT-($\cdot$) denotes different retinal thicknesses. And the better results are **bold-faced**.

Method	Modality	K-fold($K = 5$)			Test		
		RMSE ($\downarrow$)	MAE ($\downarrow$)	SMAPE ($\downarrow$)	RMSE ($\downarrow$)	MAE ($\downarrow$)	SMAPE ($\downarrow$)
Regression	RT-(a)	4.94 ± 0.23	3.12 ± 0.05	13.47 ± 0.16	-	-	-
Regression	RT-(b)	4.80 ± 0.17	3.04 ± 0.12	13.21 ± 0.36	-	-	-
Regression	RT-(c)	4.86 ± 0.22	3.13 ± 0.18	13.42 ± 0.57	-	-	-
Regression	Fundus	4.62 ± 0.07	2.95 ± 0.07	12.94 ± 0.32	4.28 ± 0.03	2.89 ± 0.06	12.13 ± 0.30
Ours(λ=0.1)	Fundus	**4.44 ± 0.27**	**2.78 ± 0.10**	**12.50 ± 0.26**	**3.69 ± 0.03**	**2.41 ± 0.04**	**11.38 ± 0.14**

4.1   The Studied Data

The studied data comes from a HM population, including 75 patients, each with diagnosis information for both eyes. For each eye, there are one fundus, VF, and MM severity category. Specifically, the fundus is captured in colorful mode, the VF is measured in the 24-2 mode with 52 effective points, and MM category is labeled by registered ophthalmologists. Besides, 34 patients (i.e., 68 eyes) have SD-OCT scans in the macular region. For these SD-OCT scans, we extract the retinal thickness with the pre-trained model [20] in order to compare our method to conventional regression using retinal thickness. According to whether the eye has SD-OCT scans or not, we divide the whole data into a training set and a test set. Specifically, the training and test data contain 68 eyes (from 34 patients) and 82 eyes (from 41 patients), respectively. It is worth mentioning that the training data and test data do not have the same patient. Besides, in the following K-fold cross-validation experiments, we split the training data based on the patient's ID to ensure that there is no information leakage.

4.2   Experimental Setup

Data Pre-processing. We choose the *left* eye pattern as our base. For fundus, VF and retinal thickness are not in the *left* eye pattern, we convert them using the horizontal flip.

Data Augmentation. Following [1], we consolidate a set of data augmentations for both fundus and retinal thickness, respectively. The details are reported in the supplementary material. Different from applying *all* [1] augmentations during training, we utilize the *TrivialAugment* [15] instead, which randomly selects one from the given data augmentations, generating more diverse augmented data.

Evaluation Methods. For quantitative evaluation, we utilize three metrics [3, 4,18,29,33]: RMSE, MAE and SMAPE. For qualitative evaluation, we visualize two representative predictions on the test set, and more visualized results are presented in the supplementary material.

Table 2. Ablation study on main components. OC denotes the ordinal classification baseline. MFF denotes multi-scale feature fusion. AUX denotes the auxiliary task. MNT denotes mitigating negative transfer from Eq. (5).

OC	MFF	AUX	MNT	RMSE ($\downarrow$)	MAE ($\downarrow$)
✓	✓	✓	✓	**3.69 ± 0.03**	**2.41 ± 0.04**
✓	✓	✓		3.74 ± 0.02	2.46 ± 0.03
✓	✓			3.73 ± 0.04	2.45 ± 0.02
✓				3.77 ± 0.02	2.49 ± 0.03

Baseline Methods. We mainly compare our method to conventional regression that estimates VF from fundus. Besides, for a more comprehensive comparison, we also compare our approach to conventional regression using different retinal thicknesses. In detail, we consider three variants: (a) the combination of GCIPL, RNFL and RCL [33], (b) the combination of GCIPL and RNFL [18], (c) only RNFL [4]. Due to the limited data, we compare our method to conventional regression using the above thickness by K-fold cross-validation on training data.

Implementation Details. We utilize the ResNet-18 [8] as the backbone. For the regression baseline, we use only one *linear* layer at last. For our method, we use the combination of *Conv2D*, *BatchNorm2D* and *ReLU* as the classification head for $\mathcal{T}_{\mathrm{pri}}$. For the MFF, we utilize the above classification head to aggregate features from different blocks. Note that the features from earlier blocks have relatively large features, thus we use *AdaptiveAvgPooling2D* to perform downsampling first. For $\mathcal{T}_{\mathrm{aux}}$, we use only one *linear* layer as the classifier. For a fair comparison, we train all methods with the same training configurations. Specifically, we train the models with 80 training epochs and the SGD optimizer, where the batch size is set to 32, the learning rate is set to 0.01, momentum is set to 0.9 and L2 weight decay is set to $1e^{-4}$. Besides, we utilize a cosine learning rate decay [9] to adjust the learning rate per epoch. Finally, we fix all input resolutions to 384 × 384 for both training and evaluation. All experiments are run independently with four seeds: 0, 1, 2, and 3. As for hyper-parameters, we search them on training data with K-fold cross-validation.

4.3 Experimental Results

Main Results. Table 1 reports the performance of our method and baselines. In general, our method achieves the best performance compared to these baselines. Specifically, compared to conventional regression using fundus, our method outperforms it by 13.79% and 16.61% according to the RMSE and MAE metric on test data. Besides, our method achieves better performance than baselines using different retinal thicknesses.

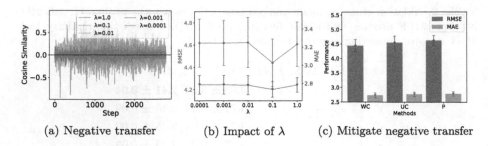

(a) Negative transfer   (b) Impact of λ (c) Mitigate negative transfer

Fig. 3. Visualization of (a) Negative transfer when optimizing Eq. (1) directly, (b) Impact of hyper-parameter λ, and (c) Different methods for mitigating the negative transfer.

Visualization of Predictions. As shown in Fig. 1, we visualize predictions from methods using fundus on two representative cases. Specifically, conventional regression fails to predict local vision loss, as its predictions share a similar and consistent pattern for both cases. In contrast, predictions from our method are more precise, revealing the local vision loss. More visualized results are presented in the supplementary material.

4.4 Ablation Study

To get a better understanding of the effectiveness of the main components in our proposed method, we conduct a series of ablation studies.

Effectiveness of Main Components. We first examine the effectiveness of the main components by ablating them. The results are reported in Table 2. In general, we can observe that all components can improve performance except AUX. Specifically, AUX denotes solely introducing the auxiliary task, which brings a degradation, because of the existence of negative transfer. Meanwhile, with the help of Eq. (5), the negative transfer can be mitigated. Besides, we observe these main components allow the model to learn high-entropy feature representations, thereby improving the model's performance [31]. More details are reported in the supplementary material.

Impact of Hyper-parameter λ. We study the impact of the hyper-parameter λ with K-fold cross validation on training data. We choose $\lambda \in \{1.0, 0.1, 0.01, 0.001, 0.0001\}$. According to the results shown in Fig. 3b, we observe that $\lambda = 0.1$ achieves the best performance.

Different Methods for Mitigating the Negative Transfer. We consider three alternatives to refine the auxiliary gradient for mitigating the negative transfer: (1) weighted cosine (WC) similarity [6] (2) unweighted cosine (UC)

similarity [6] (3) projection (P) [22]. For a fair comparison, we set $\lambda = 0.1$, then conduct experiments on training data with K-fold cross-validation. As shown in Fig. 3c, and we observe that (1) WC achieves the best performance.

5   Conclusion

In this work, we propose VF-HM for estimating VF from fundus for HM, which is the first work for VF estimation in HM; and it provides a more convenient and cost-effective way to detect HM-related vision loss. The major limitations include: first, our sample size is limited; second, we utilize both eyes from one patient as two independent inputs, which ignores their similarity; third, we only include the MM severity as the auxiliary information. Future work could be conducted as follows. First, collecting more data from different clinical sites. Second, modeling the relationship between both eyes from the same patient [33]. Third, exploring more auxiliary information. Besides, studying how to adapt our method to different domains is a crucial problem [24], as we seek to improve the generalizability. In addition, exploring VF prediction with the missing modalities [23]: either fundus or thickness is another interesting direction.

Acknowledgements. This work was supported by the Research Institute for Artificial Intelligence of Things, The Hong Kong Polytechnic University, HK RGC Research Impact Fund No. R5060-19; and the Centre for Myopia Research, School of Optometry; the Research Centre for SHARP Vision (RCSV), The Hong Kong Polytechnic University; and Centre for Eye and Vision Research (CEVR), InnoHK CEVR Project 1.5, 17W Hong Kong Science Park, HKSAR. We thank Drs Rita Sum and Vincent Ng for their guidance on data analysis of clinical population; and Prof. Ruihua Wei for external validation of the model.

References

1. Bar-David, D., Bar-David, L., Soudry, S., Fischer, A.: Impact of data augmentation on retinal oct image segmentation for diabetic macular edema analysis. In: MICCAI, pp. 148–158 (2021)
2. Cao, W., Mirjalili, V., Raschka, S.: Rank consistent ordinal regression for neural networks with application to age estimation. Pattern Recogn. Lett. **140**, 325–331 (2020)
3. Christopher, M., et al.: Deep learning approaches predict glaucomatous visual field damage from OCT optic nerve head EN face images and retinal nerve fiber layer thickness maps. Ophthalmology **127**(3), 346–356 (2020)
4. Datta, S., Mariottoni, E.B., Dov, D., Jammal, A.A., Carin, L., Medeiros, F.A.: Retinervenet: using recursive deep learning to estimate pointwise 24–2 visual field data based on retinal structure. Sci. Rep. **11**(1), 1–10 (2021)
5. Dery, L.M., Dauphin, Y.N., Grangier, D.: Auxiliary task update decomposition: The good, the bad and the neutral. In: ICLR (2021)
6. Du, Y., Czarnecki, W.M., Jayakumar, S.M., Farajtabar, M., Pascanu, R., Lakshminarayanan, B.: Adapting auxiliary losses using gradient similarity. arXiv preprint arXiv:1812.02224 (2018)

7. Hayashi, K., et al.: Long-term pattern of progression of myopic maculopathy: a natural history study. Ophthalmology **117**(8), 1595–1611 (2010)
8. He, K., Zhang, X., Ren, S., Sun, J.: Deep residual learning for image recognition. In: CVPR, pp. 770–778 (2016)
9. He, T., Zhang, Z., Zhang, H., Zhang, Z., Xie, J., Li, M.: Bag of tricks for image classification with convolutional neural networks. In: CVPR, pp. 558–567 (2019)
10. Holden, B.A., et al.: Global prevalence of myopia and high myopia and temporal trends from 2000 through 2050. Ophthalmology **123**(5), 1036–1042 (2016)
11. Lee, J., et al.: Estimating visual field loss from monoscopic optic disc photography using deep learning model. Sci. Rep. **10**(1), 1–10 (2020)
12. Lewis, R.A., Johnson, C.A., Keltner, J.L., Labermeier, P.K.: Variability of quantitative automated perimetry in normal observers. Ophthalmology **93**(7), 878–881 (1986)
13. Li, L., Lin, H.: Ordinal regression by extended binary classification. In: NeurIPS, pp. 865–872 (2006)
14. Lin, F., et al.: Classification of visual field abnormalities in highly myopic eyes without pathologic change. Ophthalmology **129**(7), 803–812 (2022)
15. Müller, S.G., Hutter, F.: Trivialaugment: tuning-free yet state-of-the-art data augmentation. In: ICCV, pp. 754–762 (2021)
16. Ohno-Matsui, K., et al.: IMI pathologic myopia. Investigative Ophthalmol. Visual Sci. **62**(5), 5–5 (2021)
17. Panwar, N., et al.: Fundus photography in the 21st century-a review of recent technological advances and their implications for worldwide healthcare. Telemedicine and e-Health **22**(3), 198–208 (2016)
18. Park, K., Kim, J., Lee, J.: A deep learning approach to predict visual field using optical coherence tomography. PLoS ONE **15**(7), 1–19 (2020)
19. Phu, J., Khuu, S.K., Yapp, M., Assaad, N., Hennessy, M.P., Kalloniatis, M.: The value of visual field testing in the era of advanced imaging: clinical and psychophysical perspectives. Clin. Exp. Optom. **100**(4), 313–332 (2017)
20. Roy, A.G., et al.: Relaynet: retinal layer and fluid segmentation of macular optical coherence tomography using fully convolutional networks. Biomed. Opt. Express **8**(8), 3627–3642 (2017)
21. Silva, R.: Myopic maculopathy: a review. Ophthalmologica **228**(4), 197–213 (2012)
22. Vivien: Learning through auxiliary tasks. https://vivien000.github.io/blog/journal/learning-though-auxiliary_tasks.html
23. Wang, S., Yan, Z., Zhang, D., Wei, H., Li, Z., Li, R.: Prototype knowledge distillation for medical segmentation with missing modality. In: ICASSP (2023)
24. Wang, S., Zhang, D., Yan, Z., Zhang, J., Li, R.: Feature alignment and uniformity for test time adaptation. In: CVPR (2023)
25. Wong, T.Y., Ferreira, A., Hughes, R., Carter, G., Mitchell, P.: Epidemiology and disease burden of pathologic myopia and myopic choroidal neovascularization: an evidence-based systematic review. Am. J. Ophthalmol. **157**(1), 9-25.e12 (2014)
26. Xiao, O., et al.: Distribution and severity of myopic maculopathy among highly myopic eyes. Investigative Ophthalmol. Visual Sci. **59**(12), 4880–4885 (2018)
27. Xie, S., et al.: Structural abnormalities in the papillary and peripapillary areas and corresponding visual field defects in eyes with pathologic myopia. Investigative Ophthal. Visual Sci. **63**(4), 13–13 (2022)
28. Xu, L., et al.: Predicting the glaucomatous central 10-degree visual field from optical coherence tomography using deep learning and tensor regression. Am. J. Ophthalmol. **218**, 304–313 (2020)

29. Xu, L., Asaoka, R., Kiwaki, T., Murata, H., Fujino, Y., Yamanishi, K.: Pami: a computational module for joint estimation and progression prediction of glaucoma. In: KDD, pp. 3826–3834 (2021)
30. Xu, L., et al.: Improving visual field trend analysis with oct and deeply regularized latent-space linear regression. Ophthalmol. Glaucoma 4(1), 78–88 (2021)
31. Zhang, S., Yang, L., Mi, M.B., Zheng, X., Yao, A.: Improving deep regression with ordinal entropy. In: ICLR (2023)
32. Zhao, X., et al.: Morphological characteristics and visual acuity of highly myopic eyes with different severities of myopic maculopathy. Retina 40(3), 461–467 (2020)
33. Zheng, Y., et al.: Glaucoma progression prediction using retinal thickness via latent space linear regression. In: KDD, pp. 2278–2286 (2019)

Content-Preserving Diffusion Model for Unsupervised AS-OCT Image Despeckling

Sanqian Li[1], Risa Higashita[1,2(✉)], Huazhu Fu[3], Heng Li[1], Jingxuan Niu[1], and Jiang Liu[1(✉)]

[1] Research Institute of Trustworthy Autonomous Systems and Department of Computer Science and Engineering, Southern University of Science and Technology, Shenzhen, China
{risa,liuj}@mail.sustech.edu.cn
[2] Tomey Corporation, Nagoya, Japan
[3] Institute of High-Performance Computing, Agency for Science, Technology and Research, Singapore, Singapore

Abstract. Anterior segment optical coherence tomography (AS-OCT) is a non-invasive imaging technique that is highly valuable for ophthalmic diagnosis. However, speckles in AS-OCT images can often degrade the image quality and affect clinical analysis. As a result, removing speckles in AS-OCT images can greatly benefit automatic ophthalmology analysis. Unfortunately, challenges still exist in deploying effective AS-OCT image denoising algorithms, including collecting sufficient paired training data and the requirement to preserve consistent content in medical images. To address these practical issues, we propose an unsupervised AS-OCT despeckling algorithm via Content Preserving Diffusion Model (CPDM) with statistical knowledge. At the training stage, a Markov chain transforms clean images to white Gaussian noise by repeatedly adding random noise and removes the predicted noise in a reverse procedure. At the inference stage, we first analyze the statistical distribution of speckles and convert it into a Gaussian distribution, aiming to match the fast truncated reverse diffusion process. We then explore the posterior distribution of observed images as a fidelity term to ensure content consistency in the iterative procedure. Our experimental results show that CPDM significantly improves image quality compared to competitive methods. Furthermore, we validate the benefits of CPDM for subsequent clinical analysis, including ciliary muscle (CM) segmentation and scleral spur (SS) localization.

Keywords: ASOCT · Unsupervised despeckling · Diffusion model

Supplementary Information The online version contains supplementary material available at https://doi.org/10.1007/978-3-031-43990-2_62.

1   Introduction

Anterior segment optical coherence tomography (AS-OCT) is a widely used non-invasive imaging modality for ocular disease [1,2]. It produces high-resolution views of superficial anterior segment structures, such as the cornea, iris, and ciliary body. However, speckle noise inherently exists in AS-OCT imaging systems [3], which can introduce uncertainty in clinical observations and increase the risk of misdiagnosis. AS-OCT despeckling has become an urgent pre-processing task that can benefit clinical studies.

To suppress speckle noise in AS-OCT images, commercial scanners [4] generally average repeated scans at the same location. However, this approach can result in artifacts due to uncontrollable movement. As a result, several post-processing denoising approaches have been developed to reduce speckles, such as wavelet-modified block-matching and 3D filters [5], anisotropic non-local means filters [6], and complex wavelets combined with the K-SVD method [7]. However, these algorithms can lead to edge distortion depending on the aggregation of similar patches. Deep learning has recently been employed for medical image processing, especially, with promising performance for image denoising tasks [8–10]. To overcome the limitations caused by the requirement for vast supervised paired data, unsupervised algorithms explore some promising stages to loosen the paired clinical data collection, including cycle consistency loss [11], contrast learning strategies [12], simulated schemes [13], or the Bayesian model [14]. Alternatively, the denoising diffusion probabilistic model (DDPM) can use the averaged image of repeated collections to train the model with excellent performance due to its focus on the noise pattern rather than the signal [15]. Given the prominent pixel-level representational ability for low-level tasks, diffusion models have also been introduced to medical image denoising based on the Gaussian assumption of the noise pattern [16,17].

Although previous studies have achieved outstanding performances, deploying AS-OCT despeckling algorithms remains challenging due to several reasons: (1). Gathering massive paired data for supervised learning is difficult because clinical data acquisition is time-consuming and expensive. (2). Speckle noise in AS-OCT images strongly correlates with the real signal, making the additive Gaussian assumption on the speckle pattern to remove noise impractically. (3). Unsupervised algorithms can easily miss inherent content, and structural content consistency are vital for clinical intervention in AS-OCT [18,19]. (4). Existing algorithms focus on suppressing speckles while ignoring the performance improvement of clinical analysis from despeckling results.

To address these challenges, we propose a Content-Preserving Diffusion Model for AS-OCT despeckling, named **CPDM**, which removes speckle noise in AS-OCT images while preserving the inherent content simultaneously. **Firstly**, we efficiently remove noise via a conditioned noise predictor by truncated diffusion model [16] in the absence of supervised data. We convert the speckle noise into an additive Gaussian pattern by considering the statistical distribution of speckles in AS-OCT to adapt to the reverse diffusion procedure. **Secondly**, we incorporate the posterior probability distribution in observed AS-OCT images

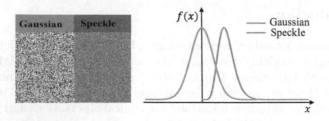

Fig. 1. Distributions of Gaussian and speckle noises.

into an iterative reverse stage to avoid getting trapped in artificial artifacts and preserve consistent content. The posterior distribution is regarded as a data fidelity term to constrain the iterative reverse procedure for despeckling. **Finally**, experiments on the AS-Casia and CM-Casia datasets demonstrate the effectiveness of CPDM compared to state-of-the-art (SOTA) algorithms. Further experiments on ciliary muscle (CM) segmentation and scleral spur (SS) localization verify that the CPDM can benefit clinical analysis.

2   The Statistical Characteristic of Speckles

Speckle noise is inherent in coherent imaging systems [3], as it results from the destructive interference of multiple-scattered waves. As shown in Fig. 1, unlike the additive Gaussian noise $Y_i = x_i + N_i$ $(i = 1, ..., n)$, the multiplicative speckle noise is modeled as $Y_i = x_i N_i$ [20], where Y denotes the noisy image, x is the noise-free image, N is the speckle noise, and i is the pixel index. Moreover, N consists of independent and identically distributed random variables with unit mean, following a gamma probability density function p_N [21]:

$$p_N(n) = \frac{M^M}{\Gamma(M)} n^{M-1} e^{-nM}, \tag{1}$$

where $\Gamma(\cdot)$ is the Gamma function and M is the number of multilook [21]. To transform the multiplicative noise into an additive one, logarithmic transform [22] is employed on both sides of Eq. 1, as: $\underbrace{\log Y}_{G} = \underbrace{\log x}_{z} + \underbrace{\log N}_{W}$. Therefore, the density of the random variable $W = \log N$ is $p_W(w) = p_N(e^w)e^w = \frac{M^M}{\Gamma(M)} e^{Mw} e^{-e^w M}$. According to the central limit theorem and analyzing the statistical distribution of transformed one in [23,24], W approximately follows a Gaussian distribution. Besides, we can obtain the prior distribution:

$$p_{G|z}(g \,|z) = p_W(g - z). \tag{2}$$

3   Content Preserving Diffusion Model

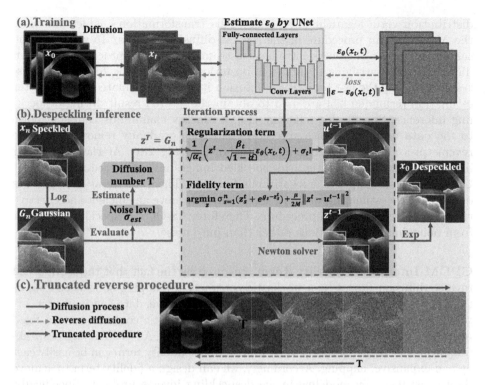

Fig. 2. Illustration of proposed CPDM algorithm. CPDM follows the training network in block(a), and learns the regularization knowledge from the trained network for image despeckling shown in block(b). Moreover, we adopt the truncated strategy shown in block(c) into the despeckling process.

Diffusion Model. The diffusion model can subtly capture the semantic knowledge of the input image and prevails in the pixel-level representation [15]. As shown in Fig. 2(a), it defines a Markov chain that transforms an image x_0 to white Gaussian noise $x_T \sim \mathcal{N}(0,1)$ by adding random noise in T steps. During inference, a random noise x_T is sampled and gradually denoised until it reaches the desired image x_0. To perfectly recover the image in the reverse sampling procedure, a practicable constraint $D_{KL}(q(x_{t-1}|\mathrm{x}_t,x_0)\|p_\theta(x_{t-1}|\mathrm{x}_t))$ was proposed to minimize the distance between $p_\theta(x_{t-1}|\mathrm{x}_t)$ and $q(x_t|\mathrm{x}_{t-1})$ [15]. Thus x_{t-1} can be sampled as follows:

$$\mathrm{x}_{t-1} = \frac{1}{\sqrt{\alpha_t}}(x_t - \frac{\beta_t}{\sqrt{1-\bar{\alpha}_t}}\varepsilon_\theta(x_t,t)) + \sigma_t I, \tag{3}$$

where ε_θ is an approximator intended to predict noise ε from x_t and $I \sim \mathcal{N}(0,1)$.

Truncated Diffusion Model. As mentioned in the previous section, speckle noise follows a gamma distribution and can be transformed into a Gaussian distribution via a logarithmic function. This transformation enables matching the Markov chain procedure in the reverse diffusion process. To speed up the sampling process, this work introduces a truncated reverse procedure that can directly obtain satisfying results from posterior sampling [16]. Figure 2(c) illustrates that only the last few reverse diffusion iterations calculated by parameter estimation technique [25] are used to obtain the desired result during despeckling inference. Specifically, following [15], a Markov chain adds Gaussian noise to the data until it becomes pure noise and then gradually removes it by the reverse procedure at the training stage shown in Fig. 2(a). At the despeckling inference stage shown in Fig. 2(b), speckled images are converted into additive Gaussian ones by applying a logarithmic function. Then, the iteration number is determined by estimating the noise levels [16] to achieve an efficient and effective truncated reverse diffusion procedure. Therefore, AS-OCT despeckling can start from noisy image distributions rather than pure noise.

CPDM Integrated Fidelity Term. Inspired by the fact that the score-based reverse diffusion process is a stochastic contraction mapping so that as long as the data consistency imposing mapping is non-expansive, data consistency incorporated into the reverse diffusion results in a stochastic contraction to a fixed point [26]. This work adopts the theory into the inverse AS-OCT image despeckling problems, as the iteration steps which impose fidelity term can be easily cast as non-expansive mapping. Accordingly, we can design a fidelity term to achieve data consistency by modeling image despeckling inverse problem. Specifically, invoking the conditional independence assumption, the prior distribution with Eq. 2 can be rewritten as:

$$\log p_{G|z}(g\,|z) = \sum_{s=1}^{n} \log pW(g_s - z_s) = C - M\sum_{s=1}^{n}(z_s + e^{g_s - z_s}). \tag{4}$$

The Bayesian maximum a posteriori (MAP) formulation leads to the image despeckling optimization with data fidelity and regularization terms.

$$\arg\min_{z} M\sum_{s=1}^{n}(z_s + e^{g_s - z_s}) + \lambda R(z), \tag{5}$$

where $R()$ is the regularization term, and λ is the regularization parameter. The unconstrained minimization optimization problem can be defined as a constrained formulation by variable splitting method [27]:

$$(\hat{z}, \hat{u}) = \arg\min_{z,u} M\sum_{s=1}^{n}(z_s + e^{g_s - z_s}) + \lambda R(u) \;\; \text{s.t.} \;\; z = u. \tag{6}$$

Motivated by the iterative restoration methods with prior information to tackle various tasks become mainstream, we explore the fidelity term Eq. 4 from the posterior distribution of observed images into the iterative reverse diffusion procedure. The fidelity can guarantee data consistency with original images and avoid falling into artificial artifacts. Moreover, we learn reasonable prior from DDPM reverse recover procedure, which can ensure the flexibility with iterative fidelity term incorporated into the loop of prior generation procedure. As shown in Fig. 2(b), the recovery result obtained from the reverse sampling of DDPM (Eq. 3) can be considered as regularization information of the image despeckling optimization model, and the fidelity term in Eq. 5 can ensure the consistency of the reverse diffusion process with the original image content. Therefore, we can achieve AS-OCT image despeckling by solving Eq. 6 with the ADMM method using variable splitting technique [21]:

$$u^{t-1} = \frac{1}{\sqrt{\alpha_t}}(z^{t+1} - \frac{\beta_t}{\sqrt{1-\bar{\alpha}_t}}\varepsilon_\theta(z^{t+1},t)) + \sigma_t I, \tag{7}$$

$$z^{t-1} \leftarrow \arg\min_z \sum_{s=1}^{n}(z_s^t + e^{g_s - z_s^t}) + \frac{\mu}{2M}\|z^t - u^{t-1}\|^2, \tag{8}$$

where the hyperparameter u control the degree of freedom. It is worth mentioning that Eq. 7 is obtained with the trained CPDM model, and Eq. 8 can be solved by the Newton method [28]. Finally, we design an AS-OCT image despeckling scheme by adopting a fidelity term integrated statistical priors to preserve content in the iterative reverse procedure.

4   Experiment

To evaluate the performance of the proposed CPDM for AS-OCT image despeckling, we conduct the comparative experiment and a ablation study in despeckling three evaluations, including despeckling evaluation, subsequent CM segmentation or SS localization.

Dataset Preparation. A series of unsupervised methods including generative adversarial networks (GAN) and diffusion models aim at learning the noise distribution rather than the signal. Therefore, we collect images by averaging 16 repeated B-scans as noisy-free data collected from AS-OCT, the CASIA2 (Tomey, Japan). This study obeyed the tenets of the Declaration of Helsinki and was approved by the local ethics committee.

AS-Casia dataset contains 432 noisy image and 400 unpaired clean image with the size of 2131×1600, which are views of the AS structure, including lens, cornea, and iris. 400 noisy data and 400 clean images were used for training, and the rest were for testing. The SS location in the noisy image is annotated by ophthalmologists.

CM-Casia dataset consists of 184 noisy images and 184 unpaired clean data with the size of 1065×1465 that show the scope of CM tissue. 160 noisy images and 160 clean data are utilized for training network, with the remaining images reserved for testing. Moreover, ophthalmologists annotated the CM regions on the noisy images.

Table 1. Quantitative evaluation of different methods.

Dataset	AS-Casia				CM-Casia				
Task	Despeckling			Localization	Despeckling			Segmentation	
Method	CNR↑	ENL↑	NIQE↓	ED↓ (um)	CNR↑	ENL↑	NIQE↓	F1↑	IoU↑
Noisy	0.52	5.12	7.05	57.09	−6.66	2.50	11.50	0.579	0.424
WBM3D [5]	1.15	6.74	6.31	56.57	−3.25	3.28	6.80	0.602	0.447
NLM [29]	1.76	22.94	6.54	96.85	−0.54	42.37	7.39	0.657	0.508
ANLM [6]	1.64	10.14	6.63	91.97	−2.18	4.18	6.52	0.627	0.474
WKSVD [7]	1.05	6.70	7.94	79.60	−4.98	5.36	8.33	0.681	0.531
UINT [30]	2.14	6.45	9.04	121.98	−1.60	12.98	9.03	0.641	0.492
CUT [12]	1.94	5.47	6.23	83.05	−4.61	5.99	6.92	0.553	0.404
CycleGAN [11]	1.82	5.13	5.58	65.42	−3.12	11.17	7.44	0.667	0.516
Speckle2void [14]	0.51	5.07	7.06	59.79	−5.47	4.71	7.86	0.665	0.514
DRDM [16]	1.28	21.23	5.86	37.96	−4.98	33.09	9.18	0.670	0.524
ODDM [17]	0.14	31.33	6.11	38.18	−7.06	91.50	9.70	0.330	0.224
LogDM	1.63	21.16	5.27	38.04	−2.08	139.25	7.70	0.679	0.535
CPDM	**2.16**	**143.68**	**4.84**	**37.43**	**−0.53**	**396.35**	**6.42**	**0.703**	**0.561**

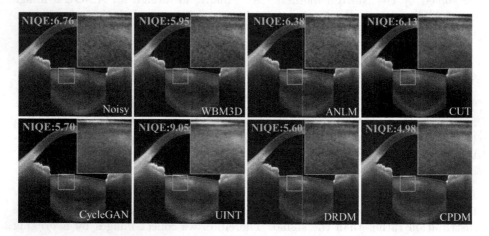

Fig. 3. The visual comparison of image despeckling results (Color figure online)

Implementation Settings. The backbone of our model is a simplified version of that in [15]. The CPDM network was trained on an NVIDIA RTX 2080TI 48GB GPU for 500 epochs, with a batch size of 2, using Adam optimizer. The variance schedule is set to linearly increase from 10^{-4} to 6^{-3} in $T = 1000$ steps and the starting learning rate is 10^{-4} and decay by half every 5 epochs. All training images were resized to 512×512 and normalized to [0,1]. In the reverse procedure, we analyzed the hyperparameters and set $\lambda = \mu/2M = 0.2$, and the iteration number was evaluated as $T = 4$. The parameters of classical blind denoising methods were tuned to reach the best performance, in which the noise-level of NLM, ANLM, WBM3D, and WK-SVD were conducted by [25]. The recent unsupervised methods, UINT, CUT, and CycleGAN are conducted by the unpaired dataset with the default setting. The methods based on diffusion

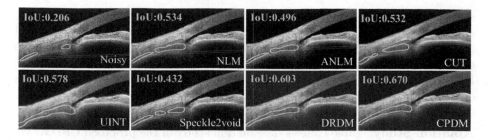

Fig. 4. Comparisons of CM segmented results (Color figure online)

models only are trained on clean data while Speckle2void is implemented only on noisy data with the default setting.

Comparison on AS-Casia Dataset. We first evaluate the despeckling performance by parameterless index, including contrast-to-noise ratio (CNR) [8], the equivalent number of looks (ENL) [8], and natural image quality evaluator (NIQE) [31]. Then we compare the despeckling results with the SOTA methods by using the SS localization task with trained models in [32]. Concretely, we calculate a euclidean distance (ED) value between the reference and the predicted SS position with despeckled images via trained models. As shown in Table 1, the proposed CPDM achieves promising despeckling results in terms of the best CNR, ENL, NIQE values and the minimum ED error in the SS localization task among all approaches. The visual comparison for denoised images with competing approaches is shown in Fig. 3: the green region has been enlarged to highlight the structure of the anterior lens capsule, which can assist in diagnosing congenital cataracts. It can be observed that the CUT and CycleGAN models oversmooths structures close to flat, the UINT method results in ringing effects while the WBM3D, ANLM and DRDM algorithms retain speckles in the lens structure. Obviously, the proposed CPDM acquires satisfactory quality with fine structure details and apparent grain.

Comparison on CM-Casia Dataset. We conduct the experiment of image despeckling and the following CM segmentation task to validate the clinical benefit with CPDM. Specifically, we train a U-Net segmentation model [33] on the CM-Casia dataset and then test the despeckled images of various methods. F1-Score and intersection over union (IoU) index for segmentation were calculated between the despeckled images and reference as reported in Table 1. It can be seen that the proposed CPDM achieves the superior despeckling performance by the highest CNR, ENL, NIQE values and segmentation metrics. Moreover, the segmented CM example of competitive methods is depicted in Fig. 4, in which the CM boundaries reference with the red line, and the yellow line means the segmented results. We can see that NLM, ANLM, CUT, and Speckle2void methods fail to the continuous segmentation results due to insufficient speckle suppression or excessive content loss while the CPDM captures a distinct CM boundary and obtains the highest IoU score. Notably, as a type of smooth muscle, CM has ambiguous boundaries, which are easily affected by speckles, resulting in difficulty distinguishing CM from the adjacent sclera and negative CNR values. Despite these challenges, the proposed CPDM can achieve the best segmentation owing to the speckle reduction while preserving the inconspicuous edge content.

Ablation Study. Table 1 shows the ablation study of the proposed CPDM. We compare our method with two variants: ODDM [17] and logDM. The ODDM only considers removing the speckles by hijacking the reverse diffusion process with the Gaussian assumption on speckles. Based on the ODDM, the logDM further transforms speckles to Gaussian distribution by analyzing the statistical characteristics of speckles. Additionally, the CPDM adopts the data fidelity term to regulate the despeckling reverse process by integrating content consistency. From Table 1, we can see that both the logarithmic function and data fidelity term can improve the quality of despeckled images and benefit the subsequent clinical analysis. Consequently, a prominent unsupervised CPDM to AS-OCT image despeckling is acquired with the proposed strategies.

5   Conclusions

Due to the impact of speckles in AS-OCT images, monitoring and analyzing the anterior segment structure is challenging. To improve the quality of AS-OCT images and overcome the difficulty of supervised data acquisition, we propose a content-preserving diffusion model to achieve unsupervised AS-OCT image despeckling. We first analyze the statistical characteristic of speckles and transform it into Gaussian distribution to match the reverse diffusion procedure. Then the posterior distribution knowledge of AS-OCT image is designed as a fidelity term and incorporated into the iterative despeckling process to guarantee data consistency. Our experiments show that the proposed CPDM can efficiently suppress the speckles and preserve content superior to the competing methods. Furthermore, we validate that the CPDM algorithm can benefit medical image analysis based on subsequent CM segmentation and SS localization task.

Acknowledgments. This work was supported in part by General Program of National Natural Science Foundation of China (Grant No. 82272086), Guangdong Provincial Department of Education (Grant No. 2020ZDZX3043), Shenzhen Natural Science Fund (JCYJ20200109140820699 and the Stable Support Plan Program 20200925174052004), A*STAR Advanced Manufacturing and Engineering (AME) Programmatic Fund (A20H4b0141) and A*STAR Central Research Fund (CRF) "Robust and Trustworthy AI system for Multi-modality Healthcare".

References

1. Radhakrishnan, S., et al.: Real-time optical coherence tomography of the anterior segment at 1310 nm. Arch. Ophthalmol. **119**(8), 1179–1185 (2001)

2. Leung, C.K.S., Weinreb, R.N.: Anterior chamber angle imaging with optical coherence tomography. Eye **25**(3), 261–267 (2012)
3. Schmitt, J.M., Xiang, S.H., Yung, K.M.: Speckle in optical coherence tomography. J. Biomed. Opt. **4**(1), 95–105 (1999)
4. Tajmirriahi, M., Amini, Z., Hamidi, A., Zam, A., Rabbani, H.: Modeling of retinal optical coherence tomography based on stochastic differential equations: application to denoising. IEEE Trans. Med. Imaging **40**(8), 2129–2141 (2021)
5. Chong, B., Zhu, Y.-K.: Speckle reduction in optical coherence tomography images of human finger skin by wavelet modified BM3D filter. Opt. Commun. **291**, 461–469 (2013)
6. Aum, J., Kim, J., Jeong, J.: Effective speckle noise suppression in optical coherence tomography images using nonlocal means denoising filter with double gaussian anisotropic kernels. Appl. Opt. **54**(13), D43–D50 (2015)
7. Kafieh, R., Rabbani, H., Selesnick, I.: Three dimensional data-driven multi scale atomic representation of optical coherence tomography. IEEE Trans. Med. Imaging **34**(5), 1042–1062 (2015)
8. Ma, Y., Chen, X., Zhu, W., Cheng, X., Xiang, D., Shi, F.: Speckle noise reduction in optical coherence tomography images based on edge-sensitive cGAN. Biomed. Opt. Express **9**(11), 5129–5146 (2018)
9. Li, H., et al.: An annotation-free restoration network for cataractous fundus images. IEEE Trans. Med. Imaging **41**(7), 1699–1710 (2022)
10. Li, S., Zhou, J., Liang, D., Liu, Q.: MRI denoising using progressively distribution-based neural network. Magn. Reson. Imaging **71**, 55–68 (2020)
11. Zhu, J.-Y., Park, T., Isola, P., Efros, A.A.: Unpaired image-to-image translation using cycle-consistent adversarial networks. In: Proceedings of the IEEE International Conference on Computer Vision, pp. 2223–2232 (2017)
12. Park, T., Efros, A.A., Zhang, R., Zhu, J.-Y.: Contrastive learning for unpaired image-to-image translation. In: Vedaldi, A., Bischof, H., Brox, T., Frahm, J.-M. (eds.) ECCV 2020. LNCS, vol. 12354, pp. 319–345. Springer, Cham (2020). https://doi.org/10.1007/978-3-030-58545-7_19
13. Göbl, R., Hennersperger, C., Navab, N.: Speckle2speckle: Unsupervised learning of ultrasound speckle filtering without clean data (2022)
14. Molini, A.B., Valsesia, D., Fracastoro, G., Magli, E.: Speckle2void: deep self-supervised sar despeckling with blind-spot convolutional neural networks. IEEE Trans. Geosci. Remote Sens. **60**, 1–17 (2021)
15. Ho, J., Jain, A., Abbeel, P.: Denoising diffusion probabilistic models. Adv. Neural. Inf. Process. Syst. **33**, 6840–6851 (2020)
16. Chung, H., Lee, E.S., Ye, J.C.: MR image denoising and super-resolution using regularized reverse diffusion. IEEE Trans. Med. imaging **42**(4), 922–934 (2023)
17. Hu, D., Tao, Y.K., Oguz, I.: Unsupervised denoising of retinal oct with diffusion probabilistic model. In: Medical Imaging 2022: Image Processing, vol. 12032, pp. 25–34 (2022)
18. Zhang, X., et al.: Attention to region: region-based integration-and-recalibration networks for nuclear cataract classification using as-oct images. Med. Image Anal. **80**, 102499 (2022)
19. Ang, M., et al.: Anterior segment optical coherence tomography. Prog. Retinal Eye Res. **66**, 132–156 (2018)
20. Amini, Z., Rabbani, H.: Statistical modeling of retinal optical coherence tomography. IEEE Trans. Med. Imaging **35**(6), 1544–1554 (2016)

21. Bioucas-Dias, J.M., Figueiredo, M.A.T.: Multiplicative noise removal using variable splitting and constrained optimization. IEEE Trans. Image Process. **19**(7), 1720–1730 (2010)
22. Forouzanfar, M., Moghaddam, H.A.: A directional multiscale approach for speckle reduction in optical coherence tomography images. In: 2007 International Conference on Electrical Engineering, pp. 1–6 (2007)
23. Boyer, K.L., Herzog, A., Roberts, C.: Automatic recovery of the optic nervehead geometry in optical coherence tomography. IEEE Trans. Med. Imaging **25**(5), 553–570 (2006)
24. Dubose, T.B., Cunefare, D., Cole, E., Milanfar, P., Izatt, J.A., Farsiu, S.: Statistical models of signal and noise and fundamental limits of segmentation accuracy in retinal optical coherence tomography. IEEE Trans. Med. Imaging **37**(9), 1978–1988 (2017)
25. Chen, G., Zhu, F., Heng, P.A.: An efficient statistical method for image noise level estimation. In: 2015 IEEE International Conference on Computer Vision (ICCV), pp. 477–485 (2015)
26. Chung, H., Sim, B., Ye, J.C.: Come-closer-diffuse-faster: Accelerating conditional diffusion models for inverse problems through stochastic contraction. In: 2022 IEEE/CVF Conference on Computer Vision and Pattern Recognition (CVPR), pp. 12413–12422 (2023)
27. Hestenes, M.R.: Multiplier and gradient methods. J. Optim. Theor. Appl. **4**(5), 303–320 (1969)
28. van Bree, S.E.H.M., Rokoš, O., Peerlings, R.H.J., Doškář, M., Geers, M.G.D.: A newton solver for micromorphic computational homogenization enabling multiscale buckling analysis of pattern-transforming metamaterials. Comput. Methods Appl. Mech. Eng. **372**, 113333 (2020)
29. Buades, A., Coll, B., Morel, J.-M.: A non-local algorithm for image denoising. In: 2005 IEEE computer society conference on computer vision and pattern recognition (CVPR 2005), vol. 2, pp. 60–65 (2005)
30. Liu, M.-Y., Breuel, T., Kautz, J.: Unsupervised image-to-image translation networks. In: Advances in neural information processing systems, vol. 30 (2017)
31. Mittal, A., Soundararajan, R., Bovik, A.C.: Making a "completely blind" image quality analyzer. IEEE Signal Process. Lett. **20**(3), 209–212 (2012)
32. Liu, Peng, et al.: Reproducibility of deep learning based scleral spur localisation and anterior chamber angle measurements from anterior segment optical coherence tomography images. Br. J. Ophthalmol. **107**(6), 802–808 (2023)
33. Ronneberger, O., Fischer, P., Brox, T.: U-net: convolutional networks for biomedical image segmentation. Pattern Recognit. Image Process. **9351**, 234–241 (2015)

Lesion-Aware Contrastive Learning for Diabetic Retinopathy Diagnosis

Shuai Cheng[1,2], Qingshan Hou[1,2], Peng Cao[1,2,3(✉)], Jinzhu Yang[1,2,3(✉)],
Xiaoli Liu[4], and Osmar R. Zaiane[5]

[1] Computer Science and Engineering, Northeastern University, Shenyang, China
[2] Key Laboratory of Intelligent Computing in Medical Image of Ministry
of Education, Northeastern University, Shenyang, China
caopeng@mail.neu.edu.cn, yangjinzhu@cse.neu.edu.cn
[3] National Frontiers Science Center for Industrial Intelligence and Systems
Optimization, Shenyang 110819, China
[4] DAMO Academy, Alibaba Group, Hangzhou, China
[5] Alberta Machine Intelligence Institute, University of Alberta, Edmonton, Canada

Abstract. Early diagnosis and screening of diabetic retinopathy are
critical in reducing the risk of vision loss in patients. However, in a real
clinical situation, manual annotation of lesion regions in fundus images is
time-consuming. Contrastive learning(CL) has recently shown its strong
ability for self-supervised representation learning due to its ability of
learning the invariant representation without any extra labelled data.
In this study, we aim to investigate how CL can be applied to extract
lesion features in medical images. However, can the direct introduction
of CL into the deep learning framework enhance the representation ability of lesion characteristics? We show that the answer is no. Due to
the lesion-specific regions being insignificant in medical images, directly
introducing CL would inevitably lead to the effects of false negatives,
limiting the ability of the discriminative representation learning. Essen-
tially, two key issues should be considered: (1) How to construct positives and negatives to avoid the problem of false negatives? (2) How to
exploit the hard negatives for promoting the representation quality of
lesions? In this work, we present a lesion-aware CL framework for DR
grading. Specifically, we design a new generating positives and negatives
strategy to overcome the false negatives problem in fundus images. Furthermore, a dynamic hard negatives mining method based on knowledge
distillation is proposed in order to improve the quality of the learned
embeddings. Extensive experimental results show that our method significantly advances state-of-the-art DR grading methods to a considerable 88.0%ACC/86.8% Kappa on the EyePACS benchmark dataset. Our
code is available at https://github.com/IntelliDAL/Image.

S. Cheng and Q. Ho—Contribute equally to this work.

Supplementary Information The online version contains supplementary material
available at https://doi.org/10.1007/978-3-031-43990-2_63.

H. Greenspan et al. (Eds.): MICCAI 2023, LNCS 14226, pp. 671–681, 2023.
https://doi.org/10.1007/978-3-031-43990-2_63

Keywords: Diabetic Retinopathy · Contrastive Learning · Hard Negative Mining · Knowledge Distillation

1   Introduction

Diabetic retinopathy(DR) is a common long-term complication of diabetes that can lead to impaired vision and even blindness as the disease worsen [13,14]. Hence, conducting large-scale screening for early DR is an essential step to prevent visual impairment in patients. Screening fundus images by the ophthalmologist alone is not sufficient to prevent DR on a large scale, and the diagnosis of DR heavily relies on the experience of the ophthalmologist [1]. Therefore, the automatic DR diagnosis on retinal fundus images is urgently needed [3,25]. Recently, in light of the powerful feature extraction and representation capabilities of convolutional neural networks, deep learning technology has developed rapidly in medical image analysis [5,22]. However, leveraging only the image-level grading annotation hinders deep learning algorithms from extracting features of suspicious lesion regions, which further affects the diagnosis of diseases. For these reasons, some previous work [17,19] considers the introduction of pixel-level lesion annotation to improve the model's feature extraction capability for lesion regions. Despite the methods have achieved promising results, the large-scale pixel-level annotation process is time-intensive and error-prone which imposes a heavy burden on the ophthalmologist. To address this problem, contrastive learning(CL) [10,11,23] has received a great deal of attention in medical images, but how to harness the power of CL in the medical applications remains unclear.

The challenges mainly lie in: (I) The diagnosis of fundus diseases relies more on local pathological features (haemorrhages, microaneurysms, etc.) than on the global information. How can contrastive learning enable models to extract features of lesion information more effectively on the large datasets with only image-level annotation? (II) The false negatives tend to disrupt the feature extraction of contrastive learning [26], resulting in the issue of inaccurate alignment of feature distributions [18] (i.e. similar samples have dissimilar features). How to address the issue of false negatives caused by introducing contrastive learning into automatic disease diagnosis? (III) The performance of contrastive learning benefits from the hard negatives [2,16]. How to effectively exploit hard negatives for improving the quality of the learned feature embeddings?

To address the aforementioned issues, we propose the lesion-aware CL framework for DR grading. Specifically, to eliminate false negatives during contrastive learning introduced in automatic disease diagnosis and ensure that samples having similar semantic information stay close in the joint embedding space, we first capture lesion regions in fundus images using a pre-trained lesion detector. Based on the detected regions, we construct a lesion patch set and a healthy patch set, respectively. Then, we develop an encoder and a momentum encoder [6] for extracting the features of positives (lesion patches) and negatives (healthy patches). The introduced momentum encoder enables the contrastive learning to maintain consistency in critical features while creating different perspectives for the positive samples. Secondly, considering the critical role of hard negatives in the contrastive learning, we formulate a two-stage scheme based on knowledge distillation [8,21] to dynamically exploit hard negatives, which further enhances the lesion-aware capability of the diagnosis models, and further improves the quality of the learned feature embeddings. Finally, we fine-tune the proposed framework in the DR grading task to demonstrate its effectiveness.

To the best of our knowledge, this is the first work to rethink the potential issues of contrastive learning for medical image analysis. In summary, our contributions can be summarized as follows. (1) A new scheme of constructing positives and negatives is proposed to prevent false negatives from disrupting the extraction of lesion features. This design can be easily extended to other types of medical images with less prominent physiological features to achieve better lesion representation. (2) To enhance the capability of CL in extracting lesion features for medical fundus image analysis and improve the quality of learned feature embeddings, a lesion-aware CL framework is proposed for sufficiently exploiting hard negatives. (3) We evaluate our framework on the large-scale EyePACS dataset for DR grading. The experimental results indicate the proposed method leads to a performance boost over the state-of-the-art DR grading methods.

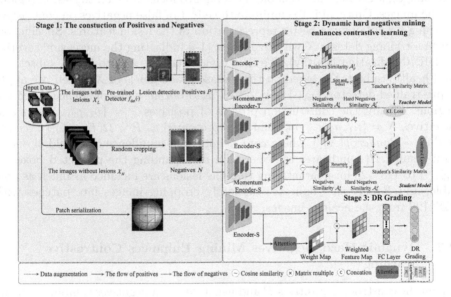

Fig. 1. The overall architecture of the proposed framework. **Stage 1:** The construction of positives and negatives based on the pre-trained lesion detector. **Stage 2:** Dynamic hard negatives mining enhances contrastive learning. **Stage 3:** Fine-tuning our model on the downstream diabetic retinopathy grading task.

2   Methodology

Fig. 1 shows the illustration of the proposed framework. In stage 1, we construct positives and negatives based on a pre-trained lesion detector pre-trained on a auxiliary dataset (IDRiD [15]) with pixel lesion annotation, to avoid the effect of false negatives on the learned feature embeddings while aligning samples with similar semantic features. In stage 2, a dynamically sampling method is developed based on knowledge distillation to effectively exploit hard negatives and improve the quality of the learned feature embeddings. In the last stage, we fine-tune our model on the downstream DR grading task. Remarkably, to bridge the gap between local patches in the pretext task and global images in the downstream task, we introduce an attention mechanism on the fragmented patches to highlight the contributions of different patches on the grading results.

2.1   Construction of Positives and Negatives

In this section, we provide a detailed description regarding the construction of positives and negatives. As opposed to traditional CL working on the whole medical images, it is essential to enable the model to focus more on the lesion regions in the images. Our goal is to eliminate the effect of false negatives on contrastive learning for obtaining a better representation of the lesion features. Specifically, given a training dataset X with five labels (1-4 indicating the increasing severity of DR, 0 indicating healthy). We first divide dataset X into lesion subset X_L and healthy subset X_H based on the disease grade labels of X. Then, we apply a pre-trained detector $f_{det}(\cdot)$ only on X_L and obtain high-confidence detection regions. Finally, the construction process of positives $P = \{p_1, p_2, \ldots, p_j\}$ and negatives $N = \{n_1, n_2, \ldots, n_k\}$ can be represented as $P = \Omega(f_{det}(X_L) > conf)$ and $N = Randcrop(X_H)$, where $conf$ denotes the confidence threshold of detection results, $\Omega(\cdot)$ indicates the operation of expanding the predicted boxes of $f_{det}(\cdot)$ to 128*128 for guaranteeing that the lesions are included as much as possible, and $Randcrop(\cdot)$ indicates randomly cropping images into patches with 128*128 from the healthy images.

2.2   Dynamic Hard Negatives Mining Enhances Contrastive Learning

Given the constructed positives P and negatives N, a negatives sampling scheme based on offline knowledge distillation is developed to enable contrastive learning to dynamically exploit hard negatives, and we adjust the update mechanism of the negatives queue(i.e. only enqueue and dequeue N to avoid confusion with P) to better adapt contrastive learning to the medical image analysis task.

Training the Teacher Network. With the positives P, we obtain two views $\tilde{P} = \{\tilde{p}_1, \tilde{p}_2, \tilde{p}_3 \ldots \tilde{p}_j\}$ and $\tilde{P}' = \{\tilde{p}'_1, \tilde{p}'_2, \tilde{p}'_3 \ldots \tilde{p}'_j\}$ by data augmentation(i.e. color distortion rotation, cropping followed by resize). Correspondingly, with the negatives N, to increase the diversity of the negatives, we apply a similar data augmentation strategy to obtain the augmented negatives $\tilde{N} = \{\tilde{n}_1, \tilde{n}_2, \tilde{n}_3 ... \tilde{n}_k\}$ (where $k \gg j$). We feed $\tilde{P}$ and $\tilde{P}' + \tilde{N}$ to the encoder $En(\cdot)$ and the momentum encoder $MoEn(\cdot)$ to obtain their embeddings $Z = \{z_1, ..., z_j | z_j = En(\tilde{p}_j)\}$, $Z' = \{z'_1, ..., z'_j | z'_j = MoEn(\tilde{p}'_j)\}$ and $\tilde{Z} = \{\tilde{z}_1, ..., \tilde{z}_k | \tilde{z}_k = MoEn(\tilde{n}_k)\}$. Then, we calculate the positive and negative similarity matrix by the samples of Z, Z' and $\tilde{Z}$. According to the similarity matrix, the contrastive loss $L_{\text{cl-t}}$ of the teacher model training process can be defined as:

$$
\begin{aligned}
L_{\text{cl-t}} &= -\sum \log \left(\frac{\exp\left(\mathcal{A}_p^t / \tau\right)}{\exp(\mathcal{A}_p^t / \tau) + \sum \exp\left(\mathcal{A}_n^t / \tau\right)} \right) \\
&= -\sum_j \log \left(\frac{\exp\left(\text{sim}\left(z_j, z'_j\right) / \tau\right)}{\exp(\text{sim}(z_j, z'_j) / \tau) + \sum_k \exp\left(\text{sim}\left(z_j, \tilde{z}_k\right) / \tau\right)} \right),
\end{aligned}
\tag{1}
$$

where $\text{sim}\left(z_j, z'_j / \tilde{z}_k\right) = \frac{\text{dot}\left(z_j, z'_j / \tilde{z}_k\right)}{\|z_j\|_2 \|z'_j / \tilde{z}_k\|_2}$, τ denotes a temperature parameter, $\mathcal{A}_p^t$ and $\mathcal{A}_n^t$ represent the similarity matrix of positives and negatives, respectively. In order to create a positive sample view different from that of $En(\cdot)$, it should be noted that the parameters θ_q of $En(\cdot)$ are updated using gradient descent, while $MoEn(\cdot)$ introduces an extra momentum coefficient $m = 0.99$ to update its parameters $\theta_k \leftarrow m\theta_k + (1 - m)\theta_q$.

Training the Student Network. Previous works [2,16] reveal that not all negatives are useful for the contrastive learning. Moreover, the hard negatives may exhibit more semantically similar to the positives than the normal negatives, indicating that hard negatives provide more potentially useful information for facilitating the following DR grading. Meanwhile, the number of hard negatives significantly affects the difficulty of training the model, in other words, the network should be capable of dynamically adjust the optimisation process by controlling the number of hard negatives. In light of the above two points, we formulate and introduce a well-balanced strategy of hard negatives during the training phase of the student model. Specifically, based on the trained teacher model, we first input P and N into both the teacher and student models to generate similarity matrices A_p^t, A_n^t and A_p^s, A_n^s, respectively. According to the negative similarity matrix A_n^t produced by the teacher model, we prioritise the negatives that are likely to be confused with the positives in descending order and only select the top δ samples for distillation learning during the student model's training phase. For each negative $\tilde{z}_k$ in A_n^t, the resampled negative set $\mathcal{A}_n^{t'}$ can be defined as:

$$
\mathcal{A}_n^{t'} = \left\{\tilde{z}_k \mid \tilde{z}_k \in Sort(\mathcal{A}_n^t), \text{sim}\left(z_j, \tilde{z}_k\right) \geq \text{sim}(z_j, \tilde{z}_\gamma)\right\},
\tag{2}
$$

where $\gamma = \delta / (cos(\frac{\pi s}{2S}) + 1)$ represents the number of the current hard negatives, s and S denotes the current and maximum training step, respectively. As s

increases during the training process, we dynamically adjust the number of hard negatives such that the difficulty of distillation learning proceeds from easy to hard. Based on the index in $A_n^{t'}$, the elements at the corresponding positions in A_n^s are obtained and a resampled negatives similarity matrix $A_n^{s'}$ is constructed. Hence, the CL loss $L_{\text{cl-s}}$ in training process of student can be formulated as:

$$L_{\text{cl-s}} = -\sum \log \left(\frac{\exp\left(A_p^s/\tau\right)}{\exp(A_p^s/\tau) + \sum_k \exp\left(A_n^{s'}/\tau\right)} \right), \tag{3}$$

In addition, to improve the quality of embeddings learned by the student model, we leverage the generated similarity matrices to facilitate the richer knowledge distilled from the teacher to the student. Formally, the KL-divergence loss L_{kd} between $\mathcal{A}_p^t$, $\mathcal{A}_n^{t'}$ and $\mathcal{A}_p^s$, $\mathcal{A}_n^{s'}$ is represented as follows.

$$L_{\text{kd}} = -\tau^2 \sum C(\mathcal{A}_p^t, \mathcal{A}_n^{t'}) \log \left(C(\mathcal{A}_p^s, \mathcal{A}_n^{s'}) \right), \tag{4}$$

where $C(\cdot)$ denotes the matrix concatenation. The final loss of the student model is $L = L_{\text{cl-s}} + \lambda_1 L_{\text{kd}}$, where the λ_1 is a positive parameter controlling the weight of the knowledge distillation loss L_{kd}.

2.3   DR Grading Task

To evaluate the effectiveness of the proposed method, we take the encoder of the pre-trained student model as a backbone and fine-tune it for the downstream DR grading task. Considering that the proposed contrastive learning framework is trained with patches, whereas the downstream grading task relies on entire fundus images, an additional attention mechanism is incorporated to break the gap between the inputs of pretext and downstream tasks. Specifically, we first fragment the entire fundus image into patches $x = \{x_{p1}, \ldots, x_{p_i}\}$. Then, feature embedding v_i of x_{p_i} is generated by the encoder. Meanwhile, an attention module with two linear layers is utilized in the DR grading task to obtain the attention weight α_i of each patch x_{p_i}.

$$\alpha_i = softmax\left(W_2^T max\left(LayerNorm\left(W_1 v_i^T\right), 0\right)\right), \tag{5}$$

where W_1, W_2 are the parameters of the two linear layers , $LayerNorm$ is the layernorm function. Finally, α_i is assigned to the corresponding patch's embedding v_i to highlight the contribution of patch x_{pi}, and the predicted results of DR obtained by $\hat{y} = W_3^T \cdot \sum_{i=1}^N \alpha_i v_i$, and W_3^T is parameter of the grading layer.

3   Experiments

3.1   Datasets and Implementation Details

EyePACS [4]. EyePACS is the largest public fundus dataset which contains 35,126 training images and 53,576 testing images with only image-level DR grading labels. According to the severity of DR, images are classified into five grades: 0 (normal), 1 (mild), 2 (moderate), 3 (severe), and 4 (prolifera-tive).

Implementation Details. The proposed framework is implemented by Pytorch on two Tesla T4 Tensor core GPUs. We employ the IDRiD dataset [15] for the pre-training of the lesion detector $f_{det}(\cdot)$. During the sample construction stage, considering the diversity of sizes of the original fundus images, all images are resized to 768×768, and the data enhancement strategies include random rotation, flipping and color distortion. During the phase of dynamically mining hard negatives, the Adam optimizer with momentum 0.9 is applied to train and update the parameters of the framework with 800 epochs, the initial learning rate of 1×10^{-3} and the batch size of 400. The designed hyper-parameter δ is set to be 4,000 after extra experiments (*please refer to the supplementary materials for more details*). In the downstream DR grading task, we fine-tune the encoder(i.e. ResNet50) for 25 epochs with an initial learning rate of 1×10^{-4} and a batch size of 32. In addition to the normal classification accuracy, we also introduce the quadratic weighted kappa metric to reflect the performance of the proposed method and a range of comparable methods.

3.2   Comparison with the State-of-the-Art

In this section, we provide qualitative and quantitative comparisons with various DR grading methods and demonstrate the effectiveness of the proposed method. As shown in Table 1, we conduct a comprehensive comparison of the proposed method with three types of comparable methods: covering the popular backbone network [7], the top two places of Kaggle challenge [4] and the current SOTA DR grading methods [9, 10, 12, 19, 20, 24].

Table 1. The comparison between our method and the SOTA methods in DR grading task on EyePACS dataset

Model	Kappa	Accuracy	Model	Kappa	Accuracy
Resnet50 [7]	0.823	0.845	AFN(2019) [12]	0.859	–
Min-pooling [4]	0.849	–	DeepMT-DR(2021) [19]	0.839	0.857
o_O [4]	0.844	–	CL-DR(2021) [10]	0.832	0.848
Zoom-in-Net(2017) [20]	0.854	0.873	CLEAQ-DR(2022) [9]	0.863	–
MMCNN(2018) [24]	0.841	0.862	**Lesion-aware CL (Ours)**	**0.868**	**0.880**

From the Table 1, it can be observed that our method consistently achieves the best results with respect to both the Kappa and Accuracy. The results show that our framework presents a notably better DR grading performance than the SOTA methods due to improve quality of the learned lesion embeddings by eliminating the false negatives and dynamically mining hard negatives, and in turn enhancing the lesion-awareness of CL, which is beneficial for DR grading.

Table 2. The ablation experiment results of the proposed framework on EyePACS

Method	Kappa	Accuracy
CL	0.844	0.858
Lesion-aware CL *w/o* CPN	0.853(**+0.9%**)	0.864(**+0.6%**)
Lesion-aware CL *w/o* DHM	0.857 (**+1.3%**)	0.871(**+1.3%**)
Lesion-aware CL	0.868(**+2.4%**)	0.880(**+2.2%**)

3.3   Ablation Study

To more comprehensively evaluate the Lesion-aware CL, we conduct ablation studies to analyze the correlation among DR grading, the construction of positives and negatives(CPN) and dynamic hard negatives mining(DHM). We compare the proposed method with its several variants. (1) CL: the proposed model is trained without CPN and DHM, it indicates a basic CL method. (2) Lesion-aware CL *w/o* CPN: the model is trained without CPN. (3) Lesion-aware CL *w/o* DHM: the model is trained without DHM.

The results of ablation study are reported in Table 2. We can draw conclusions from several aspects: (1) CL shows the worst performance and the performance of Lesion-aware CL *w/o* CPN is obviously degraded compared to Lesion-aware CL (i.e. kappa reduces 1.5%). The results suggest that CPN is critical for improving the performance when contrastive learning is introduced in fundus images. Without false negatives disrupting the feature extraction procedure of lesions, the model is able to extract a better representation for the regions of lesions and thus achieve better DR grading performance. (2) Lesion-aware CL *w/o* DHM performs worse than Lesion-aware CL. As opposed to the common CL methods which uses all negative samples, our model takes into account the difference of the negatives with difficulty level. The teacher network is able to dynamically exploit the hard negatives and transfer the learned knowledge to the student, thereby improving the quality of the feature embeddings in subsequent

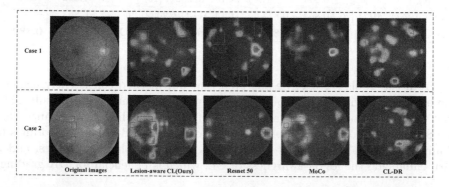

Fig. 2. Visualization results from GradCAM between the four representative methods

contrastive learning. Figure 2 shows the visualization results from GradCAM of four representative methods including Resnet50, the common CL methods (MoCo, CL-DR), and the Lesion-aware CL. Two cases with proliferate DR(DR-4) are visualized by four representative models. The intensity of the heatmap indicates the importance of each pixel in the corresponding image for making the prediction. In case 1, both Resnet and typical CL methods focus on the optic disc where has obvious physiological characteristics, while our method focuses more on the lesion regions and less on the structural aspects of the fundus image. In case 2, our method provides a promising perception of the lesion regions than other methods, suggesting that our approach allows the DR grading model to learn better representation of lesion and thus be sensitive to the DR grading.

4   Conclusion

In this paper, we propose a novel lesion-aware CL framework for DR grading. The proposed method first overcomes the false negatives problem by reconstructing positives and negatives. Then, to improve the quality of learned feature embeddings and enhance the awareness for lesion regions, we design the dynamic hard negatives mining scheme based on knowledge distillation. The experimental results demonstrate that the proposed framework significantly improves the latest results of DR grading on the benchmark dataset. Furthermore, our approaches are migratable and can be easily applied to other medical image analysis tasks.

Acknowledgments. This research was supported by the National Natural Science Foundation of China (No.62076059), the Science Project of Liaoning province under Grant (2021-MS-105) and the 111 Project (B16009).

References

1. Ayhan, M.S., Kühlewein, L., Aliyeva, G., Inhoffen, W., Ziemssen, F., Berens, P.: Expert-validated estimation of diagnostic uncertainty for deep neural networks in diabetic retinopathy detection. Med. Image Anal. **64**, 101724 (2020)
2. Cai, T.T., Frankle, J., Schwab, D.J., Morcos, A.S.: Are all negatives created equal in contrastive instance discrimination? arXiv preprint arXiv:2010.06682 (2020)
3. Cao, P., Hou, Q., Song, R., Wang, H., Zaiane, O.: Collaborative learning of weakly-supervised domain adaptation for diabetic retinopathy grading on retinal images. Comput. Biol. Med. **144**, 105341 (2022)
4. Emma Dugas, Jared, J.W.C.: Diabetic retinopathy detection (2015). https://kaggle.com/competitions/diabetic-retinopathy-detection
5. Fu, H., et al.: Evaluation of retinal image quality assessment networks in different color-spaces. In: Shen, D., et al. (eds.) MICCAI 2019. LNCS, vol. 11764, pp. 48–56. Springer, Cham (2019). https://doi.org/10.1007/978-3-030-32239-7_6
6. He, K., Fan, H., Wu, Y., Xie, S., Girshick, R.: Momentum contrast for unsupervised visual representation learning. In: Proceedings of the IEEE/CVF Conference on Computer Vision and Pattern Recognition, pp. 9729–9738 (2020)

7. He, K., Zhang, X., Ren, S., Sun, J.: Deep residual learning for image recognition. In: Proceedings of the IEEE Conference on Computer Vision and Pattern Recognition, pp. 770–778 (2016)
8. Hinton, G., Vinyals, O., Dean, J.: Distilling the knowledge in a neural network. Comput. Sci. 14(7), 38–39 (2015)
9. Hou, Q., Cao, P., Jia, L., Chen, L., Yang, J., Zaiane, O.R.: Image quality assessment guided collaborative learning of image enhancement and classification for diabetic retinopathy grading. IEEE J. Biomed. Health Inform. 1–12 (2022). https://doi.org/10.1109/JBHI.2022.3231276
10. Huang, Y., Lin, L., Cheng, P., Lyu, J., Tang, X.: Lesion-based contrastive learning for diabetic retinopathy grading from fundus images. In: de Bruijne, M., et al. (eds.) MICCAI 2021. LNCS, vol. 12902, pp. 113–123. Springer, Cham (2021). https://doi.org/10.1007/978-3-030-87196-3_11
11. Li, X., Jia, M., Islam, M.T., Yu, L., Xing, L.: Self-supervised feature learning via exploiting multi-modal data for retinal disease diagnosis. IEEE Trans. Med. Imaging 39(12), 4023–4033 (2020)
12. Lin, Z., et al.: A framework for identifying diabetic retinopathy based on anti-noise detection and attention-based fusion. In: Frangi, A.F., Schnabel, J.A., Davatzikos, C., Alberola-López, C., Fichtinger, G. (eds.) MICCAI 2018. LNCS, vol. 11071, pp. 74–82. Springer, Cham (2018). https://doi.org/10.1007/978-3-030-00934-2_9
13. Liu, X., et al.: A comparison of deep learning performance against health-care professionals in detecting diseases from medical imaging: a systematic review and meta-analysis. Lancet Digit. Health 1(6), e271–e297 (2019)
14. Ogurtsova, K., et al.: IDF diabetes atlas: global estimates of undiagnosed diabetes in adults for 2021. Diabetes Res. Clin. Pract. 183, 109118 (2022)
15. Porwal, P., Pachade, S., Kamble, R., Kokare, M., Meriaudeau, F.: Indian diabetic retinopathy image dataset (idrid): a database for diabetic retinopathy screening research. Data 3(3), 25 (2018)
16. Robinson, J., Chuang, C.Y., Sra, S., Jegelka, S.: Contrastive learning with hard negative samples. In: International Conference on Learning Representations (ICLR) (2021)
17. Shen, Z., Fu, H., Shen, J., Shao, L.: Modeling and enhancing low-quality retinal fundus images. IEEE Trans. Med. Imaging 40(3), 996–1006 (2020)
18. Tian, Y., Sun, C., Poole, B., Krishnan, D., Schmid, C., Isola, P.: What makes for good views for contrastive learning? Adv. Neural. Inf. Process. Syst. 33, 6827–6839 (2020)
19. Wang, X., Xu, M., Zhang, J., Jiang, L., Li, L.: Deep multi-task learning for diabetic retinopathy grading in fundus images. In: Proceedings of the AAAI Conference on Artificial Intelligence, vol. 35, pp. 2826–2834 (2021)
20. Wang, Z., Yin, Y., Shi, J., Fang, W., Li, H., Wang, X.: Zoom-in-Net: deep mining lesions for diabetic retinopathy detection. In: Descoteaux, M., Maier-Hein, L., Franz, A., Jannin, P., Collins, D.L., Duchesne, S. (eds.) MICCAI 2017. LNCS, vol. 10435, pp. 267–275. Springer, Cham (2017). https://doi.org/10.1007/978-3-319-66179-7_31
21. Xu, G., Liu, Z., Li, X., Loy, C.C.: Knowledge distillation meets self-supervision. In: Vedaldi, A., Bischof, H., Brox, T., Frahm, J.-M. (eds.) ECCV 2020. LNCS, vol. 12354, pp. 588–604. Springer, Cham (2020). https://doi.org/10.1007/978-3-030-58545-7_34

22. Yu, S., et al.: MIL-VT: multiple instance learning enhanced vision transformer for fundus image classification. In: de Bruijne, M., et al. (eds.) MICCAI 2021. LNCS, vol. 12908, pp. 45–54. Springer, Cham (2021). https://doi.org/10.1007/978-3-030-87237-3_5

23. Zeng, X., Chen, H., Luo, Y., Ye, W.: Automated detection of diabetic retinopathy using a binocular Siamese-like convolutional network. In: 2019 IEEE International Symposium on Circuits and Systems (ISCAS), pp. 1–5. IEEE (2019)

24. Zhou, K., et al.: Multi-cell multi-task convolutional neural networks for diabetic retinopathy grading. In: 2018 40th Annual International Conference of the IEEE Engineering in Medicine and Biology Society (EMBC), pp. 2724–2727. IEEE (2018)

25. Zhou, Y., Wang, B., Huang, L., Cui, S., Shao, L.: A benchmark for studying diabetic retinopathy: segmentation, grading, and transferability. IEEE Trans. Med. Imaging **40**(3), 818–828 (2020)

26. Zolfaghari, M., Zhu, Y., Gehler, P., Brox, T.: CrossCLR: cross-modal contrastive learning for multi-modal video representations. In: Proceedings of the IEEE/CVF International Conference on Computer Vision, pp. 1450–1459 (2021)

Efficient Spatiotemporal Learning of Microscopic Video for Augmented Reality-Guided Phacoemulsification Cataract Surgery

Puxun Tu[1], Hongfei Ye[2], Jeff Young[3], Meng Xie[2], Ce Zheng[2(✉)], and Xiaojun Chen[1,4(✉)]

[1] Institute of Biomedical Manufacturing and Life Quality Engineering, School of Mechanical Engineering, Shanghai Jiao Tong University, Shanghai, China
xiaojunchen@sjtu.edu.cn
[2] Department of Ophthalmology, Xinhua Hospital Affiliated to Shanghai Jiao Tong University School of Medicine, Shanghai, China
zhengce@xinhuamed.com.cn
[3] Department of Bioengineering, University of Texas at Dallas, Richardson, USA
[4] Institute of Medical Robotics, Shanghai Jiao Tong University, Shanghai, China

Abstract. Phacoemulsification cataract surgery (PCS) is typically performed under a surgical microscope and adhering to standard procedures. The success of this surgery depends heavily on the seniority and experience of the ophthalmologist performing it. In this study, we developed an augmented reality (AR) guidance system to enhance the intraoperative skills of ophthalmologists by proposing a two-stage spatiotemporal learning network for surgical microscope video recognition. In the first stage, we designed a multi-task network that recognizes surgical phases and segments the limbus region to extract limbus-focused spatial features. In the second stage, we developed a temporal pyramid-based spatiotemporal feature aggregation (TP-SFA) module that uses causal and dilated temporal convolution for smooth and online surgical phase recognition. To provide phase-specific AR guidance, we designed several intraoperative visual cues based on the parameters of the fitted limbus ellipse and the recognized surgical phase. The comparison experiments results indicate that our method outperforms several strong baselines in surgical phase recognition. Furthermore, ablation experiments show the positive effects of the multi-task feature extractor and TP-SFA module. Our developed system has the potential for clinical application in PCS to provide real-time intraoperative AR guidance.

Keywords: Cataract surgery · Augmented reality · Surgical phase recognition · Spatiotemporal learning

Supplementary Information The online version contains supplementary material available at https://doi.org/10.1007/978-3-031-43990-2_64.

1   Introduction

Cataract is the leading cause of blindness worldwide, and cataract surgery is one of the most common operations in health care. Among different cataract surgery techniques, phacoemulsification cataract surgery (PCS) is the standard of care [5,23]. PCS consists of the manual opening of the crystalline lens anterior capsule with forceps, removal of the opacified lens, and implantation of an intraocular lens (IOL) in the remaining capsular envelope to restore visual function.

Phacoemulsification needs microsurgical skills, which depend on numerous variables, including the amount of practice, inherent manual dexterity, and previous experience. Statistical analyses have demonstrated significant differences in completion and complication rates among ophthalmologists, with variations observed based on factors such as seniority and experience [14]. During phacoemulsification, a surgical microscope with an integrated video camera is routinely used, providing rich spatiotemporal information [15]. This information presents an excellent opportunity to develop surgical video recognition methods to extract valuable intraoperative information and overlay it on a 2D/3D screen or the microscopic eyepiece, thereby creating an augmented reality (AR) scene.

To bridge the experience gap among ophthalmologists, several intraoperative AR-guided systems have been developed. Zhai et al. [25] used two convolutional neural networks (CNNs) to segment the limbus and track the eye's rotation, and subsequently developed an intraoperative guide system for positioning and aligning the IOL. In [16], a multi-task CNN was designed to locate the pupil and recognize the surgical phase in each frame of the surgical microscope video. Nespolo et al. [17] utilized a deep CNN-based method for processing surgical videos, allowing for detecting surgical instruments and tissue boundaries to guide the ophthalmic surgery. Despite the potential of AR-guided phacoemulsification systems, limitations still hinder their clinical implementation. Firstly, the current systems process surgical videos in a frame-wise manner, enabling real-time processing but leading to lost temporal information and incoherent surgical scene recognition. Secondly, the overlaid information during surgery is not categorized by surgical phase, causing visual redundancy for ophthalmologists.

Advancements in video spatiotemporal learning, particularly in surgical phase recognition, present a promising opportunity to switch AR scenes to the current surgical phase automatically. Early attempts used a 2D CNN to extract spatial features to predict each video frame's surgical phase [18,22]. However, the lack of temporal information leads to unsatisfactory recognition accuracy. Other studies [11,24] use spatial feature maps of neighboring frames, extracted from CNNs, as an input to Gated Recurrent Unit (GRU) [2] or Long Short-Term Memory (LSTM) [10] to model the temporal dependencies and predict the surgical phase. However, these methods suffer from limited temporal reception field and non-parallel, slow inference. Recent studies focus on modeling long-range temporal relationships. Czempiel et al. [3] introduced a multi-stage temporal convolutional network (TCN) for surgical phase recognition, leveraging causal and dilated convolutions to enable global reception field and online deployment. The current state-of-the-art methods utilize transformer-based models for aggre-

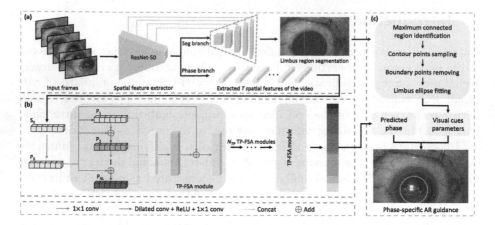

Fig. 1. Overview of our AR-guided system for PCS. (a) Limbus region-focused spatial feature extraction. (b) TP-SFA module-based spatiotemporal aggregation. (c) Phase-specific intraoperative AR guidance.

gating spatial [4,27] and spatiotemporal features [8,12]. However, focusing solely on global features may result in losing important local temporal dependencies and lead to inaccurate recognition of challenging frames.

In this study, we developed a novel intraoperative AR-guide system for PCS. Our contributions are two-fold: (1) We propose an efficient spatiotemporal network for surgical microscope video recognition, consisting of two stages: a multi-task learning stage for limbus segmentation and spatial feature extraction, and a temporal pyramid-based spatiotemporal feature aggregation (TP-SFA) module for online surgical phase recognition. (2) We use the limbus and surgical phase information to design phase-specific visual cues that offer real-time intraoperative AR guidance while avoiding distracting the ophthalmologist's attention.

2   Methods

Figure 1 presents an overview of our developed AR-guided system for PCS, which acquires intraoperative video streams from microscope and processes them using the proposed two-stage spatiotemporal network to obtain limbus and surgical phase information. Parameters of intraoperative visual cues are computed using the fitted limbus elliptic parameters and updated according to the recognized surgical phase, providing automatic AR scene switching for ophthalmologists.

2.1   Spatiotemporal Network for Microscopic Video Recognition

Limbus Segmentation and Spatial Feature Extraction. We observe that the region within the limbus displays distinguishable appearances at different phases in surgical microscope videos, whereas other regions like the sclera exhibit similar appearances. We argue that using a limbus region-focused spatial feature extraction network can improve spatiotemporal aggregation. This led us to develop a multi-task network for limbus segmentation and phase recognition in the first stage. As shown in Fig. 1(a), we employ ResNet-50 [9] as the shared backbone to extract spatial features, which are then fed into both the surgical phase recognition and limbus segmentation branches.

The surgical phase recognition branch involves a fully connected layer that is directly connected to the global average pooling layer, followed by a softmax layer. To train this branch, we use cross-entropy loss, which is defined as

$$L_{phase} = -(\sum_{s=1}^{N_s} g_s \log p_s)/N_s, \qquad (1)$$

where g_s is the ground truth binary indicator of phase s, p_s is the probability of the input frame belonging to phase s.

The limbus segmentation branch incorporates a decoder with upsampling and concatenation, resembling the U-net [20] architecture. To train this branch, we employ a hybrid loss of cross-entropy and dice, which is defined as

$$L_{seg} = -\frac{1}{N \times C} \sum_{c=1}^{C} \sum_{i=1}^{N} y_i^c \log p_i^c + \alpha(1 - \frac{2 \sum_{c=1}^{C} \sum_{i=1}^{N} y_i^c p_i^c}{\sum_{c=1}^{C} \sum_{i=1}^{N} y_i^c + \sum_{c=1}^{C} \sum_{i=1}^{N} p_i^c}), \qquad (2)$$

where y_i^c and p_i^c are the pixel-level ground truth and prediction result respectively, α is a hyper-parameter to balance the loss. The final loss function for training the first stage is defined as

$$L_{spatial} = L_{phase} + \beta L_{seg}, \qquad (3)$$

where β is a hyper-parameter to balance the loss. After training the first stage, we obtain the spatial feature $s_t \in \mathbb{R}^{2048}$ for frame t by outputting the average pooling layer of the spatial feature extractor.

Spatiotemporal Features Aggregation. We argue that the surgical phase recognition method used for intraoperative AR should fulfill the following requirements: 1) online recognition for real-time intraoperative guidance, and 2) sufficient stability to avoid distracting ophthalmologists with incorrect phase recognition. As shown in Fig. 1(b), we employ our proposed TP-SFA module in the second stage, which uses multi-scale, causal, and dilated temporal convolutions to model the temporal relationships. We denote the input spatial feature as $S_0 \in \mathbb{R}^{2048 \times T}$, where T is the sequence length of the video. The first layer of the TP-SFA module is a 1×1 convolutional layer that reduces the dimension of S_0 and outputs $P_0 \in \mathbb{R}^{32 \times T}$. To obtain temporal features with different reception fields, we apply N_L dilated layers with different dilation factors on P_0. Layer k consists of a dilated convolution with a dilation factor of 2^k, followed by a ReLU activation and a 1×1 convolution. This can be described as

$$P_k = W_2 * \text{ReLU}(W_1 * (P_{k-1} + P_0) + b_1) + b_2, \qquad (4)$$

Table 1. AR scene at different phases with different combinations of visual cues for PCS. The color of visual cues in both the text and AR scene figure is consistent.

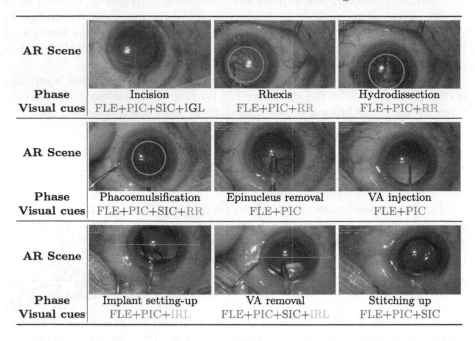

AR Scene			
Phase	Incision	Rhexis	Hydrodissection
Visual cues	FLE+PIC+SIC+IGL	FLE+PIC+RR	FLE+PIC+RR
Phase	Phacoemulsification	Epinucleus removal	VA injection
Visual cues	FLE+PIC+SIC+RR	FLE+PIC	FLE+PIC
Phase	Implant setting-up	VA removal	Stitching up
Visual cues	FLE+PIC+IRL	FLE+PIC+SIC+IRL	FLE+PIC+SIC

where $*$ denotes the convolutional operator, W_1 is the dilated convolution weights, W_2 is the weights of the 1×1 convolution, and b_1 and b_2 are bias vectors. Finally, we concatenate all $P_k(k = 1, \cdots, N_L)$ together over the temporal dimension, followed by a 1×1 convolution, a residual connection with P_0 and another 1×1 convolution to adjust the output dimension. This can be described as

$$S_1 = W_4 * (W_3 * concat(P_1, \ldots, P_{N_L}) + b_3 + P_0) + b_4, \tag{5}$$

where W_3 and W_4 are the weights of the 1×1 convolution, and b_3 and b_4 are bias vectors. Inspired by [3,7], we connect N_{tp} TP-SFA modules together and use a weighted cross-entropy loss to train the second stage. This can be described as

$$L_{temp} = -\frac{1}{N_{tp} \times N_s} \sum_{n=1}^{N_{tp}} \sum_{s=1}^{N_s} w_s g_s^n \log p_s^n, \tag{6}$$

where w_s is the weight and is inversely proportional to the surgical phase frequencies [6].

2.2   Phase-Specific AR Guidance in PCS

The proposed spatiotemporal learning network enables real-time limbus segmentation and surgical phase, facilitating the development of our intraoperative AR guidance system for PCS. The limbus contour can be fitted as an ellipse [25,26]. To accomplish this, we follow the steps shown in Fig. 1(c), including: 1) identifying the maximum connected region to remove possible mis-segmented regions; 2) extracting the contour of the maximum connected region and sampling the contour points; 3) removing contour points near the video boundaries; 4) fitting the remaining contour points to an ellipse and outputting the length and rotation of the long and short axes of the ellipse.

We segment PCS into nine phases [19]: incision, rhexis, hydrodissection, phacoemulsification, epinucleus removal, viscous agent (VA) injection, implant setting-up, VA removal, and stitching up. For intraoperative AR guidance, we designed six visual cues, including: 1) fitted limbus ellipse (FLE), extracted from the segmentation results; 2) primary incision curve (PIC), defined as an arc with a length equal to the maximum diameter of the primary incision knife; 3) secondary incision curve (SIC), defined as an arc with a length equal to the maximum diameter of the secondary incision knife; 4) incision guide lines (IGL), with included angles of 95^{circ} for BIC and 173^{circ} for SIC, respectively, relative to the reference line; 5) rhexis region (RR), with a diameter equal to half the length of the long axis of the fitted ellipse; and 6) implant reference line (IRL), defined by a horizontal line. Table 1 lists different combinations of intraoperative visual cues that are automatically updated according to the recognized surgical phase.

3   Experiments and Results

3.1   Dataset and Implementation Details

Dataset. We evaluate our methods on CATARACTS [1], a publicly available dataset for cataract surgery. It contains 50 videos with a frame rate of 30 frames-per-second (fps) and a total duration of over nine hours. All videos have been subsampled to 1 fps. Each frame has a resolution of 1920×1080 pixels and has been annotated in nineteen surgical steps. For the sake of intraoperative guidance convenience, we have reorganized the nineteen fine-grained surgical steps into nine standard phases [19]. Additionally, the limbus region of each frame has been manually delineated by two non-M.D. experts. The dataset is split into 25 cases for training, 5 cases for validation, and 20 cases for test, following [1,12].

Implementation Details. Our network was implemented in PyTorch using two NVIDIA GeForce GTX 3090 GPUs. We initialized the ResNet-50 backbone with weights trained on the ImageNet [21] and implemented random horizontal flip, random crop, random rotation ($\pm 20^{circ}$), and color jitter for data augmentation. The first stage was trained for 100 epochs using Adam optimizer with a learning rate of 5e-5 for the backbone and 5e-4 for the fully connected layer and decoder. The second stage was trained for 50 epochs using Adam optimizer with a learning rate of 1e-4. For hyper-parameters, we set $\alpha = 0.6$, $\beta = 0.5$ and $N_L = 8$.

Table 2. Quantitative comparison results with strong baselines in online surgical phase recognition. Each metric is reported as mean (%) and standard deviation (±).

Methods	Accuracy	Precision	Recall	Jaccard
ResNet-50 [9]	81.1 ± 6.4	78.4 ± 8.0	77.9 ± 7.1	63.7 ± 9.4
SV-RCNet [11]	84.8 ± 6.2	81.1 ± 7.3	81.8 ± 6.8	69.1 ± 8.5
TeCNO [3]	86.1 ± 5.5	81.6 ± 6.7	83.5 ± 6.3	70.8 ± 7.1
Trans-SVNet [8]	86.8 ± 5.9	82.7 ± 7.1	83.6 ± 6.5	71.6 ± 7.0
Ours	**87.9 ± 5.4**	**83.7 ± 6.6**	**84.5 ± 5.8**	**73.3 ± 6.3**

Table 3. Phase recognition and segmentation results of multi-task and single task.

Method	Accuracy		Dice
	First stage	Second stage	
Single phase task	81.1 ± 6.4	85.6 ± 5.9	—
Single segmentation task	—	—	94.0 ± 2.3
Phase+Segmentation	**82.6 ± 5.8**	**87.9 ± 5.4**	**94.6 ± 2.7**

3.2 Comparisons with Strong Baselines

We compare our method with several strong baselines in surgical phase recognition: 1) ResNet-50 [9], a deep residual network that predicts surgical phase in a frame-wise manner; 2) SV-RCNet [11], an end-to-end architecture that utilizes LSTM to learn temporal features; 3) TeCNO [3], a multi-stage temporal convolutional network that models global temporal features; and 4) Trans-SVNet [8], a transformer-based spatiotemporal features aggregation network. Note that we only included comparison methods that support online surgical phase recognition and excluded those that do not. For fair comparison, we use the proposed multi-task learning network in the first stage as the spatial feature extraction backbone for all methods except ResNet-50 [9]. The quantitative comparison results are listed in Table 2, which indicates that our method achieved the best performance among all compared methods. We show the quantitative results with color-coded ribbons in Fig. 2(a). The results indicate that our method can produce a smoother phase prediction compared to ResNet-50 [9] and SV-RCNet [11]. Additionally, our approach surpasses the performance of global feature aggregation-based methods [3,8] in challenging local frames.

3.3 Ablation Study

Effect of Multi-task Feature Extractor We performed experiments to evaluate the effect of the multi-task feature extractor. ResNet-50 [9] served as the backbone for all methods. We compared the accuracy of the first stage and the second stage for the single phase recognition task with that of the multi-task

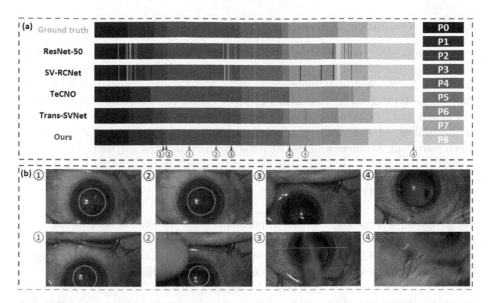

Fig. 2. (a) The comparison results with color-coded ribbons. (b) Some typical failed AR scenes. Markers in red and green represent failed scenes caused by mis-recognition of the surgical phase and mis-segmentation of the limbus, respectively. (Color figure online)

Table 4. The effect of different number of TP-SFA modules.

N_{TP}	Accuracy
1	85.7 ± 6.9
2	**87.9 ± 5.4**
3	87.6 ± 5.8
4	87.3 ± 5.2

Table 5. The effect of different combinations of TP-SFA and causal TCN.

First module		Second module		Accuracy
TP-SFA	TCN	TP-SFA	TCN	
✓		✓		**87.9 ± 5.4**
	✓		✓	86.0 ± 5.7
✓			✓	87.2 ± 5.2
	✓	✓		86.4 ± 6.3

approach. Moreover, we compared the segmentation Dice score of the single segmentation task with that of the multi-task approach. Table 3 shows the results, indicating that the multi-task feature extractor enhances both segmentation and phase recognition performance compared to the single-task approach.

Effect of the TP-SFA Module We evaluate the number of the connected TP-SFA modules. The quantitative results are listed in Table 4, which indicates that the second stage achieves the best accuracy when two TP-SFA modules are used. Furthermore, we explore different combinations of the TP-SFA module and a typical causal TCN module [13] and present the results in Table 5. Results show that two connected TP-SFA modules achieve the best performance.

690   P. Tu et al.

3.4 AR Guidance Evaluation

Our method achieves real-time intraoperative processing at a speed of 36 fps. This makes it suitable for meeting the demands of online intraoperative AR guidance, as the acquired microscope video stream has a speed of 30 fps. We show some typical failed scenes in Fig. 2(b). Failed intraoperative AR guidance can result from both mis-recognition of surgical phases and mis-segmentation of the limbus. Mis-recognition of surgical phase may introduce continuous AR scene switching problem and distract the ophthalmologist's attention.

4 Conclusion

We proposed a two-stage spatiotemporal network for online microscope video recognition. Furthermore, we developed a phase-specific intraoperative AR guidance system for PCS. Our developed system has the potential for clinical applications to enhance ophthalmologists' intraoperative skills.

Acknowledgements. This work was supported by grants from the National Natural Science Foundation of China (81971709; M-0019; 82011530141), the Foundation of Science and Technology Commission of Shanghai Municipality (20490740700; 22Y11911700), Shanghai Jiao Tong University Foundation on Medical and Technological Joint Science Research (YG2021ZD21; YG2021QN72; YG2022QN056; YG2023ZD19; YG2023ZD15), Hospital Funded Clinical Research, Xinhua Hospital Affiliated to Shanghai Jiao Tong University School of Medicine (21XJMR02), and the Funding of Xiamen Science and Technology Bureau (No. 3502Z20221012).

References

1. Al Hajj, H., et al.: CATARACTS: challenge on automatic tool annotation for cataract surgery. Med. Image Anal. **52**, 24–41 (2019)
2. Chung, J., Gulcehre, C., Cho, K., Bengio, Y.: Empirical evaluation of gated recurrent neural networks on sequence modeling. arXiv preprint arXiv:1412.3555 (2014)
3. Czempiel, T., et al.: TeCNO: surgical phase recognition with multi-stage temporal convolutional networks. In: Martel, A.L., et al. (eds.) MICCAI 2020. LNCS, vol. 12263, pp. 343–352. Springer, Cham (2020). https://doi.org/10.1007/978-3-030-59716-0_33
4. Czempiel, T., Paschali, M., Ostler, D., Kim, S.T., Busam, B., Navab, N.: OperA: attention-regularized transformers for surgical phase recognition. In: de Bruijne, M., et al. (eds.) MICCAI 2021. LNCS, vol. 12904, pp. 604–614. Springer, Cham (2021). https://doi.org/10.1007/978-3-030-87202-1_58
5. Day, A.C., Gore, D.M., Bunce, C., Evans, J.R.: Laser-assisted cataract surgery versus standard ultrasound phacoemulsification cataract surgery. Cochrane Database of Systematic Reviews (7) (2016)
6. Eigen, D., Fergus, R.: Predicting depth, surface normals and semantic labels with a common multi-scale convolutional architecture. In: Proceedings of the IEEE International Conference on Computer Vision, pp. 2650–2658 (2015)

7. Farha, Y.A., Gall, J.: MS-TCN: multi-stage temporal convolutional network for action segmentation. In: Proceedings of the IEEE/CVF Conference on Computer Vision and Pattern Recognition, pp. 3575–3584 (2019)

8. Gao, X., Jin, Y., Long, Y., Dou, Q., Heng, P.-A.: Trans-SVNet: accurate phase recognition from surgical videos via hybrid embedding aggregation transformer. In: de Bruijne, M., et al. (eds.) MICCAI 2021. LNCS, vol. 12904, pp. 593–603. Springer, Cham (2021). https://doi.org/10.1007/978-3-030-87202-1_57

9. He, K., Zhang, X., Ren, S., Sun, J.: Deep residual learning for image recognition. In: Proceedings of the IEEE Conference on Computer Vision and Pattern Recognition, pp. 770–778 (2016)

10. Hochreiter, S., Schmidhuber, J.: Long short-term memory. Neural Comput. $9(8)$, 1735–1780 (1997)

11. Jin, Y., et al.: SV-RCNet: workflow recognition from surgical videos using recurrent convolutional network. IEEE Trans. Med. Imaging $37(5)$, 1114–1126 (2017)

12. Jin, Y., Long, Y., Gao, X., Stoyanov, D., Dou, Q., Heng, P.A.: Trans-SVNet: hybrid embedding aggregation transformer for surgical workflow analysis. Int. J. Comput. Assist. Radiol. Surg. $17(12)$, 2193–2202 (2022)

13. Lea, C., Vidal, R., Reiter, A., Hager, G.D.: Temporal convolutional networks: a unified approach to action segmentation. In: Hua, G., Jégou, H. (eds.) ECCV 2016. LNCS, vol. 9915, pp. 47–54. Springer, Cham (2016). https://doi.org/10.1007/978-3-319-49409-8_7

14. Lee, J.S., Hou, C.H., Lin, K.K.: Surgical results of phacoemulsification performed by residents: a time-trend analysis in a teaching hospital from 2005 to 2021. J. Ophthalmol. **2022** (2022)

15. Ma, L., Fei, B.: Comprehensive review of surgical microscopes: technology development and medical applications. J. Biomed. Opt. $26(1)$, 010901–010901 (2021)

16. Nespolo, R.G., Yi, D., Cole, E., Valikodath, N., Luciano, C., Leiderman, Y.I.: Evaluation of artificial intelligence-based intraoperative guidance tools for phacoemulsification cataract surgery. JAMA Ophthalmol. **140**(2), 170–177 (2022)

17. Nespolo, R.G., Yi, D., Cole, E., Wang, D., Warren, A., Leiderman, Y.I.: Feature tracking and segmentation in real time via deep learning in vitreoretinal surgery-a platform for artificial intelligence-mediated surgical guidance. Ophthalmol. Retina **7**(3), 236–242 (2022)

18. Primus, M.J.: Frame-based classification of operation phases in cataract surgery videos. In: Schoeffmann, K., et al. (eds.) MMM 2018. LNCS, vol. 10704, pp. 241–253. Springer, Cham (2018). https://doi.org/10.1007/978-3-319-73603-7_20

19. Quellec, G., Lamard, M., Cochener, B., Cazuguel, G.: Real-time task recognition in cataract surgery videos using adaptive spatiotemporal polynomials. IEEE Trans. Med. Imaging **34**(4), 877–887 (2014)

20. Ronneberger, O., Fischer, P., Brox, T.: U-Net: convolutional networks for biomedical image segmentation. In: Navab, N., Hornegger, J., Wells, W.M., Frangi, A.F. (eds.) MICCAI 2015. LNCS, vol. 9351, pp. 234–241. Springer, Cham (2015). https://doi.org/10.1007/978-3-319-24574-4_28

21. Russakovsky, O.: ImageNet large scale visual recognition challenge. Int. J. Comput. Vis. **115**, 211–252 (2015)

22. Twinanda, A.P., Shehata, S., Mutter, D., Marescaux, J., De Mathelin, M., Padoy, N.: EndoNet: a deep architecture for recognition tasks on laparoscopic videos. IEEE Trans. Med. Imaging **36**(1), 86–97 (2016)

23. Wang, W., et al.: Cataract surgical rate and socioeconomics: a global study. Invest. Ophthalmol. Vis. Sci. **57**(14), 5872–5881 (2016)

24. Yi, F., Yang, Y., Jiang, T.: Not end-to-end: explore multi-stage architecture for online surgical phase recognition. In: Proceedings of the Asian Conference on Computer Vision, pp. 2613–2628 (2022)
25. Zhai, Y., et al.: Computer-aided intraoperative toric intraocular lens positioning and alignment during cataract surgery. IEEE J. Biomed. Health Inform. **25**(10), 3921–3932 (2021)
26. Zhao, W., Zhang, Z., Wang, Z., Guo, Y., Xie, J., Xu, X.: ECLNet: center localization of eye structures based on adaptive gaussian ellipse heatmap. Comput. Biol. Med. **153**, 106485 (2023)
27. Zou, X., Liu, W., Wang, J., Tao, R., Zheng, G.: ARST: auto-regressive surgical transformer for phase recognition from laparoscopic videos. Comput. Meth. Biomech. Biomed. Eng. Imaging Visual. **11**, 1012–1018 (2022)

Revolutionizing Space Health (Swin-FSR): Advancing Super-Resolution of Fundus Images for SANS Visual Assessment Technology

Khondker Fariha Hossain[1]([✉]), Sharif Amit Kamran[1], Joshua Ong[2], Andrew G. Lee[3], and Alireza Tavakkoli[1]

[1] Dept. of Computer Science & Engineering, University of Nevada, Reno, USA
khondkerfarihah@nevada.unr.edu
[2] Michigan Medicine, University of Michigan, Ann Arbor, USA
[3] Blanton Eye Institute, Houston Methodist Hospital, Houston, USA

Abstract. The rapid accessibility of portable and affordable retinal imaging devices has made early differential diagnosis easier. For example, color funduscopy imaging is readily available in remote villages, which can help to identify diseases like age-related macular degeneration (AMD), glaucoma, or pathological myopia (PM). On the other hand, astronauts at the International Space Station utilize this camera for identifying spaceflight-associated neuro-ocular syndrome (SANS). However, due to the unavailability of experts in these locations, the data has to be transferred to an urban healthcare facility (AMD and glaucoma) or a terrestrial station (e.g., SANS) for more precise disease identification. Moreover, due to low bandwidth limits, the imaging data has to be compressed for transfer between these two places. Different super-resolution algorithms have been proposed throughout the years to address this. Furthermore, with the advent of deep learning, the field has advanced so much that ×2 and ×4 compressed images can be decompressed to their original form without losing spatial information. In this paper, we introduce a novel model called Swin-FSR that utilizes Swin Transformer with spatial and depth-wise attention for fundus image super-resolution. Our architecture achieves Peak signal-to-noise-ratio (PSNR) of 47.89, 49.00 and 45.32 on three public datasets, namely iChallenge-AMD, iChallenge-PM, and G1020. Additionally, we tested the model's effectiveness on a privately held dataset for SANS and achieved comparable results against previous architectures.

1 Introduction

Color fundus imaging can detect and monitor various ocular diseases, including age-related macular degeneration (AMD), glaucoma, and pathological myopia (PM) [29]. In remote and under-developed areas, color funduscopy imaging has become increasingly accessible, allowing healthcare professionals to identify and manage these ocular diseases before they progress to irreversible stages. However, interpreting these images can be challenging for inexperienced or untrained personnel, necessitating the transfer

Supplementary Information The online version contains supplementary material available at https://doi.org/10.1007/978-3-031-43990-2_65.

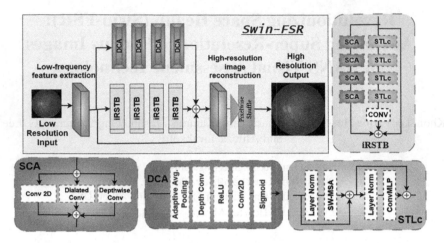

Fig. 1. An overview of the proposed Swin-FSR consisting of a low-frequency feature extraction module, improved Residual Swin-transformer BLock (iRSTB), Depth-wise Channel Attention (DCA) Block, and High-resolution image reconstruction block. Furthermore, iRSTB block consists of three parallel branches, i) Swin-Transformer with ConvMLP block (STLc), ii) Spatial and Channel Attention (SCA) Block, and iii) an identity mapping of the input, which is added together.

of data to urban healthcare facilities where specialists can make more accurate diagnoses [3,21]. Compressing and decompressing the data without losing spatial information can be utilized in this scenario by the super-resolution algorithm.

In a similar manner, color funduscopy imaging has found applications beyond the confines of the planet. For example, astronauts onboard the International Space Station (ISS) utilize this imaging to identify spaceflight-associated neuro-ocular syndrome (SANS) [10]. This condition can occur due to prolonged exposure to microgravity. It affects astronauts during long-duration spaceflight missions and can present with asymmetric/unilateral/bilateral optic disc edema and choroidal folds, which are easily identifiable by Color fundus images [11]. With the low bandwidth communication between ISS and terrestrial station [19], it becomes harder for experts to conduct early diagnosis and take preventive measures. So, super-resolution techniques can be vital in this adverse scenario. Medical experts can visualize and analyze these changes and act accordingly [18].

Image super-resolution and compression using different upsampling filters [6,20, 23], has been a staple in image processing for a long time. Yet, those conventional approaches required manually designing convolving filters, which couldn't adapt to learning spatial and depth features and had less artifact removal ability. With the advent of deep learning, convolutional neural network-based super-resolution has paved the way for fast and low-computation image reconstructions with less error [4,25,26]. A few years back, attention-based architectures [17,22,27,28] were the state-of-the-art for any image super-resolution tasks. However, with the introduction of shifted window-based transformer models [8,13], the accuracy of these models superseded attention-based ones with regard to different reconstruction metrics. Although swin-transformer-based models are good at extracting features of local patches, they lose the overall global spatial and depth context while upsampling with window-based patch merging operations.

Our Contributions: Considering all the relevant factors, we introduce the novel Swin-FSR architecture. It incorporates low-frequency feature extraction, deep feature extraction, and high-quality image reconstruction modules. The low-frequency feature extraction module employs a convolution layer to extract low-level features and is then directly passed to the reconstruction module to preserve low-frequency information. Our novelty is introduced in the deep feature extraction where we incorporated Depthwise Channel Attention block (DCA), improved Residual Swin-transformer Block - (iRSTB), and Spatial and Channel Attention block (SCA). To validate our work, we compare three different SR architectures for four Fundus datasets: iChallenge-AMD [5], iChallenge-PALM [7], G1020 [1] and SANS. From Fig. 2 and Fig. 3 , it is apparent that our architecture reconstructs images with high PSNR and SSIM.

2   Methodology

2.1   Overall Architecture

As shown in Fig. 1, SwinFSR consists of four modules: low-frequency feature extraction, deep patch-level feature extraction, deep depth-wise channel attention, and high-quality (HQ) image reconstruction modules. Given a low-resolution (LR) image $I_{LR} \epsilon \mathbb{R}^{H \times W \times C}$. Here, H = height, W = width, c = channel of the image. For low-frequency feature extraction, we utilize a 3×3 convolution with stride= 1 and is denoted as H_{LF}, and it extracts a feature $F_{LF} \epsilon \mathbb{R}^{H \times W \times C_o ut}$ with which illustrated in Eq. 1.

$$F_{LF} = H_{LF}(I_{LR}) \tag{1}$$

It has been reported that the convolution layer helps with better spatial feature extraction and early visual processing, guiding to more steady optimization in transformers [24]. Next, we have two parallel branches of outputs as denoted in Eq. 2. Here, H_{DCA} and H_{iRSTB} are two new blocks that we propose in this study, and they both take the low-feature vector as F_{LF} input and generate two new features namely, F_{SF} and F_{PLF}. We elaborate this two blocks in Subsect. 2.2 and 2.3

$$F_{SF} = H_{DCA}(F_{LF})$$
$$F_{PLF} = H_{iRSTB}(F_{LF}) \tag{2}$$

Finally, we combine all three features from our previous modules, namely, F_{LF}, F_{SF}, and F_{PLF} and apply a final high-quality (HQ) image reconstruction module to generate a high-quality image I_{HQ} as given in Eq. 3.

$$I_{HQ} = H_{REC}(F_{LF} + F_{SF} + F_{PLF}) \tag{3}$$

where H_{REC} is the function of the reconstruction block. Our low-frequency block extracts shallow features, whereas the two parallel depth-wise channel-attention and improved residual swin-transformer blocks extract spatially and channel-wise dense features extracting lost high-frequencies. With these three parallel residual connections, SwinFSR can propagate and combine the high and low-frequency information to the reconstruction module for better super-resolution results. It should be noted that the reconstruction module consists of a 1×1 convolution followed by a Pixel-shuffle layer to upsample the features.

2.2   Depth-Wise Channel Attention Block

For super-resolution architectures, channel-attention [2,15,28] is an essential robust feature extraction module that helps these architectures achieve high accuracy and more visually realistic results. In contrast, recent shifted-window-based transformers architecture [12,13] for super-resolution do not incorporate this module. However, a recent work [8] utilized a cross-attention module after the repetitive swin-transformer layers. One of the most significant drawbacks of the transformer layer is it works on patch-level tokens where the spatial dimensions are transformed into a linear feature. To retain the spatial information intact and learning dense features effectively, we propose depth-wise channel attention given in Eq. 4.

$$
\begin{aligned}
x &= AdaptiveAvgPool(x_{in}) \\
x &= \delta(Depthwise\_Conv(x)) \\
x_{out} &= \phi(Conv(x))
\end{aligned}
\tag{4}
$$

Here, δ is ReLU activation, and ϕ is Sigmoid activation functions. The regular channel-attention utilizes a 2D Conv with $1 \times 1 \times C$ weight vector C times to create output features $1 \times 1 \times C$. Given that adaptive average pooling already transforms the dimension to 1×1, utilizing a spatial convolution is redundant and shoots up computation time. To make it more efficient, we utilize depth-wise attention, with 1×1 weight vector applied on each of the C features separately, and then the output is concatenated to get our final output $1 \times 1 \times C$. We use four **DCA** blocks in our architecture as given in Fig. 1.

2.3   Improved Residual Swin-Transformer Block

Swin Transformer [16] incorporates shifted windows self-attention (SW-MSA), which builds hierarchical regional feature maps and has linear computation complexity compared to vision transformers with quadratic computation complexity. Recently, Swin-IR [13] adopted a modified swin-transformer block for different image enhancement tasks such as super-resolution, JPEG compression, and denoising while achieving high PSNR and SSIM scores. The most significant disadvantage of this block is the Multi-layer perceptron module (MLP) after the post-normalization layer, which has two linear (dense) layers. As a result, it becomes computationally more expensive than a traditional 1D convolution layer. For example, a linear feature output from a swin-transformer layer having depth D, and input channel, X_{in} and output channel, X_{out} will have a total number of parameters, $D \times X_{in} \times X_{out}$. Contrastly, a 1D convolution with kernel size, $K = 1$, with the same input and output will have less number of parameters,

$1 \times X_{in} \times X_{out}$. Here, we assign bias, $b = 0$. So, the proposed swin-transformer block can be defined as Eq. 5 and is illustrated in Fig. 1 as **STLc** block.

$$x^1 = SW\text{-}MSA(\sigma(d^l)) + x$$
$$x^2 = ConvMLP(\sigma(x^1)) + x^1 \qquad (5)$$

Here, σ is Layer-normalization and ConvMLP has two 1D convolution followed by GELU activation. To capture spatial local contexts for patch-level features we utilize a patch-unmerging layer in parallel path and incorporate **SCA** (spatial and channel attention) block. The block consists of a convolution ($k = 1, s = 1$), a dilated convolution ($k = 3, d = 2, s = 1$) and a depth-wise convolution ($k = 1, s = 1$) layer. Here, $k=$ kernel, $d =$ dilation and $s =$ stride. Moreover all these features are combined to get the final ouptut. By combining repetitive SCA, $STLc$ blocks and a identity mapping we create our improved reisdual swin-transformer block (**iRSTB**) illustrated in Fig. 1. In Swin-FSR, we incorporate four iRSTB blocks.

2.4 Loss Function

For image super resolution task, we utilize the L1 loss function given in Eq. 6. Here, I_{RHQ} is the reconstructed output of SwinFSR and I_{HQ} is the original high-quality image.

$$L = \|I_{RHQ} - I_{HQ}\| \qquad (6)$$

3 Experiments

3.1 Dataset

To assess the performance of our super-resolution models, we employ three distinct public fundus datasets: AMD [5], G1020 [1], PALM [7, 14], and one private dataset: SANS. The datasets comprise .jpg, .tif, and .png formats with a high resolution. For training purposes, we used Bicubic Interpolation to resize the images into (512×512) and converted all the images into .png format. The AMD, G1020, PALM, and SANS datasets yield 400, 1020, 400, and 276 images, respectively. We split every dataset in 80% train and 20% test set, so we end up having 320 and 80 images for AMD, 816 and 204 images for G1020, 320 and 80 images for PALM, and 220 and 56 images for SANS. We use 5-fold cross-validation to train our networks.

Data use Declaration and Acknowledgment: The AMD and PALM dataset were released as part of REFUGE Challenge, PALM Challenge. The G1020 was published as technical report and benchmark [1]. The authors instructed to cite their work [5, 7, 14] for usage. The SANS data is privately held and is provided by the National Aeronautics and Space Administration(NASA) with Data use agreement 80NSSC20K1831.

3.2 Hyper-parameter

We utilized L1 loss for training our models for the super-resolution task. For optimizer, we used Adam [9], with learning rate $\alpha = 0.0002$, $\beta_1 = 0.9$ and $\beta_2 = 0.999$. The batch size was $b = 2$, and we trained for 200 epochs for 8 h with NVIDIA A30 GPU. We utilize PyTorch and MONAI library monai.io for data transformation, training and testing our model. The code repository is provided in this link.

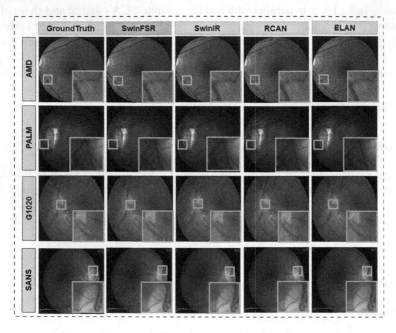

Fig. 2. Qualitative comparison of ($\times2$) image reconstruction using different SR methods on AMD, PALM, G1020 and SANS dataset. The green rectangle is the zoomed-in region. The rows are for the AMD, PALM and SANS datasets. Whereas, the column is for each different models: SwinFSR, SwinIR, RCAN and ELAN. (Color figure online)

3.3 Qualitative Evaluation

We compared our architecture with some best-performing CNN and Transformer based SR models, including RCAN [28], ELAN [27], and SwinIR [13] as illustrated in Fig. 2 and Fig. 3. We trained and evaluated all four architectures using their publicly available source code on the four datasets. SwinIR utilizes residual swin-transformer blocks with identity mapping for dense feature extractions. In contrast, the RCAN utilizes repetitive channel-attention blocks for depth-wise dense feature retrieval. Similarly, ELAN combines multi-scale and long-range attention with convolution filters to extract spatial and depth features. In Fig. 2, we illustrate $\times2$ reconstruction results for all four architectures. By observing, we can see that our model's vessel reconstruction is more realistic for $\times2$ factor samples than other methods. Specifically for AMD and SANS, the degeneration is noticeable. In contrast, ELAN and RCAN fail to accurately reconstruct thinner and smaller vessels.

In the second experiment, we show results for ×4 reconstruction for all SR models in Fig. 3. It is apparent from the figure that our model's reconstruction is more realist than other transformer and CNN-based architectures, and the vessel boundary is sharp, containing more degeneration than SwinIR, ELAN and RCAN.. Especially for AMD , G1020 and PALM, the vessel edges are finer and sharper making it easily differentiable. In contrast, ELAN and RCAN generate pseudo vessels whereas SwinIR fails to generate some smaller ones. For SANS images, the reconstruction is much noticable for the ×4 than ×2.

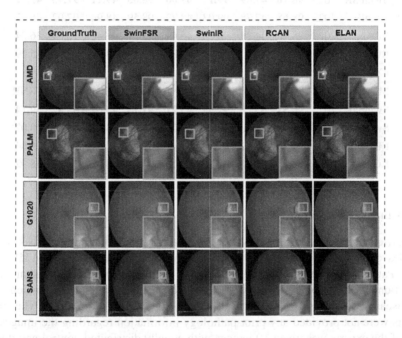

Fig. 3. Qualitative comparison of (×4) image reconstruction using different SR methods on AMD, PALM, G1020 and SANS dataset. The green rectangle is the zoomed-in region. The rows are for the AMD, PALM and SANS datasets. Whereas, the column is for each different models: SwinFSR, SwinIR, RCAN and ELAN. (Color figure online)

3.4 Quantitative Bench-Marking

For quantitative evaluation, we utilize Peak Signal-to-Noise-Ratio (PSNR) and Structural Similarity Index Metric (SSIM), which has been previously employed for measuring similarity between original and reconstructed images in super-resolution tasks [13,27,28]. We illustrate quantitative performance in Table. 1 between SwinFSR and other state-of-the-art methods: SwinIR [13], RCAN [28], and ELAN [27]. Table. 1 shows that SwinFSR's overall SSIM and PSNR are superior to other transformer and

CNN-based approaches. For ×2 scale reconstruction, SwinIR achieves the second-best performance. Contrastly, for ×4 scale reconstruction, RCAN outperforms SwinIR while scoring lower than our SwinFSR model for PSNR and SSIM.

Table 1. Quantitative comparison on AMD [5], PALM [7,14], G1020 [1], & SANS.

Dataset		AMD		PALM		G1020		SANS	
2X									
Model	Year	SSIM	PSNR	SSIM	PSNR	SSIM	PSNR	SSIM	PSNR
SwinFSR	2023	**98.70**	**47.89**	**99.11**	**49.00**	**98.65**	**49.11**	**97.93**	**45.32**
SwinIR	2022	98.68	47.78	99.03	48.73	98.59	48.94	97.92	45.17
ELAN	2022	98.18	44.21	98.80	46.49	98.37	47.48	96.89	36.84
RCAN	2018	98.62	47.76	99.04	48.83	98.53	48.29	97.91	45.29
4X									
Model	Year	SSIM	PSNR	SSIM	PSNR	SSIM	PSNR	SSIM	PSNR
SwinFSR	2023	**96.51**	**43.28**	**97.34**	**43.27**	**97.13**	**44.67**	**95.82**	**39.14**
SwinIR	2022	96.40	42.98	97.27	43.07	97.06	44.44	95.80	39.02
ELAN	2022	94.76	39.12	97.03	42.47	97.11	44.39	95.16	35.84
RCAN	2018	96.20	42.62	97.38	43.18	97.05	44.37	95.73	38.92

3.5 Clinical Assessment

We carried out a diagnostic assessment with two expert ophthalmologists and test samples of 80 fundus images (20 fundus images per disease classes: AMD, Glaucoma, Pathological Myopia and SANS for both original x2 and x4 images, and super-resolution enhanced images). Half of the 20 fundus images were control patients without disease pathologies; the other half contained disease pathologies. The clinical experts were not provided any prior pathology information regarding the images. And each of the experts was given 10 images with equally distributed control and diseased images for each disease category.

The accuracy and F1-score for original x4 images are as follows, 70.0% and 82.3% (AMD), 75% and 85.7% (Glaucoma), 60.0% and 74.9% (Palm), and 55% and 70.9% (SANS). The accuracy and F1-score for original x2 are as follows, 80.0% and 88.8% (AMD), 80.0% and 88.8% (Glaucoma), 70.0% and 82.1% (Palm), and 65% and 77.4% (SANS). The accuracy and F1-score for our model Swin-FSR's output from ×4 images are as follows, 90.0% and 93.3% (AMD), 90.0% and 93.7% (Glaucoma), 75.0% and 82.7% (Palm), and 75% and 81.4% (SANS). The accuracy and F1-score for Swin-FSR's output from ×2 images are as follows, 90.0% and 93.3% (AMD), 90.0% and 93.7% (Glaucoma), 80.0% and 85.7% (Palm), and 80% and 85.7% (SANS).

We also tested SWIN-IR, ELAN, and RCAN models for diagnostic assessment, out of which SWIN-IR upsampled images got the best results. For x4 images, the model's accuracy and F-1 score are 80% and 87.5% (AMD), 85.0% and 90.3% (Glaucoma), 70.0% and 80.0% (Palm), and 70% and 76.9% (SANS). For x2 images, the model's accuracy and F-1 score are 80% and 87.5% (AMD), 80% and 88.8% (Glaucoma), 70.0% and 80.0% (Palm), and 75% and 81.4% (SANS). Based on the above observations, our model-generated images achieves the best result.

3.6 Ablation Study

Effects of iRSTB, DCA, and SCA Number: We illustrate the impacts of iRSTB, DCA, and SCA numbers on the model's performance in Supplementary Fig. 1 (a), (b), and (C). We can see that the PSNR and SSIM become saturated with an increase in any of these three hyperparameters. One drawback is that the total number of parameters grows linearly with each additional block. Therefore, we choose four blocks for iRSTB, DCA, and SCA to achieve the optimum performance with low computation cost.

Presence and Absence of iRSTB, DCA, and SCA Blocks: Additionally, we provide a comprehensive benchmark of our model's performance with and without the novel blocks incorporated in Supplementary Table 1. Specifically, we show the performance gains with the usage of an improved residual swin-transformer block (iRSTB) and depth-wise channel attention (DCA). As the results illustrate, by comprising these blocks, the PSNR and SSIM reach higher scores.

4 Conclusion

In this paper, we proposed Swin-FSR by combining novel DCA, iRSTB, and SCA blocks which extract depth and low features, spatial information, and aggregate in image reconstruction. The architecture reconstructs the precise venular structure of the fundus image with high confidence scores for two relevant metrics. As a result, we can efficiently employ this architecture in various ophthalmology applications emphasizing the Space station. This model is well-suited for the analysis of retinal degenerative diseases and for monitoring future prognosis. Our goal is to expand the scope of this work to include other data modalities.

Acknowledgement. Research reported in this publication was supported in part by the National Science Foundation by grant numbers [OAC-2201599],[OIA-2148788] and by NASA grant no 80NSSC20K1831.

References

1. Bajwa, M.N., Singh, G.A.P., Neumeier, W., Malik, M.I., Dengel, A., Ahmed, S.: G1020: a benchmark retinal fundus image dataset for computer-aided glaucoma detection. In: 2020 International Joint Conference on Neural Networks (IJCNN), pp. 1–7. IEEE (2020)
2. Dai, T., Cai, J., Zhang, Y., Xia, S.T., Zhang, L.: Second-order attention network for single image super-resolution. In: Proceedings of the IEEE/CVF Conference on Computer Vision and Pattern Recognition, pp. 11065–11074 (2019)

3. Das, V., Dandapat, S., Bora, P.K.: A novel diagnostic information based framework for super-resolution of retinal fundus images. Comput. Med. Imaging Graph. **72**, 22–33 (2019)
4. Dong, C., Loy, C.C., He, K., Tang, X.: Image super-resolution using deep convolutional networks. IEEE Trans. Pattern Anal. Mach. Intell. **38**(2), 295–307 (2015)
5. Fu, H., et al.: Adam: automatic detection challenge on age-related macular degeneration. In: IEEE Dataport (2020)
6. Hardie, R.: A fast image super-resolution algorithm using an adaptive wiener filter. IEEE Trans. Image Process. **16**(12), 2953–2964 (2007)
7. Huazhu, F., Fei, L., José, I.: PALM: pathologic myopia challenge. Comput. Vis. Med. Imaging (2019)
8. Jin, K., et al.: SwiniPASSR: Swin transformer based parallax attention network for stereo image super-resolution. In: Proceedings of the IEEE/CVF Conference on Computer Vision and Pattern Recognition, pp. 920–929 (2022)
9. Kingma, D.P., Ba, J.: Adam: a method for stochastic optimization. arXiv preprint arXiv:1412.6980 (2014)
10. Lee, A.G., Mader, T.H., Gibson, C.R., Brunstetter, T.J., Tarver, W.J.: Space flight-associated neuro-ocular syndrome (SANS). Eye **32**(7), 1164–1167 (2018)
11. Lee, A.G., et al.: Spaceflight associated neuro-ocular syndrome (SANS) and the neuro-ophthalmologic effects of microgravity: a review and an update. NPJ Microgravity **6**(1), 7 (2020)
12. Li, B., Li, X., Lu, Y., Liu, S., Feng, R., Chen, Z.: HST: hierarchical swin transformer for compressed image super-resolution. In: Karlinsky, L., Michaeli, T., Nishino, K. (eds.) ECCV 2022. LNCS, vol. 13802, pp. 651–668. Springer, Cham (2023). https://doi.org/10.1007/978-3-031-25063-7_41
13. Liang, J., Cao, J., Sun, G., Zhang, K., Van Gool, L., Timofte, R.: SwinIR: image restoration using swin transformer. In: Proceedings of the IEEE/CVF International Conference on Computer Vision, pp. 1833–1844 (2021)
14. Lin, F., et al.: Longitudinal changes in macular optical coherence tomography angiography metrics in primary open-angle glaucoma with high myopia: a prospective study. Invest. Ophthalmol. Vis. Sci. **62**(1), 30–30 (2021)
15. Lin, Z., et al.: Revisiting RCAN: improved training for image super-resolution. arXiv preprint arXiv:2201.11279 (2022)
16. Liu, Z., et al.: Swin transformer: hierarchical vision transformer using shifted windows. In: Proceedings of the IEEE/CVF International Conference on Computer Vision, pp. 10012–10022 (2021)
17. Niu, B., et al.: Single Image Super-Resolution via a Holistic Attention Network. In: Vedaldi, A., Bischof, H., Brox, T., Frahm, J.-M. (eds.) ECCV 2020. LNCS, vol. 12357, pp. 191–207. Springer, Cham (2020). https://doi.org/10.1007/978-3-030-58610-2_12
18. Ong, J., et al.: Neuro-ophthalmic imaging and visual assessment technology for spaceflight associated neuro-ocular syndrome (SANS). Survey Ophthalmol. **67**, 1443–1466 (2022)
19. Seas, A., Robinson, B., Shih, T., Khatri, F., Brumfield, M.: Optical communications systems for NASA's human space flight missions. In: International Conference on Space Optics-ICSO 2018, vol. 11180, pp. 182–191. SPIE (2019)
20. Sen, P., Darabi, S.: Compressive image super-resolution. In: 2009 Conference Record of the Forty-Third Asilomar Conference on Signals, Systems and Computers, pp. 1235–1242. IEEE (2009)
21. Sengupta, S., Wong, A., Singh, A., Zelek, J., Lakshminarayanan, V.: DeSupGAN: multi-scale feature averaging generative adversarial network for simultaneous de-blurring and super-resolution of retinal fundus images. In: Fu, H., Garvin, M.K., MacGillivray, T., Xu, Y., Zheng, Y. (eds.) OMIA 2020. LNCS, vol. 12069, pp. 32–41. Springer, Cham (2020). https://doi.org/10.1007/978-3-030-63419-3_4

22. Song, X., et al.: Channel attention based iterative residual learning for depth map super-resolution. In: Proceedings of the IEEE/CVF Conference on Computer Vision and Pattern Recognition, pp. 5631–5640 (2020)

23. Van Ouwerkerk, J.: Image super-resolution survey. Image Vis. Comput. **24**(10), 1039–1052 (2006)

24. Xiao, T., Singh, M., Mintun, E., Darrell, T., Dollár, P., Girshick, R.: Early convolutions help transformers see better. In: Advances in Neural Information Processing Systems, vol. 34, pp. 30392–30400 (2021)

25. Zhang, K., Zuo, W., Chen, Y., Meng, D., Zhang, L.: Beyond a gaussian denoiser: residual learning of deep CNN for image denoising. IEEE Trans. Image Process. **26**(7), 3142–3155 (2017)

26. Zhang, K., Zuo, W., Zhang, L.: Learning a single convolutional super-resolution network for multiple degradations. In: Proceedings of the IEEE Conference on Computer Vision and Pattern Recognition, pp. 3262–3271 (2018)

27. Zhang, X., Zeng, H., Guo, S., Zhang, L.: Efficient long-range attention network for image super-resolution. In: Avidan, S., Brostow, G., Cissé, M., Farinella, G.M., Hassner, T. (eds.) ECCV 2022. LNCS, vol. 13677, pp. 649–667. Springer, Cham (2022)

28. Zhang, Y., Li, K., Li, K., Wang, L., Zhong, B., Fu, Y.: Image super-resolution using very deep residual channel attention networks. In: Proceedings of the European Conference on Computer Vision (ECCV), pp. 286–301 (2018)

29. Zhang, Z., et al.: A survey on computer aided diagnosis for ocular diseases. BMC Med. Inform. Decis. Mak. **14**(1), 1–29 (2014)

Representation, Alignment, Fusion: A Generic Transformer-Based Framework for Multi-modal Glaucoma Recognition

You Zhou[1], Gang Yang[1,2]([✉]), Yang Zhou[3], Dayong Ding[3], and Jianchun Zhao[3]

[1] School of Information, Renmin University of China, Beijing, China
[2] MOE Key Lab of DEKE, Renmin University of China, Beijing, China
yanggang@ruc.edu.cn
[3] Vistel AI Lab, Visionary Intelligence Ltd., Beijing, China

Abstract. Early glaucoma can be diagnosed with various modalities based on morphological features. However, most existing automated solutions rely on single-modality, such as Color Fundus Photography (CFP) which lacks 3D structural information, or Optical Coherence Tomography (OCT) which suffers from insufficient specificity for glaucoma. To effectively detect glaucoma with CFP and OCT, we propose a generic multi-modal Transformer-based framework for glaucoma, MM-RAF. Our framework is implemented with pure self-attention mechanisms and consists of three simple and effective modules: Bilateral Contrastive Alignment (BCA) aligns both modalities into the same semantic space to bridge the semantic gap; Multiple Instance Learning Representation (MILR) aggregates multiple OCT B-scans into a semantic structure and downsizes the scale of the OCT branch; Hierarchical Attention Fusion (HAF) enhances the cross-modality interaction capability with spatial information. By incorporating three modules, our framework can effectively handle cross-modality interaction between different modalities with huge disparity. The experimental results demonstrate that the framework outperforms the existing multi-modal methods of this task and is robust even with a clinical small dataset. Moreover, by visualizing, OCT can reveal the subtle abnormalities in CFP, indicating that the relationship between various modalities is captured. Our code is available at https://github.com/YouZhouRUC/MM-RAF.

Keywords: Glaucoma recognition · Multi-modal learning · Multiple instance learning · Contrastive learning

1 Introduction

Glaucoma is the second leading ophthalmic blindness disease, with nearly 70 million patients worldwide. Early glaucoma can be detected by color fundus

The computer resources were provided by Public Computing Cloud Platform of Renmin University of China.

Supplementary Information The online version contains supplementary material available at https://doi.org/10.1007/978-3-031-43990-2_66.

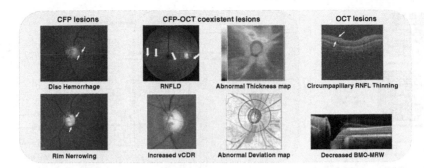

Fig. 1. Some Lesions of glaucoma in clinical examination. Lesions in the blue box appear in CFP. Yellow for OCT. (Color figure online)

photography (CFP) and optical coherence tomography (OCT). The gold standard of glaucoma in CFP includes Cup to Disk Ratio (CDR) enlargement and Retinal Nerve Fiber Layer Defects (RNFLD). Diagnosing with CFP has earned superb performance, but it only captures flat information. OCT scans the fine-grain 3D structure of the fundus, and the quantitative analysis conducted by OCT can help diagnosis for junior clinicians. Many crucial lesions coexist in both CFP and OCT images. We list some critical lesions correlated to glaucoma in Fig. 1.

Early research on automated glaucoma recognition focuses on CFP. Based on the ISNT rule [8], the vertical Cup-to-Disc rate (vCDR) is a feature with high specificity [15]. Besides, the RNFLD is an important clinical diagnostic evidence in glaucoma [6]. Previous research on OCT mainly discusses the RNFL thickness map [2], GCL thickness map [2,9], and cube scan [16]. Combining CFP and OCT for multi-modal diagnosis shows promise in providing accurate diagnostic performance and additional stereo information on retinal structure. However, multi-modal methods on glaucoma recognition have rarely been investigated. Simply combining CFP and the thickness map from the OCT report [1] will waste most information in OCT. An image pair contains merely one CFP image and up to 256 OCT B-scan frames. Interaction between two modalities with unbalanced amounts poses challenges for the existing methods. COROLLA [3] introduces supervised contrastive learning and uses a dual-stream network without modality interaction. Mehta et al. [13] use two different convolutional branches to extract the features of OCT cube and CFP respectively and then ensembles to make the diagnosis. MM-MIL [11] implements two ResNet50 to extract the features from different modalities and adopts Multiple Instance Learning to automatically discover crucial instances. However, the fusion stage is restricted by the Global Average Pooling operation, which limits the spatial information interaction.

To address the problems discussed above, We propose MM-RAF. By following the paradigm of multi-modal learning, i.e., representation, alignment and fusion, we construct three effective modules. To alleviate the semantic gap, BCA introduces bilateral contrastive loss to improve the intra- and inter-modal alignment

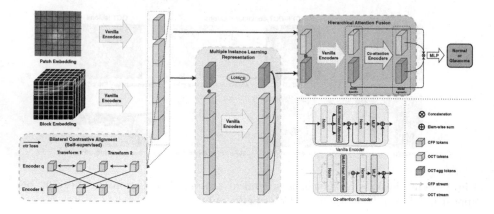

Fig. 2. Overview of our proposed MM-RAF framework.

capability. In the representation stage, MILR extracts the glaucoma-relevant OCT B-scans(instance) from the OCT volume (bag) to assemble a saturated and semantic structure and reduce the scale of the OCT branch. To solve the inability of cross-modality interaction and loss of spatial information, Hierarchical Attention Fusion (HAF) utilizes two strategies to extract modal-specific features and modal-agnostic features to make a final diagnosis. Two classical Transformer encoders are implemented throughout our framework.

Through the experiments on the private dataset, we illustrate that MM-RAF outperforms the existing multi-modal methods on glaucoma recognition. Also, it is demonstrated that MILR and HAF effectively enhance performance in the ablation study. Moreover, extensive experiments on GAMMA dataset [19] prove that the BCA module can promote robustness. We implement relevance-based methods [4] for visualization. The heatmaps illustrate that the framework manifests indistinguishable lesions like RNFLD.

2   Method

As shown in Fig. 2, MM-RAF is a two-phase framework consisting of three modules, i.e., BCA, MILR, and HAF. Inspired by Vision Transformer [7] for modeling long-range dependency, we incorporate two classical transformer encoders throughout our framework to make comprehensive interaction between different modalities. In the contrastive phase, pretraining on unlabeled multi-modal data enables BCA to align the features from different modalities into the same semantic space, diminishing the semantic gap. In the following phase, MILR employs Multiple Instance Learning to refine the cumbersome OCT branch to a semantic structure. Then, with balanced streams from two modalities, HAF renders two cross-modality interaction strategies to make inter- and intra-modal diagnoses. In the following sections, we will clarify each module specifically.

Vanilla Encoder and Co-attention Encoder. Two basic encoders enable effective intra- and inter-modal interaction. As shown in Eqs. 1–2, the Vanilla encoder duplicates the input stream into query, key and value (Q_m, K_m and V_m) components and concentrates on intra-modal interaction, while the Co-attention receives two input streams from different modalities and focuses on inter-modal interaction, with the primary stream acting as query, Q_m, and the subordinate stream replicated as key and value ($K_{\bar{m}}$, $V_{\bar{m}}$). Due to the high computational cost of the self-attention mechanism in the OCT branch, we propose block embedding, partitioning the OCT volume into n OCT blocks and each block will be embedded as $T_{O,blk}^k \in \mathbb{R}^{N \times dim}$ (N is 196, dim is 768, $k \in \{1, \ldots, n\}$).

$$Vanilla\ Encoder(Q_m, K_m, V_m) = softmax\left(\frac{Q_m K_m^T}{\sqrt{d_k}}\right) V_m \qquad (1)$$

$$Co\text{-}attention\ Encoder(Q_m, K_{\bar{m}}, V_{\bar{m}}) = softmax\left(\frac{Q_m K_{\bar{m}}^T}{\sqrt{d_k}}\right) V_{\bar{m}} \qquad (2)$$

where m and $\bar{m}$ denote different modalities, CFP or OCT in this task. d_k denotes dimension of self-attention.

2.1 Bilateral Contrastive Alignment

The BCA module aims to align the extracted features with the self-supervised strategy before interaction. The semantic gap between CFP and OCT is huge, and direct interaction between different modalities without alignment will lead to a mismatch. ALBEF [10] adopts a similar strategy with the momentum encoders and negative queues, but the weakly-correlated phenomenon and strongly-correlated negative sample in the medical area are so common that ALBEF can hardly reach convergence. To simplify the proxy task, BCA employs the theory of MoCov3 [5] and redesigns the "ctr loss" to adapt to multi-modal tasks. Considering preserving the stereo information of OCT, each block-level token, $T_{O\_blk}^k$, is averaged in the token dimension before being projected. To equally align both branches, 4 projectors are followed symmetrically in both branches to map the tokens to contrastive space. Different augmentation will cause huge discrepancies, especially for OCT images. Therefore, as shown in Fig. 2 and Eq. 3 (Bilateral loss), to mitigate the alignment difficulty for multi-modal tasks, we align both modalities by concentrating on the same modality m with different augmentation $\bar{u}$ and different modalities $\bar{m}$ with the same augmentation u.

$$\mathcal{L} = \sum_{m \in \{CFP, OCT\}} \sum_{u=1}^{2} (\mathcal{L}_{inter,m,u} + \mathcal{L}_{intra,m,u}); \qquad (3)$$

$$\mathcal{L}_{intra,m,u} = \mathbb{I}_{u \neq \bar{u}}\ ctr(preditor_{m,m,u}(q_{m,u}), k_{m,\bar{u}}); \qquad (4)$$

$$\mathcal{L}_{inter,m,u} = \mathbb{I}_{m \neq \bar{m}} ctr(preditor_{m,\bar{m},u}(q_{m,u}), q_{\bar{m},u}); \qquad (5)$$

where $q_{m,u} \in \mathbb{R}^{dim}$ (dim is 256 by default) denotes modality m with augmentation u in contrastive space with the query encoder. $k_{m,\bar{u}}$ denote the vectors from the key encoder. $preditor_{m,\bar{m},u}$ denotes the predictor to map modality m to another modality $\bar{m}$. $ctr()$ denotes "ctr loss" [5]. Intuitively, each token behaves like a "query", and BCA aligns the corresponding "values" by projection.

2.2   Multiple Instance Learning Representation

Direct interaction between different modalities with unbalanced amounts is computational-consuming, and the cross-modality relationship is difficult to build. Therefore, we conjecture that the OCT block-level tokens(features) can be formulated as an embedding-level MIL problem in two aspects: (1) In an OCT volume, only certain salient slices are related to glaucoma. (2) High-level embedding features after the BCA module are more distinguishable. In MILR, by defining the ith OCT block, namely $T^i_{O,blk}$, as an embedding instance, the integral OCT volume is taken as the bag $\mathbb{B} = \left\{ T^i_{O,blk} | 1 \leq i \leq n \right\}$. Then we concatenate the OCT bag, $\mathbb{B}$, with an aggregated tokens $T_{O,agg} \in \mathbb{R}^{N \times dim}$. As shown in Fig. 2, several Vanilla encoders render the interaction among $\mathbb{B}$ and $T_{O,agg}$, and eventually, $T^i_{O,agg}$ get semantic information from OCT block instances to form a bag prediction. To ensure that MILR aggregates glaucoma-related instances, a supervised signal is incorporated. For efficiency and effectiveness, we only pass semantic $T^{depth}_{O,agg}$ to the subsequent fusion module. MILR can be formulated as:

$$\left\{ T^{i+1}_{O,agg} | \mathbb{B}^{i+1} \right\} = Vanilla_i \left(T^i_{O,agg} \otimes \mathbb{B}^i \right), i \in \{1, \ldots, depth\} \tag{6}$$

where $depth$ is the depth of the Vanilla Encoder in MILR.

2.3   Hierarchical Attention Fusion

Before HAF, each modality interacts within its internal modality. To extract modal-specific features and modal-agnostic features respectively, HAF implements a mid-fusion strategy consisting of two fusion stages, Merged-attention and Cross-attention. As shown in Fig. 2, in the Merged-attention blocks, CFP tokens and $T^{depth}_{O,agg}$ will be concatenated, $T_{all} \in \mathbb{R}^{2*N \times dim}$, to pass through several Vanilla encoders. Except for interaction within their own modality, the salient area will also engage mildly with the other modality, e.g., CFP will leverage intra-modal (CFP) information and inter-modal (OCT) refined knowledge to reinforce the modal-specific features in CFP. Co-attention encoders in the Cross-attention stage render the CFP tokens to interact solely with OCT and thus extract the modal-agnostic features related in both modalities. Eventually, the modal-specific and modal-agnostic features from CFP and OCT will be fed into a projector for joint diagnosis. We will mention the HAF module again in the interpretability experiments in Sect. 3.4.

3 Experiments

3.1 Datasets

For multi-modal glaucoma recognition, the existing public dataset, GAMMA [19] on glaucoma grading, includes **macular** OCT and CFP. GAMMA dataset consists of 100 accessible labeled cases and another 100 unlabeled cases as the benchmark. As the dataset is limited in size for Transformer-based models, we construct a new dataset. 872 multi-modal cases are collected using Topcon Maestro-1 at the outpatient clinic in the Department of Ophthalmology of a state hospital from July 2020 to January 2021. The scan mode is 3D **Optic Disc** with one CFP image and 128 horizontal OCT B-scan images obtained simultaneously. Due to the expensive human annotations, we acquire pseudo labels of CFP by our advanced ensemble model for training. To build a trust-worthy test set for evaluation, we first split the dataset in train/val/test in 6:2:2. A clinician relabels the test set (172 cases) as GON/normal by considering CFP and OCT thickness map. The performance is evaluated on both private test set and GAMMA.

3.2 Experiment Details

Due to the high computational cost, the fixed sampling interval technique is employed to extract 32 OCT images. To avoid over-fitting, we reduce the depths of the encoders to 3 layers. For a fair comparison between all models in this study, we use the following standard setup: initializing with the pre-trained weight of ViT-Base-16 on ImageNet. In the first experiment, the existing multi-modal methods and classical baselines are compared with MM-RAF on the private test set. The baseline includes ResNet, ViT, DeiT [17], and Swin-Transformer [12] (pre-trained weight from timm [18]) with single-modal or early-fusion multi-modal experiments. Multi-modal methods include COROLLA [3], MBT [14]and MM-MIL [11]. The robustness is evaluated on the GAMMA dataset by comparing it with CNNs. The metrics are averaged over three runs. All experiments are implemented in Python 3.7 and Pytorch 1.7 with four NVIDIA TITAN X GPUs and the training configuration is included in Supplementary Material.

3.3 Experimental Results

Compared with Single-Modal and Multi-modal Solutions. As shown in Table 1, ResNet50 for CFP attains the best AP score in single modality, proving that CFP is more sensitive than OCT for glaucoma diagnosis. Besides, the transformer-based method is inferior to ResNet in CFP but surpasses CNN in OCT modality, indicating that the transformer-based methods need sufficient data to learn inductive bias. MM-ViT outperforms CFP-ViT and OCT-ViT, exemplifying that Transformer can benefit from multi-modal learning. Our framework, MM-RAF, outperformed MM-ViT with a 6% improvement in AP score and achieved **SOTA** with F1, AP, and AUC metrics in this study.

Table 1. Results of baseline and existing works comparison in the private test set. In each column, bold text denotes the best results.

lMethod		Metrics				
		Sen	Spe	F1	AP	AUC
Single-modal	CFP-ResNet50	**0.9333**	0.8078	0.8515	0.8925	0.9584
	CFP-ViT	0.6286	0.9756	0.7565	0.8483	0.9585
	CFP-DeiT	0.7348	0.9512	0.8291	0.8531	0.9512
	CFP-Swin-Transformer	0.6335	0.892	0.7729	0.8212	0.9241
	OCT-ResNet50	0.4571	0.9854	0.5994	0.8605	0.9363
	OCT-ViT	0.7333	**0.9854**	0.8397	0.8892	0.944
	OCT-DeiT	0.6286	0.9756	0.7565	0.8897	0.9585
	OCT-Swin-Transformer	0.6111	0.9281	0.7309	0.7925	0.911
Multi-modal	MM-ViT	0.8334	0.8873	0.8579	0.8983	0.934
	MM-DeiT	0.7904	0.9513	0.8629	0.8982	0.9514
	MM-Swin-Transformer	0.6111	0.7961	0.6853	0.5771	0.7639
	MM-MIL [11]	0.781	0.9416	0.8453	0.8837	0.9608
	MBT [14]	0.6667	0.9640	0.7859	0.8287	0.9384
	COROLLA [3]	0.6667	0.9376	0.7735	0.8318	0.942
	MM-RAF	0.9238	0.9027	**0.9081**	**0.9584**	**0.9855**

Table 2. Robustness experiments on GAMMA dataset. The glaucoma grading task (normal/early-glaucoma/progressive-glaucoma) is evaluated by *kappa*.

Method	*kappa*
Dual-ResNet50	0.7352
MM-MIL w/o transfer learning	0.8502
MM-MIL w/ transfer learning	**0.8562**
Ours w/o transfer learning, BCA	0.5289
Ours w/ transfer learning; w/o BCA	0.6277
Ours w/ BCA; w/o transfer learning	0.8072
Ours w/ transfer learning, BCA	**0.8467**

Robustness. Due to the limited size of GAMMA(100 cases), our transformer-based method is likely to get overfitting if training from scratch. To this end, pre-training on our private mid-scale dataset which captures **optic disc** OCT can gain enhancements even when transferring to the cross-domain dataset (GAMMA scans **macular** area). Cohen's *kappa* coefficient is implemented as metrics. As shown in Table 2, our method has better cross-domain generalization after applying BCA and transfer learning strategy. Furthermore, MM-RAF achieves results comparable to CNNs, highlighting our framework's robustness to learn inductive biases from images even with a limited dataset. When providing more domain-related data, our approach has the potential to perform marginally better than CNNs for compensating the lack of inductive bias.

Ablation Study. The ablation study on the private dataset examines the contribution of three modules, the order of HAF, and the depth of each module. From Table.3, MILR and HAF modules bring 0.03 and 0.02 AP increases, respec-

Table 3. Ablation study on private dataset. HAF-R denotes to reverse the order of two stages in HAF modules. Depth controls the encoder depth in all modules.

BCA	MILR	HAF-Merged	HAF-Cross	HAF-R	depth	Sen	Spe	F1	AP	AUC
		✓			3	0.7714	0.8759	0.8195	0.8348	0.9311
		✓	✓		3	0.8476	0.9537	0.8938	0.9275	0.9718
	✓	✓	✓		3	0.8473	**0.9803**	0.9074	0.9554	**0.9862**
✓	✓	✓	✓	✓	3	0.6574	0.904	0.7147	0.8261	0.9401
✓	✓	✓	✓		1	0.8381	0.9538	0.8906	0.9187	0.9678
✓	✓	✓	✓		6	0.8095	0.9732	0.8803	0.938	0.9793
✓	✓	✓	✓		3	**0.9238**	0.9027	**0.9081**	**0.9584**	0.9855

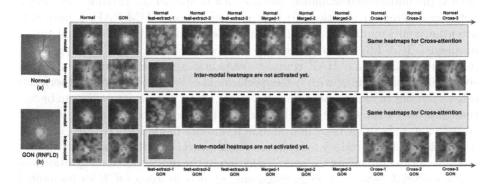

Fig. 3. Visualization. Case (a): Normal; Case (b): GON (Glaucomatous optic neuropathy) with RNFLD. Each column denotes a different class decision or a different stage in our framework. "Merged" denotes Merged-attention, "Cross" denotes Cross-attention. It is recommended to zoom in to view this figure.

tively. Reversing the order of the HAF module brings a decrease, which indicates that the modal-agnostic features should be extracted after the Merged-attention.

3.4   Visualization

For visualization, we employ a class-dependent relevance-based method [4] that captures inter- and intra-modal relevance. Since MILR has aggregated the OCT into high-level $T_{O,agg}^{depth}$ features which are complex to visualize, we choose CFP images to interpret the mechanism of how different modalities interact. For each case presented in Fig. 3, intra-modal and inter-modal heatmaps are calculated by CFP tokens and OCT tokens, respectively. In intra-modal maps, the salient area is centralized on the optic disc, while the inter-modal maps are sensitive to the temporal region where OCT can provide fine-grain stereo information of the optic nerve, indicating that our framework incorporates multi-modal information. The sparsely distributed situation, e.g., case(a) GON's inter-modal view, may be attributed to the lack of significant lesions in the image, causing a random selection of tokens. Also, we visualize how the framework considers the correct prediction with the network going deeper. In case (b), deeper layers concentrate intra- and inter-modal features on lesion areas. The intra-modal Merged-attention focuses on the optic disc(modal-specific feature), and the inter-modal Cross-attention maps are more sensitive to the temporal region (modal-agnostic feature), demonstrating the effectiveness of HAF in extracting and combining modal-agnostic and modal-specific features to make the multi-modal decision.

4   Conclusion

The challenges in multi-modal glaucoma recognition include the huge discrepancies, the unbalanced amounts, and the lack of spatial information interaction between different modalities. To this end, we propose MM-RAF, a pure self-attention multi-modal framework consisting of three modules dedicated to the problems: BCA fills the semantic gap between CFP and OCT and promotes robustness. MILR and HAF complete semantic aggregation and comprehensive relationship probing with better performance. While MM-RAF outperforms other solutions in multi-modal glaucoma recognition, the performance can be further improved with sufficient data. Our next direction is to utilize a lightweight transformer to leverage more information from both modalities. Besides, addressing the issue of uncertainty measurement and preventing the bias of any specific modality from influencing the overall decision in the multi-modal recognition scenario is crucial, especially when diagnosing glaucoma using OCT for its limited specificity. Cross-modal uncertainty measurement is also our further research direction.

References

1. An, G., et al.: Glaucoma diagnosis with machine learning based on optical coherence tomography and color fundus images. J. Healthcare Eng. **2019** (2019)
2. Asaoka, R., et al.: Using deep learning and transfer learning to accurately diagnose early-onset glaucoma from macular optical coherence tomography images. Am. J. Ophthalmol. **198**, 136–145 (2019)
3. Cai, Z., Lin, L., He, H., Tang, X.: Corolla: an efficient multi-modality fusion framework with supervised contrastive learning for glaucoma grading. In: 2022 IEEE 19th International Symposium on Biomedical Imaging (ISBI), pp. 1–4. IEEE (2022)
4. Chefer, H., Gur, S., Wolf, L.: Generic attention-model explainability for interpreting bi-modal and encoder-decoder transformers. In: Proceedings of the IEEE/CVF International Conference on Computer Vision, pp. 397–406 (2021)
5. Chen, X., Xie, S., He, K.: An empirical study of training self-supervised vision transformers. In: Proceedings of the IEEE/CVF International Conference on Computer Vision, pp. 9640–9649 (2021)
6. Ding, F., Yang, G., Ding, D., Cheng, G.: Retinal nerve fiber layer defect detection with position guidance. In: Martel, A.L., et al. (eds.) MICCAI 2020. LNCS, vol. 12265, pp. 745–754. Springer, Cham (2020). https://doi.org/10.1007/978-3-030-59722-1_72
7. Dosovitskiy, A., et al.: An image is worth 16x16 words: transformers for image recognition at scale. In: ICLR (2021)

8. Harizman, N., et al.: The isnt rule and differentiation of normal from glaucomatous eyes. Arch. Ophthalmol. **124**(11), 1579–1583 (2006)

9. Lee, J., Kim, Y.K., Park, K.H., Jeoung, J.W.: Diagnosing glaucoma with spectral-domain optical coherence tomography using deep learning classifier. J. Glaucoma **29**(4), 287–294 (2020)

10. Li, J., Selvaraju, R., Gotmare, A., Joty, S., Xiong, C., Hoi, S.C.H.: Align before fuse: vision and language representation learning with momentum distillation. Adv. Neural. Inf. Process. Syst. **34**, 9694–9705 (2021)

11. Li, X., et al.: Multi-modal multi-instance learning for retinal disease recognition. In: Proceedings of the 29th ACM International Conference on Multimedia, pp. 2474–2482 (2021)

12. Liu, Z., et al.: Swin transformer: hierarchical vision transformer using shifted windows. In: Proceedings of the IEEE/CVF International Conference on Computer Vision (ICCV) (2021)

13. Mehta, P., et al.: Automated detection of glaucoma with interpretable machine learning using clinical data and multimodal retinal images. Am. J. Ophthalmol. **231**, 154–169 (2021)

14. Nagrani, A., Yang, S., Arnab, A., Jansen, A., Schmid, C., Sun, C.: Attention bottlenecks for multimodal fusion. Adv. Neural. Inf. Process. Syst. **34**, 14200–14213 (2021)

15. Raghavendra, U., Bhandary, S.V., Gudigar, A., Acharya, U.R.: Novel expert system for glaucoma identification using non-parametric spatial envelope energy spectrum with fundus images. Biocybernetics Biomed. Eng. **38**(1), 170–180 (2018)

16. Ran, A.R., et al.: Detection of glaucomatous optic neuropathy with spectral-domain optical coherence tomography: a retrospective training and validation deep-learning analysis. The Lancet Digital Health **1**(4), e172–e182 (2019)

17. Touvron, H., Cord, M., Douze, M., Massa, F., Sablayrolles, A., Jegou, H.: Training data-efficient image transformers; distillation through attention. In: International Conference on Machine Learning, vol. 139, pp. 10347–10357, July 2021

18. Wightman, R.: Pytorch image models (2019). https://github.com/rwightman/pytorch-image-models. https://doi.org/10.5281/zenodo.4414861

19. Wu, J., et al.: Gamma challenge: Glaucoma grAding from Multi-Modality imAges (2022)

A Modulatory Elongated Model for Delineating Retinal Microvasculature in OCTA Images

Mohsin Challoob[✉], Yongsheng Gao, Andrew Busch, and Weichuan Zhang

Institute for Integrated and Intelligent Systems, Griffith University, Brisbane, Australia
mohsin.challoob@griffithuni.edu.au, {yongsheng.gao,
a.busch}@griffith.edu.au, zwc2003@163.com

Abstract. Robust delineation of retinal microvasculature in optical coherence tomography angiography (OCTA) images remains a challenging task, particularly in handling the weak continuity of vessels, low visibility of capillaries, and significant noise interferences. This paper introduces a modulatory elongated model to overcome these difficulties by exploiting the facilitatory and inhibitory interactions exhibited by the contextual influences for neurons in the primary visual cortex. We construct the receptive field of the neurons by an elongated representation, which encodes the underlying profile of vasculature structures, elongated-like patterns, in an anisotropic neighborhood. An annular function is formed to capture the contextual influences presented in the surrounding region outside the neuron support and provide an automatic tuning of contextual information. The proposed modulatory method incorporates the elongated responses with the contextual influences to produce spatial coherent responses for delineating microvasculature features more distinctively from their background regions. Experimental evaluation on clinical retinal OCTA images shows the effectiveness of the proposed model in attaining a promising performance, outperforming the state-of-the-art vessel delineation methods.

Keyword: Retinal vasculature · OCTA · Elongated responses · Contextual information · Modulatory influences · Vessel features

1 Introduction

Optical coherence tomography angiography (OCTA) is a noninvasive ophthalmic imaging modality that captures the thin vessels and capillaries, named as microvasculature, present around the fovea and parafovea regions at various retinal depths. The 2D *en face* OCTA images have been increasingly used in clinical investigations for inspecting retinal eye diseases and systemic conditions at the capillary level resolution. More specifically, the morphological changes of retinal vasculature distributed within parafovea regions are associated with diseases such as diabetic retinopathy, early-stage glaucomatous optic neuropathy, macular telangiectasia type 2, uveitis, and age-related macular degeneration [1–3]. In addition, analyzing retinal microvasculature at different depth layers can offer new pathological features that have not been reported before for clinically related

© The Author(s), under exclusive license to Springer Nature Switzerland AG 2023
H. Greenspan et al. (Eds.): MICCAI 2023, LNCS 14226, pp. 714–723, 2023.
https://doi.org/10.1007/978-3-031-43990-2_67

findings. For instance, recent studies [4, 5] have manifested that the variations in the vascular morphology exhibited in OCTA images are associated with Alzheimer's disease, mild cognitive impairment, and chronic kidney disease. Therefore, the extraction of microvasculature from OCTA image is of a great interest. The reliability of phenotypes calculated for diagnosing retinal vascular-related diseases depend on the quality of segmented vascular trees. Since retinal vascular structures appear as a wire mesh-like network that is associated with numerous branching and fusing, and considerable variations in contrast, the manual annotation of vasculature network is a labor-intensive, time-consuming, and error-prone procedure.

The automated delineation of retinal vasculature from OCTA images encounters several problems such as low signal-to-noise ratio (SNR), projection and motion artifacts, and inhomogeneous image background. A few methodologies have been used in the literature for detecting vessel trees in OCTA images. The majority of the vessel delineation approaches have been presented for color fundus images. Generally, the existing methods for vessel extraction from OCTA images can be categorized into supervised [2, 6, 7] and unsupervised approaches [3, 8, 9]. Li et al. [6] proposed an image projection network that takes 3D OCTA data as input and produces 2D vessel segmentation results. A channel and spatial attention network was introduced by Mou et al. [7] for extracting fine vessels from OCTA images. Recently, an OCTA-Net model by Ma et al. [2] was presented to segment fine and coarse vessels separately. In spite of the popularity of deep learning models in the vessel detection task, these algorithms typically need a large amount of annotated samples used in intensive training processes to achieve vessel extraction task. Apart from supervised learning, Yousefi et al. [8] developed a filtering approach based on multi-scale Hessian filter and morphological operations for the detection of microvascular structures in OCTA images. Zhang et al. [3] combined curvelet denoising and optimally oriented flux algorithms to effectively enhance OCTA microvasculature. Gao et al. [9] introduced a reflectance-adjusted thresholding method for binarizing superficial vascular complexes of en face retinal OCTA images. Further, the delineation methods [10–15] have attained encouraging results for vessel delineation in fundus color images. An operator inspired by the push-pull inhibition in the visual cortex was introduced by Strisciuglio et al. [10] to increase the robustness of vessel detection against noise. Regularized volumed ratio [13] considered vessels as rounded structures to produce strong responses for vessels with low contrast and preserve various vessel features. A bowler-hat transform [15] was proposed to detect the inherent features of vessel-like structures. The existing algorithms have achieved a great progress for vessel enhancement and extraction. However, the following problems remain unsolved and need to be overcome for a reliable vasculature delineation in OCTA images. (i) Some vessels and capillaries suffer from the problem of weak continuity, which leads to a disjoint detection in segmented vascular trees. (ii) Due to the poor SNR of OCTA images, some capillaries are presented with an inadequate contrast, causing difficulty in differentiating them from inhomogeneous background. (iii) OCTA images are associated with high noise level, which significantly interferes with vessel structures and makes their boundary irregular.

This paper introduces a new modulatory elongated model to advance the delineation methodology of retinal microvasculature in OCTA images via addressing the above problems. The contributions of this work are summarized as follows. (1) A modulatory function is proposed to include two simultaneous facilitatory and inhibitory delineation processes that distinguish vascular trees more conspicuously from background. (2) The responses of the elongated representation encode the intrinsic profile of the vessels and capillaries, elongated-like shape, which retains the subtle intensity changes of vessels and capillaries for solving the continuity issue. (3) The proposed method disambiguates the region surrounding vessel structures for addressing the disturbance of noise at vascular regions. (4) Our method achieves the best quantitative and qualitative results over the state-of-the-art vessel delineation benchmarks.

2 Methodology

The primary visual cortex (V1) has an essential role in the perception of objects in the visual system. The physiological studies revealed that the response of a V1 neuron to a stimulus in its receptive field can be either increased or decreased when more stimuli are added in the region surrounding the receptive field. The stimuli falling outside the classical receptive field (CRF) can exert modulatory effects on the activities of neurons in V1. The surrounding area beyond the CRF is named as a non-classical receptive filed (nCRF) modulation region, which captures long-range contextual information for modulating the neuron responses [16–18]. The modulatory effects resulted from the interaction between the CRF and nCRF extend the cortical area and allow visual cortex neurons to incorporate a broader range of visual field information to identify complex scenes. Based on the above neurophysiological evidences, we propose our modulatory elongated model as follows:

Responses of V1 Neurons. We propose to describe the responses of orientation-selectivity V1 neurons to the stimuli placed within the classical receptive field (CRF) by an elongated representation. The elongated kernels [19] can effectively simulate neuron responses at various characteristics such as preferred orientation, spatial scale and profile shape. The elongated responses are defined as:

$$\Psi_{\sigma,\rho}^{\theta}(x, y) = \frac{\rho^2}{\sigma^2}\left(\frac{\rho^2}{\sigma^2}(x\cos\theta + y\sin\theta)^2 - 1\right)g_{\sigma,\rho}^{\theta}(x, y) \tag{1}$$

$$g_{\sigma,\rho}^{\theta}(x, y) = \frac{1}{2\pi\sigma^2}\exp\left(-\frac{1}{2\sigma^2}[x, y]R_{-\theta}\begin{bmatrix}\rho^2 & 0 \\ 0 & \rho^{-2}\end{bmatrix}R_{\theta}[x, y]^T\right) \tag{2}$$

where $\sigma > 0$ is the scaling factor that determines the width of the kernel and $\rho > 1$ represents the anisotropic index that controls the elongation of kernel profiles. T is the matrix transpose and R_{θ} denotes the rotation matrix with angle θ while (x, y) is a point location. In our implementation, we set $\sigma \in [1 : 0.5 : 2.5]$, $\rho \in (1 : 0.1 : 1.5]$ and $\theta \in \left\{\frac{\pi}{16}, \frac{2\pi}{16}, \frac{3\pi}{16}, \ldots, \frac{15\pi}{16}, \pi\right\}$. $\Psi_{\sigma,\rho}^{\theta}(x, y)$ forms a pool of neuron responses for processing the local stimuli within the CRF. For an input OCTA image $\mathbb{I}(x, y)$, the final CRF response of a V1 neuron is obtained as:

$$\psi(x, y) = \max_{\sigma,\rho,\theta}\left(\left(\Psi_{\sigma,\rho}^{\theta}\circledast\mathbb{I}\right)(x, y)\right) \tag{3}$$

where the notation $\circledast$ represents the convolution operation in the spatial domain.

Contextual Influences for V1 Neurons. Physiological findings [20–22] illustrated that the region outside the CRF, which is termed as nCRF, of V1 neurons is alone unresponsive to visual stimuli, but it can manifest contextual influences on the neural responses to stimuli within the CRF. The contextual influences perform as a modulatory process that allows neurons in V1 to accumulate information from relatively large parts of visual space for participating in complex perceptual tasks. Also, the majority of modulatory influences are inhibitory while the reminders are facilitatory. Statistical data [20] exhibited that around 80% of the orientation-selectivity neurons in V1 manifest the inhibitory effect. About 40% of these neurons show the inhibition regardless of the relative orientation between the surrounding stimuli and the optimal stimulus. Further, it has been indicated that the modulatory strength from the nCRF decays exponentially with increasing the distance from the center of the CRF [20–22]. Therefore, the intrinsic connections between the neurons and the region around them are distance related influences. Accordingly, we consider an isotropic modulatory behavior, in which contextual influences are independent of the orientation of surrounding patterns, but they take into account the distance to the surroundings. Then, an annular function ($\mathbb{H}_i$) that captures contextual influences ($\mathbb{H}_i$) for modulating V1 neurons is expressed as:

$$\mathbb{H}_i(x, y) = \| F^{-1}((F(\mathbb{W}) * F(\psi))(x, y)) \|_2 \tag{4}$$

$$\mathbb{W}(x, y) = \frac{\lfloor \mathbb{D}o\mathbb{G}(x, y) \rfloor}{\| \lfloor \mathbb{D}o\mathbb{G}(x, y) \rfloor \|_1} \tag{5}$$

$$\mathbb{D}o\mathbb{G}(x, y) = \frac{1}{2\pi (k\sigma_w)^2} \exp\left(-\left(\frac{x^2 + y^2}{2(k\sigma_w)^2}\right)\right) - \frac{1}{2\pi \sigma_w^2} \exp\left(-\left(\frac{x^2 + y^2}{2\sigma_w^2}\right)\right) \tag{6}$$

where F and F^{-1} represent forward and inverse discrete Fourier transforms, respectively while $\|.\|_1$ and $\|.\|_2$ indicate L_1 norm and L_2 norm, respectively. $\lfloor.\rfloor$ denotes a truncated function that replaces negative values with zero. $\mathbb{W}(x, y)$ is a distance weighting function that is calculated by a 2D non-negative difference of Gaussian function $\mathbb{D}o\mathbb{G}(x, y)$ [20] with a standard deviation $\sigma_w = 7$, which is utilized to simulate the strengths of neuron connection in distance. The factor k calculates the size of the nCRF region, and we set $k = 4$ to be consistent with the neurophysiological evidence [21] that the extent of the nCRF is 2 to 5 times larger than that of CRF.

Contrast Influence on Contextual Information. Physiological experiments [22] have shown that contextual influences can vary with contrast attributes. For example, the inhibitory strength is decreased by lowering the contrast of stimulus within CRF, but when increasing the contrast, the inhibition grows relatively stronger. Thus, we introduce a contrastive operator ($\mathbb{S}_i$), as defined below, to control the modulatory strength of the nCRF so that the contextual influences are adaptively varied with contrast at each location.

$$\mathbb{S}_i(x, y) = \frac{C_l^2(x, y)}{(C_l^2 + S_d^2)(x, y)(1 - 1/N^2)} \tag{7}$$

where C_l is a local range measure that returns the range value (maximum value-minimum value) using a patch of $N \times N$ neighborhood around the pixel (x, y) in the response of ψ.

S_d computes the local standard deviation of the $N \times N$ patch around the corresponding input pixel of ψ response. We use $N = 5$ in our implementation.

Modulatory Function. The essence of the proposed modulatory model is that the response of the elongated representation at a specific point is modulated by the response of the representation presented in the area outside the region of the representation interest. We integrate the elongated responses with the contextual influences at each pixel to produce spatial coherent responses that enhance vessel structures more conspicuously from their background while reducing noise disturbances with vascular regions. The proposed modulatory model ($\mathbb{M}$) is defined as:

$$\mathbb{M}(x, y) = \frac{(\psi - \mathbb{S}_i.\mathbb{H}_i)(x, y)}{(\mathbb{S}_i + \mathbb{H}_i)(x, y) + \epsilon} \tag{8}$$

where ϵ is a small value to avoid division by zero. When there are no spurious signals in the region surrounding the elongated responses, the modulatory influence from ($\mathbb{H}_i$ and $\mathbb{S}_i$) produces a weak response and the numerator ($\psi - \mathbb{S}_i.\mathbb{H}_i$) of $\mathbb{M}$ becomes almost equal to the ψ. As a result, the term ($\mathbb{H}_i + \mathbb{S}_i$) in the denominator together performs a faciliatory process for improving the responses of ψ. However, the influence of ($\mathbb{H}_i$ and $\mathbb{S}_i$) has a strong response when containing spurious signals in the surroundings. Then, the ($\mathbb{H}_i$ and $\mathbb{S}_i$) behave as an inhibitory process in both the numerator and denominator, which drops off the contribution of ψ to almost zero response.

3 Experimental Results

Datasets and Metrics. The proposed method is assessed on the ROSE-1 database [2], which is captured from 39 participants that consist of 26 subjects with Alzheimer disease and 13 healthy controls. We perform the experiments on the superficial vascular complexes (SVC) dataset and the inner retinal vascular plexus that includes both SVC and deep vascular complexes (DVC)-termed as (SVC + DVC) dataset. Vessel delineation results are only obtained on test set images of the ROSE-1 datasets. Also, we examine our method on OCTA500_6mm dataset (300 images) [6] using the maximum projection between the internal limiting membrane (ILM) layer and the outer plexiform layer (OPL), named as an ILM_ OPL projection map. Most of OCTA500_6mm subjects (69.7%) were taken from a population with various retinal diseases such as age-related macular degeneration, choroidal neovascularization, central serous chorioretinopathy, diabetic retinopathy and retinal vein occlusion. For the used datasets, the pixel-level annotations of human experts are employed as the ground truths in our calculations. The following metrics are used to measure the quality of the delineation results in comparison with human annotations: precision-recall curve, accuracy (ACC), false discovery rate (FDR), geometric mean value (P_{VR}) of the positive predictive rate and the negative predictive rate.

Results and Comparisons. Figure 1 illustrates some examples of vascular delineation results produced by the proposed method and their corresponding manual annotations on OCTA images from the ROSE-1 (SVC and SVC + DVC) and OCTA500_6mm datasets.

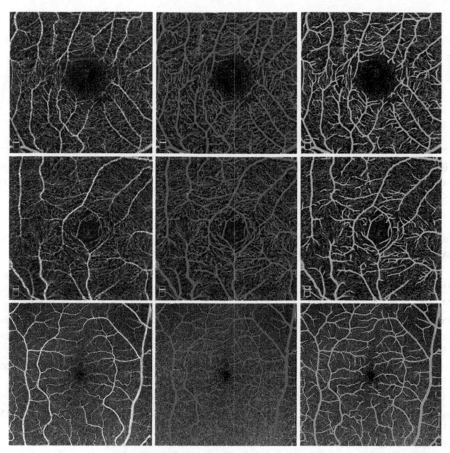

Fig. 1. Illustrative results obtained by the proposed method on example images from ROSE1-SVC (first row), ROSE1-(SVC + DVC) (second row), and OCTA500_6mm (ILM_ OPL) (last row) datasets. From left to right: original images, manual annotation results (red), and detection results by the proposed method (orange) respectively.

The performance of the proposed method on the used datasets is quantified using the precision-recall curve and compared with seven state-of-the-art vessel delineation benchmarks, as exhibited in Fig. 2. The benchmark approaches are robust inhibition-augmented curvilinear operator (RUSTICO) [10], morphological bowler-hat transform (MBT) [15], regularized volume ratio (RVR) [13], probabilistic fractional tensor (PFT) [14], scale and curvature invariant ridge detector (SCIRD) [12], phase congruency tensor (PCT) [11], and the second order generalized Gaussian directional derivative (SOGGDD) filter [19]. Table 1 reports the delineation results of all methods from the precision-recall curve by selecting the best threshold that returns the highest average ACC on each dataset. As shown in Fig. 2, the proposed method yields encouraging delineation results on the ROSE-1 and OCTA500_6mm images, validating its ability in extracting vascular structures from background interferences more effectively than the benchmark

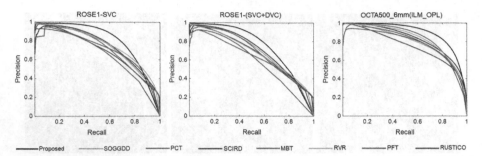

Fig. 2. Precision-Recall curves for the performance comparison of the proposed method and the state-of-the-art vessel delineation benchmarks on ROSE-1 (SVC and SVC + DVC) and OCTA500_6mm (ILM_ OPL) datasets.

Table 1. Delineation results of the proposed method compared to the state-of-the-art benchmarks on ROSE-1 and OCTA500_6mm (ILM_ OPL) datasets. Bold values indicate the best results.

Dataset	ROSE-1 (SVC)				ROSE-1 (SVC + DVC)				OCTA500_6 mm			
Method	$ACC \uparrow$	$P_{VR} \uparrow$	$FDR \downarrow$	$T_s \downarrow$	$ACC \uparrow$	$P_{VR} \uparrow$	$FDR \downarrow$	$T_s \downarrow$	$ACC \uparrow$	$P_{VR} \uparrow$	$FDR \downarrow$	$T_s \downarrow$
Proposed	**0.910**	**0.883**	**0.149**	2.80	**0.893**	**0.868**	**0.163**	2.80	**0.966**	**0.915**	**0.141**	3.86
SOGGDD [19]	0.888	0.844	0.209	2.68	0.876	0.834	0.215	2.68	0.949	0.857	0.238	3.75
RUSTICO [10]	0.874	0.810	0.265	1.50	0.868	0.810	0.256	1.50	0.941	0.835	0.270	2.08
PCT [11]	0.873	0.817	0.248	1.12	0.861	0.808	0.251	1.12	0.949	0.862	0.226	1.50
SCIRD [12]	0.890	0.849	0.202	1.40	0.878	0.839	0.208	1.40	0.952	0.876	0.203	1.85
PFT [14]	0.880	0.815	0.263	0.90	0.857	0.789	0.288	0.90	0.950	0.870	0.212	1.32
MBT [15]	0.884	0.851	0.188	**0.32**	0.874	0.839	0.201	**0.32**	0.957	0.898	0.165	**0.40**
RVR [13]	0.889	0.836	0.230	0.33	0.867	0.808	0.259	0.33	0.954	0.884	0.188	0.42

approaches. Our method outperforms all benchmarks, as reported in Table 1, by achieving the best results of ACC, FDR, and P_{VR} on the used datasets. Our delineation scores show large margins of improvement compared with benchmarks. For example, the proposed method attains the best scores of $ACC = 0.910$, $P_{VR}= 0.883$, and $FDR = 0.149$ on ROSE-1 (SVC) dataset whilst the second-best scores obtained by benchmarks are $ACC = 0.890$ [12], $P_{VR} = 0.851$[15], and $FDR = 0.188$[15]. However, the proposed method consumes in average a computational time T_s of 2.80 s and 3.86 s for processing an image of ROSE-1 and OCTA500_6mm datasets, respectively, which is longer than the benchmarks.

Performance on Challenging Cases. In Fig. 3, we investigate the delineation performance of the proposed method and comparative benchmarks on some OCTA patch images that have the challenging cases of vessels and capillaries with weak continuity

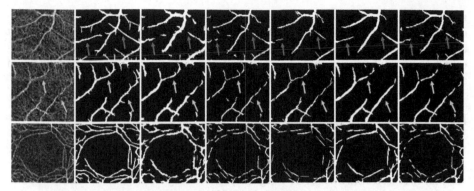

Fig. 3. Delineation results in some difficult cases. From left to right: input image, the responses of the proposed method, SCIRD, PFT, MBT, RUSTICO, and RVR benchmarks, respectively. Red arrows (first row) indicate the vessels and capillaries presented with weak continuity while yellow arrows (second row) refer to where noise signals damage vessel features. The third row includes a foveal avascular zone, in which the majority of capillaries have low contrast.

(first row), noise interferences with vessel structures (second row), and inadequate contrast of capillaries in a foveal avascular zone (third row). As illustrated in Fig. 3-first row, the proposed method attains a superior delineation performance over the benchmarks for sustaining the spatial continuity of capillaries and vessels (red arrows) in resolving the disconnections in extracted vessel trees. Also, our method produces better results over the benchmarks in overcoming the interferences of noise that damage the spatial intensity of vessel structures, as shown in Fig. 3-second row (yellow arrows). Figure 3-third row shows the encouraging ability of the proposed method in addressing the challenge of the capillaries with low visibility, which are much better extracted than the benchmark approaches.

4 Conclusion

This paper presents a modulatory elongated model that exploits contextual modulatory effects presented for tuning the responses of neurons in V1. The proposed method incorporates the elongated responses, neuron responses, at a certain location with either a facilitatory or inhibitory process, contextual information, to reliably enable the delineation of vasculatures in OCTA images and the suppression of background inhomogeneities simultaneously. The validation phase on clinically relevant OCTA images illustrates the effectiveness of our method in producing promising vessel delineation results. The proposed method not only obtains the better quantitative results over the state-of-the-art benchmarks, but also produces an encouraging performance for retaining the continuity of vessels and capillaries, detecting the low-visibility capillaries, and handling spurious signals of noise and artifacts that interfere with vessel structures. Encouraging experimental results demonstrate the effectiveness of the proposed modulatory elongated model in improving the vessel delineation performance. Further, the proposed model is a non-learning approach that does not require any annotated samples or training procedures for performing vasculature detection. As the proposed method is not only responsive to the intrinsic profile of vessels, but also is sensitive to the region surrounding vessel structures, it can be a better alternative to the existing vessel delineation approaches that depend on surroundings-unaware operators.

References

1. Kashani, A.H., et al.: Optical coherence tomography angiography: a comprehensive review of current methods and clinical applications. Prog. Retin. Eye Res. **60**, 66–100 (2017)
2. Ma, Y., et al.: ROSE: a retinal OCT-angiography vessel segmentation dataset and new model. IEEE Trans. Med. Imaging **40**(3), 928–939 (2021)
3. Zhang, J., et al.: 3D shape modeling and analysis of retinal microvasculature in OCT-angiography images. IEEE Trans. Med. Imaging **39**(5), 1335–1346 (2020)
4. Yoon, S.P., et al.: Retinal microvascular and neurodegenerative changes in Alzheimer's disease and mild cognitive impairment compared with control participants. Ophthalmol. Retina **3**(6), 489–499 (2019)
5. Vadalà, M., Castellucci, M., Guarrasi, G., et al.: Retinal and choroidal vasculature changes associated with chronic kidney disease. Graefe's Arch. Clin. Exp. Ophthalmol. **257**, 1687–1698 (2019)
6. Li, M., et al.: Image projection network: 3D to 2D image segmentation in OCTA images. IEEE Trans. Med. Imaging **39**(11), 3343–3354 (2020)
7. Mou, L., et al.: CS-Net: channel and spatial attention network for curvilinear structure segmentation. In: Shen, D., et al. (eds.) MICCAI 2019. LNCS, vol. 11764, pp. 721–730. Springer, Cham (2019). https://doi.org/10.1007/978-3-030-32239-780
8. Yousefi, S., Liu, T., Wang, R.K.: Segmentation and quantification of blood vessels for OCT-based micro-angiograms using hybrid shape/intensity compounding. Microvasc. Res. **97**, 37–46 (2015)
9. Gao, S.S., et al.: Compensation for reflectance variation in vessel density quantification by optical coherence tomography angiography. Invest. Ophthalmol. Vis. Sci. **57**(10), 4485–4492 (2016)
10. Strisciuglio, N., Azzopardi, G., Petkov, N.: Robust inhibition-augmented operator for delineation of curvilinear structures. IEEE Trans. Image Process. **28**(12), 5852–5866 (2019)
11. Obara, B., Fricker, M., Gavaghan, D., Grau, V.: Contrast-independent curvilinear structure detection in biomedical images. IEEE Trans. Image Process. **21**(5), 2572–2581 (2012)
12. Annunziata, R., Kheirkhah, A., Hamrah, P., Trucco, E.: Scale and curvature invariant ridge detector for tortuous and fragmented structures. In: Navab, N., Hornegger, J., Wells, W.M., Frangi, A.F. (eds.) MICCAI 2015. LNCS, vol. 9351, pp. 588–595. Springer, Cham (2015). https://doi.org/10.1007/978-3-319-24574-4_70
13. Jerman, T., Pernuš, F., Likar, B., Špiclin, Ž: Enhancement of vascular structures in 3D and 2D angiographic images. IEEE Trans. Med. Imaging **35**(9), 2107–2118 (2016)
14. Alhasson, H.F., Alharbi, S.S., Obara, B.: 2D and 3D vascular structures enhancement via multiscale fractional anisotropy tensor. In: Leal-Taixé, L., Roth, S. (eds.) ECCV 2018. LNCS, vol. 11134, pp. 365–374. Springer, Cham (2019). https://doi.org/10.1007/978-3-030-11024-6_26
15. Sazak, Ç., Nelson, C.J., Obara, B.: The multiscale bowler-hat transform for blood vessel enhancement in retinal images. Pattern Recogn. **88**, 739–750 (2019)
16. Zipser, K., Lamme, V., Schiller, P.: Contextual modulation in primary visual cortex. J. Neurosci. **16**(22), 7376–7389 (1996)

17. Li, C., Li, W.: Extensive integration field beyond the classical receptive field of cat's striate cortical neurons – classification and tuning properties. Vis. Res. **34**(18), 2337–2355 (1994)

18. Rossi, A., Desimone, R., Ungerleider, L.: Contextual modulation in primary visual cortex of macaques. J. Neurosci. **21**(5), 1698–1709 (2001)

19. Zhang, W., Sun, C.: Corner detection using second-order generalized Gaussian directional derivative representations. IEEE Trans. Pattern Anal. Mach. Intell. **43**(4), 1213–1224 (2021)

20. Grigorescu, C., Petkov, N., Westenberg, M.A.: Contour and boundary detection improved by surround suppression of texture edges. Image Vis. Comput. **22**(8), 609–622 (2004)

21. Kapadia, M., Westheimer, G., Gilbert, C.: Spatial distribution of contextual interactions in primary visual cortex and in visual perception. J. Neurophysiol. **84**(4), 2048–2062 (2000)

22. Wang, C., Bardy, C., Huang, J., FitzGibbon, T., Dreher, B.: Contrast dependence of center and surround integration in primary visual cortex of the cat. J. Vis. **9**(1), 1–15 (2009)

Clustering Disease Trajectories in Contrastive Feature Space for Biomarker Proposal in Age-Related Macular Degeneration

Robbie Holland[1]([✉]), Oliver Leingang[2], Christopher Holmes[3], Philipp Anders[4], Rebecca Kaye[6], Sophie Riedl[2], Johannes C. Paetzold[1], Ivan Ezhov[7], Hrvoje Bogunović[2], Ursula Schmidt-Erfurth[2], Hendrik P. N. Scholl[4,5], Sobha Sivaprasad[3], Andrew J. Lotery[6], Daniel Rueckert[1,7], and Martin J. Menten[1,7]

[1] BioMedIA, Imperial College London, London, UK
robert.holland15@imperial.ac.uk
[2] Laboratory for Ophthalmic Image Analysis, Medical University of Vienna, Vienna, Austria
[3] Moorfields Eye Hospital NHS Foundation Trust, London, UK
[4] Institute of Molecular and Clinical Ophthalmology Basel, Basel, Switzerland
[5] Department of Ophthalmology, Universitat Basel, Basel, Switzerland
[6] Clinical and Experimental Sciences, University of Southampton, Southampton, UK
[7] Technical University of Munich, Munich, Germany

Abstract. Age-related macular degeneration (AMD) is the leading cause of blindness in the elderly. Current grading systems based on imaging biomarkers only coarsely group disease stages into broad categories that lack prognostic value for future disease progression. It is widely believed that this is due to their focus on a single point in time, disregarding the dynamic nature of the disease. In this work, we present the first method to automatically propose biomarkers that capture temporal dynamics of disease progression. Our method represents patient time series as trajectories in a latent feature space built with contrastive learning. Then, individual trajectories are partitioned into atomic subsequences that encode transitions between disease states. These are clustered using a newly introduced distance metric. In quantitative experiments we found our method yields temporal biomarkers that are predictive of conversion to late AMD. Furthermore, these clusters were highly interpretable to ophthalmologists who confirmed that many of the clusters represent dynamics that have previously been linked to the progression of AMD, even though they are currently not included in any clinical grading system.

Supplementary Information The online version contains supplementary material available at https://doi.org/10.1007/978-3-031-43990-2_68.

Keywords: Contrastive learning · Biomarker discovery · Clustering · Disease trajectories · Age-related macular degeneration

1   Introduction

Age-related macular degeneration (AMD) is the leading cause of blindness in the elderly, affecting nearly 200 million people worldwide [24]. Patients with early stages of the disease exhibit few symptoms until suddenly converting to the late stage, at which point their central vision rapidly deteriorates [12]. Clinicians currently diagnose AMD, and stratify patients, using biomarkers derived from optical coherence tomography (OCT), which provides high-resolution images of

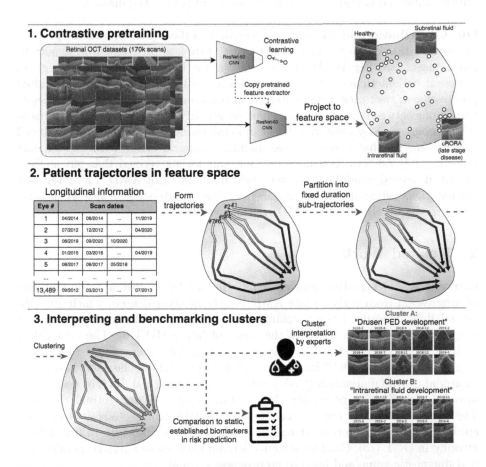

Fig. 1. Our method finds common patterns of disease progression in datasets of longitudinal images. We partition time series into sub-trajectories before introducing a clinically motivated distance function to cluster the sub-trajectories in feature space. The clusters are then assessed by ophthalmologists on their interpretability and ability to capture the progression of AMD.

the retina. However, the widely adopted AMD grading system [7,13], which coarsely groups patients into broad categories for early and intermediate AMD, only has limited prognostic value for late AMD. Clinicians suspect that this is due to the grading system's reliance on static biomarkers that are unable to capture temporal dynamics which contain critical information for assessing progression risk.

In their search for new biomarkers, clinicians have annotated known biomarkers in longitudinal datasets that monitor patients over time and mapped them against disease progression [2,16,19]. This approach is resource-intensive and requires biomarkers to be known *a priori*. Others have proposed deep-learning-based methods to discover new biomarkers at scale by clustering OCT images or detecting anomalous features [17,18,23]. However, these approaches neglect temporal relationships between images and the obtained biomarkers are by definition static and cannot capture the dynamic nature of the disease.

Our Contribution: In this work, we present a method to automatically propose biomarkers that capture temporal dynamics of disease progression in longitudinal datasets (see Fig. 1). At the core of our method is the novel strategy to represent patient time series as trajectories in a latent feature space. Individual progression trajectories are partitioned into atomic sub-sequences that encode transitions between disease states. Then, we identify population-level patterns of AMD progression by clustering these sub-trajectories using a newly introduced distance metric that encodes three distinct temporal criteria. In experiments involving 160,558 retinal scans, four ophthalmologists verified that our method identified several candidates for temporal biomarkers of AMD. Moreover, our clusters demonstrated greater prognostic value for late-stage AMD when compared to the widely adopted AMD grading system.

2   Related Work

Current AMD Grading Systems: Ophthalmologists' current understanding of progression from early to late AMD largely involves drusen, which are subretinal lipid deposits. Drusen volume increases until suddenly stagnating and regressing, which often precedes the onset of late AMD [16]. The established AMD grading system stratifies early and intermediate stages solely by the size of drusen in a single OCT image [1,6,7,10]. Late AMD is classified into either choroidal neovascularisation (CNV), identified by subretinal fluid, or geographic atrophy, signalled by progressive loss of photoreceptors and retinal thinning. The degree of atrophy can be staged using cRORA (complete retinal pigment epithelium and outer retinal atrophy), which measures the width in μm of focal atrophy in OCT [13]. Grading systems derived from these biomarkers offer limited diagnostic value and little to no prognostic capability.

Tracking Evolution of Known Biomarkers: Few research efforts have aimed at quantifying and tracking known AMD biomarkers, mostly drusen, over time

[16,19]. More work has explored the disease progression of Alzheimer's disease (AD), which offers a greater array of quantitative imaging biomarkers, such as levels of tau protein and hippocampal volume. Young et al. [25] fit an event-based model that rediscovers the order in which these biomarkers become anomalous as AD progresses. Vogel et al. [21] find four distinct spatiotemporal trajectories for tau pathology in the brain. However, this only works if biomarkers are known a priori and requires manual annotation of entire time series.

Automated Discovery of Unknown Biomarkers: Prior work for automated biomarker discovery in AMD explores the latent feature space of encoders trained for image reconstruction [18,23], segmentation [27] or generative adversarial networks [17]. However, these neural networks are prone to overfit to their specific task and lose semantic information regarding the disease. Contrastive methods [3,8,26] encode invariance to a set of image transformations, which are uncorrelated with disease features, resulting in a more expressive feature space.

However, all aforementioned methods group single images acquired at one point in time, and in doing so neglect temporal dynamics. The one work that tackles this challenge, and the most related to ours, categorises the time-dependent response of cancer cells to different drugs, measured by the changing distance in contrastive feature space from healthy controls [5].

3   Materials and Methods

3.1   OCT Image Datasets

We use two retinal OCT datasets curated in the scope of the PINNACLE study [20]. We first design and test our method on a *Development dataset*, which was collected from the Southampton Eye Unit. Afterwards, we test our method on a second independent *Unseen dataset*, which was obtained from Moorfields Eye Hospital. All images were acquired using Topcon 3D OCT devices (Topcon Corporation, Tokyo, Japan). After strict quality control, the *Development dataset* consists of 46,496 scans of 6,236 eyes from 3,456 patients. Eyes were scanned 7.7 times over 1.9 years on average at irregular time intervals. The *Unseen dataset* is larger, containing 114,062 scans of 7,253 eyes from 3,819 patients. Eyes were scanned 16.6 times over 3.5 years on average.

A subset of 1,031 longitudes was labelled using the established AMD grading protocols derived from known imaging biomarkers. Early AMD was characterised by small drusen between 63–125μm in diameter. We also recorded CNV, cRORA ($\geq$ 250μm and <1000μm), cRORA ($\geq$ 1000μm) [13] and healthy cases with no visible biomarkers. Visual acuity scores, which measured the patient's functional quality of vision using a LogMAR chart, are available at 83,964 time points.

3.2 Self-supervised Feature Space Using Contrastive Learning

We use the BYOL contrastive loss [8] in Eq. 1 to train a ResNet50 (4x) model f over each batch of twice transformed images x

$$\mathcal{L}(x) = 2 - 2 \frac{\langle f(x), f'(x) \rangle}{||f(x)||_2 \cdot ||f'(x)||_2} \tag{1}$$

where the output of the momentum updated 'teacher' network f' is passed through a stop-gradient, so that only the student network f is updated. As several of the contrastive transformations designed for natural images are inapplicable to medical images, such as solarisation, colour shift and greyscale, we use the set tailored for retinal OCT images by Holland et al. [9]. Models were trained on the entire dataset for 120,000 steps using the Adam optimiser with a momentum of 0.9, weight decay of $1.5 \cdot 10^{-6}$ and a learning rate of $5 \cdot 10^{-4}$. After training f, we first remove the final linear layer before projecting all labelled images to the feature space of 2048 dimensions.

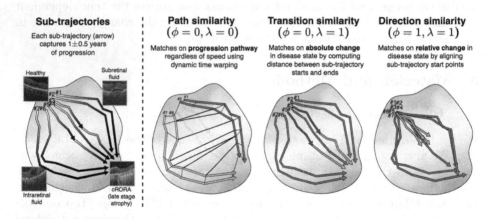

Fig. 2. Illustration of sub-trajectory distance functions which each encode temporal criteria for similarity (see Eq. 4). We illustrate clusters assignments, denoted by colour, resulting from three combinations of ϕ and λ.

3.3 Extracting Sub-trajectories via Partitioning

Naively clustering whole time series of patients ignores two characteristics of longitudinal data. Firstly, individual time series are not directly comparable as patients enter and leave the study at different stages of their overall progression. Secondly, longer time series can record multiple successive transitions in disease stage. Inspired by TRACLUS [11], the state of the art in trajectory clustering, we adapt their *partition-and-group* framework by assuming that trajectories can be partitioned into a common set of *sub-trajectories* that capture singular transitions between progressive states of the disease.

For each eye, we first form piecewise-linear trajectories by linking points in feature space that were derived from consecutively acquired OCT images. We then extract sub-trajectories by finding all sequences of images spanning 1.0 ± 0.5 years of elapsed time within each trajectory. Next, to avoid oversampling trajectories with a shorter time interval between images, we randomly sample at most one sub-trajectory in every 0.5-year time interval.

3.4 Sub-trajectory Distance Functions and Clustering

In order to find common patterns of disease progression among sub-trajectories we cluster them. To this end we introduce a new distance function between sub-trajectories that incorporates three distinct temporal criteria (see Fig. 2). The first, formulated in Eq. 2, matches two sub-trajectories, U and V, of patients who progress between the same start and end states:

$$D_{transition}(U,V) = \|U_{start} - V_{start}\|_2 + \|U_{end} - V_{end}\|_2 . \tag{2}$$

Since all sub-trajectories cover a similar temporal duration, $D_{transition}$ also differentiates between fast and slow progressors and stable periods of no progression. However, by ignoring intermediary images, this metric does not respect the disease pathway along which patients progress. To incorporate this, we include a second metric that measures path dissimilarity, calculated using dynamic time warping (DTW) [4,14,15]. DTW finds the optimal temporal alignment between two time series before computing their distance. This re-alignment allows us to match sub-trajectories that traverse the same disease states in the same order, irrespective of the rate of change between states. We combine $D_{transition}$ with DTW using a $\lambda \in \mathbb{R}, 0 \le \lambda \le 1$ coefficient so the overall distance between U and V is

$$D_{path}(U,V) = \lambda \, D_{transition}(U,V) + (1 - \lambda) \, \text{DTW}(U,V). \tag{3}$$

The third and final temporal criteria is to match time series that progress in the same relative direction, regardless of absolute disease states. We weight the contribution of this with $\phi \in \mathbb{R}, 0 \le \phi \le 1$ in Eq. 4:

$$D_{subtraj}(U,V) = \phi \, D_{path}(U - U_{start}, V - V_{start}) + (1 - \phi) \, D_{path}(U,V). \tag{4}$$

Spectral Clustering: As the non-linearity of $D_{subtraj}$ prohibits the use of k-means for clustering, we instead use spectral clustering [22] to group similar sub-trajectories. Hereby, we construct an affinity matrix $\mathcal{A}$ encoding the negative of the distance $D_{subtraj}$ between all pairs of sub-trajectories. Using $\mathcal{A}$, we group sub-trajectories into K clusters.

Development dataset   **Unseen dataset**

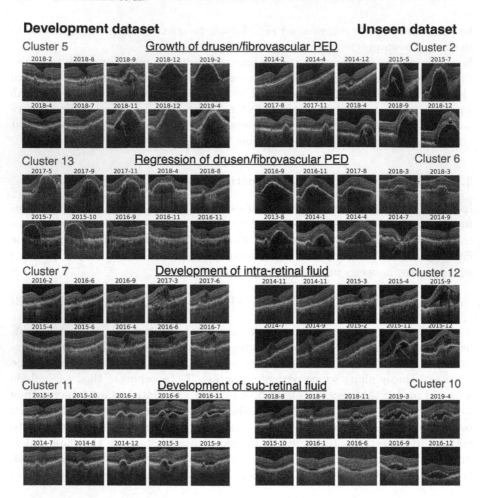

Fig. 3. We show four clusters from the *Development dataset* (left half) and the equivalent clusters in the *Unseen dataset* (right half). Ophthalmologists identified clusters capturing the same progression dynamics in both datasets, providing clinical interpretations (underlined). Clusters show two representative sub-trajectories originating from different patients, each containing five longitudinal images with the time and location of greatest progression marked by arrows.

3.5   Qualitative and Quantitative Evaluation of Clusters

Initially, we tune the hyperparameters, λ, ϕ and K, on the *Development dataset* by heuristically selecting values that result in higher uniformity between subtrajectories within each cluster. Two teams of two ophthalmologists then review 20 sub-trajectories from distinct patients in each cluster, interpreting and summarising any consistently observed temporal dynamics. Next, using the same hyperparameters we apply the method directly to the *Unseen dataset*. The ophthalmologists then review these clusters and confirm whether they capture the

same temporal biomarkers observed in the *Development dataset*.
In addition to the qualitative evaluation, we also validate the utility of our clusters as biomarkers that stratify risk of disease progression. We test this by predicting the time until conversion to late AMD and its subtypes, CNV and cRORA. Additionally, we predict current visual acuity. For prediction, each sub-trajectory is characterised by a vector of size K that encodes proportional similarity to each cluster. This vector is then used by a Lasso linear regression model. Similarly, we fit an equivalent linear regression model to the static biomarkers from the established grading system detailed in Sect. 3.1. We also include a demographic baseline using age and sex. We also add a temporally agnostic baseline that clusters only single time points. Finally, to demonstrate the performance gap between our interpretable approach and black-box supervised learning algorithms, we include a fully supervised deep learning baseline by fitting an SVR directly to the feature space. Each experiment uses 10-fold cross validation on random 80/20 partitions, while ensuring a patient-wise split. Finally, we repeat the entire method, starting from sub-trajectory extraction, followed by clustering and then regression experiments, using 7 random seeds and report the means and standard deviations.

Table 1. Sub-trajectory clusters were comparable to the established clinical grading systems for AMD in predicting future disease, shown by reduced MAE in years for Late AMD, CNV and cRORA and MAE in LogMAR for visual acuity.

	Development dataset			
	Time to Late AMD ↓	Time to CNV ↓	Time to cRORA ↓	Current visual acuity ↓
Demographic	0.756±0.01	0.822±0.012	0.703±0.028	0.381±0.007
Current grading system	0.757±0.01	0.819±0.012	0.685±0.035	0.367±0.008
Single timepoint clusters	0.747±0.013	0.776±0.015	0.630±0.05	0.230±0.005
Sub-trajectory clusters	0.739±0.01	0.748±0.011	0.636±0.031	0.375±0.007
Fully supervised	0.709±0.015	0.726±0.012	0.609±0.033	0.199±0.004
	Unseen dataset			
Demographic	1.343±0.027	1.241±0.017	1.216±0.062	0.188±0.007
Current grading system	1.308±0.018	1.244±0.022	1.286±0.053	0.177±0.008
Single timepoint clusters	1.325±0.049	1.341±0.080	1.297±0.096	0.136±0.005
Sub-trajectory clusters	1.322±0.029	1.235±0.027	1.257±0.056	0.188±0.006
Fully supervised	1.301±0.044	1.298±0.08	1.255±0.097	0.135±0.006

4   Experiments and Results

Sub-trajectory Clusters are Candidate Temporal Biomarkers: By first applying our method to the *Development dataset* we found that using $\lambda = 0.75$, $\phi = 0.75$ and $K = 30$ resulted in the most uniform and homogeneous clusters while still limiting the total number of clusters to a reasonable amount. Achieving the same cluster quality with smaller values of ϕ required many more clusters in order to encode all combinations of possible start and end disease states. The expert ophthalmologists remarked that many of the identified clusters capture dynamics that have already been linked to the progression of AMD, even though they are not currently included in any clinical grading system. Using the same hyperparameters our method generalised to the *Unseen dataset* which yielded clusters with equivalent dynamics and quality (see Fig. 3).

Ophthalmologists identified clusters capturing the same variants of temporal progression in both datasets. They named these as '*rapid growth of drusen pigment epithelial detachments (PED)*', '*regression of drusen PED*', '*development of subretinal fluid*', '*development of intraretinal fluid*', '*development of hypertransmission*' and '*stable state*' (no signs of progression at each disease state).

Sub-trajectory Clusters Predict Conversion to Late AMD: Next, we validated that our clusters are predictive of progression to late AMD. Our sub-trajectory clusters were comparable to, and in some cases outperformed, the current widely adopted grading system in predicting risk of conversion (see Table 1). In all tasks the standard biomarkers are only marginally more indicative of risk than the patient's age and sex. This experiment confirms that our clusters are related to disease progression.

5   Discussion and Conclusion

Motivated to improve inadequate grading systems for AMD that do not incorporate temporal dynamics we developed a method to automatically propose biomarkers that are time-dependent, interpretable, and predictive of conversion to late-stage AMD. We applied our method to two large longitudinal datasets, cataloguing 3,218 total years of disease progression. The found time-dependent clusters were subsequently interpreted by four ophthalmologists. They found them to capture distinct patterns of disease progression that have been previously linked to AMD, but are not currently included in clinical grading systems. Furthermore, we experimentally demonstrated that the found clusters predict conversion to late-stage AMD on par with the established grading system.

In the future, biomarkers identified by our method can be further refined by clinicians. We will also use the full volumetric image to model progression dynamics outside the macular. As late stage patients were overrepresented in our datasets, we also intend to apply our method to datasets with greater numbers of patients progressing from earlier disease stages. Ultimately, we envision that proposals from our method may inform the next generation of grading systems for AMD that incorporate the temporal dimension intrinsic to this dynamic disease.

Acknowledgements. The PINNACLE study is funded by a Wellcome Trust Collaborative Award (ref. 210572/Z/18/Z). This work is also funded by the Munich Center for Machine Learning.

References

1. Bird, A.C., et al.: An international classification and grading system for age-related maculopathy and age-related macular degeneration. Surv. Ophthalmol. **39**(5), 367–374 (1995)
2. Chen, K.G., et al.: Longitudinal study of dark adaptation as a functional outcome measure for age-related macular degeneration. Ophthalmology **126**(6) (2019)
3. Chen, T., Kornblith, S., Norouzi, M., Hinton, G.: A simple framework for contrastive learning of visual representations. In: ICML, pp. 1597–1607. PMLR (2020)
4. Cuturi, M., Blondel, M.: Soft-dtw: a differentiable loss function for time-series. In: International Conference on Machine Learning, pp. 894–903. PMLR (2017)
5. Dmitrenko, A., et al.: Self-supervised learning for analysis of temporal and morphological drug effects in cancer cell imaging data. In: Medical Imaging with Deep Learning (2021)
6. Ferris, F.L., et al.: A simplified severity scale for age-related macular degeneration. Arch. Ophthalmol. **123**(11), 1570–1574 (2005)
7. Ferris III, F.L., Wilkinson, C., Bird, A., Chakravarthy, U., Chew, E., Csaky, K., Sadda, S.R., for Macular Research Classification Committee, B.I., et al.: Clinical classification of age-related macular degeneration. Ophthalmology 120(4) (2013)
8. Grill, J.B., et al.: Bootstrap your own latent-a new approach to self-supervised learning. NeurIPS **33**, 21271–21284 (2020)
9. Holland, R., et al.: Metadata-enhanced contrastive learning from retinal optical coherence tomography images. CoRR abs/2208.02529 (2022)
10. Klein, R., et al.: Harmonizing the classification of age-related macular degeneration in the three-continent AMD consortium **21**(1), 14–23 (2014)
11. Lee, J.G., et al.: Trajectory clustering: a partition-and-group framework. In: ACM SIGMOD, pp. 593–604 (2007)
12. Mitchell, P., et al.: Age-related macular degeneration. The Lancet **392**(10153), 1147–1159 (2018)
13. Sadda, S.R., et al.: Consensus definition for atrophy associated with age-related macular degeneration on oct: classification of atrophy report 3. Ophthalmology **125**(4), 537–548 (2018)
14. Sakoe, H.: Dynamic-programming approach to continuous speech recognition. In: 1971 Proc. the International Congress of Acoustics, Budapest (1971)
15. Sakoe, H., Chiba, S.: Dynamic programming algorithm optimization for spoken word recognition. IEEE Trans. Acoust. Speech Signal Process. **26**(1), 43–49 (1978)
16. Schlanitz, F.G., et al.: Drusen volume development over time and its relevance to the course of age-related macular degeneration. BJO **101**(2), 198–203 (2017)
17. Schlegl, T., et al.: f-anogan: fast unsupervised anomaly detection with generative adversarial networks. Med. Image Anal. **54**, 30–44 (2019)
18. Seeböck, P., et al.: Unsupervised identification of disease marker candidates in retinal oct imaging data. IEEE TMI **38**(4), 1037–1047 (2018)
19. Steinberg, J.S., et al.: Longitudinal analysis of reticular drusen associated with geographic atrophy in age-related macular degeneration. IOVS **54**(6), 4054–4060 (2013)
20. Sutton, J., et al.: Developing and validating a multivariable prediction model which predicts progression of intermediate to late age-related macular degeneration-the pinnacle trial protocol. Eye, pp. 1–9 (2022)
21. Vogel, J.W., et al.: Four distinct trajectories of tau deposition identified in Alzheimer's disease. Nat. Med. **27**(5), 871–881 (2021)

22. Von Luxburg, U.: A tutorial on spectral clustering. Stat. Comput. **17**, 395–416 (2007)
23. Waldstein, S.M., et al.: Unbiased identification of novel subclinical imaging biomarkers using unsupervised deep learning. Sci. Rep. **10**(1), 1–9 (2020)
24. Wong, W.L., et al.: Global prevalence of age-related macular degeneration and disease burden projection for 2020 and 2040: a systematic review and meta-analysis. Lancet Glob. Health **2**(2), e106–e116 (2014)
25. Young, A.L., et al.: A data-driven model of biomarker changes in sporadic Alzheimer's disease. Brain **137**(9), 2564–2577 (2014)
26. Zhao, A., et al.: Prognostic imaging biomarker discovery in survival analysis for idiopathic pulmonary fibrosis. In: MICCAI Proceedings, pp. 223–233. Springer (2022). https://doi.org/10.1007/978-3-031-16449-1_22
27. Zheng, Q., et al.: Pathological cluster identification by unsupervised analysis in 3,822 UK biobank cardiac MRIS. Front. Cardiovasc. Med **7**, 539788 (2020)

Incomplete Multimodal Learning for Visual Acuity Prediction After Cataract Surgery Using Masked Self-Attention

Qian Zhou[1], Hua Zou[1(✉)], Haifeng Jiang[2], and Yong Wang[2]

[1] School of Computer Science, Wuhan University, Wuhan, China
[2] Aier Eye Hospital of Wuhan University, Wuhan University, Wuhan, China
zouhua@whu.edu.cn

Abstract. As the primary treatment option for cataracts, it is estimated that millions of cataract surgeries are performed each year globally. Predicting the Best Corrected Visual Acuity (BCVA) in cataract patients is crucial before surgeries to avoid medical disputes. However, accurate prediction remains a challenge in clinical practice. Traditional methods based on patient characteristics and surgical parameters have limited accuracy and often underestimate postoperative visual acuity. In this paper, we propose a novel framework for predicting visual acuity after cataract surgery using masked self-attention. Especially different from existing methods, which are based on monomodal data, our proposed method takes preoperative images and patient demographic data as input to leverage multimodal information. Furthermore, we expand our method to a more complex and challenging clinical scenario, *i.e.*, the incomplete multimodal data. Firstly, we apply efficient Transformers to extract modality-specific features. Then, an attentional fusion network is utilized to fuse the multimodal information. To address the modality-missing problem, an attention mask mechanism is proposed to improve the robustness. We evaluate our method on a collected dataset of 1960 patients who underwent cataract surgery and compare its performance with other state-of-the-art approaches. The results show that our proposed method outperforms other methods and achieves a mean absolute error of 0.122 logMAR. The percentages of the prediction errors within $\pm$ 0.10 logMAR are 94.3%. Besides, extensive experiments are conducted to investigate the effectiveness of each component in predicting visual acuity. Codes will be available at https://github.com/liyiersan/MSA.

Keywords: Incomplete Multimodal Learning · Visual Acuity Prediction · Self-Attention

Supplementary Information The online version contains supplementary material available at https://doi.org/10.1007/978-3-031-43990-2_69.

1   Introduction

Cataract has been the leading cause of vision loss. As the only treatment option, cataract surgery is one of the most commonly performed surgeries worldwide. Nevertheless, not all patients achieve complete visual recovery after surgery, which can be due to various factors, such as pre-existing eye conditions, surgical complications, and postoperative inflammation. The ability to accurately predict visual recovery can help clinicians identify high-risk patients and provide appropriate interventions to improve their visual outcomes.

Over the past decades, many efforts have been made to predict the Best Corrected Visual Acuity (BCVA) after cataract surgeries. Most of them are based on traditional approaches like retinometer [8,11,13] or visual electrophysiology [2,5,14]. These methods require specialized expertise to perform and are subject to significant variability in their results. Even though some computer-aided approaches have been proposed, most of them use traditional machine learning algorithms [1] and focus on single-modal data [9,15,16]. While they can be effective to some extent, they often have limited predictive power due to the reliance on a single source of information. In addition, traditional machine learning methods (*e.g.*, linear regression, decision trees, and support vector machines) may not be able to capture complex relationships between different modalities, such as clinical data and imaging data. What's more, clinical scenarios are more complex. For example, the multimodal data may be incomplete due to medical conditions, as shown in Fig. 1. Therefore, there is a need for more sophisticated computer-aided techniques that can integrate multiple sources of data and leverage the strengths of each to improve predictive accuracy as well as address the challenging modality-missing problem.

Transformers [12], which are a type of neural network architecture originally developed for natural language processing tasks, have recently shown

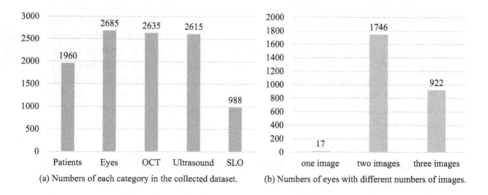

(a) Numbers of each category in the collected dataset.   (b) Numbers of eyes with different numbers of images.

Fig. 1. Statistics for the collected dataset. (a) Number of patients, eyes, and three image modalities. OCT refers to Optical Coherence Tomography, and SLO stands for Scanning Laser Ophthalmoscopy. (b) Number of multimodal or monomodal samples. For instance, "two images" denotes the samples with two image modalities. We can see that only one-third of cases have complete multimodal images.

great promise in computer vision applications, such as image classification [4], object detection [3], and segmentation [20]. Transformers have the ability to learn from large amounts of data and can capture complex patterns and relationships in the data, which makes them well-suited for analyzing multimodal learning. Motivated by the tremendous success of Transformers in multimodal learning [10,17,18], we propose a new framework that utilizes incomplete multimodal data for predicting BCVA after cataract surgeries. In particular, our framework contains three stages: modality-specific feature extraction, attentional feature fusion, and visual acuity prediction. Firstly, for each input modality, a pre-trained efficient transformer will be used to extract features. To better leverage the clinical diagnosis keywords, an auxiliary classification loss is added to the image transformer. And to extract text features more efficiently, we apply a CLIP-like [10] input to combine discrete clinical words (*e.g.*, age, sex, and preoperative visual acuity) into sentences. Secondly, we apply an attentional transformer to fuse multimodal features. Specifically, to address the issue of missing modalities, we introduce modality embeddings and attentional masks to prevent the interference of missing modalities with the remaining modalities. Finally, a prediction head takes the fused features as input to predict the BCVA with the Mean Square Error (MSE) loss.

The main contributions of this work can be summarized as follows: (1) We develop a novel framework that uses multimodal data to predict BCVA as well as tackle the complex modality-missing issue. (2) An auxiliary classification loss is adopted to extract more comprehensive pathological features for images. Also, discrete textual words are combined into sentences to better fit the text transformer input. (3) Extensive experiments are conducted to prove the effectiveness of our method. The compared methods include incomplete multimodal learning approaches and monomodal BCVA prediction methods. We also analyze the importance of each component of our method.

2   Method

2.1   Framework Overview

As shown in Fig. 2, our framework contains three parts: modality-specific encoder, multimodal fusion network, and BCVA prediction head. During feature extraction, we take pre-trained transformers as the backbone, *i.e.*, ViT [4] as the image encoder, and CLIP [10] as the text encoder. After that, a cross-modal transformer is used to fuse features from multiple modalities, in which an attentional mask is added to tackle the missing modalities. Finally, a fully connected (FC) layer is used as the prediction head.

2.2   Monomodal Feature Extraction

Text Encoder. The text encoder is a pre-trained CLIP [10] model. The discrete physiological information is combined into a sentence format that benefits the

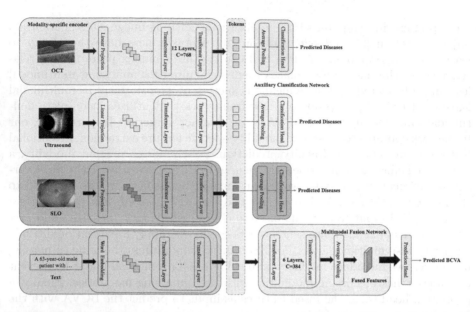

Fig. 2. Pipeline of the proposed framework. The modality-specific encoders utilize vanilla multi-head self-attention. In contrast, the multimodal fusion network employs masked multi-head self-attention. Notably, the fusion network takes features (*i.e.*, tokens) from all modalities as input, whereas each auxiliary classification network receives features from a single modality as input.

text encoder. For instance, the text "male, 67 years old, preoperative visual acuity 0.52 logMAR" will be combined into the sentence "A 67-year-old male patient with preoperative visual acuity of 0.52 logMAR". By combining the text in this way, the physiological texts of all patients are fed into the text transformer in a unified manner. Compared with directly concatenating the texts and inputting them into the model, the combined sentences are more in line with the real scene and are easier to be understood by the model to extract key semantic information. Moreover, the CLIP text encoder is trained on a large text corpus, enabling excellent generalization performance. Thus, during the training of the overall model, the weights of the text encoder can be fixed.

Image Encoder. The image encoder adopts ViT [4]. Since ViT is trained on natural images and may not be directly applicable to medical images, it can not be fixed during the training. However, the pre-trained weights can be utilized to expedite convergence. In addition, there are diagnostic keywords given by ophthalmologists for each image in the dataset. It is not appropriate to directly use the CLIP text encoder to extract medical features from these keywords since CLIP is trained on natural language texts. To this end, we introduce an auxiliary classification loss in the image encoder. Specifically, for each input image, a multi-label classification network is incorporated after the image encoder to predict the diseases contained in the image. The classification network is composed of an average pooling layer and a fully connected layer. For simplicity, we adopt the binary cross-entropy loss as the auxiliary classification loss as Eq. (1).

$$L_{CLS} = \sum W^i L^i_{BCE} \tag{1}$$

where W^i equals 1 if the i-th modality is available and equals 0, otherwise.

By adding the classification loss, the final loss to train the whole model is:

$$L = L_{MSE} + \alpha L_{CLS} \tag{2}$$

where L_{MSE} is the mean square error loss between the predicted BCVA and the ground truth, α is a hyper-parameter and set to 0.5.

2.3 Multimodal Feature Fusion

We use a cross-modal Transformer as the multimodal fusion network. The obtained modality-specific tokens are projected into the same dimension and concatenated into an input sequence. In contrast to the modality-specific Transformer, besides adding positional embeddings to all input tokens, we also add learnable modality-specific embeddings to all tokens indicating the modality information.

Complete Multimodal Learning. In the collected dataset, all cases have corresponding text modality, and almost all cases have corresponding OCT modality. Therefore, we built a complete multimodal prediction model based on the text modality and OCT modality. In such a situation, only the OCT encoder and text encoder will be preserved, while the SLO encoder and Ultrasound encoder will be dropped. Since transformers can handle sequences of any input length, we don't need to make any changes to the fusion network. Using complete multimodal learning, we can compare our methods to other BCVA prediction approaches which do not consider the incomplete multimodal scenario.

Incomplete Multimodal Learning. Not all cases have complete modalities for the three image modalities (*i.e.*, OCT, SLO, and Ultrasound). For the missing modalities, one possible way is to simply represent them by 0 values [18]. However, the 0 values will be regarded as noise by the model as they do not contain any useful information. To avoid model degradation, we add attentional masks in the vanilla self-attention to exclude the interactions among missing modalities and available modalities.

The proposed attentional masks are easy to implement. As shown in Eq. (3), self-attention in transformers is mainly matrix multiplication.

$$Attn(Q, K, V) = softmax(\frac{QK^T}{\sqrt{d_z}})V \tag{3}$$

in which, Q, K, and V are queries, keys, and values obtained from tokens, respectively. d_z is the projection dimension. To avoid interactions between irrelevant tokens, the masked self-attention is computed as:

$$Attn_{Mask}(Q, K, V) = softmax(\frac{QK^T}{\sqrt{d_z}} + M)V \tag{4}$$

in which, M is the mask matrix. For each element in M, it will be 0 if the interactions should be included, or it will be negative infinity to avoid unnecessary interactions. By adding negative infinity, the results of softmax will be very close to 0. Figure 3 shows an example of masks when modalities are missing.

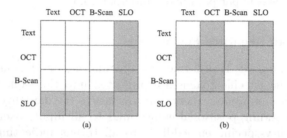

Fig. 3. An example of the attentional mask. (a) means only SLO is missing; (b) represents both OCT and SLO are missing. White cells represent a value of 0, and gray cells represent a value of negative infinity. (Color figure online)

2.4   Implementation Details

To validate the effectiveness of the proposed method, we have conducted extensive experiments implemented with Pytorch and 8×RTX 3090 GPUs. The input images are resized to 224 × 224. The Adam optimizer is adopted with an initial learning rate of 0.001 and $\beta_1 = 0.9$, $\beta_2 = 0.99$. The mini-batch size is set to 32. We train the model for 100 epochs in total, and the learning rate will be decayed by 0.1 every 20 epochs. Besides, we randomly split all samples to 80% for training and 20% for testing. All experiments are conducted with 5-fold cross-validation to produce more solid results.

3   Experiments

3.1   Datasets and Evaluation Metrics

The collected dataset consists of 1960 patients (2685 eyes) having cataract surgeries at Aier eye hospital of Wuhan University. The collected modalities are texts and images. The images contain 2635 Optical Coherence Tomography (OCT), 2615 Ultrasound, and 988 Scanning Laser Ophthalmoscopy (SLO). The textual information includes sex, age, preoperative and postoperative visual acuity. For each image, three ophthalmologists will label it to obtain the diagnosis (*i.e.*, clinical diagnosis keywords) of 14 retinal diseases. The retinal diseases include normal, vitreous opacity, posterior staphyloma, stellate vitreous degeneration, pathological myopia changes, retinal atrophy, macular degeneration, epiretinal membrane, ellipsoid band partially missing, retinoschisis, retinal hemorrhage, macular edema, macular hole, and retinitis pigmentosa. We use Mean Absolute Error (MAE), Symmetric Mean Absolute Percentage Error (SMAPE), and prediction accuracy as the metrics. For simplicity, we consider predictions to be accurate if the prediction errors are within $\pm$ 0.10 logMAR.

3.2   Quantitative Performance

We have compared the results on the collected dataset with other approaches. The compared methods include state-of-the-art methods which aim to predict BCVA using OCT like CTT-Net [15], Wei *et al.* [16], and other algorithms considering incomplete multimodal learning such as Huang *et al.* [6], Ma *et al.* [7], and Zhao *et al.* [19]. For the former methods, we compare our method with them using the complete data. As for the latter approaches, they are not proposed for BCVA prediction. Therefore, we finetune them so that they can be applied to incomplete BCVA prediction data.

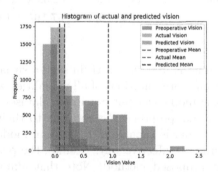

Fig. 4. Distribution of preoperative, predictive, and postoperative visual acuity. Actual vision means the actual postoperative visual acuity.

From Table 1, we can see that our proposed framework achieves the best performance. Specifically, we have improved CTT-Net [15] and shown a sharp rise compared to Wei *et al.* [16] on the complete dataset. Even though CCT-Net uses both text and oct modalities, the utilization of text is still limited. Wei *et al.* [16] directly ignores the textual information and only uses some simple frameworks to predict BCVA, thus achieving the worst results. When using incomplete data, the performance is also improved greatly, and we still achieve the best performance. Huang *et al.* [6] try to apply image synthesis to solve the modality-missing problem. However, the collected images in our dataset are not

Table 1. Quantitative prediction results on the collected dataset. Complete means text and OCT modalities are available and complete. Incomplete means text and three image modalities are available, but the image modalities may be incomplete.

Dataset	Methods	MAE (↓)	SMAPE (↓)	Accuracy (↑)
Complete	CTT-Net [15] (OCT+Text)	0.168 ± 0.014	85.236 ± 3.277	0.887 ± 0.022
	CTT-Net [15] (OCT)	0.174 ± 0.013	89.635 ± 2.881	0.872 ± 0.016
	Wei et al. [16] (OCT)	0.237 ± 0.093	93.587 ± 3.236	0.723 ± 0.056
	Ours (OCT)	0.153 ± 0.012	65.615 ± 1.690	0.901 ± 0.018
	Ours (OCT + Text)	0.142 ± 0.009	62.550 ± 1.668	0.923 ± 0.014
Incomplete	Huang et al. [6]	0.176 ± 0.054	88.672 ± 3.051	0.854 ± 0.017
	Ma et al. [7]	0.139 ± 0.013	61.722 ± 2.007	0.917 ± 0.015
	Zhao et al. [19]	0.133 ± 0.021	59.673 ± 2.362	0.921 ± 0.021
	Ours	**0.122 ± 0.007**	**57.165 ± 1.610**	**0.943 ± 0.012**

Table 2. Ablation study results on incomplete datasets.

Model	MAE (↓)	SMAPE (↓)	Accuracy (↑)
Baseline	0.176 ± 0.014	87.642 ± 3.023	0.874 ± 0.014
Baseline + combined sentences	0.163 ± 0.012	78.932 ± 3.672	0.893 ± 0.011
Baseline + cls loss	0.157 ± 0.009	71.023 ± 2.346	0.912 ± 0.015
Baseline + attentional mask	0.145 ± 0.010	62.328 ± 2.064	0.925 ± 0.013
Baseline + all	**0.122 ± 0.007**	**57.165 ± 1.610**	**0.943 ± 0.012**

aligned, and image synthesis may not work in such a situation. Zhao et al. [19] propose to learn common representations of all modalities, and this idea works in cases of minorly missing modalities but not in our dataset. Ma et al. [7] achieve similar performance to our proposed method due to their robust design. The results show that our model is capable of extracting modality-specific features as well as fusing them in an effective way. Besides, the performance also shows the robustness of our proposed method. Note that almost all results on the incomplete multimodal dataset outperform results on the complete multimodal dataset. This is due to the incomplete multimodal dataset will always contain OCT and text modalities, thus having more information in the input. Figure 4 shows the distribution of preoperative, predictive, and postoperative visual acuity. We can see that the predicted visual acuity largely overlaps with the true postoperative visual acuity, demonstrating the effectiveness of our method.

3.3   Ablation Study

As shown in Table 2, the proposed framework mainly benefits from the auxiliary classification loss and attentional fusion mask. Our analysis is as follows: Images provide more valuable information than text, making the auxiliary classification loss more effective than text combining. In missing multimodal learning, feature fusion takes precedence, and the masked self-attention mechanism contributes the most. Additionally, we conducted experiments to evaluate each modality's effectiveness. The results can be seen in the supplementary material.

4 Conclusion

In this paper, we present a new framework for BCVA prediction on the collected incomplete multimodal dataset. We take full advantage of multimodal information through our framework. The text modality is better utilized through the combination of the words. Moreover, image modality is explored effectively by the auxiliary classification loss. The attentional mask addresses the modality-missing issue. Extensive experiments have proved the effectiveness and superiority of our method.

Acknowledgements. This work is partially supported by Bingtuan Science and Technology Program (No. 2022DB005 and 2019BC008) and Key Research and Development Program of Hubei Province (2022BCA009).

References

1. Alexeeff, S.E., et al.: Development and validation of machine learning models: electronic health record data to predict visual acuity after cataract surgery. Perm. J. **25**, 188 (2021)
2. An, J., Zhang, L., Wang, Y., Zhang, Z.: The success of cataract surgery and the preoperative measurement of retinal function by electrophysiological techniques. J. Ophthalmol. **2015**, 401281 (2015)
3. Carion, N., Massa, F., Synnaeve, G., Usunier, N., Kirillov, A., Zagoruyko, S.: End-to-end object detection with transformers. In: Vedaldi, A., Bischof, H., Brox, T., Frahm, J.-M. (eds.) ECCV 2020. LNCS, vol. 12346, pp. 213–229. Springer, Cham (2020). https://doi.org/10.1007/978-3-030-58452-8_13
4. Dosovitskiy, A., et al.: An image is worth 16x16 words: transformers for image recognition at scale. arXiv preprint arXiv:2010.11929 (2020)
5. Forshaw, T.R.J., Ahmed, H.J., Kjær, T.W., Andréasson, S., Sørensen, T.L.: Full-field electroretinography in age-related macular degeneration: can retinal electro-physiology predict the subjective visual outcome of cataract surgery? Acta Oph-thalmol. **98**(7), 693–700 (2020)
6. Huang, Z., Lin, L., Cheng, P., Peng, L., Tang, X.: Multi-modal brain tumor segmen-tation via missing modality synthesis and modality-level attention fusion. arXiv preprint arXiv:2203.04586 (2022)
7. Ma, M., Ren, J., Zhao, L., Testuggine, D., Peng, X.: Are multimodal transformers robust to missing modality? In: Proceedings of the IEEE/CVF Conference on Computer Vision and Pattern Recognition, pp. 18177–18186 (2022)
8. Mimouni, M., Shapira, Y., Jadon, J., Frenkel, S., Blumenthal, E.Z.: Assessing visual function behind cataract: preoperative predictive value of the Heine lambda 100 Retinometer. Eur. J. Ophthalmol. **27**(5), 559–564 (2017)
9. Obata, S., et al.: Prediction of postoperative visual acuity after vitrectomy for macular hole using deep learning-based artificial intelligence. Graefe's Archive for Clinical and Experimental Ophthalmology, pp. 1–11 (2021)

10. Radford, A., et al.: Learning transferable visual models from natural language supervision. In: International Conference on Machine Learning, pp. 8748–8763. PMLR (2021)
11. Tharp, A., Cantor, L., Yung, C.W., Shoemaker, J.: Prospective comparison of the Heine Retinometer with the mentor Guyton-Minkowski potential acuity meter for the assessment of potential visual acuity before cataract surgery (1994)
12. Vaswani, A., et al.: Attention is all you need. In: Advances in Neural Information Processing Systems 30 (2017)
13. Wald, C.S., Unterlauft, J.D., Rehak, M., Girbardt, C.: Retinometer predicts visual outcome in Descemet membrane endothelial keratoplasty. Graefes Arch. Clin. Exp. Ophthalmol. **260**(7), 2283–2290 (2022)
14. Wang, H., et al.: Electrophysiology as a prognostic indicator of visual recovery in diabetic patients undergoing cataract surgery. Graefes Arch. Clin. Exp. Ophthalmol. **259**, 1879–1887 (2021)
15. Wang, J., et al.: CTT-Net: a multi-view cross-token transformer for cataract postoperative visual acuity prediction. In: 2022 IEEE International Conference on Bioinformatics and Biomedicine (BIBM), pp. 835–839. IEEE (2022)
16. WeiL, L., et al.: An optical coherence tomography-based deep learning algorithm for visual acuity prediction of highly myopic eyes after cataract surgery. Front. Cell Develop. Biol. **9**, 652848 (2021)
17. Xu, J., et al.: GroupViT: semantic segmentation emerges from text supervision. In: Proceedings of the IEEE/CVF Conference on Computer Vision and Pattern Recognition, pp. 18134–18144 (2022)
18. Zhang, Y., et al.: mmFormer: multimodal medical transformer for incomplete multimodal learning of brain tumor segmentation. In: Wang, L., Dou, Q., Fletcher, P.T., Speidel, S., Li, S. (eds.) Medical Image Computing and Computer Assisted Intervention – MICCAI 2022. MICCAI 2022. LNCS, vol. 13435. Springer, Cham (2022). https://doi.org/10.1007/978-3-031-16443-9_11
19. Zhao, J., Li, R., Jin, Q.: Missing modality imagination network for emotion recognition with uncertain missing modalities. In: Proceedings of the 59th Annual Meeting of the Association for Computational Linguistics and the 11th International Joint Conference on Natural Language Processing (Volume 1: Long Papers), pp. 2608–2618 (2021)
20. Zheng, S., et al.: Rethinking semantic segmentation from a sequence-to-sequence perspective with transformers. In: Proceedings of the IEEE/CVF Conference on Computer Vision and Pattern Recognition, pp. 6881–6890 (2021)

UWAT-GAN: Fundus Fluorescein Angiography Synthesis via Ultra-Wide-Angle Transformation Multi-scale GAN

Zhaojie Fang[1], Zhanghao Chen[1], Pengxue Wei[3], Wangting Li[3],
Shaochong Zhang[3], Ahmed Elazab[4], Gangyong Jia[2], Ruiquan Ge[2(✉)],
and Changmiao Wang[5(✉)]

[1] HDU-ITMO Joint Institute, Hangzhou Dianzi University, Hangzhou 310018, China
[2] School of Computer Science and Technology, Hangzhou Dianzi University,
Hangzhou 310018, China
gespring@hdu.edu.cn
[3] Shenzhen Eye Hospital, Jinan University, Shenzhen 518040, China
[4] School of Biomedical Engineering, Shenzhen University, Shenzhen 518060, China
[5] Medical Big Data Lab, Shenzhen Research Institure of Big Data, Shenzhen 518172,
China
cmwangalbert@gmail.com

Abstract. Fundus photography is an essential examination for clinical
and differential diagnosis of fundus diseases. Recently, Ultra-Wide-angle
Fundus (UWF) techniques, UWF Fluorescein Angiography (UWF-FA)
and UWF Scanning Laser Ophthalmoscopy (UWF-SLO) have been gradually put into use. However, Fluorescein Angiography (FA) and UWF-FA
require injecting sodium fluorescein which may have detrimental influences. To avoid negative impacts, cross-modality medical image generation algorithms have been proposed. Nevertheless, current methods in
fundus imaging could not produce high-resolution images and are unable
to capture tiny vascular lesion areas. This paper proposes a novel conditional generative adversarial network (UWAT-GAN) to synthesize UWF-
FA from UWF-SLO. Using multi-scale generators and a fusion module
patch to better extract global and local information, our model can generate high-resolution images. Moreover, an attention transmit module is
proposed to help the decoder learn effectively. Besides, a supervised approach is used to train the network using multiple new weighted losses on
different scales of data. Experiments on an in-house UWF image dataset
demonstrate the superiority of the UWAT-GAN over the state-of-the-art
methods. The source code is available at: https://github.com/Tinysqua/
UWAT-GAN.

Keywords: Fluorescein Angiography · Cross-modality Image
Generation · Ultra-Wide-angle Fundus Imaging · Conditional
Generative Adversarial Network

Supplementary Information The online version contains supplementary material
available at https://doi.org/10.1007/978-3-031-43990-2_70.

1   Introduction

Fluorescein Angiography (FA) is a commonly utilized imaging modality for detecting and diagnosing fundus diseases. It is widely used to image vascular structures and dynamically observe the circulation and leakage of contrast agents in blood vessels. Recently, the emergence of Ultra-Wide-angle Fundus (UWF) imaging has enabled its combination with FA and Scanning Laser Ophthalmoscopy (SLO), namely UWF-FA and UWF-SLO. The UWF-FA imaging enables simultaneous and high-contrast angiographic images of all 360°C of the mid and peripheral retina [1,4,21]. However, both FA and UWF-FA require injecting a fluorescent dye (i.e., sodium fluorescein) into the anterior vein of the patient's hand or elbow, which then passes through the blood circulation to the fundus blood vessels. Some patients may experience adverse reactions such as vomiting and nausea during or after the examination. Moreover, it is not suitable for patients with serious cardiovascular and other systemic diseases.

Cross-modality medical image generation provides a new method for solving the aforementioned problems. Multi-scale feature maps from different input modalities usually have similar structures. Hence, different contrasts can be merged to generate target images based on multimodal deep learning to provide more information for diagnosis [15]. Recently, the generative adversarial networks (GANs) [5] and their variants have made breakthroughs in this field. The idea of PatchGAN [12] was proposed to synthesize clearer images. Liu et al. [14] proposed an end-to-end multi-input and multi-output deep adversarial learning network for MR image synthesis. Xiao et al. [22] proposed a Transfer-GAN model by combining transfer learning and GANs to generate CT high-resolution images. By merging multi-scale generators, these networks can explore fine and coarse features from images [10]. Kamran et al. [9] proposed a semi-supervised model called VTGAN introducing transformer module into discriminators, helping the synthesis of vivid images. However, previous methods yielded lower-resolution situations and most discriminators can only take squared inputs (width equal to height) [18]. Moreover, misaligned data and lower attention on disease-related regions as well as the correctness of synthesized lesions remain significant issues. Furthermore, the highly non-linear relationship between different modalities makes the mapping from one modality to another difficult to learn [23].

In this paper, we present the Ultra-Wide-Angle Transformation GAN (UWAT-GAN), a supervised conditional GAN capable of generating UWF-FA from UWF-SLO. To address the image misalignment issue, we employ an automated image registration method and integrate the idea of pix2pixHD [20] to use multi-scale discriminators. In addition, we use the multi-scale generators to synthesize high-resolution images as well as improve the ability to capture tiny vascular lesion areas and employ multiple new weighted losses on different scales of data to optimize model training. For evaluation metrics, we use Fréchet Inception Distance (FID) [6], Kernel Inception Distance (KID) [11], Inception Score (IS) [2] and Learned Perceptual Image Patch Similarity (LPIPS) [24] to quantify the quality of images. Finally, we compare UWAT-GAN with the state-of-the-art

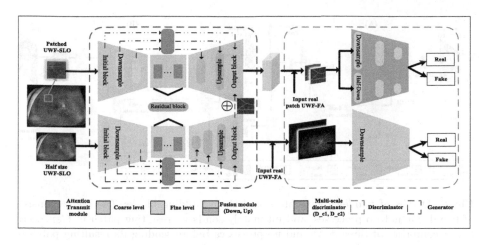

Fig. 1. The overall framework of UWAT-GAN.

image synthesis frameworks [3,7,20] for qualitative assessment and conduct an ablation study. Our main contributions are summarised as follows:

1). To the best of our knowledge, we present the first study to synthesize UWF-FA from UWF-SLO, overcoming the limitations of UWF-FA imaging.

2). We propose a novel UWAT-GAN utilizing multi-scale generators and multiple new weighted losses on different data scales to synthesize high-resolution images with the ability to capture tiny vascular lesion areas.

3). We assess the performance of the UWAT-GAN on a clinical in-house dataset and adopt an effective preprocessing method for image sharpening and registration to enhance the clarity of vascular regions and tackle the misalignment problem.

4). We demonstrate the superiority of the proposed UWAT-GAN against the state-of-the-art models through extensive experiments, comparisons, and ablation studies.

2   Methodology

We propose a supervised conditional GAN for synthesizing UWF-FA from UWF-SLO images, as illustrated in Fig. 1. In order to achieve the desired outcome, we propose the concept of a fine-coarse level generator in whole architecture (Sect. 2.1) and a fusion module works on result of both level generators (Sect. 2.2). Then the Attention Transmit Module is put forward to improve the U-Net-like architecture (Sect. 2.3). Additionally, we provide a comprehensive description of up-down sampling process and architecture of multi-scaled discriminators (Sec. 2.4). Eventually, we discuss the proposed loss function terms and their impacts in detail (Sect. 2.5).

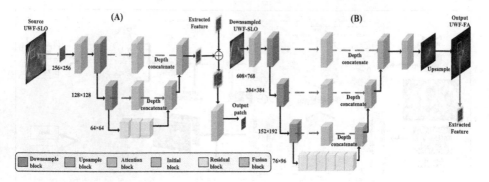

Fig. 2. Details of two-level generators and their building blocks. (A) Gen_F consists of two down and up processes, and ultimately sends its resulting patch to Gen_C; (B) Gen_C comprises of three down and up processes before sending its resulting patch to Gen_F.

2.1   Overall Architecture

UWF-FA has global features such as eyeballs and long-thick blood vessels, as well as local features such as small lesions and capillary blood vessels. However, generating images with both global and local features using a single generator is challenging. To address this issue, we devise two different levels of generators. The coarse generator Gen_C extracts global information and generates a result based on this information, while the fine generator Gen_F extracts local information. The results of global and local information can be used, alternately, as a reference for each other. Hence, this allows the extraction and utilization of both global and local information. In Fig. 1, the original image is fed into the entire model. After down-sampling, the image is passed into Gen_C. Then, we extract a patch from the original image as an input to Gen_F as described in Sect. 2.3. Both generators share the down-sample residual block and attention concatenated modules. Note that both generators generate a UWF-FA image and pass it to the discriminators. However, the output of Gen_F is the one we considered in the later experiments.

2.2   Patch and Fusion Module

In Fig. 1, Gen_F receives a randomly cropped patch as the input instead of the original image. This is because directly inputting the original image would occupy a large amount of memory and significantly reduce the training speed. Therefore, we only feed one cropped patch of the original image to Gen_F in each step. In Fig. 2(A), a fusion block takes both patches from Gen_F and Gen_C. To get the same region from Gen_F and Gen_C, we resized the images to the same size, and cropped the patches from the same position. These two patches are concatenated at the depth level and passed into a two-layer fusion operation.

2.3   Attention Transmit Module

In Fig. 2, the attention module is designed based on U-Net-like [16] architecture, whose sampling process can provide more information to the decoder. Whereas, when synthesizing UWF-FA from UWF-SLO, the information density of the source image is low. For instance, the eye sockets in the periphery of the image are sparsely distributed blood vessels in some areas. Therefore, completely passing the graphs from the down-sampling process to the up-sampling process is not appropriate. Subsequently, images that pass through Attention Transmit Module can first extract useful information so that the decoder uses this information, efficiently. The multi-head attention [19] and the CNN-Attention blocks are shown in Fig. 3.

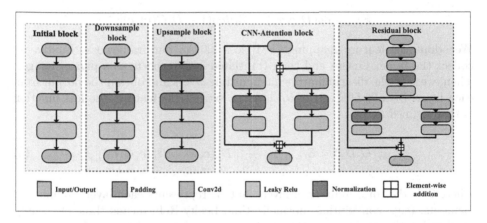

Fig. 3. Different blocks of the proposed generator. From left to right; the initial block, downsampling block, upsampling block, CNN-Attention block, and residual block.

2.4   Generator and Discriminator Architectures

After conducting multiple experiments, we choose three down-sample layers for Gen_F and two down-sample layers for Gen_C. In addition, each generator includes an initial block, down-sample block, up-sample block, residual block, and attention transmit module, which are shown in Fig. 3. The initial block contains a reflection padding, a 2D convolution layer, and the Leaky-Relu activation function. The down/up-sampling blocks consist of a 2D convolution/transposed layer and the activation function combined with normalization. Additionally, the multi-scale discriminator in pix2pixHD [20] is employed to evaluate the output of the generator. For generator Gen_X, the first discriminator D_{X1} takes the original and generated image P_1 while the second discriminator D_{X2} takes the down-sampled version of P_1. Although theoretically, a multi-level discriminator can be applied by generating an image pyramid for an image, we use D_{C1} and D_{C2} for Gen_C, and D_F for Gen_F in our framework.

2.5 Proposed Loss Function

Denote the two generators Gen_C and Gen_F as G_C and G_F, the three discriminators as D_{C1}, D_{C2}, and D_F, and the paired variables $\{(c_i, x_i)\}$, where c represents the distribution of original input as a condition and x represents the distribution of ground truth (i.e., real UWF-FA image). Given the conditional distribution c, we aim to maximize the loss of D_{C1}, D_{C2}, and D_F while minimizing the loss of Gen_C and Gen_F using the following objective function:

$$\min_{G_C} \max_{D_{C1}, D_{C2}} \sum_{k=1,2} \mathcal{L}_{cGAN}(G_C, D_{Ck}) + \min_{Gen_F} \max_{D_F} \mathcal{L}_{cGAN}(G_F, D_F), \qquad (1)$$

where $\mathcal{L}_{cGAN}$ is given by:

$$\mathbb{E}_{(c,x)}[log(D(c,x))] + \mathbb{E}_c[log(1 - D(G(c), c)]. \qquad (2)$$

We adopt the feature mapping (FM) loss [20] in our framework. Firstly, we collect the target images and their translated counterparts as a pair of images. Then, we split the discriminators into multiple layers and obtain the output from each layer. Denote $D^{(i)}$ as the ith-layer to extract the feature, the loss function is then defined as:

$$\mathcal{L}_{FM}(G.D_k) = \mathbb{E}_{(c,x)} \sum_{i=1}^{T} \frac{1}{N_i} [\|D_k^i(c,x) - D_k^i(c, G(c))\|_1], \qquad (3)$$

where T is the total number of layers and N_i represents each layer's number of elements. (e.g., convolution, normalization, Leaky-Relu means three elements). Minimizing this loss ensures that each layer can extract the same features from the paired images. Additionally, we use the perceptual loss [8] in our framework, which is utilized by a pretrained VGG19 network [17], to extract the features from the paired images and it is defined as:

$$\mathcal{L}_{VGG}(G.D_k) = \sum_{i=1}^{N} \frac{1}{M_i} [\|V^i(c,x) - V^i(c, G(c))\|_1], \qquad (4)$$

where N represents the total number of layers, M_i denotes the elements in each layer, and V^i is the ith-layer of the VGG19 network. The final cost function is as follows:

$$\min_{G_C}(\max_{D_{C1}, D_{C2}} \sum_{k=1,2} \mathcal{L}_{cGAN}(G_C, D_{Ck}) + \lambda_{FMC} \sum_{k=1,2} \mathcal{L}_{FM}(G_C, D_{Ck})$$
$$+ \lambda_{VGGC} \sum_{k=1,2} \mathcal{L}_{VGG}(G_C, D_{Ck})) + \min_{G_F}(\max_{D_F} \mathcal{L}_{cGAN}(G_F, D_F) \qquad (5)$$
$$+ \lambda_{FMF}\mathcal{L}_{FM}(G_F, D_F) + \lambda_{VGGF}\mathcal{L}_{VGG}(G_F, D_F)).$$

where $\lambda_{FMC}, \lambda_{VGGC}, \lambda_{FMF}, \lambda_{VGGF}$ indicate adjustable weight parameters.

3   Experiments and Results

3.1   Data Preparation and Preprocess

In our experiments, we utilized an in-house dataset of UWF images obtained from a local hospital, comprising UWF-FA and UWF-SLO images. The UWF-SLO are in 3-channel RGB format, whereas the UWF-FA images are in 1-channel format. Each image pair was collected from a unique patient. However, from a clinical perspective, images taken with an interval of more than one day or those with noticeable fresh bleeding were excluded. Additionally, images that contain numerous interfering factors affecting their quality were also discarded. After the quality check, we have 70 paired images with the size of 3900×3072, of which 70% were randomly allocated for training and 30% for testing, respectively.

Furthermore, we employed image sharpening through histogram equalization to enhance the clarity of images. We then utilized automated image registration software, i2k Retina Pro, to register each pair of images which changed the image size. To standardize the size of each image, we resized the registered images to 2432×3702. Subsequently, we randomly cropped the resized images with a size of 608×768 into different patches. And 50 patches could be obtained for each image. Finally, we adopted data augmentation using random flip and rotation to increase the number of training images from 49 pairs to 1960 pairs.

3.2   Implementation Details

All our experiments were conducted on the PyTorch 1.12 framework and carried out on two Nvidia RTX 3090Ti GPUs. Our model was trained from scratch to 200 epochs. The parameters were optimized by the Adam optimizer algorithm [13] with learning rate $\alpha = 0.0002$, $\beta_1 = 0.5$ and $\beta_2 = 0.999$. We used a batch size of 2 to train our model and set $\lambda_{FMF} = \lambda_{FMC} = \lambda_{VGGF} = \lambda_{VGGC} = 10$ (Eq. 5).

3.3   Comparisons

We first compared the performance of our model with some state-of-the-art GAN-based models including: Pix2pix [7], Pix2pixHD [20] and StarGAN-v2 [3]. For a fair comparison, we took the default parameters of the open-source codes of the competing methods, ensuring that the data volume matched the number of training cycles. We used the $FID(\downarrow), KID(\downarrow), LPIPS(\downarrow)$ and $IS(\uparrow)$ to evaluate the generated UWF-FA. Table 1 shows the generation performance of different methods. Overall, our method achieves the best in all metrics compared to other models. The Pix2Pix attained the worst performance in all evaluation metrics while PixPixHD and StarGAN had comparable performance. In general, our method outperformed the competing methods and improved FID, KID, IS, and LPIPS by at least 24.47%, 39.95%, 3.59%, and 14.04%, respectively. Although StarGAN-v2 yielded the second-best performance, it is still less comparable with the proposed UWAT-GAN due to the lesion generation module which could capture tiny image details and improve overall performance (see Fig. 4).

Table 1. Comparison with the state-of-the-art methods using 4 evaluation metrics. The * means that the official code hasn't provided the way to measure it.

Methods	FID($\downarrow$)	KID($\downarrow$)	IS($\uparrow$)	LPIPS($\downarrow$)
Pix2Pix [7]	135.4038	0.1094	1.2772	0.4575
Pix2PixHD [20]	76.76	0.0491	1.0602	0.4451
StarGAN-v2 [3]	74.38	0.0433	*	0.4577
UWAT-GAN M_{NA}	67.96	0.0308	1.2757	0.4086
UWAT-GAN	**55.59**	**0.0260**	**1.323**	**0.3826**

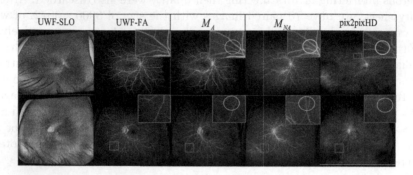

Fig. 4. Visualization of original and generated images. From left to right: source UWF-SLO, UWF-FA, the proposed framework with and without the attention transmit module, and the pix2pixHD, respectively.

3.4   Ablation Study

To evaluate the significance of the attention transmit module proposed in UWAT-GAN, we trained our model with and without this module, namely M_A and M_{NA}, respectively. Unlike the generated images of M_A, we found that M_{NA} was not so distinctive as some vessels were missing and some interference of eyelashes was incorrectly considered as vessels. In Fig. 4, we showed the original pair of UWF-SLO and UWF-FA images and the generated images with M_A and M_{NA}. It is clear that the proposed method can generate good images and preserve small details. It becomes more distinctive when the attention module was used, as shown in the enlarged view of the red rectangle. It is also obvious that the FID and KID scores were improved by 22.25% and 18.46%, respectively.

4   Discussion and Conclusion

To address the potential adverse effects of fluorescein injection during FA, we propose UWAT-GAN to synthesize UWF-FA from UWF-SLO. Our method can generate high-resolution images and enhance the ability to capture small vascular lesions. Comparison and ablation study on an in-house dataset demonstrate the superiority and effectiveness of our proposed method. However, our model still has a few limitations. First, not every pair of images can be registered since some paired images may have fewer available features, making registration difficult. Second, our model's accuracy in synthesizing very tiny lesions is not optimal, as some lesions cannot be well generalized. Third, the limited size of our dataset is relatively small and may affect the model performance. In the future, we aim to expand the size of our dataset and explore the use of the object detection model, especially for small targets, to push our model pay more attention to some lesions. After further validation, we aim to adopt this method as an auxiliary tool to diagnose and detect fundus diseases.

Acknowledgements. This work was supported by the National Natural Science Foundation of China (No. U20A20386, U22A2033), Zhejiang Provincial Natural Science Foundation of China (No. LY21F020017), Chinese Key-Area Research and Development Program of Guangdong Province (2020B0101350001), GuangDong Basic and Applied Basic Research Foundation (No. 2022A1515110570), Innovation teams of youth innovation in science and technology of high education institutions of Shandong province (No. 2021KJ088), the Shenzhen Science and Technology Program (JCYJ20220818103001002), and the Guangdong Provincial Key Laboratory of Big Data Computing, The Chinese University of Hong Kong, Shenzhen.

Data statement. Dataset used in this work is privately collected from our collaborative hospital after ethical approval and data usage permission. Data maybe be available upon the permission of the related authority and adequate request to the corresponding author(s).

References

1. Ashraf, M., Shokrollahi, S., Salongcay, R.P., Aiello, L.P., Silva, P.S.: Diabetic retinopathy and ultrawide field imaging. In: Seminars in Ophthalmology, vol. 35, pp. 56–65. Taylor & Francis (2020)
2. Barratt, S., Sharma, R.: A note on the inception score. arXiv preprint arXiv:1801.01973 (2018). http://arxiv.org/abs/1801.01973
3. Choi, Y., Uh, Y., Yoo, J., Ha, J.W.: Stargan v2: diverse image synthesis for multiple domains. In: Proceedings of the IEEE/CVF Conference on Computer Vision and Pattern Recognition, pp. 8188–8197 (2020)
4. Ehlers, J.P., Jiang, A.C., Boss, J.D., Hu, M., Figueiredo, N., Babiuch, A., Talcott, K., Sharma, S., Hach, J., Le, T., et al.: Quantitative ultra-widefield angiography and diabetic retinopathy severity: an assessment of panretinal leakage index, ischemic index and microaneurysm count. Ophthalmology **126**(11), 1527–1532 (2019)
5. Goodfellow, I., et al.: Generative adversarial networks. Commun. ACM **63**(11), 139–144 (2020)
6. Heusel, M., Ramsauer, H., Unterthiner, T., Nessler, B., Hochreiter, S.: Gans trained by a two time-scale update rule converge to a local nash equilibrium. Advances in neural information processing systems 30 (2017)
7. Isola, P., Zhu, J.Y., Zhou, T., Efros, A.A.: Image-to-image translation with conditional adversarial networks. In: Proceedings of the IEEE Conference on Computer Vision and Pattern Recognition, pp. 1125–1134 (2017)

8. Johnson, J., Alahi, A., Fei-Fei, L.: Perceptual losses for real-time style transfer and super-resolution. In: Computer Vision-ECCV 2016: 14th European Conference, Amsterdam, The Netherlands, October 11–14, 2016, Proceedings, Part II 14. pp. 694–711. Springer, Cham (2016). https://doi.org/10.1007/978-3-319-46475-6_43

9. Kamran, S.A., Hossain, K.F., Tavakkoli, A., Zuckerbrod, S.L., Baker, S.A.: Vtgan: Semi-supervised retinal image synthesis and disease prediction using vision transformers. In: Proceedings of the IEEE/CVF International Conference on Computer Vision, pp. 3235–3245 (2021)

10. Kamran, S.A., Hossain, K.F., Tavakkoli, A., Zuckerbrod, S.L., Sanders, K.M., Baker, S.A.: Rv-gan: Segmenting retinal vascular structure in fundus photographs using a novel multi-scale generative adversarial network. In: Medical Image Computing and Computer Assisted Intervention-MICCAI 2021: 24th International Conference, Strasbourg, France, September 27-October 1, 2021, Proceedings, Part VIII 24. pp. 34–44. Springer (2021)

11. Knop, S., Mazur, M., Spurek, P., Tabor, J., Podolak, I.: Generative models with kernel distance in data space. Neurocomputing **487**, 119–129 (2022)

12. Li, C., Wand, M.: Precomputed real-time texture synthesis with Markovian generative adversarial networks. In: Leibe, B., Matas, J., Sebe, N., Welling, M. (eds.) ECCV 2016. LNCS, vol. 9907, pp. 702–716. Springer, Cham (2016). https://doi.org/10.1007/978-3-319-46487-9_43

13. Lihua, L.: Simulation physics-informed deep neural network by adaptive adam optimization method to perform a comparative study of the system. Eng. Comput. **38**(Suppl 2), 1111–1130 (2022)

14. Liu, X., Yu, A., Wei, X., Pan, Z., Tang, J.: Multimodal MR image synthesis using gradient prior and adversarial learning. IEEE J. Sel. Top. Signal Process. **14**(6), 1176–1188 (2020)

15. Luo, S.: A survey on multimodal deep learning for image synthesis: applications, methods, datasets, evaluation metrics, and results comparison. In: 2021 the 5th International Conference on Innovation in Artificial Intelligence, pp. 108–120 (2021)

16. Ronneberger, O., Fischer, P., Brox, T.: U-Net: convolutional networks for biomedical image segmentation. In: Navab, N., Hornegger, J., Wells, W.M., Frangi, A.F. (eds.) MICCAI 2015. LNCS, vol. 9351, pp. 234–241. Springer, Cham (2015). https://doi.org/10.1007/978-3-319-24574-4_28

17. Simonyan, K., Zisserman, A.: Very deep convolutional networks for large-scale image recognition. arXiv preprint arXiv:1409.1556 (2014). http://arxiv.org/abs/1409.1556

18. Tavakkoli, A., Kamran, S.A., Hossain, K.F., Zuckerbrod, S.L.: A novel deep learning conditional generative adversarial network for producing angiography images from retinal fundus photographs. Sci. Rep. **10**(1), 21580 (2020)

19. Vaswani, A., et al.: Attention is all you need. Advances in neural information processing systems 30 (2017)

20. Wang, T.C., Liu, M.Y., Zhu, J.Y., Tao, A., Kautz, J., Catanzaro, B.: High-resolution image synthesis and semantic manipulation with conditional gans. In: Proceedings of the IEEE Conference on Computer Vision and Pattern Recognition, pp. 8798–8807 (2018)

21. Wang, X., et al.: Automated grading of diabetic retinopathy with ultra-widefield fluorescein angiography and deep learning. J. Diabetes Res. **2021** (2021)

22. Xiao, Y., et al.: Transfer-gan: multimodal CT image super-resolution via transfer generative adversarial networks. In: 2020 IEEE 17th International Symposium on Biomedical Imaging (ISBI), pp. 195–198. IEEE (2020)

23. Yang, Q., Li, N., Zhao, Z., Fan, X., Chang, E.I.C., Xu, Y.: Mri cross-modality image-to-image translation. Sci. Rep. **10**(1), 3753 (2020)
24. Zhang, R., Isola, P., Efros, A.A., Shechtman, E., Wang, O.: The unreasonable effectiveness of deep features as a perceptual metric. In: Proceedings of the IEEE Conference on Computer Vision and Pattern Recognition, pp. 586–595 (2018)

Clinical Applications – Vascular

Clinical Applications – Vascular

Topology-Preserving Automatic Labeling of Coronary Arteries via Anatomy-Aware Connection Classifier

Zhixing Zhang[1], Ziwei Zhao[1,6], Dong Wang[2], Shishuang Zhao[3], Yuhang Liu[3], Jia Liu[4], and Liwei Wang[2,5(✉)]

[1] Center for Data Science, Peking University, Beijing, China
zhangzhixing@stu.pku.edu.cn
[2] National Key Laboratory of General Artificial Intelligence, School of Intelligence Science and Technology, Peking University, Beijing, China
[3] Yizhun Medical AI Co., Ltd., Beijing, China
[4] Peking University First Hospital, Beijing, China
[5] Center for Machine Learning Research, Peking University, Beijing, China
wanglw@pku.edu.cn
[6] Pazhou Lab, Guangzhou, China

Abstract. Automatic labeling of coronary arteries is an essential task in the practical diagnosis process of cardiovascular diseases. For experienced radiologists, the anatomically predetermined connections are important for labeling the artery segments accurately, while this prior knowledge is barely explored in previous studies. In this paper, we present a new framework called TopoLab which incorporates the anatomical connections into the network design explicitly. Specifically, the strategies of intra-segment feature aggregation and inter-segment feature interaction are introduced for hierarchical segment feature extraction. Moreover, we propose the anatomy-aware connection classifier to enable classification for each connected segment pair, which effectively exploits the prior topology among the arteries with different categories. To validate the effectiveness of our method, we contribute high-quality annotations of artery labeling to the public orCaScore dataset. The experimental results on both the orCaScore dataset and an in-house dataset show that our TopoLab has achieved state-of-the-art performance.

Keywords: Automatic Labeling · Coronary Arteries · Topology-Preserving

---

Z. Zhang and Z. Zhao—These authors contributed equally to this work.

Supplementary Information The online version contains supplementary material available at https://doi.org/10.1007/978-3-031-43990-2_71.

1   Introduction

Coronary Computerized Tomography Angiography (CCTA) is a commonly used non-invasive approach for the diagnosis of potential coronary artery diseases [11]. In clinical practice, accurate labeling of coronary artery segments (see Fig. 1(a)) is a crucial step toward the subsequent diagnosis and analysis of the image. However, the vast variability of coronary artery anatomy across individuals makes it challenging to achieve precise and automatic labeling.

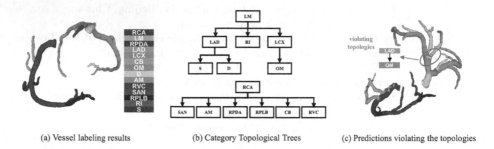

(a) Vessel labeling results   (b) Category Topological Trees   (c) Predictions violating the topologies

Fig. 1. (a) The task of coronary artery labeling aims to assign each vessel segment an anatomical category. (b) The anatomically predetermined connections among different categories of coronary arteries form a tree structure – *category topological trees*. (c) shows an example containing obvious mistakes violating the prior topology made by previous methods.

Previous studies on labeling coronary arteries using deep learning-based methods [20,22–24] have shown promising results by introducing graph convolutional networks and point cloud analysis. However, these approaches have overlooked the essential prior knowledge that **different categories of coronary arteries have anatomically predetermined connections** [3]. For instance, LAD, LCX, and RI originate from LM, while S and D arise from LAD. All of the anatomical connections form a tree structure with prior topology, as shown in Fig. 1(b). We call this structure *category topological trees*. Due to lacking the utilization of the *category topological trees*, existing methods [20,22–24] often make some obvious mistakes that violate the prior topology as illustrated in Fig. 1(c). We argue that incorporating the *category topological trees* into the network explicitly is the key to improving automatic labeling performance, especially in reducing topology-violation labeling errors.

In this paper, we propose a novel framework called TopoLab to perform topology-preserving automatic labeling of coronary arteries. Our model mainly contains two components: the hierarchical feature extraction module and the anatomy-aware connection classifier. The hierarchical feature extraction module introduces the segment query to achieve intra-segment feature aggregation via Transformer [17] and relies on graph convolutional network [8] to establish inter-segment feature interactions. Moreover, to incorporate the *category topological trees* into the network explicitly, we further propose the anatomy-aware connection classifier (AC-Classifier). Unlike previous methods that classify each segment independently, AC-Classifier performs classification for every connected segment pair. Specifically, all of the connections derived from the *category topological trees* are used to construct the ground truth connection templates, and each connected segment pair is categorized into one of these templates. Since the connection templates inherently conform to the topology, AC-Classifier has effectively prioritized the anatomically predetermined connections and the network is enabled to preserve the topology by design.

To the best of our knowledge, there is currently no publicly available dataset with annotations for artery labeling. In this work, we contribute high-quality annotations to the orCaScore dataset [19]. The experimental results on the public dataset orCaScore and an in-house dataset have demonstrated that our TopoLab outperforms previous state-of-the-art methods, especially in the topology-related metrics.

Our contributions can be summarized as follows. (1) We are the first to incorporate the *category topological trees* into the deep learning models for automatic labeling of coronary arteries by introducing AC-Classifier. (2) We propose a novel hierarchical feature extraction module to achieve intra- and inter-segment feature aggregations. (3) Our approach achieves state-of-the-art performance on both public and in-house datasets. (4) We provide high-quality annotations of artery labeling for the public dataset orCaScore, which is available at https:// github.com/zutsusemi/MICCAI2023-TopoLab-Labels/.

2 Related Work

Traditional methods [1,21] for automatic labeling of coronary arteries usually align the extracted artery trees with a 3D coronary artery tree model which provides the anatomical connections as prior knowledge. However, these works rely heavily on logical rules, which can not always capture the complexities of the anatomical structure. To overcome the limitations, deep learning has been introduced in this area [20,22–24] with its great success in medical imaging [7,14–16,18]. For instance, TreeLab-Net [20] uses bidirectional tree structural LSTM [6] to model the coronary artery trees, while CPR-GCN [22] constructs a vessel graph by treating each segment as a node and leverages GCN [8] to aggregate segment features. CorLab-Net [24] regards spatial and anatomical dependencies as the explicit guidance for artery labeling based on point cloud networks [13]. Nevertheless, all of these deep learning based methods ignore the important priority about the predetermined anatomical structure – *category topological trees*. In this paper, we aim to incorporate this priority into the network design explicitly for developing the topology-preserving models.

3 Methodology

3.1 Overview

We start by extracting centerlines from the vessel segmentation annotations in a CCTA image, using a traditional 3D thinning algorithm [9]. Next, we use the minimum spanning tree algorithm [5] to construct two coronary artery trees, one for the left domain (LD) and one for the right domain (RD). Following the branch bifurcation rules in [22] (details can be found in supplementary materials), the coronary artery trees are split into several segments, denoted by $\mathcal{S} = \{S_i\}_{i=1}^{N}$, where N is the number of segments, $S_i \in \mathbb{R}^{L_i \times 3}$ denotes the i-th segment comprised of the 3D positions of L_i centerline points. Our model takes as input the vessel segments $\mathcal{S}$ and their connections $\mathcal{C} = \{(i_1, j_1), (i_2, j_2), ..., (i_{N_c}, j_{N_c})\}$, where (i_k, j_k) indicates that the i_k-th segment is connected with the j_k-th segment. The objective is to predict the classes of the vessel segments.

As illustrated in Fig. 2, we design a novel framework named TopoLab for the automatic labeling of coronary arteries. In Sect. 3.2, we will introduce the hierarchical feature extraction module comprised of intra-segment feature aggregation and inter-segment feature interaction. In Sect. 3.3, we will elaborate on the details of AC-Classifier which exploits the *category topological trees* effectively.

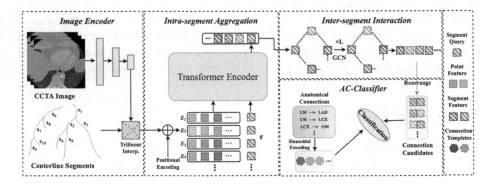

Fig. 2. The pipeline of our proposed TopoLab.

3.2   Hierarchical Feature Extraction

We first feed the CCTA image $X \in \mathbb{R}^{H \times W \times D}$ into an image encoder (*e.g.* U-Net [2]) to obtain the downscaled feature map $F \in \mathbb{R}^{H/4 \times W/4 \times D/4 \times C}$, where C is the channel size.

For each segment S_i, trilinear interpolation in the downsampled feature map F is adopted for the centerline point $v \in S_i$ to obtain the corresponding point features $f(v) = \text{Tri}(v, F) \in \mathbb{R}^C$. The segment features are denoted as $\{f(v)|v \in S_i\}$, simplified as $E_i \in \mathbb{R}^{L_i \times C}$.

Intra-Segment Feature Aggregation. To extract the segment-level features for labeling, aggregation of sequential point features belonging to the same segment is required. Previous studies [20,22] employ Bidirectional LSTM [6] to summarize the tubular sequential features. However, due to the weak representative ability of LSTM for the vessel segments with huge length variability, the performance of these methods is limited.

We introduce Transformer [17] as our intra-segment feature aggregator for its strong capability to model the relationships among the sequences with varying lengths. Concretely, we set a learnable embedding $q \in \mathbb{R}^C$ called segment query to aggregate intra-segment point features. The segment query is concatenated with the segment features E_i to obtain the new tensor $\tilde{E}_i \in \mathbb{R}^{(L_i+1) \times C}$. Following [4], we feed $\tilde{E}_i$ augmented by the learnable 3D positional encodings into a Transformer encoder containing several standard sub-blocks. Each standard sub-block has the same architecture as in [17], which consists of multi-head attention, feed-forward network, layer normalization and ReLU activation. The state of the segment query q at the output of the Transformer encoder serves as the aggregated segment representation $\hat{E}_i \in \mathbb{R}^C$.

Inter-Segment Feature Interaction. The branching structure of coronary arteries is inherently graph-like, with each segment serving as a node and the connections between segments serving as edges. Thus, we leverage graph convolutional network [8] (GCN) to capture the interactions among different segments.

Specifically, let $\hat{E} \in \mathbb{R}^{N \times C}$ denotes the aggregated segment features where the i-th item of $\hat{E}$ is $\hat{E}_i$, and $A \in \mathbb{R}^{N \times N}$ is the adjacency matrix for the vessel segment graph derived from the segment connections $\mathcal{C}$. The process of GCN layers is as follows:

$$\hat{E}_{l+1} = \sigma(A\hat{E}_l W_l), \tag{1}$$

where σ is the ReLU activation, $W_l \in \mathbb{R}^{C \times C}$ is the learnable parameters for the l-th GCN layer, the input for the first layer is $\hat{E}_0 = \hat{E}$.

Finally, we fuse the input segment features $\hat{E}$ and the output of the final GCN layer $\hat{E}_f$ to obtain $\overline{E} = [\hat{E}_f, \hat{E}]W_f \in \mathbb{R}^{N \times C}$ with the parameters $W_f \in \mathbb{R}^{(C+C) \times C}$. The enhanced segment features $\{\overline{E}_1, \overline{E}_2, ..., \overline{E}_N\}$ are forwarded to the classifier for segment labeling, where $\overline{E}_i \in \mathbb{R}^C$ is the i-th item of $\overline{E}$.

3.3   Anatomy-Aware Connection Classifier

The direct approach for labeling the coronary arteries is to use a linear layer to classify each segment independently as in previous methods. To incorporate the *category topological trees* into the classifier design, we propose to conduct the classification task for every connected segment pair.

We begin by defining the ground truth segment connections which are composed of the topology-conforming connections derived from the *category topological trees* (like LM→LAD, LCX→OM, etc.). Note that the self-connections (*e.g.* RCA→RCA) are also considered as the ground truth connections. The ground truth segment connections have the corresponding template embeddings for classification, which are represented by $G \in \mathbb{R}^{N_g \times 2C}$, where N_g is the number of ground truth segment connections. Denote $g_i = \text{Concate}(\text{Enc}(x), \text{Enc}(y)) \in \mathbb{R}^{2C}$ as the i-th item of G, where x and y stand for the segment classes indexed by the i-th ground truth connection, Enc denotes the sinusoidal encoding as in [17].

The connection templates G are used to enable classification for the connected segment pairs.

Then, given the segment features $\{\overline{E}_i\}_{i=1}^N$, and all of the connected segment pairs $\mathcal{C} = \{(i_1, j_1), (i_2, j_2), ..., (i_{N_c}, j_{N_c})\}$, the connection features $P \in \mathbb{R}^{N_c \times 2C}$ are obtained by rearrangement of the segment features, where the k-th item of P is $\mathrm{Concate}(\overline{E}_{i_k}, \overline{E}_{j_k}) \in \mathbb{R}^{2C}$. We use an MLP layer to further fuse the features $\hat{P} = \mathrm{MLP}(P) \in \mathbb{R}^{N_c \times 2C}$.

Training Loss. The loss function can be written as:

$$\mathcal{L} = \sum_i^{N_c} \mathcal{L}_{\mathrm{cls}}(\hat{p}_i, y_i), \tag{2}$$

$$\mathcal{L}_{\mathrm{cls}} = -\log \frac{\exp(\mathrm{sim}(\hat{p}_i, g_{y_i})/\tau)}{\sum_{j=1}^{N_g} \exp(\mathrm{sim}(\hat{p}_i, g_j)/\tau)}, \tag{3}$$

where y_i is the ground truth of the i-th segment connection, $\hat{p}_i$ denotes the i-th item of $\hat{P}$. sim in Eq. 3 stand for the cosine similarity, $\mathrm{sim}(x, y) = \frac{x^T y}{||x||_2 ||y||_2}$, and the temperature τ is a hyperparameter which is set as 0.05 by default.

Inference. During the inference stage, for each segment, we first select the connection with the largest confidence score among all segment connections that have covered the given segment. And then the corresponding category indexed by the selected connection serves as the prediction of the specific segment.

4   Experiments

4.1   Setup

Datasets. We train and evaluate our method on two datasets. The **orCaS-core** [19] MICCAI 2014 Challenge contains 72 contrast-enhanced CTA images and non-contrast enhanced CT scans. As the original dataset only contains the labels of calcifications, we have annotated vessel segmentation and anatomical categories for coronary arteries with experienced radiologists. Considering the small amount of data, we randomly split the dataset into five folds to perform cross-validation. The mean values of cross-validation results are reported. We also collect an **In-house Dataset** containing 1200 CTA scans which have been annotated by at least two experts. The dataset is collected in compliance with the terms of the licensing agreement and ethical certification. We randomly split the dataset into train, validation and test set with 800, 200 and 200 scans respectively. The annotations for both datasets adhere to the same standard, which includes 14 classes of coronary artery segments (see Fig. 1(b)). It is a challenging task, surpassing the scope of studies like TreeLab-Net [20] and CPR-GCN [22], which only consider 10 and 11 categories, respectively.

Evaluation Metrics. Following previous methods [20,22,24], we adopt the mean metrics of all categories of the segments including recall, precision and

F1. Note that the mean metric is the weighted average based on the number of segments of different categories. To further evaluate the topological accuracy of connected segments, we propose two new metrics: viola and viola^c, which reflect the segment-level topological accuracy and the case-level topological accuracy, respectively. Specifically, viola is calculated as the ratio of the number of connections violating the topology to the total number of connections, while viola^c is calculated as the ratio of the number of test cases containing any topology-violating connection to the total number of test cases.

Table 1. Comparison with other methods on orCaScore dataset(%). Each entry of the table shows the average value of 5 folds with the standard deviation in subscript. "Recall, precision, and F1" of each fold are the weighted averages of the 14 vessel categories.

Method	Recall	Precison	F1	Viola	Viola^c
TaG-Net [23]	$82.29_{2.07}$	$83.41_{2.28}$	$82.14_{2.32}$	$10.87_{2.49}$	$87.62_{4.85}$
CorLab-Net [24]	$82.09_{1.03}$	$83.83_{1.53}$	$82.15_{1.18}$	$9.01_{1.09}$	$75.05_{10.01}$
TreeLab-Net [20]	$83.35_{2.80}$	$84.90_{2.11}$	$83.12_{2.96}$	$3.75_{1.08}$	$55.52_{17.26}$
CPR-GCN [22]	$82.88_{1.44}$	$83.61_{1.59}$	$82.72_{1.55}$	$4.94_{1.49}$	$53.91_{9.53}$
TopoLab(Ours)	$\mathbf{87.13}_{1.03}$	$\mathbf{88.31}_{1.60}$	$\mathbf{87.23}_{1.29}$	$\mathbf{1.52}_{0.73}$	$\mathbf{22.29}_{8.30}$

4.2 Implementation Details

3D ResUNet [25] is employed as the image encoder for feature extraction with channel dimension $C = 64$. The transformer encoder has 3 standard blocks and the number of graph convolution layers is set to 4. We train the network using AdamW optimizer [10] with a base learning rate of 5e-4 and the cosine learning rate schedule during the training stage. The batch size is set to 4. All networks are implemented by Pytorch [12] and trained on four NVIDIA GeForce RTX 3090 GPUs. We train our model on the orCaScore dataset for 3.5k iterations and in-house dataset for 12.5k iterations.

4.3 Comparison with Other Methods

Quantitative Results. We compare TopoLab with other deep learning based approaches including TaG-Net [23], CPR-GCN [22], TreeLab-Net [20], and CorLab-Net [24] which are implemented by ourselves with the same training configurations for a fair comparison. Note that for the point cloud-based methods [23,24], we use intra-segment voting to transform the point-level predictions into segment-level predictions. From the results on the orCaScore dataset in Table 1 and the in-house dataset in Table 2, we can conclude that the proposed TopoLab outperforms all existing methods by a large margin. The performance gains are more significant in the topology-related metrics, which demonstrates the effectiveness of the utilization of prior knowledge. More detailed results on each category of coronary arteries can be found in the supplementary materials.

Qualitative Results. In Fig. 3, we present a qualitative comparison of TopoLab with other methods. Consider the first case as an example, where our approach successfully avoids topology-violating errors, whereas other methods incorrectly classify RI as D, leading to the RI→D, D→RI, or LM→D connections that violate topology. These visualizations demonstrate the effectiveness of the usage of *category topological trees*. More qualitative results can be found in supplementary materials.

Table 2. Comparison with other methods on in-house dataset (%). Each entry of the table shows the average value of 5 trials with the standard deviation in subscript. "Recall, precision, and F1" of each trial are the weighted averages of the 14 vessel categories.

Method	Recall	Precision	F1	Viola	Viola^c
TaG-Net [23]	$88.26_{0.30}$	$88.47_{0.34}$	$88.23_{0.34}$	$6.46_{0.28}$	$65.90_{1.98}$
CorLab-Net [24]	$88.85_{0.35}$	$88.96_{0.34}$	$88.83_{0.35}$	$4.75_{0.31}$	$57.10_{1.46}$
TreeLab-Net [20]	$88.58_{0.43}$	$88.56_{0.44}$	$88.52_{0.44}$	$2.34_{0.09}$	$32.70_{0.51}$
CPR-GCN [22]	$90.92_{0.15}$	$90.84_{0.13}$	$90.84_{0.14}$	$2.89_{0.17}$	$38.70_{3.93}$
TopoLab(Ours)	$\mathbf{92.23}_{0.23}$	$\mathbf{92.21}_{0.25}$	$\mathbf{92.19}_{0.23}$	$\mathbf{0.77}_{0.13}$	$\mathbf{9.40}_{1.69}$

Table 3. Ablation Study on orCaScore Dataset(%). Each entry of the table shows the average value of 5 folds. "Recall, precision, and F1" of each fold are the weighted averages of the 14 vessel categories.

Method	Recall	Precision	F1	Viola	Viola^c
w/o IFA	$79.61_{3.38}$	$81.47_{2.71}$	$79.29_{3.35}$	$4.00_{1.82}$	$48.86_{20.39}$
w/o IFI	$85.79_{2.61}$	$86.74_{2.77}$	$85.66_{2.83}$	$2.98_{1.54}$	$41.43_{15.55}$
w/o ACC	$86.88_{1.39}$	$88.15_{1.74}$	$86.95_{1.57}$	$2.85_{0.85}$	$37.71_{10.26}$
TopoLab	$\mathbf{87.13}_{1.03}$	$\mathbf{88.31}_{1.60}$	$\mathbf{87.23}_{1.29}$	$\mathbf{1.52}_{0.73}$	$\mathbf{22.29}_{8.30}$

4.4   Ablation Study

In this subsection, we explore the effectiveness of different components in TopoLab on orCaScore dataset, as shown in Table 3.

Intra-segment Feature Aggregation (IFA). Transformer [17] is leveraged to achieve the intra-segment feature aggregation in our model. To validate its benefits, we replace it with the Bi-LSTM used in CPR-GCN. From the results in the first line of Table 3, our method surpasses Bi-LSTM by 7.94% in F1 score and 26.57% in Viola^c.

Inter-segment Feature Interaction (IFI). We use GCN to establish inter-segment feature interactions due to the natural graph structure of coronary arteries. When directly removing this module, the performance drops by 1.57% in F1 score and 19.14% in Viola^c as illustrated in the second line of Table 3.

Anatomy-Aware Connection Classifier (ACC). AC-Classifier which exploits the prior knowledge from *category topological trees* is adopted to classify each connected segment pair. We replace it with a commonly used linear layer to enable classification for single segments as in previous methods, and the performance drops by 1.33% in Viola and 15.42% in Viola^c, which effectively demonstrates the superiority of the proposed method.

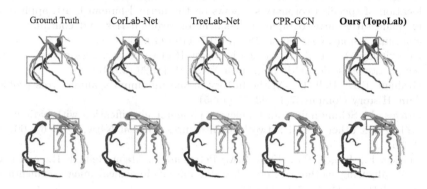

Fig. 3. Visual comparison between our TopoLab and other methods. Errors violating topology are highlighted in red boxes, while errors that do not violate topology are marked in orange boxes. The corresponding positions in the ground truth and TopoLab predictions are indicated by blue boxes. (Color figure online)

5   Conclusion

In this study, we review the essential task of coronary artery labeling and exploit the prior knowledge of the predetermined anatomical connections. The proposed strategies of intra- and inter-segment feature aggregation guarantee effective feature extraction, while the AC-Classifier preserves the clinical logic in the network design. The extensive experiments on orCaScore dataset and in-house dataset reveal that the proposed TopoLab has achieved new state-of-the-art performance. We hope our paper could encourage the community to explore the use of clinical priority to facilitate the design of more effective algorithms.

Acknowledgements. We would like to thank Dr. Nianxi Liao for valuable discussions. This work is supported by National Key R&D Program of China (2022ZD0114900) and National Science Foundation of China (NSFC62276005).

768   Z. Zhang et al.

References

1. Cao, Q., et al.: Automatic identification of coronary tree anatomy in coronary computed tomography angiography. Int. J. Cardiovasc. Imaging **33**, 1809–1819 (2017)
2. Çiçek, Ö., Abdulkadir, A., Lienkamp, S.S., Brox, T., Ronneberger, O.: 3D U-Net: learning dense volumetric segmentation from sparse annotation. In: Ourselin, S., Joskowicz, L., Sabuncu, M.R., Unal, G., Wells, W. (eds.) MICCAI 2016. LNCS, vol. 9901, pp. 424–432. Springer, Cham (2016). https://doi.org/10.1007/978-3-319-46723-8_49
3. Dodge Jr, J.T., Brown, B.G., Bolson, E.L., Dodge, H.T.: Intrathoracic spatial location of specified coronary segments on the normal human heart. applications in quantitative arteriography, assessment of regional risk and contraction, and anatomic display. Circulation **78**(5), 1167–1180 (1988)
4. Dosovitskiy, A., et al.: An image is worth 16x16 words: transformers for image recognition at scale. arXiv preprint arXiv:2010.11929 (2020)
5. Graham, R.L., Hell, P.: On the history of the minimum spanning tree problem. Ann. History Comput. **7**(1), 43–57 (1985)
6. Graves, A., Schmidhuber, J.: Framewise phoneme classification with bidirectional LSTM and other neural network architectures. Neural Netw. **18**(5–6), 602–610 (2005)
7. Isensee, F., Jaeger, P.F., Kohl, S.A., Petersen, J., Maier-Hein, K.H.: nnu-net: a self-configuring method for deep learning-based biomedical image segmentation. Nat. Methods **18**(2), 203–211 (2021)
8. Kipf, T.N., Welling, M.: Semi-supervised classification with graph convolutional networks. arXiv preprint arXiv:1609.02907 (2016)
9. Lee, T.C., Kashyap, R.L., Chu, C.N.: Building skeleton models via 3-d medial surface axis thinning algorithms. CVGIP: Graph. Models Image Process. **56**(6), 462–478 (1994)
10. Loshchilov, I., Hutter, F.: Decoupled weight decay regularization. arXiv preprint arXiv:1711.05101 (2017)
11. Mowatt, G., et al.: 64-slice computed tomography angiography in the diagnosis and assessment of coronary artery disease: systematic review and meta-analysis. Heart **94**(11), 1386–1393 (2008)
12. Paszke, A., et al.: Pytorch: An imperative style, high-performance deep learning library. In: Wallach, H.M., Larochelle, H., Beygelzimer, A., d'Alché-Buc, F., Fox, E.B., Garnett, R. (eds.) NeurIPS (2019)
13. Qi, C.R., Yi, L., Su, H., Guibas, L.J.: Pointnet++: Deep hierarchical feature learning on point sets in a metric space. In: Advances in neural information processing systems 30 (2017)
14. Ronneberger, O., Fischer, P., Brox, T.: U-Net: convolutional networks for biomedical image segmentation. In: Navab, N., Hornegger, J., Wells, W.M., Frangi, A.F. (eds.) MICCAI 2015. LNCS, vol. 9351, pp. 234–241. Springer, Cham (2015). https://doi.org/10.1007/978-3-319-24574-4_28
15. Shen, D., Wu, G., Suk, H.I.: Deep learning in medical image analysis. Annu. Rev. Biomed. Eng. **19**, 221–248 (2017)
16. Shit, S., et al.: cldice-a novel topology-preserving loss function for tubular structure segmentation. In: Proceedings of the IEEE/CVF Conference on Computer Vision and Pattern Recognition, pp. 16560–16569 (2021)

17. Vaswani, A., et al.: Attention is all you need. Advances in neural information processing systems 30 (2017)
18. Wang, D., Zhang, Z., Zhao, Z., Liu, Y., Chen, Y., Wang, L.: Pointscatter: Point set representation for tubular structure extraction. In: Computer Vision-ECCV 2022: 17th European Conference, Tel Aviv, Israel, October 23–27, 2022, Proceedings, Part XXI, pp. 366–383. Springer, Cham (2022). https://doi.org/10.1007/978-3-031-19803-8_22
19. Wolterink, J.M., et al.: An evaluation of automatic coronary artery calcium scoring methods with cardiac CT using the orcascore framework. Med. Phys. **43**(5), 2361–2373 (2016)
20. Wu, D., et al.: Automated anatomical labeling of coronary arteries via bidirectional tree LSTMs. Int. J. Comput. Assist. Radiol. Surg. **14**, 271–280 (2019)
21. Yang, G., et al.: Automatic coronary artery tree labeling in coronary computed tomographic angiography datasets. In: 2011 Computing in Cardiology, pp. 109–112. IEEE (2011)
22. Yang, H., Zhen, X., Chi, Y., Zhang, L., Hua, X.S.: Cpr-gcn: conditional partial-residual graph convolutional network in automated anatomical labeling of coronary arteries. In: Proceedings of the IEEE/CVF Conference on Computer Vision and Pattern Recognition, pp. 3803–3811 (2020)
23. Yao, L., et al.: Tag-net: topology-aware graph network for centerline-based vessel labeling. IEEE Trans. Med. Imaging (2023)
24. Zhang, X., Cui, Z., Feng, J., Song, Y., Wu, D., Shen, D.: CorLab-Net: anatomical dependency-aware point-cloud learning for automatic labeling of coronary arteries. In: Lian, C., Cao, X., Rekik, I., Xu, X., Yan, P. (eds.) MLMI 2021. LNCS, vol. 12966, pp. 576–585. Springer, Cham (2021). https://doi.org/10.1007/978-3-030-87589-3_59
25. Zhang, Z., Liu, Q., Wang, Y.: Road extraction by deep residual u-net. IEEE Geosci. Remote Sens. Lett. **15**(5), 749–753 (2018)

AngioMoCo: Learning-Based Motion Correction in Cerebral Digital Subtraction Angiography

Ruisheng Su[1,6]($\boxtimes$)(iD), Matthijs van der Sluijs[1](iD), Sandra Cornelissen[1](iD),
Wim van Zwam[2](iD), Aad van der Lugt[1](iD), Wiro Niessen[1,3](iD),
Danny Ruijters[4](iD), Theo van Walsum[1](iD), and Adrian Dalca[5,6](iD)

[1] Erasmus University Medical Center, Rotterdam, The Netherlands
r.su@erasmusmc.nl
[2] Maastricht University Medical Center, Maastricht, The Netherlands
[3] Delft University of Technology, Delft, The Netherlands
[4] Philips Healthcare, Best, The Netherlands
[5] Massachusetts Institute of Technology, Boston, USA
[6] Massachusetts General Hospital, Harvard Medical School, Boston, USA

Abstract. Cerebral X-ray digital subtraction angiography (DSA) is the standard imaging technique for visualizing blood flow and guiding endovascular treatments. The quality of DSA is often negatively impacted by body motion during acquisition, leading to decreased diagnostic value. Traditional methods address motion correction based on non-rigid registration and employ sparse key points and non-rigidity penalties to limit vessel distortion, which is time-consuming. Recent methods alleviate subtraction artifacts by predicting the subtracted frame from the corresponding unsubtracted frame, but do not explicitly compensate for motion-induced misalignment between frames. This hinders the serial evaluation of blood flow, and often causes undesired vasculature and contrast flow alterations, leading to impeded usability in clinical practice. To address these limitations, we present AngioMoCo, a learning-based framework that generates motion-compensated DSA sequences from X-ray angiography. AngioMoCo integrates contrast extraction and motion correction, enabling differentiation between patient motion and intensity changes caused by contrast flow. This strategy improves registration quality while being orders of magnitude faster than iterative elastix-based methods. We demonstrate AngioMoCo on a large national multi-center dataset (MR CLEAN Registry) of clinically acquired angiographic images through comprehensive qualitative and quantitative analyses. AngioMoCo produces high-quality motion-compensated DSA, removing while preserving contrast flow. Code is publicly available at https://github.com/RuishengSu/AngioMoCo.

Keywords: Angiography · X-Rays · Registration · Motion Artifacts

Supplementary Information The online version contains supplementary material available at https://doi.org/10.1007/978-3-031-43990-2_72.

1   Introduction

Cerebral X-ray digital subtraction angiography (DSA) is a widely used imaging modality in interventional radiology for blood flow visualization and therapeutic guidance in endovascular treatments [25]. It is a 2D+T image series obtained by subtracting an initial pre-contrast image from subsequent post-contrast frames, leaving only the contrast-filled vessels visible. The injection of contrast medium and the subtraction process effectively eliminate soft tissue and bone, enabling high-resolution visualization of the vessels and the blood flow. However, this subtraction technique assumes the absence of motion between frames during exposure. In clinical practice, this premise is often violated. Involuntary motions, caused by swallowing, coughing, stroke, or endovascular procedures, are nearly inevitable. Body motion results in undesired artifacts in subtracted images, leading to decreased image quality and impaired interpretability of DSA (Fig. 1).

Over the last three decades, various motion correction techniques have been proposed to mitigate the impact of body motion retrospectively [18]. Registration algorithms typically employ template matching with corresponding control points or landmarks to align images [3,4,6–10,16,17,19,22,26–28]. These algorithms rely on features based on vessels [8], edges [9,17,19,28], corners [30], textures [20], temporal correspondence [3], and non-uniform grids [27]. To capture both local and global transformations, multi-resolution search [21,31], block

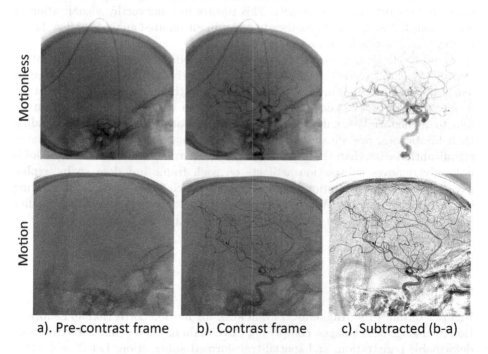

a). Pre-contrast frame   b). Contrast frame   c). Subtracted (b-a)

Fig. 1. Illustration of motion artifacts in DSA: a) the pre-contrast frame; b) a subsequent post-contrast frame; c) subtracted frame (b-a).

matching [9], and iterative estimations [20, 30] have been proposed. To limit undesirable vessel distortions, sparse key points [19] and non-rigidity penalties [26] have been used. Although these methods are effective in motion compensation, they require time-consuming iterative computation for each frame, limiting their clinical applicability.

Recent generative models, such as pix2pix [13], have been adapted to address subtraction artifacts without registration [11, 12, 29]. These models leverage deep learning techniques to predict a subtraction image from an input post-contrast image by discerning foreground contrast from the body background, resulting in reduced artifacts. However, these models do not explicitly compensate for motion-induced misalignment between frames. More importantly, they may cause hallucinations or modification of contrast and vessels, and lack interpretability as there is no subtraction. Consequently, these shortcomings hinder the serial evaluation of blood flow and impede the diagnostic utility of DSA.

To overcome these limitations, we introduce AngioMoCo, a fast learning-based motion correction method for DSA that avoids severe contrast distortion. We employ a supervised CNN module that distinguishes between motion displacement and contrast intensity change. The output contrast-removed image and the pre-contrast image are then input to a subsequent self-supervised learning-based registration model for deformable registration, where a deformation regularization loss limits the local irregularity. By excluding contrast enhancements from the deformation learning processing, AngioMoCo avoids undesired distortion of the vessels. This results in trustworthy visualization of continuous blood flow and promises to assist in automated analysis of flow-based biomarkers relevant to endovascular treatments.

Overall, classical non-rigid registration methods use various regularization strategies to limit vessel distortion, but are prohibitively time-consuming. Recent learning-based methods are fast, but do not explicitly model the motion between frames, and as a result can negatively distort the very clinical information we aim to highlight. We build on the strengths of both directions while avoiding their limitations. Specifically, we propose a novel learning-based strategy that is significantly faster than traditional non-rigid registration methods. AngioMoCo not only removes subtraction artifacts on each frame but does so by explicitly compensating for motion between frames, which is not available in existing image-to-image models. We demonstrate that AngioMoCo achieves high-quality registration while avoiding undesirable contrast reduction or vessel erasure.

2   Method

2.1   Model

Figure 2 outlines the AngioMoCo framework for motion correction and subtraction in angiographic images, comprising three main modules: contrast extraction, deformable registration, and spatial-transformed subtraction. Let $\mathcal{X} = \{x_t\}_{t=0}^{T}$ be the 2D+T DSA series of a patient, where x_0 is the pre-contrast frame and $\{x_t\}_{t=1}^{T}$ are the post-contrast frames.

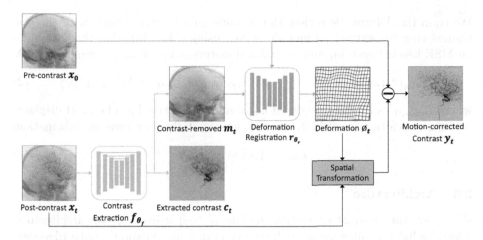

Fig. 2. Overview. The proposed framework takes a pre-contrast image x_0 and a post-contrast image x_t as input. The contrast extraction module $f_{\theta_f}(\cdot)$ splits x_t into contrast c_t and contrast-removed m_t. Next, m_t and x_0 are registered using network $r_{\theta_r}(\cdot, \cdot)$, which outputs a deformation field ϕ_t. Subsequently, ϕ_t is applied to the post-contrast image x_t to obtain the final output subtracted image y_t, which corrects misalignment between frames.

We define a contrast extraction module $f_{\theta_f}(x_t) = c_t$ with parameters θ_f that takes as input a post-contrast frame x_t. This function separates x_t into a contrast image c_t and a contrast-removed image m_t where $m_t = x_t - c_t$. The values in c_t are within $[-1, 0]$ as the injected contrast medium can only lead to a decrease in pixel intensity relative to the input image with an intensity range of $[0, 1]$. The contrast extraction module aims to reduce contrast discrepancies between the pre- and post-contrast frames. Such image-to-image modules can lead to hallucination and may not fully capture distal vessels, relatively less contrasted vessels, and vessels behind bone structures. Therefore, in AngioMoCo, we only employ this module to enable easier registration of the frame x_t to the pre-contrast x_0 using the intermediate contrast-extracted m_t image.

We define a registration function $r_{\theta_r}(x_0, m_t) = \phi_t$ with parameters θ_r to estimate the deformation ϕ_t. We obtain the motionless subtraction angiography y_t by subtracting the pre-contrast frame x_0 from the warped post-contrast frame w_t: $y_t = w_t - x_0 = x_t \circ \phi_t - x_0$, where $\circ$ defines a spatial warp.

2.2 Training

We train the contrast extraction $f_{\theta_f}(\cdot)$ and deformable registration $r_{\theta_r}(\cdot, \cdot)$ modules separately. The contrast extraction module is trained on a motionless subset of data with an MSE loss between the ground truth contrast, estimated via subtraction between post- and pre-contrast frames $(x_t - x_0)$, and the predicted c_t:

$$\mathcal{L}_{\text{ext}}(\theta_r; x_t) = \mathcal{L}_{\text{MSE}}(x_t - x_0, f_{\theta_f}(x_t)). \tag{1}$$

We train the deformable registration module on a motion subset, with the pre-trained contrast extraction module frozen, using a loss function that combines an MSE loss between m_t and x_0 and a smoothness loss $\mathcal{L}_{\text{smooth}}$, weighted by λ:

$$\mathcal{L}_{\text{reg}}(\theta_f; x_0, m_t \circ \phi_t) = (1 - \lambda)\mathcal{L}_{\text{MSE}}(x_0, m_t \circ \phi_t) + \lambda\mathcal{L}_{\text{smooth}}(\phi_t), \qquad (2)$$

where $\mathcal{L}_{\text{smooth}}$ is the mean squared horizontal and vertical gradients of displacement u_t in deformation field ϕ_t, that enforces spatial smoothness of deformation:

$$\mathcal{L}_{\text{smooth}}(\phi_t) = \|\nabla u_t\|^2. \qquad (3)$$

2.3   Architecture

We design the contrast extraction module $f_{\theta_f}(\cdot, \cdot)$ using a U-Net architecture, which includes a contracting path (encoder) and an expanding path (decoder) connected by skip connections. The encoder stage comprises eight convolutional and max-pooling layers with the number of channels being 8, 16, 32, 64, 128, 256, 512, and 512 respectively. The convolutions operate with a 3×3 kernel size and a stride of 2. Similarly, the decoding path employs eight upsampling, 3×3 convolution, and concatenation operations with 32 feature maps per layer to restore the spatial dimension up to the input size. Each convolution is accompanied by an instance normalization and a LeakyReLU activation layer. We also use three additional 3×3 convolutions. The final convolution employs a negative sigmoid activation, confining the output pixel intensity to $[-1, 0]$.

We employ a deformable registration module $r_{\theta_r}(\cdot, \cdot)$ based on VoxelMorph to learn motion correction in DSA [2]. We add instance normalization between the convolution layers of the encoder and decoder. We utilize this deformable registration module to predict bi-directional dense deformation fields using diffeomorphism that allows to spatially transform either pre- or post-contrast frames.

3   Experiments

We assess AngioMoCo in terms of vessel contrast preservation, artifact removal, and computation efficiency compared to existing approaches.

3.1   Experimental Setup

Data. We identified 272 patients with unsubtracted cerebral angiographic images available from MR CLEAN registry [14], an ongoing prospective observational multi-center registry of patients with acute ischemic stroke who underwent endovascular thrombectomy (EVT). This comprised 788 angiographic series, consisting of 16,641 frames in total, acquired between attempts of thrombus retrieval. The DSA series were acquired using various imaging systems, including Philips, GE, and Siemens, and had a size of 1024×1024 pixels. The series had varying lengths, ranging from 10 to 50 frames, and temporal resolutions between 0.5 and 4 frames per second (fps). We performed image resizing to 512×512

pixels and min-max intensity normalization to obtain intensity values within the range of $[0,1]$. To ensure the coherency of the intensity along the series, the maximum intensity is calculated on the series level based on the stored bits in the DICOM header.

Based on visual assessment, we categorized the dataset into two subsets: motionless and motion. We use the motionless subset, consisting of 107 series (1933 frames) from 21 patients, for pre-training the contrast extraction module. The motion subset, which contains 681 series (14708 frames) from 251 patients, is used for overall training and evaluation. We split data on the patient level independently on the motionless and motion subsets, with a ratio of 50%, 20%, and 30% for training, validation, and testing, respectively.

Baselines. We compare AngioMoCo with two widely used image registration approaches, elastix-based affine registration and VoxelMorph [1,2], and an image-to-image approach employing a U-Net [24] architecture. We followed the implementation of [2] for VoxelMorph with deformation regularization $\lambda = 0.01$. For the U-Net, we employed the same architecture as the contrast extraction module $f_{\theta_f}(\cdot, \cdot)$ with the same preprocessing and augmentations. We trained the U-Net using the motionless subset and used mean squared error (MSE) as the optimizing objective. We implemented the methods using Python 3.10.6 and PyTorch [23].

Training Details. We use an NVIDIA 2080 Ti GPU (11 GB), the Adam optimizer [15] and the ReduceLROnPlateau scheduler with an initial learning rate of 0.001, a patience of 300 epochs, and a decay of 0.1. We set the batch size to 8 and applied early stopping with a patience of 500 epochs. We selected these optimization parameters based on validation performance using a grid search. We applied data augmentations using Albumentations [5], including *HorizontalFlip*, *ShiftScaleRotate*, and *RandomSizedCrop*, each with a probability of 0.5.

Evaluation. We carry out both qualitative and quantitative analyses on the hold-out test set of the motion subset. A key challenge is to minimize motion and subtraction artifacts while retaining clinically important features. We use mean squared intensity (MSI) as a proxy to quantify the preservation of contrast intensity within vessels and the ability of motion correction outside vessels. As ground truth deformations are not available for image sequences with motion, we manually segment the blood vessels in post-contrast frames (Supplemental Fig. 6), and use the resulting masks to quantify MSI inside and outside blood vessels. We used paired t-tests for statistical significance.

3.2 Results

Quantitative Analysis. The optimal outcome is represented by the top left corner of Fig. 3, indicating high vessel contrast preservation and complete artifact removal (Supplemental Table 1). Compared to elastix affine registration,

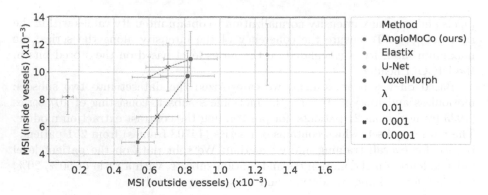

Fig. 3. Mean squared intensity (MSI) on the test set. Better methods will preserve the MSI (i.e., vessel contrast) inside vessels (↑, y-axis) while minimizing the MSI (i.e., artifacts) outside vessels (←, x-axis), moving towards the top left of the graph.

AngioMoCo($\lambda = 0.001$) achieves similar vessel preservation ($P = 0.2$), while substantially decreasing the MSI outside vessels (by about half). Compared to VoxelMorph, AngioMoCo demonstrates substantial improvement, with higher vessel preservation and better (more to the left) artifact removal. While the image-to-image U-Net yields the lowest MSI outside vessels, it sacrifices a substantial amount (30%) of contrast inside vessels, harming the precise clinical signal we are interested in.

Qualitative Analysis. Figure 4 presents visual comparisons of the methods through three representative examples. The image-to-image U-Net generates images with fewer motion artifacts than other methods, but it often fails to capture vessel contrast behind bone structures (Row 1), distal vessels (Row 1), and loses high-frequency spatial features, leading to blurry images (Row 2). These errors can have substantial negative effects on downstream clinical applications. VoxelMorph operates on pre- and post-contrast images, which can cause considerable modifications in the vessel contrast flow. For example, the motion-corrected image of VoxelMorph in Row 3 has lighter vessel contrast than its counterparts. In contrast, AngioMoCo overcomes these limitations of U-Net and VoxelMorph by learning to disentangle contrast flow from motion.

Runtime. Compared to iterative registration methods, deep-learning-based registration methods, including AngioMoCo, require orders of magnitude less time. For example, AngioMoCo takes less than a second to process a series on GPU, while iterative registration methods are mostly implemented on CPU where they require minutes.

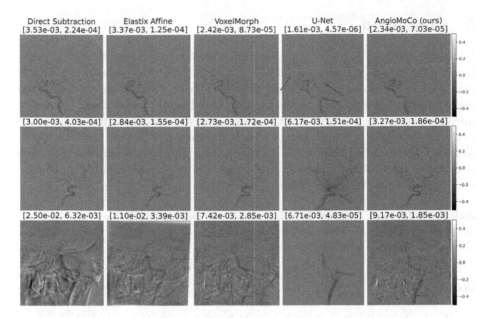

Fig. 4. Representative visual comparisons. We report MSI values inside (left) and outside (right) vessels in brackets. Red arrows point to undesired vessel contrast erasure or modifications. AngioMoCo achieves better background artifact removal and vessel enhancement than other methods. The UNet achieves excellent artifact removal, but it comes at the cost of severe damage to the vessels of interest, making it clinically less useful.

4 Discussion

We find that AngioMoCo achieves high-quality motion correction in DSA, while preserving vessel details, which is of critical clinical importance. While the image-to-image U-Net resulted in fewer artifacts, it substantially degrades the vessel contrast, harming its usability in clinical usefulness.

These results suggest that AngioMoCo is clinically relevant for endovascular applications, enhancing the utility of DSA in diagnosis and treatment planning. The tool can extract contrast flow while outputting smooth bi-directional deformation fields that provide interpretability. Unlike image-to-image models, the contrast flow visualization is driven by motion-compensation of the post-contrast frames to the pre-contrast image, and hence avoids undesirable hallucinations and modifications of vessel contrast.

We also examined the end-to-end training strategy of AngioMoCo, which did not yield superior results to VoxelMorph or the modularly trained AngioMoCo (Supplemental Fig. 5). To further enhance registration accuracy, future research may explore the integration of 3D spatio-temporal CNN and the utilization of vessel masks as auxiliary supervision.

5   Conclusion

We have presented AngioMoCo, a deep learning-based strategy towards motion-free digital subtraction angiography. The approach leverages a contrast extraction module to disentangle contrast flow from body motion and a deformable registration module to concentrate on motion-induced deformations. The experimental results on a large clinical dataset demonstrate that AngioMoCo outperforms iterative affine registration, learning-based VoxelMorph, and image-to-image U-Net. Overall, AngioMoCo achieves high registration accuracy while preserving vascular features, improving the quality and clinical utility of DSA for diagnosis and treatment planning in endovascular procedures.

Acknowledgment. This work is done during a visit at MGH. The visit was made possible in part by the Academy Van Leersum grant of the Academy Medical Sciences Fund of the Royal Netherlands Academy of Arts & Sciences (KNAW).

References

1. Balakrishnan, G., Zhao, A., Sabuncu, M.R., Guttag, J., Dalca, A.V.: An unsupervised learning model for deformable medical image registration. In: Proceedings of the IEEE Conference on Computer Vision and Pattern Recognition, pp. 9252–9260 (2018)
2. Balakrishnan, G., Zhao, A., Sabuncu, M.R., Guttag, J., Dalca, A.V.: VoxelMorph: a learning framework for deformable medical image registration. IEEE Trans. Med. Imaging **38**(8), 1788–1800 (2019)
3. Bentoutou, Y., Taleb, N.: A 3-D space-time motion detection for an invariant image registration approach in digital subtraction angiography. Comput. Vis. Image Underst. **97**(1), 30–50 (2005)
4. Bentoutou, Y., Taleb, N., El Mezouar, M.C., Taleb, M., Jetto, L.: An invariant approach for image registration in digital subtraction angiography. Pattern Recogn. **35**(12), 2853–2865 (2002)
5. Buslaev, A., Iglovikov, V.I., Khvedchenya, E., Parinov, A., Druzhinin, M., Kalinin, A.A.: Albumentations: fast and flexible image augmentations. Information **11**(2), 125 (2020)
6. Buzug, T.M., Weese, J.: Image registration for DSA quality enhancement. Comput. Med. Imaging Graph. **22**(2), 103–113 (1998)
7. Buzug, T.M., Weese, J., Fassnacht, C., Lorenz, C.: Using an entropy similarity measure to enhance the quality of DSA images with an algorithm based on template matching. In: Höhne, K.H., Kikinis, R. (eds.) Visualization in Biomedical Computing: 4th International Conference, VBC 1996 Hamburg, Germamy, 22–25 September 1996, Proceedings, pp. 235–240. Springer, Cham (2006). https://doi.org/10.1007/BFb0046959
8. Cao, Z., Liu, X., Peng, B., Moon, Y.S.: DSA image registration based on multiscale Gabor filters and mutual information. In: 2005 IEEE International Conference on Information Acquisition, pp. 6-pp. IEEE (2005)
9. Chu, Y., Bai, N., Ji, Z., Chen, S., Mou, X.: Registration for DSA image using triangle grid and spatial transformation based on stretching. In: 2006 8th international Conference on Signal Processing, vol. 2. IEEE (2006)

10. Cox, G.S., de Jager, G.: Automatic registration of temporal image pairs for digital subtraction angiography. In: Medical Imaging 1994: Image Processing, vol. 2167, pp. 188–199. SPIE (1994)
11. Crabb, B.T., et al.: Deep learning subtraction angiography: improved generalizability with transfer learning. J. Vasc. Intervent. Radiol. **34**, 409-419.e2 (2022)
12. Gao, Y., et al.: Deep learning-based digital subtraction angiography image generation. Int. J. Comput. Assist. Radiol. Surg. **14**, 1775–1784 (2019)
13. Isola, P., Zhu, J.Y., Zhou, T., Efros, A.A.: Image-to-image translation with conditional adversarial networks. In: Proceedings of the IEEE Conference on Computer Vision and Pattern Recognition, pp. 1125–1134 (2017)
14. Jansen, I.G., Mulder, M.J., Goldhoorn, R.J.B.: Endovascular treatment for acute ischaemic stroke in routine clinical practice: prospective, observational cohort study (MR CLEAN Registry). BMJ **360**, k949 (2018)
15. Kingma, D.P., Ba, J.: Adam: a method for stochastic optimization. arXiv preprint arXiv:1412.6980 (2014)
16. Liu, B., Zhao, Q., Dong, J., Jia, X., Yue, Z.: A stretching transform-based automatic nonrigid registration system for cerebrovascular digital subtraction angiography images. Int. J. Imaging Syst. Technol. **23**(2), 171–187 (2013)
17. Meijering, E.H., et al.: Reduction of patient motion artifacts in digital subtraction angiography: evaluation of a fast and fully automatic technique. Radiology **219**(1), 288–293 (2001)
18. Meijering, E.H., Niessen, W.J., Viegever, M.: Retrospective motion correction in digital subtraction angiography: a review. IEEE Trans. Med. Imaging **18**(1), 2–21 (1999)
19. Meijering, E.H., Zuiderveld, K.J., Viergever, M.A.: Image registration for digital subtraction angiography. Int. J. Comput. Vision **31**, 227–246 (1999)
20. Nejati, M., Amirfattahi, R., Sadri, S.: A fast image registration algorithm for digital subtraction angiography. In: 2010 17th Iranian Conference of Biomedical Engineering (ICBME), pp. 1–4. IEEE (2010)
21. Nejati, M., Pourghassem, H.: Multiresolution image registration in digital X-ray angiography with intensity variation modeling. J. Med. Syst. **38**, 1–10 (2014)
22. Nejati, M., Sadri, S., Amirfattahi, R.: Nonrigid image registration in digital subtraction angiography using multilevel B-spline. BioMed Res. Int. **2013**, 236315 (2013)
23. Paszke, A., et al.: PyTorch: an imperative style, high-performance deep learning library. In: Advances in Neural Information Processing Systems, pp. 8026–8037 (2019)
24. Ronneberger, O., Fischer, P., Brox, T.: U-Net: convolutional networks for biomedical image segmentation. In: Navab, N., Hornegger, J., Wells, W.M., Frangi, A.F. (eds.) MICCAI 2015, Part III. LNCS, vol. 9351, pp. 234–241. Springer, Cham (2015). https://doi.org/10.1007/978-3-319-24574-4_28
25. Shaban, S., et al.: Digital subtraction angiography in cerebrovascular disease: current practice and perspectives on diagnosis, acute treatment and prognosis. Acta Neurologica Belgica **122**(3), 763–780 (2021)
26. Staring, M., Klein, S., Pluim, J.P.: A rigidity penalty term for nonrigid registration. Med. Phys. **34**(11), 4098–4108 (2007)
27. Sundarapandian, M., Kalpathi, R., Manason, V.D.: DSA image registration using non-uniform MRF model and pivotal control points. Comput. Med. Imaging Graph. **37**(4), 323–336 (2013)
28. Taleb, N., Jetto, L.: Image registration for applications in digital subtraction angiography. Control. Eng. Pract. **6**(2), 227–238 (1998)

29. Ueda, D., et al.: Deep learning-based angiogram generation model for cerebral angiography without misregistration artifacts. Radiology **299**(3), 675–681 (2021)
30. Wang, J., Zhang, J.: An iterative refinement DSA image registration algorithm using structural image quality measure. In: 2009 Fifth International Conference on Intelligent Information Hiding and Multimedia Signal Processing, pp. 973–976. IEEE (2009)
31. Yang, J., Wang, Y., Tang, S., Zhou, S., Liu, Y., Chen, W.: Multiresolution elastic registration of X-ray angiography images using thin-plate spline. IEEE Trans. Nucl. Sci. **54**(1), 152–166 (2007)

CenterlinePointNet++: A New Point Cloud Based Architecture for Coronary Artery Pressure Drop and vFFR Estimation

Patryk Rygiel[1], Paweł Płuszka[1], Maciej Zięba[2], and Tomasz Konopczyński[1(✉)]

[1] Hemolens Diagnostics Sp. z o.o., Wrocław, Poland
{patryk.rygiel,pawel.pluszka,tomasz.konopczynski}@hemolens.eu
[2] Wrocław University of Science and Technology, Wrocław, Poland

Abstract. Estimation of patient-specific hemodynamic features, and in particular fractional flow reserve (FFR) in coronary arteries is an essential step in providing personalized and accurate diagnosis of coronary artery disease (CAD). In recent years, in the domain of computed tomography angiography (CTA), a virtual FFR (vFFR) derived from coronary CTA using computational fluid dynamics (CFD), has been used as a compelling, non-invasive, *in-silico* replacement for invasive diagnostic techniques. Unfortunately, the time and computational demands of CFD are major obstacles to introducing vFFR from CT as a commonly used prophylactic tool. In this work, we propose a novel geometric-based artificial deep learning (DL) architecture, CenterlinePointNet++, which acts as a surrogate for CFD engines for the task of hemodynamic features estimation of the coronary arteries. Our architecture works directly on the vessel geometry represented as a surface point cloud and a centerline graph. As a result of that, it utilizes implicit geometry embedding without the need for hand-crafted features to estimate directly hemodynamic features. We evaluate our approach on the task of pressure drops and vFFR estimation for a synthetically generated dataset of coronary arteries and showcase significant improvement over commonly used geometry-based approaches.

Keywords: deep learning · geometric deep learning · point clouds · CAD · FFR · hemodynamics

1 Introduction

Estimation of patient-specific hemodynamic features in coronary arteries is an essential step in providing personalized and accurate diagnosis and treatment of CAD which is one of the main causes of death in the world [15]. In the assessment of CAD, one of the most important biomarkers is a Fractional Flow Reserve

---

Supplementary Information The online version contains supplementary material available at https://doi.org/10.1007/978-3-031-43990-2_73.

H. Greenspan et al. (Eds.): MICCAI 2023, LNCS 14226, pp. 781–790, 2023.
https://doi.org/10.1007/978-3-031-43990-2_73

(FFR [3]), measured during an invasive coronary angiography procedure. With developments in CT and CFD, non-invasive approaches for accurate hemodynamic features estimation such as vFFR in patients with suspected ischemic heart disease became possible [9]. Unfortunately, although CFD offers a compelling *in-silico* replacement for invasive coronagraphy procedures, prolonged vFFR computation times, over many hours, have been a major concern [14].

To solve the computation time drawback, AI-based solutions have been proposed to lower the estimation time of vFFR down to minutes, or even seconds at the cost of accuracy [1,22]. Itu et al. [8] propose to compute a set of local and global hand-crafted features from the coronary anatomy to create a representation of a local stenotic segment. Based on this representation, a multilayer perceptron (MLP) model is trained to perform an FFR regression in the stenotic areas. They experiment on a dataset of 12,000 synthetically generated coronary geometries and evaluated the model on 125 patient-specific anatomical models extracted from CTA, reporting a correlation of 0.729 with invasive FFR on real anatomical models. Wang et al. [24] propose a similar method that utilizes hand-crafted features to feed a recurrent neural network (RNN) to estimate the FFR along the coronary artery. Authors experiment with training on 71 patient-specific anatomies extracted from CTA and report a correlation of 0.686 with invasive FFR measurement. Both approaches report compiling results of the vFFR estimation and showcase a drastic improvement in time consumption: 2.4 s for Itu et al. [8] and 120 s for Wang et al. [24]. The main drawback in both cases is the utilization of hand-crafted features. Additionally, authors do not share details of their features making their work impossible to reproduce [20]. A natural improvement over these methods would be an implicit feature learning method that could be performed on a vessel surface modelled as a mesh or a point cloud. However, there is no such method known to the authors that would tackle the problem of vFFR estimation for coronary arteries from CTA. For a similar task, Li et al. [11] propose a PointNet-like [17] architecture to predict velocity and pressure fields on point cloud geometries containing aorta, coronary arteries and bypass graft. Similarly, Suk et al. [21] utilizes mesh-based neural networks [5,6,23] to estimate wall shear stress along the vessels. Both approaches would be unsuitable for the vFFR estimation from CTA, as it would require the processing of more complex, elongated geometries.

In this work, we tackle the problem of estimating hemodynamic features such as pressure drops and vFFR along the coronary arteries extracted from CTA. We propose a novel CenterlinePointNet++ architecture that is tailored towards the processing of complex, elongated structures such as coronary arteries, that can be represented as a surface point cloud and a centerline graph along branches. In our approach, implicit feature extraction is guided by the proposed centerline grouping aggregation. It is a replacement for the commonly used topology-agnostic [18], centerline-agnostic [7,19] or connectivity-based [25] aggregation strategies. To our best knowledge, this is the first method that is tackling the problem of pressure drops and vFFR estimation utilizing a point cloud neural network. We train and evaluate our approach on different scenarios

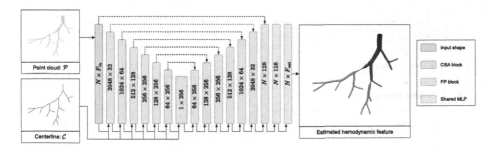

Fig. 1. Diagram of CenterlinePointNet++ architecture. The model incorporates an encoder-decoder structure seen in PointNet++. The encoder branch consists of seven CSA blocks that process the surface point cloud and centerline graph. The decoder branch is built out of seven FP blocks and topped with the head consisting of two Shared MLP layers.

on the dataset of $1,700$ synthetically generated coronary arteries. We test the model capability for two tasks: estimation of pressure drop and vFFR under different GT CFD settings. For each setting, we report an improvement over Euclidean distance based grouping in terms of Mean Absolute Error (MAE) and Normalized Absolute Error (NMAE). We further test the approach performance of vFFR estimation and report a correlation with CFD vFFR of 0.93.

2 Method

In this section, we provide a detailed description of the proposed CenterlinePoint-Net++ architecture (see Fig. 1) for the problem of estimating hemodynamic features. We improve upon the well-known PointNet++ architecture [18] by taking into account *a priori* knowledge of coronary arteries geometries. We extend its input to accept an additional channel with a centerline graph and change its encoder blocks to the proposed Centerline-Set-Abstraction (CSA) blocks that utilize the centerline graph and the centerline grouping method which facilitates geodesic metric. We refer to the input vessel mesh as $\mathcal{M}$ and the point cloud representing its surface that is sampled by extracting all its vertices is denoted as $\mathcal{P}$. The centerline graph, which can be extracted from the mesh with an algorithm of choice is denoted as $\mathcal{C} = (V_{\mathcal{C}}, E_{\mathcal{C}})$, where $V_{\mathcal{C}}$ is the set of nodes and $E_{\mathcal{C}}$ a set of undirected edges. Since the centerline graph is a 3D structure in Euclidean space, we can represent each $v \in V_{\mathcal{C}}$ as a point with 3D coordinates. The architecture takes as an input a point cloud $\mathcal{P}$ and centerline graph $\mathcal{C}$ and returns per-point hemodynamic features of choice. The details of the method are described in the following sections.

2.1 Encoder

The encoder is built out of the proposed CSA blocks in a hierarchical manner as it is shown in Fig. 1. It consists of n number of CSA blocks down to the

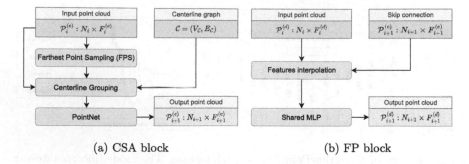

(a) CSA block   (b) FP block

Fig. 2. CenterlinePointNet++ blocks. The CSA block performs feature extraction by downsampling point cloud with the FPS algorithm, aggregating surface points in a centerline-guided manner and processing them with PointNet. The FP block performs the upsampling procedure by interpolating features onto the new points passed via skip-connection and processing them with the Shared MLP layer.

bottleneck. Each CSA_{i+1} takes as an input a point cloud $\mathcal{P}_i^{(e)}$ of dimensions $N_i \times F_i^{(e)}$ from the previous block and a centerline graph $\mathcal{C}$, where N stands for the number of points, F stands for the number of features, $i + 1$ is the index of the current CSA block, i is the index of the previous block, and (e) stands for "encoding". The CSA block is described in Fig. 2a. In the first step of a CSA block, a representative sampling procedure in form of Farthest Point Sampling [4] (FPS) is utilized to downsample $\mathcal{P}_i^{(e)}$ and create a representative point cloud $\mathcal{R} \subseteq \mathcal{P}$.

For each point $r \in \mathcal{R}$, a *centerline grouping* procedure $g(r, \mathcal{C})$ is performed to extract their point neighborhoods $\mathcal{G}_r \subseteq \mathcal{P}$. In the last step, all extracted neighbourhoods are processed independently with a PointNet [17] to construct neighbourhood feature vectors for all the points in $\mathcal{R}$.

Centerline Grouping: We propose a novel centerline-guided point grouping scheme for point cloud structures. Commonly used Euclidean distance based grouping strategies do not take into account an underlying surface manifold formed by the points. Thus they tend to fail when the point cloud topology is complex, by considering points in the close Euclidean distance to be neighbours while their distance along the formed manifold is much larger (see Fig. 3a). By utilizing a centerline-guided point grouping, the geodesic metric is facilitated, and thus the point neighbourhoods are constructed along the manifold (see Fig. 3b). Additionally, grouping along the centerline allows for a robust sequential embedding of the complex vessel trees and thus facilitates the learning of hemodynamic features.

For a given point cloud $\mathcal{P}$, its representation $\mathcal{R}$ and its centerline $\mathcal{C}$, the centerline grouping procedure $g(r, \mathcal{C})$ is performed independently for each $r \in \mathcal{R}$. We define a mapping function $M : \mathcal{P} \mapsto V_{\mathcal{C}}$, which assigns a closest, in Euclidean distance sense, centerline node $v \in V_{\mathcal{C}}$ to each point $p \in \mathcal{P}$. Since a

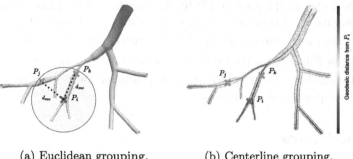

(a) Euclidean grouping.   (b) Centerline grouping.

Fig. 3. Grouping comparison. Points P_j and P_k are considered to be in the same size neighborhood of P_i when grouped with Euclidean distance, even though their geodesic distances are substantially different. Grouping along the centerline solves this issue by facilitating geodesic metric.

representative point cloud $\mathcal{R}$ is a subset of point cloud $\mathcal{P}$, the mapping is also defined for each point $r \in \mathcal{R}$. Due to the utilization of a mapping function M, the grouping procedure can be performed directly on the centerline graph $\mathcal{C}$ where the topological structure of the vessel tree can be more accurately represented. Based on the 3D coordinates of centerline nodes $V_{\mathcal{C}}$, the weights of the edges $E_{\mathcal{C}}$ are calculated as distances between the connected nodes - since the centerline graph is not-mutable during training, the weights can be pre-computed in the pre-processing stage.

Having a mapping function M we can directly work on the centerline graph itself and thus we define a centerline neighbourhood $\mathcal{Q}_v$ which is computed independently for each $v \in M[\mathcal{R}]$ and can be expressed with the following equation:

$$\mathcal{Q}_{v_0} = \{v \in V_c : d_{\mathcal{C}}(v, v_0) \leq t\}, \tag{1}$$

where $d_{\mathcal{C}}$ is a centerline distance function which returns the length of the weighted shortest path between two given nodes in the centerline $\mathcal{C}$ and t is the distance threshold which marks whether nodes should be considered neighbours or not. Having centerline neighbourhoods Q extracted we need to map them back onto the point cloud $\mathcal{P}$ to obtain the point neighbourhoods $\mathcal{G}$ for each $r \in \mathcal{R}$:

$$\mathcal{G}_r = \{p \in P : M(p) \in \mathcal{Q}_{M(r)}\}. \tag{2}$$

2.2   Decoder

The decoder is built out of the Feature-Propagation (FP) blocks in a similar manner as in PointNet++ [18]. Its architecture is shown in Fig. 2b The FP takes as an input a point cloud $\mathcal{P}_i^{(d)}$ from the previous block and a reference point cloud $\mathcal{P}_{i+1}^{(e)}$ passed via skip-connection from the respective encoder block, where

$i + 1$ is the index of the current FP block, i is the index of the previous block and (d) stands for "decoding". In the first step, the features from the point cloud $\mathcal{P}_i^{(d)}$ are interpolated onto the $\mathcal{P}_{i+1}^{(e)}$ by averaging features of three nearest neighbours in Euclidean distance sense. The resulting point cloud $\mathcal{P}_{i+1}^{(d)}$ is then processed with shared MLP which is applied independently to each $p \in \mathcal{P}_{i+1}^{(d)}$ to form new per-point feature vectors.

3   Experiments

In this section, we describe our evaluation of the proposed model architecture in the tasks of pressure drops and vFFR estimation of a synthetic coronary artery geometry with respect to different stenosis severity grades and different biologically relevant ranges of blood flow characteristics. We test our architecture against standard PointNet++ grouping with the same number of adequate layers as a reference due to the impossibility of reproduction [20] of the other two known relevant methods [8,24]. We train CenterlinePointNet++ and PointNet++ with the same set of hyperparameters. Both need approximately 15 s for inference on the RTX 3090 24 GB graphic card. The models are trained for 500 epochs with the batch size of 8 and Adam optimizer with the default constant learning rate of 0.001. For the loss function we use the Mean Squared Error (MSE) which is the mean of per point squared errors.

3.1   Dataset

We utilize a dataset of 1,700 synthetically generated coronary arteries to train and evaluate the model's performance. We split the data to train, validation and test set of sizes 1,500, 100 and 100, respectively.

We generate synthetic coronary arteries using ranges of geometric quantities such as the radii of branches, bifurcation angles, and degree of tapering with distributions similar to other studies [8,13]. On top of that, we model stenotic areas based on Coronary Artery Disease - Reporting and Data System (CAD-RADS) [16] stenoses percentage intervals. Our point clouds are further preprocessed such that each point is described by the position in the 3D space, Euclidean distance to the centerline and geodesic distance to the inlet (first point on the centerline).

For the ground-truth (GT) labels generation we use a commercial CFD engine of choice [10] designed for vFFR calculation.

The considered simulations for GT are stationary and take up to two hours on a CPU with 16 processes per synthetic coronary artery. We generate labels from a biologically relevant range that aim to simulate a patient under rest, mild exercise and high-intensity exercise conditions [12]. We experiment with three values of input flow Q_{in}: three, five and seven ml/s, and three values of inlet pressure p_{in}: 80, 100 and 120 mmHg.

Table 1. Comparison of PointNet++ (PN) and CenterlinePointNet++ (CPN) in the task of pressure drop and vFFR estimation with various BC of CFD GT simulation. Metrics are computed on the centerline-projected surface predictions.

	Q_{in} (ml/s)	Pressure drops						vFFR ($p_{in}=80$ mmHg)			vFFR ($p_{in}=100$ mmHg)			vFFR ($p_{in}=120$ mmHg)		
		NMAE (%)			MAE (m^2/s^2)			MAE ($\times 10^{-2}$)			MAE ($\times 10^{-2}$)			MAE ($\times 10^{-2}$)		
		mean	median	75th	mean	median	75th	mean	median	75th	mean	median	75th	mean	median	75th
PN	3	2.01	0.74	1.36	2.23	0.80	1.52	2.83	1.04	1.92	2.18	0.80	1.47	1.89	0.69	1.28
CPN		1.57	0.55	1.32	1.75	0.64	1.52	2.21	0.77	1.86	1.70	0.60	1.43	1.47	0.52	1.24
PN	5	1.68	0.63	1.31	4.31	1.59	3.35	5.36	2.01	4.17	4.13	1.55	3.21	3.58	1.34	2.78
CPN		1.58	0.56	1.12	3.99	1.44	2.79	5.03	1.80	3.59	3.87	1.38	2.75	3.35	1.20	2.39
PN	7	1.74	0.60	1.34	7.74	2.63	5.90	9.70	3.34	7.44	7.46	2.57	5.73	6.47	2.23	4.96
CPN		1.54	0.50	1.17	6.86	2.23	5.18	8.56	2.80	6.50	6.58	2.16	5.00	5.71	1.87	4.33

In our experimental setup, one coronary artery is composed of a surface point cloud with 100,000–200,000 points and a centerline graph with 300–600 vertices, depending on the complexity of the vascular tree. The surface point clouds are uniformly downsampled to 20,000 points during the training process. In the inference setting the per-point results are obtained via a mean aggregation over multiple inference runs, five in our case, of different random splits of a point cloud into chunks of 20,000 points. The process of projecting the surface features onto the centerline graph is done by averaging the estimated features of the faces' vertices which intersect with the orthogonal plane to the centerline placed for a given centerline node (Table 1).

3.2 Results

We showcase the comparison between PointNet++ and proposed Centerline-PointNet++ in the task of pressure drop and vFFR estimation for the synthetic dataset under different biologically relevant boundary conditions (BC) in Fig. 4. We report the MAE and NMAE which is MAE normalized by the absolute largest pressure drop in the testing set. The metrics are computed for projected results on a centerline for the evaluation to be more informative. Centerline-PointNet++ outperforms PointNet++ in every combination of input flow Q_{in} and inlet pressure p_{in} for both pressure drop and vFFR estimation.

We show a visual comparison between the PointNet++ and CenterlinePoint-Net++ capabilities in the task of the vFFR in Fig. 4 for two example synthetic simulations of a patient under rest with input flow Q_{in} of 3 ml/s and inlet pressure p_{in} of 80 mmHg. Both models tend to achieve good results on non-stenotic and mild-stenotic segments. However, when it comes to larger stenoses, Point-Net++ lacks correct assessment of the stenosis impact, while CenterlinePoint-Net++ estimates correct values. We attribute this robustness towards sequential embedding done along the centerline which follows the natural blood flow in the coronary artery. Both examples in Fig. 4 showcase the common problem of Point-Net++ for vessel trees with close branches. Since the aggregation is done via Euclidean distance, the branches close to each other in this space are considered neighbours. Due to that, the predicted vFFR on one branch tends to spill on

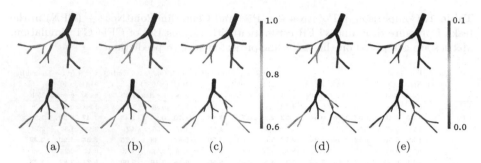

Fig. 4. Qualitative comparison of vFFR between PointNet++ and CenterlinePoint-Net++. (a) CFD GT vFFR, (b) PointNet++ vFFR, (c) CenterlinePointNet++ vFFR, (d) PointNet++ vFFR MAE, (e) CenterlinePointNet++ vFFR MAE on two examples. Due to the lack of sequential embedding, PointNet++ tends to be less robust in assessing the impact of stenosis on pressure drops.

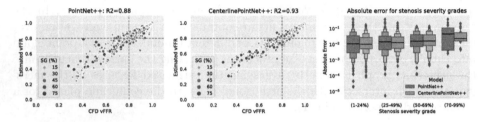

Fig. 5. Evaluation on vFFR estimation with respect to different stenosis severity grades. The vFFR is estimated at stenotic segments with the maximum percentage of radius change reported as a severity grade (SG). We report a correlation of 0.88 and 0.93 for PointNet++ and CenterlinePointNet++, respectively. Dashed lines at 0.8 indicate the clinical threshold.

the second branch as seen on the whole left side of the sample in the first row Fig. 4.

We evaluate estimated vFFR with respect to the stenosis severity grade and group stenosis grades into the relevant intervals based on CAD-RADS scale [16]. Figure 5 showcases scatter plots for PointNet++ vFFR and CenterlinePoint-Net++ vFFR estimation against GT CFD vFFR with respect to the stenosis severity grade. We report the correlation of 0.88 and 0.93 for PointNet++ and CenterlinePointNet++ vFFR estimation, respectively. Performance in correct stenosis impact assessment seems to deteriorate with the increase of stenosis severity grade, however, only outliers achieve the absolute error above 0.1. CenterlinePointNet++ vFFR estimations achieve smaller variations than Point-Net++ over all classes, and significantly lower MAE for the stenosis severity grade (70–90%). We showcase that for the stenosis severity grade $\leq$ 70% the median of CenterlinePointNet++ estimated vFFR absolute error is kept under 0.01. Additionally, we evaluate the clinical viability of the proposed method by computing the accuracy of whether to perform intervention or not based on the predicted vFFR. We set the vFFR threshold to the clinically accepted one of 0.8 [2], and report the accuracy of 94.11% and 95.11% for PointNet++ and CenterlinePointNet++, respectively.

4 Conclusions

In this study, we propose a novel point cloud based neural network architecture CenterlinePointNet++ which is tailored towards the analysis of complex vessel trees by incorporating multi-modal input of surface point cloud and centerline graph. We show an improvement in the vFFR time estimation from approx. two hours for a CFD simulation to around 15 s per synthetic coronary artery. Our centerline grouping approach is confronted with PointNet++ in the task of pressure drop and vFFR estimation in synthetic coronary arteries and achieves better results in NMAE and MAE for every set of values of the input flow and pressure of CFD simulation. The evaluation of FFR showcases a correlation of 0.93 with the CFD vFFR. One of the limitations of the method is the fact that the model is trained for the specified set of boundary conditions of underlying CFD simulation. In the future, we plan to expand the approach by incorporating boundary conditions as an additional input to the network. We also aim to conduct a comprehensive study using real patients' geometries for both training and evaluation, while comparing the results with invasive FFR.

References

1. Arzani, A., Wang, J.X., Sacks, M.S., Shadden, S.C.: Machine learning for cardiovascular biomechanics modeling: challenges and beyond. Ann. Biomed. Eng. **50**(6), 615–627 (2022)
2. Corcoran, D., Hennigan, B., Berry, C.: Fractional flow reserve: a clinical perspective. Int. J. Cardiovasc. Imaging **33**, 961–974 (2017)
3. De Bruyne, B., Sarma, J.: Fractional flow reserve: a review. Heart **94**(7), 949–959 (2008)
4. Eldar, Y., Lindenbaum, M., Porat, M., Zeevi, Y.: The farthest point strategy for progressive image sampling. IEEE Trans. Image Process. **6**(9), 1305–1315 (1997). https://doi.org/10.1109/83.623193
5. de Haan, P., Weiler, M., Cohen, T., Welling, M.: Gauge equivariant mesh CNNs: anisotropic convolutions on geometric graphs. CoRR abs/2003.05425 (2020). https://arxiv.org/abs/2003.05425
6. Hamilton, W.L., Ying, R., Leskovec, J.: Inductive representation learning on large graphs. CoRR abs/1706.02216 (2017). https://arxiv.org/abs/1706.02216
7. He, J., et al.: Learning hybrid representations for automatic 3D vessel centerline extraction. In: Martel, A.L., et al. (eds.) Medical Image Computing and Computer Assisted Intervention – MICCAI 2020, pp. 24–34. Springer, Cham (2020). https://doi.org/10.1007/978-3-030-59725-2_3
8. Itu, L.M., et al.: A machine learning approach for computation of fractional flow reserve from coronary computed tomography. J. Appl. Physiol. (Bethesda, Md.: 1985) **121**, jap.00752.2015 (2016). https://doi.org/10.1152/japplphysiol.00752.2015

9. Ko, B.S., et al.: Noninvasive CT-derived FFR based on structural and fluid analysis: a comparison with invasive FFR for detection of functionally significant stenosis. JACC Cardiovasc. Imaging **10**(6), 663–673 (2017)
10. Kosior, A., Mirota, K., Tarnawski, W.: Patient-specific modeling of hemodynamic parameters in coronary arteries (20 June 2019), US Patent App. 16/217,328 (2019)
11. Li, G., et al.: Prediction of 3D cardiovascular hemodynamics before and after coronary artery bypass surgery via deep learning. Commun. Biol. **4**, 99 (2021). https://doi.org/10.1038/s42003-020-01638-1
12. Małota, Z., et al.: The comparative method based on coronary computed tomography angiography for assessing the hemodynamic significance of coronary artery stenosis. Cardiovasc. Eng. Technol. **14**, 364–379 (2023)
13. Medrano-Gracia, P., et al.: A study of coronary bifurcation shape in a normal population. J. Cardiovasc. Transl. Res. **10**(1), 82–90 (2017)
14. Morris, P.D., van de Vosse, F.N., Lawford, P.V., Hose, D.R., Gunn, J.P.: "virtual" (computed) fractional flow reserve: current challenges and limitations. JACC Cardiovasc. Interv. **8**(8), 1009–1017 (2015)
15. Nabel, E.G.: Cardiovascular disease. N. Engl. J. Med. **349**(1), 60–72 (2003)
16. Nikolaev, A., Feger, J., Weerakkody, Y., et al.: Coronary artery disease - reporting and data system. Reference article, Radiopaedia.org. https://doi.org/10.53347/rID-56786. Accessed 14 Feb. 2023
17. Qi, C.R., Su, H., Mo, K., Guibas, L.J.: PointNet: deep learning on point sets for 3D classification and segmentation. CoRR abs/1612.00593 (2016). https://arxiv.org/abs/1612.00593
18. Qi, C.R., Yi, L., Su, H., Guibas, L.J.: PointNet++: deep hierarchical feature learning on point sets in a metric space. CoRR abs/1706.02413 (2017). https://arxiv.org/abs/1706.02413
19. Rygiel, P., Zieba, M., Konopczynski, T.: Eigenvector grouping for point cloud vessel labeling. In: Bekkers, E., Wolterink, J.M., Aviles-Rivero, A. (eds.) Proceedings of the First International Workshop on Geometric Deep Learning in Medical Image Analysis. Proceedings of Machine Learning Research, vol. 194, pp. 72–84. PMLR, 18 November 2022. https://proceedings.mlr.press/v194/rygiel22a.html
20. Sklet, V.: Exploring the capabilities of machine learning (ML) for 1D blood flow: application to coronary flow. Master's thesis, Norwegian University of Science and Technology, Trondheim, Norway (2018)
21. Suk, J., Haan, P., Lippe, P., Brune, C., Wolterink, J.: Mesh convolutional neural networks for wall shear stress estimation in 3D artery models, September 2021
22. Taebi, A.: Deep learning for computational hemodynamics: a brief review of recent advances. Fluids **7**(6), 197 (2022). https://doi.org/10.3390/fluids7060197, https://www.mdpi.com/2311-5521/7/6/197
23. Verma, N., Boyer, E., Verbeek, J.: Dynamic filters in graph convolutional networks. CoRR abs/1706.05206 (2017). https://arxiv.org/abs/1706.05206
24. Wang, Z., et al.: Diagnostic accuracy of a deep learning approach to calculate FFR from coronary CT angiography. J. Geriatr. Cardiol. JGC **16**, 42–48 (2019)
25. Yao, L., et al.: TaG-Net: topology-aware graph network for vessel labeling. In: Manfredi, L., et al. (eds.) Imaging Systems for GI Endoscopy, and Graphs in Biomedical Image Analysis, pp. 108–117. Springer, Cham (2022). https://doi.org/10.1007/978-3-031-21083-9_11

SPR-Net: Structural Points Based Registration for Coronary Arteries Across Systolic and Diastolic Phases

Xiao Zhang[1,2], Feihong Liu[1,2(✉)], Yuning Gu[2], Xiaosong Xiong[2],
Caiwen Jiang[2], Jun Feng[1], and Dinggang Shen[2,3(✉)]

[1] School of Information Science and Technology, Northwest University, Xi'an, China
`fhliu@nwu.edu.cn`
[2] School of Biomedical Engineering, ShanghaiTech University, Shanghai, China
`dgshen@shanghaitech.edu.cn`
[3] Shanghai United Imaging Intelligence Co., Ltd., Shanghai, China

Abstract. Systolic and diastolic registration of coronary arteries is a critical yet challenging step in coronary artery disease analysis. Most existing methods ignore the important relationship between vascular geometric shape and image contextual information in the two phases, leading to limited performance. In this paper, we propose a novel structural point registration network, which comprehensively captures both point-level geometric features and image-level semantic features as enriched feature representations to assist coronary registration. Specifically, given the systolic and diastolic CCTA images, our method improves coronary artery registration from three aspects. First, the point cloud encoder learns the spatial geometric features of the points in the 3D coronary mask to effectively capture the vascular shape representation. Second, a vision transformer (ViT) is employed to extract the image semantic information as a complementary condition of the geometric features to identify the bi-phasic correspondence of different vascular branches. Third, we design a transformer module to fuse the features across points and images to obtain the corresponding structural points in the two phases and then use structural points to guide the coronary artery registration via the thin-plate spline (TPS) method. We evaluated our method on a real-clinical dataset. Extensive experiments show that our proposed method significantly outperforms the state-of-the-art methods in coronary artery registration.

Keywords: Coronary artery registration · Intrinsic structural points learning · transformer · 3D Point cloud

1 Introduction

Coronary artery disease (CAD) is one of the most prevalent critical cardiovascular diseases with up to 32% mortality rate [18]. The CAD diagnosis necessitates reconstructing a 3D coronary artery tree, *e.g.*, from CCTA images, so that

H. Greenspan et al. (Eds.): MICCAI 2023, LNCS 14226, pp. 791–801, 2023.
https://doi.org/10.1007/978-3-031-43990-2_74

the diagnosis decision could be finalized according to the vascular anatomical information, *e.g.*, annotations of vascular branches and vascular morphological properties [7]. However, conventional reconstruction methods merely exploit the images obtained from the diastolic phase that only reveals partial coronary arteries [1,15,22], which potentially makes vessel lesions invisible, *i.e.*, misdiagnosis.

In fact, a cardiac cycle has two phases, *i.e.*, diastole and systole. The reconstructed arteries in the two-phased CCTA images are incomplete coronary trees, but they complement each other. By accurately aligning the arteries in both phases, the complete coronary tree can be reconstructed. Nevertheless, there are three challenges for successful coronary reconstruction. 1) Since the heart beats vigorously, its surrounding arteries can be squeezed by heart chambers and become invisible in one of the phases, easily causing the misalignment of a significant number of arteries in the two-phased images (short for component variation), as pointed by yellow (visible in diastole only) and cyan (visible in systole only) arrows in Fig. 1(a). 2) Arteries deform along with heartbeats, their shape, size, and location may vary significantly across the two phases, causing difficulties in alignment, as demonstrated in Fig. 1(b). 3) Arteries are tiny tubular tissues, which only occupy a very small part ($\leq 0.5\%$) of the whole CCTA image (Fig. 1(c)), causing imbalance issues for image-based registration methods.

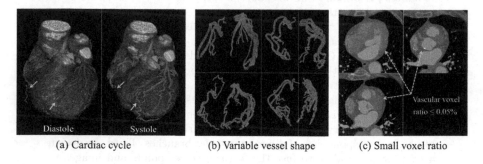

    (a) Cardiac cycle   (b) Variable vessel shape   (c) Small voxel ratio

Fig. 1. Three challenges of coronary artery registration.

For vessel registration, there are mainly three main branches of methods, *i.e.*, image-based, point-cloud-based, and hybrid-based registration. Image-based methods utilize image features to register the entire volume, and the obtained deformation field is then used to align vessels to the target space. Those methods have been extensively applied to the registration of coronary arteries [14,16], pulmonary vessels [13,17], cerebral vessel [10], heart chamber [11], etc. Although those methods demonstrate promising performance on the whole image scale, the vessels are not necessarily well-aligned and cannot be employed to reconstruct the complete coronary tree. By contrast, point-cloud-based registration directly aligns the vessels, which are firstly labeled or segmented from CCTA images and then modeled as point clouds for registration. For example, point-cloud networks [20,21] or graph convolutional networks [24] commonly exploit geometric features of the vascular point-cloud, which are more flexible and accurate than those image-based methods. The limitation of those point-cloud-based methods mainly involves the disability in geometric feature representation to distinguish the arteries, because different arteries or artery branches can share very similar morphology [12]. Similarly, the hybrid-based methods [4,8] also extract the vessel masks in the images for registration, but the lack of effective image information limits its performance. Integrating the advantages of both domains (image and point cloud) may produce improved outcomes, but has not yet been explored.

In this paper, we propose a structural point registration network (SPR-Net) to align coronary arteries from the systolic and diastolic phases. The SPR-Net is designed to exploit both image-based and point-cloud-based features, in which the image and point cloud are encoded as intrinsic features. Additionally, we propose a transformer-based feature fusion module to fully exploit the obtained intrinsic features in extracting structural points, *i.e.*, key points that delineate the anatomical morphology of arteries across the two phases and are solely used to compute the deformation field. For those obtained structural points, a simple thin-plate spline [5] method is employed to align coronary arteries of systole and diastole. Extensive experiment results demonstrate the superiority of our method over eight methods (Fig. 2).

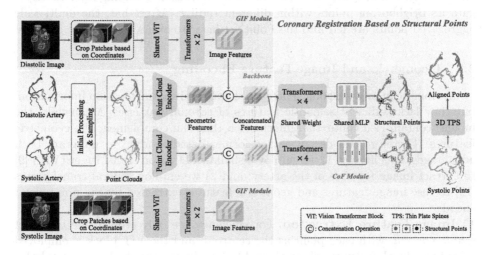

Fig. 2. Overview of the proposed framework consisting of four components: 1) Geometric feature learning module for point cloud; 2) Image semantic feature extraction by ViT modules; 3) Geometric and image semantic feature encoding by transformers; 4) TPS-based dense deformation field interpolation.

2   Method

We propose the SPR-Net method, which simultaneously utilizes geometric features extracted from point clouds and image features extracted from CCTA images with the goal of generating structural points to align arteries across systole and diastole, with shape, location, and component variations. In this section, we first introduce the extraction of geometric features (Sect. 2.1), then the extraction of image features (Sect. 2.2), next the extraction of structural points and their usage in registration procedures (Sect. 2.3), and finally the loss function (Sect. 2.4).

2.1   Geometric Feature Learning

The coronary arteries share a tubular structural shape. The point cloud network has the advantages of effectively learning the spatial geometric shape of arteries and providing accurate relative positional relationships of points [24], so that the obtained point features are more discriminative. Inspired by [6], we employ the point cloud encoder, with the same structure as [6] that composes three layers (*i.e.*, sampling layer, multi-scale grouping layer, and PointNet layer), to extract the geometric features of each point.

Given the input diastolic and systolic point clouds P and Q, we first use a sampling layer in the point cloud encoder to obtain the down-sampled points $\bar{P} = \{\bar{p}_1, \bar{p}_2, \cdots, \bar{p}_m\}$ with $\bar{p}_i \in R^3$ and $\bar{Q} = \{\bar{q}_1, \bar{q}_2, \cdots, \bar{q}_m\}$ with $\bar{q}_i \in R^3$, respectively. $\bar{P}$ and $\bar{Q}$ are then filled into the multi-scale grouping layer to aggregate its neighboring points within different radii r. After that, the multi-scale aggregated points are fed into the PointNet layer to extract geometric features.

2.2   Geometric and Image Feature Encoding

Point clouds can provide good geometric shapes and spatial location information, but they lack sufficient semantic features of coronary arteries. Meanwhile, the images contain rich contextual information that can complement the geometric features. Therefore, we design a transformer-based module to integrate both advantages. Specifically, 1) we employ a shallow 3D vision transformer (ViT) [9] to extract image features of the artery; and 2) we employ general transformers [19] to fuse image features and geometric features extracted by the point-cloud encoder.

1) Image Feature Extraction. For efficiency, we only crop image blocks of size $h \times w \times d$, with each point as the centroid, and the ViT block is employed to extract local features. Since these blocks are extracted along the tubular structures, the extracted local features reveal intrinsic relationships. To exploit their correlations, we employ a self-attention mechanism-based transformer. The coordinates of each point serve as the position encoding, which is added to its local image feature as the input to the following transformer blocks.

$$f_i^{img}(a, b, c) = E_i(a, b, c) + I_i(a, b, c) \tag{1}$$

where E_i and I_i respectively indicate the position encoding and image features for the i-th rectangular volume. $(a, b, c) \in R^3$ is the point coordinates. $f_i^{img} \in R^l$ is the self-attention input of transformer layer, and l is the feature dimension.

2) Geometry and Image Co-embedding. Given concatenated features of pointwise and image features, four transformer layers are employed to further explore comprehensive contextual features between the two phases. The transformer layer incorporates an encoder and decoder block, which are based on a multi-head attention mechanism. We use the concatenated features of the diastolic phase as input to the transformer encoder and decoder respectively, and the opposite for the systolic phase, to learn the feature dependencies between the two phases.

2.3 Registration via Structural Point Correspondences

1) Integration of Structural Points. The input of MLP is the contextual features extracted by the transformer, and the output is the probability of each point. Specifically, given the sampled points $\bar{P}$ with the fused features F_P from diastole, we input the features into the shared MLP to generate the probability maps $V_p = \{v_1, v_2, \cdots, v_k\}$ with $v_i \in R^m$. Thus, the diastolic structural points S_p can be calculated as follows:

$$s_i^p = \sum_{j=1}^{m} \bar{p}_j v_i^j \quad \text{with} \quad \sum_{j=1}^{m} v_i^j = 1 \quad \text{for each } i \qquad (2)$$

Note that, the systolic structural points S_q are calculated in the same way as the diastolic structural points.

2) Structural Points based Registration using TPS. Based on the correspondence established between the structural points S_p and S_q in the two phases, we apply a simple but effective idea of the TPS method to interpolate the dense deformation field. For the two sets of structural points, S_p and S_q, the nearest projection from structural points S_q to the S_p is calculated, and the S_q is warped to the S_p in the diastolic phase. Eventually, each systolic point is re-meshed by the closest point to the structural point and further warped to the original points Q using the estimated dense deformation field.

2.4 Loss Function

We design a structure-constrained registration loss for SPR-Net,

$$L_{total} = L_{rec}(S_p, P) + L_{rec}(S_q, Q) + L_{rec}(S_p, S_q) \qquad (3)$$

where,

$$L_{rec}(X, Y) = \frac{1}{|X|} \sum_{x_i \in X} \min_{y_j \in Y} \|x_i - y_j\|_2^2 + \frac{1}{|Y|} \sum_{y_j \in Y} \min_{x_i \in X} \|y_j - x_i\|_2^2, \qquad (4)$$

Here L_{rec} is chamfer distance, and X and Y denote two point clouds respectively. The first part $L_{rec}(S_p, P)$, and the second part $L_{rec}(S_q, Q)$ assure the predicted structural points in two different phases are close to their corresponding original point clouds. The third part $L_{rec}(S_p, S_q)$ encourages an accurate alignment of structural points between the two phases, ensuring that structural points with the same semantics align on the same vessel branch.

3 Experiments and Results

3.1 Dataset and Evaluation Metrics

Data Processing. In our experiments, we collected 58 pairs of CCTA images with both diastolic and systolic phases. All coronary artery masks are first extracted using [25] and refined by three experts. Then, the annotated arteries were down-sampled and modeled as 3D point clouds; meanwhile, their coordinates were normalized to the range of [0,1]. We choose the five-fold cross-validation evaluation strategy, with 40 training subjects and 18 testing subjects. **Evaluation Metrics.** Since the artery branches of systole and diastole only partially overlap, *i.e.*, some coronary branches only appear in one phase, we define a common Dice coefficient (CoDice) to accurately evaluate the results.

$$\text{CoDice}(P_o, Q_o) = \frac{2|P_o \cap Q_o|}{|P_o| + |Q_o|} \tag{5}$$

where P_o and Q_o denote the set of coronary branches common to diastolic and systolic phases, respectively. Moreover, the Dice coefficient (Dice), Chamfer distance (CD), and Hausdorff distance (HD) are also employed for evaluation.

3.2 Implementation Details

The initial inputs of the SPR-Net contain 4096 point clouds for each phase, and a volume size of $16 \times 16 \times 8$ is cropped around each point. The point cloud encoder consists of two set abstraction blocks with 1024 and 256 grouping centers respectively. In each set abstraction block, we utilize the grouping layer with two scales r to combine the multi-scale features, containing scales (0.1, 0.2, 0.4) and (0.2, 0.4, 0.8) respectively. The transformer blocks we used are composed of vanilla transformer layers. The outputs of the point cloud encoder and ViT have 512-D and 128-D features, respectively, which are concatenated together to form 640-D contextual features. The configuration of the MLP block in the structural point integration depends on the number of structural points. All experiments were implemented using Pytorch on 1 NVIDIA Tesla A100 GPU. We trained the networks using Adam optimizer with an initial learning rate of 10^{-4}, epoch of 600, and batch size of 8.

3.3 Comparison with State-of-the-Art Methods

Our SPR-Net was quantitatively and qualitatively evaluated, compared with eight SOTA registration methods, which belong to three categories:1) image-based registration, including SyN [2], VoxelMorph [3], and DiffuseMorph [11]; 2) hybrid-based registration, TMM [8]; 3) point-cloud based registration, including Go-ICP [23], DCP [20], STORM [21], and ISRP [6]. **Quantitative Results.** The quantitative results are listed in Table 1. We can find the superiority of point cloud-based methods if compared to image-based methods, which supports the previous conclusion about the limitation of image-based methods. We can also find that our proposed method significantly out-performs other methods since SPR-Net fully encodes and fuses features of the images and point clouds. Notably, SPR-Net achieves significantly better perfor-mance than ISRP, the closest competing method, with an improvement of 10% (*i.e.*, increasing Dice from 58.31% to 68.58%).

Table 1. Results of comparison experiments.

Method	Dice (%)↑	CoDice (%)↑	CD↓	HD↓
Benchmark	25.60	27.59	20.56	19.78
SyN [2]	30.60 ± 3.59	33.59 ± 2.84	14.26 ± 3.17	15.05 ± 2.42
VoxelMorph [3]	32.78 ± 3.71	35.65 ± 2.44	11.70 ± 2.82	12.15 ± 2.15
DiffuseMorph [11]	37.55 ± 2.89	42.57 ± 2.06	9.28 ± 2.31	9.57 ± 2.26
Go-ICP [23]	37.89 ± 2.85	41.60 ± 2.39	11.29 ± 2.44	12.14 ± 2.15
DCP [20]	48.23 ± 2.30	54.58 ± 2.28	5.76 ± 2.21	6.08 ± 1.87
TMM [8]	54.76 ± 1.98	60.21 ± 2.14	4.11 ± 1.87	3.35 ± 1.41
STORM [21]	56.04 ± 2.24	61.21 ± 2.08	3.76 ± 1.82	3.42 ± 1.55
ISRP [6]	58.31 ± 1.46	63.45 ± 1.75	2.21 ± 1.29	2.55 ± 1.34
Ours	**68.58 ± 1.14**	**72.24 ± 0.93**	**1.46 ± 0.75**	**1.73 ± 0.62**

Table 2. Quantitative results of ablation analysis of different components.

Method	CoF	GIF	Number-SP	Dice (%)↑	CoDice (%)↑
# 1			768	58.31 ± 1.46	63.45 ± 1.75
# 2	✓		768	60.82 ± 1.62	66.27 ± 1.51
# 3		✓	768	63.02 ± 1.54	68.96 ± 1.69
# 4	✓	✓	256	60.73 ± 1.49	65.27 ± 1.51
# 5	✓	✓	384	62.87 ± 1.62	68.38 ± 1.55
# 6	✓	✓	512	66.95 ± 1.33	70.19 ± 1.32
# 7	✓	✓	768	**68.58 ± 1.14**	**72.24 ± 0.93**
# 8	✓	✓	1024	66.45 ± 1.22	69.70 ± 1.28

Qualitative Visualization. Since the correspondence of structural points is vital for registration, we show the structural points (colored) in systole (green) and diastole (red) in Fig. 3 for demonstrating their correspondence. Those struc-tural points with correspondence to the same vascular branch are marked by the same color denoted by the dashed boxes in the 2nd column of Fig. 3. Notably, we can find that the structural points are distributed at positions such as the end-points or bifurcation points, as shown in the 1st and 2nd columns, which properly delineate the morphology of the point clouds when the number of structural points is small. With increased number, structural points do not only locate at endpoints or bifurcation points but also diffuse along the vessel branches, forming the vessel skeleton, as shown in the 3rd and 4th columns of Fig. 3. In the 5th column of Fig. 3, a complete coronary tree is obtained by exploiting the registration ($K = 768$).

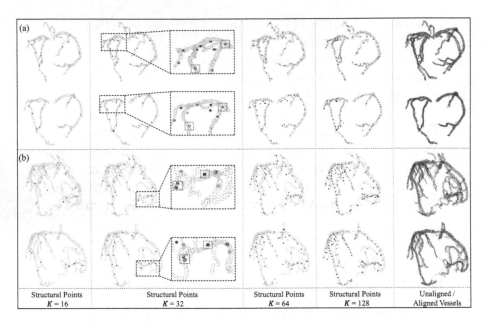

| Structural Points $K = 16$ | Structural Points $K = 32$ | Structural Points $K = 64$ | Structural Points $K = 128$ | Unaligned / Aligned Vessels |

Fig. 3. Structural points and registration results of two subjects (a and b), and point clouds in green and red denote systole and diastole phase, respectively. From the 1st to 4th columns, the number of structural points generated increases, and the colors of structural points denote the correspondence across the two phases. The last column shows the results before and after registration according to the correspondence of the structural points. (Color figure online)

3.4 Ablation Study

We also conduct the ablation studies with the same backbone point cloud encoder by following three groups of configurations: 1) Whether using the four transformer layers, denoted as CoF, to encode and fuse the systolic and diastolic geometry. 2) Whether fusing the geometry features of point cloud with image-level semantic features, denoted GIF. 3) Testing the network on different numbers of structural points (Number-SP). Table 2 summarizes the ablation study results.

If without employing CoF and GIF, only the backbone encoder is used to generate structural points. 1) With the same 768 structural points, we can find the individual modules of CoF and GIF can both improve the Dice performance. Meanwhile, combining the two modules lead to the best performance, which may suggest the importance of fusing the two different aspects of features. 2) By equipping both CoF and GIF, we can find that SPR-Net's performance has been improved when the structural points number increases from 256 to 768. However, the performance decreases when it is further increased to 1024, indicating that dense structural points negatively affect the results, which is probably caused by the increasing number of outlier points. It can also be found that SPR-Net demonstrates inferior performance than both backbone+CoF and backbone+GIF when using 256 structural points, which is probably caused by the sparsity of structural points that are largely located at the endpoints and bifurcation positions, which cannot well delineate the morphology of vessel tree. Therefore, the number of structural points is a key parameter that affects registration performance. Through extensive experiments, we determine the optimal number of structural points to ensure one-to-one correspondences between diastole and systole (Table 2).

4   Conclusion

In this paper, we have proposed an intrinsic structural point learning-based framework for systolic and diastolic coronary artery registration. The framework identifies structural points in the arteries across the two different phases using both the spatial geometric features extracted by the point cloud network and the complementary image semantic information extracted by ViT. By strategically fusing the image and point geometric features through a transformer, structural points with strong correlations in two different phases are extracted and used to guide the registration process. Compared with the existing image-based registration methods and point cloud-based methods, our integrated method achieves superior performance and outperforms the state-of-the-art methods by a large margin, which suggests the potential applicability of our framework in real-world clinical scenarios for CAD diagnosis.

Acknowledgment. This work was supported in part by National Natural Science Foundation of China (grant number 62131015, 62073260, 62203355), and Science and Technology Commission of Shanghai Municipality (STCSM) (grant number 21010502600).

References

1. Achenbach, S., et al.: Influence of heart rate and phase of the cardiac cycle on the occurrence of motion artifact in dual-source CT angiography of the coronary arteries. J. Cardiovasc. Comput. Tomogr. **6**(2), 91–98 (2012)

2. Avants, B.B., Epstein, C.L., Grossman, M., Gee, J.C.: Symmetric diffeomorphic image registration with cross-correlation: evaluating automated labeling of elderly and neurodegenerative brain. Med. Image Anal. **12**(1), 26–41 (2008)
3. Balakrishnan, G., Zhao, A., Sabuncu, M.R., Guttag, J., Dalca, A.V.: VoxelMorph: a learning framework for deformable medical image registration. IEEE Trans. Med. Imaging **38**(8), 1788–1800 (2019)
4. Bayer, S., et al.: Intraoperative brain shift compensation using a hybrid mixture model. In: Frangi, A.F., Schnabel, J.A., Davatzikos, C., Alberola-López, C., Fichtinger, G. (eds.) MICCAI 2018. LNCS, vol. 11073, pp. 116–124. Springer, Cham (2018). https://doi.org/10.1007/978-3-030-00937-3_14
5. Bookstein, F.L.: Principal warps: thin-plate splines and the decomposition of deformations. IEEE Trans. Pattern Anal. Mach. Intell. **11**(6), 567–585 (1989)
6. Chen, N., et al.: Unsupervised learning of intrinsic structural representation points. In: Proceedings of the IEEE/CVF Conference on Computer Vision and Pattern Recognition, pp. 9121–9130 (2020)
7. Çimen, S., Gooya, A., Grass, M., Frangi, A.F.: Reconstruction of coronary arteries from X-ray angiography: a review. Med. Image Anal. **32**, 46–68 (2016)
8. Çimen, S., Gooya, A., Ravikumar, N., Taylor, Z.A., Frangi, A.F.: Reconstruction of coronary artery centrelines from X-Ray angiography using a mixture of student's t-Distributions. In: Ourselin, S., Joskowicz, L., Sabuncu, M.R., Unal, G., Wells, W. (eds.) MICCAI 2016. LNCS, vol. 9902, pp. 291–299. Springer, Cham (2016). https://doi.org/10.1007/978-3-319-46726-9_34
9. Dosovitskiy, A., et al.: An image is worth 16×16 words: transformers for image recognition at scale. arXiv preprint arXiv:2010.11929 (2020)
10. Fu, K., Liu, Y., Wang, M.: Global registration of 3D cerebral vessels to its 2D projections by a new branch-and-bound algorithm. IEEE Trans. Med. Robot. Bionics **3**(1), 115–124 (2021)
11. Kim, B., Ye, J.C.: Diffusion deformable model for 4D temporal medical image generation. In: Wang, L., Dou, Q., Fletcher, P.T., Speidel, S., Li, S. (eds.) International Conference on Medical Image Computing and Computer-Assisted Intervention, pp. 539–548. Springer, Cham (2022). https://doi.org/10.1007/978-3-031-16431-6_51
12. Li, Y., Harada, T.: Lepard: learning partial point cloud matching in rigid and deformable scenes. In: Proceedings of the IEEE/CVF Conference on Computer Vision and Pattern Recognition, pp. 5554–5564 (2022)
13. Pan, Y., Christensen, G.E., Durumeric, O.C., Gerard, S.E., Reinhardt, J.M., Hugo, G.D.: Current-and varifold-based registration of lung vessel and airway trees. In: Proceedings of the IEEE Conference on Computer Vision and Pattern Recognition Workshops, pp. 126–133 (2016)
14. Pang, J., et al.: High efficiency coronary MR angiography with nonrigid cardiac motion correction. Magn. Reson. Med. **76**(5), 1345–1353 (2016)
15. Schroeder, S., et al.: Influence of heart rate on vessel visibility in noninvasive coronary angiography using new multislice computed tomography: experience in 94 patients. Clin. Imaging **26**(2), 106–111 (2002)
16. Shechter, G., Resar, J.R., McVeigh, E.R.: Rest period duration of the coronary arteries: implications for magnetic resonance coronary angiography. Med. Phys. **32**(1), 255–262 (2005)
17. Smeets, D., Bruyninckx, P., Keustermans, J., Vandermeulen, D., Suetens, P.: Robust matching of 3D lung vessel trees. In: MICCAI Workshop on Pulmonary Image Analysis, vol. 2, pp. 61–70 (2010)
18. Timmis, A., et al.: European society of cardiology: cardiovascular disease statistics 2021. Eur. Heart J. **43**(8), 716–799 (2022)

19. Vaswani, A., et al.: Attention is all you need. In: Advances in Neural Information Processing Systems, vol. 30 (2017)
20. Wang, Y., Solomon, J.M.: Deep closest point: learning representations for point cloud registration. In: Proceedings of the IEEE/CVF International Conference on Computer Vision, pp. 3523–3532 (2019)
21. Wang, Y., Yan, C., Feng, Y., Du, S., Dai, Q., Gao, Y.: STORM: structure-based overlap matching for partial point cloud registration. IEEE Trans. Pattern Anal. Mach. Intell. 45(1), 1135–1149 (2022)
22. Weissman, N.J., Palacios, I.F., Weyman, A.E.: Dynamic expansion of the coronary arteries: implications for intravascular ultrasound measurements. Am. Heart J. 130(1), 46–51 (1995)
23. Yang, J., Li, H., Campbell, D., Jia, Y.: Go-ICP: a globally optimal solution to 3D ICP point-set registration. IEEE Trans. Pattern Anal. Mach. Intell. 38(11), 2241–2254 (2015)
24. Yao, L., et al.: TaG-Net: topology-aware graph network for centerline-based vessel labeling. IEEE Trans. Med. Imaging (2023)
25. Zhang, X., et al.: Progressive deep segmentation of coronary artery via hierarchical topology learning. In: Wang, L., Dou, Q., Fletcher, P.T., Speidel, S., Li, S. (eds.) International Conference on Medical Image Computing and Computer-Assisted Intervention, pp. 391–400. Springer, Cham (2022). https://doi.org/10.1007/978-3-031-16443-9_38

WarpEM: Dynamic Time Warping for Accurate Catheter Registration in EM-Guided Procedures

Ardit Ramadani[1,2,3]($\boxtimes$), Peter Ewert[2], Heribert Schunkert[2,3], and Nassir Navab[1,4,5]

[1] Computer Aided Medical Procedures, Technical University of Munich, Munich, Germany
ardit.ramadani@tum.de
[2] German Heart Center Munich, Munich, Germany
[3] German Centre for Cardiovascular Research, Munich Heart Alliance, Munich, Germany
[4] Munich Institute of Robotics and Machine Intelligence, Technical University of Munich, Munich, Germany
[5] Computer Aided Medical Procedures, Johns Hopkins University, Baltimore, USA

Abstract. Accurate catheter tracking is crucial during Minimally Invasive Endovascular Procedures (MIEP), and Electromagnetic (EM) tracking is a widely used technology that serves this purpose. However, registration between preoperative images and the EM tracking system is often challenging. Existing registration methods typically require manual interactions, which can be time-consuming, increase the risk of errors and change the procedural workflow. Although several registration methods are available for catheter tracking, such as marker-based and path-based approaches, their limitations can impact the accuracy of the resulting tracking solution, consequently, the outcome of the medical procedure.

This paper introduces a novel automated catheter registration method for EM-guided MIEP. The method utilizes 3D signal temporal analysis, such as Dynamic Time Warping (DTW) algorithms, to improve registration accuracy and reliability compared to existing methods. DTW can accurately warp and match EM-tracked paths to the vessel's centerline, making it particularly suitable for registration. The introduced registration method is evaluated for accuracy in a vascular phantom using a marker-based registration as the ground truth. The results indicate that the DTW method yields accurate and reliable registration outcomes, with a mean error of 2.22 mm. The introduced registration method presents several advantages over state-of-the-art methods, such as high registration accuracy, no initialization required, and increased automation.

Supplementary Information The online version contains supplementary material available at https://doi.org/10.1007/978-3-031-43990-2_75.

Keywords: Electromagnetic Catheter Tracking · Dynamic Time
Warping for Registration · Minimally Invasive Endovascular Procedures

1   Introduction

Minimally Invasive Endovascular Procedures (MIEP) are becoming increasingly
popular due to their non-invasive nature and quicker recovery time compared to
traditional open surgeries. MIEP encompasses various applications, from peripheral artery disease treatments to complex procedures in kidneys, liver, brain,
aorta, and heart. Catheter tracking is an essential component of these procedures, allowing for accurate guidance of the catheter through the vasculature [2].
   Electromagnetic (EM) tracking is a widely used technology for catheter tracking in MIEP. However, accurate registration between preoperative images and
the EM tracking system remains a challenge [4,15]. Existing registration methods often require manual interaction, which can be time-consuming and may
alter the procedural workflow. Achieving accurate and effective registration is
essential for a seamless and fast integration of EM tracking, preoperative data,
and the patient into the same intraoperative coordinate space.

1.1   Related Work

Electromagnetic Tracking is a popular tracking technology that uses integrated sensors in the catheter tip and a field generator to enable localization of
the sensor's pose in 3D. EM tracking does not require line-of-sight, which makes
it particularly advantageous for MIEP [20]. The sensors come in various shapes
and sizes and can be tracked in translation and orientation, with respect to the
generated EM field. Therefore, EM tracking is typically used in conjunction with
intra- or preoperative images to guide the procedure [4,15].

Registration is an essential step that enables intraoperative guidance of procedures through tracking technologies such as EM. Numerous methods have been
developed and presented for EM tracking registration, particularly in vascular
procedures [6,10]. One method commonly used for registering all involved components in the same intraoperative coordinate space involves the use of external
markers or fiducials, which must remain affixed to the patient's body throughout
the preoperative and intraoperative phases of the procedure. In order to achieve
this registration, surgeons are required to match multiple physical marker points
with their corresponding preoperative positions in the image, generating a set of
point correspondences that can then be used to calculate the registration transformation. Some of the works using the marker-based registration method are
presented in [7,9,11,19]. However, this method is subject to several disadvantages, including changes to the procedural workflow, the necessity of ensuring
marker stability between the pre- and intraoperative phase, and sensitivity to
natural changes in the patient's anatomy.

Other registration methods include the use of the EM-tracked catheter paths and registration to the vessel's centerline, using Iterative Closest Point (ICP) algorithms, as described in [8,13], while others explored the registration of EM paths to points within the vasculature, as reported in [5]. Extensive research has been conducted on these methods, which have demonstrated accurate and reliable registration results. However, these methods are also subject to some drawbacks, such as limitations in accurately capturing the movement of the catheter, requiring a good initialization for accurate registration, and reduced accuracy when missing data segments or limited number of data points.

Dynamic Time Warping is a widely-used technique in signal processing to compare and align two time-dependent sequences. Dynamic Time Warping (DTW) calculates the similarity between the sequences by optimally aligning them in a nonlinear fashion, taking into account time differences and sequence sampling rates [1,3]. The warping function enables the matching of corresponding features in the sequences, allowing for accurate alignment even when there are temporal differences or missing data points. DTW has many applications, including biomedical signal processing, speech, and gesture recognition [1,3].

Despite the well-known advantages of DTW as a technique for signal alignment, its application as a registration method in EM-tracked MIEP is still unexplored. Previous works have reported using DTW and EM tracking; however, these works have primarily used DTW for data processing and evaluation due to its advantages in interpreting intravariability, rather than for registration purposes [17,21]. This paper introduces a novel approach to registering EM tracking systems and preoperative images using DTW. To the best of the authors' knowledge, this is also the first work that explores the temporal analysis of 3D EM catheter paths, which are subsequently used to register two systems within a single coordinate space. The introduced method utilizes DTW to warp the time-dependent 3D EM-tracked catheter path to the vasculature's centerline, creating corresponding points between the two. These points are then filtered based on the minimal cost in the DTW algorithm to generate a set of correspondence points that are used for registration between the EM path and centerline.

2   Method

This paper introduces a novel approach for automatically registering EM tracking systems to preoperative images using DTW. The introduced method includes a preoperative phase, during which the targeted vasculature is segmented from preoperative images such as Magnetic Resonance Imaging (MRI), computed Tomography (CT), or CT Angiography (CTA), and the centerlines are extracted using the SlicerVMTK toolkit. In the intraoperative phase, the catheter with an integrated EM sensor in its tip is guided through the respective branch of the vascular tree and records a 3D EM path. In this particular implementation, a catheter-shaped EM sensor is used instead. The recorded EM path is then processed using DTW and warped to the centerline, providing point correspondences between the EM path and the centerline. These correspondences are used to perform a closed-form solution using Coherent Point Drift (CPD) algorithm, translating the EM path to the centerline, resulting in the registration of the two coordinate spaces. A detailed overview of the method is presented in Fig. 2.

2.1 EM Tracking System

The tracking technology used in this paper is the Aurora tracking system from Northern Digital Inc. - NDI (Waterloo, Ontario, Canada). It comprises of a system control unit, sensor interface unit, and a tabletop field generator, which allows for tracking of EM sensors in a $420 \times 600 \times 600$ mm space. This tracking system is capable of recording data with a frequency of up to 40 Hz. The Aurora 5DOF FlexTube, which has a 1 mm diameter, was the catheter-shaped EM sensor used in this paper. It is highly versatile in applications since it can be navigated independently or integrated into catheters. The EM sensor was used without a catheter and was navigated in the phantom through its long cable. The tracking data is recorded using ImFusion Suite software (ImFusion GmbH, Munich, Germany), running on a laptop computer with the following specifications: Windows 11Pro, Intel Core i7-8565U CPU, 16 GB RAM, and Intel UHD 620 Graphics. Figure 1 provides a detailed overview of the experimental setup, including the EM tracking system and the phantom utilized in this paper.

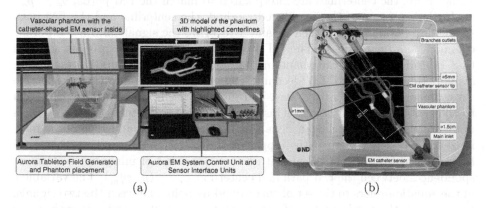

(a)   (b)

Fig. 1. Full system setup including: (a) EM tracking system with the vascular phantom, EM catheter-shaped sensor, and phantom model visualization on the screen; and (b) Top-view of the vascular phantom used in the experiments, featuring labeled branches for easy identification and color-matching with the results.

2.2 Phantom

For EM tracking data acquisition purposes of this paper, an STL model obtained from [18] was utilized. The STL model was simplified while preserving six of the primary branches that represent natural vasculature features, including bifurcations, stenosis, and curvatures. Additionally, the model was resized to dimensions suitable for catheterization, with vessel diameters ranging from 1.5 cm to 5 mm and a length of 22 cm. In order to enable visibility of the catheter-shaped EM sensor from the outside, the phantom was 3D printed using rigid transparent Polylactic Acid (PLA) material. The phantom was then rigidly fixed in a box and positioned on top of the EM tracking field generator to conduct the experiments and record EM tracking data. Figure 1b provides a detailed illustration of the phantom and the catheter-shaped EM sensor utilized in this paper.

2.3   Dynamic Time Warping Registration

During the intraoperative phase, two preoperative components are used, namely the segmented vascular model and its corresponding branch centerline points. The centerline points are referred to as $p_c^n \in \mathbb{P}_{preop}^3$, where n represents the number of points, and $\mathbb{P}_{preop}^3$ represents the 3D preoperative coordinate space. Firstly, the EM catheter-shaped sensor is guided through the vascular phantom, and the resulting EM path is recorded. The EM path points are referred to as $p_{em}^m \in \mathbb{P}_{em}^3$, where m represents the number of points, and $\mathbb{P}_{em}^3$ represents the 3D coordinate space of the EM tracker. In order to use DTW with 3D signals, the number of points must be consistent across the signals; therefore, the signal with fewer points is linearly interpolated to match the number of points in the other signal. Here, the centerlines are interpolated to match the EM paths, $p_c^n \rightarrow p_c^m$.

In the next step of the introduced method, the centerline and the EM path signals are normalized between -1 and 1 to bring the signals into a temporary common coordinate space. The DTW algorithm registration process assumes that the orientation of the phantom (patient) and the preoperative model are similar. Here forwards, DTW decomposes both 3D signals into their respective axes, namely x, y, and z over time, and matches each point from the EM path to the centerline's counterpart. This iterative process stretches the two signals until the sum of the euclidean distances between corresponding points is minimized. The output warp paths represent the corresponding indices that have been warped from the DTW algorithm, creating a set of minimum cost correspondences between the EM path and centerline, $c_{i,j}^u = (p_c^i, p_{em}^j)$. The variable c in the equation refers to the set of corresponding points between the two signals, u represents the total number of correspondences, while i and j represent the indices of the corresponding matched points. For each point in the EM path, there exist one or more points in the centerline that have been warped together and vice versa. For further reading on the DTW algorithm used in this paper, please refer to the following works with more implementation details [14,16].

In order to register the two signals, we leverage the point correspondences generated by the DTW algorithm to select three sets of equally distributed points from three equal segments of the signals, $c_{i,j}^3 = (p_c^i, p_{em}^j)$. The correspondences in each signal segment are selected based on the minimum cost return function of the DTW algorithm, where the sum of the euclidean distances between corresponding points is minimal. Utilizing the three segments facilitates the equitable distribution of point matching across the entirety of the signal. This step is crucial to ensure high confidence matching and produce a reliable registration that does not rely entirely on one part of the vasculature. The selected correspondence points are then used to find the rigid transformation between them using the CPD algorithm, as described in [12]. The introduced solution is a closed-form algorithm that produces a transformation $T = (R, t)$ between the set of correspondence points while minimizing the distance between them. In the transformation matrix T, R represents the 3×3 rotation matrix, and t represents the 3×1 translation vector. Finally, the registration transformation is applied to the recorded EM path, registering the two systems in the same coordinate space, $p_c^m, p_{em}^m \in \mathbb{P}_{intraop}^3$, which represents the intraoperative 3D coordinate space.

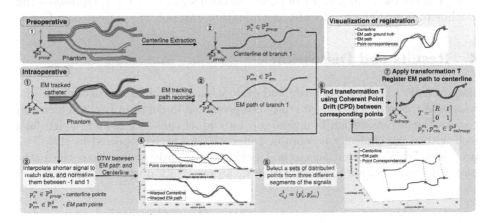

Fig. 2. Overview of the introduced DTW registration method for EM guided MIEP.

3   Evaluation

In this study, the introduced DTW method for registration of EM-guided MIEP is evaluated using the mean registration error criterion. The method is compared to path-based ICP registration methods from state-of-the-art. To evaluate the DTW method for registration accuracy, we conducted experiments by recording EM-tracked paths while navigating through each of the six branches of the phantom five times. For each run, the EM catheter-shaped sensor was manually pulled from the main inlet of the phantom towards the outlets of each branch, with an average speed of 1–2 cm/s. Subsequently, the recorded EM paths were registered to the phantom's centerline using the introduced DTW method.

The ground truth registration used in this study is a marker-based method, which is considered a benchmark in the literature. Ten unique easily-identifiable landmarks throughout the phantom are used to register the preoperative model to the EM tracking system for calculating the ground truth.

3.1   Mean Registration Error

This criterion is employed to assess the accuracy of the introduced DTW method compared to the ground truth registration. The mean registration error is computed by summing the euclidean distance of each DTW-registered EM path point to its closest point in the ground truth EM-registered path. Equation (1) represents the mathematical expression employed for this computation.

$$\text{Mean error} = \frac{1}{m} \sum_{i=1}^{m} \min_{j} ||p_{em}^i - p_{gt}^j|| \tag{1}$$

The p_{em}^i represents an EM path point transformed by the introduced DTW registration method, p_{gt}^j represents the closest point from ground truth to p_{em}^i, m represents the total number of points in both signals, and $|| \cdot ||$ represents the Euclidean norm. The results of this evaluation criterion are presented in Fig. 3.

4   Results and Discussion

Based on the experimental setup and evaluation criterion mentioned above, the results of this proof-of-concept study are presented in detail in Fig. 3. The mean registration error over all branches and individual runs is 2.22 mm, which falls well within the clinically acceptable range of <5 mm, as reported in [13]. The proposed DTW registration method performed slightly better than the path-based ICP registration method, with a mean registration error of 2.86 mm.

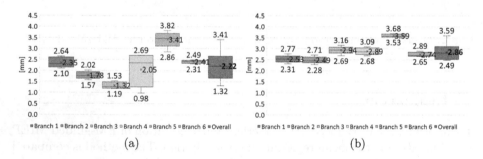

(a)   (b)

Fig. 3. Mean registration error in millimeters across all branches, and overall accuracy (bars represent the standard deviation, while whiskers indicate the minimum and maximum error). (a) DTW registration of EM paths to centerlines, and (b) ICP registration of EM paths to centerlines.

The results presented in this paper demonstrate that the introduced DTW registration method achieves accurate and reliable registration. The method outperforms the established path-based registration methods that use ICP in all reported branch results. The variability of registration accuracy among different runs is higher in mostly straight vessels, when the automatic selection of registration feature points follows a straight line. This variability is mostly noticeable in results from Branch 4 and 5, where the translation is correctly matched, but the orientation of the signals is not perfectly aligned, and the standard deviation in the results exceed that of ICP.

The introduced methods offers the advantage of automation in the registration process, with no changes in the intraoperative workflow. Unlike marker-based registration methods, which require manual interactions to match point correspondences, the DTW method automatically warps the signals, selects corresponding points, and performs registration. Additionally, the DTW method is not dependent on initialization compared to other path-based registration methods, which may fail to provide a transformation if not correctly initialized. In this paper, the ICP method was consistently initialized from the registered position of the DTW method. This step was necessary as the direct application of ICP registration would not converge to provide a transformation due to the significant distance between the signals.

Furthermore, unlike other path-based registration methods, the DTW registration method does not rely on registering the entire EM path to the centerline. In ICP-based methods, the aim is to minimize the difference between all points to be registered, which means that any deformations in one part of the signal would affect the entire registration. In contrast, the DTW method matches points between signals beforehand and employs algorithms to check the confidence of the matched correspondences, ensuring that the set of points to be registered would result in a reliable registration.

In comparison to prior studies, the proposed method aligns well with other state-of-the-art approaches in the research community. Marker-based techniques outlined in [9] report registration errors of 1.28 mm in phantom studies and 4.18 mm in in-vivo studies. Similarly, two additional studies employ path-based registration methods with ICP and report registration accuracies of 3.75 mm and 4.50 mm in [13] and [8] respectively. Another ICP registration approach proposed in [5] reports a mean registration accuracy of 1.30 mm. While these comparisons are relative due to differences in tracking data utilized, they demonstrate acceptable registration accuracy, which the introduced DTW method achieves.

Future research directions for the introduced DTW registration include conducting additional evaluations with various phantoms and potentially in-vivo studies, which may provide more evidence of the method's effectiveness in clinical settings. In addition to exploring the impact of catheter movements on DTW matching, another research direction involves improving the approach to registering EM paths to centerlines. In time-dependent series, alternating forward or backward catheter movement can change the signal appearance and result in incorrect matching. To overcome this, one solution is to use the 3D localization and motion-capturing abilities of EM to detect forward and backward movements and then backward warp the signal when the catheter direction changes. Last, advanced algorithms that more accurately depict catheter motion dynamics could be implemented to improve registration accuracy. Current solutions still exhibit significant variability between ground truth catheter movement and the centerline.

5   Conclusion

Accurate catheter tracking is essential in MIEP, and this paper introduces a novel catheter registration method using DTW. As far as the authors know, this study represents the first attempt at using temporal analysis of 3D EM signals for catheter registration. The method is evaluated on a vascular phantom with marker-based registration as the ground truth. The results indicate that the DTW registration method achieves accurate and reliable registration, outperforming path-based ICP registration methods. Furthermore, it provides several advantages compared to existing solutions, including high registration accuracy, registration process automation, preservation of procedural workflow, and elimination of the need for initialization. The method introduced in this paper is a proof-of-concept study, and further experiments by the research community are necessary to establish its applicability and effectiveness in clinical settings. Overall, the introduced DTW registration method has the potential to enhance the accuracy and reliability of catheter tracking in MIEP.

Acknowledgment. The project was funded by the Bavarian State Ministry of Science and Arts within the framework of the *"Digitaler Herz-OP"* project under grant number 1530/891 02. We also thank ImFusion GmbH and BrainLab AG for their software support and valuable interactions.

References

1. Müller, M.: Dynamic time warping. In: Information Retrieval for Music and Motion, pp. 69–84. Springer, Heidelberg (2007). https://doi.org/10.1007/978-3-540-74048-3_4
2. Abi-Jaoudeh, N., et al.: Multimodality image fusion-guided procedures: technique, accuracy, and applications. Cardiovasc. Intervent. Radiol. **35**(5), 986–998 (2012). https://doi.org/10.1007/s00270-012-0446-5
3. Efrat, A., Fan, Q., Venkatasubramanian, S.: Curve matching, time warping, and light fields: new algorithms for computing similarity between curves. J. Math. Imaging Vis. **27**(3), 203–216 (2007). https://doi.org/10.1007/s10851-006-0647-0
4. Franz, A.M., Haidegger, T., Birkfellner, W., Cleary, K., Peters, T.M., Maier-Hein, L.: Electromagnetic tracking in medicine-a review of technology, validation, and applications. IEEE Trans. Med. Imaging **33**(8), 1702–1725 (2014). https://doi.org/10.1109/TMI.2014.2321777
5. de Lambert, A., Esneault, S., Lucas, A., Haigron, P., Cinquin, P., Magne, J.L.: Electromagnetic tracking for registration and navigation in endovascular aneurysm repair: a phantom study. Eur. J. Vasc. Endovasc. Surg. **43**(6), 684–689 (2012). https://doi.org/10.1016/j.ejvs.2012.03.007
6. Liao, R., Zhang, L., Sun, Y., Miao, S., Chefd'Hotel, C.: A review of recent advances in registration techniques applied to minimally invasive therapy. IEEE Trans. Multimed. **15**(5), 983–1000 (2013). https://doi.org/10.1109/TMM.2013.2244869

7. Lin, Q., Yang, R., Dai, Z., Chen, H., Cai, K.: Automatic registration method using EM sensors in the IoT operating room. EURASIP J. Wirel. Commun. Netw. **2020**(1), 1–16 (2020). https://doi.org/10.1186/s13638-020-01754-w

8. Luo, X.: A bronchoscopic navigation system using bronchoscope center calibration for accurate registration of electromagnetic tracker and ct volume without markers: a bronchoscopic navigation system. Med. Phys. **41**(6 - Part1), 061913 (2014). https://doi.org/10.1118/1.4876381

9. Manstad-Hulaas, F., Tangen, G.A., Gruionu, L.G., Aadahl, P., Hernes, T.A.N.: Three-dimensional endovascular navigation with electromagnetic tracking: ex vivo and in vivo accuracy. J. Endovasc. Ther. **18**(2), 230–240 (2011). https://doi.org/10.1583/10-3301.1

10. Matl, S., Brosig, R., Baust, M., Navab, N., Demirci, S.: Vascular image registration techniques: a living review. Med. Image Anal. **35**, 1–17 (2017). https://doi.org/10.1016/j.media.2016.05.005

11. Mittmann, B.J., et al.: Reattachable fiducial skin marker for automatic multi-modality registration. Int. J. Comput. Assist. Radiol. Surg. **17**(11), 2141–2150 (2022). https://doi.org/10.1007/s11548-022-02639-7

12. Myronenko, A., Song, X.: Point set registration: coherent point drift. IEEE Trans. Pattern Anal. Mach. Intell. **32**(12), 2262–2275 (2010). https://doi.org/10.1109/TPAMI.2010.46

13. Nypan, E., Tangen, G.A., Manstad-Hulaas, F., Brekken, R.: Vessel-based rigid registration for endovascular therapy of the abdominal aorta. Minim. Invasive Ther. Allied Technol. **28**(2), 127–133 (2019). https://doi.org/10.1080/13645706.2019.1575240

14. Paliwal, K., Agarwal, A., Sinha, S.S.: A modification over Sakoe and Chiba's dynamic time warping algorithm for isolated word recognition. Signal Process. **4**(4), 329–333 (1982). https://doi.org/10.1016/0165-1684(82)90009-3

15. Ramadani, A., Bui, M., Wendler, T., Schunkert, H., Ewert, P., Navab, N.: A survey of catheter tracking concepts and methodologies. Med. Image Anal. **82**, 102584 (2022). https://doi.org/10.1016/j.media.2022.102584

16. Sakoe, H., Chiba, S.: Dynamic programming algorithm optimization for spoken word recognition. IEEE Trans. Acoust. Speech Signal Process. **26**(1), 43–49 (1978). https://doi.org/10.1109/TASSP.1978.1163055

17. Sielhorst, T., Blum, T., Navab, N.: Synchronizing 3D movements for quantitative comparison and simultaneous visualization of actions. In: Fourth IEEE and ACM International Symposium on Mixed and Augmented Reality (ISMAR'05), pp. 38–47, October 2005. https://doi.org/10.1109/ISMAR.2005.57

18. Sutton, E.E., Fuerst, B., Ghotbi, R., Cowan, N.J., Navab, N.: Biologically inspired catheter for endovascular sensing and navigation. Sci. Rep. **10**(1), 5643 (2020). https://doi.org/10.1038/s41598-020-62360-w

19. Wood, B.J., et al.: Navigation with electromagnetic tracking for interventional radiology procedures: a feasibility study. J. Vasc. Interv. Radiol. **16**(4), 493–505 (2005). https://doi.org/10.1097/01.RVI.0000148827.62296.B4

20. Zhang, H., et al.: Electromagnetic tracking for abdominal interventions in computer aided surgery. Comput. Aided Surg. **11**(3), 127–136 (2006). https://doi.org/10.3109/10929080600751399

21. Zhang, Z., Liu, Z., Singapogu, R.: Extracting subtask-specific metrics toward objective assessment of needle insertion skill for hemodialysis cannulation. J. Med. Robot. Res. **04**(03n04), 1942006 (2019). https://doi.org/10.1142/S2424905X19420066

Author Index

© The Editor(s) (if applicable) and The Author(s), under exclusive license
to Springer Nature Switzerland AG 2023
H. Greenspan et al. (Eds.): MICCAI 2023, LNCS 14226, pp. 813–818, 2023.
https://doi.org/10.1007/978-3-031-43990-2

Printed in the United States
by Baker & Taylor Publisher Services